FAQ's on Forestry – For IFS Aspirants

A Decade of Solved Papers

ABOUT THE AUTHOR

Dr. K.T. Parthiban is a professor borne at the faculty of forestry and engaged in professional forestry teaching, research and extension activities for the past 25 years. He has conceived the value chain models for Industrial Agroforestry through Technological, Organizational and Marketing interventions. The author is pioneer in Tree varietal development, Consortia Mode Tree Farming, Tree Insurance, Mini Clonal Technology, Price Support System for farm grown trees and Value Addition Technology. He is credential of establishing India's first Forest Business Incubator and leveraged technology based business entrepreneurship development.

He has handled numerous projects funded by DST, DBT, ICAR, NOVOD Board, SLUB, Department of Environment, USAID and several wood based industries. He has published 12 books and over 100 research papers in forestry research. Some of his books like Objective Forestry, Subjective Forestry to IFS Aspirants, Tree Improvement, Industrial Agroforestry and Forestry Technologies – A Value Chain Approach have found wide range of utility by the students, professional foresters and wood based industries.

Dr. K.T. Parthiban travelled widely including UK, France, Republic of Ireland, Germany, Holland, Italy, Hungary, Philippines, Malaysia, Brazil, Denmark etc. and gained skill and expertise on forestry technologies and applied them in India with suitable modifications. Above all, he is a prominent resource person for various Institutes, organizations and industries. He has received several awards including ICFRE award for Excellence in Forestry and TNAU's Award for Niche Area of Excellence in Industrial Agroforestry.

FAQ's on Forestry – For IFS Aspirants

A Decade of Solved Papers

Author

Dr. K.T.Parthiban

Professor (Forestry)

Co-Authors

B. Palanikumaran

N. Krishnakumar

R. Thirunirai Selvan

N. Kanagaraj

Forest College and Research Institute
Tamil Nadu Agricultural University
aMeMupalayamaaaMMMaMMM

Published by

SCIENTIFIC PUBLISHERS (INDIA)
5 A, New Pali Road, P.O. Box 91
Jodhpur 342 001 (INDIA)
E-mail: info@scientificpub.com
Website: http://www.scientificpubonline.com

ISBN: 978-93-92590-06-1

E-ISBN: 978-93-92590-09-2

Printed in India

PREFACE

The professional manpower needs of the country are growing at an alarming rate in order to protect and manage the natural resources on a sustainable basis. Forestry is one such profession where the demand is very critical and is generally met through various competitive examinations. Among several competitive examinations conducted in the country, the Indian Forest Service (IFS) is one of the attractive professions among the graduates for the reason that they are all intrinsically attracted towards forests and nature. Annually several thousands of graduates are appearing for Indian Forest Service (IFS) examinations with success and failure. Based on the discussion and consultations with the aspiring and successful graduates, it was found that the major obstacle towards preparation for the examination is unavailability of comprehensive reference books with questions and answers. Wide range of objective type books are available for forestry but there is no single subjective question and answer type books is available for reference which demanded such a compilation.

Accordingly, it has been planned to develop a comprehensive reference book incorporating questions and answers of IFS examinations. A decade of Indian Forest Service (IFS) question papers (2005 to 2016) have been collected, reviewed, analyzed and suitable answers were collected from the existing text books, scientific reports from government institutions and discussion with various Indian Forest Service officers and scientists working across the country.

This book has been prepared by referring over 150 textbooks, scientific documents and furnished in an easy language for the readers to follow. Hence the book will find profound utility by all sections of graduates aspiring for Indian Forest Service (IFS). The compilation of the book will also find utility for other state service and Agricultural research service examinations.

The book has included as far as possible nearly correct answers and solutions to the questions indicated; however there may be some errors, short of answers etc., which would have happened due to oversight in a huge volume of compilation. Hence the readers are most welcome to offer comments and suggestions for improvement of this book in the future editions.

AUTHORS

ACKNOWLEDGEMENT

The senior author is profoundly thankful to

- The Vice-Chancellor, TNAU, Coimbatore for his sustained encouragements in writing competitive examination books.
- The Authors and Publishers of various books and publications which act as a source material for such a voluminous compilation.
- Sri. Rajsekar, I.F.S., Conservator of Forests, Punjab for answering questions related to field level application
- Sri. P. Kathirvel, I.F.S., Deputy Conservator of Forests, for extending suitable answers to the questions related to wild life, policy, law and JFM.
- Sri. S.Narendra Babu, I.F.S., Divisional Forest Officer, Munar, Kerala for answering the questions related to working plan.
- Dr. R. Revathi, Professor and Head (Forest Products and Utilization), Dr.K.Sivakumar, Asst. Professor (Soil Science and Agricultural Chemistry) Dr.M.Tilak, Asst. Professor (Microbiology), Dr.R.Ravi Kumar, Asst. Professor (Mathematics), Dr.K.Baranidharan, Asst. Professor (Forestry), Dr.V.Karthik, Assistant Professor (Agrl. Economics), Dr.P.Sudha, Asst. Professor (Food Processing and Engineering), Dr.P.Renganayaki, Professor (Seed Science and Technology) Dr.M.Uma Devi, Asst. Professor (Plant Breeding and Genetics), Dr.P.G.Kavitha, Asst. Professor (Nematology) are gratefully acknowledged for extending suitable answers specific to questions.
- My students M. Karthick, K. Karthikeyan, T.M. Akhil Raj, S.Mahesh, M.V. Jawahar Vishnu and M.Sathya are gratefully acknowledged for their contribution towards collection of wide range of books and literatures.
- Thanks are also due to Mrs. R.D. Sharmila for her patience and excellently executing the compilation continuously over a period of 12 months.

BOOK REVIEWERS

- Sri. P. Kathirvel, I.F.S.,
- Sri. N. Vijay, I.F.S., Rajasthan, currently on deputation as Conservator of Forests, Central Academy for State Forest Service (CASFOS), Coimbatore.

CONTENTS

Chapter 1

Silviculture

1. What are the conditions necessary for establishment of a tree nursery? (2005)

The establishment of a nursery is very essential to produce quality planting stock in a required number and within the stipulated period in order to meet the seedling requirements of both afforestation and reforestation programme. The following factors and conditions are very essential for establishment of a nursery.

(i) **Location:** The area chosen for nursery should be centrally located for transportation and communication facility as near as possible to forest establishments to facilitate frequents supervision. The location should also be amenable for lifting and transportation of seedlings easily from the nursery.

(ii) **Light:** This is an important factor to be considered in selecting an area for raising seedlings. The nursery area should be free from permanent shade. The seedlings raised under shade lead to failure in the establishment under field conditions due to scorching sunlight.

(iii) **Slope and aspect:** Nursery should be gentle and flat to ensure good drainage facilities for proper growth and establishment of the seedlings. Generally sloppy areas can be avoided for nursery establishment.

- North aspects are cooler and more humid in lower elevation
- Southern aspects are warmer and suitable for 1200m elevation and above.

(iv) **Size:** The size of the nursery should be decided based on the seedling demand and catering needs. Choosing a large area for a small demand is unmanageable and uneconomical. A minimum nursery site of 1 hectare is preferred but depending on the quantity, the size of the nursery can be up to 4 hectare (10 acres).

(v) **Soil:** The nursery should be located on the best available soil preferably on deep sandy loam or loamy sand. Physical characters are more important than fertility. Soil should be porous and evenly distribute soil water. To improve soil conditions, addition of organic matter is essential. For any nursery seedling production, sand is preferable to heavy clay. The soil pH between 6.5 and 7.5 is preferred for production of quality planting materials.

(vi) **Water:** The site chosen for nursery should have a good source of water for watering the seedlings particularly during the driest months. The pH of the water should be around 7.0 and the water should be free from high concentrations of carbonates of calcium, magnesium and potassium. Similarly the water should be free from high concentration of chlorides of sodium and potassium, sulphate and phosphates of calcium to avoid drying of plants and to ensure quality planting material. Site near the river, pond, stream, lakes or reservoir ensure availability of water round the year.

(vii) **Labor:** The nursery activities demand huge volume of labor requirement to cater the needs of input transportation, loading and unloading, mother bed formation, sowing, bag filling, transplantation, watering, pest and disease management etc. All these activities demand continuous availability of labor and hence the nursery should be located near to a village or an area where sustainable work force is available for successful establishment and management of the nursery. Now due to mechanization the labor requirement has been drastically reduced. Nurseries should invest on necessary infrastructure particularly the micro irrigation facility, mechanized filling operations etc. to manage the labor problem during unforeseen conditions.

(viii) **Flood:** The site should not be located in the areas amenable for periodic flooding. The site with heavy clay soil is amenable for water logging and flooding and hence this cites may be avoided.

(ix) **Frost:** The nursery site should be avoided when the area is subjected to frost and snow.

(x) **Wind:** The areas subjected to strong winds should also be avoided. Alternatively the area should be protected by establishing one or two rows of wind breaks or three to five rows of shelter belts depending on the wind speed.

2. a. Classify the methods of regeneration of forests. How will you select the trees for seed collection? (2007)

b. Artificial regeneration and Natural regeneration. (2011)

c. Discuss in details the objectives of artificial regeneration. (2014)

There are two main methods of regeneration of forest crops but in practice, a combination of the two methods is also sometimes adopted. Thus, the methods of regeneration of forest crops are

(i) Natural regeneration
(ii) Artificial regeneration and
(iii) Natural regeneration supplemented by artificial regeneration

(i) Natural Regeneration: Natural regeneration is defined as 'the renewal of a forest crop by self-sown seed or by coppice or root suckers'. It also refers to the crop so obtained. Thus, the natural regeneration may be obtained from the following two main sources:

(a) From seed and
(b) From vegetative parts like coppice shoots and root suckers.

When regeneration obtained from seed forms a crop, it is called a seedling crop. When this seedling crop grows into a forest, it is called a high forest. When regeneration obtained by coppice forms a crop, it is called coppice crop and when it develops into a forest, it is called coppice forest to differentiate it from the high forest. Root suckers are, however, not used for large scale regeneration operation.

Condition for Natural regeneration

- Parent tree with sufficient quantity of seeds
- Soil condition suitable for growth of seedlings
- Shelter to safeguard young saplings
- Sufficient sunlight for the growth of sapling

(ii) Artificial regeneration: Artificial regeneration is defined as 'the renewal of a forest crop by sowing, planting or other artificial methods. It also refers to the crop so obtained.' Normally such a crop is called by another term 'plantation' which is defined as 'a forest crop raised artificially, either by sowing or planting.' Sowing refers to direct sowing which is defined as the 'sowing of seed directly on an area where a crop is to be raised as opposed to sowing in a nursery'.

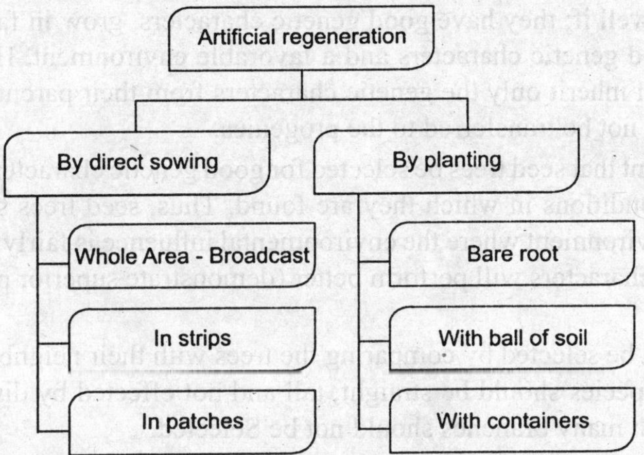

Artificial regeneration is mainly carried out for the following two objects:

(a) Reforestation (b) Afforestation.

Reforestation may be defined as the 'restocking of felled or otherwise cleared woodland' by artificial means. In other words, reforestation is the raising of a forest artificially in an area which had forest vegetation before. On the other hand, afforestation is the 'establishment of a forest by artificial means on an area from which forest vegetation has always or long been absent.'

Objectives of artificial regeneration

(a) To assist or supplement natural regeneration
(b) To change the composition of the existing tree crop
(c) To develop plantations for domestic and industrial utility
(d) To rehabilitate barren and wastelands
(e) To develop agro and farm forestry plantations
(f) To introduce exotics

(iii) Natural regeneration supplemented by artificial regeneration: It is a process which is practiced in forests where natural regeneration is a failure. Under such conditions, natural regeneration is adequately replanted either by sowing or planting by artificial means.

Example: The Sal forest of moist deciduous forest in Central and North India.

The teak forest of dry deciduous forest in Madhya Pradesh, Maharashtra and Andhra Pradesh.

Selection of trees for seed collection

For seed collection, the trees have to be identified and selected to collect seeds of fully matured, fertile and good seeds towards deploying them in nursery practices. The following guidelines are to be followed for selection of trees towards seed collection.

- Proper tree selection will help maintain and improve the quality of the seed collected.
- Seed trees should be selected from the best stand available on a homogenous (uniform) population.
- Tree performance (phenotype) is determined by the genotype (genetic characters), the environment and their interaction.
- Trees will grow well if: they have good genetic characters, grow in favorable environments, or have both good genetic characters and a favorable environment. However, the progenies of these trees will inherit only the genetic characters from their parents. The influence of the environment will not be transferred to the progenies.
- Thus it is important that seed trees be selected for good genetic characters, not for the favorable environmental conditions in which they are found. Thus, seed trees should be selected in a homogeneous environment where the environmental influence is fairly uniform and trees with superior genetic characters will perform better (demonstrate superior phenotype) and be easy to identify
- Seed trees should be selected by comparing the trees with their neighbors. For instance, seed trees for timber species should be straight, tall and not effected by disease or insects. Small crooked trees with many branches should not be Selected.
- Do not select isolated trees, those that are separated from other trees of the same species by a distance of 100 m or more. Isolated trees produce seed by inbreeding which often results in growth depressions and susceptibility to pests and diseases in progeny.
- The selection of trees for seed collection also depends on the purpose of management. For example timber purposes the seeds should be collected from straight, fast growth and thinned branched tree.
- In case of fodder purposes the seeds should be selected from fast growth, multiple branches, more trunk and quick leaf after harvesting
- The mother trees are identified with desired characteristics like vigorous growth, free from disease and insect nearly mature, desired quality of seed bearer, growing in the same species stand

3. Impact of deforestation on ecological aspects? (2007)

Deforestation has many negative effects on the environment. The most dramatic impact is a loss of habitat for millions of species. Seventy percent of Earth's animals and plants live in forests and many cannot survive the deforestation and destroys their habitats. The following are the impact of deforestation in terms of ecological effects.

E-ects of deforestation

(a) **Deterioration and degradation of soil:** The first and foremost impact of deforestation is the damage to soil and the associated ecology. The deforested areas are exposed to all natural calamities like sun and rainfall which causes rapid evaporation and drying up of soil and thus alters the soil structure and texture. Similarly the opened area when exposed to rainfall have more chances of being eroded resulting in loss of top soil. These activities due to deforestation will deteoriate the soil and changes the soil ecosystem.

(b) Impact on water cycle: When forests are destroyed, the atmosphere, water bodies and the water table are all affected. Trees absorb and retain water in their roots. A large part of the water that circulates in the ecosystem of rainforests remains inside the plants. Some of this moisture is transpired into the atmosphere. When this process is broken, the atmosphere and water bodies begin to dry out. The watershed potential is compromised and less water will run through the rivers. Smaller lakes and streams that take water from these larger water bodies dry up.

(c) Loss of Biodiversity: Many wonderful species of plants and animals have been lost, and many others remain endangered. More than 80% of the world's species remain in the Tropical Rainforest. It is estimated that about 50 to 100 species of animals are being lost each day as a result of destruction of their habitats. Many creatures, both plants and animals have vanished from the face of the earth.

(d) Climate Change: Plants absorb Carbon Dioxide CO_2 (a greenhouse gas) from the atmosphere and uses it to produce food (carbohydrates, fats, and proteins that make up trees). In return, it gives off Oxygen. Destroying the forests mean CO_2 will remain in the atmosphere and in addition, destroyed vegetation will give off more CO_2 stored in them as they decompose. This will alter the climate of that region. Cool climates may get a lot hotter and hot places may get a lot cooler. The consequences like global warming and acid raining will be at an alarming rate because forests are identified as one of the potential sink for carbondioxide.

(e) Impact on environment: The deforestation will have serious impact on the environment as a whole thereby changes the ecosystem and the environmental issues. Trees in general are significant for their aesthetic and recreational benefits which will be lost due to deforestation. The revenue generated from ecotourism will be a major threat due to deforestation.

(f) Impact on cultural values: The deforestation may affect the livelihood activities of tribal community who are recognized as saviors of ecology and the ecosystem. Hence deforestation may lead to loss of several tree species which may be of livelihood system coupled with recreational and religious values.

4. a. Why thinning is necessary in forest crops? Discuss in brief the various methods of thinning? (2007)

b. Whatisordinarythinning?Explainbrieflydi-eren tgradesofordinarythinning? (2006)

c. Di-erent grades of ordinary thinning (2009)

d. Describe important objectives of thinning. Di-erentiate crown thinning from ordinary thinning. Write grades of ordinary thinning? (2016)

Thinning is a felling made in an immature stand for the purpose of improving the growth and form of the trees that remain, without permanently breaking the canopy. Thinning is a tending operation carried out in a crop beyond the sapling stage and up to the beginning of regeneration period.

Thinning is necessary for forestry crop due to the following reasons

- Thinning helps to improve the hygiene of forest crop, by removing dead, dying and diseased trees.
- Trees which have lost in the struggle of life are usually weak and very much susceptible to be attacked by diseases and insects. Such trees are likely to be a source of infection for those which are fit in the struggle. Thinning helps to remove such trees which are infected with pest and diseases.
- Thinning helps to improve the growth and development of the retained trees due to availability of more space resulting in less competition and more availability of natural resources.
- Thinning helps to improve the horizontal growth of tree species thereby helps to achieve the diameter increment.
- Thinning helps to develop uniform and regularly managed crop suitable for varied domestic and industrial utility.
- Thinning helps to bring interim benefits through the harvest and sale of thinned materials.

Methods of thinning

The following methods of thinning are normally practiced in all forestry activities

I. Mechanical thinning
II. Ordinary thinning
III. Crown thinning
IV. Free thinning
V. Craib's advance thinning
VI. Numerical thinning

I. Mechanical thinning

This type of thinning is usually applied in young crops or young plantations before the crown differentiation has taken place. There are two general systems followed in this type of thinning *viz.*

(i) **Row-thinning:** Thinning in which trees are cut in line or rows throughout the area.

(ii) **Spacing thinning:** Thinning in which trees at fixed intervals of distance are selected tor retention and all other are cut. The mechanical thinning since applied in the younger crops, it is not possible to identify the vigorous trees because most of them grow with more or less same vigour at an early age.

Example: In teak plantations of Kerala, first and second thinnings carried out at the age of 4 and 8 years are usually mechanical thinning and consists of the removal of alternate diagonal lines or rows of trees reducing the stocking from 2,500 trees per ha to 1250 trees per ha, after the first mechanical thinning. Then the population is reduced to 625 trees per hectare after second thinning with spacing increased from 2m x 2m to about 2.85 m x 2.85m and then ultimately to 4m x 4m after second thinning. The thinning practices followed are shown in the fig 1.

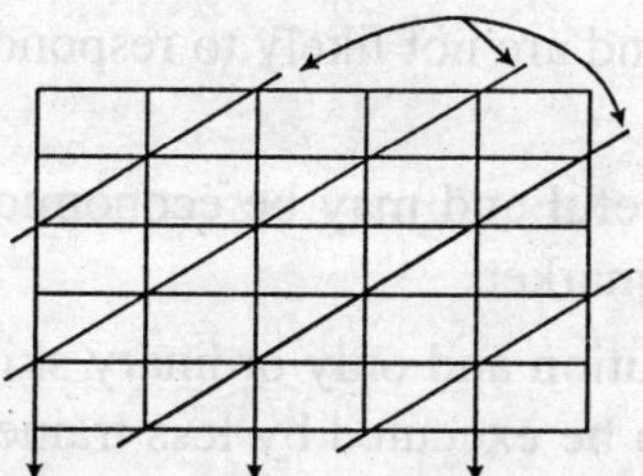

Fig 1. Spacing after the first and second mechanical thinning

Practice of mechanical thinning even in plantations becomes difficult in areas where casualties are more and already gaps exist in the area. The application of row thinning or space thinning is very much limited and depends upon the extent of natural regeneration, as species with dense natural regeneration can be thinned by either of these methods in earlier stages.

Some thumb rule methods have been developed to facilitate in execution by less technical staff. These are based on the relation between the average diameter of trees and the average spacing between the trees.

The relationship between these two is usually a linear relation and can be represented by

$$d = aD + b$$

Where

a and *b* are the two constants,

d the spacing in feet

D the average diameter of trees in inches.

Disadvantages of Mechanical thinning

(i) The main disadvantage of this method is arbitrary selection of the trees in rows or lines which may result in removal of good trees and leaving of poor ones in a crop.

(ii) The method is good for areas with relatively uniform productivity alone.

(iii) However, if the productivity or area is varying, it may result in the removal of the trees which may be growing on better place and are likely to produce more yield. Removal of wolf trees from natural stands or plantations along with thinning is called wolfing.

II. Ordinary thinning

This is also called as 'low thinning' or 'German thinning' or 'thinning from the below'. In ordinary thinning, the trees are removed from lower crown classes. The trees which are removed in this thinning belong to diseased trees of all classes, suppressed, dominated and some dominants depending upon the grades of the thinning. Ordinary thinning has been devised to follow the nature i.e. those trees, which have been unsuccessful in the struggle of existence are removed first.

Advantages and applications of ordinary thinning

(i) It is most suited for light demander species, Eg: chir, teak, sal, sissoo, semal, gamhari, etc. because in light demander species, the suppressed and dominated individuals do not usually retain power of recuperation and are not likely to respond to improved conditions of growth due to thinning.

(ii) This method of thinning is useful and may be economically applied , in species and areas where small size timber has a market.

(iii) The method is simple in execution and only ordinary skill is required for selection of trees, therefore, where thinning is to be executed by less trained persons, it is better to adopt this thinning.

(iv) Removal of lower crown classes includes several diseases and insect infected individuals. Thus, this thinning practice improves the hygiene of the crop.

(v) The method is suitable for areas where there is no danger of soil deterioration as a result of removal of smaller sizes trees.

(vi) The method is useful in forests infested with climbers and where there is risk of crown fire

(vii) In this type of thinning, trees of lower crown classes are removed to accelerate the growth of remaining trees. It is reported in case of several species that growth in dominant trees is stimulated only when some gap is created and root competition is considerably reduced by removing some trees in the top canopy and removal of smaller trees does not significantly affect the growth. Only heavier grades of thinning are likely to produce desired results. Therefore, this method of thinning with proper grade has to be selected taking into consideration the growth, soil loss, etc.

(viii) Cutting of lower crown classes at later stages of development helps to obtain natural regeneration.

Disadvantages of ordinary thinning

(i) This method of thinning rests on removal of trees of lower crown classes. In several areas, troublesome and desirable trees and shrubs in the understorey may develop. In sal, it is reported that dense understorey develops in areas where heavy opening are created in lower crowns.

(ii) This thinning is designed to be executed when trees have been in competition for sufficiently a long time to develop crown differentiation. This means that whole crop including the dominants has already suffered due to adverse effect of competition.

(iii) Sometimes, lower crown classes may be useful particularly when predominants and dominants are not able to utilize the site perfectly. The surplus of the nutrients store is utilized by lower crown classes and is kept in cycling.

(iv) If there is no demand of small trees, it is often difficult to utilize the thinned material economically. When thinned material remains undisposed, it increases fire hazard and does not improve the hygiene of the crop

(v) The removal of trees of lower crown classes may expose the soil. This increases soil erosion particularly in slopy areas.

Grades of ordinary thinning

The following grades of ordinary thinning are practiced.

(i) **Light thinning (A grade):** This is limited to the removal of dead, dying, diseased arid suppressed trees i.e. classes V, IV and III. This grade is of no practical use, but forms a convenient initial stage, specially in comparative research on the effect of thinning on the increment, volume, etc.

(ii) **Moderate thinning (B grade):** This consists in the further removal of defective dominated stems and whips. Branchy advance growth, which is impracticable or not desirable to prune or lop may also be taken out i.e. classes V, IV, III, II (b), I (d) and an occasional I (c). This grade is also of little use in ordinary practice, having little influence on the increment of the remaining stems.

(iii) **Heavy thinning (C grade):** This consists in the further removal of the remaining dominated stems and such of the dominated stems and such of the defective co- dominants as can be removed without making lasting gaps in the canopy, i.e. classes V, IV, III, II and I (b), I (c) and I (d). This is the standard grade in vogue for yield table computation.

(iv) **Very heavy thinning (D grade):** The distinguishing feature of this grade is that it also takes some of the dominants, subject to the same condition of not making any lasting break in the canopy. Trees selected for removal are such that the remaining crop consists of as far as possible trees with good boles and crowns. The trees are well and evenly distributed over the area. The trees which are removed are of class V, IV, III, II, 1(b), 1(c), 1(d) and some 1(a).

(v) **Extremely heavy thinning (E grade):** This is the heaviest thinning that can be done in a crop without making permanent gaps in the canopy. More of the dominant stems even of class 1(a) are removed. This is chiefly used for the research purposes.

In grades C to E of ordinary thinning, the tree classes e.g. V, IV and III classes may be left if removal is of no economic or hygienic value. Intermediate grades of thinning between C and D and between D and E are also recognized.

III. Crown thinning

Crown thinning is a method of thinning which is primarily directed to the dominant trees in a regular crop, the less promising ones being removed in the interest of the best individuals; the dominated and suppressed stems are retained unless they are dead, drying or diseased. This method of thinning is known as French thinning because of its origin in France. This is also called thinning from the above or high thinning to differentiate it from ordinary thinning.

The beneficial effects of removal of lower crown classes on the growth of trees of top crown classes is doubtful and some of the trees of lower crown classes need to be retained for protection of the site and checking invasion of undesired under storey. This method of thinning therefore, aims at removing the inferior trees from among the dominant class. The bulk of trees of lower crown classes, e.g. dominated and suppressed, are ignored and allowed to grow after each thinning. In this method of thinning, trees from upper crown classes are removed in order to favour the growth and development of most promising frees of the same class. The selection of trees to be retained after thinning depends primarily on the health, vigour, form and species. If all the trees are of good health, vigour, form and species, codominants are removed which are interfering with the growth of predominant trees. If there is a choice between relatively poor predominant and promising codominant, the codominants are favoured for retention. In crown thinning, the object is to promote the growth of trees which show signs of success in the competition and promise a good value as a future crop.

Crown thinning is suited for a number of conditions,

(i) It is best suited for moderately shade tolerant species, e.g. sal, deodar, etc. because, in such species, dominated and suppressed trees retain the power of recuperation for a long period.

(ii) It is suited for areas where there is no demand for small sized timer.

(iii) It is suitable for forests where considerable damage due to frost, snow, wind, etc. occur. The vacancy created by the death of dominant trees due to these factors is filled by the trees of lower crown classes.

(iv) The method is especially suited for area where there is danger of deterioration due to removal of trees of lower crown classes.

Advantages of crown thinning

(i) Crown thinning provides better environment for growth and development of retained dominant trees, they grow faster than in ordinary thinning. The production of timber is of big size and of better quality.

(ii) The trees of lower crown classes are not removed under this method of thinning. They keep weeds and shrub growth under control.

(iii) The pressure of trees of lower crown classes results better pruning of side branches. This helps in production of better timber.

(iv) This method of thinning helps in protection of the site and reduces the damages due to frost snow, wind, etc.

Disadvantages of Crown thinning

(i) There is higher root competition for moisture and nutrients adversely affecting the growth of dominants when growth factors are limiting.

(ii) Crown thinning presents problems and obstruction in felling, logging and extraction of thinned material.

(iii) Crown thinning is more flexible method than ordinary thinning. It requires greater skill in execution. It requires more strict numerical check on the trees to be removed.

(iv) The trees of lower crown classes, if, diseased or insect infested, serve as a nucleus of infection in the retained crop. Therefore, closer look on suppressed and dominated trees would be necessary.

Grades of crown thinning

The following two grades of crown thinning are recognized.

(i) **Light crown thinning (L.C. grade):** This consists of the removal of dead, dying and diseased trees with such of the defective, and after them the better dominants, as are necessary to leave room for the further development of the best available trees, evenly distributed over the area. Trees of classes V, IV, 1(d), 1(c) and manyof 1(b) and a few of 1(a) are removed, but not III and II.

(ii) **Heavy crown thinning (H.C. grade):** This grade of thinning pays more attention in favouring the selected best stems by removing all the remaining 1(b) trees, which can be taken without creating permanent gaps, and more of 1(a). The trees removed belong to crown classes V, IV, 1(d), 1(c), most of 1(b) some of 1(a), but not of III and II.

IV. Free thinning

Free thinning is a modification of crown thinning and is also called 'single stem silviculture', or 'elite thinning".

In this method, 'elites' are first selected in number appropriate to the size or age of the crop with special regard to their stem form and to their even spacing. In this method, instead of treating the crop as a whole, attention is concentrated for the provision of optimum growing space for future stems throughout, the growing period of the crop.

The required number of future stems corresponding to the rotation age are first selected and retained more or less uniformly distributed all over the area. In selection of future stems, co-dominants or even dominated trees of perfect stem form and standing in the appropriate position relative to neighbouring elites stems, may be preferred to quite fair but less acceptable dominants. The elites may be marked for easy identification. Once the elites are identified, the balance crop is considered from the point of view of their effect on the elites. If there removal is likely to help the elites, they are cut otherwise they are retained for the protection of the site.

V. Advance thinning

In this method, thinning is carried out before the competition among individual trees has set in; the competition is never allowed to take place. This may cause loss in the total volume production, but ensures maximum growth of the elite trees. This method was suggested by Craib (1939) for plantations in South Africa and usually known as Craib advance thinning.

In the earlier stages of development, there is an intense competition among the plants for water, nutrients, light and space. This competition is regulated by removing the surplus individuals and diverting the resources to the residual trees. Sometimes the trees which are freed take 3 to 4 years for starting faster rate of growth. This leads to the conclusion that greatest increment is possible only when no suppression has yet set in the crop.

VI. Numerical thinning

Various factors such as number of stems per ha and diameter at breast height relationship, N/H^2 (number of stems per ha and height relationship), tree area ratio, bole area, basal area, etc., can be very good guide for the optimum density of the crop at a particular age and site and can be used as a numerical check on thinning,

(i) N/D relationship: The relationship between number of stems per ha and diameter for several Indian timber species has not been found independent of site quality.

Stand density index is the number of trees corresponding to a given standard average diameter. Such numerical parameter on the basis of relationships between number of trees and average diameter are suggested on the basis of standard relationship of number of stems per hectare and average diameter of the crop.

(ii) N/H^2 relationship: For a given site quality, the optimum number of trees per unit area is found to be inversely proportional to the square of the height of the stand i.e. N/H^2. However this method is not used in Indian forests.

5. What is succession? Describe primary and secondary succession along with examples. How the succession is studied in natural forest? (2007)

Succession

Succession is the gradual replacement of one plant community by the other community in the development of vegetation towards a climax is termed as succession. Succession is the gradual supplanting of one community of plants by another, the sequence of communities being termed as sere as each stage seral.

(a) Primary Succession: It is the succession that takes place in an area which never supported any type of vegetation. This succession is otherwise termed as autogenic succession

(b) Secondary succession: It is the succession that takes place in an area after destruction of whole or part of the previous vegetation of the site.

Causes of succession

Succession is a continuous process and many factors and agents are involved in the process. They are categorized into two types.

(i) Primary or Initial causes: The factors responsible for initiating the succession to occur.

(ii) Secondary or continuing causes: The factors responsible for continuous development of plant communities.

Causes	Primary succession	Secondary succession
	Erosion	Climate
Primary or initial causes	Physiography	Topography
	Elevation	Biotic factors
		Migration
		Establishment
Secondary or continuing causes		Aggregation
		Competition
		Reaction

I. Initial causes for primary succession

(i) Erosion: Any primary succession occurs in an area where there is no previous vegetation. This is possible due to sedimentation and deposition of alluvium in the area of succession development. The new alluvium deposit attracts regeneration of grasses and other plant communities. Hence erosion is the primary cause for any succession.

(ii) Physiography: The primary succession depends on the configuration of the land which supports the possible erosion and transfers soil from one area to other area and lead to succession development.

(iii) Elevation: It also plays a vital role in new soil formation due to soil disturbances from high elevated areas to low elevated areas.

Examples of primary succession

(i) Riverain succession

- *Saccharum spontanum*, *Tamarix dioica*
- *Acacia catechu*, *Dalbergia sissoo*
- *Acacia catechu*, *Holoptelia*, *Adina*, *Albizzia procera*.
- *Holoptelia*, *Adina*, *Lagestroemia parviflora*, *Bombax*, *Terminliya belerica*
- *Adina*, *Lagerstromia parviflora*, *Terminalias*, *Shorea*
- *Shorea*, *Lagerstromia*, *Terminalias*, *Adina*.
- If soil is moist, *Trewia nudiflora* and *Toona ciliata* appear in stage
- *Terminalia tomentosa* in stages
- *Syzygium cumini* below the top canopy

(ii) Coniferous succession:

- Shrub association
- *Pinuswallichiana*
- Mixed forest of deodar, spruce and blue pine
- Mixed forest of spruce, fir and deodar

(iii) Estuarine succession

- Mangrove scrub
- Mangrove trees
- Slow growing salt water Heritiera
- Finer brackish water Heritiera forest
- Fresh water swamp forest without Heritiera
- Evergreen or semi – evergreen forest

II. Initial causes in secondary succession

The secondary succession occurs in an area where there is already established vegetation and the vegetation changes due to prevailing climate, existing topography and the biotic factors including man. All these three factors play a critical role in changing or replacing the plant communities.

Continuing causes

Secondary or continuing causes like migration, establishment, aggregation, competition, reaction occurs in areas where vegetation started occurring or already occurred and creates changes in plant community.

(i) **Migration**: The movement of plant or animal communities in mass from one location to other is a major cause for succession in any ecosystem.

(ii) **Establishment**: The regeneration and the associated establishment in a locality is another factor responsible for succession. The migrated seeds and seedlings through factors like wind, water, animals will have to establish in the new area and get stabilized towards succession formation.

(iii) **Aggregation:** The migrated species in the form of seeds and seedlings will get colonized in the new area and create the new vegetation.

(iv) **Competition:** The aggregated or colonized or established vegetation compete themselves for all natural resources like land, water, solar radiation etc. and only the dominant survivors will lead the succession.

(v) **Reaction:** The established plant communities in an ecosystem will react towards climatic, edaphic and biotic factors which thus lead to changes in plant communities or vegetation. The extreme conditions of water availability, wind, pest and diseases etc. create major changes in plant establishment and adaptation and act as a serious factor in succession development.

Examples of secondary succession

- Macaranga spp in moist forest
- In tropical evergreen forest, secondary succession starts with the arrival of *Acrocarpus fraxinifolius,* Callicarpa, Trema and Macaranga.
- In hilly area, secondary succession is often initiated by shrub growth, *Woodfordia fruticosa, Indigofera pulchella* and *Berberis lycium* depending on aspect and altitude.

Succession in natural forests

Succession in natural forests is studied through the changes observe in an ecological community following perturbation that opens up a relatively large space. The earlier studies described the sequence of species that successively invade a site. However the more recent studies have described in other characteristics such as biomass, productivity, diversity, niche breadth etc. To understand the actual mechanism of succession in natural forests can be studied by using both through the changes in species composition and also through the studies on biomass, productivity, diversity etc. The studies already carried out on succession of natural forests in several ecosystem have identified and confirmed the following process of succession.

THE PROCESS OF FOREST SUCCESSION			
⇨	⇨	⇨⇨⇨⇨	**Climax** tree species (very tolerant)
Grass	Pioneer tree species (very tolerant)	**Seral** tree species (increasing in tolerance as replacement continues)	↻
	Stage 1, Stand re-initiation	**Stage 2, Stem exclusion..........Stage 3, Understorey re-initiation**	**Stage 4, Old-growth**
⇦	⇦	⇦ ⇦	⇦
.... Disturbance sets succession back to an earlier stage			

6. Di-erentiate between site quality and site index. (2010)

Site quality

Site quality is the relative productive capacity of a site for particular a species.

Applications of site quality

- Inter relationships can help utilize its growth potential of a site to its optimum level.
- Used to evaluate the site factors / vegetation.
- Site Quality is used as an indicator of site when a specific management objective is being pursued. It may be defined as: An indicator of the productive capacity of a specific area of land for a particular species.
- In the context of timber management, site quality can be defined as:
  The potential of the site to produce timber given a particular species or forest type. The factors of the site and the plants themselves are interacting and interdependent making it difficult to assign cause and affect relationships. There is no single measure of site quality that has been found to be entirely satisfactory.

Site quality can influence

- Species composition (through competition)
- Response to disturbances
- The ease or difficulty of getting desired regeneration
- Cultural practices
- The growth and yield of stand

Site index

Site index is the relative measure of forest site quality based on height of the dominant trees at a specific age.

Applications of site index

- Construct yield table includes age of stand and total height of each tree.
- Average height of dominants used to determine site index.
- Site Index- Height to which a tree will grow under forest conditions in 100 yrs.
- Site index is expressed as height of a dominant tree at a particular diameter.
- More sensitive indicator of site quality than yield table.
- Represent full height attainable.
- Applied in a mixed-uneven aged- poorly stocked stand.
- More economical and quick way of estimation.

Differences between site quality and site index

Site Quality (SQ)	Site Index (SI)
SQ is a broader term	SI is a narrow term
SQ take into consideration edaphic and climatic factors	SI takes into consideration the eadphic factors
SQ is not species specific	SI is species specific
SQ is assessed by CVP index, evaluation of vegetation and indicator species.	SI is assessed by height age correlation of particular species in a site or by site index method

7. Di-erentiate between Net primary productivity and yield (2010)

Di3erences between Net primary productivity and Yield

Net primary productivity	Yield
It is the amount of energy or organic matter fixed or stored at tropic level I.	It refers to the output/harvest from a forest made at a particular time which is measured either in volume or value units
This excludes the energy lost through respiration	It includes all photosynthetic and respiration process.
The productivity is measured in gram/m^2/day or year.	The yield is measured in metric tonne or m^3 / ha.

8. Di-erentiate between gregarious flowering and sporadic flowering in bamboo. (2010)

Gregarious flowering	Sporadic flowering
It is the general flowering, within one or a few years and over considerable areas, of all or most of the individuals of certain species that do not flower annually.	It is the flowering of one or a few culms in a clump or a few clumps in a locality.

After gregarious flowering most plants die	Plants do not die after sporadic flowering.
Example: *Bambusa bambus* *Dendrocalamus strictus*	**Example:** *Arundinaria wightiana, Bambusa lineata, Ochlandra stridula*, etc.

9. Di-erentiate between growth and development of trees. (2012)

Various stages of growth and development of a plant are designated as follows:

(i) **Seedling** – Seedling is a plant grown from seed till it attains a height of about one metre, *i.e.*, before it reaches the sapling stage.

(ii) **Sapling** – Sapling is defined as a young tree from the time when it reaches about one metre (3 feet) in height till the lower branches begin to fall. A sapling is characterized by the absence of dead bark and its vigorous height growth.

(iii) **Pole** – Pole is defined as a young tree from the time when the lower branches begin to fall off to the time when the rate of height growth begins to slow down and crown expansion becomes marked.

(iv) **Tree** – Tree is the stage of growth beyond the pole stage when the rate of height growth begins to slow down and crown expansion becomes marked.

Development of trees

The tree starts its life as a small seedling which grows by increase in length and diameter of its shoot and root. As the shoot grows upwards, it develops branches and foliage. The root grows downward and develops lateral roots and its branches. Thus the seedling grows not only by increase in the size of its shoot and root but also by the formation of new organs. The increase in size is commonly referred to as growth or increment and the formation of new organs is referred to as development. Thus both growth and development are responsible for the change that takes place in a small seedling growing into a tree.

10. Write notes on Grading of seedling (2009)

Grading of seedlings is very essential in order to establish successful plantations. Before shifting the seedlings for plantation establishment, the seedlings have to be graded for their quality and fitness towards plantation development. Generally the seedlings are based on the following factors.

(i) Age
(ii) Height
(iii) Root-shoot ratio or
(iv) Health and vigour of the plant.

(i) **Age:** Age of the seedling is very essential in order to produce quality seedlings. The seedlings are over aged or under aged have to be summarily rejected in order to ensure ideal planting stock. Eg. For teak, stump preparation one year old seedlings are optimal. For Casuarinas and Eucalyptus clonal plants 90-120 days aged planting stock is optimal.

(ii) **Height of the seedling:** Height is one of the important factors, which should be considered while grading the seedlings. The height of the plants depends on age as well as containers used for growing seedlings. For example 30-40 cm height is ideal for the seedling of Melia, Ailanthus, Silver oak etc. for commercial plantations. For urban and road side planting the seedling with 6-8 feet height are raised and graded.

(iii) Root shoot ratio: The optimal root shoot ratio is one of the important factors for grading of seedlings. The higher shoot root ratio or root shoot ratio leads to imbalanced growth and ultimately affect survival and growth increment during plantation operation. Higher shoot root ratio leads to root coiling which may affect intermodal growth and development.

(iv) Health and vigour: The seedlings have to be graded based on health particularly from pest and diseases and vigour through their external appearance. This will ensure high survival and rapid development.

Objectives of the grading

(a) To eliminate sub standard seedlings

(b) To eliminate culled, diseased and damaged seedlings

(c) To segregate the plantable stock into two or three grades so that good seedlings may be planted at important locations. The seedlings should have well proportioned or balanced root-shoot ratio. Medium sized stock between 50 and 60 cm height with a woody root collar show better survival than smaller plants. The grading for root-shoot cannot be done all at one time.

The quality of seedlings can be measured by two factors viz.,

(i) Genetic make-up of the parent stock: The seedlings have to be raised deploying the seeds collected from superior genetic stocks or preferably from seedling and clonal seed orchards in order to gain maximum productivity improvement.

(ii) Physical growth of plant: It is very essential because improper shoot and root growth may lead to stunted growth and problem in survival. The plants with ideal shoot and root ratio will ensure excellent survival and the associated growth and development.

In a holistic perspective grading of seedlings is very essential which decides the success and failure of plantations.

11. a. Why do forest plantations fail? Cite relevant examples. (2010)

b. Failure of forest plantation. (2012)

The failure of forest plantation is the common feature nowadays due to increasing interest on plantations. There are several reasons for a plantation failure which are furnished below.

(i) Site Selection: Site is one of the critical factor for successful establishment of plantations. But in many cases, the site quality is not assessed for its suitability towards various plantation establishment. This resulted in failure of plantations. In tropical countries like India the site exhibited significant variability in terms of structure, texture and nutrient composition which decide the success of any plantations.

(ii) Choice of species: The correct choice of species for the locality is very essential phenomena which decide the success or failure of a plantation. One of the classical example is several industries introduced agar wood (*Aqualaria malachensis*) which is a native species in northeastern states and its introduction in South India has failed due to mismatching of species and climate. Similarly introduction of Pawlounia plantations in Tamil Nadu is another failure due to mismatching of species and factors of locality.

(iii) Lack of proven genetic resources: Another major problem in the failure of plantation is due to the absence of improved genetic stocks suitable for varied agroclimatic regions. In fact country like India wherein a few clones are commercially promoted for establishment of plantations across agro-climatic zones which resulted in failure of plantations. For example, the Eucalyptus ITC 286 and ITC 413 are very productive in Andhra Pradesh but its introduction in Tamil Nadu found to be poor producer and ultimately the plantation failed. Since trees are long rotation species systematic breeding programme has not been carried out for wide range of species which resulted in mismatch of varieties and the locality and lead to failure of plantation. The classical example is non-flowering nature of tamarind plantations in several parts of Tamil Nadu which resulted in uprooting of the plantations. This failure is due to poor genetic resource.

(iv) Adaptation and allocation: The plantations grown tree species are inventorized from various forest species and domesticated without any organized evaluation programme. The growth and development in wild ecosystem is different from its introduction in farm lands due to poor adaptation. This poor adaptation is due to poor allocation of photosythetates which resulted in poor growth and failure of plantations.

(v) Epidemic status of pest and diseases: The monoculturebased plantation establishment has attracted new pests and diseases which sometimes become epidemic in nature. This epidemic nature of pest and diseases kill the tree and ultimately the plantations is a failure.

Example: Eucalyptus gallwasp (*Leptocybe invasa*) is a major pest in Eucalyptus and has turned into an epidemic status during 2008-10. Due to this epidemic status several acres of Eucalyptus were under threat and the plantations failed in terms of poor productivity. Similarly the stem wilt disease in Casuarina caused by *Trichosporium vesiculosam* is a major disease which killed wide range of Casuarina plantations.

(vi) Forest fire occurrences: The occurrence of fire due to both natural and man-made is also one of the reasons for a failure of plantations across the country. The classical example is failure of the Casuarina is due to fire caused during riots and fire in bamboo plantations due to lightening are the evidences for failure of plantations.

(vii) Failure due to Grazing and Browsing: Some of the plantations are failed due to continuous grazing and browsing of both domestic and wild animals. The establishment of *Melia dubia* plantations nearby forest area have attracted both domestic and wild animals towards grazing and ultimately the plantation is failure.

(viii) Failure due to natural calamities: Natural calamities like wind, cyclone, typhoon and Tsunami causes wide range of threat to coastal plantations. There are significant evidences for the complete failure of Casuarina, Eucalyptus and Teak based plantations in coastal districts due to cyclones.

12. What are the di-erent types of containers used in forest nurseries? (2010)

The forest nurseries in the country deploy the following types of containers depending on the locality and the commercial utilization.

(i) **Brick Containers:** They are made of unbaked bricks comprising of clay, farmyard manure and sand in ratio 1:1:1 - A paste is made by adding water and semi- viscous mixture is put in the mould. Planting-bricks are broad at the base and narrowed towards the top. The usual dimension is: Base = (15 cm x 15 cm): top = 10 cm x 10 cm; height = 25 to 30 cm. A small hole having 2 cm diameter and 8 cm depth is made in the centre of the top of the brick. After drying, the hole is filled with usual soil mixture and seeds are sown.

(ii) **Leaf containers**: It is one of the traditional containers which are used in the nursery for production of seedlings. These containers are otherwise called as donna containers and are prepared by using various tree leaves. Eg: the leaves of FICUS, Mango, bauhinia, *butea monosperma*, teak, teleospermum etc are used to grow primary seedlings.

(iii) **Poly bags:** The cheap and convenient method for raising large scale forestry plantations are by using polythene bag plants. They are also easy to keep, handle and transport. It is normally white transparent but black coloured polythene bags are also used sometimes. Before filling polythene bags with light textured loamy mixture of soil, sand and farmyard manure, the various holes are made by punching it. This gives good aeration and proper drainage. The algae and moss are developed inside the polypots due to action of light therefore, it is advisable to keep polypots buried inside the soil. At the same time they need to be shifted twice or thrice to prevent roots going into the soil.

Size of polythene containers: The polythene containers of various sizes viz., 12.5 cm × 20 cm, 10 cm × 15 cm, 20 cm × 30 cm, 10 cm × 20 cm, 10 cm × 15 cm, 12.5 cm × 30 cm, 20 cm × 40 cm, 12 cm × 20 cm etc. are available at different thickness viz., 100, 200, 300 and 400 gauge. Based on the need, suitability and durability the size and the gauge can be selected by the forester.

(iv) **Root trainers:** The various shapes of plastic or metal containers can be designed to prevent root coiling. They can be cubical, cylindrical or of any desired shape. The basic principle is to make the walls corrugated when moulding it in the desired shape and size. The corrugated walls do not allow coiling of roots. The roots are trained along the corrugation. They have holes at the bottom for aerial pruning of tap root, aeration and drainage. These plastic containers are usually made of black plastic. The root are now commercially preferred over other containers due to their advantage of preventing root coiling, the durability of root trainer and the easiness in logistic operation besides acting as a sanitized containers.

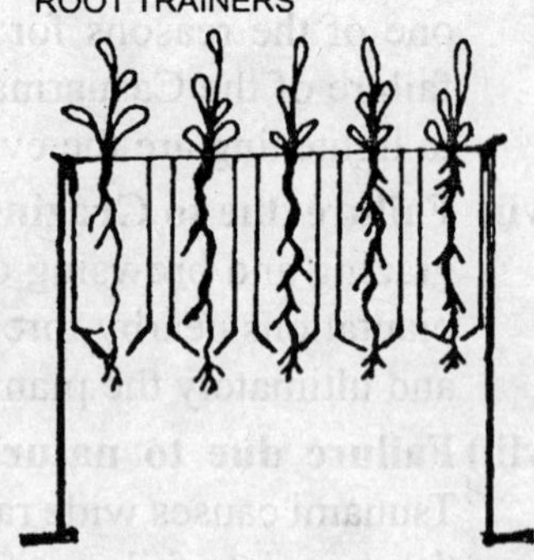

(v) **Ji3y pots:** These are type of containers made from compressed peat and held together with biodegradable net. The advantage of jiffy pots is the plant roots can easily be penetrated and there is no root coiling.

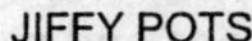

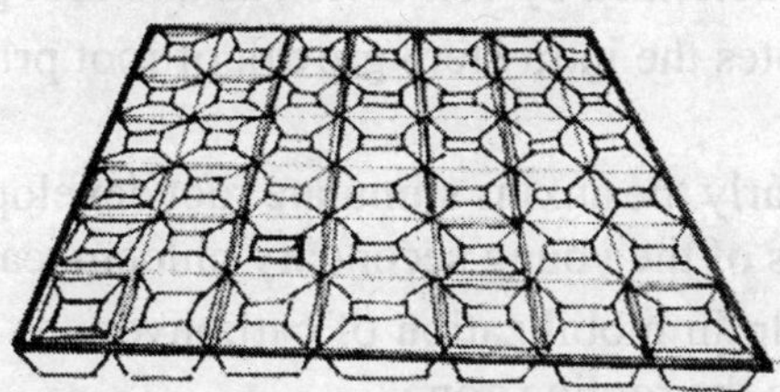

(vi) Clay or concrete pots: One of the traditional containers used in forest nursery is clay or concrete pot which are used to young plants. The containers are heavy and porous hence loose moisture readily. The clay pots are rarely used but concrete pots are used in raising ornamental plants.

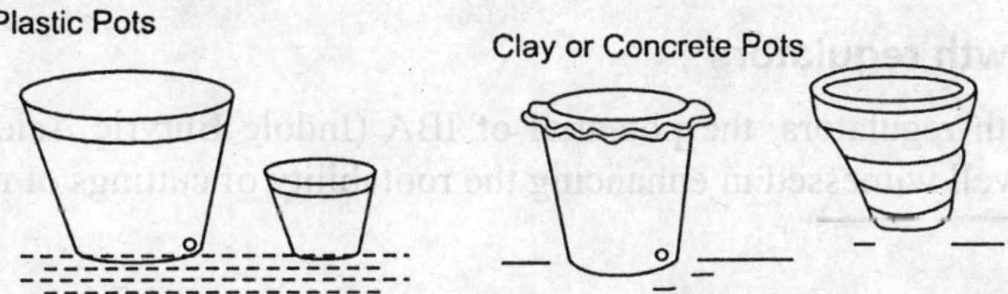

(vii) Fiber pots: these pots are made from wood fibers and coconut fibers which are added with fertilizers and plant protection chemicals. These pots are biodegradable and they are set in the soil along with the plants.

(viii) Paper containers: several ornamental and tree seedlings are successfully raised in paper pots or paper tube containers. These pots are highly amenable for mechanical filling, application of automatic seeders and facilitate air pruning of root system. These containers are biodegradable and gaining significance in commercial forestry operations.

13. Explain the role of growth regulators in rooting of cutting (2010)

Plant growth regulators may be defined as any organic compounds, which are active at low concentrations (1-10 mg/ml) in promoting, inhibiting or modifying growth and development.

The naturally occurring (endogenous) growth substances are commonly known as plant hormones, while the synthetic ones are called growth regulator.

Plant hormone is an organic compound synthesized in one part of the plant and translocated to another part, where in very low concentrations it causes a physiological response. The plant hormones are identified as promoters (auxins, gibberellin and cytokinin), inhibitors (abscissic acid and ethylene) and other hypothetical growth substance (florigen, flowering hormone, *etc*).

Role growth regulators in rooting of cuttings

- Application of growth regulators to cuttings causes metabolic changes during the adventitious root formation, which consists of three successive but independent phases, namely induction, initiation and expression.
  - **Induction phase** comprises of molecular and biochemical events without visible changes.

- **Initiation phase** is characterized by cell divisions and root primordial organization.
- **Expression phase** denotes the intra-stem growth of root primordial and the emergence of roots.

- Growth regulators particularly the auxins stimulate root development by inducing root initials that differentiate from cells of the young secondary phloem, cambium and pith cells.
- Growth regulators also help in mobilization of carbohydrates in leaves and upper stem and accelerate their transport to the rooting zone.
- The oxidative enzymes widely distributed in higher plants have special significance during the rooting.
- Growth regulators change the pattern of IAA-oxidase and peroxidase (POX) activities and have been proposed as being the biochemical markers for the successive rooting phases.

Root inducing growth regulators

Among several growth regulators, the potential of IBA (Indole Butyric Acid) and IAA (Indole Acetic Acid) is very well witnessed in enhancing the rootability of cuttings of most forest trees.

14. Write short notes on seed coating and pelleting (2010)

(a) Seed Coating: It is the process of coating the seeds with polymers of different colours along with nutrients and plant protectants to increase the aesthetic values of the seed with required benefits.

Methodology: Coat the seeds with polykote (3g + 5ml water / kg) after proper dilution. Mix fungicide (Carbendazim 2g) and pesticide (Imidachloprid 2 ml / Kg) with the polykote to increase the resistance to the pest and diseases. Shade the seed before using / storing.

Advantages of seed coating

- Polykote gives uniform colour and shape to the seeds and also increases 5 to 10 percent of germination % and vigour index.
- Improved seedling emergence and vigour.
- Protection against pest and disease during storage.
- Free from environment pollution
- Avoid wastage of seed treating chemicals.

(b) Seed Pelleting: In seed pelleting treatments, the surface of the seeds is coated with some inert material, the sticker, to which chemicals of various kinds may be added.

Advantages of Seed Pelleting

- The incorporation of fertilizer material in the pellet furnishes the young germinated seedlings with needed nourishment.
- Plant growth regulators or stimulants may promote rooting or hasten emergence of the seedlings.

- Fungicides and insecticides are more effective when in direct contact with the seeds.
- Seeds may be protected against rodents by adding unpalatable, repellent or poisonous substances.
- Small seeds become larger and heavier and this improves the scatter pattern in aerial seeding.

Pelleting can also be used to incorporate a brightly coloured substance with the seed, which renders it more visible against the soil and thus facilitates uniform sowing rates. Coating with an anti-desiccant material such as alginate can preserve the seed against drying out, until adequate rain has fallen to permit safe germination.

15. Explain the following point in relation to nursery management: site selection and layoutSelection of Site (2011)

Selection of site plays a vital role in nursery management. The following points should be considered while selecting a site for nursery.

(a) Nursery should be situated as centrally as possible with reference to the area to be planted.

(b) It should preferably be near a Ranger's or forester's quarter for close and regular supervision.

(c) It should be close to the source of supply of labour to do various works as and when required.

(d) In order to have good irrigation facility, it should be situated near of slightly below the source of adequate water supply.

(e) Soil should be well-drained sandy loam. Clayey soils should not be selected as their aeration and drainage is poor and they are liable to crack during summer.

(f) As far as possible, the nursery should be made in a newly cleared forest land with no overhead shade except in hot dry localities where it is beneficial. Some side shade from the south is, however, usually advantageous.

(g) It should never be located in a natural blank.

(h) In the hilly areas a gentle slope not exceeding 5° should be selected for establishment of nursery. If the slope is steeper, cost of terracing increases.

(i) In the western Himalayas, northerly aspect should be preferred but towards higher limits of the altitudinal zone of the species, southern aspect may be preferred.

(j) In the eastern Himalayas, northerly aspect are suitable only up to an altitude of 1200 m; beyond that they become too cold and wet and so westerly or south westerly aspects are preferred.

(k) Altitudinal zone of the species should also be kept in view while selecting site for nursery in hills.

Layout

Lay out nursery is very essential in order to house and locate the various needs of nursery establishment.

(a) Layout of nursery should be done with demarcation. After demarcation, nursery should be divided into rectangular blocks by permanent paths, about 1.8m wide. These blocks are further divided by paths 0.9m wide into smaller blocks which are subdivided into rectangular nursery beds by paths 0.5 m wide.

(b) While laying out a nursery, location of irrigation channels should also be kept in view so that they run along paths. In the hills, nursery beds are made after terracing the area. The width of terraces should be sufficient to accommodate at least on bed in width and paths on both sides of it. It should thus be atleast 2m wide. The beds should be laid out with their lengths east to west so that they can be shaded against frost and sun without difficulty.

(c) In the nurseries made for extensive mechanized plantations, there should be provision for wide roads on which tractors with trailors can be operated. Thus each nursery bed should have a tractor road on one side of its length and an irrigation channel on the other. In other words, there should be a tractor road after every two beds so that polythene bag plants can be loaded from the bed directly on the trailor without any difficulty.

(d) The layout should appropriately locate mother bed, greenhouse, hardening chamber and open nursery if they are planned for clonal seedling production.

Points to be considered for a good nursery layout

(i) Well defined and well protected boundary by fencing, gates, clear areas and fire buffers.
(ii) Roads and inspection paths.
(iii) Store room, office building, labour huts, garage etc.
(iv) Platforms for natural weathering of seeds and other seed treatment facilities.
(v) Seed germination and central growing area as nursery beds & additional area set aside for expansion if demand increases.
(vi) Irrigation and sprinkler systems.
(vii) Drainage systems.
(viii) Compost pits.

16. Explain the following point in relation to nursery management: soil working (2011)

Soil working is very essential in nursery management as it is the growing media which decides the growth and development of plants. The following soil working principles need to be considered for efficient nursery management.

(a) The soil should be very clear and all stumps, bolders, large pebbles and grass roots are to be removed.

(b) The best soil medium would be a mixture of loam and sand. Suppose in the nursery locality the available soil is not good means the loam and sand should have to be brought from outside and mixed with the soil.

(c) The clayey soils should be avoided as it affects the drainage and aeration which will affect the growth and development of nursery seedlings.

(d) The soil texture should be management by efficiently mixing sand and fully rotten farmyard manure mix.

(e) The soil should be prepared in such a way that it is not attracted by ants and termites. In case, if they are present application of 5% dursban should be added.

(f) The pH of the soil can be corrected by adding charcoal dust and ash.

(g) The soil should be solarized for protecting them against insects and microbes.

(h) The soil should be prepared and mixed with wood ash, farmyard manure and compost in order to improve the structure, texture and quality.

(i) The fertility of the soil can be increased by adding farmyard manure, fertilizers and other coir composites.

(j) The soil acidity can be managed by adding lime.

(k) The addition of bio-fertilizers like VAM improves the soil fertility.

(l) Soil working for raised bed, sunken bed, containers filling, transplant beds, drainages, flooding pattern and soil erosion control measures.

17. Explain the following point in relation to nursery management: Methods of raising nursery stocks (2011)

The nursery stocks are raised through the following approaches.

(i) Preparation of naked seedling in the bed

(ii) Raising polythene container based nursery stock production

(iii) Root trainer based clonal nursery stock production

(iv) Tissue culture based nursery stock production

(i) Preparation of naked seedling in the bed: There are species like Casuarina nursery stocks are produced predominantly through naked seedling production. For this purpose the sunken beds of size 10 x 1 m are dug up and filled with soil, sand and farmyard manure. Then ½ kilo seeds are sown in the nursery bed and the seedlings are raised. Weeding, hoeing and irrigation are done regularly in order to produce healthy seedlings. These seedlings are pricked out either with ball of earth or naked root for planting out.

(ii) Raising polythene container based nursery stock production:

(a) Sowing in the mother bed and transplantation : This method is used predominantly for producing seedlings of all major tree species like Teak, Terminalia, Neem, Dalbergia, Eucalyptus, Melia, Silk cotton etc. In this method, the seeds are sown in the mother bed during April – May and the germinated seedlings are transplanted in the polythene containers filled with soil and farmyard manure. The transplanting is to be done carefully, so that the roots cannot get damage. After transplanting the seedlings have to be watered regularly and applied with fertilizers and plant protection measures upto a period of six months for planting out in the field.

(b) Direct sowing in the container: Several tree species like Acacia, Prosopis, Subabul, Amla, Sterculia etc. are large sized seeds and they can be sown directly in the polythene bag after application of necessary seed treatment process. The seeds started germinating in about of 15 days and the seedlings have to be maintained for the period of 6-12 months depending on the objective and need. During this period the seedlings have to be irrigated regularly and protected with necessary plant protection measures. Periodic shifting of bag is necessary to avoid root penetrating into the earth.

(iii) **Root trainer based clonal nursery stock production:** Recently due to increasing attention on the promotion of industrial forestry and agroforestry plantations there is a need to produce massive nursery stock production with genetically improved planting material. For this purpose, trees such as Casuarina, Eucalyptus, Poplars, Bamboos, Melia, Subabul etc. are mass multiplied through clonal technology using apical shoot buds or nodal cuttings. These vegetative propagules are obtained from clonal mother garden and after due disinfection they are planted in root trainers filled with coir compost and grown under mist chamber with controlled environmental conditions. The rooting of cuttings will be induced within 21 days and the rooted plants are shifted to hardening chamber and then to open nursery for secondary hardening and planted out in the field.

(iv) **Tissue culture based nursery stock production:** The recent developments in science and technology have attracted attention for production of nursery stock through tissue culture technology for mass multiplication of superior, endangered and endemic genetic resources. Through this technology the explants (any vegetative plant part) can be used to induce large number of shoots which in turn sub cultured in an artificial medium (predominantly modified MS medium) and grown under controlled environmental conditions (2000 – 4000 lux light intensity and 60 – 80% RH). The invitro grown plants are further hardened both in hardening chamber and open nursery and then planting out in the field. Species like Teak, Sandal, Eucalyptus and Bamboos are commercially growth through this approach.

18. Explain the following point in relation to nursery management: plant protection measures. (2011)

The primary aim of forest nursery is to produce healthy planting stock and this can be obtained by

(a) **Better nursery management practices:** The best management practices include providing adequate sanitation facilities coupled with cultural practices in order to avoid the housing of pathogens and to create spic and span in the nursery areas.

(b) **Control of biological organisms:** The biological organisms are the most damaging factor causes large scale mortality and are poor growth of seedlings. These include insect, pests such as grubs, cut worms, crickets, termites, leaf feeders, sap suckers and non insect pests like Nematodes and rodents. Hence these biological determinants need to be controlled through adequate plant protection measures.

The following are the plant protection measures for management of a nursery.

(i) Better sanitation of the nursery area

(ii) Adoption of suitable cultural practices

(iii) Integrated pest and disease management

Similar to nursery pests, the nursery seedlings are also affected with microorganisms. The populations of bacteria and fungi are increased due to improper soil management, poor soil fertility, too much of watering are large scale introduction of soil organisms. Therefore the nursery soil requires skillful management as a pre-requisite for controlling diseases.

Diseases like damping off, wilt, root rot, colar rot, twig blight and other foliar diseases such as leaf spot, leaf blight, powdery meldew are very common and causes considerable damage to the seedlings. The pathogens may be seed borne, soil borne or air borne in nature. The pre-treatment of seeds with fungicides such as captan (0.2%) or emisan (0.1%) can control the diseases.

Alternatively the grown up seedlings can be surveyed for the presence of pest and diseases periodically and suitable control measures through fungicidal application can be practiced in order to control pest and diseases. Regular treatment of bavistin at 0.1 to 0.2% will help to control the nursery diseases and spraying monochrotophos will help to control the nursery pests.

19. Explain the following point in relation to nursery management: Nursery register (2011)

Nursery Register

It is a journal containing all information related to nursery activities. For any nursery maintenance of register is the mandatory in order to produce healthy seedlings in a required number and in a required period.

Each nursery should have a nursery register giving full information about the nursery. In the beginning it should give the general particulars about the nursery, viz., name of the nursery, location, year of formation, are in hectares, soil, method of irrigation, type fencing with initial cost, soil working done at the time of initial formation with cost, manuring, its details and cost, rent if any.

The nursery register should have the following information for easy management, inspection and monitoring.

- The nursery register should clearly demarcate the location, area, map of nursery, etc.
- The year of establishment of the nursery with the capacity in terms of area and size
- The expenditure made towards establishment of nursery in the form of clearing, leveling, soil working, fencing, etc. should have to be mentioned.
- The nursery register should have details of all the fixed assets like buildings, borewell, vehicles, machineries, watering arrangement etc.
- The nursery should contain yearwise and itemwise – fixed and variable expenditure made on annual schedule of operations.
- Records of seed collected and purchased, number of beds, polybags, number of stocks produced and their use is recorded.
- Details of watering, manuring, weeding, diseases, insecticides, etc. applications have to be mentioned with date and details.
- Name of the staff in-charge of nursery, number of workers including watch and ward have to be maintained in the nursery.

- It should have a provision for writing inspection notes, comments and their compliance.

The counting of nursery stock should be made on 1st of January and 1st of June every year.

20. Write short notes on Cultural operation (2011)

Cultural operation is defined as 'the operation, as a rule not directly remunerative, undertaken to assist or complete existing regeneration, to promote the proper development of the crop or to minimize the after-effects of felling damage.

It, therefore, includes subsidiary felling, weeding, cleaning, unremunerated improvement fellings, and thinning in groups of advance growth, girdling or poisoning of unwanted growth, climber cutting and even pilling of felling debris, and controlled-burning but usually not other ground operations nor pruning. It is generally associated with silvicultural systems relying primarily on natural regeneration.

Major cultural operations

(i) Removal of marked trees left unfilled provided their felling is still considered necessary.

(ii) Removal of trees so badly damaged in felling that their retention is of no use.

(iii) Cutting back malformed and ill-developed advances growth if it can give a good coppice shoot.

(iv) Removal of inferior species or weed growth where it is interfering with the growth of the economically more important species and where advance growth has been coppiced to admit light for the coppice to develop.

(v) Thinning of even-aged groups of poles if it has not been done with felling.

(vi) Climber cutting is also one of the important cultural operation which will help to improve the growth and development of a tree.

21. How snow a-ect the forest vegetation (2013)

Snow

At higher altitudes, the decrease in temperature results in precipitation taking the form of snow. The amount of snowfall and the period during which it remains on the round depends upon temperature and the amount of winter precipitation. Snow occasionally falls down to 1200 m in north western and central Himalayas but it stays only above 2000 m. In the eastern Himalayas and the south, the altitude to which snow falls, is higher.

Injurious E-ects of Snow

(i) Snowfall results in the mechanical bending of stems of trees. Snow gets accumulated on the uphill side of the young saplings and poles and causes them to bend outward at the base. This curve is maintained even when the poles develop into trees and can be seen on all mature trees. The bend makes this portion of the tree unfit for utilization. As this portion of the stem has maximum diameter, there is a great loss in volume.

(ii) Accumulation of snow on the crowns of trees results in breaking of branches and tops of trees. Kail is most susceptible to snow break and deodar comes next.

(iii) Sliding snow not only causes erosion but also uproots trees. Often snow slides wipe out strips of forests along their course. The folds of the hills, in which snow slides regularly, are devoid of vegetation and thus a fair amount of area is without tree growth.

(iv) Snow shortens the period of vegetative growth.

(v) Snow is reported to favour the growth of certain fungi, e.g. *Fomes* and *Trametes*.

Beneficial E-ects of Snow

(i) Snow influences the distribution of deodar, fir and spruce and their best forests are found in places of heavy snowfall. Heavy water snowfall is essential for satisfactory natural regeneration of deodar.

(ii) Snow is the source of water in streams and rivers. All perennial rivers and streams have their origin from snow glaciers.

(iii) Snow acts as a blanket, prevents further drop in temperature and thus protects seedlings and other vegetation from the damaging effect of excessive cold and frost.

22. Explain elite thinning are often di-cult to execute (2009)

Free thinning or elite thinning

Elite thinning is defined as a method of thinning in which attention is concentrated on evenly spaced selected stems (called elites or alpha-stems) which are retained until maturity or till the last thinning or two, thinning being directed to the removal of other stems hindering their optimum development. It is applied to regular crops.

This method of thinning has been applied in India with slight modifications under various names, such as, Elite thinning of Kerala, Tamil Nadu and U.P.

Elite thinning is difficult to execute because of the following reasons.

(i) It is not possible to select trees at an early stage as differentiation of growth and crown height, in several cases, does not set in.

(ii) The plantation during the sapling and pole stages is usually subjected to heavy damage by wild animals, particularly elephants and because of the conspicuous paint ring put on the elites, they attract the attention of the curious elephants. Thus, in young plantations, the elites were extensively damaged.

(iii) The heavier thinning around the elites compared to non-elites caused the trees to grow into 'wolf tree'.

(iv) As one elite was to be selected from every quadrant (10 m x 10 m), it sometimes become obligatory to select a bad elite because the whole lot consists of bad individuals.

(v) Between the thinning interval, elite may be attacked by some disease or may be damaged by wind and there may be many suitable substitute near to it.

(vi) The thinning in the 'buffer stock' between elites is likely to be neglected, which may result in islands of congested stock, growing in suppression, which ultimately affects the intermediate yield.

(vii) The practice is more time consuming and costly.

23. Explain the specific problem of costal land and hot deserts (2014)

Coastal land

Soils having higher proportion of soluble salts tend to affect adversely the growth of plants.

The salt level in saline soils exceeds a limit of 4.0 dSm^{-1}. Mostly these soils are dominant with chloride and sulphate salts. These salts are neutral salts and hence the pH of these soils may not exceed 8.5.

Saline soils are formed through a soil forming process called salinization in semi arid and arid zones. Salinization refers to accumulation of soluble salts in the soil surface horizons.

E-ects of soil salinity

The characteristics feature of saline soil is white encrustation on surface of soils due to evaporation of water to atmosphere leaving the salts on surface of soils. Presence of salts leads to alteration of osmotic potential of the soil solution. Consequently water intake by plants restricted and there by nutrients uptake by plants are also reduced. In these soils due to high salt levels microbial activity is reduced. Reduced microbial activity results in slower decomposition of organic matter. Slow decomposition leads to slow nutrient availability particularly nitrogen and sulphur. Due to osmotic potential alteration water from plants cells moves to soil and plants are affected due to dehydration. As a result drying of leaves and finally death of plants are commonly seen in saline soils. Apart from above effects specific ion effects on plants are also seen due to toxicity of ions like chloride, sulphate etc.

Hot deserts

Hot deserts are the areas where the seasons are generally warm throughout the year and very hot in the summer. The winters usually bring little rainfall. Temperatures exhibit daily extremes because the atmosphere contains little humidity to block the Sun's rays. Desert surfaces receive a little more than twice the solar radiation received by humid regions and lose almost twice as much heat at night. Many mean annual temperatures range from 20-25° C. The extreme maximum ranges from 43.5-49° C. Minimum temperatures sometimes drop to -18° C.

Rainfall is usually very low and/or concentrated in short bursts between long rainless periods. Evaporation rates regularly exceed rainfall rates. Sometimes rain starts falling and evaporates before reaching the ground. Rainfall is lowest on the Atacama Desert of Chile, where it averages less than 1.5 cm. Some years are even rainless. Inland Sahara also receives less than 1.5 cm a year. Rainfall in American deserts is higher – almost 28 cm a year.

Soils are course-textured, shallow, rocky or gravely with good drainage and have no subsurface water. They are coarse because there is less chemical weathering. The finer dust and sand particles are blown elsewhere, leaving heavier pieces behind.

Canopy in most deserts is very rare. Plants are mainly ground-hugging shrubs and short woody trees. Leaves are "replete" (fully supported with nutrients) with water-conserving characteristics. They tend to be small, thick and covered with a thick cuticle (outer layer). In the cacti, the leaves are much-reduced (to spines) and photosynthetic activity is restricted to the stems. Some plants open their stomata (microscopic openings in the epidermis of leaves that allow for gas exchange) only at night when evaporation rates are lowest. These plants include: yuccas, ocotillo, turpentine bush, prickly pears, false mesquite, sotol, ephedras, agaves and brittlebush.

24. Why does height of a tree considered better criteria for site selection than its diameter? Discuss. (2014)

Height of the tree is considered better criteria for site selection than its diameter because height is perhaps the single parameter best related to productivity of site of a given species. Though it is dependent on other factors but still it is one of the robust factors to indicate the site quality rather than diameter. Several studies have indicated a close correlation between variations in height growth due to variation in site quality and advocated the use of height as an indicator of site quality.

In India, the site quality classes of various tree species are determined and established based on the top height of trees. The tree height is influenced by the meristematic cells at the tip of the shoot which ensures the vertical growth and leads to maximum height which is directly influenced by the local site quality. Though site the tree height growth depends on other factors like crop density and genetic factors but the major influence is due to site quality. Hence height is considered as one of the important factor for assessing site quality than the diameter.

It is an easy measure which decides the various management issues to improve the factor to achieve desired growth of a stand. Height has a minimum impact on stand density and composition compared to diameter. It does not have major impact due to thinning.

It is a measure of productivity of site to particular species. It will be difficult to get the volume immediately for a stand and also the diameter. But the height is more convenient measure to get some reliable level of the site index and site quality is easy to adopt in practical way than other measures.

For a given site and the site factors the tree height reaches maximum at earlier stage than other parameter such as diameter and volume. For easy practical management purpose, forester uses the height to decide the site quality and it helps him as an easy and practically approachable method.

Height growth of trees is attained maximum at earlier stage and it became less influenced in the production / yield growth of volume at the end of the growth period. During the later stage of growth and development, the height increment is very low and only diameter increment is active and hence influence of height on overall volume increment is less.

25. Explain the importance of snow in regeneration of *Cedrus deodara*. (2014)

- Snow influences the distribution of deodar particularly in the higher altitudes. The best deodar forests are found in places with heavy snow fall. The heavy winter snow fall is very essential for satisfactory natural regeneration deodar due to the following actions.

- The snow cover is a source of water and melts during summer and increases the soil moisture regime and the same is utilized by deodar for its survival growth and development.
- Snow acts as a blanket and prevents further drop in temperature and thus protect the young regenerants.
- It also acts as a barrier against excessive cold and frost and helps to enhance the germination and protects the already germinated seedlings.

Hence snow is a very important climatic factor helps in regeneration of *Cedrus deodara* and plays a vital role in silvicultural management of this important timber species.

26. Describe regeneration survey (2014)

The natural regeneration obtained in the forest under the various systems is generally hidden in grass and shrubs covering the forest floor and therefore its ocular estimates are generally inaccurate. In the absence of 100% enumeration, reasonably accurate assessment of natural regeneration can be obtained by regeneration survey.

Regeneration survey is 'a survey for the assessment of established and unestablished regeneration generally by sample enumeration.'

The regeneration survey may be carried out at the time of revision of working plans with the following objects:

(i) To compare natural regeneration in any regeneration area at the end of the working plan with that in the beginning to evaluate the effects of operations carried out during the working plan period.

(ii) To prepare a stock map of any area proposed to be regenerated to prescribe correct Silvicultural treatment for various parts on the basis of the status of the regeneration in them as well as to serve as a basis for comparison the end of the plan.

Sampling patterns and intensity

Regeneration survey can be done along with the enumeration or independently. In the flat terrain, the regeneration survey may be carried out in linear strips or line plot surveys while in the mountainous terrain, it can be carried out by topographical units. At present, however, regeneration surveys are carried out in flatter terrain found in forests managed under shelterwood system. The intensity is generally from 2 to 4%.

Methodology

A base line is drawn on the base of the map and parallel survey lines are drawn at right angles to the base line 200m or 100 m apart after selecting the position of the first line by random sample.

Regeneration survey is then done in 2m x 2m plots on either side of the survey line. In this survey, it is sufficient to record the occurrence of the most promising category of regeneration only and regeneration survey followed in sal forest is furnished.

Categories of Regeneration survey followed in Sal

	Category of regeneration (1)	Symbol (2)	Explanation (3)
1.	Established regeneration whose height should be 2.5 m or more and d. b. h. 10 cm	e	Symbol e indicates that at least one such plant is present, which is sufficient to stock the quadrat.
2.	Woody shoot, which is not established but which is large and vigorous and so expected to become established early	w	In the absence of e, w indicates that there is at least one woody shoot which is sufficient to stock the quadrat.
3.	Woody shoot which has been browsed	w^+	-
4.	Whippy unestablished seedling whose height is more than 50 cm	u^+	In the absence of e and windicates that there are more than one unestablished whippy seedlings more than 50 cm height in the quadrat.
5.	Whippy unestablished seedlings whose height is more than 50 cm	u	In the absence of e, w andu$^+$, u indicates that there is one unestablished whippy seedling more than 50 cm in height, in the quadrat.
6.	Sub-whippy unestablishedseedling whose height is less than 50 cm	s^+	In the absence of other symbols (i.e., e, w, u^+, u), s^+ indicates that there are more than one sub-whippy seedlings in the quadrat.
7.	Sub-whippy unestablished seedlings whose height is less than 50 cm.	s	In the absence of other symbols, s indicates that there is one subwhippy seedling in the quadrat.
8.	Recruit (Current year's seedling).	r	In the absence of other symbols, r indicates that there are only current year's seedlings in the quadrat.
9.	Blank	o	o indicates that there is no regeneration in the quadrat.

27. Comment upon the dieback phenomenon in *Shorea robusta*, it is a problem or an adaption? (2015)

Dying Back of Sal Seedlings

Die back is an adaptation mechanism which has been experienced over years. Generally die back means dying usually backwards from the tip of any position of plant leading to death of shoots and roots remain alive.

Normally the die back occurs in sal after good seed year with commencement of rain. Immediately after rain sal seeds germinates steadily and huge volume of natural regenerated seedlings can be seen in the forest floor. After this gregarious regeneration, the seedling start dying especially during the rainy season duc to dampness followed by wilting. Some of the seedlings dry during winter season.

Due to the influence of cold and some die during summer due to the influence of high temperature.

These adverse climatic factors leads to death of whole shoot as only a part of shoot but root position will be still active. In the following year one or two wheepy shoots are produced and they also experience the same effects if the above factor continues.

In each and every year the root portion starts developing new shoots and if the natural calamities are absent in one particular year the shoots develop into a fully matured stage and there after they do not die. Sometimes the development process takes 40 years to establish thc full tree. Hence die back seems to be a problem for sal establishment but it is the natural process of sal regeneration get adopted to it.

Factors responsible for die back

The sal die back is caused due to

(i) Dense overhead canopy
(ii) Inadequate light and other growth factors
(iii) Competition due to weeds
(iv) Heavy accumulation of leaf litter
(v) Accumulation of carbon-di-oxide
(vi) Adverse climatic factors
(vii) Drip
(viii) Intensive grazing and browsing

28. Describe Raunkiaer's Life forms (2012)

A Danish ecologist, Christen Raunkiaer (1876-1960), proposed a classification of plants based on the position of these parenting organs in relation to the soil surface. This classification is known as Raunkiaer's Life Forms. The classification by Raunkiaer is as follows:

Phanerophytes

A tall, woody or herbaceous perennial with resting buds more than 25cms above soil level, e.g. deciduous trees and shrubs.

Chamaephytes

Perennating buds or shoot apices are borne very close to the ground, or a small, woody herbaceous plant having resting buds not more than 25cms above soil level, e.g. rockrose and bilberry.

Hemicryptophytes

Plants with resting buds at or near the level of the soil, e.g. daisy.

Cryptophytes

Plants with resting buds lying either beneath the surface of the ground as a rhizome, bulb, corm, etc., or a resting bud submerged under water.

Cryptophytes are sometimes divided up into one of the following:

- **Geophytes** resting in dry land, e.g. Crocus, Lily-of-the-Valley.
- **Helophytes** resting in marshy, lake or pond edges, e.g. Reed Mace.
- **Hydrophytes** resting by being submerged under water, e.g. Frog bit. Some over-winter as turions.

Some aquatic plants develop special winter food storing buds, known as Turions. A turion is an underground bud or shoot which develops into an aerial stem; a winter bud, characteristic of many aquatic plants, that contains food material and becomes detached from the parent plant, either floating or resting at the bottom of the water until favorable conditions stimulate its development and growth into a new plant.

29. What are the ecological aspects for selecting the tree species? Discuss (2012)

The term environment denotes the circumstances or surroundings of the region. In essence, everything external to an organism is its environment. Environment of a tree means everything other than the particular tree. Therefore, environment is the whole sum of the surroundings, external conditions within which an organism, a community or an object exists. Environment of a forest means all those factors and surroundings which affect the forest. Forest environment is very complex in nature. It can be understood by the influences which it produces on the forests.

Factors of Site

Factors of site may be called as habitat factors or locality factors. These factors affect the vegetation considerably. These factors control species composition, density structure, rate of growth, etc. Some of these factors affect the forests directly while others have indirect effects. For example, rainfall affects the forest directly. It also affects the development (They interact together and affect the forest of a particular locality). These factors of site or locality may be grouped as follows:

(i) Climatic factors
(ii) Edaphic factors
(iii) Physiographic factors
(iv) Biotic factors

(i) Climatic factors: The climatic factors which influence the vegetation are solar radiation, light, air temperature, precipitation, atmospheric humidity, wind, lightening and atmospheric injuries. The climatic site factors are very important as these factors determine the distribution of vegetation. These factors determine the climatic regions and these climatic regions have different vegetation types. The distribution of forest in a climatic region is further controlled by other site factors, e.g. edaphic, physiographic and biotic.

(ii) Edaphic factors: Edaphic factors relate to soil conditions. Some of the important factors relating to soil are depth, parent material, soil composition, soil texture, structure, porosity, soil moisture, soil reaction, drainage, humus, soil temperature, CN ratio, micro-organisms, etc.

(iii) Physiographic factors: The physiographic factors include the form and structure of the land surface, altitude, configuration, topography, slope and exposure. Physiographic factors affect climate and soil and thus affect the development of vegetation.

(iv) **Biotic factors:** Biotic factors include conditions which relate directly or indirectly to plant or animal agencies. The factors of site relating to climate, soil and physiography are responsible for the composition, form, growth and the development of forests. Biotic factors can retard, arrest or change or enhance the course of development of forests. The important biotic factors are plant, animal and man.

30. Enumerate the factors which decide the choice of species for plantation? (2013)

The choice of species for plantation establishment depends on the following factors.

(i) **Climate and micro climate:** For any plantation development climate and micro climate play a vital role in deciding the choice of species. The climate in terms of temperature, rainfall and humidity play a vital role in selecting the species choice. For example, Casuarina is more adoptable to coastal climate than to the inland climate. Similarly Silver oak experiences rapid growth under cool climate than under hot climate.

(ii) **Soil factor:** Site is another major factor which decides the choice of species for a plantation. The soil based on its structure, texture, nutrient availability decides the success of any species of a plantation establishment. The following species are selected based on the site.

Soil	**Choice of species**
Black cotton soil	*Acacia nilotica, Prosopis juliflora, Leucaena leucocaephala, Albizzia lebbeck*
Laterite soil	*Acaica auriculiformis, Bamboos, Eucalyptus, Teak*
Saline Alkaline soil	*Prosopis, Neem, Thespesia*
Alluvium soil	*Teak, Terminalia, Pterocarpus*
Sandy Loam	*Casuarina, Melia, Kadam*

(iii) **Objective of management:** The choice of species depends on the objective of management like timber production, plywood production, pulpwood production, fuel wood production, oil seed production etc. The following are the choice of species for each objective.

Management objective	**Choice of species**
Timber production	*Teak, Dalbergia, Pterocarpus, Terminalia*
Pulpwood production	*Casuarina, Eucalyptus, Bamboos*
Plywood production	*Poplar, Eucalyptus, Melia*
Matchwood production	*Ailanthus, Albizzia, Poplar*
Fuel wood production	*Prosopis, Acacia*

(iv) **Market demand:** The choice of species depends on the market demand. In India, the demand for wood and wood products are increasing at an alarming rate and there are over 25 species are commercially cultivated across the country. However not all the species are getting adequate market demand which leads to failure of few plantations. Hence successful plantation establishment depends on the species which has adequate market demand.

(v) Rate of growth: The choice of species for establishing plantation depends on the rate of growth. Based on growth rate the trees are classified into short duration, medium duration and long duration crop. The stakeholders depending on their interest decide the tree crop

Duration	Choice of species
Long duration	*Teak, Dalbergia, Pterocarpus, Terminalia*
Medium duration	*Poplar, Eucalyptus, Melia*
Short duration	*Casuarina, Eucalyptus, Bamboos*

(vi) Ease of establishment: The choice of species depends on the easiness in terms of establishment. For this purpose, several trees are preferred commercially due to their easiness in establishment, survival, growth rate and the associated management options.

(vii) Utility: The choice of species depends on the utility of crop like timber, pulp, paper, ply, fuel wood, match splints, sports goods, etc.

(viii) Cost and returns: The success of any plantation depends on the species which involve low cost and maximum returns. The species which accrue the cost benefit ratio of 1: 3.5 are highly preferred in today's plantation establishment programme.

31. Explain why after gregarious flowering bamboo always dies. (2014)

Some of the Bamboo species are manocarpic in nature which results in gregarious flowering followed by death.

The science behind death after gregarious flowering is not well understood and there are no proven evidences for this cause. But generally the gregarious flowering requires huge volume of anagy and the bamboos divert all its energy for the following phase which leads sessation of vegetative growth.

This may be one of the possible reasons for the death after gregarious flowering.

However, there are two school of hypothesis for the death of bamboo after gregarious flowering.

(i) Predation satiation hypothesis: This hypothesis indicates that the death of adult clone is due to resource extortion as it would be more effective for the parent plants to divert all resources to create a large seed crop than to whole back energy for their own regeneration.

(ii) Fire cycle hypothesis: This hypothesis indicates that flowering followed by death of adult plants has evolved as a mechanism to create disturbance in the habitat then providing the seedlings with the gap to grow. It argues that the dead culm creates large fuel load and also a large target for lightening strikes, increasing the likelihood of wildfire. Since bamboos are aggressive at its early stages the seedlings would be able to outsit other plants and take over the space left by their parents.

However, these two hypotheses have been disputed for several reasons and it is a nature or adoptive mechanism of bamboos which lead to death after gregarious flowering.

32. Mixed forest stands o-er complete utilization of land. Comment. (2014)

Yes, the mixed forest stand utilizes the land completely due to the following reasons.

(i) The mixed stand occupies each and every portion of the sites and fully utilizes the site due to its adoption to the microsite of the locality.

(ii) It mixed stands are arranged properly taking into considerations the species mix which result in complete and most efficient utilization of the land.

(iii) Mixed forest stands are considered better for their positive, productive and ameliorative effect on the sites but the pure forests of several species are reported to exhaust the nutrients and create adverse effect on the site.

(iv) The mixed forest stands are adopted to wide range of site including the poor sites. The mixed crop balances its soil properties much better than the pure crop.

(v) The mixed forest stand can draw nutrients from different depth of soil and utilizes the land fully compared to pure crops which make constant and unvarying demand on the site.

(vi) Similarly the mixed forest stand drains water from all parts of the site and water use efficiency of mixed stands are better than the pure stands which results in the maximum utilization of land.

(vii) The mixed forest stand will have both shallow and deep rooted plant which not only utilizes the site fully but also protects them from all natural calamities like wind, water, snow, frost etc. compared to pure stands.

(viii) The mixed stands are not only utilizes the land fully but also it avoids the epidemic status of pest and diseases compared to monoculture.

(ix) Above all the mixed stands are able to produce diverse or multifarious forest products by judicial utilization of land and meet the growing demands of wood and wood products.

33. Discuss the natural regeneration in soil. Give steps recommended for ensuring its successful regeneration. (2014)

The natural regeneration is the renewal of forest crop by natural means. The natural regeneration is possible through seeds, coppice and / or root suckers.

The natural regeneration is influenced by the following factors.

(i) Seed Production and Dispersal: The natural regeneration depends on the seeds produced by the trees which are dependent on the species, location, size, age, crown structure etc. Several species produce seeds every year and some may produce seeds periodically. Even some species produce seeds only once during its left the like bamboos.

The natural regeneration depend not only seed production but also its dispersal to the site of regeneration. The seeds are dispersed by wide range of agents wind, water, animal which again depends on the species.

(ii) Seed Germination: Successful natural regeneration is possible only with profused seed germination. However, germination depends on internal and external factors. The internal factors include permeability of seeds to water and air, embryo development, embryo maturity, viability and germinablity of the seeds. Tree species differ significantly for all the above factors. The external factors include all biotic and abiotic factors of the locality which include light, temperature, humidity and site quality.

(iii) **Seedling Establishment:** Germinated seeds should develop into a seedling for successful completion of the regeneration process. Seedling establishment is very essential and it depends on the favorable growing conditions in a locality. The growing seedlings have to be protected from the biotic and abiotic components of the locality.

Natural Regeneration in Soil

- Natural regeneration is successful in soils with good physiochemical properties. The soil should be moist and free from weeds, leaf litter and inhibitory factors.
- Natural regeneration can be seen in abundant cultivation areas, rough terrain and in clear fell areas. Natural regeneration is robust in the fire occurred areas.
- In contract the acidic soil with low pH value, saline soil and alkaline soil with high pH do not encourage natural regeneration.
- Natural regeneration is profuse if the soil deep and porous. If the soil is heavy, the roots of seedlings may not be able to penetrate into the soil and lead to death of seedlings.
- Soils with good porosity leads to enhanced regeneration compared to waterlogged soil which lead to carbon-di-oixde build up resulting in root respiration. This may cause death of the plants.
- The forest floor should be free from humus accumulation which may create acidity and affect the natural regeneration process. The sal regeneration process is affected due to heavy litter accumulation.
- The successful natural regeneration suitable soil condition along with necessary light which enhances the growth and development of new recruits. The well rooted seedlings adopt to the locality by efficiently utilizing moisture and other nutrients which will help them to compete the surrounding environment.
- Natural regeneration can also be enhanced due to adequate soil working in regular forest the soil may be heavy sometime with clay and canker formation which will not encourage germination. Hence soil working is recommend for ensure successful regeneration.

34. Comment 'Forest has moderating influences on soil and air temperature. (2016)

It is true that, forest exercises a moderating influence on air temperature. The influence of forest on temperature reduction depends on density and type of forest. Higher density leads to more reduction than lower density. It has been well proven that forest lowered the temperature as much as 391°F and this temperature reduction is less fluctuated in daily extremes. Among the type of forest the maximum moderating influence on air temperature was observed in beach forest and scotch pine forest.

Similarly, in Pine and Bamboo forest the temperature than in the open. These observations indicated the following conclusion on the moderating effect on air temperature.

(i) Forest lower the daily mean temperature in summer and raises slightly in winter.

(ii) The forest reduces the daily maximum temperature and increases the daily minimum temperature.

(iii) Forest reduces the range of variation in temperature.

(iv) The dense evergreen forests have more influences.

(v) The forest not only influences the temperature inside its ecosystem but also it influences the surrounding locality.

Forest influences on soil temperature

The influence of forest in moderating soil temperature is also similar to influence on air temperature but the reduction in soil temperature is much greater than air temperature.

(i) The tree canopy reduces the insolation and the radiation which results in reduction in soil temperature.

(ii) The leaf litter and the thick humus cover on the forest floor has influenced significance reduction in soil temperature.

(iii) Due to the influences of canopy and the humus layer, soil summer temperatures are lower and soil winter temperatures are higher compared to open land.

(iv) The reduction in soil temperature during summer protects recruits and seedlings from higher temperature and helps in their growth and development.

(v) Similarly the increase in winter temperature protects the recruits and regenerates from frost and severe cold.

35. Describe the role of light in obtaining high quality timber. (2009)

Light is an important factor for production of high quality timber, because light is essential for photosynthetic activity which results in formation of carbohydrates and used by the plants for their growth and development.

- Light affects the form and development of trees. Based on the requirement of light the trees are classified as light demanders and shade demanders. The light demanders are generally tall and dominant trees which lead to development of straight cylindrical bole and ultimately quality timber.
- The low light intensities and infrared radiation tend to stimulate the growth in height while white light of high intensity increases the root growth. Height growth is retarded under intense light conditions. The deficiency of light in forest is responsible for death of lower branches due to natural pruning. This develops stems with longer bole and cylindrical form lead to the production of quality timber.
- The crown competition in the upper reaches restricts its expansion which leads to the development of cylindrical bole and produce quality timber.
- The state of growth of trees is enhanced if the trees are exposed to favourable light conditions. The trees utilizes the light fully due to thinning and heavy opening, the lights available to the retained tree is very maximum and add more volume increment. The process increasing volume increment through light is called light increment.
- Growth and development: The photo chemical reaction play by the light is very essential for growth and development of trees through height, diameter and volume increment.

Hence light is an important and essential factor for obtaining high quality timbers.

36. a. Describe the e-ect of low temperature on plant growth. (2009)

b. Briefly discuss low temperature injuries in forest trees. (2010)

Low temperature has significant impact on plant growth. The temperature reduction below 5°C may cause chilling injury due to the excess of transpiration over absorption. The low temperature also leads to excess of respiration over photosynthesis which results in breaking down more proteins than its synthesis. Further reduction in temperature leads to frost and snow which causes the following serious damages to the plant.

(i) **Killing of young plants and their parts:** The young plants are killed due to excessive transpiration because roots are not able to supply moisture because of frozen ice. The water in the soil gets frozen and the roots are not able to trap the water.

(ii) **Death of plants due to cell damage:** Due to low temperature the water in the cell wall and intercellular spaces of the plant gets frozen which withdraws water from the interiors of the cells which results increasingly concentration of salts and dehydration of cells. This lead to coagulation and precipitation of cell colcoid resulting in death of plants.

(iii) **Injuries to the crown:** The low temperature leads to accumulation of intensive frost and the tree canopy and causes injuries to the crown this is followed by infestation by heart rot and affects the trees.

(iv) **Formation of frost cracks in wood:** Due to low temperature the moisture inside the wood particularly the free water gets frozen and the water from bound water moves out which lead to drying and shrinkages in wood. The outerwood shrinks rapidly than the inner which may create longitudinal cracks in the stem which may affect the quality of timber and death of plants.

(v) **Canker formation:** The low temperature may cause accumulation of frost and the crown which will result in damages to branches and ultimately branches die. The dead branches may follow leaving the cut end which will divide and produce callus. This callus on exposure no frost may die. Subsequently, this repeated effect of frost may create cancerous growth and affect the tree growth and development.

Injurious E-ects of Snow

(i) Snowfall results in the mechanical bending of stems of trees. Snow gets accumulated on the uphill side of the young saplings and poles and causes them to bend outward at the base. This curve is maintained even when the poles develop into trees and can be seen on all mature trees. The bend makes this portion of the tree unfit for utilization. As this portion of the stem has maximum diameter, there is a great loss in volume.

(ii) Accumulation of snow on the crowns of trees results in breaking of branches and tops of trees. Kail is most susceptible to snow break and deodar comes next.

(iii) Sliding snow not only causes erosion but also uproots trees. Often snow slides wipe out strips of forests along their course. The folds of the hills, in which snow slides regularly, are devoid of vegetation and thus a fair amount of area is without tree growth.

(iv) Snow shortens the period of vegetative growth

(v) Snow is reported to favour the growth of certain fungi. Eg. *Fomes* and *Trametes.*

37. What do you understand by the term locality factor? How these a-ect the decision of plantation undertaken by the Silviculturist? (2009)

Locality factors

The term locality factors are the factors which decide the growth and development of vegetation. The locality factors control species composition, density, structure, rate of growth, incidence of pest and diseases, presence of wild animals etc. The locality factors affect the forest both directly and indirectly by influencing the local climate, soil and the associated biological ecosystem. The success of vegetation depends on the locality factors.

- The silviculture is before initiating any plantation programme, he will assess the factors of locality for establishing correct plantation in a correct site because the forester may perform various silvicultural operations which may influence the factors of locality and this may or may not be good for growth of trees.
- Several site factors viz., light, temperature, soil moistures, water table, rate of decomposition, biotic facts can be modified by the forester through silvicultural practices.

Hence for locality factor may influence, or affect the decision of the silviculturist. For example soil moisture is one of the most important factors for growth and development. Proper stand density and stand composition may be required for soil conservation. Higher stand density require more moisture than lower density. In areas with low moisture, deciduous species may be given preference. Similarly with higher moisture regime evergreen species preferred.

Drought is the common phenomenon and the decision of a forester is significantly influenced by locality factors. For example the forester is planning for establishment of Terminalia and Teak in a locality which faces severe drought. Under such situation, Terminalia and Teak may not survive or even if survives, the increment will be very low. Under such circumstances, the locality factors prefer selection of drought tolerance species against the species selected by the Silviculturist.

Similarly the climatic factors significantly affect the decision of the forester. For example in evergreen and semi evergreen forest the vegetation growth is very dense and any plantation development within the forest requires manipulation of canopy in order to invite more light manipulation of top canopy by executing seedling felling in teak, sal and deodar forest is done with the view to improve light conditions and avoid overhead competitions. If light demanding species are planted, if their forest without canopy management, the factors of locality do not allow successful crop.

Another major locality factors is biotic factor. Suppose the forester plans for establishment of multipurpose tree species without ensuing the fodder value of the tree species. Under such circumstances the trees planted may be heavily browsed by local animals which may result in failure of the plantation.

The topographic factors particularly the altitude, slopes, aspect and exposure play a critical role by affecting the decision of the forester. Species preference towards altitude is highly variable. If a forester wants to establish Silver oak in lower altitude and Casuarina in higher altitude, the locality factors may severely affects the decision as both the species are not suitable for this locality.

Hence, locality factors of a site are very essential for successful development of plantation and these factors may or may not affect the decision of plantation undertaken by the Silviculturist.

38. Why LAI important in deciding the productivity of forest tress? Explain the concept of optimum LAI and how it varies with type of forest and climate. (2010)

Leaf is the principle organ in which the primary production is taking place. Leaves due to the presence of chlorophyll, synthesizes its energy by utilizing light, water and carbon-di-oxide from the atmosphere. The productivity of leaf is refers to the production in a unit area in a unit time and for this purpose leaf area is very important which decides the productivity of a plantation. This leaf area is estimated by LAI method. The area of a leaf in relation to a ground area is of vital significance towards productivity.

In most of the forest ecosystem the standing crop biomass of green plants is over 98% of the living biomass. But in actual sense, the primary productivity is the rate of energy conversion from the radiant form to the chemical bonds of the organic substances which lead to increase in weight and biomass. As indicated above, this energy trapping process is called photosynthesis and it is measured through primary productivity.

The Leaf Area Index is the ratio of total surface of leaf of a plant to the ground area available to that plant. It is the sum of total area of leaf per unit area of the ground and expressed as follows.

$$\text{LAI} = \frac{\text{TLA}}{\text{TGA}}$$

Where LAI – Leaf Area Index

TLA – Total Leaf Area

TGA – Total ground Area

39. Distinguish between Tending operation and cultural operation. (2010)

In a forest, competition among plants can be increased with increasing age. In order to obtain a maximum yield the competition among the plants can be optimally managed by reducing the number of stems per unit area with the progress of age. All these can be covered under the tending operations

	Tending	Cultural operations
1.	Tending is carried out for the benefit of forest crop by creating better environment	It is done for assisting regeneration, promote development and to decrease felling damage
2.	This is not aims at natural regeneration	This aims at natural regeneration
3.	It includes pruning but not controlled burning	It includes controlled burning but not pruning
4.	Tending carried out from seedling stage	Cultural operations carried out after felling
5.	It is carried out in both natural as well as artificial regeneration	This is carried out in silvicultural system relying primarily on natural regeneration

40. Define micro climate giving suitable example? (2011)

Microclimate

Microclimate is defined as the climate of small area which for some reasons differ significantly from the general climate of the area; more particularly the climate under the plant or other vegetative cover, differing in extremes of temperature moisture from the climate outside that cover.

Forest play a vital role in moderating the microclimate of the region particularly the solar radiation, temperature, rainfall, humidity, wind speed, etc. The factors which decide development of microclimate are altitude, aspect, slope, drainage condition and the vegetation.

The influence of forest on climatic factors like temperature precipitation, humidity and wind movement etc. may lead to some changes in macro climate or create microclimate in zone of its occurrence and the surrounding area. This microclimate may influence the locality factors.

(i) Forest influences the local microclimate of the region which leads to increased precipitation. The studies indicate that forest influences the local precipitation because evapo-transpiration from the forest is more rapid and a large amount of moisture available for precipitation in that locality.

(ii) The temperature of the microclimate is moderated or influenced by forests, it lowers the daily mean maximum temperature and increases the daily minimum temperature. In the microclimate region, the daily range of variation in temperature is very low. The dense evergreen forests have more influence on air temperature than deciduous forest.

(iii) The forest reduces the wind velocity by obstructing or deflecting the moving wind currents. Dense forests with heavy foliage have the greatest effect in reducing wind velocity in the microclimate region. Trees in general, trees reduces wind velocity not only in the microclimatic region but it extends upto the distance of 25 times the height of the tree.

(iv) The humidity in the microclimate is very significant. The evapotranpiration from the forest area is large which releases large amount of water vapour in the atmosphere. Therefore the microclimate under the forest canopy exhibited higher relative humidity compared to other areas.

41. Explain the various factors a-ecting the choice between artificial and natural regeneration with reasoning? (2013)

The choice between natural and artificial regeneration is decided based on the following factors.

(i) Ecological considerations and Biodiversity

(ii) Soil productivity

(iii) Pest and diseases

(iv) Cost and economics

(v) Economic valuc

(vi) Income and employment generation

(vii) Genetical considerations

(viii) Stocking

(ix) Flexibility

(x) Effectiveness.

(i) **Ecological considerations and Biodiversity:** If the propose of regeneration is to enriched biodiversity and maintain the ecological balance means the choice is towards natural regeneration over artificial regeneration. The natural regeneration enhances species richness and biodiversity compared to artificial regeneration where it will be concentrated on single or double species.

(ii) **Soil productivity:** Natural regeneration always enhances the soil productivity due to species richness and the associated accumulation of more litterfall. This litterfall addition enhances the decomposition and improves the soil fertility. In case of artificial regeneration, the accumulation of litterfall is low due to minimal number of species and there are evidences that artificial regeneration with teak has depleted soil fertility. Hence, if the objectives is to improve the soil fertility, then the preference towards natural regeneration.

(iii) **Pest and disease:** The natural regeneration is housed with mixed of species and the possibility of epidemic status of pest and disease are low. But there are enough evidences to prove that plantations through artificial regeneration have suffered from epidemic outbreaks of pest and diseases. This outbreak is due to reduced diversity, high density and the practices of silviculture. The occurrence of teak defoliator, sal borer and defoliator in Ailanthus are examples of epidemic status of pests. Casuarina wilt, die back in sisham are examples of epidemic disease due to artificial regeneration. Hence to control pest and diseases, it is preferred to choose natural regeneration than artificial regeneration.

(iv) **Cost and economics:** Generally natural regeneration process is very cheap compared to artificial regeneration. The establishment of 1 ha of artificial regeneration required about Rs. 75,000/- compared to less than Rs. 10,000/- in natural regeneration process. If the investment is not available and infrastructure is not adequate then natural regeneration preferred over artificial regeneration.

(v) **Economic value:** It is one of the important factors which decide the choice of regeneration. If the objective is to produce more timber of high quality then artificial regeneration is preferred over natural regeneration. This is because artificial regeneration involves implementation of necessary silvicultural practice which will ensure quality timber production. Similarly, if the timber is planned for sustained yield then artificial regeneration is preferred over natural regeneration

(vi) **Income and employment generation:** Natural regeneration do not support adequate income and employment generation activities as the major objective is towards conservation. Under such circumstances, artificial regeneration is preferred over natural regeneration.

(vii) **Genetical considerations:** Natural regeneration helps to conserve wide range of genetic diversity but artificial regeneration aims to develop one or two species. Hence if the objective is to encourage and conserve more diversity then natural regeneration is preferred over artificial regeneration.

(viii) Stocking: Stocking is an important factor which decides the choice of regeneration. Full and over stocking demands natural regeneration and under stocking demands artificial regeneration. Based on the objective and the stocking density regeneration can be decided.

(ix) Flexibility: The flexibility of operation is one of the major important factors which decide the choice of regeneration. Artificial regeneration is more flexible as it involves only one species and the tending operations are easy and flexible. But natural regeneration has more number of species which have complex tending operations specific to species. Hence for easy and flexibility in operations, artificial regeneration is preferred.

(x) E3ectiveness: The effectiveness depends on the productivity of plantations raised through either natural or artificial regeneration. The productivity of natural forests is less than $1 m^3$/ha/annum and the productivity of artificially regenerated plantations have recorded over $20 m^3$/ha/annum. Hence considering the effectiveness, artificial regeneration is preferred over natural regeneration.

42. What do you mean by plantation schedule? Give in detail the factors which decide the success of planting program? (2014)

Plantation schedule

Plantation schedule is the schedule of operations from site preparation upto harvesting of plantation through a planned and timely operations.

The schedule consists of various activities in plantation viz., the layout, demarcation, marking, pitting, planting, weeding, pruning, thinning, pest and disease management etc. Similarly, the plantation schedule should incorporate collection of seeds, sowing, seeding production, seedling management etc. Plantation schedule should also incorporate periodic removal, felling and transportation of wood. The schedules of the operations have to be planned well in advance in order to plan and execute the plantation development.

Example: Sequential Plantation schedule operation for *Melia dubia*

S.No.	Plantation operations	Period
i	Clone development	3 months
ii	Land preparation	
iii	Ploughing	0-3 months
iv	Layout and marking	
v	Pitting	3-6 months
vi	Planting	
vii	Irrigation	Once in every 5 days throughout the lifecycle except monsoon season
viii	Fertilization	Once in every 3 months from planting upto harvesting
ix	Pest and Disease management	Based on occurrence
x.	Pruning	Annual pruning during April – May

xi	Thinning	2,4 and 6 years under high density plantation One thinning at third year in medium density plantation Silvicultural thinning under low density plantation
xii	Harvesting	2 and 4th year for pulp wood 6 and 8th year for plywood
xiii	Transportation	At the time of harvest

Factors deciding the success of planting programme:

(i) Quality of planting material: The success of planting programme depends on the quality of the planting material. Inferior seedlings with poor genetic back up will result in failure of plantations. Alternatively improved seed sources from seed orchards or clones developed through elite genetic resources will ensure successful planting programme.

(ii) Site quality: The success of planting programme depends on the correct site quality. Mismatching the species with site quality resulting in failure of a plantation irrespective of the quality of the planting materials.

(iii) Time of planting: The success of planting programme depends on the timely planting operations. In India, most forest operations are carried out during monsoon season which results in success of planting programme. The failure of monsoon or planting during non monsoon season may result in failure of plantation.

(iv) Post planting operations: Immediately after planting programme, necessary soil working coupled with application of plant protection measures will help to ensure successful establishment of plantations.

43. Describe the methods of pre-sowing treatment of seeds for raising nursery. (2015)

Tree seeds are commonly endowed with many kinds of dormancy. Hence pre-sowing treatment for tree seeds is a must to enhance rapid and uniform germination.

(i) Mechanical scarification: Large quantities are hard coated seeds can be mechanically scarified using concrete mixture with sharp gravel or sand. This treatment mechanically injures the hard seed coat and enhances germination. Care should be taken to avoid over treatment to prevent damage to the embryo.

Eg: *Acacia, Albizzia, Cassia, Delonix regia* etc.

(ii) Water soaking: Soaking in cold water for a period of few hours to several days is a safe and simple treatment. This treatment is intended to soften hard seeds and to leach out chemical inhibitors. Leguminous seeds respond well to immersion in boiling water for few minutes and soaking in cold water for a period ranging from 2 to 48 hours.

Eg: Teak, Red Sanders, *Santalum album*, Lannea and Terminalia

(iii) Acid scarification: Soaking in concentrated sulphuric acid is the most common method of treatment to soften the seed coat. To scarify the seeds, pour the acid until all the seeds are uniformly wetted. Stir them for few minutes depending upon the species and wash them in running tap water for 5-10 minutes to remove all traces of acid.

Eg: Acacia, Albizia, Prosopis and Subabul

(iv) **Dry heat treatment:** Spreading the light inflammable litter over the fruits and then ignite it. Adjust the heat of the fire to achieve maximum effect on the pericarp without damaging the embryo.

Eg: *Acacia mangium* and *Tectona grandis*

(v) **Warm stratification:** Thistreatment helps to overcome mechanical and morphological dormancy. Soak the seeds in several times their volume of cold water to approximately 3-5°C for 48 hours. Drain of the water and mix the seeds with 2-4 times their volume of moist and water retained medium such as sand, peat or vermiculate. Store at a warm temperature occasionally open the container and mix well for uniform moisturing.

Eg: Abies and Oak

(vi) **Cold stratification:** This treatment is recommended to remove physiological dormancy. Seeds are soaked in cold water 3-5°C for 48 hours. After soaking the water is drained and moist seeds are stored in 3-5°C for the period appropriate to each species. They may be stored with medium or without medium. The combination of high moisture and low temperature appears to trigger off the changes in cell energy relationships leading to biochemical activity which transforms complex food substance into simpler forms utilized by the embryo when it renews growth at germination.

Eg: Abies, Acer, Primus, Persica and Fagus

(vii) **Biological methods:** In nature, animals and microorganisms play an important role in the breaking the impermeability of seeds. It is difficult to use these organisms in controlled environment. Successful results in germination are proven when certain tree seeds passed through the bodies of animals.

Eg: *Santalum album*, Acacia, Coffee etc.

The combination of warmth, moisture and chemical action of enzymes inside the digestive tract of animals enhance germination.

44. Explain the term of hardening o-. What are the internal factors a-ecting frost resistance? (2015)

Hardening off is the natural process by which plants become adopted to drought, cold or heat.

Internal factors a-ecting frost resistance:

(i) **Size of the cell:** Cell size is very important for frost resistance and the tree with small tree cell size are generally are more frost hardy than trees with large sized cells.

(ii) **Cell water content:** Water content in the cell is very critical factor in frost resistance. If the water content more in the cells then the water gets freezed and ice formation occurs both in the intra and intercellular regions. Hence trees with low intercellular water content are more frost resistant than trees with high intra and intercellular water content.

(iii) Osmatic concentration: If the osmatic concentration within the cells are high then it will extent greater resistant to frost. This resistant is due to lower freezing point of cell sap which results in lower internal ice formation.

(iv) Permeability to water: The permeability water and the frost resistance are directly proportional. Higher the permeability of protoplasm to water greater is the frost resistance.

(v) Water binding colloids: It is also directly proportional to frost resistance high water binding colloids in cells, greater in the frost resistance. This resistance is due to

(a) Reducing the amount of free water that can be frozen

(b) Low internal ice formation

(c) Reducing the amount of water that can be removed on external ice formation.

These are the internal factors which lead to frost resistance.

45. Di-erentiate clearly between natural and artific ial regeneration of forest. Describe the manner in which regeneration of Teak, Sal and deodal takes places? (2012)

S.No.	Particulars / Factors	Natural Regeneration	Artificial Regeneration
1.	Nature of regeneration	It occurs through artificial means	It occurs by natural means through seeds, coppice or root sucker
2.	Species composition	Artificial regeneration is raised for one or two species in given area based on the utility or human interest.	Natural regeneration contain hundreds of trees and shrubs belonging to the various species composition
3.	Ecological consideration	Artificial regeneration is leads to uniform age and variation among them is very less.	All age classes are present in NR
4.	Canopy	Single layer of canopy	Multiple layer of canopy
5.	Change in richness and diversity	It devoid of species richness and diversity	There is a tendency to produce mixed crop which are having high diversity and richness.
6.	Reduction in soil productivity	The AR is raised by removing the existing plantation which leads to exposition of the sites. which reduce the productivity of forest	In NR the loss of soil productivity is very less due to the accumulation of leaf litter and decomposition matters.
7.	Natural balance	The natural ecosystem is best for soil protection and climate amclioration.	The NR takes full advantages of sites by giving due place to light demanded as well as to shade bearer species.

8.	Economic value of the crop	High	Low
9.	Employment	AR is labour intensive and can generate enough employment to the rural poor people.	NR is depend on nature and can provide little employment to the poor people.

Regeneration of Sal, Teak and Deodar

S.No	species	Natural regeneration	Artificial regeneration
1	*Shorea robusta* (sal)	• Seeding, germination and establishment are the factors influencing natural regeneration • Dyeing back, influence of dead leaves, light, frost, grazing, fire, heavy soil, bad drainage are also influencing natural regeneration	• Difficult due to short viability • Fresh seeds have to be used germination starts in 7-10 days and transplanted to poly bags. • 1 year old seedlings are used for planting • Timely weeding and irrigation ensures artificial regeneration.
2	*Tectona grandis* (Teak)	• Seed production, germination, establishment, moisture, temperature and soil aeration and mild grazing are the key factors influencing regeneration.	• Mature seeds with alternate wetting and drying is sown in the mother bed. • Seed rate: 1kg/m² • Germination starts in 15 days and completes in 60 days • 9-12 months old seedlings are lifted and stump prepared • Stump size: 2.5cm shoot • 22.5cm root • 1-2 cm collar dia • Spacing: 2m×2m • Watering and weeding should be done to ensure regeneration
3	*Cedrus deodar*	• The major factors which decides natural regeneration are seeding condition, germination, weather and soil conditions • Canopy management followed by regeneration felling are important	• Seeds raised in nursery beds should be protected from rodents • Seedlings should be pricked and planted in poly bags • Weeding, lopping and pruning is essential. • Spacing: 2m×2m or 3m×3m

46. Define a forest types.Discuss the di-erent forest types found along tidal swamp forests with their species composition. Give a note on how *Rhizophoraracemosa is managed* in mangrove forests in Sunderbans? (2011)

Forest type

Forest type is a category of forest defined with reference to its geographical location, climatic and edaphic features, composition and condition.

Champion and Seth define it as a unit of vegetation which possesses (broad) characteristics in physiognomy and structure sufficiently pronounced to permit of its differentiation from other such units. This is irrespective of physiographic, edaphic or biotic factors. It is selected in the first place subjectively from the ever-varying cover of vegetation, with boundaries arbitrarily imposed on what are in fact gradual changes.

Tidal Swamp forests

The tidal swamp forests are classified within the littoral and swamp forest of champ & seth classification. These forests are occurring on swampy areas and occur predominantly in Ganga and Brahmaputra. The temperature ranged between 23 to 30°C Rainfall between 100 mm to 3000 mm.

Species composition

(i) Heritiera species
(ii) Braguiera species
(iii) Ceriops spp
(iv) Rhizophora spp

This Tidal swamp forest is further divided into 4 subtypes and they are

(i) Mangrove scrub
(ii) Mangrove forest
(iii) Salt water mixed forest Heritiera
(iv) Palm swamp

S.No	Forest type	Species composition
1.	Mangrove scrub	*Ceriops roxburghiana, Avicennia alba, Aegialitis rotundifolia, Phoenix paludosa, Sunneratia apetala* etc.
2.	Mangrove forest	*Rhizophora candelaria, Kandelia candel, Avicennia alba, Bruguiera conjugatus, Ceriops tagal, Sonneratia apetala, etc*
3.	Salt water mixed forest Heritiera	*Heritiera minor, Excoecaria agallocha, Ceriops roxburghiana, Bruguiera conjugate, etc*
4.	Palm swamp	*Phoenix paludosa*

47. What do you understand aerial seeding? Discuss briefly its success and failures in the Indian context? (2005)

Aerial seedling is the process of broadcasting the seeds from the atmosphere using through aircrafts / helicopter.

Purpose of aerial seedling

- The main purpose is to vegetate the remote, rugged and inaccessible areas
- Afforesting denuded and degraded hillocks
- Regeneration of riverines

Experiences in India

In India, aerial seedling has been tried in several states and resulted in both success and failures. Several trees species like *Acacia nilotica, Catachu, Tortalis*, *Prosopis juliflora*, *Dalbergia sissoo*, *Dendrocalamus spp*. etc. have been sown through aerial seedling. In India, sambal ravine in Madhya Pradesh, Uttar Pradesh, Western ghats of Maharastra, arid regions of Rajasthan have been attempted with aerial seedling. In Tamil Nadu, the entire palani hills were aerial seeded with *Acacia nilotica* and *Acacia planifrons*. Though aerial seedling was attempted its successfulness has not been evidenced fully barring few exceptions.

48. a. Explain in brief di-erent methods of vegetative p ropagation in tree species (2006)

b. What are the di-erent grafting their ecological adaptation methods used for forest tree species? (2005)

Methods of propagation

Vegetative propagation is possible by the following two methods.

I. Macro vegetative propagation

(i) **By cuttings:** The term cutting refers to any portion removed from the parent plant which is meant for regeneration. The following cuttings are generally employed for vegetative propagation viz., Stem cuttings, Root cuttings, Leaf cuttings, Leaf bud cuttings, Softwood cuttings, Semi hardwood cuttings, Hardwood cuttings.
Cuttings are collected from the mother tree or from hedge garden or from mini garden, treated properly to avoid fungal infection and on treatment with growth regulators, the cuttings are planted in containers and kept for rooting under mist chamber conditions.

(ii) **By grafting:** Grafting is an art of joining plant parts together in such a way that they will readily unite and continue to grow as one plant. A graft has two parts viz., scion and root stock. The scion refers to the upper or top portion of the plant collected from identified mother tree and the roots stock is the lower portion of the graft combination which is raised predominantly through seeds. For successful union of graft, it is very essential that the cambium of scion be placed in close contact with the cambium of the stock.

Methods of grafting

(A) **Splice or Whip Grafting:** First the scion and stock plants are selected. Then the stock is cut back to the height where grafting is to be done and a slanting cut of about 3-5 cm long is made and a similar cut of the same length and slope is given to the scion plants which ensures perfect match. These two cut surfaces are placed together and wrapped tightly with polythene grafting tape.

Whip and tongue grafting

(B) Whip and Tongue Grafting: It is another grafting method which is applicable when the stock and scion are of similar thickness. In general seedlings, rooted layers or cuttings are whip grafted in late winter or early spring.

The scion and stock are prepared as in the case of whip grafting. On each of these cut surfaces, a reverse cut is made. It is started downward at a point about one third of the distance from the tip, and should be about one-half the length of the first cut. The scion is then slipped into the stock so that they interlock.

- Diagonal cut at stock and scion with tongue
- The graft is tied and wrapped.

(C) Saddle Grafting: The stock and the scion should be nearly equal in size for good cambial contact and complete graft. The rootstock is prepared in the form of a wedge. A vertical cut is then given at the base of the scion and is inserted to the rootstock so that they fix well. The components are then tied and coated with grafting wax.

(D) Cleft or Wedge Grafting: It is done in early spring when the tree is nearing dormancy. The shoot tip of the stock plant is removed and a vertical cut is made through the centre of pith. The scion is prepared by making tapering cuts in the form of a wedge. It is then inserted into the split so that the cambial layers of both the scion and the stock are well matched. The union is tied and covered with grafting wax to avoid water entry.

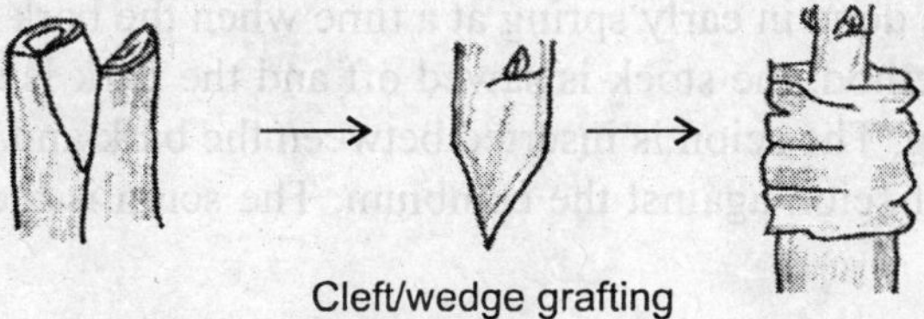

Cleft/wedge grafting

Side Grafting: An oblique or slanting cut is made in the side of the stock for about 3.5 cm. The scion is prepared with sloping cuts, opposite to each other on the base. Then the rootstock is slightly bent such that it allows sufficient space for scion insertion. The scion should match the cambial layers of the stock. Then the graft is tied with polythene tape to seal the entire cut area. After graft union, the top of the stock is cut and painted with tree paint.

(F) Veneer Grafting: The base of the scion is cut diagonally and a wedge shaped patch of bark of the same size is removed from the side of the stock. The scion and stock are bound together with cambial layers matching. After healing of stock, the upper portion of stock is removed.

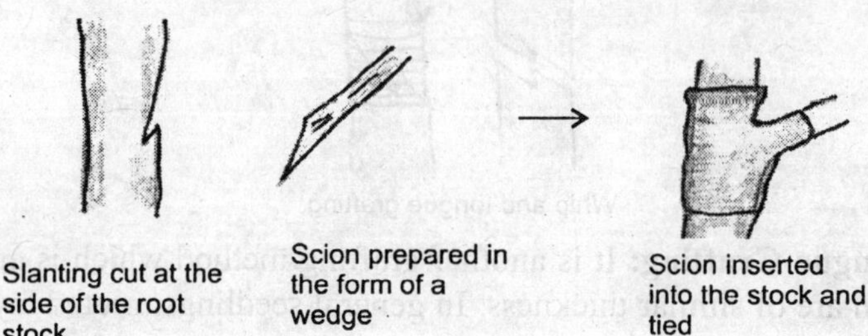

(G) Approach Grafting: Two independent self sustaining plants are grafted together. A cut is made through the cambium and slightly into the wood is given on both stock and the scion. The cut surfaces are pressed firmly and tied with polythene tape. After the union is achieved the scion is severed below the union and the stock above the union.

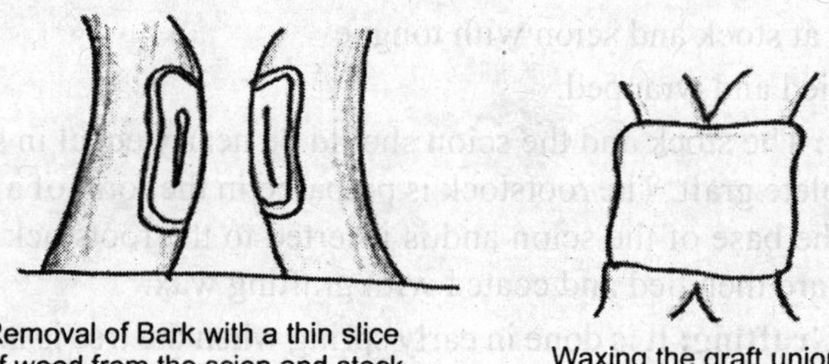

- Removal of a thin slice of wood and bark from both stock and scion.
- Positioning and tying the stock and scion.
- Waxing the graft union.

(H) Bark Grafting: It is done in early spring at a time when the bark separates from the wood with ease. In this method, the stock is sawed off and the bark is split downward from the apex about 5 cm long. The scion is inserted between the bark and the wood of the stock by placing longer cut of scion against the cambium. The scion is then tied and covered with grafting wax.

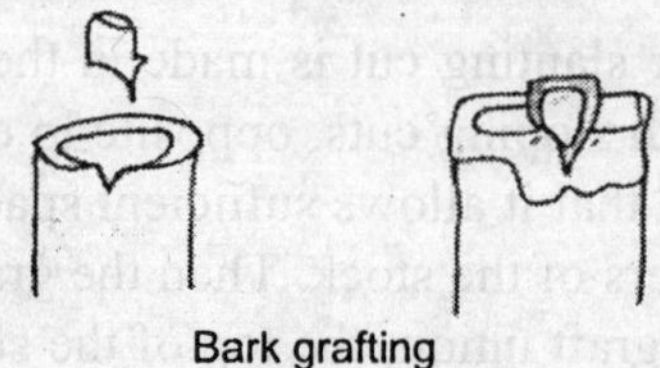
Bark grafting

(I) Root Grafting: The rootstock seedling or rooted cutting or layered plant is dug up, and the roots are used as the stock for the graft. The entire root system may be used, or the roots may be cut into small pieces and each piece is used as a stock. As the roots used are relatively small, whip grafting method is generally followed.

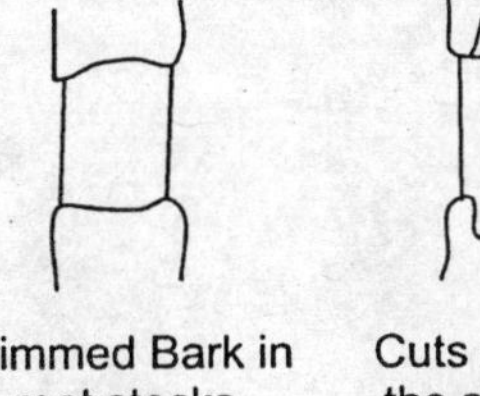

Trimmed Bark in root stocks

Cuts made in the stock for scion and insertion

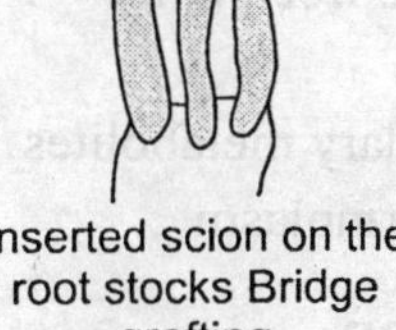

Inserted scion on the root stocks Bridge grafting

(J) Bridge Grafting: It is a form of grafting followed when there is injury to the trunk. The wounded area is trimmed. The scion should be taken from dormant one year old shot in the form of strips of uniform size. It is then inserted around the injured section for every 5-7.5 cm and connected to both the upper and lower ends into live undamaged bark.

Among all the grafting methods, as far as forest tree species are concerned due to their ecological requirement adaptation, cleft grafting has been found to be more useful.

(ii) By layering: It is an auto vegetative propagation method and is popularly known as Chinese layering or marcottage. This method is practiced in a wide range of tree species due to its simplicity. There are two types of layering viz.

(a) Air layering: Large shoot of one o two years old are most suitable for making air layers. First the leaves are removed from the selected shoot. Then the stem is girdled by removing a ring of bark about two to three cm wide. The girdled/ cut portion is covered with *sphagnum* moss and wrapped with polythene film tightly to retain the moisture and allows gaseous exchange. This portion will induce roots in four to eight weeks and the rooted portion of the branch is separated out and planted in containers for hardening and further planting in the field.

(b) Ground layering: In this method, the girdled stem is buried inside the soil. As soon as the rooting occurs, the plant is separated out and planted in containers for hardening and further planting in the field.

II. Micro vegetative propagation

This is one of the methods of propagation wherein micro shoots viz.; explants (shoot tip, node, internode, meristem, inflorescence, leaf and leaf bits) are cultured under *in-vitro* conditions. Micropropagation leads to mass multiplication from a single explant.

Stages of micropropagation

- Establishment stage
- Shoot multiplication stage
- Root multiplication stage
- Hardening stage

Advantages of micropropagation

- Mass multiplication
- Production of uniform plants
- Production of disease free plants
- *In-vitro* conservation
- Production of secondary metabolites
- Easy exchange of germplasm

Mini Clonal Propagation

A mini clonal technology has been developed for casuarinas and Melia which is one of the pioneering attempts in the country for these industrial wood species. Under this technology, the superior clonal plants were planted in a mini clonal garden and are provided with regular irrigation and fertilization in order to enhance shoot multiplication. In this method, the mother plants are planted at 10 cm x 10 cm spacing and 60 days old plants are ready for collection of cuttings.

Clonal Garden establishment

The clonal garden can be established at a size of 10 x 1 x 0.6 m or 5 x 1 x 0.6 m or 3 x 1x 0.6m using cement trough or GI trough. The bed should be filled with 20 mm stones upto 25 cm and over which finely sewed river sand can be filled. The trough should have facility for drainage. The raised beds can be covered with 100 micron UV stabilized polyethylene film on the top and covered with insect proof mesh to protect the plants and to ensure its freeness from pest and diseases.

Fertigation

The clonal garden should be maintained with irrigation at an interval of every one hour and supplemented with the following nutrients.

- Urea – 300-400 g/m^2
- SSP – 150-175 g/m^2
- KCL – 175-250 g/m^2
- Micro nutrient mixture – 100 g/m^2

The nutrients can be applied twice or thrice depending on the rate of growth of plants.

Clonal Management

The plants are allowed to grow upto 60 days by applying the required nutrient composition. After 60 days the plants are pruned at required size preferably at half of the plant to induce new shoots. With continuous irrigation and nutrient management the cuttings will start producing shoots from 8-10 days onwards and after 15-20 days the cuttings can be collected and treated with 2% carbendazim solution.

Clonal Treatment

The newly induced shoots were separated from the plants and are treated with or without 1500 ppm IBA (liquid formulation) and planted in 90 cc root trainers filled with decomposed coir pith. The rooting started in 15 days and 25 days old rooted plants are ready for hardening.

Green House Conditions

The root trainers are kept under green house conditions with a temperature regime of 32–35°C and a relative humidity of 85-95%. Periodical watering once in every 30 minutes is preferred.

Acclimatization and Hardening

The rooted plants are hardened in a shade house condition with 50% shading for 7-15 days and maintained with adequate irrigations. After hardening chamber, the plants are lifted to open nursery for 30 days. Watering is done 2 times a day and the fertilizer of all 19 (N:P:K) can be applied at the rate of 5g/plant. During this hardening, application of carbendazim (2g/l) or triazophos (2ml/lr) is recommended based on the incidence of diseases and pests.

49. What are the factors that adversely a-ect natural regeneration of the species? (2006)

Natural regeneration of the species depends on the germination capacity and the behaviour of the intended species. The natural regeneration is inhibited by the following factors viz., Seed production and dispersal, seed germination and seedling establishment and genetic factors

(i) Seed production and dispersal: Seed production in a tree species is a major factor which affects the natural regeneration. The quantity and quality of seed depends upon the species, age of the trees, crown size, apical dominance and the associated biotic and abiotic factors. All the tree species do not produce seeds annually. Based on seed production either annually or periodically, years are classified into good seed year, moderate seed year or bad seed year. Natural regeneration is very good in good seed year and very poor in bad seed year. Hence, according to species and seed production capacity, natural regeneration varies from site to site and species to species. Not only seed production is essential but also its dispersal to ideal locations is significantly important for the success of natural regeneration. Good seed production in wrong locality and bad seed production in good locality may lead to poor regeneration. The seeds produced have to be dispersed safely by all means to ensure and complete the natural regeneration process.

Seed dispersal process in most tree crops is facilitated by the following factors.

Wind	:	Cas*uarina, Terminalia, Poplar, Pterocarpus, Acacia, Dipterocarpus*
Water	:	Most mangroves and riverine species
Gravity	:	*Quercus, Juglans, Aesculus*
Birds	:	*Prunus*, *Mulberry*, *Diospyros*
Animals	:	*Prosopis, Acacia*

After abundant seed production, if the seeds are not dispersed, this may affect the natural regeneration process.

(ii) Seed germination: Natural regeneration is affected due to improper seed germination and behaviour. Germination depends on internal and external factors. The internal factors include permeability of seeds to water and air, embryo development, embryo maturity, viability and germinablity of the seeds. Tree species differ significantly for all the above factors.

(a) **Permeability to water and oxygen:** Moisture is very critical to germination and will adversely affect the regeneration. The presence of hard seed coats will prevent the entry of moisture and oxygen thereby germination gets affected which in turn affects the natural regeneration.

(b) **Development of embryos:** Embryo development is very critical factor which decides the natural regeneration. If the embryo is not fully developed it will affect the germination and the associated natural regeneration process. Even if the embryo is developed it may not be germinable due to the presence of chemical substances which will delay the germination and the natural regeneration process.
Example: *Fraxinous floribunda, Juniperus macropoda* and *Tectona grandis*.

(c) **Seed viability:** It is the capacity of the seed germinates and depends predominantly on the natural factors like rain. If the monsoon is delayed the seeds may lose viability and there won't be any natural regeneration. The external factors include all biotic and abiotic factors of the locality which include light, temperature, humidity and site quality. The increase in temperature reduces the germination. Similarly light is another critical factor which rapidly influences regeneration process. The biotic influence particularly the damage caused by grazing and browsing animals are also one of the factors affecting natural regeneration process.

(iii) **Seedling establishment:** Germinated seeds should develop into a seedling for successful completion of the regeneration process. Seedling establishment is very essential and it depends on the favorable growing conditions in a locality. The growing seedlings have to be protected from the biotic and abiotic components of the locality. If the growing seedling is not established then natural regeneration process is a failure.

(iv) **Genetic factors:** Natural regeneration also depends on the genetic factors. Seeds from inferior, lean and emaciated population may adversely affect the natural regeneration process.

50. Explain how the knowledge about the silviculture characteristics of a tree enable us to manage the species in a better way? (2015)

The management of species depends on the silvicultural characteristics of tree species. The following silvicultural characteristics and the knowledge on these aspects will help to manage the species very effectively.

(a) **Phenology and reproductive biology:** The phenology of the species particularly its architecture, growth pattern, flowering behaviour, fruit and seed production are very essential towards managing the tree. If the phenology of the species is not known and the tree species and the associated plantations may not be effectively managed.

(b) **Climatic requirements:** One of the major silvicultural characters of any species is its climatic requirement particularly the temperature. Some species may thrive on extreme weather conditions and several species may suffer due to extreme climatic conditions. Hence the knowledge on silvicultural character especially its climatic requirement is very important to manage the tree.

(c) **Soil requirements:** Another major important silvicultural requirement of the tree is its soil requirement. Mismatching the species with the soil lead to failure of plantations. Hence the knowledge on soil requirement of the species will help to establish and manage the species very effectively.

(d) **Growth characteristics:** The growth characteristics of species particularly its height and diameter growth and the demand for light play a vital role in managing the species. Several tropical species are strong light demanders, some are shade bearers and some are shade demanders. Based on these light requirements the growth characteristics are variable and hence the knowledge on growth characteristics is very essential.

(e) **Water and fertilizer requirement:** The management of tree depends on its demand on water and the nutritional requirement. Hence the knowledge on water and nutritional requirement of species is very significant towards successful management of a species.

(f) **Utilization:** The use of the tree is very significant towards its management. Hence multifarious utilization characters of the species need to be understand thoroughly in order to manage them for timber production, pulpwood production, plywood production, fuel wood production, fodder and other non timber values.

Hence the knowledge on silvicultural characteristics of the species is very essential towards managing the tree species.

51. Write a short note on Hi-tech nursery? (2007)

The Hi-tech nursery is the nursery established using modern tools and technologies in order to produce genetically identical and improved planting materials in a required volume in a specified time.

Components of Hi-tech nursery

Any Hi-tech nursery will have the following components.

(i) **Tissue culture lab:** It is very essential to produce the mother plants of superior genetic origin. Since the trees are highly recalcitrant towards vegetative propagation, the identified trees are to be multiplied initially through tissue approach and forwarded to mother garden.

(ii) **Mini clonal mother garden:** The clones screened have to be planted in the sand bed trough for mass multiplication. The sand bed trough can be of various sizes viz., 10 x 1 x 0.6 m or 5 x 1 x 0.6 m or 3 x 1 x 0.6 m depending on the annual planting material requirement and the associated management technologies. The trough of the required sizes can be constructed using cement concretes or fabricated through GI sheets. Once the bed size is designed and established, the bottom of the trough is filled with 20 mm stones upto the height of 20 to 25 cm from the base upwards. Then the portion above stone filling should be filled with fine sewed river sand. The trough should have facility for proper drainage at both the ends. The trough can be covered with 100 micron UV stabilized poly ethylene film on the top and covered with insect proof mesh along the sides in order to protect the plants from pest and diseases.

(iii) **Propagation chamber (Mist chamber or green houses):** This chamber is for multiplication of clonal plants. This chamber has misting facility coupled with automated temperature and humidity control. Incorporation of cool seal pad system will help to control the growing environment within the chamber. The green house growing conditions are with a temperature regime of 32 – 35°C and a relative humidity of 85-95%. Periodical watering once in every 30 minutes will ensure early, easy and uniform rooting process.

Ideal size of the structure (25000 plants 30 x 60 ft)

Larger size (40 x 100 ft)

(iv) **Hardening chamber (Shade house)**: This is most predominantly pronounced as shade house which are established using 30% / 50% / 70% agro shade net depending on the species in order to filter the intensity of light towards hardening the clonally multiplied plants.

Size : (40 x 100 ft, 60 x 100 ft, 80 x 120 ft)

(v) **Open nursery with facilities for sprinkler irrigation**: The open nursery should have facility for good water coupled with irrigation through automated sprinkler system. This will ensure uniform growth and development and assures higher survival.

Other requirement of Hi-tech nursery

Containers: All the Hi-tech nurseries are now equipped with wide range of containers particularly the root trainers which will facilitate easy rooting and avoid root coiling. This also ensures contamination free cultures and facilitates easy transportation.

Media: The Hi-tech nursery should have improved potting media resources like decomposed coir compost, saw dust, vermiculte, perlite, Styrofoam etc. in order to facilitate good water holding capacity, porosity and ensures rapid root development.

Other tools: The scissors, secateurs, pruning saws, sprayers, digging forks etc. are essential for handling the cuttings. The mobile trolleys are also very essential in Hi-tech nursery towards handling and shifting the clonal plants.

52. Explain the role of fire in silviculture of *Shorea robusta*? (2011)

Sal is one of the promising timber species belong to the family Dipeterocarpaceae with the scientific name of *Shorea robusta*. Normally Sal is light demander but behaves as a shade bearer. Sal is regenerated through both natural as well as artificial. Among them natural regeneration of sal has been attracting the attention of foresters for a long time. The sal natural regeneration is facilitated or augmented by the fires. This is possible due to the burning of leaf litter under the sal forests. In dry deciduous forests the accumulation of leaf biomass hinders the germination of fallen seeds. Hence to enhance germination of sal controlled burning is practiced to drive away the loose biomass and allow the seeds to germinate. In case of very moist types burning or cutting down weeds and climbers annually for some years can produce good seed bed suitable for germination.

In sal forests with a broken or open canopy subject to annual or periodic fires plenty of regeneration comes in and persist for a long time in the whippy form where protection can only produce conditions for the growth of seedlings. In sal forests a fire is generally good preparation for regeneration, but is unfavourable to the existing regeneration. Hence fire is a pre-requisite for germination of sal forests but it should be managed efficiently in order to protect the growing regenerants.

Fire has been considered one of the main factors affecting both positive as well as negative towards sal stand development. Controlled burning of leaf litter before seeding was used to ensure good regeneration.

Fire was extensively and intentionally used to promote regeneration and maintain the sal forest as a climax type in wet sal forests regions of India.

Studies are also available that sal trees can resist fire once they have passed the sapling stage and hence it is regarded as fire tolerant or resistant species.

53. What is a regeneration stock map? How it is prepared (2011)

Regeneration stock map is a map which indicates an accurate delineation of the extent of young regeneration and advance growth is specially significant in case of crops managed under the conversion, uniform or the group system.

The regeneration stock maps are usually prepared in the scale range of 1: 15000 to 1:5000 scale.

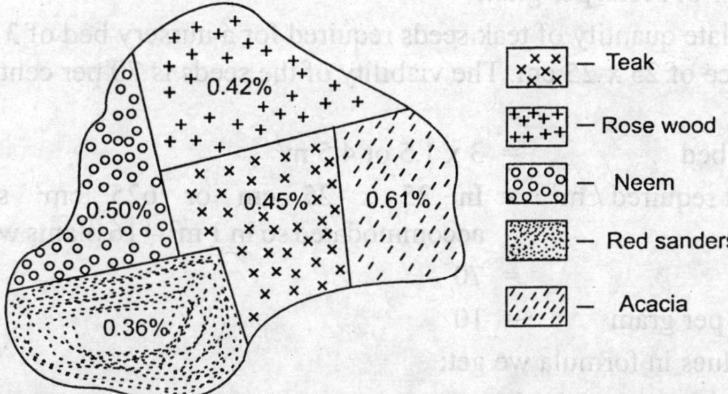

Method of recording

S.No.	Symbol	Category of Regeneration	Remarks
1.	e	Woody shoots of estt. Ht. or over. (Weightage Factor-W.F=5)	*e* symbol indicates presence of atleast one such plant sufficient to stock the quadrat.
2.	w	Woody shoot, less than estt. Ht., healthy and vigorous to become estd.	In the absence of (1), symbol *w* indicates at least one such plant to stock the quadrat completely.
3.	u+	Non-woody (whippy) old, unestd. Regeneration (W.F.=2)	In the absence of (1) and (2), symbol *u*+ indicates more than one such plant.
4.	u	-do- -do –do-(W.F.=1)	-do- -do- one such plant only

5.	r	Recruitment (W.F.=0.1)	In the absence of 1, 2, 3 and 4, *r* may be recorded.
6.	o	No seedlings: blank	-

54. How do we calculate the seed requirement of a species while raising nursery? Also explain the method of calculating the number of plants require / ha of plantation (2011)

A rough estimate of the quantity of seeds that will be required for sowing in a specified bed can be calculated with the following formula:

$$W = \frac{A \times D}{P \times N}$$

Where

W = Weight of seeds required,

A = Area of nursery bed (m^2)

D = Number of plants required per m^2,

P = Plant per cent of species

N = Number of seeds per gram

Example: Calculate quantity of teak seeds required for a nursery bed of 3 x 1.5 m, which are to be sown at a distance of 25 x 25 cm. The viability of the seeds is 70 per cent and number of seeds per gram is 10.

Area of nursery bed = 3 x 1.5 or 4.5 m^2

Number of plant required / m^2 = In 25 x 25 cm or 625 cm^2 space 4 plants are accommodated so in 1 m^2 – 16 plants will be accommodated.

Plant per cent = 70

Number of seed per gram = 10

Putting these values in formula we get:

$$W = \frac{4.5 \times 16}{70 \times 10} \times 100$$

$$W = 10.28 \text{ g (approx)}$$

Method of calculating number of plants / ha and spacing in di3erent patterns of planting

Pattern of planting	Number of plants / ha	Spacing (if number of plants / ha is known)
Rectangular	$\frac{100 \times 100}{\text{Plant – Plant} \times \text{Row – Row}}$	$\frac{100 \times 100}{\text{No. of plants ha} \times \text{distance between row-row}}$
Square	$\frac{100 \times 100}{(\text{Planting Distance})^2}$	$\sqrt{\frac{100 \times 100}{\text{No. of plants/ha}}}$

Triangular	$\frac{100\times100\times1.155}{(\text{Planting Distance})^2}$	$\sqrt{\frac{100\times100\times1.155}{\text{No. of plants/ha}}}$
Quincunx	$\frac{2\times100\times100}{\text{Distance between plant in row or column}}$	$\sqrt{\frac{2\times100\times100}{\text{No. of plants/ha}}}$

Value obtained for spacing of square, triangular and quincunx is expressed in m x m and of rectangular method in m. To obtain number of plants in more then one hectare or less than one hectare, the value obtained per hectare is multiplied with the size of the area.

55. What are the traditional practices and reason advances in tropical silvicultural research? (2012)

Silviculture play a critical and significant role in managing the forests and producing the useful products towards the meeting the domestic and industrial requirement. Traditionally the silviculture followed natural system of management with a main focus on controlled and regulated harvesting and protective measures to sustain or increase the commercial value of subsequent stands based primarily on the natural regeneration of native species. But the major problem and issues indicated are selective harvesting reduces the number of good phenotypes and may even eliminate their population totally. The traditional forest management has never take into consideration the recent development forestry science and technology that led to poor productivity of less than 1m^3/ha/annum. This low forest cover coupled with poor productivity has ushered in a total mismatch between demand and supply besides closing the natural forests for production purposes due to policy and legal regulations. This necessitated introduction and implementation of modern silvicultural research leading to productivity improvement on a sustainable basis.

Recent trends in silvicultural research

(i) **Development of high yield varieties for plantation forestry:** The recent trends in tree research have started with development high yielding varieties in order meet the growing demand. Accordingly several species like Poplars, Eucalyptus, Casuarina, Teak, Melia, Bamboos etc. have been researched and high yielding varieties have developed and introduced as a high yielders with a productive potential of over 20m^3/ha/annum.

(ii) **Development of short duration varieties:** The research has now been directed towards generating short duration varieties which are amenable for harvest in less than 5 years for soft wood species and less than 20 years for hard wood species. This has generated wide range of species amenable for harvest in short period viz., Casuarina (3 years), Eucalyptus (3 – 5 years), Melia (3-6 years), Poplars (3-8 years), Bamboos (Annual harvest) which has attracted the profitability of plantation forestry in the country.

(iii) **Inventory and identification of alternate industrial wood species:** The recent science and technology development has also attracted silvicultural research towards development of new products like composite woods or engineered products as a replacement to the existing timber based products. This resulted in identification of new species amenable for varied industrial utility like Kadam for plywood, Melia for plywood and particle board, Acrocarpus for plywood and timber, Chuckrassia for timber and other medium density products.

(iv) **Precision silvicultural research:** The silvicultural research has gained significant attraction due to the advancements made in precision agriculture and horticulture. Similar research approach has been directed towards silvicultural research which developed technologies for rapid growth and development amenable for varied industrial utility.

(v) **Fertigation research:** The silvicultural research leading to optimization of irrigation and fertilizer is of recent origin and attracted several industries and research organizations to adopt this research as a replacement to traditional silvicultural management.

(vi) **Industrial agroforestry research:** The policy and legal directions from state and central government have compelled several wood based industries towards development industrial agroforestry plantations. However for want of suitable research advancements the achievement in this front was very low till the recent past. But during the last one decade the advancements made in this sector have resolved issues in production, processing and value chain management.

(vii) **Value addition and value chain management:** Yet another advancement made in silvicultural research is value addition of plantation residues and augmenting the associated value chain management. The value addition of plantation residues in the form of briquettes, pellets, activated carbon, biomass power generation, ethanol generation etc., have attracted several industries towards silviculture.

Hence the recent advancements made in silvicultural research have replaced the traditional management system and improved the productivity and profitability of plantation sector on a sustainable basis besides protecting the natural forests.

56. Ectomycorhizae and Endomycorhizae. (2011, 2013 and 2015)

Botanically, the mycorrhiza is a mutualistic association (non-pathogenic association) between soil borne fungi and the root of the higher plants. The term mycorrhiza (Gr.Mykes = Fungus or Mushroom; Rhiza=Root) i.e. "Fungal Root" was coined by the German Pathologist Frank in the year 1885, to describe the union of two different beings to form a single morphological organ in which the plant nourishes the fungus and fungus does the same for the plant.

Ectomycorrhizas

Ectomycorrhizas are mutualistic symbiotic associations formed between roots of higher plants with certain soil fungi. The mycorrhizal fungi benefit associated host trees in a number of ways although the most important is by enhancing nutrient mobilization particularly for elements with low mobility in the soil (N and P) and several micronutrients, release of nutrients from mineral particles or rock surfaces via weathering, effects on carbon cycling, interactions with mycoheterotrophic plants, mediation of plant responses to stress factors such as drought, soil acidification, toxic metals, and plant pathogens, and rehabilitation and regeneration of degraded forest ecosystems, as well as a range of possible interactions with groups of other soil microorganisms .The fungi in return get benefited by receiving 30–60% of the net photosynthate produced by the host . Worldwide ectomycorrhizal fungi are represented by 343 genera including 11,950 species, of which 252 genera belong to Basidiomycota, 84 to Ascomycota, and 5 to Zygomycota.

Endomycorrhiza

An Endomycorrhiza arbuscular mycorrhiza (plural mycorrhizae or mycorrhizas) is a type of mycorrhiza in which the fungus penetrates the cortical cells of the roots of a vascular plant. Arbuscular mycorrhizae (AMs) are characterized by the formation of unique structures such as arbuscules and vesicles by fungi of the phylum Glomeromycota (AM fungi). AM fungi help plants to capture nutrients such as phosphorus and micronutrients from the soil. It is believed that the development of the arbuscular mycorrhizal symbiosis played a crucial role in the initial colonisation of land by plants and in the evolution of the vascular plants.

Endomycorrhiza	Ectomycorrhiza
The most important type of mycorrhizal association found in all tropical Zones is known as endomycorrhizae.	The most important type of mycorrhizal association found in temperate regions is known as Ectomycorrhizae
The endomycorrhizae characteristic of all colonize are present in the tropical forest trees.	The ectomycorrhizae characteristic of all colonize temperate to boreal forest trees, are also present in the tropical forest trees.
Most of the endomycorrhizal fungi belong to Zygomycota	Most of the ectomycorrhizal fungi belong to Basidiomycetes; some also belong to Ascomycetes
Arbuscular mycorrhiza (plural mycorrhizae or mycorrhizas) is a type of mycorrhiza in which the fungus penetrates the cortical cells of the roots of a vascular plant.	Fungus grows into the root of the host plant and the hyphae penetrate between outer cortical cells, forming a typical structure called the "Hartig net".
They form Arbuscules and Vesicles	On the root surface the fungus forms a mantle or sheath, a structure, typical of ectomycorrhizae.

57. Treatment for breaking of Exogenous dormancy. (2009)

It is otherwise called as seed coat dormancy which is due to hard or impermeable seed coat or pericarp. This dormancy is further classified into physical, chemical and mechanical dormancy.

(i) **Physical dormancy**: This dormancy is predominantly due to impermeable seed coat by the presence of thick walled, cutinized palisade like macroscleroid cells layer.

Eg: *Acacia, Prosopis, Albizia* spp.

(ii) **Chemical dormancy**: This is due to the presence of chemical inhibitors present on the seed coat.

Eg: *Betula* Spp.

(iii) **Mechanical dormancy**: This dormancy is due to the presence of hard woody fruit wall especially endocarp and sometimes mesocarp. The endocarp acts as a mechanical barrier to a germinating embryo.

Eg: Species of *Terminalia, Pterocarpus*

Seed treatments to resolve dormancy

Different types of dormancy can be treated differently to induce and enhance germination. The following seed treatment methods will help to improve the germination.

(a) **Mechanical scarification:** It is one of the widely practiced method of seed treatment to resolve dormancy in hard coated seeds. Large quantities of hard coated seeds like *Acacia, Albizia,* Sandalwood etc. can be mechanically scarified mixed with gravel or sand to break the seed coat dormancy.

(b) **Water treatment**: Water treatment is intended both for softening the seed coat and to leach out the chemical inhibitors present on the seed coat. Water soaking is normally done for a few hours to few days depending on the species and the demand for treatment.

For example, Ailanthus is treated in cold water for 2-3 days continuously changing the water daily.

Acacia species are treated with 2-48 hours of soaking depending the requirement.

(c) **Acid scarification**: Concentrated sulphuric acid is predominantly used in this scarification to soften the seed coat. The seeds are scarified by pouring and undiluted acid until all the seeds are wetted. The seeds with acid are continuously stirred using glass rod or a wooden stick for 5-10 minutes to 30-60 minutes depending on the species. The scarified seeds are then washed with running tape water for 5-10 minutes to remove all traces of acid. Seeds like *Prosopis juliflora*, *Leucaena leucocephala*, *Acacia nilotica* are scarified in this method.

(d) **Dry heat treatment**: This method is used to remove the pericarp without damaging it. In this method, the inflammable materials are spread over the fruits and the inflammable materials are fired until the pericarp is removed without damaging the seed embryo.

For example, species like *Acacia mangium* at 100°C and *Tectona grandis* at 50°C found effective in germination.

(e) **Warm stratification**: In this method, seeds are soaked in water for several hours and then seeds are mixed with sand, peat or vermiculate and stored at temperature ranged between 20 and 30 °C alternatively. This method is found effective in many species.

(f) **Cold stratification**: Some species experiences physiological dormancy which is overcome by cold stratification. In this method, seeds are soaked in water for 48 hours at 3-5° C. After that the water is drained and the moist seeds are stored in 3-5°C for few hours to several days depending on the species. Several species are found amenable for this method.

(g) **Biological scarification**: Several animals and micro organisms are found responsible for breaking the seed hard coat and improving the permeability. The animals after feeding the fruits or pods deliver the seeds through their excreta. These seeds are found amenable for easy germination due to the combined action of moisture, warm and chemical reaction of digestive juices which softens the seed coat. The species like *Santalum album*, *Prosopis juliflora* and *Melia dubia* expressed high germination from the seeds dropped by the animals.

(h) Hormonal treatment: It is found that several growth hormones like gibberlins, auxins, ethylene, cytokinnis etc. are found to break the seed dormancy in several tree species.

(i) Chemical treatment: Similar to hormonal treatment some of the chemicals like potassium nitrate, hydrogen peroxide and thiourea are found very effective in breaking the dormancy and improving the germination.

58. Explain the evolution of the concept of plant succession. (2013)

Succession

Succession refers to the gradual replacement of one set of biotic communities or biota by another set of different nature. The species and composition will change in due course of time. The phenomenon, thus, taking place is known as succession. The development of the biological community, beginning with unoccupied sites and proceeding without disturbance is called primary succession. Succession, which is subsequent to a disturbance, which destroys existing biotic community either partially or totally is called secondary succession.

Evolution of the Concept of Plant Succession

- The concept of plant succession has been developed during the last century by observation of changes in isolated cases. Starting with Richard Peters who in 1806 described the departure of southern pine timber as a proof of the tendency in the nature to a change in products on the same soil.
- Dawson (1847) and Thoreau (1863) described the changes in localized places. It was Thoreau who, first, used the term forest succession.
- Later Cawles (1899) described the change in vegetation on sand dunes.
- It was, however, Clements (1916) who developed the idea and elaborated the theory of plant succession. While the American foresters were working on this concept, foresters of England and the continent were also working on the idea. However, it is the Clements theory of plant succession that is widely accepted.

The basic features of the concept of plant succession are:

(i) There is a continual change in the vegetation as a result of interaction of plant community and the habitat factors

(ii) The succession is inherently and inevitably progressive and the end product is the climax and

(iii) The succession is the progressive development of vegetation on the same site in course of time

59. Describe the tangible and intangible benefits of forests? (2011)

Forests play a vital role by extending tangible and intangible benefits. The tangible benefits include supply of wood and non wood products coupled with meeting other direct demands of the society. The intangible benefits include the role of forests in protection and conservation of environment, soil, habitat, water, biodiversity, wild animals etc. The tangible and intangible benefits are discussed elaborately under the following heads.

Tangible benefits

(i) **Fuel wood:** Wood is used as fuel for thousands of years, until the advent of coal, oil, gas, electricity, etc. Wood constitutes as chief source of fuel. Even today more than half of the total world consumption of wood is for fuelwood. Wood remains the major source of domestic fuel in India. Approximately 175 million m^3 of wood is used as fuel in the country. It is estimated that by 2010, most of the 3 billion people who depend on it for their daily living will find it hard to obtain. Already rural families spend precious hours in collecting, firewood instead of other productive work that causes losses to the tune of US$ 50 billion to the world economy

The important fuel wood species are *Acacia spp.*, *Casuarina equisetifolia*, Prosopis, *Azadirachta indica, Leucaena leucocephala*, etc.

(ii) **Timber:** Forests act as a resource for supply of timber needs of the society till the recent past. However after the Supreme Court order 1996, only timber from approved working plan areas alone is available. Currently the trees grown outside the forests in the form of agroforestry is able to provide over 70% of the timber demands of the country.

Major timber species: Teak, Sal, Rosewood, Pterocarpus, Terminalias, Cedar

(iii) **Non timber resources:** Forests provide wide range of non timber products like tendu leaves, tamarind, amla, gums, resins, dyes, honey, medicinal plants, grasses, TBOs etc. which provide direct benefit to the society through income generation activity. The NTFP contributes over 60% of the country's forestry export and thus it is one of the major tangible benefit.

(iv) **Fodder:** Forests extend fodder to the cattle due to their richness in protein, fiber and fat. The forest is India meet about 30% of the fodder requirement through provision of grazing and browsing facility. It also supplements the fodder needs of the cattle during lean season.

Fodder trees: Acacia, Albizia, Hardwickia, Pterocarpus, Robinia etc.

(v) **Industrial raw material:** Forests play a vital role in supplying raw material to the wood based industries like pulp and paper, plywood, veneer, pencil, sports goods, packing, construction industry and agriculture implements. It has been estimated that the country needs over 153 million m^3 of wood as a raw material requirement of various wood based industries and forests are the major resources.

Pulp and paper	:	Bamboos, Eucalyptus, Casuarina
Plywood	:	Teak, Rose wood, Terminalia
Packing cases	:	*Pinus* sp., Silver oak, Fir
Matchwood	:	Ailanthus, Simaruba, Bombax
Toys	:	Adina, Red sanders, Rose wood

(vi) **Dendro power:** Forests also contributes towards power generation. It is estimated that 1.5 tonnes of wood is able to generate one MW power which will help to generate decentralized power production and help to resolve the energy crisis.

Example: Prosopis, Casuarina, Eucalyptus, Subabul, Acacia

Intangible benefits

(i) **Forest and Hydrological cycle:** Forests are vital resources for watersheds. Water and fertile soil are most important pre-requisite for food security. The forests due to their thick humus layer and the soil retention capacity of tree roots are very important towards maintaining and regulating water flows and sub soil water regimes. Almost all the water bodies like rivers are originated predominantly from forest lands and forest derived water tables.

(ii) **Forest and ecological security:** Forests are the habitats for diverse species and sustain the ecological diversity. The viable populations of flora and fauna housed within the forests, help to maintain the ecological food chain and maintains the ecosystem sustainably. This will help to protect the flora and fauna.

(iii) **Forest and ecotourism:** The concept of ecotourism is gaining significant attraction due to urbanization and population explosion. For this purpose the forests and other protected areas are acting as a hub for ecotourism activities. Forests indirectly help to create recreational benefits coupled with aesthetic values which are very essential for sustenance of human life.

(iv) **Forest and pollution:** Forests are excellent sink for combating various pollutants. They absorb more carbon-di-oxide and release oxygen through photo synthetic process and help to balance carbon concentration in the atmosphere. Due to this process, the forests help to reduce the global warming and mitigate acid raining. Forests are one of the best land use system to mitigate and adapt to growing climate change associated risk.

(v) **Forest and climate:** Forests help to moderate the climate due to its multifarious actions. The interception of solar light by the tree canopy helps to augment the micro climate by reducing the temperature and the associated evapo transpiration. The increasing humidity and the increase in precipitation are also essential intangible benefits extended by the forests.

(vi) **Forest and soil:** One of the major intangible benefits extended by the forests is its role in soil conservation. The tree intercepts soil and water erosion and improves the soil physical, chemical and biological properties through the addition of leaf litter followed by decomposition. The soil structure is enhanced which lead to increased permeability and arrests soil and water erosion thereby protect the soil.

Chapter

2

Silvicultural System

1. What is the di-erence between coppice with standards system and the shelter wood coppice system? (2005)

Difference between Coppice-with-Standards System and Shelter Wood Coppice System

S. No.	Coppice-with-Standards System	Shelter Wood Coppice System
i.	Standards are seedling origin and the age will be multiples of coppice rotation. These are the permanent feature throughout the life of the crop	Some standards are retained as over wood against frost-protection.
iii.	The main objectives of standards in are to produce large-sized timber, to protect the regeneration against frost, to enrich the coppice growth and to augment the vigor of coppice	Standards are retained for frost-protection
iii.	It has two rotations. One for the coppice and the other for standards	Applied in frosty localities and long rotation species.

Coppice-with-Standards System

Coppice-with-Standard system is a silvicultural system where the regeneration is based on coppice origin and the overwood of standards are usually of seedling origin and composed of trees of various ages which are kept over coppice over periods. The age of standards will be multiples of coppice rotation. Thus this system has following two peculiarities.

A. **Constitution of crop:** The crop worked under coppice-with-standards system is composed of two distinct parts;

(i) A lower storey which consist of even-aged coppice crop, treated on principles of simple coppice, and usually grown for firewood or poles; and

(ii) An over-wood consist of scattered trees, preferably of seedling origin and of different age-classes, treated as a high forest on principles of selection system, usually grown for larger-sized timber.

B. **Rotation:** This system has two rotations, one for the coppice and the other for the standards. The standards are kept in this system to

(i) Supply of large-sized timber

(ii) Protect against frost

(iii) Enrich the coppice growth

(iv) Provide seedling regeneration to maintain vigour of the coppice

(v) Increase the revenue.

The Shelter Wood Coppice System

The modification of the simple coppice system is Shelter Wood Coppice System wherein the first clear felling some Shelter wood are retained towards protection from the frost. This system was developed by Trevor, suitable for frosty localities.

2. Describe the Indian irregular shelter wood system with examples of its application in tropical rain forest? (2005)

The Indian Irregular Shelter Wood System

The Indian Irregular Shelter wood System is a silvicultural system in which the crop to be regenerated is opened up irregularly and the resultant crop is uneven-aged. It is modification of the uniform system, where large quantities of advance growth of various ages and sizes, and trees even up to 40 cm (16 inches) diameter, are retained as part of the future crop. It is also stated that this system had been described by Glover as the Punjab Shelter Wood System in case of deodar forests. As the uniform character of the crop is lost by retention of trees up to 40 cm diameter, the uneven-aged and irregular nature of the crop necessitates separate treatment of the system.

This system has two special characteristics which differentiates it from the uniform system

- It facilitates the retention of groups of well-grown poles and mature trees as part of the future crop
- It permits the adoption of selection fellings on steep and rugged portions of compartment being worked under uniform system.

Pattern of felling

The felling of the crop to be regenerated is irregular and does not follow any regular pattern.

Nature of crop produced

As the immature trees of the old crop are retained as part of the future crop, and as regeneration appears slowly over a long period, character of the resultant crop is uneven-aged.

Application of irregular Shelterwood System in tropical rain forest:

Irregular Shelterwood System followed in the areas where the number of mature and over-mature trees is proportionately higher, the surplus growing stock of such trees should be removed in the first one or two felling cycles to facilitate the application of Irregular Shelterwood System.

(a) Tropical rain forest in Arunachal Pradesh: Tropical rain forests are being worked under Indian Irregular Shelterwood System. Under this system, all trees above 60 cm diameter of the top canopy are removed in one operation, while those below this limit are retained as a part of the future crop. The trees of the middle storey are not felled, to control weed growth. The regeneration of principal species, *Dipterocarpus macrocarpus* and *Shorea assamica* present in seedling and sapling stages, is tended and supplemented. For this 1.2 m wide strips spaced 10 m apart are cleared. The regeneration in these strips is weeded and cleaned by removing all unwanted weed growth and miscellaneous species. The gaps in the regeneration are filled either by sowing 6 to 8 seeds of *Dipterocarpus macrocarpus* in thalies1.8m apart or by planting polythene bag plants of *Shorea assamica.* The tending operations are carried out for 3 years successively, and thereafter in 5th, 7th and 10th years when regeneration in the areas gets established.

(b) Tropical rain forest in Andamans: The evergreen and deciduous forests of Andamans are also worked under Indian Irregular Shelterwood System though it is locally called as 'Andamans Canopy-lifting Shelterwood System'. Under this system, all saleable trees above a prescribed g.b.h., which is 150 cm for hard-woods, 120 cm for soft-woods and 180cm for others, are felled and their timber is expeditiously removed. All sound trees of commercial species, below the prescribed g.b.h. mentioned above, are left as advance growth. Closely following the extraction of timber, canopy lifting is done by:

(a) Felling all poles of non-commercial species and under growth upto 10 m height

(b) Girdling trees of 10 m to 20 m height not needed as seed bearers or as a part of the future crop.

3. Explain the selection system and its application to Sal forest? (2006)

Selection system

Selection system is defined as the silvicultural system where felling and regeneration are distributed over the whole of the area and the resultant crop is uneven-aged. In selection system, regeneration operation is carried out throughout the life of crop.

Application of selection system in Sal Forests

The moist and dry sal forests, occasionally moist-bhabar and teraisal, plains sal, and considerable areas of peninsular sal forests are worked under a system of selection. The chief aim is to obtain maximum sustained yield of timber from trees of exploitable diameter as far as silvicultural conditions permit. In addition, trees below the exploitable diameter are preserved under best possible conditions of growth by thinning. Thus the following two operations are carried out at the time of main felling:

(a) Removal of trees of and above the exploitable diameter: Trees of and above the exploitable diameter are marked upto the numerical limit fixed by the Working plan provided they are silviculturally available. A selection tree is considered silviculturally available it is removal.

(i) Does not create a permanent gap in the canopy

(ii) Tends to improve the remaining crop in terms of production.

(iii) Helps the growth of groups of young trees down to the sapling stage of any of the valuable species.

(b) **Thinning trees below the exploitable diameter:** In order to nurse up all trees below the exploitable diameter under best conditions of growth, thinning is carried out therein.

In the year following the main felling, usual cultural operations are carried out to minimize the after effects of felling. According to some recent working plans, along with cultural operations, the regeneration is also attended by carrying out the following:

(i) Cutting back all badly grown saplings and suppressed poles if there is reasonable hope of obtaining good coppice shoots

(ii) Cleaning in groups of sal saplings

(iii) Removal of undergrowth interfering with the proper growth of sal regeneration in woody or large-leaved whippy stage

(iv) Fencingof compact groups of young regeneration over about 10 hectares or more in accessible areas.

Merits and Demerits

- Risk of soil deterioration and erosion is minimum.
- Marking and felling of trees of the over wood are simpler than the selection system.
- Vulnerability due to invasion of other species like weeds and grasses is minimum.
- The young recruits are well protected against adverse climatic factors.
- The new crop is superior since regeneration is obtained from the seeds of best selected trees.
- The system is suitable for regenerating both light-demanders as well as shade-bearers.
- The success of regeneration and control over the distribution of age classes is fairly clear.
- It reduces the risk of multiplication of injurious insects.
- The average length of rotation is shortened as compared to clear felling system.
- The growing space is effectively utilized.
- This system produces even-aged crop of trees with long clean cylindrical boles.
- Supervision and control of various operation are easy.
- Aesthetic point of view, it is preferable over the clear-felling system.

Demerits

- As the over wood is removed in series of operation, there is much damage to the young crop.
- Manipulation of canopy requires skill in case of mixed forests with the species having different light requirements.
- The isolated seed bearers are susceptible to wind damage
- There is possibility of invasion by weeds after seeding felling and regeneration may be affected.
- For the species with longer regeneration period, weeding and cleaning have to be done for longer period and natural regeneration becomes costly.

4. Explain the kinds of pattern of fellings in uniform system? (2006)

Uniform system is silvicultural system "wherein the canopy is uniformly opened up over the whole area of a compartment to obtain uniform regeneration. The uniform system is otherwise called shelter wood compartment system or, compartment system".

Pattern of felling

(i) Seeding Felling
(ii) Secondary felling
(iii) Final felling

(i) Seeding Felling: Seeding felling is the opening of mature stand canopy in order to provide favorable conditions like soil moisture and light etc. for obtaining natural regeneration from the seeds of trees retained for the purpose. This is the first stage of regeneration fellings under a shelter wood system.

(ii) Secondary felling: Secondary felling is a regeneration felling carried out between the seeding felling and the final felling under a shelter wood system. During the secondary felling, shelter woods are removed gradually so as to admit more light to the regenerated crop. If the regeneration requires more light, some of the seed bearers retained at the time of seeding felling are also removed. The secondary fellings not only improves the light availability but also helps in manipulation of species composition.

(iii) Final felling: In the final felling, the last seed or shelter trees after the successful regeneration areremoved. It is the final stage in the regeneration fellings. Final felling is carried out when the whole regeneration area is completely stocked with established regeneration and no longer requires any shelter against adverse condition.

Regeneration

The regeneration is predominantly obtained through natural, or sometimes it is supplemented by artificial regeneration

Tending

After the seeding fellings, the slash left on the forest floor is cut and collected away and burnt during winter. Once the regeneration has started, weeding and cleaning are done regularly till the entire area is completely regenerated.

Nature of the crop

The crop produced in this system is even-aged.

Regeneration period

Regeneration period is the time period required to regenerate the whole periodic block.

Periodic blocks

Periodic block defined as the part or parts of a forest set aside to be regenerated or otherwise treated during a specified period.

Length of the Regeneration Period

The following are some important factors which affect the length of the regeneration period to regenerate an area or a periodic block

(i) Frequency of seed years
(ii) Light requirement
(iii) Other climatic factors
(iv) Soil conditions
(v) Condition of grass and other competing weed growth
(vi) Incidence of grazing and browsing
(vii) Fire

5. a. How are forests managed under the clear-felling system? (2007)
b. Briefly discuss a Silvicultural system in which equal or equi-productive areas of mature crops are successfully felled. (2010)

Clear felling is the silvicultural system in which equal or equi-productive areas are sequentially clear felled is one operation and regenerated mostly artificial and some time naturally also.

The process of clear felling

The process of clear felling is normally every year and the area to be clear felled is arrived at 1/n of the total area allotted to the system. Where n is number of years in the rotation and it is usually referred as annual coups. The coupes are divided into equi-productive areas and felling is done every year generally is clear felling system and the felling and regeneration are carried out throughout year and this system ensure normal series of age-gradation.

Forest management under clear felling system

The regeneration and felling operation are the major activities in clear felling systems which need to be managed effectively. In general the clear felling areas are to be mapped with clear boundaries.

The felling of trees is executed through auctioning and the bidder who offered maximum money will fell the tree and clears the area at one time harvest.

Regeneration of clear felled areas

Regeneration is normally carried out by following two methods

(i) Departmental plantations
(ii) Taungya plantation

(i) Management under Departmental plantations: After clear felling the area is surveyed and inspected and soil map should be prepared. After this treatment of suitability map is to be established. Once it is over then the area has to be fenced and other preparation works like land preparation, pitting, planting, irrigation management, application of fertilizer and beating out operations are carried the forest department towards the establishment of plantations.

The plantation further managed by forest department through periodic weed, protection against fire in order to ensure satisfactory growth and development.

(ii) **Taungya Plantation:** It is the methods of raising forest plantation along with agricultural crop and based on the management the taungya are classified in to departmental, leased and village taungya.

(a) **Departmental Taungya:** In this method, the forest department raise forest plantations and co-cultivate agricultural crops. The main objective of raising agricultural crops is to reduce the weed growth and to get supplementary revenue.

(b) **Leased Taungya:** In this method, the forest department raises the forest plantations and the area is leased to the person who offers maximum lease money. In this case, the area is leased immediately after clear felling for a period of two years and one year after planting trees. This type of management ensures full protection to the tree due to the involvement of local people.

(c) **Village Taungya:** This method is done by forest settlers or local village and the entire management is vested with the local communities. The raising of forest plantations, tending operations and protection operations are all carried out by the settlers at their own cost.

Method of regeneration

In most cases, under clear felled area the method of regeneration is artificial and rarely natural.

Nature of the new crop

The crop is even-aged and with the normal series of age gradations.

Merits

- It is the simplest silvicultural system and not requires a high degree of skill.
- The cost of felling and extraction is less and introduction of mechanization in conversion and extraction is possible.
- Introducing fast growing exotics and regulating species composition through artificial regeneration is advantageous.
- Supervision of all operations is easy.
- Damage to the new crop by felling is nil.
- Production of trees with relatively cleaner and more cylindrical boles is possible in clear felling system.
- Cost of regeneration operations and length of rotation are minimum.
- The coupe can be opened up for grazing soon due to the presence of advance growth.
- Establishment of normal forest can be achieved easily through clear felling system.

Demerits

- Among the various systems, it is the most artificial one.
- Danger of soil deterioration and erosion is high.
- Threats due to weeds, grass invasion and fire hazard is high.
- The produced even aged crop is less resistant to damage by wind and snow.
- The resultant pure crop is more vulnerable to the damages caused by insects, plant parasites and pathogens.
- During initial period, it sacrifices all the immature trees which are under active growing period.
- This system fails to consider the site factors.
- Aesthetic point of the forests this system is undesirable.

6. Discuss the salient features of shelter wood system of forest management with merits and demerits. (2007)

Shelter wood system

Shelter wood system is the silvicultural system aiming at concentrated regeneration through gradual removal of over wood in two or more successive fellings depending on the success of regeneration. The primary objective of this system is to protect and provide shelter to the young recruits.

Salient features of shelter wood systems

- In shelterwood system the vigorously growing trees are retained with the objective of providing shelter to the new regenerates.
- The vigorously growing trees are also retained for the purpose of seed production to facilitate natural regeneration
- This system ensures rapid diameter and volume increment of the retained trees
- This system ensures soil protection against all natural deterioration
- The trees retained in this system will be large, dominant, wind firm, desirable species coupled with desirable physical characters

Process

This system involves gradual removal of the entire stand in two or more successive fellings which spread over a part of the rotation. During the regeneration felling the opening up of canopy will facilitate the start of natural regeneration under the shelter of the retained over wood. In the progress of natural regeneration, the over wood is gradually removed and the entire shelter is opened when no longer the protection is required and allowing the new crop to make full use of the growing space.

Merits

- Risk of soil deterioration and erosion is minimum.
- Marking and felling of trees of the over wood are simpler than the selection system.
- Vulnerability due to invasion of other species like weeds and grasses is minimum.
- The young recruits are well protected against adverse climatic factors.
- The new crop is superior since regeneration is obtained from the seeds of best selected trees.
- The system is suitable for regenerating both light-demanders as well as shade-bearers.
- The success of regeneration and control over the distribution of age classes is fairly clear.
- It reduces the risk of multiplication of injurious insects.
- The average length of rotation is shortened as compared to clear felling system.
- The growing space is effectively utilized.
- This system produces even-aged crop of trees with long clean cylindrical boles.
- Supervision and control of various operations are easy.
- Aesthetic point of view, it is preferable over the clear-felling system.

Demerits

- As the over wood is removed in series of operation, there is much damage to the young crop.
- Manipulation of canopy requires skill in case of mixed forests with the species having different light requirements.
- The isolated seed bearers are susceptible to wind damage
- There is possibility of invasion by weeds after seeding felling and regeneration may be affected.
- For the species with longer regeneration period, weeding and cleaning have to be done for longer period and natural regeneration becomes costly.

7. Define Clear-Felling System. How is Sarandasal (*Shorea robusta*) forest managed under clear felling by natural regeneration? (2016)

Definition

The Clear-felling system is defined as a Silvicultural system in which equal or equi-productive areas of mature crop are successively clear-felled in one operation to be regenerated, most frequently, artificially but sometimes naturally also.

Clear-felling technique for Sarandasal

Sal is managed under clear felling system followed by natural regeneration from advanced growth. The advance growth is the established seedlings, saplings and young pole of the sal already present in the coupe before clear felling.

This advance growth of seedlings and saplings have survived and stagnated under suppression from the old trees and also these young regenerates are partly damaged due to clear felling operations

The damaged advance growth seedlings and saplings have to be cut in order to induce vigorous coppice shoot and they are pronounced as seedling coppices. If it is done in pole stage then it is referred as coppice shoots.

The dependence of coppice shoot does not alter the character of the system (i.e) from high forest to coppice forest. In regular coppice system, the coppices are included from the stools of felled main crop but in this clear felling system, coppices have been obtained from stools of the damaged trees. Hence the coppice shoot which will form the part of the new crop are derived from advance growth and not from clear felled trees. This system is followed in sal forests of Bihar, Madhya Pradesh in the clear felled area. However, this advance growth is alone not sufficient to restock the clear felled area and hence this natural regeneration is supplemented with artificial regeneration in case of sal forests.

8. Write notes on permanent and floating periodic blocks. (2009)

Periodic block

Periodic block is the part or parts of a forest set aside to be regenerated or otherwise treated during a specific period.

(i) **Permanent Periodic Block:** This periodic block may be a permanent periodic block one or floating periodic block. Permanent periodic block is the periodic block which retain their territorial identity at each working plan revision

The area allotted to periodic block is determined by the following formulae.

$$\text{Area of Periodic Block} = \text{F.S} \times \frac{P}{R}$$

Where, F.S. is the area of felling series

P is the Number of years

R is the rotation

This periodic block may be located entirely in one place as a compact unit and/or sub compartments. Hence based on the location the permanent periodic block is further classified into

(i) Self contained Periodic Block

(ii) Scattered periodic block

The self-contained periodic blocks is advantageous because all the operations can be carried out in a concentrated manner and makes their supervision easy and reduces the cost. It also facilitates deployment of mechanization in silvicultural operations. At the same time it may not be possible to obtain regeneration uniformly in fixed regeneration period either due to refractory nature of the species or due to unfavourable site factors. These drawbacks can be minimized by using scattered periodic blocks, in which each compartment or sub-compartment is allotted to appropriate periodic block on the basis of its crop

(ii) **Floating periodic block:** Floating periodic block consisting of mature forest areas taken up for regeneration during the working plan period, without any predetermined limit on the size of regeneration block or the period in which it is to be regenerated. At each revision of working plan, fully regenerated compartments or sub-compartments are excluded and fresh area needed for regeneration operations are added. Thus, at each revision of the working plan, the floating periodic block consists of following two categories of areas.

(i) Compartments partly regenerated during the expiring wording plan

(ii) New compartments requiring regeneration operations

The calculation of area and regulation of its allotment to periodic block is a function of forest management, and so it has been only described briefly here.

$$\text{Area of the Periodic Block} = \text{Area of the felling series} \times \frac{\text{Estimated regeneration peirod}}{\text{Rotation}}$$

This method was first introduced in France where the floating periodic block was called *Quartier Blue* because the area allotted to the regeneration Periodic Block is blue coloured on the map, while the rest of the area of the working circle is left uncoloured and called as *Quartier Blank*.

9. Comment on the following statement – "Conversion is an accepted Silvicultural system". (2010)

Yes, conversion is an accepted silvicultural system because of the following reasons and features

(i) To meet the growing demands of domestic and industrial utility

(ii) Yield improvement due to changes in composition

(iii) To protect the soil and other associated natural resources

The conversion system is the change from one silvicultural system or one set of silvicultural system to another. This involves change in crop composition and the associated silvicultural system by which the crops are regenerated and replaced by new crops of distinctive form.

Necessity for conversion system

(i) Advantage of particular system

(ii) Failure of an existing system

(iii) Advances in precision silviculture

(iv) Growing demand for wood and wood products

(i) Advantage of particular system: Suppose if one particular silvicultural system known to offer greater advantages over other systems in terms of higher yield, simplicity of work and improvement of growing stock then adopting the system by removing the old system is preferred. The classical example is in several species selection system replaces by uniform system and coppice with reserve system

(ii) Failure of an existing system: If the introduced silvicultural system apply to species fail to regenerate then it has to be replaced with new system. For example uniform system followed in fir and spruce has been changed to selection system as well as clear felling system. Teak managed under uniform system was converted into clear felling system followed natural or artificial regeneration.

(iii) **Advances in precision silviculture:** Due to the advancement made in science and technology which has led to the development of precision silviculture. Hence there is a need to change the silvicultural system by adopting the newer technology. For example traditionally eucalyptus was managed under coppice system but due to advancement in clonal technology coppice system is replaced with clear felling followed by artificial regeneration

(v) **Growing demand for wood and wood products:** Increasing demand for wood and wood products has completed the productivity improvement of several species. Hence, there is a need to change the silviculture system from the traditional to modern.

10. (i) Successful regeneration in a forest depends upon Silvicultural system. (2012)

(ii) Successful regeneration in a forest stand depends upon judicious choice of a Silvicultural system. Comment. (2010)

The success of any silvicultural system depends wholly on the success of regeneration. The failure of regeneration leads to the failure of silvicultural system. Several species like Teak, Casuarina are worked under clear felling system followed by artificial regeneration. This has resulted in successful regeneration due to completion of artificial regeneration process in a shortest period and hence the silvicultural system is successful. Again it depends on species for example light demanders prefer clear felling system and shade bearer wood prefers shelter wood or selection system.

Suppose if the seed production of the main crop occurs is long interval in fir (*Abies spp*) it is not advisable to adopt shelterwood system with short regeneration period. Alternatively fir regeneration has been achieved through selection system. In certain species natural regeneration is highest, uncertain and erratic due to internal and external seed dormancy. Under such cases, it is better to raise the seedling in nursery and plant them in clear felled areas with or without resources depend light requirement of the species.

Similarly the species which have excellent coppicing ability may not be preferred other silvicultural system. For example Eucalyptus with simple coppice system highly prefer selection and shelter wood system. Some species may produce seeds profusely and the natural regeneration is very high. Under such circumstances, these species has worked under clear felling system followed by natural regeneration.

In some species both natural and artificial regeneration are not possible due to various reasons. For this species it is advisable to follow either Indian irregular system or selection system in order to have successful regeneration.

Hence the successful regeneration of forests depends on the silvicultural system followed.

11. a. Describe the major factors which a-ect the length of regeneration period in a periodic block? (2015)

b. Explain the factors which a-ect the length of regeneration period in a periodic block. (2013)

The length of regeneration period in a periodic block is influenced by several factors and important factors are listed below.

(i) Site and soil condition
(ii) Light intensity
(iii) Requirement at climatic factors
(iv) Development of roots
(v) Presence of under growth
(vi) Grazing and browsing
(vii) Fire
(viii) Drip
(ix) Composition of crop.

(i) **Site and soil condition:** Soil plays a vital role in regeneration and the length of regeneration period. The small seedling has to depend upon the soil for its food, moisture and air and its establishment, depends upon favourable soil conditions. Soil moisture is very important and excess or its deficiency is both deleterious to regeneration and alters the regeneration period. The presence of thick humus layer also affects the regeneration. soil aeration is the another important factor which play a critical role in regeneration and regeneration period. The presence of nutrient in soil is also very important in influencing the regeneration period in periodic block.

(ii) **Light intensity:** Light is a very important factor in seedling regeneration and regeneration period because it varies from species to species and even in same species it varies due to locality and age. For example in moist deciduous forest regeneration of teak is influenced by light, but in dry locality, light is the problem. This light not only influence regeneration but also influences the regeneration period of the species.

(iii) **Requirement at climatic factors:** Extremely high or extremely low temperature are both harmful for seedling establishment. In extremely high temperature, seedlings are killed due to insolation while in extremely low temperature they are killed by frost. For seedling establishment, only adequate rainfall is not essential but its proper seasonal distribution is also essential. Otherwise, the long dry season after the monsoonic rains, kills most of the seedlings.

(iv) **Development of roots:** Growth and development of seedling depends upon the food resources of cotyledon and after that and it has depends on own resources. For this purpose, development of well established root system is very essential to absorb and transfer moisture and nutrients. If the root development does not reach required stage in first growing season, the seedling may be killed by drought and the regeneration period may extend. Hence sufficient root development is very important which besides length of the regeneration period.

(v) **Presence of under growth:** The regeneration and the associated regeneration period is influenced by the growth of grasses and competing weeds. The density of weed and grasses growth has influenced on establishment of regeneration. The dense growth of grass is very harmful and extends the regeneration periods. Only very few species like teak can survive in the dense weedy growth. The dense growth of Strobilanthus and Petalidium is very harmful for teak regeneration. Similarly the dense growth of Clerodendron and Parratia are harmful to sal and deodar respectively. This dense growth of grasses and weeds affect the regeneration and extend the regeneration period. Sometimes the dense growth is advantageous for regeneration of some species particularly the deodar and kail regeneration is high under dense vegetation. Hence the undergrowth plays a significant role in affecting the regeneration period.

(vi) **Grazing and browsing:** Uncontrolled grazing and browsing is very harmful in regeneration establishment and influencing the regeneration period. Sometimes the grazing and **browsing** reduces the growth of seeds and destroys the under composed organic matter which may favour rapid growth in seedling.

(vii) **Fire:** Fire is one of the injurious factors which decide the regeneration and the regeneration period for several species fire is deleterious and for some species like sal, fire is advantageous to enhance the regeneration.

(viii) **Drip:** Drip from the large leaves of species such as sal, teak is very harmful for seedling establishment because it removes soil from the roots of the tiny seedlings in splash erosion thereby exposing the roots resulting in the death of plants. The splashed soil covers the shoot of the tiny seedling and besides preventing it to perform photosynthesis, also kills it by rotting of shoot.

(ix) **Composition of crop:** The composition of the crop affects soil conditions and therefore affects the establishment. A mixed crop is believed to create more favorable condition for seedling establishment than pure crops.

12. Comment upon the variations practiced in the Silvicultural system of *Shorea robusta* in the state of UP, Bihar, Odisha and West Bengal? (2015)

Particulars	Uttar Pradesh	Bihar	Orissa	West Bengal
Silviculture system	Uniform system followed by Indian irregular shelterwood system	Clear felling system	Uniform system	Uniform system
Regeneration	Natural and artificial Regeneration	Artificial Regeneration	Natural regeneration	Natural regeneration
Area divided into	Floating periodic block	Periodic block	Periodic block	Periodic Block
Nature of the crop	Uneven-aged	Even-aged	Even-aged	Even-aged

13. Highlights salient features of (i) Aerial seeding (ii) Stump planting. (2012)

(i) **Aerial seeding:** Aerial seedling is the process of broadcasting the seeds from the atmosphere using through aircrafts / helicopter.

Purpose of aerial seedling

- The main purpose is to vegetate the remote, rugged and inaccessible areas.

- Afforesting denuded and degraded hillocks.
- Regeneration of riverines.

Experiences in India

In India, aerial seedling has been tried in several states and resulted in both success and failures. Several trees species like *Acacia nilotica, Catachu, Tortalis, Prosopis juliflora, Dalbergia sissoo, Dendrocalamus spp*. etc. have been sown through aerial seedling. In India, sambal ravine in Madhya Pradesh, Uttar Pradesh, Western ghats of Maharastra, arid regions of Rajasthan have been attempted with aerial seedling.

In Tamil Nadu, the entire palani hills was aerial seeded with *Acacia nilotica* and *Acacia planifrons*. Though aerial seedling was attempted its successfulness has not been evidenced fully barring few exceptions.

(ii) Stump planting: Stump planting is the method of artificial regeneration which is the planting of cuttings are root shoot ratio cuttings. Stump planting is normally practiced in species like *Tectona grandis, Dalbergia sissoo, Terminalia sp. Pterocarpus marsuipium*, etc.

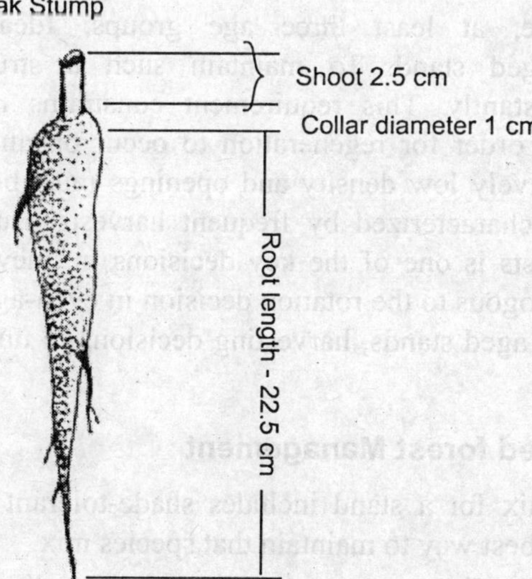

Method of stump preparation

Normally stumps are prepared from the well grown seedlings raised under mother bed system. For example in teak stumps are prepared from 12- 15 months old seedlings grown under mother bed.

For this purpose, the teak seeds after necessary seed treatments are sown in the raised bed and growth for a period of 12 -15 months. Once the seedlings attained the collar diameter ranged between 1 – 2 cm then they are eligible for stump preparation.

The seedlings with 1-2 cm collar diameter are pricked out and the root and shoot portion are cut to a standard sizes. The shoot portion is cut 2.5 cm above collar region and root portion is cut 22.5 cm below the collar region. This shoot root cuttings of 25 cm length in teak is called as Stump.

Similar stump preparation method is applicable to other trees with variability in collar diameter, shoot and root portion.

- It is easy and cheap method of planting stock preparation
- It is very easy and involves cheap transportation with large stock volume
- It requires easy soil preparation
- It requires less labor and cost of product is cheap.
- It is best suited method particularly for in accessible areas
- Stump planting ensure early and better root development and hence assures good survival.

14. Give a short account on the management of uneven-aged forest? (2014)

An uneven-aged forest is a group of trees that differ significantly in ages, by convention, the spread of ages exceeds 25% of the planned life span for an age class. In uneven-aged management, we remove mature trees, or groups of them, leaving gaps and young trees to grow, allocating a portion of the growing space to regeneration.

Uneven-aged stands have, at least three age groups. Ideally, all age groups are represented in an uneven-aged stand. To maintain such a structure, regeneration must occur frequently, if not constantly. This requirement constrains many of the decisions in uneven-aged management. In order for regeneration to occur on an on-going basis, the stand must be maintained at a relatively low density and openings must be created frequently. Thus, uneven-aged management is characterized by frequent harvests usually every 5 to 15 years. The frequency of these harvests is one of the key decisions in uneven-aged management, and this decision is somewhat analogous to the rotation decision in even-aged management. However, in contrast to harvesting even-aged stands, harvesting decisions in uneven-aged stands are quite complex.

Advantages of uneven-aged forest Management

- If the desired species mix for a stand includes shade-tolerant tree species, uneven-aged management may be the best way to maintain that species mix.
- Uneven-aged stands have a diverse structure, with small, medium and large trees providing a multi-layered canopy. This forest structure provides needed or preferred habitat for many plant and wildlife species.
- Many people are offended by the sight of a clear cut. With uneven-aged management, the stand is never clear cut. Thus, an uneven-aged management system may socially or aesthetically beneficial than an even-aged management system.
- Uneven-aged stands provide continuous cover on a site, reducing problems with erosion and excessive run-off after heavy rains.
- Uneven-aged management may be more compatible with many owners' financial constraints. It provides more frequent cash flow, and, because it typically relies on natural regeneration, it has relatively low investment requirements.

- In theory, because the site is always occupied by trees, uneven-aged management could provide more growth than even-aged management. In practice, however, the densities required to maintain adequate regeneration are relatively low, so growth rates are typically lower with uneven-aged management than with even-aged management.

Disadvantages of Uneven-aged Management

- A key disadvantage of uneven-aged management is that it does not work well when shade-intolerant species are desired. Many of the most valuable timber species – including most pines, douglas-fir, aspen, black cherry, hickories and to a lesser degree, oaks – are relatively shade-intolerant.
- Many wildlife species prefer or require habitat provided by even-aged stands. The openings created by clear cuts and other even-aged harvesting methods and the more mature stages of even-aged stands are particularly important.
- Uneven-aged management is complex. A truly uneven-aged stand often is difficult to achieve and maintain. Some management costs are higher because more detailed information about the stand is required.
- While clear cuts admittedly have their disadvantages, the frequent, lighter harvests required for uneven-aged management also have numerous problems. First, since less volume is removed per unit of area, average woods hauling distances and logging costs will be greater. Second, it is difficult to avoid some damage to the residual stand. Third, more frequent entries into the stand increase the potential for site degradation due to compaction and rutting.

15. Explain the terms: (2009)

(i) Seeding Felling
(ii) Secondary felling
(iii) Final felling

(i) Seeding Felling: Seeding felling is defined as opening the canopy of a mature stand to provide conditions for securing regeneration from the seed of trees retained for the purpose. This is the first stage of regeneration felling under a Shelterwood System.

(i) Selection of trees to be retained,
(ii) The distance between the trees retained and their position.

The trees to be retained have to be selected as they are the source of seed. As the quality of seed depends upon the qualities of the parent tree, it is necessary that middle-aged trees with long clean cylindrical boles and well-developed crowns, free from disease and defects or, in short, genetically superior trees, should be selected as seed bearers. The distance between the seed bearers or, in other words, their number per unit area, affects the light, warmth, etc., reaching the forest floor and varies according to silvicultural requirement of the species, and climatic and other factors. The shade-bearing species, and those with heavy seed, require less opening, i.e., larger number of seed bearers than light demanders and species with light winged seeds.

(ii) Secondary felling: Secondary felling is defined as a regeneration felling carried out between the seeding felling and the final felling under a shelterwood system in order gradually to remove the shelter and admit increasing light to the regenerated crop.

When the regeneration resulting from the seeding felling progresses, it requires more light and this is achieved by removing some of the seed bearers retained at the time of seeding felling. Besides the amount of light required, there is another factor which affects the removal of trees and that is the appearance of regeneration. The seeding felling may result in the appearance of regeneration at certain places, while some may still be devoid of it. In such cases, the seed bearers retained at the time of seeding felling cannot be removed from places where the regeneration has not come up. Thus, while some parts of the coupe still require the continuance of conditions of the seeding felling stage others, in which regeneration has progressed, require the removal of some seed bearers to admit more light for the proper development of regeneration. In other words, the removal of trees in secondary felling is governed by the progress of regeneration and its light requirement.

Secondary felling also helps in the manipulation of mixture of species in a young mixed crop. As the requirements of light of different" constituent species vary, these can be favored or eliminated by regulating light over their young regeneration. The number of secondary felling depends upon the species and the progress of its regeneration. For species in which a larger number of seed bearers are retained at the time off seeding felling, the number of secondary felling would be more than for species in which the number of seed bearers retained was less. Thus, while in silver fir and spruce forests, there may be two or more secondary felling, in chir there may be none as the number of seed bearers is only 12 to 18 per hectare, and these can be removed in one operation when the regeneration is established, without the necessity of any intermediate felling. Similarly, in the slow-growing sensitive shade bearers, the number of secondary felling will be more than in comparatively fast- growing light-demanding pine species. Felling, logging and extraction of trees marked for secondary felling do a lot of damage to the young regeneration, particularly when the species does not coppice. Therefore, though, theoretically, two or more secondary felling may be necessary, in practice, either their number is kept to an absolute minimum to minimize the damage to / regeneration or they are not carried out at all.

(iii) Final felling: Final felling is defined as the removal of the last seed or shelter trees after regeneration has been effected under a Shelterwood System. It is thus the final stage in regeneration felling it is carried out when the coupe or regeneration area is more or less completely stocked with established regeneration which no longer requires any shelter against frost, isolation and/or adverse weed competition.

The condition of established regeneration, when considered to be no longer in need of shelter wood, varies with species and climatic and biotic factors of the locality in which it grows. For example, the young crop of chir is considered fit for final felling when the regeneration has attained a height of 2 m to 6 or 7 m on ordinary sites and has been control-burnt at least thrice.

The regeneration interval, which is defined as the interval between the seeding felling and final felling on a particular area, under one of the shelterwood systems, is sometimes increased, *i.e.,* the final felling are deferred to enable the young crop attain a height when logging and conversion can be carried out below its canopy. Though final felling means that all the shelter wood is removed, it is usual to keep 3 to 5 seed bearers per hectare in chir even after final felling, mostly as an insurance against devastating fires and, sometimes, to allow them to put on increment and meet the demands of the local people.

16. Describe the impact of felling height and felling intensity in restocking of bamboo species. (2009)

Bamboos are worked under felling cycles of 2 to 4 years and the felling period exhibited wider variations from state to state. The felling of bamboo is very significant in terms of felling height and felling intensity.

Felling height

The regulation of height at which bamboo should be cut is very important factor in bamboo felling. The minimal height at which bamboos are cut is generally 15 cm above ground level with the condition that at least one node should be left. This will ensure coppice regeneration from the nodal segments besides rhizome regeneration. Normally the felling height varies from place to place and it is 25 cm in Uttar Pradesh, 45 cm in Madhya Pradesh and nearly 1 feet in Tamil Nadu. This felling height helps to include new shoots and ensure restocking of bamboos. The new shoots developed from the cut portion also help to induce the rhizome growth which in turn helps in restocking of bamboo vegetation.

Felling intensity

Though bamboo working is over 100 years old, it experiences different felling intensity. Felling intensity is the planned number of fellings carried out during one particular year is termed as felling intensity.

Bamboos are generally felled at 2 year cycle, 3 year cycle and 4 year cycle depending on its growth and development.

The felling intensity of bamboo varies from site quality to site quality. In site quality I, a minimum of 20 culms should be retained and others can be removed. Similarly in site quality II – 15 culms and in site quality III – 10 culms are to be retained and the remaining can be harvested. The retention of these culms will help in regeneration of rhizomes and restock the bamboo plantation.

17. Briefly describe the selection system with particular reference to the following? (2011)

(i) Character of the Crop Produced
(ii) Felling cycle
(iii) Tending
(iv) Mode of regeneration
(v) Advantages and Disadvantages

The selection system is defined as a Silvicultural system in which felling and regeneration are distributed over the whorl of the area and the resultant crop is so uneven aged that trees of all ages are found mixed together over every part of area.

(i) **Character of the Crop Produced:** In the typical Selection System, the crop is absolutely uneven-aged with all age-classes mixed together on each unit of area.

(ii) **Felling cycle:** Felling cycle is defined as the time that elapses between successive main fellings on the same area. The length of the felling cycle affects the silviculture of species, exploitation of forests and the natural of crop produced.

(iii) **Tending:** Weeding and cleanings should be carried out for two or three years to enable the young regeneration to grow up, in addition to the usual cultural operations carried out after a felling.

(iv) **Mode of regeneration:** In the Ideal or Periodic Selection System, it is expected that natural regeneration will come up in the gaps created by felling of trees. Thus, natural regeneration is mostly relied upon. In case, however, natural regeneration does not follow, artificial regeneration has to be attempted because the essence of a silvicultural system is that it should provide for satisfactory regeneration of the area. It may, however, be conceded that artificial regeneration under a shelter wood is possible only in case of shade-bearing species. Attempts made with light demanding species have not been successful unless the canopy was fairly open.

(v) **Advantages of selection system**

- The system creates or maintains uneven aged stand and site factors are fully utilized.
- It provides excellent site protection.
- The forest produced is most resistant to insect pests and adverse climate factors.
- It prevents invasion of grass and weeds.
- Natural regeneration success-fully achieved due to continuous seed supply from over storey and the protection.
- This system facilitated in producing large-sized trees per unit area.
- This system provides a continuous flow of high quality timber.
- The resultant forest is biologically superior as well as aesthetically pleasing.
- Excellent system to promote natural regeneration of shade-tolerant species.
- It provides capital returns at short intervals.
- The vegetation maintains provide continuous cover and food supply for wildlife.
- It provides a stable environment, prevents erosion and landslides as well as rapid runoff.

Disadvantages of selection system

- Considerable skill is required for execution of this system.
- Mature trees to be removed are scattered leads to high cost of logging.
- Felling, logging and extraction can cause damage to the young crop.
- Seed from bad trees can cause genetic deterioration of future crop
- Damage to regeneration by grazing is high.
- Strict fire protection is difficult to achieve.

- Success or failure of regeneration is difficult to assess.
- The growing stock progressively degenerates with every felling when inferior trees dominate the superior one.
- Limited applicability to shade intolerant species.

18. a. Describethemethodfollowedtointroducetheuniformshelterwoodsystemina forest "*Cedrusdeodara*" and "*Pinuswallichiana*" (2013)

b. How can a forest with shade bearer and light demander tree species be managed under uniform shelter wood system? (2015)

In Himachal Pradesh the deodar and kail forest havc bcen worked under uniform shelter wood system. In these two species, deodar is a shade bearer and kail is the light demander and hence seeding felling is first carried out to cater to the needs of deodar regeneration

Once deodar regeneration has come out, the overwood light demanding kail has to be opened up to enable jail regeneration to come up in the blank and complete the regeneration.

Seeding felling

The seedling felling usually carried out in such a way so as to leave one crown width between seed bearer which means that atleast 45-50 seed bearers per hectare in deodar and 25 -30 seed bearer per hectare in kail have to be retained. In both the species, the well grown advance growth may be retained.

Treatment of a deodar and kail forest

- Once the felling are over, the slash is cut and collected in heaps away from the shade bearer and advance growth and these heaps are burn before the onset of snow fall.
- Grazing is allowed till a good seeding crop is achieved.
- In deodar and kail forest there are always accumulation of thick layer of un-decomposed needles and hence a contour strips are cleared before seed fall.
- Then secondary felling can be carried out to remove the over wood gradually as per the regeneration process and usually secondary fellings can be carried out in 2 – 3 stages.
- Since deodar, kail forest is a complex system, the regeneration is not uniform throughout the area and hence it must be supplemented with artificial regeneration. Sometimes the pole trees and small trees of advance growth are also retained as a future crop. This will create uneven-aged system and hence the name of uniform system may not apply here.

19. Discuss the uniform system with artificial regeneration. (2014)

Uniform System

The uniform system is also aiming atconcentrated regeneration, wherein thecanopy is uniformly opened up over thewhole area of a compartment to obtainuniform regeneration. The uniform systemis otherwise called shelter woodcompartment system or, compartmentsystem.

Artificial Regeneration

The regeneration is predominantlyobtained through natural, or sometimes itis supplemented by artificial regeneration.

- Wherever the natural regeneration is not successful in uniform system as in the case of sal the regeneration must be supplement with artificial regeneration
- For artificial regeneration seedlings have to be revived from known seed sources and are kept ready for planting
- Then area is demarked and seeding felling has to be initiated. All the debris and the undergrowth present in the demarked area need to be cleared by controlled burning.
- Then the area is laid out, markings are made, pitting are dug and seedlings are planted in the demarked area.
- The area should be protected from grazing and browsing animal along with fire control
- Once artificial regeneration progresses then secondary felling can be initiated by removing some of the overwood on order to allow more light.
- Cleanings and thinning have to be followed in the artificial regenerated area
- The resultant crop is uniform and even-aged

20. Discuss two-storied high forest and explain its advantages. (2014)

Two-storied High Forest System is an accessory Silvicultural system which results in the formation of a two-storied forest, *i. e.,* a crop of trees in which the canopy can be differentiated into two strata, in each of which the dominant species is usually different. The crop in each storey is approximately even-aged, and is of seedling origin. The lower storey may be obtained by natural regeneration by seed brought from outside, but usually by under-planting which is done for one of the following objects:

(i) For protection of soil: When the upper storey is of strong light demanding species like teak, some evergreen species is planted under it to cover the soil. Planting of Evodia, Mesua, Swietenia, and bamboo under teak are examples of this object.

(ii) **For increasing the proportion of valuable species in the moist deciduous and semi-evergreen forests:** When the proportion of valuable species in the upper canopy is low, or when the natural regeneration of the valuable species in the upper canopy cannot be assured, some other valuable species is planted under the main crop. This is also calledenrichment planting. Planting of teak in the moist sal forests of U. P., Hopea in the mixed deciduous forests in Tamil Nadu and Kerala and gurjan in the Gmelina forests of South Bengal are examples of this object.

(iii) **For propagation of species which cannot be raised in theopen:** When a valuable species cannot be raised in the open, it is planted under the canopy of some other species; e.g., planting of mulberry under sissu.

Advantages of the Two-storied High Forest System

- To protect soil with the lower storey crop when the upperstorey crop is incapable of doing so;

- To increase production by growing two crops on the same area.
- To propagate shade-bearing or frost-tender species which cannot be grown without protection;
- To change the species gradually.
- To provide for a vertical mixture in species composition; and (vi) To obtain early returns.

Disadvantages of the system

- Under-planting is difficult operation and, unless done carefully and painstakingly, it is likely to fail.
- The under-storey is likely to be damaged during thinnings or fellings in the upper-storey.
- ❖ The under-storey crop may affect the growth of the upper storey crop.

21. How is *Tectona grandis* forest managed under clear felling followed by artificial regeneration? (2015)

Teak (*Tectona grandis*) is worked under clear felling followed by artificial regeneration, which is predominantly carried by either departmental plantation or taungyas. This method of clear felling followed by artificial regeneration has been followed by allapalli in Kerela termed as allapalli Technique for Teak.

Management Techniques

- Though clear felling is followed by artificial regeneration but still the technique relied on advanced natural regeneration growth.
- The areas containing advanced growth of teak in the form of established saplings, numbering about 750 per hectare selected for natural regeneration.
- The felling area is year marked and a treatment map with workable and unworkable to be marked.
- In this workable area, the area to be regenerated naturally is demarcated separately with boundary billers.
- In the unworkable area which includes 20 m strip along nallas and very steep slope in which only dead and dying trees are marked for felling.
- In other areas of the coupe are trees are marked for felling.
- The working of each coupe is spread over two years before planting is taken up. In first year of valuable timbers are extracted followed by remaining timber and firewood in second year
- The felling operations are carried by department and logging and extraction work completed before the end of December.
- In areas intended for artificial regeneration, all the trees are felled and the felling debris like leaf litter, branches, bark, root bits etc. are burned during march- April. In order to carried out artificial regeneration easily.
- Then the area is laid out and demarking is made at 2m x 2m spacing.
- Then the stumps of size 25 cm (2.5 cm shoot portion, 22.5 cm root, 1-2 cm collar diameter)

are used for planting after the onset of monsoon. The stumps started sprouting and rapid growth and development takes place during early years.

- In order to accelerate the growth and development of teak, three weeding are done in first year, two weeding in second year and one weeding in third year onwards.
- The weeding should be followed by clearing operation across the coupe areas in order to remove the multiple shoot and leaving one leading shoot.
- Thinning is established in management of teak and both mechanical and silvicultural thinning is processed in Teak management under clear felling system followed by artificial regeneration.

22. How will you classify a Silvicultural system? Discuss important features of uniform system with reference to *Pinus roxburghii* and give it Merits and Demerits? (2012)

Silvicultural Systems have been classified in a variety of ways but, in India, the most commonly used classification is based primarily on the mode of regeneration, and this is further classified according to the pattern of felling carried out in the crop. According to mode of regeneration, Silvicultural Systems are classified into following two main categories or groups:

(i) High Forest Systems and

(ii) Coppice Systems.

(I) High Forest Systems

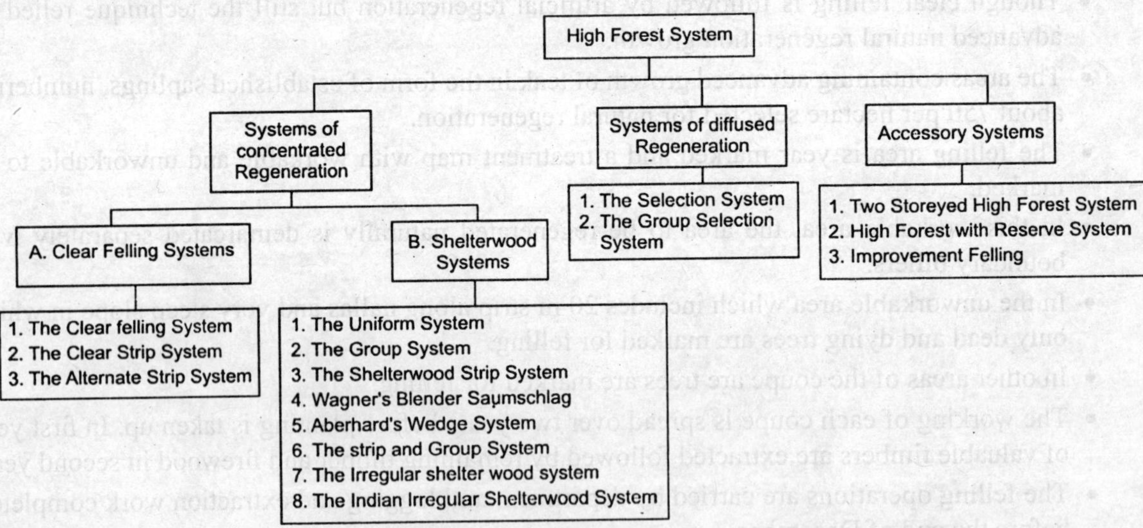

High Forest Systems are those *silvicultural* systems in which the regeneration is normally of seedling origin, either natural or artificial (or a combination of both) and where the rotation is generally long (Fr. *Futaie*; Ger. *Hochwald*). These are further classified on the basis of pattern of felling which, in turn, affects the concentration or diffusion of regeneration and the form or character of the new crop so produced. Thus the High Forest Systems are further classified s shown in the following figure:

(II) Coppice Systems

Coppice Systems are those Silvicultural systems in which the crop originates mainly from coppice and where the rotation of the coppice is short. This main group is further differentiated into the following Silvicultural systems on the basis of pattern of felling:

(i) The Simple Coppice System
(ii) The Coppice of the Two-rotation System
(iii) The Shelterwood Coppice System
(iv) The Coppice-with-Standards System
(v) The Coppice-with-Reserves System
(vi) The Coppice Selection System
(vii) The Pollard System

Pinus roxburghii (Chir pine)

In its optimum zone, chir regenerates naturally easily and is therefore, admirably suited for being worked under the Uniform System The number of seed bearers retained at the time of seeding felling varies from locality to locality, and in the same locality with aspect.

For example, the number of seed bearers retained per hectare in Palampui (H.P.) is 15 to 18, more in dry localities; in Chakrata (U.P.), it is 12 to 15 on hotter aspects the number is increased to 20 to 25 per hectare.

After retaining the required number of seed bearers, all other growth is usually felled but, in certain places, it is usual to retain advance growth of vigorously growing saplings and poles upto 20 cm in diameter, found in the regeneration area, and to remove all overwood from above them.

After the felling is over, all slash left in the area is cut and collected away from the seed bearers and groups of poles, and burial during winter. The area is control-burnt annually and allowed to be grazed till a good seedling crop is obtained as a result of good seed year. After this, the area is closed to grazing and protected against fire.

Weeding's and cleanings are done as and when necessary. Protection is continued till the regeneration is 1 to 2 m high on gentle slopes and 2.5 to 3 m high on steeper slopes after which control-burning is started.

If the entire area does not have uniform growth, the portions with smaller regeneration are fire-protected by making a fire trace roundthem. Grazing is allowed as soon as this can be done without harm because it reduces fire hazard. Final felling are carried out when the regeneration is established, the height when the regeneration is considered established varying from 2 m to 6 or 7 m, sometimes even 10 in, depending on the risks to which the young crop is exposed, In areas where destructive fires occur frequently, it is usual to retain 3 to 5 trees per hectare as insurance against fire. If, as a result of difficulty in enforcing effective closure, natural regeneration is patchy, it is supplemented by artificial regeneration, mostly by planting polythene bag plants.

Merits

- In this system, the soil is not completely denuded so there is little risk of soil deterioration and erosion, to which it is exposed in clear-felling system. The soil is adequately protected in this system.
- The marking and felling of trees of the over wood are simpler in execution, than in the other shelter wood system as well as the Selection System.
- As the regeneration operations are carried out under the shelter of older crop, there is little danger of invasion of the area by weeds, as happens in the clear felling system.
- The young crop is protected against adverse climatic factors such as frost, cold winds, insolation, etc.
- As the regeneration is obtained from the seeds of the best selected trees, the new crop is superior from the point of view of hereditary characters.
- The system is fairly flexible with the result that it is used for regeneration of both light-demanders as well as shade-bearers. Even in mixed forests of light-demanders and sensitive shade-bearers, as are found in Himalayan moist temperate western mixed coniferous forest (type 13/C1d) it controls the admixture of species by regulation of light reaching the forest floor.
- The success of regeneration operations and the distribution of age-classes, though not so apparent as in Clear-felling System, is fairly clear and is not obscured as in the Selection System.
- The system is superior to clear felling system as it reduces the risk of multiplication of injurious insects commonly found in clear-felled areas.
- As the new crop appears before the old one is harvested, the average length of rotation is shortened as compared to that in the Clear-felling system.
- The growing space is more fully utilized as the regeneration grows under the shelter of older trees, so there is little loss of potential wood production. On the other hand, the rapid diameter increment in the selected seed bearers after seeding felling, results in producing larger-sized trees and larger volume per unit area as compared to clear-felling system, in which crop remains closed till the end.
- The system produces even-aged crop of trees with long clean cylindrical boles.
- As the work is fairly concentrated, it makes supervision and control of various operations easy.
- From the aesthetic point of view, it is preferable to clear felling system which produces a desolate look.

Demerits

- As the over wood is removed in more than one operation, the removal of trees, after the regeneration has come up, damages it to a great extent.
- In mixed forests with species having different light requirement, the manipulation of canopy requires skill and knowledge of silvicultural requirements of species composing the mixture, so that the desired proportion could be attained in the new crop.

- The isolated seed bearers are very susceptible to be thrown over by wind, particularly if the species is shallow-rooted. Even when they are deep-rooted, they are generally broken by wind at the middle.
- In the species, in which interval between seed years is long. The Seeding Felling, particularly when it is on the heavier side, results in the invasion of grass and weeds in the openings, thus making the conditions unfavorable for regeneration to appear later.

23. a. Define "Indian irregular shelter wood system". Describe the application of this system to "sal" forest of this country. What are the di-erence between the selection system and Indian irregular shelter wood system? (2009)

b. Describe the Silvicultural system which is introduced to manage *Shorea robusta* after the failure of Uniform system? (2015)

Indian irregular shelterwood system

The Indian Irregular Shelter wood System is a silvicultural system in which the crop to be regenerated is opened up irregularly and the resultant crop is uneven-aged.

Silviculture system in sal

Sal (*Shorea robusta*) has worked under uniform system with various states which resulted in excellent growth of even aged sal forest though it was successful in some areas, in other areas, the uniform system ran into difficulties as the regeneration did not follow the seedling felling. This is due to failure in establishment of sal seedling even though the regeneration was very profuse. Due to this problem the uniform system is gradually replaced by irregular shelterwood system.

Indian irregular shelterwood system for sal

The Indian irregular shelterwood system applied in sal forest of Uttar Pradesh due to the failure of uniform system. The uniform system does not result in regeneration over large area within the expected regeneration period. Hence, the Indian irregular shelterwood system was adopted.

This system was found scattered over almost the whole area of the working circle in patches of varying sizes and in all stages of regeneration. It includes, the recruits, sub-whippy population and population with pole stage since the regeneration is distributed / scattered all over the area, the felling area included floating periodic block.

In this following periodic block, the areas are allotted into two categories,

(i) Areas which have regenerated in part are contained some woody or whipy regenerates.

(ii) Areas which are deficient in regeneration but may respond to soil working and seed broadcasting.

Treatment of area

(i) No felling is prescribed in areas where regeneration is absent or exists in the form of recruits. In these areas, only dead, dry and diseased trees are felled.

(ii) In areas with sub-whippy regeneration, thinning of heavy foliaged tree or under storey trees like mellotes, Ehneagratia are done. In areas with woody stage regeneration sea or degrade thinning is carried out. In addition mature and over mature trees are felled in preference to sound and vigorously growing trees.

(iii) In areas with well established regeneration stage, the old canopy may be completely removed except the young trees.

(iv) The healthy and well grown trees up to 40 cm diameter are retained as a part of future crop and all facilitated by necessary thinning.

(v) The top canopy should have a mixture crop up to 10% light crown miscellaneous species.

(vi) Species like mango, Mahua and other oil yielding trees can be retained unless they are diseased or died.

Other operations

- After the felling is over, the undergrowth and undesirable species which interfere the growth of sal seedlings have to be removed to facilitate sal growth.
- The regeneration area should be adequately fenced in units of 12-15 acres.
- Annual shrub cutting should be done during the rainy season to facilitate regeneration
- Controlled burning should be done between the patches in order to simulate the growth of stagnating whippy regeneration; it should be protected from fire.

Di3erence between Indian irregular shelterwood system and Selection system

S.No	Indian irregular shelterwood system	Selection system
1.	In this system, regeneration felling continued with in the compartment or sub-compartment	In selection system, regeneration is spread over the entire area
2.	In this system, Regeneration is completed in certain years	Regeneration is a continuing process throughout the life of the crop
3.	It is applicable to tropical rain forests	It is applicable to sal and deodar forests.

24. Define conversion. Discuss the importance of conversion in forest management. Write di-erences between coppice-with-standards system and coppice of two rotations system? (2009)

Conversion is defined as a change from one Silvicultural system or one set of species to another. Thus, the concept of conversion involves a change in crop composition and/or the Silvicultural system by which the crops are regenerated and replaced by new crops of distinctive form.

Importance of conversion

(i) Advantage of particular system

(ii) Failure of an existing system

(iii) Advances in precision silviculture

(iv) Growing demand for wood and wood products

(i) **Advantage of particular system:** Suppose if one particular silvicultural system known to offer greater advantages over other systems in terms of higher yield, simplicity of work and improvement of growing stock then adopting the system by removing the old system is preferred. The classical example is in several species selection system replaces by uniform system and coppice with reserve system

(ii) **Failure of an existing system:** If the introduced silvicultural system apply to species fail to regenerate then it has to be replaced with new system. For example uniform system followed in fir and spruce has been changed to selection system as well as clear felling system. Teak managed under uniform system was converted into clear felling system followed natural or artificial regeneration.

(iii) **Advances in precision silviculture:** Due to the advancement made in science and technology which has led to the development of precision silviculture. Hence there is a need to change the silvicultural system by adopting the newer technology. For example traditionally eucalyptus was managed under coppice system but due to advancement in clonal technology coppice system is replaced with clear felling followed by artificial regeneration

(iv) **Growing demand for wood and wood products:** Increasing demand for wood and wood products has completed the productivity improvement of several species. Hence, there is a need to change the silvicultural system from the traditional to modern.

Di3erences between coppice-with-standards system and coppice of two rotations system

S.No.	Coppice-with-Standards System	Coppice of Two Rotations Systems
1.	Standards are seedling origin and the age will be multiples of coppice rotation. These are the permanent feature throughout life of the crop	The standards are selected at the beginning of the second rotation and retained only for one more rotation. At the end, the retained standards are felled and fresh standards are selected.
2.	The main objectives of standards in are to produce large-sized timber, to protect the regeneration against frost, to enrich the coppice growth and to augment the vigor of coppice.	Standards are retained for production of large-sized timber.
3.	It has two rotations. One for the coppice and the other for standards	It has only one rotation

25. Explain the forest conversion process of coppice system to uniform broad leaved high forest by natural regeneration? (2014)

Conversion

Conversion is defined as a change from one silvicultural system or one set of species to another. Thus, conversion involves change in crop composition, silvicultural system by which the crops are regenerated and replaced by new crops of distinctive form.

This system was evolved in France and carried out with the preparatory period of rest or intensive preservation.

(a) Preparatory period of rest: In this approach, the entire working circle into a number of periodic blocks taking into consideration the rotation and the regeneration period. For example if the rotation is 80 years and the period to regenerate is 20 years. Then the areas are divided into 4 periodic blocks. In periodic block one, the large and over mature stands are removed and the vigorously growing stands to meet the seed requirement are retained. The coppice crop is allowed to grow during the period of rest.

The rest of the working circle, coppiced an usual after the preparatory period of rest, regeneration fellings are carried out in periodic block one and another quarter of the forest area can be selected for allotment to periodic Block one.

In 20 years, the periodic block 1 gains regeneration and the uniform crop is tended by cleanings and thinning for rest of the rotation. Similarly, Periodic block two will take another 20 years and there after felling and thinning carried out. Similar exercise can be done for periodic block three and four to complete entire working circle with regeneration containing four different age classes.

(b) Method of intensive reservation: It is otherwise called as Aburt system which is designed to increase the early returns and to shorten the conversion period. In this method, 500-600 best coppice shoots and seedlings standards are retained per hectare to form the future crop by suitably removing the older crops. After 20 years thinning are started and at 60 years of are regeneration fellings can be carried out. In this method, the periodic blocks are not allotted in advance. This method was practices in India in oak shelterwood circle.

26. Discuss in brief the simple coppice system. Write di-erent types of coppice system? (2016)

The simple coppice system

Definition

The Simple Coppice System is defined as a Silvicultural system based on stool coppice, in which the old crop is clear-felled completely with no reservation for shelterwood, or any other purpose.

Mode of Regeneration

The regeneration is from stool coppice. But trees cannot keep on coppicing indefinitely and they progressively lose their coppice vigour at every felling and ultimately die. Therefore, in every coppice coupe, some stools do not coppice in each rotation. These blanks are filled up usually by planting stumps .or container plants.

Tending

In the year following the main felling or, at the latest, in the second year after it, a cleaning is done, to remove the climbers and inferior species interfering with the growth of coppice, and to reduce the number of shoots to two or three.

Thinning is based on the object of management. If the object of coppice crop is fuel production only, nothing is called for, as thinnings have no effect on total volume production. But if the object of coppice is production of poles or even timber, thinning becomes necessary. The first thinning is usually carried out in the fifth year and in this the number of shoots is reduced only to one per stool; the operation being known as the singling out operation.

Character of the Crop Produced

The crop produced under this system is even-aged.

Di-erent types of coppice system

(i) **The simple coppice system:** It is a silvicultural system where regeneration is based on stool coppice and the old crop is clear-felled completely without reservation.

(ii) **The coppice of the two-rotation system:** It is a modification of the Simple Coppice System where at the end of the first rotation of coppice, a few selected poles are left scattered singly over the coupe in order to attain bigger size in the second rotation.

(iii) **The Shelter wood coppice system:** Another modification of the simple coppice system is Shelter Wood Coppice System where in the first clear felling some Shelter wood (125 to 150 per hectare) are retained towards protection from the frost. This system was developed by Trevor, suitable for frosty localities

(iv) **The coppice-with-standards system:** Coppice-with-Standards is a silvicultural system where the regeneration is based on coppice origin and the overwood of standards are usually of seedling origin and composed of trees of various ages which is kept over coppice over periods.

(v) **The coppice-with-reserves system:** Under this system, felling is done only in suitable areas likely to benefit, after reserving all financially immature growth of principal as well as other valuable miscellaneous species, either singly or in optimally spaced groups.

(vi) **The coppice selection system:** In this system fellings are carried out on the principles of selection system but regeneration is obtained by coppice. The trees of exploitable girth or diameter are fixed and accordingly felling cycle is decided.

(vii) **The pollard system:** The tree whose stem has been cutoff in order to obtain a flush of shoots, usually above the height to which the browsing animals can reach is pollarding.

27. Write importance of soil organic matter in forests. How calculation of number of seedlings is carried under Line, square, Triangular and Quincunx methods of planting. (2016)

The transformation and movement of materials within soil organic matter is a dynamic process influenced by climate, soil type, vegetation and soil organisms. Soil organisms are responsible for the decay and cycling of both macronutrients and micronutrients, and their activity affects the structure, tilth and productivity of the soil.

Importance of soil organic matter

- Soil organic matter contains many living organisms such as bacteria, fungi, algae, actinomycetes, earthworms, nematodes, and arthropods and involved in nutrient cycling.
- All contribute to the biological and microbiological activity which results in the breakdown of plant material, roots, dead organisms and other organic residues that accumulate in soils.
- This organic material starts to decompose, soil organic matter is formed.
- Ten to fifteen percent of this organic material is readily decomposable, while the remainder takes much longer.
- Soil organic matter provides nutrients for crops as it decomposes and contributes to the cation exchange complex necessary for holding applied nutrients in the soil.
- Soil aggregation is improved by increased organic matter content and hence has a role in maintaining soil structure, drainage and aeration.
- Soil organic matter also plays a role in increasing moisture retention and consequently the drought tolerance of the crop.
- A soil with an adequate level of organic matter will be less erodible, have increased nutrient retention, and also be easier to work and plow.
- Soil organic matter is resistance to soil crusting and compaction,
- It gives greater fertility, better root growth of crops, and improved crop yields.
- A continuous soil cover of living plants, which together with the soil architecture facilitates the capture and infiltration of rainwater and protects the soil;
- A litter layer of decomposing leaves or residues providing a continuous energy source for macro- and micro-organisms;
- The roots of different plants distributed throughout the soil at different depths permit an effective uptake of nutrients and an active interaction with microorganisms;
- The major period of nutrient release by micro-organisms coincides with the major period of nutrient demand by plants;
- Nutrients recycled by deep-rooting plants and soil macrofauna and microfauna.

The following patterns of planting are followed in Indian Forestry:

- Line Planting
- Square planting
- Triangular planting
- Quincunx planting

(a) Line Planting: In line planting, plants are planted at some spacing in lines which are also some distance apart.

Line Planting

Calculate the Number of Plants Required

$$\text{Number of plants required} = \frac{100 \times 100}{\text{Distance of plants in lines} \times \text{Distane between the lines}}$$

(b) Square planting: Square planting is defined as planting in square pattern, i.e., with plants occupying the four corners of each successive square. This is achieved by planting plants in lines at the same spacing as that of the line themselves.

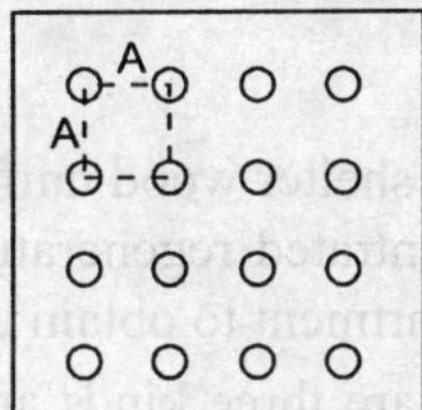

Square planting

$$\text{Number of plants required} = \frac{100 \times 100}{\text{Square of the planting distance}}$$

(c) Triangular planting: Triangular planting is defined as the planting in the pattern of equilateral triangles, *i.e.,* with plants occupying the three corners of each adjacent equilateral triangle.

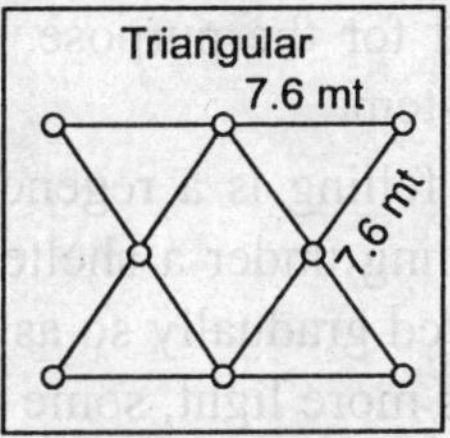

Triangular planting

$$\text{Number of plants required} = \frac{100 \times 100 \times 1.155}{\text{Square of the planting distance (i.e. the side of the try angle)}}$$

(d) Quincunx planting: Quincunx is defined as 'a group of five points, four of which from the corners of a square with the fifth as center'. Thus quincunx planting is that method of planting 'in which an extra plant is placed in the center of the each square of four plants'.

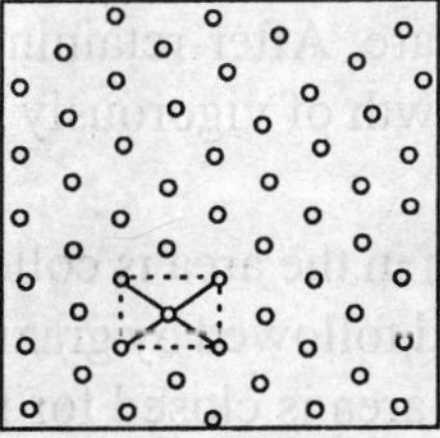

Quincunx planting

$$\text{Number of plants required} = \frac{2 \times 100 \times 100}{\text{Square of the side of the planting square}}$$

28. Describe kinds and pattern of felling followed in shelter wood uniform system. Explain uniform system which is followed to manage *Pinus roxburghii* forest of Himachal Pradesh. (2016)

Definition

Uniform system is an abbreviation of the shelter wood uniform system, which is defined as a kind of shelter wood system aiming at concentrated regeneration, in which the canopy is uniformly opened up over the whorl area of a compartment to obtain uniform regeneration

In shelterwood uniform system there are three kinds and patterns of felling are followed and they are,

(i) Seeding felling
(ii) Secondary felling
(iii) Final felling

(i) Seeding Felling: Seeding felling is the opening of mature stand canopy in order to provide favourable conditions like soil moisture and light etc. for obtaining natural regeneration from the seeds of trees retained for the purpose. This is the first stage of regeneration fellings under a shelter wood system.

(ii) Secondary Felling: Secondary felling is a regeneration felling carried out between the seeding felling and the final felling under a shelter wood system. During the secondary felling, shelter woods are removed gradually so as to admit more light to the regenerated crop. If the regeneration requires more light, some of the seed bearers retained at the time of seeding felling are also removed. The secondary fellings not only improves the light availability but also helps in manipulation of species composition.

(iii) Final Felling: In the final felling, the last seed or shelter trees after the successful regeneration are removed. It is the final stage in the regeneration fellings. Final felling is carried out when the whole regeneration area is completely stocked with established regeneration and no longer requires any shelter against adverse condition.

Application of Uniform system in *Pinus roxburghii*

In Palambur Himachal Pradesh, the chir pine is worked under uniform system. In this system, 15-18 seed bearers are retained per hectare. After retaining the seed bearers all other growth is felled except the retention of advance growth of vigorously growing saplings and pole up to 20 cm diameter.

After the felling is over the entire slash in the area is collected and burned during winter season. Annual controlled burning is also practiced followed by grazing till a good seedling crop is obtained. Once the seedling crop is established, the area is closed for felling. Final felling is carried out once the regeneration is established and the regeneration height varied between 2 m and 7m.

29. Briefly describe the clear felling silvicultural system with reference to (2010)

(i) Nature of crop produced
(ii) Felling system
(iii) Tending
iv) Regeneration
(v) Advantages and disadvantages

(i) Nature of the new crop: The crops are even-aged with normal series of age gradations.

(ii) Felling system: The mature crop will be felled and removed in one operation.

(iii) Regeneration: Regeneration is obtained mostly artificially but occasionally naturally also.

(iv) Tending: Three and two weeding may be required in the first year for the regeneration obtained through sowings as well as planting respectively. Second and third year weeding are also necessary in case of regeneration through sowings. From the 5th year onwards, cleanings and thinning are carried out on 5 or 10 year cycle.

(v) Advantages

- It is the simplest silvicultural system and not requires a high degree of skill.
- The cost of felling and extraction is less and introduction of mechanization in conversion and extraction is possible.
- Introducing fast growing exotics and regulating species composition through artificial regeneration is advantageous.
- Supervision of all operations is easy.
- Damage to the new crop by felling is nil.
- Production of trees with relatively cleaner and more cylindrical boles is possible in clear felling system.
- Cost of regeneration operations and length of rotation are minimum.
- The coupe can be opened up for grazing soon due to the presence of advance growth.
- Establishment of normal forest can be achieved easily through clear felling system.

Disadvantages

- Among the various systems, it is the most artificial one.
- Danger of soil deterioration and erosion is high.
- Threats due to weeds, grass invasion and fire hazard is high.
- The produced even aged crop is less resistant to damage by wind and snow.
- The resultant pure crop is more vulnerable to the damages caused by insects, plant parasites and pathogens.
- During initial period, it sacrifices all the immature trees which are under active growing period.
- This system fails to consider the site factors.
- Aesthetic point of the forests this system is undesirable

30. Describe the silviculture system which may be introduced to manage a plantation forest of *Azadirachta indica*. (2016)

Neem is a versatile tree of Indian origin which has a wide range of domestic and industrial utility. The tree grows throughout the greater parts of India more specially in the drier parts of India. Neem is intermingled with Indian livelihood system and meets the demands of timber, firewood, fodder, medicinal value, pesticide, fungicide besides playing a vital role in cultural religious festivals. The neem has been raised as a individual tree as a road side avenue trees, as sporadic trees in agricultural field, bund planting and now as commercial plantations. As of now, there is no standard silvicultural system followed for neem. However due to the growing interest and importance the following silviculture system may be introduced to manage neem plantations in India.

(i) Selection system: This system is more applicable particularly when neem is aimed for multifarious utility. Currently the neem is planted in 3m x 3m espacement for timber as well as for seed production. The seed production depends on the canopy size and structure which indicates that larger the canopy more will be seed production. Under such circumstances, the selection system will help to retain the superior individuals and eliminating the inferior ones. Though in selection system regeneration operations are mandated but as far as neem is concerned the regeneration is not expected in the current day management strategies and hence inferior population alone can be removed.

(ii) Clear felling system: Wherever the neem is mandated for timber production, the entire plantation needs to be felled at a time in order to generate timber both in terms of quantity as well as quality. This system is may find rare utility.

(iii) Coppice system: The neem leaves are now gaining significant attraction across the world as a source of insecticide and also as a medicinal value. Hence coppice system may be introduced to neem plantations. Though neem is not a best coppicer for wood production but it is ideal for leaf production. Hence the industries involved in neem leaf based product development can use coppice system in order to generate required quantity and quality of neem leaf on a sustainable basis.

Chapter

3

Silviculture Mangroves and Cold Desert

1. Write a short note on cold desert. (2009, 2013)

Cold desert

Cold desert is an area which experiences extreme cold weather and denuded terrain. This is also referred as temperate desert. It occurs in Ladakh, Leh and Kargil area of kashmir and Spiti valley of Himachal Pradesh. In India cold desert occupies 0.97 million ha predominantly distributed in 2 states of Jammu and Kashmir and Himachal Pradesh. The dryness of the locations is caused due to low temperature and also due to dryness of the atmosphere. The cold deserts are distributed over 3000 m altitude regions.

Characteristics of cold desert

- Temperature is less than zero degrees centigrade for most of the period which Inhibit water absorption by plant.
- Atmosphere is very dry and the mean annual rainfall is less than -100 mm which creates desert like influence.
- Snowfall is heavy and occurs between late November and early April.
- Wind erosion is more common.
- Soil is sandy to sandy loam and neutral to slight alkaline in nature.
- Soil has poor organic matter content and low in water retention capacity.
- Growing period is narrow which is mostly during the summer season.
- Because of extreme cold conditions, growth of vegetation is slow and stunted nature.

Dominant species

- Dry temperate zone: *Betula utilis*, *Salix spp*, *Juniperus recurva*
- Alpine zone: *Junipers, Birch*, Rhododendron with grasses.
- Perpetual snow zone: No vegetation due to permanently frozen soil

2. Give the zonation of land – sea interface in a typical mangrove ecosystem with its characteristics. (2010)

Zonation in Mangroves

Mangroves typically show zonation. In India this zonation may be very distinctive (east coast of India) or merging (west coast of India). The zonation in mangroves is not so simple and varies from place to place. Every species has its own level of salinity tolerance. Estuaries on east coast show distinct zonation. The high salinity range on the east coast estuaries may be the principal reason for this distinct zonation. The range and force of tidal action also play a role in creation and maintenance of zones. A very broad and general distinction is describes below:

(i) **Proximal Zone (Front mangroves):** This zone is towards sea water, subject to regular tidal effect where intensity of soil accumulation and inundation is a continuous process. The mangrove species in this zone are specially adapted with stilt roots, prop roots for stability and anchorage. Main species with these features are Rhizophora apiculata and Rhizophora mucronata. Avicennia spp., Sonneratia, Caseolaris. Both Avicennia and Sonneratia produce pneumatophores.

(ii) **Middle Zone (Mid mangroves):** This zone represents die core zone of mangroves. The main species in this zone are Bruguiera gymnorrhiza, B. cylmdrica. Lumnitzera racemosa. L littoralis, Ceriops tagal and Aegiceras comiculatum, Ceriops and Bruguiera develop a strong hold fast in the form of knee roots or bent roots as a special adoption for supporting the erect bole.

(iii) **Distal Zone (Back mangroves):** This zone is towards land region, subjected to less sea action. Towards island area mangroves like *Excoecaria agallocha*, Heritiera littoralis and Xylocarnus spp occur. Both Heritiera and Xylocarpus produce buttresses. Generally the salinity is on lower side in this zone occurring towards hill sides where run off of fresh water is for a prolonged period. The duration of tidal submersion is low in this zone compared to front mangroves.

Characteristics of mangrove ecosystem

- Roots of these plants remain submerged under water.
- The tree species can tolerate high salt concentration.
- Roots develop preumatophars or aerial root for sucking oxygen.

Climate: 1500-3000 mm of rain fall

Light: Long day, high intensity of full sun light

Coastal physiographic: Shallow shores mudflats and tidal estuarine deltas

Soil: Unripened soil, Ripening soil, Organic soil

Wind and storm: As agents of pollination and seed dissemination.

3. Comment on following statement. "Mangrove ecosystems have physiologically dry soil" (2010)

Yes, the mangrove ecosystems have physically wet soil but physiologically dry in nature. The physiological process which allows the mangrove will live in the consistently changing environment where the land meets the sea which is unique among the plant system. The mangrove species must content with drying effect on sun, wind, osmotic imbalance caused by the salinity of sea water. These species has to grow in salty, oxygen deficiency and water logged soils. Mangroves have physiological adaptive ability to with stand action of tides, exposure to fresh water and diurnal and seasonal fluctuation of temperature. The evolutionary adjustments of the varying coastal marine environment have produced some astounding biological characters within mangrove plant communities. Though these plants grow in the area of saturated water but this water cannot directly available to mangroves because of high concentration of salts in the soil and water. Thus mangroves are growing on physically wet soils but they are physiologically dry in nature.

Plants cannot absorb soil water due to high salt present in the soil. Generally the soil is deltaic alluvium and is characterized to have high saline humic and acid sulphate soils of humid tropical region. Acidic sulphate soil has high acidic pH, high EC, presence of humic horizon, dominance of sulphates and chloride salts. These soils always exhibit poor aeration and water holding capacity. Hence plants have adaptation which excludes salt by filtration in root system. Some plants store the salts and excrete them through specialized glands.

4. Illustrate the successive of mangrove vegetation in sea coast. (2010)

Succession

Succession is a process of ecological change, involving the progressive replacement of earlier biotic communities with others over time. Succession usually begins with the disturbance of a pre-existing ecosystem, followed by recovery. The succession of mangrove is dependent on the available seeds or propagules, their size or length and the tidal fluctuation.

- The mangrove ecosystem is periodically subject to catastrophic disturbance, usually associated with severe windstorms such as hurricanes.
- These can be energetic enough to uproot and kill mature trees, and to initiate ecological recovery through primary succession.
- The tidal current results in more sedimentation of mud on sandy soil and facilitates succession.
- Species of *Rhizophora* are often the primary mangrove colonist, followed by *Avicennia* and other secondary species as the ecosystem begins to stabilize and mature.
- The patterns of successional dynamics of mangrove forests are related to the environmental tolerances of the species. Usually
- The succession culminates in a mature forest of mangrove species. In some cases succession in the mangrove ecosystem can sufficiently reduce the influence of tidal waters to allow relatively fresh-water conditions to develop.
- The succession can result in the development of a relatively species rich forest that is lacking in mangrove species.

5. How are mangrove forest managed in India. (2012)

Mangroves are a salt tolerant group of tropical plants that occupy the inter tidal zones of sheltered course like estuaries and lagoons. Mangrove plants include trees, shrubs, ferns and palms. These plants are found in the tropics and sub tropics on river banks and along coastlines, being unusually adapted to anaerobic conditions of both salt and fresh water environments. These plants have adapted to muddy, shifting, saline conditions. They produce stilt roots which project above the mud and water in order to absorb oxygen.

Management of Mangroves in India

India has a long tradition of mangrove forest management and the first management plan was implemented in 1892. The Government of India has set up the National Mangrove Committee in 1976 to advise the government about mangrove conservation and development.

Mangrove conservation and management strategies

- Identification of selected mangrove areas for conservation
- Preparation of a management plan
- Promotion of research
- Adoption of a multidisciplinary approach involving state governments, universities, research institutions and local organizations.

Research and Development for Mangrove management

- Nationwide mapping of the mangrove areas, preferably by remote sensing techniques coupled with land surveys, and time series to assess the rate of degradation of the ecosystems
- Quantitative surveys of area, climatic regime, rate of growth of forest trees and seasonal variations of environmental parameters
- Assessment of suitable sites for reserve forests
- Conservation programmes
- Afforestation of degraded mangrove areas
- Study of management methods, the ecology of mangroves, their flora and fauna, their microbiology and the biochemistry of organic matter and sediments.

Mangrove management

Mangrove management has been carried out through the following activities.

(i) **Involvement of local people:** People in the local community should be involved in mangrove management programme on public lands. To management of mangroves should be included by creating necessary awareness through films, Newspapers, Magazines, workshop, Seminars, Study tour etc in the mangrove forest.

(ii) **Restoration:** The restoration are establishment of mangroves is one of the management option that is suggested to stabilize unvegetated areas or to rehabitate the existing areas. For this purpose species like Rhizophora, Aegiceras, Avicennia, Laguncularia etc. can be planted either by seeds or through established seedlings. Successful restoration has been witnessed in Sunderbands in West Bengal and Pichavaram in Tamil Nadu through planting of container raised seedlings.

(iii) **Establishment of Conservation Park:** Establishment of mangrove conservation parks will help to improve the management strategies. This conservation park should be housed with wide range of species coupled with its role and functions in ecosystem stability have to be indicated through suitable signs and boards. Celebration of mangrove conservation day will help to conserve and manage the mangrove plantation.

(iv) Incentive Programme: Incentive programme is one of the sustainable management options which will help to involve the local people and society in management and conservation issues. Extending cash incentives and awards to the people who are actively involves in the mangrove management will help to sustain these activities. It will also help to expand the area under mangroves both in public and private lands.

(v) Legal and policy issue: The existing legal and policy provisions both from forest and environment laws should be enforced in order to protect, conserve and management of mangrove forest.

(vi) Research and Development: Continued and suitable R & D activities have to be carried out through institutional linkages. This research and development should take care of restoration research, management research, protection research and utilization research which will help to manage the mangrove vegetation effectively and efficiently.

6. Describe fish-bone technique. (2014)

Fish Bone Technique

This technique is developed in Pitchavaram, Tamil Nadu. The planting is done in a fish bone or herringbone fashion that aligns the planting at an angle of 45 degree to the prevailing tide direction. Bed for planting is dug in a fish bone fashion. Inside the bunds of the channel seedlings are planted. In high tide time water enters the channel and during low tide time water drains out from the channels. In addition to that, seedlings will escape from the erosive action of tides so that it cannot be uprooted. Thus successful establishment of mangrove seedlings is ensured.

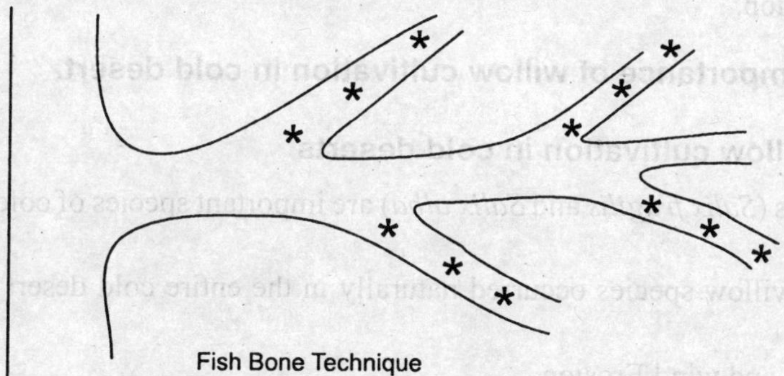

Fish Bone Technique

7. Explain the following mangrove habitat. (2013)

(i) Deltaic mangrove habitat

(ii) Coastal mangrove habitat (both east and west coast)

(i) Deltaic mangrove habitat: It is characterized by high tidal range with associated strong bi-directional tidal currents. These currents are responsible for the dispersion of sediments brought to the coasts by rivers and the main river channels, which are funnel shaped and are fed by numerous tidal creeks. It includes the major estuaries of East coast and two gulfs of West coast, Gujarat.

(ii) Coastal mangrove habitat: It is characterized by low tidal range. River discharge of fresh water and sediments lead to rapid deposition of terrigenous sands, silts and clays to form deltas. These deltas build seawards over flat offshore slopes composed of fine-grained pro-delta sediments. It includes the inter-tidal coastlines, minor river mouths, sheltered bays and backwater areas of the West coast.

(a) East coast mangrove habitats: Sundarban mangrove forest, Ganga delta-West Bengal; Subarnarekha mangrove forest, Bhitarkanika mangrove forest, Mahanadi delta-Orissa; Godavari and Krishna delta-Andhra Pradesh; Pitchavaram Estuary, Cauvery Estuary-Tamil Nadu are considered to fall under deltaic mangrove forest, situated at the mouth of respective major rivers carrying freshwater, facing Bay of Bengal by North to South and North-West to South directions. They altogether cover about 4700 km^2 or 70% of the total mangrove area, and are rich in mangrove species. In East coast habitats, SF of Indian part alone covers about 4200 km^2, while the undivided Sundarbans are the largest single mangrove blocks in the world.

(b) West coast mangrove habitats: Cochin estuary- Kerala; Kundapur, Malpe-Karnataka; Zuary estuary-Goa; Bombay mangrove creeks- Maharashtra; and Bhabnagar estuary-Gujarat, are considered to fall under coastal mangrove habitats in the inter tidal zones along with mouths of minor rivers or minor estuaries and back waters, facing Arabian sea by North-East to West and East to West directions. They altogether occupy about 850 km^2 (12% of the total Indian mangrove area). The mangrove vegetation is sparse, less extended, and confined to patches due to scanty upstream freshwater supply, excessive amount of silt-clay deposition, low average rainfall and relatively low tidal fluctuation.

8. Discuss the importance of willow cultivation in cold desert. (2014)

Importance of willow cultivation in cold deserts

- Willow species (*Salix fragilis* and *Salix alba*) are important species of cold desert agroforestry systems
- The shrubby willow species occurred naturally in the entire cold desert up to an altitude of 3.850 MSL.
- It prevent Soil and wind Erosion
- It grows in sandy to sandy loam and neutral to slight alkaline soil in cold desert.
- Willow mainly conserves the soil moisture and effectively utilizes it.
- Their ability to grow through shoot-cuttings plantations under extreme and xeric climatic conditions of cold deserts makes them ecologically suited and socially accepted for forestry programmes.
- It satisfies huge fuel wood for their domestic consumption.

- Willow species in cold deserts may combat desertification and can thrive well in these areas as compared to any other species.
- Willow species are used as subsistence resources and for socio-religious purposes.

9. Write the characteristics of cold desert. Discuss soil working and planting techniques for cold desert. (2016)

Characteristics of cold desert

- Temperature is less than zero degrees centigrade for most of the period which Inhibit water absorption by plant.
- Atmosphere is very dry and the mean annual rainfall is less than 100 mm which creates desert like influence.
- Snowfall is heavy and occurs between late November and early April.
- Wind erosion is more common.
- Soil is sandy to sandy loam and neutral to slight alkaline in nature.
- Soil has poor organic matter content and low in water retention capacity.
- Growing period is narrow which is mostly during the summer season.
- Because of extreme cold conditions, growth of vegetation is slow and stunted nature.
- Native vegetation is scarce and the important species in this area are *Pinus gevan, Daphne acuminate*, *Indigofitra spp*, *Desmodium spp, Butea minor*, *Arthuisla spp,* Common, shrubs in this area are *Berberis vulgaris*, *Rosa webbiana*, *Loliumspp*, *Poa alpine* and *Cymbopogon* are the common grasses in this zone.

Soil working and planting techniques

There are two types of soil working carried out in cold desert and they are furnished below.

(i) Trench cum Pit Type: It is useful for gentle sloping lands as well as stony areas. Contour ditches are digged staggered fashion at 10 m interval each being 4.3 m long, 0.60 m wide at bottom and 1m at top. In between this, pits of 60 x 60 x 100 cm made at a spacing of 2 x 2 m. Pits are filled with 50:50 of dug up soil and fresh silt. Mulching is also given with available grass materials.

(ii) Irrigation cum Trench Type: This method is suitable for low lying areas viz., river banks and marshy lands with high salt concentration. Planting is done in 45-50 cm crow bar hole and mounds of 45 cm are made around the plants by piling the excavated soil from ditches. Ditch 100 cm wide at top and 60 cm in bottom and 45 cm in depth. Each spaced at 12.5 cm away on either side of the planted row. Depth and width are regularly maintained. Excess water from these ditches is drained into a permanent depression or through artificially constructed channel. Irrigation is given once in a week from May to October. The beating up operation is essential in the next year. The plantation has to be protected from grazing to get a good establishment.

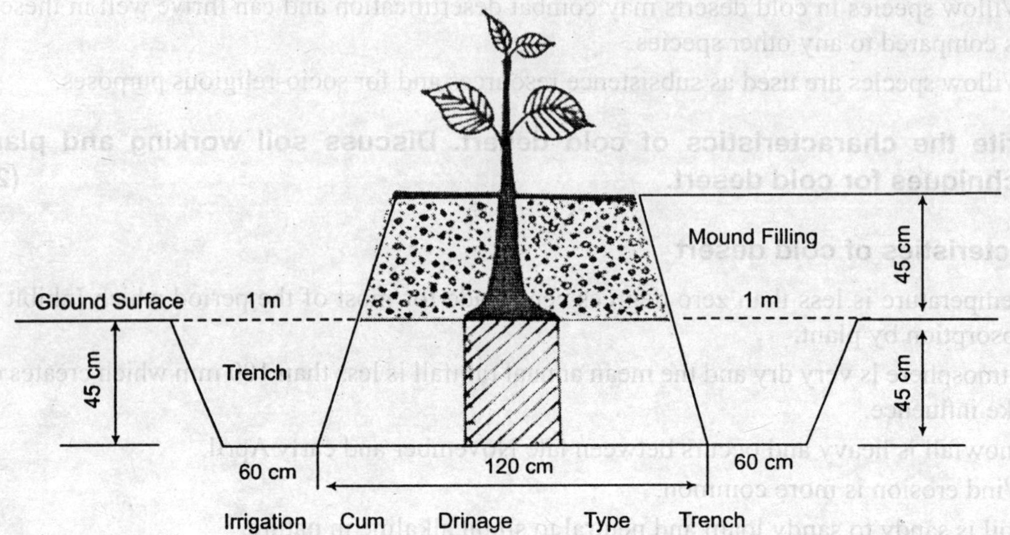

Suitable species for cold desert

Betulautilis, Juniperus recurve, Hippophae rhamnoides, Pinus gerardiana, Berberis vulgaris, Rosa webbiana etc.

10. How and why should cold desert forest ecosystem be conserved? (2010)

Cold desert are prominent in ladakh region of Jammu and Kashmir and Kinnaur and Spithi region of Himachal Pradesh. These regions are characterized by extreme cold weather, denuded terrain and the vegetation is very sparse. The ecosystem of this cold desert region is very very unique due to their sparse distribution and unique morphology specific to that region. This ecosystem based silvicultural character of cold desert species are very unique and gained significant attraction towards their conservation.

Climatic factors and plant adaptations

This region is subjected to several drastic climatic conditions not suitable for usual plant life. The region also experiences scanty or no rainfall. The extreme variations in temperature, heavy snowfall, ultraviolet rays, poor soil fertility and high wind velocity are the common features of cold deserts. Due to these factors, the plants tend to become prostrate, thick wooly, cushion forming, bushy, hardy, spiny, long roots and small leaves which necessitated conservation measures in order to protect cold desert ecosystem.

The unique feature of the vegetation of the Lahaul – Spiti valley is that south and east facing hill slopes are practically devoid of any vegetation and north and west facing slopes show growth of plants. The number of species decreases as one move towards high cold arid region due to harsh climatic factors. Hence it warrants conservation.

Species

Pinus gerardiana, Daphne acuminate, Indigofera spp., Desmodium spp., Butea minor, Artimisia spp., Juniper, Birch, Rhododenron.

Need for conservation

(i) Over grazing is a major problem which has caused serious destruction and limits the regeneration of desert plants.

(ii) Indiscriminate felling to meet the domestic demand is also a major problem which forces conservation issues.

(iii) Water is a critical resource and due to this several plants died and demand conservation measures.

(iv) The low temperature is the most damaging factor particularly when the plants are budding and growing in the spring season.

(v) The desiccation due to climatic changes also causes death of plants. Winter kill is a serious problem which causes serious severe damage.

Due to the above factors, the cold desert plants are under serious damaged conditions and demand conservation.

Conservation of cold desert

(i) The cold desert plants can be conserved using a species known to grow the same climate with similar photo periodicity and similar soil.

(ii) Improving the soil condition for surface stability, water retention and nutritional status which will help to improve the vegetation and ultimately conserve them.

(iii) Practicing regress protection measures particularly for protection against extreme weather conditions to minimize the damages.

(iv) Adopting suitable planting techniques like trench cum pit type or irrigation cum drainage ditch type in order to utilize the rains and to conserve soil and water.

(v) Wherever possible the cold desert plants have to be given life irrigation to protect the plants.

(vi) Fencing the areas with wooden poles will help to prevent the damages due to grazing and conserve the plants.

(vii) Soil working is yet another conservation approach for cold desert plants.

11. a. Suggest measures to rehabilitate degraded mangrove forest. (2011)

b. How degraded mangrove forests rehabilitated? (2006)

Causes for Mangrove Degradation

All over the tropical world, mangroves are regularly harvested by coastal communities to meet their needs but mangrove use is not confined to subsistence, small-scale harvesting. For example, before the discovery of oil in Nigeria, about 20000 m³ of mangrove wood was utilized annually in the coal mining industry as pit props. Even though the demand for coal has drastically declined, mangrove wood is still extensively used in Nigeria and the extraction of much higher volumes of wood is undertaken exclusively as selection harvesting under license agreements with the competent authorities.

Three factors are responsible for the degradation of the mangrove wetlands:

(a) Silvicultural system - Coupe felling,

(b) Reduction in freshwater flow and

(c) Over grazing

Rehabilitation of degraded mangrove forest

Regeneration of Mangroves

Mangrove forests have an efficient mechanism for natural regeneration, particularly in areas within mangrove stands where site degradation has not taken place. Because of the daily or periodic inundation by tidal water as well as the deposition of water-borne soil particles, the forest bed is soft and always suitable for regeneration and normally does not require any site preparation. Profuse natural regeneration seems to occur in areas under mangrove formations.

Restoration technique

Canal method

This method is otherwise called trench method and is largely followed to restore mangrove areas that are degraded due to clear felling in the past under coupe system of management. It is also called fish bone canal method because canals designed like a fish bone produce better results. In the mangrove reserve forests of Tamil Nadu such as Pitchavaram and Muthupet and of Andhra Pradesh such as Krishna and Godavari, mangrove forests were clear felled in large areas for revenue generation by various government agencies since the time of the British government.

This system of clear felling was called coupe felling and it was followed in 15 to 25 year rotation. Because of large scale felling, soil water in the coupe felled areas evaporated (mangrove soil contains 80% of water), which in turn caused subsidence of sediments in these areas. As a result, smooth topography of the mangrove wetland has become trough shaped and tidal water entering into these trough shaped areas stagnates. The evaporation of the stagnant water leads to hyper saline condition and that prevents natural regeneration of mangroves in the coupe felled areas.

It was found out that these degraded patches could be restored by ensuring free flow of tidal water in and out of the degraded area during the high tide and low tide by digging canals and linking them to the nearby natural creeks.

These manmade canals also facilitate flooding of the degraded area with low saline or freshwater during the monsoon season. These actions create suitable environmental condition for the regeneration of mangrove species.

Canal technique produces better results in areas, which are of about 10 to 15 ha in size and where there is stagnation of tidal water. Canals for restoration consist of main and feeder canals, which should be dug in the form of a fish bone for better movement of water in and out of the degraded area.

Planting

Planting is of two types:

(i) Direct planting (dibbling) of propagules or seeds of mangrove plants and

(ii) Planting of nursery raised saplings (pot seedlings).

Fish Bone Technique

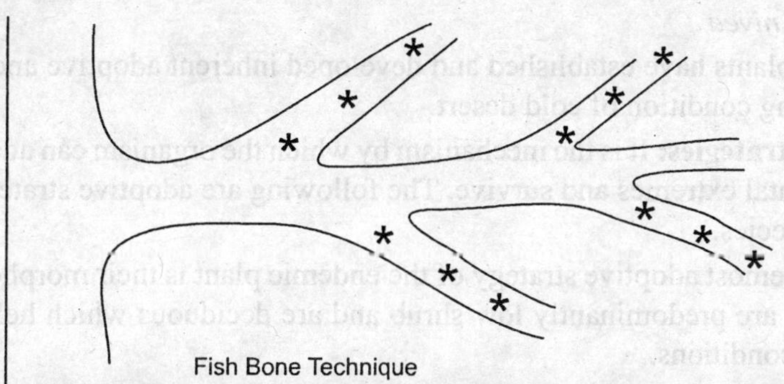

Fish Bone Technique

This technique is developed in Pitchavaram, Tamil Nadu. The planting is done in a fish bone or herringbone fashion that aligns the planting at an angle of 45 degree to the prevailing tide direction. Bed for planting is dug in a fish bone fashion. Inside the bunds of the channel seedlings are planted. In high tide time water enters the channel and during low tide time water drains out from the channels. In addition to that, seedlings will escape from the erosive action of tides so that it cannot be uprooted. Thus successful establishment of mangrove seedlings is ensured.

12. Give the scientific name of following (2015)

(i) Five tree species of mangrove forest

(ii) Five trees / shrubs species of cold desert

S.No	Cold Desert	Mangrove Forests
1.	Betulautilis	*Rhizophora racemose*
2.	Juniperus recurve	*Excoecaria agallocha*
3.	Hippophae rhamnoides	*Sonneratia apetala*
4.	Pinus gerardiana	*Avicenia marina*
5.	Berberis vulgaris	*Rhizophora mangle*
6.	Rosa webbiana	*Pellicierarhizophorae*

13. Comment upon the adaptive and survival strategies of the plant species endemic to the cold desert area of the western Himalaya. (2015)

Cold desert

Cold deserts in the Western Himalayas are one of the unique ecosystem and has wide range of endemic species. The following are the endemic species in the Western Himalayas.

(i) *Astragalus ciliolatus*
(ii) *Dianthus deltoids*
(iii) *Saussurea subulata*
(iv) *Corydalis adiantiflora*
(v) *Inula falconeri*
(vi) *Waldhemia nivea*

These endemic plants have established and developed inherent adoptive and survival strategies against the prevailing condition of cold desert.

(i) Adoptive strategies: It is the mechanism by which the organism can adapt to the prevailing environmental extremes and survive. The following are adoptive strategies evolved by the endemic species.

(a) The foremost adoptive strategy of the endemic plant is their morphology. The endemic species are predominantly low shrub and are deciduous which help to adopt the cold desert conditions.
(b) Several plants shed their leaves during extreme climatic conditions.
(c) Several cold desert plants have spiny leaves which help them to proven evaporation or protect the plant against scarce resources.
(d) The endemic plants develop shallow and widely spread root system which helps to absorb as much as water possible and help the plants to adopt.
(e) The endemic plants are short shrubs and height ranged between 6 and 48 inches which help them adapt to the locality.
(f) The cold desert plants are xerophytic and halophytic in nature which helps them adapt to the extremes of cold desert conditions.

(ii) Survival strategies: Similar to adoptive strategies the cold desert plants have established survival strategies through the following physiological mechanism.

(a) The cells of survive the freezing temperature by withstanding dehydration and avoid formation of ice crystals in the cells.
(b) The plant cells of cold desert accumulate solutes like sucrose and proline which help to depress the freezing point of water and to stabilize membranes.
(c) The cold desert plants synthesizes anti-freeze proteins which help them preventing ice crystal formation in the extra cellular and inter cellular spaces thereby the plant survives.
(d) The plants also survives by synthesizing a protein called dehydrin produced in cytoplasm which may bind water molecules and alter the collective structure of water in the cell to stabilize membranes.

(e) The plant cells for survival can alter lipid composition of cellular membranes in order to adjust the fluidity to cold temperature.

Table: Adaptive features of cold desert plants

S.No.	Characters	Mechanism
1.	Grow close to the ground	Reduce damage caused by wind and ice particles
2.	Small leaves	Conserve water which can be lost through the leaf surface
3.	Shallow root systems	Allows the plant to grow in the active layer and avoid the permafrost
4.	Grow in close proximity to one another	Each plant acts as a barrier for others for the wind and ice particles
5.	Stem, buds and leaves are covered in small hairs	Creates a layer of insulation for protection against cold temperatures
6.	Can photosynthesize in extremely cold weather conditions	Allows the plant to store energy despite a lack of sunlight for large parts of the year
7.	Develop and produce seeds in a relatively short time	Allows for germination to be possible in a small favourable climatic window of time
8.	Ability to survive on bare rock (Lichen)	Survival of a species where soil does not exist

14. a. Describe the characteristics of mangrove forest. Explain the silviculture system followed to manage the mangrove forest. (2009)

b. Discuss silviculture system for manage mangrove forests in India. (2014)

Characteristics of mangroves

- It is an evergreen ecosystem.
- Mangrove flora tolerates and flourishes under high saline mud condition.
- It produces blind roots or pneumatophores so as to overcome the anaerobic soil conditions.
- Vivipary (i.e., seeds are germinated in the tree itself) is more common.
- Most mangrove vegetation has lenticellated bark which facilitates more water loss.
- Many mangrove species produces coppices readily (Rhizophora spp).
- Leaves of many species contain salt secreting glands.
- The cell osmotic concentration is higher.

Silvicultural systems in mangroves

With the exception of Asia, in most countries with mangrove formations no systematic silviculture or management is applied to the resource. All over the tropical world, mangroves are regularly harvested by coastal communities to meet their needs but mangrove use is not confined to subsistence; small-scale, harvesting.

(i) Clear-felling system: Different adaptations of the clear-felling system are in practice in Thailand, Malaysia and Indonesia and, in the past were also used in Viet Nam. This system has been found suitable for *Rhizophora apiculata* and *R. conjugata*, which are strong, light-demanding species and can withstand competition in open areas.

(ii) Shelterwood system: In shelterwood system, two thinnings are carried out followed by a regeneration felling to enhance seedling establishment. It was observed subsequently that adequate regeneration could be achieved without a regeneration felling and, in the absence of regeneration, felling trees could grow for a few extra years and gain in volume. This resulted in the silvicultural system being changed to one which has been termed as clear-felling with the retention of standards. The practice involves the retention of seven trees per hectare at the time of final felling when all trees above the diameter of 7.5 cm are removed. A narrow 3 m belt of trees is retained adjacent to river banks or the coast to prevent erosion. In order to avoid damage to regeneration during harvest operations, extraction is mostly carried out with axes. Trees are cut into 1.6 m billets which are carried out of the forest manually to minimize damage to the remaining vegetation.

(iii) Selection system: The selection system is practiced in Pakistan, India, Bangladesh and Myanmar where trees above certain predetermined diameters are harvested from the annually stipulated coupes. The Sundarbans were managed under a selection system in 1892-1893 when the first management plan for the forest was implemented. The modified selection system has proved to be very appropriate for the sustainable management of the forest.

(iv) Selection-cum-improvement silvicultural system: The mangroves of the Sundarbans especially *Rhizophora racemosa* two years after planting are managed under a selection-cum-improvement silvicultural system. Separate annual coupes are laid out for timber, fuel wood and pulpwood extraction. All types of harvest are carried out in a 20-year cycle where a single harvest operation is carried out once in each 20-year period.

Exoecaria agallocha is exclusively used as pulpwood and matchwood. All trees above the exploitable diameter are harvested in a single operation. In the case of *Sonneratia apetala*, all trees above 30 cm in diameter are removed, provided such removal does not create any gap in the forest canopy.

15. Write in details the term girdling and pruning. Write scientific names of five trees/ shrubs each for the cold desert and mangrove forests. (2016)

Girdling

Girdling is defined as cutting through bark and other living layers of woods in a continuous incision all around the pole of a tree. This is the economical method of killing of inferior trees inside the forests. Girdling of stem restricts flow of photosynthates from leaves to roots. Thus girdling cause death of roots by starvation. This is restricted to certain species only. Girdling enhances more fire hazard inside the forests.

Girdling, also called ring-barking is the complete removal of a strip of bark (consisting of cork cambium, phloem, cambium and sometimes going into the xylem) from around the entire circumference of either a branch or trunk of a woody plant. Girdling results in the death of the area above the girdle over time. A branch completely girdled will fail and when the main trunk of a tree is girdled, the entire tree will die, if it cannot regrow from above to bridge the wound.

Objectives

(i) To kill`the trees of inferior species, where they are removal by felling is not economical.
(ii) To remove the inferior trees in order to benefit the going regeneration.
(iii) To kill the trees in order to protect the damage due to the fire.
(iv) Girdling is done to improve the growth of the retain trees without damaging the retain trees during traditional felling of inferior trees.

Pruning

Pruning is defined as the elimination of branches in order to obtain trees with clean bole.

Objective

The objective of pruning is to obtain knot free timber and it is likely to be obtained more effectively by the removal of lateral buds.

Natural pruning: The elimination of branches by physical and biotic agencies of the environment is called 'natural pruning'. Pruning occurs naturally when the crop is dense enough particularly in younger stage.

The process of natural pruning completes in three stages

(i) Killing of branches
(ii) Shedding of dead branches
(iii) Healing over of the branch stub

The killing of the branches is determined by the density of the crop. The lower branches of trees usually die when crowns of the adjacent trees are close to each other. Death in such cases occurs below the point of closer.

The shedding of branches occurs after the dead branches are attacked by saprophytic fungi, insects, etc., or broken away by winds etc. The shedding of dead branches from the main stem is followed by the occlusion of the short stubs produced by the dead branches. Natural pruning can be accelerated by manipulation of the density and composition of the crop. The rate of killing of lower branches, their shedding and healing of the branch stub depends on species and their habitat factors. The retention of under storey trees may also lead to effective natural pruning. The simplest method of obtaining natural pruning is to develop and maintain dense stocking in the main crop. This may decrease the rate of diameter increment. Therefore, it is necessary to reduce the crop density as soon as natural pruning is achieved.

Artificial pruning

Removal of branches from the seated portions of the tree by mechanical means is referred as artificial pruning. With the adoption of the technique of artificial regeneration, it sometimes becomes necessary to carry out artificial pruning. Artificial pruning results in production of clear bole trees on shorter rotation than would be required in natural pruning. Plantations cannot be raised in closer spacing because it leads to higher investment and therefore artificial pruning, sometimes may be necessary

In the artificial pruning, the moribund and lower green branches are cut off from the main stem as near to the bole as possible. The use of handsaw has been found useful than bladed instruments. The height upto which pruning should be carried out artificially depends upon species, age and local conditions. In young plantations, it should be carried out upto one half to three-fifth of the total height of the tree.

Scientific names of trees/ shrubs for the cold desert and mangrove forests

S.No	Cold Desert	Mangrove Forests
1.	*Betula utilis*	*Rhizophora racemose*
2.	*Juniperus recurve*	*Exoecaria agallocha*
3.	*Hippophae rahmnoides*	*Sonneratia apetala*
4.	*Pinus gerardiana*	*Avicenia marina*
5.	*Berberis vulgaris*	*Rhizophora mangle*
6.	*Rosa webbiana*	*Pelliciera rhizophorae*

16. How is the accrual of tangible and intangible benefits of mangrove forests being a-ected by their degradation? (2016)

The degradation of mangrove forests will affect the following tangible benefits.

(i) E3ect on Tangible benefits

(a) Wood and wood products: The mangroves supply wide range of wood and wood products to meet predominant domestic needs which will get affected due to degradation.

(b) Firewood: The local inhabitants predominantly derive their firewood from mangrove species. The degradation of mangrove will affect this tangible benefit.

(c) Nonwood forest products: The mangrove supports wide range of non wood products from edible fruits to wild fruits which are affected due to degradation.

(d) Income and employment generation: The plantation establishment and management of mangrove ecosystem help to generate wide range of income and employment generation activities and the degradation of mangrove will affect this tangible benefits.

(e) Fishing and other associated activities: The mangrove ecosystem supports wide range of fishing activities and their degradation will affect both quality and quantity of seashore and estuarine fishes.

(f) Medicinal plants: Some of the mangrove species are acting as a local medicine for the coastal people and the degradation of mangroves will seriously affect their availability.

(ii) E3ect on Intangible Benefits

(a) Mangroves moderate monsoonal tidal floods and reduce inundation of coastal low lands. The degradation of mangrove vegetation will increase the tidal floods which may lead to inundation of coastal low lands.

(b) The coastál areas are severely affected due to soil erosion especially after degradation which will affect soil health and quality.

(c) The mangroves act as a bio-shield or barrier against all natural calamities like Tsunami, Hurricane and floods. The degradation of mangroves will have a serious impact and may cause death of human beings and other coastal bio-resources.

(d) The mangroves reduces the wind velocity thereby protect the neighbouring agriculture, homes, livestock and aqua culture. The degradation may lead to increased wind speed which may affect the all livelihood opportunities in and around the coastal ecosystem.

(e) The degradation of mangroves will reduce the natural recycling process and hence soil quality will be degraded.

(f) The degradation of mangroves will affect the populations of flora and fauna and other wild animals associated with the mangrove ecosystem.

17. Discuss a-orestation of cold desert with suitable tree species. (2006)

The cold desert afforestation is a very difficult task and afforestation in these areas should take into consideration the following factors.

(i) To use a species known to grow in the same climate with similar photoperiodicity and similar soil,

(ii) To modify the site thereby ascertaining improved surface stability, water retention and nutritional status and

(iii) To apply rigorous protection measures to minimize all damage to plantations.

Species suitable for a-orestation of cold deserts

Pines, Willows, Poplars, Eucalyptus are important species suitable for afforestation in cold deserts. The following specific species will be suitable for cold desert afforestation.

Eucalyptus	
E. parviflora	Tolerates up to –18ºC temperature
E. gunii	
E. niphophile	
E. coccifera	Tolerates up to –15ºC temperature
E. johntoni	
E. urnigera	
E. aggregate	Tolerates up to –12ºC temperature
E. delegrateusis	

Poplars	
Populus euphratica	Doing well in Leh
P. alba	Successful in Leh
P. candicansa	Successful in Leh
Robinia pseudoacacia	
Shrubs	
Berberis vulgaris	Soil binders
Rosa webbiana	
Hippophae rhamnoides	
Cotoneaster spp	
Legumes	
Medicago falcate	*Nitrogen fixing legumes. These may be planted in between the pits of the trees*
Trifolium repens	
Astragalus spp	
Grasses	
Lolium spp	
Poa alpine	
Cymbopogon spp.	

(Readers can also refer question no. 1, 8 and 9)

Silviculture of Trees

1. Describe the silviculture of the following species. (2005)

a. *Anthocephalus cadamba*

I. Botanical description

It is commonly called as Kadam belongs to the family of Rubiaceae. It can grow upto a height of 45 m without branches. The crown is umbrella shaped and the branches are characteristically arranged in tiers. Leaves are simple, 13-32 cm long. Flowers are in orange colour, small and in dense and globose heads. They appear like solid, hairy orange balls. The fruits are small capsules, packed closely together to form a fleshy, yellow or orange coloured inflorescence containing approx. 8,000 seeds. The small capsules split into four parts releasing the seed at maturity. There are approximately 20,000 seeds per gram.

The tree is a strong light demander and it is sensitive to frost, drought and grazing. It coppices vigorously and tree is quite tolerance of water logged areas where the water inundates for 2-3 months.

II. Environmental Requirement

- Altitude - 300 - 800 m
- Temperature - 23°C
- Rainfall - 1600 mm
- Soil type - well drained entisols

III. Regeneration

It can be propagated through seeds as well as through cuttings.

Propagation through seeds: Seeds are extracted from fruits and mixed with sand for sowing in mother bed. Germination starts in 7-15 days. After germination, seedlings are transplanted into polybags.

- Number of seeds/Kg - 9,00,000 – 27,00,000 seeds/kg
- Pre sowing treatment - Seeds are dipped in 2% Hydrochloric acid solution for 24 hours
- Germination percentage - 70%

Vegetative propagation: *A. cadamba* predominantly propagated through cuttings. In this method, cuttings are collected from the coppice shoots and treated with IBA at 1000 ppm and planted in polybags or root trainers filled with coir compost. These containers are kept in mist chamber under controlled environmental conditions with intermittent misting. The rooting starts in

7-15 days and after 30 - 45 days old cuttings are ready for hardening under shade house conditions. Thirty days after hardening the plants are ready for planting out operation.

IV. Planting techniques

(i) **Spacing:** For pulpwood plantations 2m × 2m For plywood plantations 3m × 3m to 5m x 5m

(ii) **Pit Size:** 30 cm × 30 cm × 30 cm

(iii) **Basal application:** Application of 50 g DAP + 50 g neem cake in the pits before planting.

(iv) **Planting:** The graded seedlings are to be planted in the pits and proper stacking has to be given in windy areas. Immediately after planting, a circular basin has to be formed followed by irrigation.

(v) **Weeding:** Regular weeding is essential in order to encourage rapid early growth.

(vi) **Rate of Growth:** It is a fast growing tree and attains a height of 40 feet and a girth of 60 cm in fifth year with the productivity of 20 m^3 under well managed conditions.

(vii) **Rotation:** The growth of Kadamba is usually fast for the first 6-8 years. At the age of 4 years, it can be harvested for pulp and paper production. For plywood production, the rotation is ranged between 6 and 8 years.

V. Utilization

- The fresh leaves are excellent fodder with a high protein content of over 16%.
- The fragrant orange flowers attract honey – bees as pollinated by bats also roots on this tree and eat its fruit.
- Timber is used for plywood, light construction, pulp and paper, boxes and furniture making.
- Kadamba yields a pulp of satisfactory blightness and performance as a hand sheet.
- A yellow dye can be obtained from the root bark.
- The dried bark is used to relieve fever and as a tonic. An extract of the leaves serves as a mouth garge.
- Kadamba flowers are an important raw material used for the production of Indian perfumes with sandalwood base.
- Kadam wood is also used for manufacture of pencil.

b. *Hardwickia binata*

(i) **Botanical description:** It is a large deciduous tree and reaches the height of upto 30 m with a girth reaching about 3m. The tree is straight, cylindrical bole. The leaves are small, 2-3 cm wide, 2-6 cm long, alternate pinnate greyish green in colour and kidney shape. The flowers are small pale yellowish green in axillary and terminal lacs panicle raceme. The parts are flat and samaroid, lanceolate, coriaceous and narrowed in both ends the parts also have veins running parallel in longitudinal direction and it has one seed near apex.

(ii) Environmental Requirement

- Altitude - 0-300 m
- Temperature - 22°C to 34°C
- Rainfall - 250 - 1500 mm
- Soil type - Sand stone, quartzite, granite with sandy loam, gravel sand: tolerate acid to neutral soil.

(iii) Regeneration

(i) **Natural regeneration:** It experiences the natural regeneration in their original habitat. However it suffers due to die back. Natural regeneration occurs through coppice and root suckers.

(ii) **Artificial regeneration**

Predominantly by seeds

- Number of seeds/Kg - 4800 – 5,400 seeds
- Pre sowing treatment - Water soaking for 24 hours
- Germination percentage - 38 – 80 %
- Period - 15 days to 5 weeks

(iv) Planting techniques

(i) **Spacing:** 3 x 3 m

(ii) **Pit Size:** 30 cm x 30 cm x 30 cm or 45 cm x 45 cm x 45 cm

(iii) **Basal application:** Application of 2- 5 kgs of farmyard manure per pit is advisable.

(iv) **Planting:** 1 – 2 years old seedlings are to be planted during monsoon season. Care should be taken to avoid root injury.

(v) **Weeding:** Regular weeding is essential in order to encourage rapid early growth.

(vi) **Rate of Growth:** Under natural condition the growth is low and altitude condition it grows very rapidly.

(vii) **Rotation:** The traditional rotation is 30 years which has come down 10- 20 years with advancement in Silvicultural technology.

(v) Utilization:

For timber: Wood is heavy to very heavy with density of 737-1025 kg/m^3. Timber is fairly durable and very difficult to work. Wood is used for making wheels, plough shears, oil mils, post and beam. It is also used for making agricultural implements and railway sleepers.

Fuel wood: It is an excellent fuel wood with a calorific value of 4950 k cal/ kg. It produces a good charcoal.

Fodder: Leaves are good fodder and contain around 9% crude protein and 30% fibre.

Fiber: The tree yields fiber for rope making.

Agricultural implements: The wood is used to produce handles of agriculture implements

c. *Populus deltoides*

I. Botanical description

It is a fast growing deciduous tree and grows to a height of 50 – 80 feet tall. The twigs are yellowish and the leaves are green with coarsely toothed and gummy in buds which make different from other Populus. *Populus deltoides* is native to North America and it is an exotic to India. The trees are deciduous, male flowers are reddish and female flowers give dehiscent capsules. Bark is ridged and dark grey. It is light demander frosty hardy and good coppicer.

II. Environmental Requirement

- Altitude - Up to 1000 m
- Temperature - 8°C to 14°C
- Rainfall - 600 - 1500 mm
- Soil type - Sandy loams, and fairly stiff Clays, moist, well-drained fine sandy loams

III. Regeneration

- Poplars are regenerated predominantly by vegetative means and seed propagation is not a common practice. For production of vigorous plants, planting of cuttings, the clones have to be selected and from the clones, the cuttings of size 18 – 25 cm are collected and soaked in water before planting.
- For production of vigorous plants, cuttings are to be planted at 80 cm x 60 cm spacing after treatment with a suitable fungicide.
- The best time for planting cutting is January and February. The nursery beds are flooded with water after that quick insertion of cutting is to be carried out.
- The cuttings are vertically inserted into the soil leaving one bud above the ground and the soil is compacted gently and firmly.
- Sandy loam soil is better for rooting of *Populus deltoides* cuttings.
- Regular watering, weeding, soil working is to be carried out. The collar shoots have to be thinned and only most vigorous shoots can be allowed.
- Similarly debudding is essential process which has to be carried out regularly.
- The nursery soil should be enriched with farm yard manure and fertilizer. Application of urea is essential at early stages for robust growth of cuttings.
- The one year old cuttings are taken out from the nursery with care in order to avoid the damage. The cuttings are to be lifted by digging the furrows and the roots are wrapped with gunny bags and the plants should be kept covered during transportation period before reaching the planting site to avoid desiccation. The gunny bags are to be watered regularly. These are called Entire Transplants (ETPs) which are taken to the plantation site and placed horizontally in the pre-planned channels where a regular flow of fresh water is ensured.
- The ETPs are submerged under water for 48 hours. Then the ETPs can be used for planting during January – February at the spacing of 3-6m depending on the need and utility.

IV. Planting techniques

(i) **Spacing:** 3- 6 m depending on the purpose

(ii) **Pit Size:** Depth of pit: 70-80 cm; Width of the pit: 40 – 50 cm

(iii) **Basal application:** Application of 2 kg of farmyard manure; 50 g superphosphate and 5 g urea.

(iv) **Planting:** 1 – 2 years old plants (ETPs) are taken out from the nursery and used for planting operations. Since Poplars is planted by entire planting method, two persons are essential to hold the plants and place them vertically in the pits. After that the soil around the ETPs are compacted and it should be irrigated immediately. During planting operation, it should be ensured that the collar of the plants should be shrunken at 20 – 50 cm in the pits.

(v) **Tending:** Weeding, soil working and pruning are essential operations have to be carried out to increase the growth and development. Thinning is recommended to remove dead and diseased, lean and lanky which is of silvicultural nature. First dose of 75 g urea, second dose of 150 g urea and third dose of 250 g urea should be applied after one year. Application of fertilizer is continued to augment the growth and development. Irrigation is very essential.

(vi) **Rate of Growth:** It is a fast growing tree amenable for short rotation. The tree reaches a height of 10 meters in 3 years and amenable for harvest from fourth year onwards. The species has a potential productivity of over 25m^3/ha/annum.

(vii) **Rotation:** 4-8 years depending the utility and objective of management.

V. Utilization

- Poplar is important plywood in Northern India and it is acting as a raw material resource for entire industries located in north western region
- The wood is an excellent source for match splints and match box production due to its higher rate of recovery of splints per kg of wood, colour and fissility properties.
- It is an excellent raw material for pulp and paper industry.
- Poplars are good for dendro power generation

2. Describe silviculture of the following species (2006)

a. *Dalbergia sissoo* (Sisham)

Dalbergia sissoo

I. Botanical description

Sissam is a medium to large deciduous tree native to India and grows to a height 25 m and diameter upto 3m. Leaves are leathery, alternate and pinnately compound, flowers are whitish, fragrant, nearly sessile and in dense clusters. Parts are oblong, flat, thin and light brown in colour. These parts contain 1-5 flat green shaped seeds; trunk is often crooked with fissured bark. It's a strong light demander, frost hardy and good coppicer.

II. Environmental Requirement

- Altitude - Up to 1500 m
- Temperature - 4 °C to 49°C
- Rainfall - 760 - 4570 mm
- Soil type - Sandy and gravelly alluvium

III. Regeneration

Dalbergia sissoo is regenerated both through seeds and cuttings.

Seed Propagation: Seeds are collected from mature pods.

- Number of seeds/Kg - 53000 seeds
- Pre sowing treatment - soaking of pods in cold water for 24 hours
- Germination percentage - 90%

Clonal Propagation: Sissoo is now predominantly multiplied through clonal technology using apical shoots. The superior genetic stocks are identified, felled at 15 – 30 cm from ground level. The stumps are induces coppice and apical shoots are collected from these coppicer rooted under controlled environmental condition. These plants are planted in clonal mother garden at 10 cm x 10 cm in order to accommodate 100 plants/m². These mother plants are irrigated and fertilized method. After 30 days the mother plants are cut and new shoots will produced in about 15-20 days. These new apical shoots are collected and rooted under controlled environmental condition. This technology is now commercially deployed for mass multiplication.

IV. Planting techniques

(i) **Spacing:** 2m x 2m for biomass utility, 3m x 3m, 4m x 4m, 5m x 5m for timber purpose

(ii) **Pit Size:** 30 cm x 30 cm x 30 cm or 45 cm x 45 cm x 45 cm

(iii) **Basal application:** 50 gram DAP + 100 gram neem cake

(iv) **Planting:** The seedlings or the clones of *Dalbergia sissoo* can be planted during monsoon season. The plants or immediately compacted with the soil after planting and followed by irrigation. Application of dursban as powder formulation at the rate of 2 – 5 g per pit will help to prevent termite attack.

(v) **Tending:** Weeding is essential at early stage. Trees have pruned periodically to induce clear bole in case of low density plantations.

Thinning is essential for high density plantations particularly 2m x2m and 3m x 3m and three thinnings are followed. The first thinning is mechanical at 2-3 years and other thinnings are Silvicultural thinning at 4th and 8th year in case of timber plantations. For biomass energy plantations *Dalbergia sissoo* is managed as a short duration of 3 years crop in the state of Tamil Nadu. Application of fertilizers of Urea at 100 g and 200 g in two doses at first year, 250 g DAP in second year and 250 g all 17 in the third year is practiced for energy plantations.

(i) **Rate of Growth:** It is a fast growing tree under intensive management the sissoo plantation able to produce 100 tonnes / ha in 3 years for biomass purpose. For timber purpose the tree at the age of 5th year is able to produce 400 kg/tree.

(ii) **Rotation:** Three years for biomass and pulpwood, 8 to 15 years for timber.

V. Utilization

Timber: It is an excellent timber due to its hard heartwood. The density of species varied between 620 kg/m^3 to 820 kg/m^3. The timber is amenable for making high quality furniture and cabinet making.

Plywood: It is an excellent wood for manufacture of decorative plywood and aircraft grade plywood.

Sports goods: Wood is useful for making sports goods, tool handle and other ornamental turnery purpose.

Fuel wood: Sissoo is an excellent fuel wood, with the calorific value is over 4900 kcal/ kg.

Fodder: The leaves are good fodder with the protein content of over 12%.

Pulpwood: It is also a good pulpwood and suitable for writing and printing paper.

b. *Tectona grandis*

I. Botanical description

Teak is very popular timber in India and it is a deciduous tree attaining height of 30 – 40 m and a girth of 4-6 m. It is a large sized tree and leaves are broad and opposite. The flowers are in large, lacs cluster at the end of branches and white in colour. The fruits are drupe, spongy in nature and contain 1-4 seeds.

Silvicultural characters

Teak is a strong light demander and does not tolerate suppression and requires complete overhead light and fair amount of side room for proper development. It produces a large and deep root system. However, teak is sensitive to frost and drought. It has greater resistance to the effects of fire. Teak is a good coppicer and pollards vigorously. The coppicing power is being retained up to a considerable size. The vitality of teak stumps is remarkable.

II. Environmental Requirement

- Altitude - 0 to 1200 m
- Temperature - 2°C to 48°C
- Rainfall - 1000 - 5000 mm
- Soil type - Deep, well-drained, Fertile Alluvial-colluvial soil

III. Regeneration

Teak is regenerated predominantly through seeds. However the seeds need pretreatment like alternate wetting and drying for 7-14 days.

- Number of seeds/Kg - 1500 - 2000 seeds/kg
- Pre sowing treatment - Soaking of seeds in cold Water, Chemical treatment (20 minutes sulphuric acid), Soaking of seeds in cow dung solution
- Germination percentage - 90 %

(i) Natural regeneration: On the level ground teak seed germinates that are placed under mother trees. In the hillside seed is washed down the slopes. Sufficient degree of moisture, temperature and soil aeration is the key factors that influence the natural regeneration. Grazing, if not heavy, appears to be favorable for natural regeneration of teak by keeping down weeds and grasses.

(ii)Artificial reproduction: The Collection seeds from mature trees can be done from December to March depending on locality. The collected fresh seeds should be dried for storage otherwise it needs pretreatment like immersing the seed in hot water for a few hours, hot water treatment, immersing in cold water for a number of days, alternate soaking and drying, placing seeds in a paste of cow dung mixed with water and acid treatment. These pretreatments will break the exogenous dormancy and will facilitate in early germination and seedling vigor. The pretreated seeds can be sown on the raised mother bed at the rate of 1 kg / m^2. Regular nursery care practice like watering and weeding is followed for the production of healthier planting material. Once the seedling has attained four leaf stages, it has to be transplanted form mother bed to polythene bag which contains usual soil mix. These seedlings can be maintained in the nursery for a period of one year and screened vigorous seedlings can be utilized for main field planting.

Stump planting: Teak can also be propagated through stumps. The six month old seedlings grown on the mother bed can be pricked out and stumps can be prepared with 2.5 cm shoot and 22.5 cm root. These stumps further can be planted in the poly bags for vigorous development or in the moist areas stumps can be directly planted in the field during the monsoon period. Normally

The spacing adopted for the establishment of commercial plantations is 2m x 2m. Watering and weeding should be done during the initial period in order to get better growth.

IV. Planting techniques

(i) **Spacing:** 2m x 2m

(ii) **Pit Size:** 30 cm x 30 cm x 30 cm

(iii) **Basal application:** 2 kg of farmyard manure, 100 g DAP

(iv) **Planting:** Teak is normally planted by using stumps. The stumps are prepared from one year old seedlings with 2.5 cm shoot portion, 22.5 cm root portion and with a collar diameter of 1 – 2 cm are used for planting in the pit holes dug through crow bar. After planting the soil around the stumps are compacted and irrigated immediately.

(v) **Tending:** Teak is very sensitive to weed competition and hence weeding is essential along with climber cutting throughout the period. The pruning operations are to be carried out in order to have clean bole height. Thinning is a very important silvicultural operations and it is carried out in different years at different locations. In Kerala, thinning cycle of 4, 8, 12, 18, 26 and 36 years have been followed for 50 years rotation. In Tamil Nadu the thinning cycle of 5, 10, 18, 25 and 36 years are followed for 50 years rotation. In both the cases, the first two thinning are mechanical and the rest are silvicultural at C grade thinning.

(vi) **Rate of Growth:** Teak is a fast grower and attains a height of 15feet to 20 feet in one year under well managed condition.

(viii) **Rotation:** 20 years under well irrigated and managed condition, 40-60 years under forest site condition, 15 years under bund plantations.

V. Utilization

- Teak is a moderately strong timber with a density of 660 kg/m^3 and is preferred as a most suitable timber both for domestic and industrial utility.
- Teak is known as a renowned timber due to durability, dimensional stability, working quality and resistant to termites.
- Teak wood is used in all construction purpose such as beams, columns, doors, windows, flooring, paneling etc.
- It is one of the best timbers for furniture and cabinet making wagon and railway cadges.
- For marine construction and ship building teak is preferred due to dimensional stability.

c. *Acacia nilotica* (Babul)

I. Botanical description

A.nilotica commonly called as babul it is moderate size evergreen tree with a short trunk the crown spreading with feathery foliage. Leaves are bipinnate, flowers are golden yellow in colour and born in penduncculate heads. The parts are long, moniliform, green and tomatoes during young stage and greenish block during matures. Parts are indehiscent deeply constructed between the seeds. Normally simple pod contain 8-10 seeds.

Silvicultural character: It is strong light demander, fairly drought resistant, fairly good coppicer in south India.

II. Environmental Requirement

- Altitude - Below 450 m
- Temperature - -1° C to 48°C
- Rainfall - 200 – 1270 mm
- Soil type - Fresh alluvial soil and black Cotton Soil

III. Regeneration

A.nilotica can be raised either through direct sowing or through containerized plants.

- Number of seeds/Kg - 6600-11600 seeds/kg
- Pre sowing treatment - Soaking of seeds in cold Water (48 hours), Chemical treatment (10 to 30 minutes Sulphuric acid)
- Germination percentage - 50-90%

IV. Planting techniques

(i) **Spacing:** 3m x3m

(ii) **Pit Size:** 30 cm^3

(iii) **Basal application:** 2 kg of farmyard manure and 100 g DAP per pit

(iv) **Planting:** Planting is carried out in the month of July. Timely planting is very essential for success of the plantation.

(v) **Tending:** It is most important for establishing of *A.nilotica*. Annual soil working followed by cleaning will ensure good growth and development under high density plantation thinning is recommended at the age of 5 years.

(vi) **Rate of Growth:** It is fast growing tree and attains a height of 5m in 5years with MAI of around 5-8 m^3/ha under managed condition.

(vii) **Rotation:** 10 years under social forestry plantations. Otherwise it is ranged between 20 – 25 years.

V. Utilization

- It is a very strong and heavy timber with the density of 830 kg/m3. It is suitable for cart making, agricultural implements, boat handle and turnery etc.
- Wood is also useful for doors and windows making and also for mining purposes.
- The wood of acacia is an excellent fuel wood with the calorific value of 4700 k cal/kg.
- Wood yield is high grade charcoal with calorific value of over 6000 kcal/kg.
- The bark contains tannin which ranged between 12 to 20%.
- The leaves are excellent fodder and which in crude protein which ranged between 12 – 14%.
- The tree also yields a black gum which is used in calico printing and dyeing.

d. *Dendrocalamus strictus*

I. Botanical description

It is a medium sized gregarious and toughed bamboo which reaches the height of 10 meters and culm diameter of 8 cm and the culms are pale, green or dull green and yellowish when mature. The culms are solid in nature. The leaves are linear lanceolate and the tip is sharply acuminate with twisted points. Inflorescence is large panicles of dense globular heads and the flowering is monocarpic. The spicklets are hairy with fertile and sterile intermixed. Fruit is a caryopsis.

II. Environmental Requirement

- Altitude - 2000 m
- Temperature - -6°C to 46°C
- Rainfall - 800 – 5000 mm
- Soil type - Sand stone, Granite and Granite Gneiss

III. Regeneration

Bamboo regeneration is predominantly by seeds, rhizomes and offsets. Since the seeds are available only once in life time. Rhizomes as well as offset planting is gaining significance generally the seeds are raised in mother beds and germinated seedlings with 10-15 cm height are pricked out and planted in polybags filled with standard mixture

Six months to a year old plants are used for planting out depending on locality.

Rhizomes: In the absence of seeds rhizomes are collected from mother plants and planting in mother bed to facilitate root and shoot growth. After one year the rhizomes is dugged out and used to planting.

O3sets or culm cutting: In this method, the actively growing culms are collected and they are buried in the soil. In the culm, holes are made and IBA 2000 – 2500 ppm solution is poured inside the holes. The nodes of the culm started inducing roots. Six months after planting, culms are digged out and individual rooted culm nodes called offsets are separated out and planted in polythene bags. Again the offsets are maintained in bags for 6 months and used in planting out operation.

(i) **Artificial regeneration**

- Number of seeds/Kg - 29000 - 32000 seeds/kg
- pre sowing treatment - Soaking of seeds in cold Water (12 Hours)
- Germination percentage - 50 %
- Germination Period - Upto 21 days

IV. Planting techniques

(i) **Spacing:** 4m × 4m, 4.5m × 4.5m, 5m × 5m, 6m × 6m

(ii) **Pit Size:** $60cm^3 – 90cm^3$

(iii) **Basal application:** 5 kg of farmyard manure, 100 g Urea

(iv) **Planting:** Two year old seedlings are planted out during monsoon. One to two months old, seedlings are also recorded highest rate of success during planting out operation

(v) **Tending:** Weeding is very essential at earlier stage. Thinning of culms is necessary in order to induce more culms in a clump. Culm thinning is also practices along with harvest operation in order to induce more culms. Annual culm cleaning coupled with soil working is important to increase number of culms. In culm cutting operations, old culms are removed after which the clumps must be worked with soil around the clump to induce new shoots

(vi) **Rate of Growth:** Bamboo is one of the fast growing trees and reaches height of 12-15 m in 2-3 years. As an average a single clump under well managed condition is capable of producing 10-15 culms at young age and 25 culms at maturity.

(vii) **Rotation:** 2-4 years of felling cycle

V. Utilization

- Bamboos are known as poor man's timber and used in house constructions
- It is used as pole and scaffolding materials
- Bamboo is a principle raw material in cottage industries for mats, baskets, furniture, tool handles, designer doors and windows.
- Bamboo is an excellent raw material for pulp and paper industries due to its high fiber length.
- Bamboos are used as a potential raw material for manufacture of activated carbon.
- It has excellent calorific value of over 3800 kcal and used in biomass based power generation industries.
- Bamboo is an excellent raw material for agar bakthi industries.

3. Neem (*Azadirachta indica*) is a tree for solving global problems. Justify the statement. (10)(2007)

Yes – Neem – *Azadirachta indica* is a tree for solving global problems due to its following multifarious utility.

Neem as an insecticide

Neem has emerged as an ideal source of pesticides and insecticides. About 350 species of insect pests, 18 species of Nematodes and equal number of fungi have already been found susceptible to neem effect. Unlike conventional insecticides, bioactive components in neem comprise an array of novel chemicals that affect not only the physiological functions but rather act on different behavioural and physiological processes. The identification of Azadirachtin, Nimbin, Nimbidin etc. have found profound utility in controlling the pest and disease of several tropical and temperate species. Hence it can resolve the global problem of pest and diseases.

Neem as an oil

Neem seed yields deep yellow fatty oil which is known as margosa oil in commerce. It is used in treatments of leprosy, skin diseases, preparation of face cream, hair lotions, medicated show, tooth paste, disinfectants etc. which are finding utility across the world and hence it is a solution for global problems. The nimbidin, constituent of margosa oil heals bleeding gums and cures pyrrohoea when it is used in mouth wash and tooth paste. Hence neem as an oil has a global market and can resolve the global problems.

Neem as a fertilizer

The oil cake obtained after expelling is greatly valued as a fertilizer and repellent for insects. It contains more sulphur (1 – 1.36%) than any other cake and nitrogen content varies from 2-3%. Application of urea mixed with neem cake renders this fertilizer more efficient. Due to growing interest in organic farming across the world neem as a fertilizer gain significant attraction and can resolve the soil related problem.

Neem as a fodder

Neem is regarded as a good fodder and is heavily lopped for goats. It has high crude protein content varied between 12 and 18%, fiber between 11 and 23 % besides having excellent mineral composition. This fodder quality will resolve the issues of feed problem for animals faced across the world.

Neem as a medicine

Neem leaves are used as antidote to poisons, itches and worms; antiseptic for skin diseases. Simple preparations of nimbidin based ointments are known to be effective against several skin diseases, wounds, boils, chicken pox, measles, small pox etc. In the unani system the leaves are used in a medication for purifying the blood, seeds of piles, pimples etc. Hence neem can act as a medicine to treat the global human diseases.

Neem as a wood

Neem wood is very strong and has a density of 833 kg/m^3. The timber is coarse texture and reported to season well and durable. The wood is used for local furniture, agriculture implements and found profound utility in meeting the domestic wood needs. The wood is excellent fire wood due to high calorific value of 6900 kcal.

Hence neem has multifarious utility and may can resolve the global problems which may be to humans or animals or soil ecosystem etc and hence the statement is very true and opt.

4. Describe the silviculture of any four of the following species: (2007)(10)

(i) Pinus roxburghii (Chir)

I. Botanical description

It is a large evergreen, coniferous tree and reaches the height to a maximum of 55m and girth of around 2-4 m the bark is dark grey fissured the leaves are needle like and the ranches are 2 types long and dwarf. The long shoot in the seedling stage bears a whorl of 14-18 cotyledonary leaves, 2-6 cm long triquetrous, pale green and acicular. Dwarf shoots arise in the axis of scale leaves green in the beginning and turn into brown. Dwarf shoots bear 3 needles, triquetrous; acicular with one side rounded apex narrowing into long fine pointed needles remain on the tree for up to three years. Chir pine is a monoecious tree where in male and female parts occur on different branches on the some plant. Male cones are cylindrical, ovoid and female cones are solitary, borne at the tip of large shoots restricted to upper branches, ovoid thick and woody. Seeds are long thin with membranous wings. Strong light demander suppressed under shade, drought hardy and wind firm.

II. Environmental Requirement

- Altitude - 450 to 2300 m
- Temperature - 15 °C to 38°C
- Rainfall - 900 – 3000 mm
- Soil type - Sand stone, and quartzite

III. Regeneration

Chir pine seeds are collected from mature selected trees and separated from the cones. The cleaned seeds are placed in the water in order to separate the unsound seeds which are floating are removed and sinked seeds are further soaked in water for 24 hours. The seeds are mixed with fresh dung and wrapped in gunny bag to retain the moisture for another 10 days the seeds start sprouting and the sprouted seeds can be planted in polythene containers filled with standard potting mixture.

The seedlings have to be maintained by the application of fungicide and pesticide in order to control nursery pest and disease.

- Number of seeds/Kg - 8800-12300 seeds/kg
- Pre sowing treatment - Seeds are placed in a Container of Water, the unsound seed that float on the waster is rejected.
- Germination percentage - 60%

IV. Planting techniques

(i) **Spacing:** 3m x 3m or 2.5 x 2.5m

(ii) **Pit Size:** 30cm^3

(iii) **Basal application:** 100 g DAP per pit

(iv) **Planting:** The seedlings are transported to the site with care and planting out operation is done during rainy season.

(v) **Tending:** Weeding is very essential at earlier stage. Pruning of branches is very important tending operation where in one third of the tree height has to be pruned. Soil working, pest and disease management are very important. Fire control is very essential

(vi) **Rate of Growth:** Slow growing at initial stages and fast growing at later stages. The tree reaches 10 cm diameter in 18 years and 40 cm diameter in 67 years. Under site quality 1: the tree attains 39 m in 100 years with 50 cm diameter and is a slow growing tree.

(vii) **Rotation:** 120 years

V. Utilization

- Wood is moderately hard with density of 575 kg/m^3.
- One of the excellent raw materials for paper and pulp industry.
- The resin obtained from chir pine is very potential and is used in multi various utility. The resin is used in manufacture of turpentine and rosin. The needles are used to extract oil and are used for multi various utility.
- The needles are used in fiber board and particle board production.

(ii) Tectona grandis

Reader can refer Q. 2 (b)

(iii) Eucalyptus tereticornis

I. Botanical description

Eucalyptus is a fast growing, medium- sized to tall tree attaining 20-50m in height and upto 2m in diameter. The tree has a deep tap root system with mycorrhizal associations which increases its ability to draw nutrients and water. The tree has a smooth silvery white stem. The leaves are leathery in texture, hang obliquely or vertically are studded with glands containing aromatic oil. Flowering takes place during July-August. Flowers in bud are covered with a cup-like membrane (Hence the name of the genus, derived from the Greek 'eucalyptus' meaning- 'well covered'), which is thrown off as a lid when the flower expands. The fruiting occurs during September - October. The fruits are surrounded by a woody, cup-shaped receptacles and contain numerous minute seeds.

Silvicultural characters: It is a light demander. It produces good coppices freely and vigorously. It is a fast growing species and adaptability to a wide range of soil and climatic condition.

II. Environmental Requirement

- Altitude - 0 - 1000 m
- Temperature - 2°C to 32°C
- Rainfall - 600 – 1500 mm
- Soil type - Sandy loams, gravels and Alluvial Soil

III. Regeneration

(a) **Seed Propagation:** Planting is more common than sowing. Eucalyptus seeds starts ripen from January to March. The ripen capsules fall to the ground are collected and dried for four days to dehisce. The opened capsules are sieved to extract the small seeds. The seeds does not warrants any pre sowing treatment and sown on the raised beds filled with a mixture of soil, sand and Farm Yard Manure and the surface is smoothed well with fine soil / sand. The vigorous seeds are broadcasted on the raised bed and lightly covered with a layer of fine earth. Initially it requires shade for its germination and later the shade can be removed. The beds should be watered with sprinklers regularly.

At two to four leaf stages, the seedlings should be pricked out form the mother bed and transferred into polythene bag. Pricking out requires much care as seedlings are very sensitive to any damage to root or stem.

Field planting is done during the monsoon season along with the ball earth. In low lying or swampy ground, planting in mounds is desirable. Spacing for Eucalyptus varies from 2 m to 3 m apart.

- Number of seeds/Kg - 3,20,000 – 6,00,000 seeds/kg
- Pre sowing treatment - No treatment required for Eucalyptus
- Germination percentage - 60 %
- Germination Period - 7 – 21 days

(b) **Clonal propagation of Eucalyptus:** Currently most Eucalyptus plantations are established deploying clones through clonal technology. It can be easily propagated through vegetative propagation by cuttings (clonal propagation) using IBA based powder formulated rooting hormones at the concentration ranging from 1500 - 2000 ppm. The treated cuttings should be cultured under green house conditions with intermittent misting coupled with the humidity of 70-80%. A rooted cutting in mist chambers is also found very successful for mass multiplication of elite clones. Superior trees are normally propagated through this method of propagation are called clones.

Similarly a mini clonal garden can be established by planting 100 ramets in 100 cm^2 area which will facilitate to take cuttings from 45 days after planting. Through proper nutrition and irrigation, this method will help to take cuttings periodically for 3 to 5 years.

The cuttings after proper trimming can be treated with IBA at 1500- 2000 ppm and rooting can be achieved very easily. Soft wood cuttings expressed excellent rooting which help to produced quality seedlings.

IV. Planting techniques

(i) **Spacing:** 3m x 1.35m

(ii) **Pit Size:** 30cm x 30cm x 30cm

(iii) **Basal application:** 250g of Vermi-compost or 2kg of Farmyard manure per pit with 50-100 g DAP

(iv) **Planting:** The better planting season for Eucalyptus is June to October. After planting the soil around the plants should be packed compactly to avoid the dryness of plants. After planting the life irrigation should be done for better initial establishment. The causality replacement should be done after one month from planting.

(v) **Tending:** Initial frost protection and weeding is necessary. Coppice with standards system was tried but abandoned owing to poor yield. The clear felling followed by coppice or artificial regeneration is the best system for its regeneration.

(vi) **Rate of Growth:** Fast growing short rotation trees. It yield an average of 125 - 150 tonnes / ha in 3 years.

(vii) **Rotation:** 3 Years for Biomass,
5-6 Years for Pulpwood
6-8 Years for Ply wood

V. Utilization

- Eucalyptus is one of the fastest growing trees and is an excellent raw material for pulp and paper production due to higher pulp yield ranged between 44 and 48 percent.
- Wood is strong and used in particleboard and hardboard industries.
- It is also an excellent source of fuel wood and charcoal due to high calorific value of over 4500 kcal / kg.
- Eucalyptus wood is also used for light and heavy construction, railway sleepers, bridges, piles, poles and mining timber.
- Leaf extracts of the species have pesticidal properties and can be promoted as a biopesticide.
- The leaves of the Eucalyptus species are rich in essential oils that have many medicinal uses.
- Eucalyptus is a large ornamental tree suitable for parks and avenue plantations.
- The tree species can be effectively used for regeneration of denuded lands and prevention of soil erosions in drought affected areas.

(iv) Prosopis juliflora

I. Botanical description

Prosopis is commonly known as mesquite. It is introduced to Indian continent in 1890 and subsequently naturalized in arid and semi-arid part of country. It is a spreading scrub to moderate sized tree and reaches an average height of 1-3 meter, but in tree forms it reaches upto 12 meter. The bark is 2-3 cm thick grey or dark purple with blackish tinch. The bark splits longitudinally. The leaves are cluster on short shoots and bipinnate. The flowers are in auxiliary spike, 8-10 cm long and bearing greenish white flower that turns light yellow when mature. The tree flowers three times in a year. The pods are usually flatten and straight, but increased at the apex. The pods are 6-30 cm long and yellowish brown in colour. The endocarp may have upto 29 round rectangular segment; each carried a seed within it.

Silvicultural character: It is a strong light demander, profuse colonisers and an excellent coppicer. It withstands drought and suitable for saline and alkaline soils.

II. Environmental Requirement

- Altitude - 0-1500 m
- Temperature - 40° C to 45°C
- Rainfall - 250 - 760 mm
- Soil type - Rocky hills, saline flats and Shifting Sand dunes

III. Regeneration

Prosopis is regenerated through naturally. The flowering is abundant but the seed production is low. The seeds have hard seed and for Artificial Regeneration, the seeds to be scarified in acid or boiling in hot water, but under natural condition it is regenerate well and spread very easily due to its easy colonisation. The organized regeneration in southern part of Tamil Nadu identifies that the prosopis seeds 2-5 kg are required broad cost in 1 acre of land. The land is first ploughed and the seeds treated with hot water followed by cold water (24 hours) are broad casted. The seeds are germinated 10-15 days and regenerated fastly and colonised the area very rapidly.

- Number of seeds/Kg - 32000 – 35200 seeds/kg
- Pre sowing treatment - Soaking of seeds in cold Water (24 - 48 hours), Chemical treatment (15 – 30 minutes sulphuric acid), Soaking of seeds in cow dung solution
- Germination percentage - 60 %

IV. Planting techniques

(i) **Spacing:** 1.8 m × 1.8 m to 3m × 3m, but there is no systematic planting.

(ii) **Pit Size:** 30 cm^3 for container (Broadcasting of seeds is normally practiced).

(iii) **Basal application:** 1 kg of farmyard manure, 50 g Urea

(iv) **Planting:** Planting is carried out with 2-3 month or 1 year old seedlings at the onset of monsoon. For stabilization of shifting sand 6-9months old plants are planted without any soil working.

(v) **Tending:** There is no organized tending operation. However the species needs pruning and thinning. The regular thinning will help to induce more number of shoots for biomass production.

(vi) **Rate of Growth:** It is a fast growing tree and reaches the height of at 8 – 10 meter in 3 years. The productivity of the species ranged between 10 and 25 m^3/ha/annum.

(vii) **Rotation:** The rotation is after 3 years and with 2-3 year cycles for fuel wood plantation.

V. Utilization

- It is an excellent fire wood with higher calorific value, which recorded over 4800 K Cal.
- It is used for commercial charcoal production and activated charcoal process.
- The wood is heavy and used for light construction purposes and also in furniture industries.
- Pod is excellent feed stock for animals due to higher protein content of over 20%.
- It also gives gum for medicinal and other utility.
- Prosopis act as a major source for honey.
- It is an excellent crop for soil reclamation

(v) Treminalia tomentosa

I. Botanic description

Terminalia tomentosa is a large deciduous tree with spreading canopy. It has straight bole and reaches a height of over 30 metres. The leaves are coriaceous, elliptic or ovate in shape and it has one or two glands near the base. The flowers are auxiliary panicle and yellowish in colour. The fruits are globorous with five coriaceous wings and are brown in colour. This bark is dark grey to black in colour and longitudinal fissured and transversely cracked.

Silvicultural characteristics: It is a light demander species and sensitive to drought. The tree is excellent coppicer and grows in a wide range of deep fertile soils.

II. Environmental Requirement

- Altitude - Up to 1000 m
- Temperature - 22°C - 28°C
- Rainfall - 900mm -3000 mm
- Soil type - It grows best on fertile, loam soils with good drainage

III. Regeneration

The tree is regenerated artificially by direct sowing or by container raised seedling. In case of direct sowing, trees are dippled at 15 cm gap. In case of container raised seedling the seeds are sown after treatment in cold water for 48 hours in a nursery bed prepared by mixing soil sand and FYM at 1:1:1: ratio. The seeds are sown during March to April. Watering is essential. Germination commences after 15 to 30 days. 3 to 5 inches of seedlings are pricked out and planted in polythene bags of 15 X 20 cm. These seedlings can be maintained in another 30 days. Then regular watering, weeding and plant protection measures are needed. Six month old seedlings are ready for planting out.

IV. Planting techniques

(i) **Spacing:** 3m × 3m, In case of bund planting 2m × 2m.

(ii) **Pit Size:** 45 cm^3

(iii) **Basal application:** 2 kg of farmyard manures, 100 g DAP

(iv) **Planting:** One year old seedlings are used for the planting out operation than the stump planting or direct sowing. Planting operation is done in the rainy season.

(v) **Tending:** Weeding is very essential in early stages following by soil working around the plants. Thinning is carried out at the age of 4-5 years based on silvicultural attributes.

(vi) **Rate of Growth:** The tree growth is variable under various climatic conditions depending upon the soil. At the age of 10 years, it reaches 8 meter height in Site quality I and diameter of 14.5 cm. In 50 years, it reaches a height of 20 meter and diameter of 53 cm and it is considered as slow growing tree.

(vii) **Rotation:** 40 - 60 years. However under agroforestry system the tree has been harvested between 15 and 25 years.

V. Utilization

- The wood is dense wood with density of 817 kg/m^3 and it is used in construction industries, beams, joints, doors, window frame etc.
- It is also used for manufacturing all kinds of furniture and carpentry and mining purposes.
- It is also used as telegraphy and electric poles.
- The wood is an excellent fire wood, calorific value of over 5000 K cal/kg.
- Bark contains over 18 % tannin and is used in leather industries.
- The leaf is excellent fodder and contains 10 % of proteins.
- The leaves are used as feed stock for tasar silk production.

5. Reasons of dying *Dalbergia sissoo*: (2012)

There are several reasons behind the declining population of *Dalbergia sissoo* in the Indo-Gangetic Plain of India.

(i) The first and foremost is the sudden attack of a fungal disease known as 'die back' of sissoo caused by Deuteromycetous fungus *Fusarium solani* f. sp. Dalbergia. This threatening disease has caused large scale mortality of sissoo trees in last two decades in the region, growing on private and public lands. The *F.solani* f.sp. Dalbergia often in conjunction with soil borne Basidiomycetous fungus *Ganoderma lucidum* parasitizes the roots of *Dalbergia sissoo* tree. This results in dying of roots which lead to death of the whole tree. This disease has spread across the indo gangetic plain which resulted in complete dying of *Dalbergia sissoo* plantations in these regions.

(ii) The second most important reason for dying of *Dalbergia sissoo* is the persistence of the 'powdery mildew' disease of sissoo in the region caused by an Ascomycetous fungus *Phyllactinia corylea.* This fungi weakens the trees making them susceptible to other diseases and pests ultimately resulting into their death.

These are the two major reasons for dying of *Dalbergia sissoo* plantations in Northern and Central India.

6. Gregarious flowering is an indicator of drought in the area. Discuss briefly. (2012)

Yes. There is a belief that gregarious flowering is an indicator drought in the area. This is true in case of gregarious flowering in bamboo experienced in North eastern states. Bamboo flowering is a unique feature leading to monocarpic flowering nature and flowers once in their life cycle.

The bamboo flowering is very mysterious and the forest of bamboo separated by hundreds of kilometers will flower simultaneously. The theory is that the rhizomes of bamboo have some kind of memory or imperturbable inner clock ticking away until the present alarm goes off simultaneously. In this quit cacophony, all the bamboo plants burst into bloom simultaneously. After flowering they all die in some kind of a mass suicide. Dying after flowering is a characteristic of monocarpic flowering which is experienced by several bamboo species.

The bamboos are normally acts as a major food supplement for the people in North-East India and in other localities its food supplement for the local tribal population. Several studies indicated the bamboo seeds are livelihood opportunity and subsistence for the tribal peoples.

Under such circumstances, the gregarious flowering followed by death will affect the avail of bamboo seed for the tribal and local people for the next 20-40 years. This may be the reason for drought in the area. **(Readers can also refer the answers mentioned in question 31 in Chapter I)**

7. *Acacia catechu* and Prosopis juliflora have emerged as important species for marginal lands. (2010)

Yes *Acacia catechu* and *Prosopis juliflora* have been emerged as a important species for arid and marginal lands due their adaptation to low fertile arid and semiarid locations.

(i) ***Acacia catechu***: *Acacia catechu* commonly called catechu tree is distributed throughout India except in humid and temperate region. The species thrives the maximum temperature 40-50°C and 2.5-7.7°C. This species is essential tree of dry region due to its adaptation, where others cannot survive. It grows in a wide range of geological formation and soil. The tree is capable of grow in poor shallow soil even with the rocky areas with calcareous nature. Though the natural regeneration of the species is vary unreliably but excessive regeneration has been recorded in difficult sites, due to adaptation to heavy soils, dry localities and profused regenerations coupled with economic utility, the species is emerged as important one. The species gives the important product called cutch and katha. It gives good economic return. Besides satisfying, the timber and fire wood demand of the local people. Hence under marginal land *Acacia catechu* become very important.

(ii) ***Prosopis julifora***: *Prosopis julifora* (Seemai Karuvelam in Tamil) grows tremendously and spreads mainly due to its inbuilt mechanism to overcome adverse conditions of like drought and salt. This is facilitated by its high 'proline' content under stress conditions, which helps the plant to thrive where other species die.

With deep penetrating roots, it can draw water from deeper layers. Though the species has played a role in changing land use and the livelihood security of huge populations of the world, due to poor management practices, it has colonized many important ecosystems, creating a negative pressure on biodiversity.

Prosopis juliflora can be a very aggressive invader and replaces native vegetation and takes over rangelands. Due to this invasiveness, *Prosopis* is very important in marginal lands.

The species has been found multifarious utility as timber, fuelwood, charcoal and activated carbon production. The pods are found as an excellent fodder due to high protein. Hence the species extends wide range of income and employment opportunities to the rural poor and play a significant role in the livelihood system of marginal lands. Moreover this species tolerates the pH of upto 11 and so it is identified as one of the potential species marginal and degraded lands. Due to its colonization the species has been regarded as an excellent soil reclamation species.

Hence both *Acacia catechu* and *Prosopis juliflora* are considered as important species in the marginal lands due to their productivity and adaptability.

8. Discuss the sustainable management of mangrove forest in India. (2015)

The sustainability of mangrove forest has to be managed taken into consideration the ecological, economic and social issues.

The sustainability of mangroves will therefore that requires that the mangroves could be utilized at the rate or equal to the natural rate of regeneration. The following strategies of mangrove management system will help to protect this unique marine ecosystem.

(i) **Ecological sustainability of mangrove forests:** The mangrove forests have to be managed mainly to protect the ecological issues. This involves maintenance of the genetic diversity within the mangrove ecosystem. Mangrove forests are housed with wide range of flora and fauna and they act as resilience mechanism to maintain the diversity and protect the ecological system. Hence mangrove forests have to be managed taking into consideration the ecological issues.

(ii) **Economic sustainability of mangrove forests:** Mangrove forests are the sources of domestic timber and non timber products besides satisfying the other demands of fodder requirement and fishery activities. Continuous harvesting and exploitation may affect sustainability of these resources availability. Hence mangrove forests have to be managed systematically in order to generate the domestic requirement and the products associated with livelihood opportunities not only for the current generation but also for the future utility.

(iii) **Social sustainability of mangrove forests:** The continuous degradation of mangrove ecosystem will have serious impact on the social and cultural diversity of the region. Mangrove forest act as a livelihood opportunities for coastal dwellers and their degradation will have a serious impact on social issues. Similarly the many trees on coastal forest haves cultural and religious significance for the coastal dwellers and any damage to mangrove ecosystem will have implication on these factors. Hence local and national institutions have to be created for managing the mangrove forests on a sustainable basis in order to protect the livelihood system and social justice. For long term sustainability, ensuing full participating of the local communities through decision making will help to resolve the issues. Similarly, improvement of women through skill development programme will help sustainable management of mangrove forests. Above all the legal and policy issues promulgated by state and central government have to be judicially deployed for sustainable management of mangroves.

9. Discuss in brief the Silvicultural of the following species: (2012) (5)
(i) Shorea robusta (Sal)

I. Botanical description

It is one of the primary timbers in India and is a large deciduous tree. It attains a height of upto 30 meters and girth of 2 meters. The leaves are ovate, oblong, round or cordate at base, shortly acuminate, coriaceous when mature. The leaves are initially reddish and on maturity become dark green. The flowers are hermaphrodite.

Silvicultural character: Sal is a strong light demander under moist condition but act like a shade bearer in dry localities. It is sensitive to frost but fire resistance species. It has good coppicing ability.

II. Environmental Requirement

- Altitude - 100 m - 1500 m
- Temperature - 1°C to 47°C
- Rainfall - 1020mm - 4600 mm
- Soil type - Well-drained, Moist, Slightly Acid Sandy to clayey soils

III. Regeneration

Sal is regenerated artificially by seeds either direct sowing or stump planting or pre sprouted species or the container plants. The seeds are collected from the identified trees and can be sown directly in the field by digging out trenches. Alternatively, the seeds are raised in mother bed and stumps are prepared from the plants of 1-2 years old. The stump with collar diameter of 0.6 to 1 cm and 38 - 45 cm length are used for planting in 45 cm deep pits.

- Number of seeds/Kg - 516 – 531 seeds/kg
- Germination percentage - 90 %

IV. Planting techniques

(i) **Spacing:** 2m × 2m, 3m × 3m

(ii) **Pit Size:** 35 cm^3 - 45 cm^3

(iii) **Basal application:** 2 kg of farmyard manure, 100 g DAP

(iv) **Planting:** Nursery raise seedlings are used for planting out with ball of earth, Polythene containers, Dona containers and basket containers. For basket containers, 1-2 years old nursery stock is used.

(v) **Tending:** Cleaning and thinning is very important tending operation in sal. First thinning is mechanical at the age of 10 years; second thinning is at 20 years and there after 20 years circle. These thinnings are Silvicultural thinning conforming to C grade and sal is management under coppice system with the rotation ranged between 20 to 60 years.

(vi) Rate of Growth: Sal is a slow grower and the rate of growth varied between locations to location. In the site quality 1 the height reaches of 39 m and diameter of 54 cm but in the site quality IV in the same period, the height recorded was 18m and the associated diameter was only 33 cm.

(vii) Rotation: Rotation ranged between 20 and 60 years

V. Utilization

- The sal is known for its timber and it is used in railway sleepers, mining timber, electric and telegraph post. It is also used for making beams, columns, bridges, piles and other construction purposes like flooring, roofing, doors and windows frames etc.
- The timber is used exclusively for making agricultural implements, boat building, cart, marine piles and lorry body building, ship building and tool handle etc.
- Sal is good fire wood and the calorific value exceed over 5000 k cal/kg.
- Sal bark contains tannin of 9% and leaves with 20% and used in tannin industries.
- Sal is also used in packing cases and it produces an oleo resin called dammar which is used in medicinal purpose.
- Sal seed yields edible oil, due to its higher oil content in the seeds. The seeds contain higher palmatic acid and hence used for edible purpose.

(ii) Bamboo species

a. *Bambusa bambos*

I. Botanical description

It is a spinous bamboo and the culms are green and reach the height of upto 30m with a diameter of culm upto 18cm. The nodes are swollen and the internode are long ranged between 20 – 40cm. the leaves are lanceolate to linear acute, glabrous, leaf sheath ligulate pubescent, auricled with silvery bristles.

The inflorescence is panicle with 5 pale spikelets. The spikelets are lanceolate, acute, sessile and consist of 1 to 2 empty glumes. The flowers are normally hermaphrodite at lower level only male flowers in the upper reaches. The unique features of flowering in bamboo is monocarpic and gregarious which flowers once in 30-40 years followed by death.

II. Silvicultural characters

Bamboos are both light demanders and shade bearers and grow well on loamy, clay loamy soil with adequate moisture. The best growths are witnessed in alluvial along with streams and water courses.

III.Envlronmental Requirement

- Altitude - Up to 1250 m
- Temperature - 22-36°C

- Rainfall - 2000 - 2500 mm
- Soil type - It grows best on loamy and clay loam

IV. Regeneration

Predominantly through seeds 1 Kg seed is expected to produce over 55000 plants. The seeds are collected and sown over the raised beds of 30 cm height. The seeds are covered with thin layer of sand and irrigated daily for a week. After 7 days the mother bed should be irrigated alternate days continued for 30 days. After that depending on the need irrigation can be continued.The seed start germination within 7 days continued and completed in another 15 days. 2-5 inches seedlings are pricked out and planted in polybags filled with soil and FYM in 3:1 ratio. The seedlings are to be watered regularly upto 6 months and these seedlings are used for planting out operation.

V. Planting techniques

(i) **Spacing:** 5m x 5m, 6m x 6m

(ii) **Pit Size:** 60-90 cm^3 pits

(iii) **Basal application:** 2 kg FYM and 100 g DAP.

(iii) **Planting:** Six months old seedlings are planted with basal application of manures and fertilizers. After planting the soil around the plants are packed tightly to ensure rigid planting. It has to be followed by irrigation once in every week. The planting area should be fenced in order to protect the plantation from domestic and wild animals.

(iv) **Weeding:** The plantation are to be weeded 3 times in the first year, 2 weeding in second year and third year onwards it is annual weeding or by ploughing operation.

(v) **Rate of Growth:** it is a Fast growing tree species and are able to produce 20-40 culms within 3 years and reaches a height of 13 m and culm diameter of upto 6 cm in 3 years

(vi) **Rotation:** Bamboos are harvested in alternative years after three years

VI. Utilization and economic importance

- Bamboos are widely used for rafters, house posts, ladders, tent poles, scaffoldings and basket making etc.
- Shoots are edible which is used for making pickles
- Wood is used for pulp and paper making. The paper produced from the bamboo is highest grade due to its high fiber length.
- Leaves contain 18- 64% crude protein which intern used as fodder for livestock's besides young twigs are eaten by elephants and buffaloes
- Due to its high levels of silica content, it is widely used in Ayurveda medicine.

b. *Dendrocalamus strictus* Reader can refer Q. 2 (d)

10. Give climatic requirement, rotation age, spacing, tending operation and yield of the following forest species: (2010)

***(i) Populus deltoides* (Family: Salicaceae)**

(a) Climatic requirement: *P. deltoides* tolerates frost, heavy soil, sand, slope, and waterlogging. Because of its intolerance to competition and the absence of suitable seedbeds under existing stands, it does not usually succeed itself. It is estimated to range in forest life zones from warm temperate dry to moist through cold temperate dry to moist.

Altitude: Up to 1000 m, Mean annual temperature: 8-14°C, Mean annual rainfall: 600-1500 mm Soil type: Said to persist on infertile sands, fine, sandy loams, and fairly stiff clays, but thrives on moist, well-drained, fine, sandy loams or silts close to streams, pH of 4.5-8.

(b) Rotation age: Rotation age of *Populous deltoides* is 4 to 8 years depending on the utility.

(c) Spacing: The spacing to be adopted for *P.deltoides* will depend upon the object of management, soil fertility and survival in the plantation. The spacing for the production of matchwood or veneer logs will have to be wider than that for pulpwood production. The spacing may be a little closer on better sites than on poor sites. In India, Spacing for *P.deltoides* is 5m x 5m, whereas in Taungya 5m x 3m spacing is sufficient. In farmers field 5m x 4m spacing is more commonly used.

(d) Tending operation:

Pruning: Pruning in popular is very important to improve the stem form and timber quality. If not properly done, it may reduce growth and may stimulate epicormic branches. In India pruning has to be done at the winter season, because populars are leafless and it is easier to decide which branch needs to be removed. Under farm forestry plantations, pruning should be done during second year, and it should restricted to the lower 1/3rd part of the stem, till the trees are 4 years old. From 5th to 7th year, pruning may be taken a little 1/3rd and upto ½ of the total height depending upon the requirement of the individual tree.

Thinning: The results of both selective and systematic thinning of *P. deltoides* growing with various spacing have indicated that an increase in d.b.h growth as a result of thinning may not be possible in less than 10 years old plantations on good sites. The explanation for this may be that the released trees in 6 to 10 years age class showed very little, if any, crown growth into the opening. Therefore the initial spacing to be adopted for poplars plantations to be harvested at short rotation should be such as may not necessitate any thinning upto the final harvesting age.

(e) Yield: A rotation of 12 years was suggested for *P.deltoides* when the trees attain 30cm diameter, and can be commercially utilized. However under agroforestry system Poplars are rotated between 4 and 6 years and hence pronounced as short rotation tree. It is estimated that at 5m x 4m spacing, of 12 years age a total volume production of about 100 $m^3$of wood and 20 m^3 pulpwood per hectare.

***(ii) Casuarina equisetifolia* (Beaf wood)**

(a) Climatic requirement: *Casuarina equisetifolia* is found from Burma and Vietnam throughout Malaysia east to French Polynesia, New Caledonia, Vanuatu, and south into Australia. Along the coastal regions, Casuarina thrives in the temperature extending to 47°C. Under inland conditions, it tolerated extreme temperatures, but exhibits poor growth. It grows well both in South West and North East monsoon areas which receive rainfall between 900 to 3800 mm. Growth is poor in low rainfall areas. Casuarina grows best in loose, fine coastal sands. It can grow well under inland conditions of well drained sandy soils. It tolerates lateritic and red soils and also saline, alkaline and acidic conditions. The trees have nitrogen fixing root nodules and hence dependability on nitrogen supplement is less.

(b) Rotation age: Rotation age of Casuarina is 36 months.

(c) Spacing: The recommended espacement for *Casuraina equsetifolia* is 1.5 x 1.5m to 2 x 2 m. But this varies according to the locality.

(d) Tending operation

Weeding: Weeding is one of the important tending operation in Casuarina in order to increase the growth attributes. Two weedings per annum is very essential.

Pruning: The tree needs to be pruned once in every six months. The pruning has to be carried out 1/3 of the tree in early stages and ½ to ¾ of the tree in the later stages.

(e) Yield: Yield of *Casuarina equisetifolia* is 100 to 150 tonnes / ha in 3 years.

Describe economic importance of *Acacia nilotica*, *Terminalia belerica*, *Vitex negundo* and *Madhuca latifolia*. (2010)

(i) Economic importance of *Acacia nilotica:*

Timber: It is a very strong and heavy timber with the density of 830 kg/m^3. It is suitable for cart making, agricultural implements, boat handle and turnery etc.

Fuel Wood: The wood of acacia is an excellent fuel wood with the calorific value of 4700 k cal/kg. Wood yield is high grade charcoal with calorific value of over 6000 kcal/kg. Wood is also useful for door and windows making.

Fodder: The leaves are excellent fodder and which in crude protein which ranged between 12 – 14%.

Tannin: The bark contains tannin which ranged between 12 to 20% and used in leather industries.

Gum: The tree also yields a black gum which used in calico printing and dyeing

(ii) Economic importance of *Terminalia belerica:*

Timber: The wood of *Terminalia bellerica* is strong and moderately heavy with density of 737kg/m^3.The timber is used for construction work, planks, rafts, boats and carts etc.

Wood: The wood is also used in heavy packing cases and tea chest boards. The *Terminalia bellerica* wood is also suitable for commercial plywood and block board production. The wood is raw material for manufacturing printing and packing papers.

Fodder: The leaves are excellent fodder for domestic animals and for silk worm. The *Terminalia bellerica* is known for production of tassar silk worm.

Myrobolan: The fruit is commercially known as belleric myrobolan which contains over 20% of tannin and used in leather tanning industry. The fruit is an excellent source of medicine which forms a constituent of triphala in natural medicine. The fruit is and astringent and the kernels are used as a narcotic agent.

Oil: The kernels contain non edible oil and are used in the manufacturing of soap.

Gum: The tree yields gum and the bark contains 12- 19 % oxalic acid.

(iii) Economic importance of *Vitex negundo:*

Timber: *Vitex negundo* has the wood density of 0.67 g/cm³

Fuel: Stems are used as firewood.

Fiber: Branches and twigs are used in basket making.

Essential oil: Leaves yield an essential oil.

Medicine: All parts of the plant are commonly used in Indian medicine. They are aromatic and are smoked for relief of headache and catarrh and a decoction is employed in smoke baths for the treatment of febrile, catarrhal and rheumatic affections. The juice of the leaves is used for the treatment of foetid discharges. They show anti-inflammatory, antibacterial and anti-fungal activity. Roots are used in local medicine for dysentery and are anthelmintic, flowers are astringent and fruits are considered vermifuge.

Miscellaneous: Leaves have insecticidal properties and are laid over stored grain to ward off insects.

(vi) Economic importance of *Madhuca latifolia:*

Timber: Mahua is a heavy wood with density of 939 kg/m³.

Wood: Mahua is predominantly used in the manufacturing of wooden cars for carrying god and goddess. It is an excellent timber for building purposes as a source of beams, doors and windows etc.

Fuel wood: It is also useful for heavy work such as bridges, piles, cart wheels, ships, boats and sports goods. Mahua yields an excellent fire wood with calorific value of 4800 kcal/kg.

Fodder: The Mahua leaves are excellent fodder prepared by most of the domestic animals.

Oil: The Mahua flowers are rich source of sugar, vitamins, calcium and essential oil. Hence it is used in preparation of country made distilled liquor. Mahua seeds are excellent source of edible oil and is able to yield over 40% of oil and act as a excellent cooking oil for rural communities. The oil is used in cooking, burning lamps, manufacturing of margarine soap etc. besides acting as an adultrate in ghee production.

Oil Cake: Oil cake is used as a fertilizer due to its higher nitrogen content.

11. Briefly describe the Silvicultural characters and natural regeneration of the following species: (2015)

(i) *Dendrocalamus hamiltoni*

(a) **Silvicultural characters:** The *Dendrocalamus hamiltoni* is a strong light demander. In the initial stages it can tolerate shade to some extent. It is resistant to wind damage, drought, fire and mechanical injury. The species does not tolerate cattle damage in the initial stages. It can with stand moderate winds, however succumbs to heavy gales in the formative years. Drought, fire and mechanical injury can with stand to some extent.

(b) **Natural regeneration:** The natural regeneration of the species under natural conditions occur after gregarious flowering as the seed fallen on the ground germinate after first monsoon rains. The seedlings are tended in a proper manner for establishment of rhizomes and spacing.

(ii) *Acacia auriculiformis*

(a) **Silvicultural characters:** It is strong light demander, and intolerant to competition. It is an evergreen tree retaining considering foliage during the year. It is an excellent speed producer and can regenerate under its own shade and other species. It produces root suckers profusely. It is moderate drought resistant, but fails to with stand severe drought, and is fire tender. It is not browsed by cattle after the first year. The tree possesses superficial root system. Due to excessive branches and crooked stems, the tree is unsuitable for sawn timber.

(b) **Natural regeneration:** The species has the remarkable capacity to regenerate itself under its own shade or of other species where the seed has been wind – dispersed from other areas. Thinning out, opening up the crop or clear felling tends to seedlings to grow rapidly. The young seedling can be smothered by weeds as the tree is shade intolerant, but once the seedlings are established, they outstrip any competition.

(iii) *Dalbergia sissoo* (Sissham)

(a) **Silvicultural characters:** Shisham is a strong light demander and it is decidedly frost hardy. Only leaves affected by severely by the frost but the tree is not seriously injured. It is fairly hardy to drought but in the irrigated plantations suffer badly if water supply is cut off for any length of time. Seedlings are sensitive to drought. *Dalbergia sissoo* is a good coppicer and recovers well from mechanical damage. This species is easily browsed by cattle; goats and camels. Rats and porcupines eat the roots and bark. In its natural state, dense weed growth prevents the establishment of regeneration. Regeneration by root suckers is common.

(b) **Natural regeneration:** The matures pods are disseminated by wind in non-riveraine tracts and by wind and water in reveraine tract; reaches the ground from the beginning of December till end of April and lie until they get soaked when the monsoon rains commensurate in June – July. Early heavy thunder storms results in earlier germination. Numerous seedlings are found after rain in riveraine tracts in places well above the reach of floods.

In non-riveraine tracts, regeneration is easy on exposed ground such as on new embankments, landslips, soil worked by pigs, abandoned cultivated areas if free from weeds. Dense weed growth prevents establishment of regeneration. Fire and grazing are inimical to natural regeneration.

12. Give distribution, phenology, method of regeneration and brief Silvicultural management of two tree species belonging to each family below: (2012) (10)

(a) Meliaceae

(i) *Melia dubia*

(i) Distribution: It occurs in tropical moist deciduous forests, and yields a useful timber. It occurs in Sikkim Himalayas, North Bengal and upper Assam, Khasi hills of Orissa, N.Circars, Deccan and Western Ghats at altitudes of 1,500 – 1,800 m. It grows on variety of soils .However deep fertile sandy loam soils shows optimum growth, while shallow gravelly soils shows stunt growth. The tree is a light demander, the seedling are suppressed under shade. Seedlings tolerate some frost but severe frost kills them. It is susceptible to damage by fires and sapling suffers from browsing. The trees are sensitive in water logged conditions.

(ii) Phenology: *Melia dubia* is also called as a Mahaneem or Forest neem which is fastest growing tree species and within 6-7 years the plantation is ready to harvest. The wood is having good demand from ply wood and paper industries. A height of 20.m with a spreading crown and a cylindrical straight bole of 9 m length and 1.2 – 1.5 m girth. The bark is dark brown, exfoliating in thin, narrow strips with broad, shallow, longitudinal cracks. Leaves bi-pinnate or occasionally tri-pinnate. Leaflets ovate – Lanceolate to ovate round, entire or crenulate; flowers greenish white, fragrant, intense panicles; fruit an ovoid or ellipsoid drupe with 5 or less seeds.

(iii) Method of regeneration

Natural regeneration: Natural propagation is mostly through seeds and the germinability is less. It coppices well and produce root suckers when the roots are injured. It pollards well and clusters of new shoots are thrown out from dormant buds on stems and branches.

Artificial regeneration

Seed propagation: It can be raised either by direct sowing or planting in nursery for raising seedlings or stumps. Direct sowing is recorded to give poorer results than planting of samplings or stumps; the latter is considered the best.

Seed treatment: Seeds are collected from ripened fruits (Jan – Feb) by rubbing, washing and drying and are stored in sealed tins. The germinability of the seed is less than 25%. In nursery, the seeds are sown in raised nursery beds. The best seed treatment is treating the seeds with cow dung solution for one day. Then the treated seeds are sown over the raised nursery bed. It takes two or three months for the seeds to germinate. Irrigation should be done regularly. The seedling takes 6 months to 12 months to complete its nursery stage. Alternatively the seeds can be extracted from the fruits without any damage. The seeds are then soaked in 100 ppm GA overnight which has recorded over 50% germination.

Clonal Propagation: FC&RI has developed a mini clonal technology for mass multiplication of *Melia* clones. The technology for mass multiplication is furnished below.

Clonal Garden establishment: The clonal garden can be established at a size of 10 x 1 x 0.6 m or 5 x 1 x 0.6 m or 3 x 1x 0.6 m using cement trough or GI trough. The bed should be filled with 20 mm stones upto 25 cm and over which finely sewed river sand can be filled. The raised beds can be covered with 100 micron UV stabilized polyethylene film on the top and covered with insect proof mesh to protect the plants and to ensure its freeness from pest and diseases.

(iv) Silvicultural management

Fertigation: The clonal garden should be maintained with irrigation at an interval of every one hour and supplemented with Urea (300-400 g/m^2), SSP150 (175 g/m^2), KCL (175-250 g/m^2) and Micro nutrient mixture (100 g/m^2). The nutrients can be applied twice or thrice depending on the rate of growth of plants.

Irrigation: The tree responds well to irrigation at once in every 3-5 days. The water requirement varies from 10 liter / day to 25 liter / day depending on the age.

Fertilizer requirements: Application of organic manure (Vermicompost) @ 0.5 kg for each pit at the time of planting is advisable. Application of N, P, K mixture of 25 -50 g per tree at the beginning, two times in a year help to augment the growth. From 3rd year onwards the fertilizer level can be increased upto 100 gm/ tree. The fertilizer requirements can be scheduled on need basis depending on the growth and development of the tree.

Pruning: Straight pole fetches good price in the market.

Uses: The wood is used for packing cases, cigar boxes, ceiling planks, building purposes, agricultural implements, pencils, match boxes, splints and Kattamarams. It is employed for outriggers of boats. It is suitable for musical instruments, tea boxes and plyboard. It is a good fuel wood (Calorific value: 3,400 - 4,100 cal.)The fruit of the plant is bitter. It is considered anthelmintic. It gives positive tests with alkaloid reagents.

Yield

For pulpwood: 100-150 tonnes / ha in three years.

For Plywood: 200 tonnes / ha in six years

(ii) *Azadirachta indica* (Neem)

(i) **Distribution:** The tree is believed to be a native of upper Myanmar. Its occurrence in the Siwalik forest of Uttar Pradesh is also considered natural. In India, neem occurs in tropical dry deciduous and thin forests in drier parts upto 1500 m. It is found in almost all states of India.

(ii) **Phenology:** Neem is almost evergreen but becomes near leafless in dry localities for a short period during February-March. New leaves appear in March-April, before the old ones are shed. Flowering occurs from January-March; in the southern parts of India. In Kerala flowering starts in January, in Karnataka, Tamil Nadu and Andhra Pradesh during February-March; in Central India during the first week of April. Thus flowering is progressively delayed from South to North in the sub-Himalayan area, flowering occurs during the first week of May. Fruiting also follows the pattern of flowering; fruits ripen from June to August. The tree starts fruiting at the age of five years but economic yield of fruits is obtained at the age of 10-12 years. About 3300-4500 seeds weigh one kg and on an average, a medium sized tree produces 37-55 kg fruits.

(iii) Method of regeneration

Natural regeneration: Neem is seldom found growing gregariously. It regenerates naturally from seed, coppice and root suckers. Fruit ripening coincides with rainy season; fallen fruits germinate within a fortnight giving a thicket of seedlings under the mature trees. The seedlings establish with tolerable freedom under the protection of thorny bushes, provided it is not subjected to browsing and frost. The germination is epigeal.

Artificial regeneration: Neem can be raised easily through direct sowing, planting out of entire plants or stump planting. Direct sowing is more successful in neem. The tree can also be raised successfully in combination of with field crops by Taungya provided protection from animals assured. Late fruiting and low viability of seed are the limitation for its regeneration.

Nursery Techniques: Physiologically mature seeds with maximum germination capacity and longevity are obtained 10-12 weeks after flowering. The fruit attains peak-green weight and the embryo is fully developed. Green fruits are collected by beating the branches and heaped in the shade by mixing ash, till the seed becomes easily extractable. Seeds are dried in shade for 3-4 days and then sown or stored. Neem seeds do not require any special pre-treatment but seeds do not retain viability very long and have to be sown within 2 or 3 weeks after collection. De-pulping and cleaning of seed improve the germination per cent and seeds passed through digestive system of birds show better germination. Seeds are sown in the nursery beds 15-cm apart in rows, 25 cm apart at a depth of 2.5 cm germination normally takes one or two weeks and it may vary from 9 to 55 days. The beds should be sparingly watered and soil kept loose to prevent cutting, excessive water should be avoided.

Seedlings are pricked out to transplanting beds at 15 x 15 cm spacing or to polypots when they are about 5 cm high; neem seedlings do not require shade except during pricking out stage. In frosty localities, the plants should be provided with shelters. Seeds can also be sown in polypots directly.

(iv) Silvicultural management

Weeding and hoeing: Weedings are essential in the year of formation. Generally, two weedings in the first year, after the germination completes and the second about a month after the first are recommended. A third weeding may also be necessary just before the close of rains. Unweeded plants show poor development and generally get suppressed and killed by frost.

Thinning: First mechanical spacing is done when the crop completes the age of 3-4 years. A second thinning may be necessary at the age of 6 years. Thinning schedule can be worked out depending upon the site quality and growth.

(b) Papilionaceae

(i) *Dalbergia latifolia* (Rose wood)

(i) Distribution: Indian rosewood is reported to be scattered in the dry deciduous forests throughout the Indian peninsula. It grows in the sub-Himalayan tract from Oudh eastwards to Sikkim, Bihar, Orissa, and throughout central and southern India. Optimum growing conditions for the species are reported to be in the Bombay region. In India, it's found throughout Andhra Pradesh, in moist deciduous forests but more common along the Godavari, but nowhere abundant. Identified by full rounded crown and cylindrical, straight bole. It is drought resistant and a good coppicer. It prefers good, deep loam or clayey soil containing lime.

(ii) Phenology: *Dalbergia latifolia* can grow to a height of 20 - 40 m tall, with a trunk diameter of 1.5 - 2 m. The bark is gray, thin with irregular short cracks, peeling in fibrous longitudinal flakes. Alternately arranged leaves are odd-pinnate with 5-7 unequal sized leaflets originating from the same rachis. Leaflets are broadly blunt-tipped, dark green above and pale below. White flowers are borne in panicles, 0.5-1 cm long, in leaf axils. The brown pods are oblong-lance like, and pointed at both ends. They contain 1-4 smooth brown seeds and do not open at maturity. The species name *latifolia* means broad-leaved.

(iii) Method of regeneration

Natural regeneration: The species propagates itself freely by root suckers. It also regenerates itself from seed. Germination takes place early in the rainy season. Chief cause of mortality is drought in situations exposed to direct sun. Seedlings are found usually in places with loose moist soil with moderate growth of grass, etc. In fire protected areas where grazing also is restricted, natural regeneration is usually good under light shade.

Artificial regeneration: The best method of raising is by planting one year old stumps early in the rainy season. Direct sowing is next best and entire plants the worst. Stumps of 0.2" to 0.6" diameter give the best results in point of survivals and height growth.

In Madras artificial regeneration has been confined so far to growing in mixture with teak and lately this too has been given up owing to slower growth of rosewood in the earlier stages as compared with teak. It was tried in gap regeneration in areas subjected to selection felling but here too its slow growth in the early stage and its susceptibility to browsing has necessitated the abandonment of this method. So far as is known no concentrated regeneration of the species on any systematic and large scale has been attempted.

Seed collection – Seeds from December to May. Best time of collection is February/March.

Method of collection – Collected from trees by lopping branches. Only ripe pods are to be collected. Pods turn bark brown on ripening.

Seed treatment – Dry in the sun, not necessary to extract seed, better results by sowing pods.

(iv) Silvicultural management

- The general spacing followed for *Dalbergia latifolia* is 3m x 3m. However under agroforestry system it varied between 2m and 5 m.

- After planting or direct sowing, regular weeding is necessary until trees can withstand weed competition.
- Loosening soil around seedlings also improves growth. Weeding and soil loosening should be done before weeds become dense.
- Application of DAP 100 g – 250 g per plant depending on the age and growth is recommended to enhance the volume production.
- Trees are usually harvested in 30-40 years. *D.latifolia* is generally managed by clear felling followed by artificial regeneration.

(ii) *Dalbergia sissoo* (Sisham)

(i) **Distribution:** *Dalbergia sissoo* is native to Afghanistan, Bangladesh, Bhutan, India, Malaysia and Pakistan. It is normally distributed in the altitudes from 1000 m to 1500 m above MSL with mean annual temperature ranges between -4°C and 45°C and mean annual rainfall of 500 mm - 4500 mm. It thrives best on alluvial soil in beds of streams of rivers. It prefers porous soil and sand, pebbles and boulders. Generally it is associated with *Pinus roxburghii*, *Acacia catechu* and *Shorea robusta*. *Dalbergia sissoo* is adapted to a seasonal monsoon climate and a dry season of up to 6 months.

(ii) **Phenology:** This species starts sheds their mature brown colour leaves in the beginning of November. The young leaves appear in the 2nd half of January or in the 1st half of February. The flower buds appear along with the new leaves and opens in March or April. Young pods appear rapidly from April but they ripen towards the end of November – December.

(iii) **Method of regeneration**

Natural regeneration: The matures pods are disseminated by wind in non-riveraine tracts and by wind and water in reveraine tract; reaches the ground from the beginning of December till end of April and lie until they get soaked when the monsoon rains commensurate in June – July. Early heavy thunder storms results in earlier germination. Numerous seedlings are found after rain in riveraine tracts in places well above the reach of floods.

In non-riveraine tracts, regeneration is easy on exposed ground such as on new embankments, landslips, soil worked by pigs, abandoned cultivated areas if free from weeds. Dense weed growth prevents establishment of regeneration. Fire and grazing are inimical to natural regeneration.

Artificial regeneration: The matured pods have to be collected during December to February from the selected straight grown trees by lopping branches and from the ground. The collected pods need to be dried in the sun soon after collection. The dried pods can be stored in gunny bags and care should be taken against the damage by rats. Before sowing, soak pods in water for 2 days. Direct sowings are preferable to planting and suggested spacing between the lines is 3 m to 4 m apart. Otherwise, the pretreated seeds are dibbled in polythene bags which contains the potting mixture of soil: sand: Farm Yard Manure (3:1:1). Germination will starts within 7 to 15 days and 90% germination can be achieved. The seedlings need to water daily and weeded constantly.

The one year old healthy and vigorous seedlings is found to be suitable for field planting at the espacement of 2 m x 2 m or 3 m x 3 m. Planting of 2 year old stumps is also the best method where the competition due to grasses is severe.

Clonal Propagation: Sissoo is now predominantly multiplied through clonal technology using apical shoots. The superior genetic stocks are identified, felled at 15 – 30 cm from ground level. The stumps are induces coppice and apical shoots are collected from these coppicer rooted under controlled environmental condition. These plants are planted in clonal mother garden at 10 cm × 10 cm in order to accommodate 100 plants/m^2. These mother plants are irrigated and fertilized method. After 30 days the mother plants are cut and new shoots will produced in about 15-20 days. These new apical shoots are collected and rooted under controlled environmental condition. This technology is now commercially deployed for mass multiplication.

(iv) Silvicultural management:

- The *Dalbergia sissoo* plantations are worked under coppice with standards system, seedling regeneration being obtained both naturally by cleaning the ground and weeding.
- Fire protection in first four years and afterwards early burning can be attempted.
- The riveraine Sissoo forests are worked under selection fellings or by simple coppice for a rotation period of 20 years.
- Species regeneration measures should be adopted with regular cleanings and thinning. The most appropriate system will be clear felling with artificial regeneration.

13. Describe the silvics of *Tectona grandis* under the following heads: (2011)

(i) Distribution and morphology

(ii) Silvicultural character

(iii) Silvicultural system and management

(iv) Utilization

(i) Distribution and morphology: This species is indigenous to Burma, India, Indonesia, Laos, Japan and Java. In India, it occurs in Madhya Pradesh, Orissa, Uttar Pradesh, Tamil Nadu Kerala and Rajasthan. It is also introduced in moist deciduous forests of West Bengal, Assam, Bihar, Orissa and the Andamans. Teak is widely distributed in the tropical moist and dry deciduous forest with average annual rainfall of 1500 to 2000 mm. It grows well on the wide geological formations like basalt and other lime and base rich rocks, sandstone, quartzite and other rocks which are usually non-calcareous. It prefers well drained sandy alluvium to old red soils except laterite soils. Teak occurs in soils of pH range between 6.5 and 7.5; below 6.0 is practically absent and beyond 8.5 affects its growth. Teak is very popular timber in India and it is a deciduous tree attaining height of 30 – 40 m and a girth of 4-6 m. it is a large sized tree and leaves are broad and opposite. The flowers are in large, lacs cluster at the end of branches and white in colour. The fruits are drupe, spongy in nature and contain 1-4 seeds.

Silvicultural character: Teak is a strong light demander and does not tolerate suppression and requires complete overhead light and fair amount of side room for proper development. It produces a large and deep root system. However, teak is sensitive to frost and drought. It has greater resistance to the effects of fire. Teak is a good coppicer and pollards vigorously. The coppicing power is being retained up to a considerable size. The vitality of teak stumps is remarkable.

(ii) Silvicultural system and management

- Excepting for dry teak forest which is worked under coppice system all areas in moisture places are worked under clear felling system followed by regeneration.
- The stand density should be properly maintained as it reflects on the current and mean annual increment. Hence it is prescribed to do two mechanical thinning during 5th and 10th year and one Silvicultural thinning during 15th year.
- Necessary and timely pruning and tending operations like climber cutting etc. should be carried out in the teak plantation in order to produce the wood with better form and quality.

(iii) Utilization

- Teak is a moderately strong timber with a density of 660 kg/m^3 and is preferred as a most suitable timber both for domestic and industrial utility.
- Teak is known as a renowned timber due to durability, dimensional stability, working quality and resistant to termites.
- Teak wood is used in all construction purpose such as beams, columns, doors, windows, flooring, paneling etc.
- It is one of the best timbers for furniture and cabinet making wagon and railway cadges.
- For marine construction and ship building teak is preferred due to dimensional stability.

14. Describe the silvics of *Casuarina equisetifolia* under the following heads: (2011)

(i) Ecology and Distribution
(ii) Propagation and management
(iii) Functional uses
(iv) Pest and disease management

(i) Ecology and Distribution: *Casuarina equisetifolia* is found from Burma and Vietnam throughout Malaysia east to French Polynesia, New Caledonia, Vanuatu, and south into Australia. Along the coastal regions, *Casuarina* thrives in the temperature extending to 47^{0}C. Under inland conditions, it tolerated extreme temperatures, but exhibits poor growth. It grows well both in South West and North East monsoon areas which receive rainfall between 900 to 3800 mm. Growth is poor in low rainfall areas. *Casuarina* grows best in loose, fine coastal sands. It can grow well under inland conditions of well drained sandy soils. It tolerates lateritic and red soils and also saline, alkaline and acidic conditions. The trees have nitrogen fixing root nodules and hence dependability on nitrogen supplement is less.

(ii) Propagation and management

Seed propagation: The *Casuarina* seeds are disbursed to some extent by wind and abundant seed production leads to wider distribution. Natural regeneration can be augmented by factors viz., abundance of light, porous sandy soil free from weed growth, absence of heavy rain during germination and *early* life of the seedling, sufficient moisture in the soil to prevent mortality from drought during the first two or three years and absence of serious damage by ants and other insects as well as crabs.

Clonal Propagation: It is propagated vegetatively from lateral or side shoots by using IBA based powder formulated *rooting* hormones at the concentration ranging from 3000 – 6000 ppm. The treated cuttings should be cultured under greenhouse conditions with intermittent misting coupled with the humidity of 70-80%. Air layers and rooting of sprigs in mist chambers is also found very successful for mass multiplication of elite clones. Superior trees are normally propagated through this method of propagation and are called clones. Normally farmers adopt 0.8 to 1 m spacing which results in thin boles. Hence a spacing of 4 x 4 to 6 x 6 feet is ideal which allows intercropping in the first year and also it gives good growing space for the trees to put on excellent girth at an early state. For planting of bag plants in sandy soils, the soil is scooped to the depth and width of the bag, planted preferably with onset of rains.

If watering is done along with the planting, establishment will be faster. In hard clayey soils, a small pit of 30 cm is dug out and bags planted is done after removing the ploythene cover carefully, without disturbing the soil in the bag. Planting of naked seedlings or treated naked seedlings can be done in holes made with a crowbar, to depths of the root length. The seedling is placed with roots straight in the hole before planting the thin part of the root should be cut cleanly with a knife.

Management

Irrigation: Casuarina is an irrigated crop and the irrigation is required for once in every 10-15 days. It is also advisable to adopt drip irrigation in order to effectively manage the available water and to increase the yield.

Fertilization: The trees have nitrogen fixing root nodules (an actinomycetes called Frankia) and hence, dependability on nitrogen supplement is less. However, 40 -50 kg per ha of nitrogen can be applied in four equal splits. The Super Phosphate @ 150 kg per ha and Muriate of Potash @ 100 kg per ha can be applied in four to five equal splits.

Pruning: Branches are pruned flush to the stem of up to 1/3rd of the stem height to augment height increment and to obtain clean bole during early stage (6-12 months).

(iii) Functional uses

Pulp wood: Casuarina is a good pulpwood species. Farmers are cultivating mostly this species for pulpwood. Yield and strength properties of *Casuarina* pulp are reported satisfactory for wrapping paper and duplex paper. It makes good pulp by use of neutral sulfite semi chemical process. The material with bark is used for pulping.

Fuel wood: Casuarina is considered to be the best firewood in the world, burns even when green. Its calorific value is 4950 Cal/Kg and hence, it can be a source of bioenergy.

Timber: The sapwood is pale brown, heartwood dark reddish brown. Timber is strong and heavy (average 850 kg/m^3). It is liable to crack and split, not easy to saw and season. It is used as poles, scaffolding, transmission lines and rafters.

Medicinal: Bark of Casuarina is a tonic and astringent useful in dysentery and diarrhea. Decoction of leaves and twigs is used in colic and powdered seed in made into a paste and applied as balm for headaches. Bark contains 6-18% tannin, which is also used for dyeing wool and silk fabrics and for toughening fishermen nets. It also yields a resin. Needles of Casuarina have been used for preparing activated carbon by the zinc chloride method.

Avenue: Casuarina forms a good avenue tree and most suitable for landscaping sea beaches. It forms a good hedge plant and can be shaped to desired form.

Windbreak: Casuarina with deep taproot can withstand cyclonic storms than any other species and is very useful as a windbreak and for sand dune stabilization.

(iv) Pest and Disease

Major pest

S.No	Common pest	Control Measure
i.	Bag worm	Topical application endosulfan or chlorlpyriphos 2ml per litre.
ii.	Bark feeding caterpillar	Remove the feeding galleries and apply insecticide soaked cotton (15ml of dichlorvos) in the bore holes.
iii.	Stem borer	Inserting wire to remove feeding larvae and applying insecticide soaked cotton (15ml of dichlorvos).
iv.	Mealy bug	Topical application of methyl dematon or dimethoate (2ml per litre)
v.	Termite	Application for chlorlpyriphos 2 ml per litre

Termites also cause serious damage in young plantation by destroying the root system leading to the death of young trees. This can be arrested by soil drenching with chlorpyriphos at 0.2 per cent.

Major disease: The casuarina plantation are found vulnerable to various diseases viz., stem canker and die back, pink disease, root infection and die back and wilt. The stem canker and die back can be controlled by the application of Bavistan at 0.01 per cent active ingredients.

Wilt disease caused by Trichosporium is a serious disease and could be managed with proper soil and water management. Severely damaged trees have to be uprooted immediately to avoid further spreading.

S.No	Common disease	Control Measure
i.	Damping off	Providing proper drainage Seed treatment with captan or thiram @ 4g/kg. Soil drenching with carbendazim (Bavistin) @ 0.1 %
ii.	Stem canker	Spray with mancozeb @ 0.25 %
iii.	Die back	Removal of infected plant parts and spray with mancozeb @ 0.25 % or copper oxy chloride @ 0.25 %.

iv.	Pink disease	Removal of severely affected plant parts or scrap the infected portions and apply with Bordeaux Paste.
v.	Wilt disease	Remove the infected trees immediately.
		Digging the trenches around the infected tree.
		Scrap the infected portion and spray with copper oxy chloride @ 0.25 %.

15. Write phenology, regeneration and Silvicultural characters of *Gmelina arborea* (2005)

(i) Phenology

It is a deciduous fast growing tree, attains a height of 30 m and girth over 4 m. The tree will have straight bole and cylindrical crown. Bark grey coloured, and the leaves are greyish green. The flowers labiate type and yellow panicle with brown and it occurs in narrow branching clusters. The fruits are large, pulpy and globosely. It is a light demanders, coppicing ability is good

(ii) Regeneration

1. **Regeneration**
   - **(i) Natural regeneration:** Natural regeneration is very poor in this species & hence predominantly by artificial regeneration.
   - **(ii) Artificial regeneration**
     - Number of seeds/Kg - 1400 seeds/kg
     - Pre sowing treatment - Depulping and water soaking treatment for 24 hours
     - Germination percentage - 13 % - 90 %
     - Germination Period - 21 – 45 days
2. **Planting techniques**
   - Spacing - 3 x 3 m, 4x4 m and 5x5m
   - Pit size - 30 cm^3 to 45 cm^3
   - Planting - 10 months old containers are planted in the prepared field

(iii) Silvicultural characters

Gmelina is a strong light demander, moderately frost and drought hardy. The tree coppices well and the coppice grows vigorously. Gmelina produces root suckers also. Gmelina is propagated through seeds, cutting and root suckers. It is susceptible to grazing and browsing damage.

16. "Indian Sandalwood (*Santalum album*) is the most valuable wood in India. However, silvicultural production of this species is not satisfactory." Discuss the above relation to the phenology of Santalum album and the forest laws governing its cultivation and trade. (2010)

Indian sandalwood is a most valuable wood known for its oil which is pronounced as the most famous East Indian sandalwood oil. This oil is produced from heartwood of sandal through distillation process and it found its profound utility in perfume industries, cultural and religious and also as a source of medicine. The sandalwood is highly priced in the market and the current estimates indicated that one ton of sandalwood is Rs. 56 lakhs, which is the highest among the Indian timbers. Hence sandalwood is highly value.

Silvicultural production

The silviculture of the species though century old, but still does not delivered the expected level of productivity. Generally sandalwood grows in rainfall ranged between 500 mm and 2000 mm and it's reported that good heartwood formation occurred in rainfall ranged between 800 mm and 1000 mm. It is light demanding species and grows well in sandy clay, red lateritic, loamy and even in black cotton soil. But sandal growth is very well observing in ferruginous loam soil which is laying on metamorphic rocks.

Propagation

Sandal is propagated both naturally and artificially. The artificial sandal cultivation has experienced success and failure which resulted in problem in silvicultural production. For artificial regeneration the depulped seeds are treated with Gibberlic acid at 0.05% overnight soaking and sown in raised bed. The seeds germinate within four weeks and 4-6 leaf stage seedlings are pricked out and planted in polybags along with suitable host.

Requirement of host

Another major problem in silviculture production is, it association with other plants. Sandal is partial root parasite and it desired part of its nutritional requirement from the host species. Though several reports are available on host suitability but the exact sandal host association has not delivered significance combination which resulted poor growth and development in sandal and some cases the resulted in failure of plantations.

Planting

The sandal seeds are planted as espacement of 3m × 3m, 4m × 4m, 5m × 5m and there is no prescribed and standard spacing available for cultivation of sandalwood and the compiled with the lack of prescribed host planting leads to problem in silviculture production. The Irrigation and nutritional requirement of sandal is not available which led to slow growth and development and results in problem in silviculture production.

The sandal growth and development is very slow and there is no fixed rotation for this species even after 100 years of scientific management in Indian forestry, sandal is managed in physical rotation which is resulted in problem in silviculture production. Hence the sandal silviculture production problematic due to the lack of information and technology on suitable host on nursery and field condition, requirement of irrigation and fertilisation leads to problem in silviculture production.

Sandal phenology

Sandal flowers twice a year and the flower is cymes. Though the flower is profuse, but the seed production is only half of the flowering intensity due to cross pollination, which is predominantly by bees and ants. Another peculiar feature in sandal flowering is even in young trees of less than 5 years are able to flower and produce fruits. And the seeds experienced erratic germination behaviour and leads to failure of regeneration.

Forest laws governing cultivation and trade

Since sandal is predominantly available in two states of Karnataka and Tamil Nadu. The law varies from state to state. As far as sandal cultivation is concern there is no restriction, however if it is cultivated outside the natural forest, it should be registered in the land patta. As far as felling is concern, it comes under the Tamil Nadu sandalwood transit rules, 1967 and Tamil Nadu sandalwood tree on patta land 2008. As per this rules legally sandalwood trees are permitted for extraction, but the sandal tree grower has inform the forest department for extraction of trees by providing the certificate of ownership, the entry of chitta and adangal, sketches of the patta land and the list of sandalwood trees proposed to be sold. Based on these request the forest department verifying the sandal tree ownership and follow the process of enumeration, extraction and fixing of prices. As per the sandal rules 2008, after fixing the price, 80 % of sale price *i.e* Net sale price will be paid to the owner of the sandal tree and 20 % to the forest department. These are the laws governing sandal cultivation and trade.

17. Give an account on the management of Eucalyptus or popular plantation?

(2005)

(a) **Management of *Populus deltoides:*** Popular plantations are established predominantly through entire transplanting of clones. These popular plantation managed for the production of short rotation plywood raw material of popular plantation include the following

(i) **Plantation establishment:** 1 – 2 years old plants are taken out from the nursery and used for planting operations. Since Poplars is planted by entire planting method, two persons are essential to hold the plants and place them vertically in the pits. After that the soil around the ETPs are compacted and it should be irrigated immediately. During planting operation, it should be ensured that the collar of the plants should be shrunken at 20 – 50 cm in the pits.

(ii) **Irrigation management:** Popular plantations are managed profitably with irrigation the growing of Populus along with agricultural crop with judicial irrigation management help to be get economic yield. During the first year of plantation irrigation is done at weekly intervals particularly during the period between February and June. From July to September irrigation must be provided based on the rainfall. After September up to February two irrigation per month is required as the growth activity of popular ceases. During second year weekly irrigation to 2 irrigation per month may be given depending on the rainfall. The well irrigated plantation exhibit more uniform growth and leads to higher productivity.

(iii) **Fertilizer management:** Normally Populus are very reactive towards fertilizer application. The basal application of 50 g superphosphate, 50 g Urea with 2 Kg FYM is recommended. Application of 75 g, 150 g, 250 g Urea as three split doses as encouraged rapid growth and development at early stages. The fertilizer applications have to be continued up to harvesting period. Immediately after the application of fertilizer the plantation are to be irrigated.

(iv) **Spacing management:** Spacing of Populus depends on its utility. For pulpwood plantation closer spacing is recommended. For plywood plantation wider spacing is recommended. In U.P large scale plantation are carried out at espacement of 5m × 5m and under Agroforestry system the spacing of 5m × 4m is practiced.

(v) **Debudding:** Debudding is an essential Silvicultural operation which involves gently rubbing of newly forming buds to avoid branch formation and to ensure production of clean, straight and balanced plantation.

(vi) **Pruning:** It is necessary to improve the stem form and timber quality. Pruning is done during winter season when Populus are leafless. Pruning of leader shoots and young shoots are done based on the growth and development.

Under Agroforestry systems pruning is done up to 1/3rd of the plant height at the early stages and ½ of the plant height after five years. Careful pruning is very essential and it should not damage the bark. After pruning, application of Bordeaux paste is useful in avoiding the diseases.

(vii) **Thinning management:** Populus are normally worked under selective and systematic thinning. If planted with high density, systematic thinning done at 3 and 6 years. In order to get intermediate yields from the thinning and to allow the remaining crop to reach harvestable maturity for plywood production for long rotation and low density plantation selective thinning is essential to remove the inferior and to realize economic returns from the retained superior trees.

(viii) **Pest and disease management:** The populus are invaded by wide range of pest and disease and most serious is shoot borer caused by *Apriona cinerea.* This insect can be controlled by injection of carbon dishulphide or formalin in the borer hole and plugging the hole with cotton wool or clayey soil. Termite is another major problem which can be controlled by application of 2-5% dusban.

In case of disease occurrence, the root rot caused by Fomes and Ganoderma is very serious which can be controlled through silvicultural treatment by removing infected, diseased and dead plants.

(b) Management of Eucalyptus plantations

(i) **Site preparation:** The site preparation is one of the important aspects to be addressed carefully for establishment of eucalyptus plantations. The site preparation should be done before the onset of monsoon i.e., May/June. The site preparation activities includes the following,

(a) Bush clearing

(b) Ploughing

(c) Compartmental bund formation

(a) Bush clearing: Most of the eucalyptus plantations are established at degraded or barren land, these sites may be covered with unwanted bushes and thorny shrubs. Though the bushes availability will be lesser it is always needed to clear the vegetation for better establishment of plantation. Hence complete clearing of the sites is needed for raising eucalyptus plantations. These bushes present in the site can be clearly removed with the help of dozers and the debris should be removed from the field and may keep along the periphery of the field as boundary fence.

(b) Ploughing: Incase of hardy area the whole area should be ploughed with disc to breaks the hard pan. After this, one ploughing with tyne cultivator should be done to breaks the clots and loosening the soil to collects maximum rain water. It allows the water to percolate in the subsoil and thereby controls the runoff. Incase of sandy soils one ploughing with tyne cultivator is essential to loosen up the soil.

(c) Compartmental bund formation: After ploughing, the compartmental bunding should be done along the contour to avoid runoff and soil erosion and collect the rain water with in the field. This will ensure better success rate of plantation by means of providing good root zone and highest water penetration rate.

(ii) Planting: The better planting season for Eucalyptus is June to October. The optimum pit size should be 45cm x 4cm x 45cm for polybags plants. The root trainer contained plants may be planted in the pit size of 30cm x 30cm x 30cm. While planting application of 250g of Vermi-compost or 500g of Farmyard manure per pit will improve the establishment of plantations. After planting, the soil around the plants should be packed compactly to avoid the dryness of plants. After planting the life irrigation should be done for better initial establishment. The causality replacement should be done after one month from planting.

(iii) Weed Management: Initially the weed management is essential for better establishment and growth of eucalyptus plantations. These weeds need to be removed either by tyne ploughing or manually. During planting year, one ploughing with tyne during the month of November along with soil working will remove the weeds present in the field. The same practices need to be followed during first year of maintenance for better weed management.

(iv) Water Management

(a) Rainfed Farming: Since most of the eucalyptus plantations are raised under rainfed area the life irrigation after planting is essential for better establishment. Incase of failure of monsoon, the protective irrigation for first two years is essential. The available rain water will be harvested through ploughing activities. During planting year, the tyne ploughing at November and disc ploughing during February will improve the soil penetration properties and collect the available rain water. Also the compartmental bunds formed in the plantations will reduce the run off.

During first and second year of maintenance, two disc ploughing and one disc ploughing during third year of maintenance will also harvest the available rain water and improve the growth of plantations.

(b) Irrigated Farming: In the recent past, Eucalyptus Clonal plantations are raised under intensive management system through drip fertigation. The National Agricultural Innovation Project on Industrial Agroforestry has identified irrigation schedule for Eucalyptus Clonal plantation which is furnished below.

Month	Water Requirement (Litre /Plant /Day)		
	1st Year	2nd Year	3rd Year
June (Planting Season)	8	10	10
July	8	10	10
August	8	10	10
September	6	8	8
October	5	6	6
November	4	6	6
December	8	8	10
January	8	10	12
February	10	12	12
March	12	14	15
April	12	14	15
May	12	15	15

(v) Nutrient Management Rainfed Farming: The eucalyptus plantations growth can be improved by application of organic and chemical fertilizers during initial years. The application of fertilizers should be done based on the soil analysis. In general, one time application of 50g of Complex Fertilizer (17:17:17) per plant during planting and first year will improve the establishment and growth of plantations. Also application of 500g of bio-compost per plant will improve the growth of plantations. Application of gypsum in the salt affected area will improve the soil conditions.

(vi) Pest and disease management: The following pest and disease may occur in the Eucalyptus plantations and the management practices are indicated.

Pests / Diseases	**Management Practices**
Pests	
G$_{a}$n	Selection of gall resistance variety*Application systemic insecticide like imidacloprid (2ml / litre)
Termite	Application for Chlorphyriphos 2 ml per litre
Diseases	
Pink disease	Application of Bordeaux mixture
Stem canker	Application of Bordeaux mixture at root portion

(vii) **Harvest management:** The rotation period for eucalyptus plantation ranged between 2 and 5 years depending on the irrigation and or rainfed farming. Plantation will be harvested by adopting simple coppice system. The better season for harvest of eucalyptus plantations is rainy period which will improve the growth of new sprouts. The total area will be harvested through clear felling and the stumps should be retained for next rotation. The optimum height of the stump should be 15 cm and the cut should be done in slanting phase. The coppice shoots will be sprouted from this stump and the two superior coppice shoots should be retained and the balance shoots should be removed. This singling out operations should be done after 6 months of harvest. This coppice shoots will form the first rotation coppice plantations. Upto 3 rotations, these stumps can be retained and after that it should be removed from the field since there is loss in vigour in coppice from the third coppice onwards.

18. a. Give an account of Silvicultural management of Teak plantation? (2006)
b. Explain thinning operation with respect to teak plantation (2005)

Silvicultural characters

Teak is a strong light demander and does not tolerate suppression and requires complete overhead light and fair amount of side room for proper development. It produces a large and deep root system. However, teak is sensitive to frost and drought. It has greater resistance to the effects of fire. Teak is a good coppicer and pollards vigorously. The coppicing power is being retained up to a considerable size. The vitality of teak stumps is remarkable.

Plantation establishment

Teak plantations are established predominantly by stump planting at an espacement of 2m x 2m.

Irrigation management

Teak plantations are both under irrigated and rainfed conditions in Tamil Nadu. The teak plantations are managed under irrigation – 2 irrigation / month. Immediately after planting in absents of rainfall irrigation should be done immediately followed by another irrigation after 3-5 days. In Kerala, Teak is planted during rainy season and assisted by irrigation during the lean period.

Weeding

Teak plantations are very sensitive to weed growth especially the grasses and climb 3 weeding in1styear, 2 weeding 2ndyear and annual weeding from 3rd year onwards will help to produce quality timber. Besides weeding climber cutting, removal of double leaders *etc*will help to improve the plantation.

Fertilizer

Teak is responsive crop for fertilizer application and basal application of 2-5 kg FYM + 50g DAP will help to ensure rapid initial growth from 2nd year onwards 75 g to 150 g DAP and after 1st

thinning application of 250g DAP/plant along with 5kg of FYM/plant will augment the growth of teak.

Thinning operation in teak plantation

Thinning is an important factor in teak management plan. The thinning depends on initial spacing, site quality, weed growth and rate of growth especially during the initial period. The latter thinning depends on the merchantability of the product and the objective of management. Teak is widely grown across the country and there are variations in thinning practices followed in India which is furnished below.

Teak thinning in Kerala

Teak is widely planted in Kerala particularly the Nilambur teak is very famous for quality in Nilambur forest division. The thinning cycle of 4, 8, 12, 18, 26 and 36 years has been followed for 50 years rotation. Among these thinning, the first two thinning are mechanical and the rest four thinnings are silvicultural. In the first mechanical thinning alternative diagonal rows are removed. In the second mechanical thinning alternative planting rows are removed. In case of silvicultural thinning 'D' grade thinning is generally practiced.

Teak thinning in Tamil Nadu

In Tamil Nadu, teak is spaced at 2m x 2m and the thinning cycles are 5, 10, 18, 25 and 40 years for 60 years rotation cycle. Here also first two thinnings are mechanical and the remaining three are silvicultural thinning.

Other teak thinning regimes

Due to variation in spacing various thinning cycles are followed in India. In teak the initial spacing is 1.8 m x 1.8 m. The first mechanical thinning is done when teak attains a height of 7.5m to 9m. Second thinning is done when the crop is 10-12 m height. Third thinning is done at the age of 10-15 years and fourth thinning at 40 years.

Silvicultural treatment

Excepting for dry teak forest which is worked under coppice system all areas in moisture places with adequate advance growth, which on cutting back results in fully stocked new crops. The stand density should be properly maintained as it reflects on the current and mean annual increment.

Pest and disease management

Nature of damage	Casual organism	Control measures
Defoliation	*Hybleae purea*	Adoption of Silvicultural practices coupled with Silvicultural control which ensure removal of infected and dead trees
Skeletonizing	*Eutectona machaeralis*	

Disease

Among the disease, occurrence of powdery mildew *Uncinulla tectonae* and leaf rust *Olivia tectonae* were prominent and controlled by silvicultural management.

Control measures

Adoption of Silvicultural practices coupled with Silvicultural control which ensures removal of infected and dead trees.

Chapter

5

Agroforestry, Social Forestry, Joint Forest Management and Tribology

1. Enumerate di-erent Agro forestry systems with models. (2005)

(a) Agri-Silviculture

(b) Silvipasture

(c) Agri-Silvipasture

(d) Miscellaneoussystem

(a) Agri-Silviculture: Agri-Silviculture system refers to combined production of agriculture crop and trees including shrubs/vines and forest crops.

(i) Improved fallow species in shifting cultivation

(ii) The Taungya system

(iii) Multispecies tree gardens

(iv) Alley cropping (Hedgerow intercropping)

(v) Multipurpose trees and shrubs on farmlands

(vi) Crop combinations with plantation crops

(vii) Agroforestry fuelwood production

(viii) Shelterbelts

(ix) Windbreaks

(x) Soil Conservation hedges

(b) Silvipasture: Silvipasture system refers to combined production of trees and fodder crops/ grasses and/or animals. This system deals with combine cultivation of timber yielding trees and fodder yielding trees plus grasses and or animals.

The production of woody plants combined with pasture is referred to Silvipasture system. The trees and shrubs may be used primarily to produce fodder for livestock or they may be grown for timber, fuel wood and fruit or to improve the soil.

This system is classified in to three categories

(i) Protein bank

(ii) Live fence of fodder trees and hedges

(iii) Trees and shrubs on pasture

Protein bank: In this Silvipastural system, various multipurpose trees (protcin rich trccs) are planted on or around farmlands and range lands forcut and carry fodder production to meet the feed requirement of livestock during the fodder deficit period in winter.

Example: *Acacia nilotica, Albizia lebbeck, Azadirachta indica, Leucaena leucocephala, Gliricidia sepium* and *Sesbania grandiflora*

(ii) Livefence of fodder trees and hedges: In this system, various fodder trees and hedges are planted as live fence to protect the property from stray animals or other biotic influences.

Example: *Gliricidia sepium, Sesbania grandiflora, Erythrina sp.* and *Acacia sp.*

(iii) Trees and shrubs on pasture: In this system, various tree and shrub species are scattered irregularly or arranged according to some systemic pattern to supplement forage production.

Example: *Acacia nilotica, Acacia leucophloea, Tamarindus indica* and *Azadirachta indica.*

(c) Agri-Silvipasture: The production of woody perennial combined with annuals and pastures is referred Agrisilvopastural system.

This system is grouped into two categories.

(i) Home gardens

(ii) Woody hedgerows for browse, mulch, green manure, soil conservation

(i) Home gardens: This system is found extensively in high rainfall areas in tropical South and South eastasia. This practice finds expression in the states of Kerala and Tamil Nadu with humid tropical climates and where coconut is the main crop. Many species of trees, bushes, vegetables and other herbaceous plants are grown in dense and in random or spatial and temporal arrangements. Most home gardens also support a variety of animals. Fodder grass and legumes are also grown to meet the fodder requirement of cattle. In India, every homestead has around 0.20 to 0.50 ha land for personal production.

Home gardens represent land use systems involving deliberate management of multipurpose trees and shrubs in intimate association with annual and perennial agricultural crops and livestock within the compounds of individual houses. The whole tree- crop- animal units are being intensively managed by family labour.Home gardens can also be called as

Multitier system orMultitier cropping: Home gardens are highly productive, sustainable and very practicable. Food production is primary function of most home gardens.

Structure of Home Gardens: Home gardens are characterized by high species diversity and usually 3-4 vertical canopy strata. The layered configuration and compatible species admixture are the most conspicuous characteristics of all home gardens. Generally all home gardens consist of herbaceous layer near the ground, a tree layer at the upper levels and an intermediate layer. The lower layer can be partitioned in to two, the lowermost being at less than 1.0m in height, dominated by different vegetables and the second layer of 1.0 -3.0/m height comprising food crops such as banana, papaya and so on. The upper tree layer can also be divided into two, consisting of emergent, full grown timber and fruit trees occupying the upper most layers of 25m height and medium size trees of 10-20m occupying the next lower layer. The intermediate layer of 5-10m height is dominated by various fruit trees.

Choice of species

(a) Woody species: *Anacardium occidentale*, *Artocarpus heterophyllus, Citrus spp, Psiduim guajava, Mangifera indica, Azadirachta indica, Cocus nucifera,*

(b) Herbaceous species: Bhendi, Onion, cabbage, Pumpkin, Sweet potato, Banana, Beans, etc.

(ii) Woody Hedgerows: In this system various woody hedges, especially fast growing and coppicing fodder shrubs and trees are planted for the purpose of browse, mulch, green manure soil conservation etc. The following species *viz., Erythrina sp, Leucaena leucocephala* and *Sesbania grandiflora* are generally used.

(d) Miscellaneous system

- Understory cropping
- Home garden
- Dispersed trees
- Improved fallows
- Mixed perennials
- Alley cropping
- Vegetative strips
- Wind breaks
- Taungya system
- Shelter wood
- Park land system
- Multiple tree garden
- Trees with plantation crop
- Multipurpose trees and shrubs on farmland
- Home stead plantation

2. What is the role of tree species in watershed management? (2005)

The trees in a watershed play a significant role due to their influenes on interception, infilteration, soil moisture storage, runoff and evaporation which thereby helps to improve the sustainable watershed management activities. The following are the specific roles played by trees in a watershed.

- The improvement of watershed conditions for sustaining water resources demand sustainable forest management practices. The forested catchments supply water for domestic, agricultural, industrial and other ecological needs in down stream areas.
- The trees in the watershed will have direct impacts on the quality of water yields from the watersheds.

- The forests contribute to the control of soil erosion and consequently reduce the levels of sedimentation. It has been estimated that in India 5333 million tonnes of soil is annually detached and eroded whichi indicates a loss of 16.53 tonnes of soil per hectatre.
- Out of this soil erosion 29% is carried away by rivers and 10% is deposited in the reservoir resulting in approximately 2% loss of storage capacity. If the watershed is well managed with sufficient forests and trees soil erosion can be controlled, sedimentation can be reduced and water harvesting can be augmented.
- Trees play a vital role in controlling the soil erosion and help to conserve both soil and water. The tree adds leaf litter which gets decomposed and improves the soil physical, chemical and biological properties. This improved soil properties enhances infilteration and perculation thereby reduces the soil erosion and conserves water. The rainfall is intercepted and infilterated by the tree leaf litter which thereby helps to conserve soil and water. The tree root system protects the soil from detachment and helps to conserve soil.
- These activities will help to improve the water table not only within the watershed area but also across the farmlands located in the vissinity of watershed and improves the agricultural situation around the watershed.
- The trees in the watershed play a significant role in reducing the peak discharge and slow the flow of water and reduce the impacts.
- The forests and the associated forest management influence the downstream uses of water resources for agriculture, animal husbandry, grazing lands, industrial utility, domestic consumption etc.
- The trees in the watershed provide all the basic needs of the people in terms of timber, fuel, fodder, fiber and food thereby help to sustain the livelihood opportunities.
- The trees in the watershed help to augment the biodiversity status.
- The trees in the watershed have a significant role in combating climate change and its associated risks in climate variables through mitigation and adaptation mechanisms.
- The forests also play a vital role in the watershed in all ecological, social and economic dimensions.

3. What is the role of NGOs in JFM activities? (2005)

The NGOs play a vital role in JFM activities and they are furnished below.

- NGOs are facilitating village communities as well as the forest department in the formation of JFM Committees and have been instrumental in articulating the needs of forest-dependent communities to the state.
- NGOs have develope their own participatory forestry models based on the JFM policy directives of the government.
- During the inception of JFM in India, the forest department was skeptical of NGO involvement in assisting village communities in undertaking community forestry programmes.
- Over the last decade, however, the state of affairs has changed in favor of NGOs, which may be mainly attributed to the change in the mindset of forest department personnel towards forest management.

- Now substantial rural developmental funds earmarked by the government are routed through NGOs for JFM programmes.
- Additionally, pressure on the forest department from external aid agencies to involve NGOs in JFM programmes and to restructure the forest department accordingly, as a condition for aid in India, has also resulted in overcoming the problem between NGOs and the forest department.
- Now NGOs are a major stakeholder in forest policy formulation in the country.
- NGOs can also play a vital role in building awareness and environmental protection and other developmental programmes of the JFM areas.
- NGOs can play a role in establishing basic infrastructure facilities like roads, protected drinking water, health, sanitation, electrification and improvement of the habitate.
- The NGOs can play a vital role in integrating women activities through micro credits and other income generation activities.
- The sustainable collection, scientific process and value addition of non timber forest products can be very well played by the NGOs.
- Above all creating and strengthening local institutions will help to protect the social fencing and to sustain the JFM activities.
- The NGOs can facilitate integration of schemes operating at the various departments.
- NGOs can play a vital role in microplan preparation.
- NGO's are easily approachable and hence they can collect reliable information on socioeconomic status.

4. Describe a model of Silvipasture. (2005)

It is one of the major agroforestry systems wherein woody tree species are grown along with grasses. The tree species incorporated in silvipastoral system may be of multiple utility from fodder to timber. The silvipastoral system is further classified into three categories and they are,

(i) Trees and shrubs as pasture
(ii) Protein based fodder banks
(iii) Fodder trees, shrubs and hedges as live fence

(i) Trees and shrubs as pasture: It is one of the silvipastoral systems in which multipurpose trees and shrubs are planted sporadically with or without systematic pattern in the pasture lands. The main idea of introducing trees on pasture land is to supplement fodder availability during the lean season.

Preferred species: *Acacias, Prosopis, Tamarind, Emblica, Holoptelia, Hardwickia*, Neem.

(ii) Protein based fodder banks: It is one of the important Silvipastoralsystems in which protein rich tree species are planted in and around pasture lands. These tree species provide protein rich leaf fodder to supplement fodder production during the lean season.

Protein rich tree species are

- *Acacia* spp.

- Albizia
- Leuceana
- Gliricidia
- Sesbania
- Prosopis

(iii) Fodder trees, shrubs and hedges as live fence: In this system, several fodder trees and hedges are planted around the farm land as live fence to protect the crops within the farm land from all biotic influences. The preferred species for live fences are

- *Gliricidia sepium*
- *Sesbania grandiflora*
- *Acacia leucophloea*
- *Leuceana leucocephala*
- *Protium cordatum*
- *Euphorbia*

Kangeyam (Tamil Nadu) Silvipastural Model

The state of Tamil Nadu is known for Kangeyam silvipastural model which includes three components viz., tree, grasses and animals. This model is over 100 years old and still in existence due to its productivity and sustainability.

Tree Component: *Acacia leucophloea* is a major species which is a multipurpose nitrogen fixing trees. The tree is amenable for arid and semi-arid localities and survive even in extreme temperature. This tree produces nutritious fruit as a fodder due to its high protein content of over 16% and the fruits are available during summer season. During this period the availability of grass is very low and the people use this pod as a cattle feed. The milk obtained from this feed is known to contain good nutritional properties.

The tree is also providing local timbers to the people in order to meet their domestic needs of agricultural implements, cart wheels, small furnitures etc. The tree is a very dense wood with density of over 600 kg/m^3 and hence amenable for rough construction. The wood exhibit high calorific value of over 5000 kcal/kg and hence suitable as a fire wood.

Grass component

The grass component in this model *Cenchrus ciliaris.* This grass is fast growing and amenable for arid locations. The tree is excellent for grazing animals especially during the lean season. The grass regenerates well on its own and during the rainy season the seed disperses and develop into a dominant grass. The grass has excellent palatability and nutritional quality.

Animal Component

The animals include both cattles and sheep which feed on both trees and grasses. The animals are allowed inside the silvipastural model for the whole day for feeding on the tree and grazing on the grasses.

This silvipastural model is one of the socially acceptable model in the arid tract of tamil nadu. This model acts as a livelihood opportunity for the local farmers. This model incorporates 60-80 trees in one acre area which gives boarder canopy spread and ensures production of over 100 kilos green fodder per tree and satisfy the quality fodder requirements of the locality.

5. Urban forestry (2005)

Definition: Urban forestry is defined as 'the art, science and technology of managing trees and forest resources in and around urban community ecosystems for the physiological, sociological, economic, and aesthetic benefits trees provide society.'

Significance of Urban Forestry

The growing population coupled with increase in infrastructure development in the urban areas demanded the development of greenery to benefit the society both in terms of recreational aspects and to combat the growing environmental hazards.

For this purpose, trees are recognized as an integral part of any landscape which not only beautify the surroundings but also functionally help to improve the quality of environment. Trees in general reduce the temperature and augment the microclimate besides sequestering carbon-di-oxide and releasing oxygen to the atmosphere.

Trees reduce temperature around the area inhabited them by absorbing solar radiation and moderate the heat island a condition formed due to significant increase in temperature. Similarly, in industrial and urbanized areas trees act as a tool for combating air, water and noise pollution.

It is estimated that trees sequester carbon to the tune of 50% of the volume of wood produced. Trees also reduce noise pollution from all sources. Trees and their associated ground cover near water streams, rivers and lakes serve as a sponge, reducing the flow of pollutants such as agricultural chemicals before they reach the fresh water.

The trees in urban environment will have a major role to improve soil and water quality and the associated biodiversity improvement.

Trees accumulate large volume of litter which will create a conducive environment for both micro and macro organisms that maintain and enrich soil fertility. This litter will act as a sponge for water storages and their percolation will enrich the water table.

Urban areas are housed with multi storey buildings, hospitals, industrial parks, schools and colleges, office buildings, hotels, religious places and the human settlements. These urbanized developments create a stress in human system and may create health hazards. To overcome all these issues, trees are recommended for planting in these urbanized developments in order to increase the scenic and recreational beauty of the city.

Trees are selected for specific uses based on the purpose of growing, tree architecture, colour of foliage or flower, availability of space, soil and climatic factors. Native trees and leguminous trees, which enrich soil with nitrogen, are gaining importance in modern landscaping and bioaesthetic planning of cities and villages.

Hence trees are highly preferred in urban and semi urban landscape environment as a source of recreational and aesthetic factors. These recreational and aesthetic developmental activities require specific trees suitable for varied site conditions.

Trees suitable for various urban localities

Trees in general are characterized based on their foliage and flower characters. The foliage and flowering trees are generally preferred for urban planting as a source of aesthetic and recreational benefits. Some of the trees are preferred for their aromatic smell to eliminate the domestic sewage smell and also planting in and around the industrial premises. The trees suitable for various urban locations are furnished.

Ornamental trees suitable for town roads

Foliage trees for outer avenue	Flowering trees for inner avenue
Anthocephalus cadamba	*Cassia fistula*
Polyalthia longifolia	*Bauhinea purpurea*
Putranjiva roxburghii	*Peltophorum ferrugineum*
Sterculia alata	*Spathodea nilotica*
Pithecolobium saman	*Jacaranda mimosaefolia*
Melia azadirachta	*Lagestroemia flosreginae*
Alstonia schloris	*Grevillea robusta*
	Gliricidia maculata

Trees for narrow roads	Trees for wider roads
Calophyllum inophyllum	*Ficus lacor*
Cassia javanica	*Ficus retusa*
Casuarina equisetifolia	*Mimusops elengi*
Filicium decipiens	*Polyalthia longifolia*
Grevillea robusta	*Pongamia pinnata*
Jacaranda mimosaefolia	*Samanea saman*
Kigelia pinnata	*Simarouba glauca*
Lagestroemia flosreginae	
Melia azadirachta	
Putranjiva roxburghii	

Trees for providing colour and aesthetic loon on the road

Ornamental flowering trees for small roads	Ornamental trees for wider roads
Bauhinia acuminata	*Bauhinia triandra*
B.variegata	*Bombax ceiba*
Cassia fistula	*Erythrina indica*

jacaranda mimosifolia	*Bauhinia triandra*
Lagestroemia speciosa	*Simarouba glauca*
Saraca indica	*Ficus bengalensis*
Spathodea campanulata	

Trees suitable for di-erent localities

(i) Industrial areas and factories: Industrial areas and factory sites are known for continuous accumulation of pollution in the form of gaseous, particulate and noise pollutions. Hence trees have to be identified in order to reduce the gaseous, particulate and noise pollutions to create better living in and around the industrial premises. The following table indicates the suitable species for these sites.

Trees for reduction of gaseous pollution	Trees for reduction of noise pollution
Albizia lebbeck	*Alstonia scholaris*
Alstonia scholaris	*Butea monosperma*
Azadirachta indica	*Grevillea robusta*
Ficus religiosa	*Erythrina indica*
Lagestroemia flosreginae	*Lagestroemia flosreginae*
Polyalthia longifolia	*Pterospermum acerifolium*
Terminalia arjuna	*Terminalia arjuna*
Albizia lebbeck	*Mangifera indica*
Alstonia scholaris	
Azadirachta indica	
Ficus religiosa	

(ii) Parks and other recreation places: In urban environment, the aesthetic and recreational beauty is enhanced through the establishment of parks and other recreational sites. These parks and recreational sites have to be planted with foliage plants, flowering plants and accent plants which are furnished below.

Ornamental flowering trees	Ornamental foliage trees	As accent plants
Bauhinea purpurea	*Caesalpinea coriarii*	
Cassia fistula	*Ficus benghalensis*	*Araucaria excelsa*
Delonix regia	*Ficus retusa*	*Cassia fistula*
Jacaranda mimosaefolia	*Mimusops elengi*	*Callistemon lanceolatus*
Lagestroemia speciosa	*Polyalthia longifolia*	*Delonix regia*
Saraca indica	*Pongamia pinnata*	*Ficus krishnae*
Spathodea campanulata	*Samanea saman*	*Lagestroemia flosreginae*

(iii) Public buildings and educational institutions: Cities are known for wide range of buildings of both private and public sector ownership coupled with array of schools and colleges. In these establishments, tree cover will not only play a role of simple urban beautification but also help to educate the people and ameliorate the environment by reducing the heat, temperature, wind and improving the humidity and general environment. The following tree species will help to auger the green cover in and around the public buildings and educational institutions.

Larger gardens	Maidans, recreation grounds etc	Wind Break
Cassia renigera	*Caesalpinea coriarii*	*Melia azadirachta*
C.siamea	*Ficus benjamina*	*Millingtonia hortensis*
Kigelia pinnata	*Ficus religiosa*	*Polyalthia longifolia*
Mimusops elengi	*Kigelia pinnata*	*Pongamia glabra*
Peltophorum ferrugineum	*Mimusops elengi*	*Casuarina junghuhniana*
Polyalthia longifolia	*Polyalthia glabra*	
Pterospermum acerifolium	*Thespesia populnea*	
Spathodea campanulata		

(iv) Hospitals: One of the major development in urban environment is hospitals of small, medium and multispecialty nature. These buildings have to be planted with trees which will attract the patients and enhance the hospital ambience. The establishment of flowering trees, foliage trees and in some cases aromatic trees will help the patients for easy recovery and their comfortable living in the hospital environment.

Flowering trees	Foliage trees	Aromatic flowering trees
Bauhinia purpurea	*Araucaria excelsa*	*Alstonia scholaris*
Cassia nodosa	*Azadirachta indica*	*Couroupita guianensis*
Citharexylum spinosum	*Ficus benjamina*	*Magnolia grandiflora*
Jacaranda mimosaefolia	*F.elastica*	*Michelia champaca*
Michelia champaca	*F.reliogiosa*	*Millingtonia hortensis*
Millingtonia hortensis	*Melia azadirachta*	*Mimusops elengi*
Tabebuia rosea	*Mimusops elengi*	*Plumeria rubra*
	Polyalthia longifolia	*P.alba*

(v) Trees suitable for di3erent soil and climatic conditions: Urban environ1ment comprises wide range of soil and climatic features which demand proper planning before establishment of tree cover. The urban settlements have both dry locations and marshy areas due to continuous sewage accumulation. For these areas the trees with wide range of adaptation have to be identified and planted. The following table suggests the trees suitable for different urban locality.

For marshy areas	For dry localities	For saline soils
Casuariana equisetifolia	*Acacia auriculiformis*	*Azadirachta indica*
Largerstroemia flos-reginae	*Butea monosperma*	*Butea monosperma*
Melia azadriach	*Cassia fistula*	*Causarina Spp.*
	Erythrina indica	*Eucalyptus Spp.*
	Jacaranda mimosaefolia	*Madhuka indica*
	Kigelia pinnata	
	Melia azadirach	
	Peltophorum ferugineum	
	Plumeria alba	

For moist and humid areas	For acid soils
Bauhinia variegata	*Butea monosperma*
B.purpurea	*Cassia fistula*
Brownea grandiceps	*Casuarina equisetifolia*
Cassia marginata	*Melia azadirach*
Lagerstroemia speciosa	
Samanea saman	
Saraca indica	
Spathodea campanulata	

(vi) Trees suitable for urban sewage dumped area: The urban sewage dumped area is a major problem in the city which emit uncomfortable odour. Hence these areas are to be planted with aromatic trees which are furnished below.

Millingtonia hortensis	*Michelia chempaca*
Mimusops elengi	*Cestrum nocturnum*
Nyetanthes arbortristis	

6. ICRAF (2005)

The International Centre for Research in Agroforestry (World Agroforestry Centre) is an international institute headquartered in Nairobi, Kenya, and founded in 1978 as "International Council for Research in Agroforestry". The Centre specializes in the sustainable management, protection and regulation of tropical rainforest and natural reserves. In 2002 the Centre acquired the World Agroforestry Centre brand name, although International Centre for Research in Agroforestry remains its legal name and it continues to use the acronym ICRAF.

ICRAF's missionis to harness the multiple benefits trees provide for agriculture, livelihoods, resilience and the future of our planet, from farmers' fields through to continental scales.

ICRAF's three operating principles focus on:

- **People:** collaboration and partnership, learning, attracting, nurturing, and rewarding talent
- **Science:** quality science, communicating for accelerated impact, value for money, testing development options
- **Processes:** efficiency and effectiveness, accountability, subsidiarity and empowerment

Role of ICRAF

- Support for the development of national agroforestry policies
- Development of tree varieties
- Strengthening of rural advisory service provisions
- Standardizing and employing land health assessment methods
- Multiple scale geo-spatial analyses
- Formulation of systems science paradigms
- Providing negotiation support tools
- Stimulating environmental service rewards
- Devising climate adaptation approaches
- Developing bio-energy options
- Formulating rainbow water concepts
- Increasing decision support

7. Industrial Agro forestry (2005)

Definition

Industrial Agroforestry is the specific land use system wherein the industrial utility tree species are planted in an agroforestry system which include both agro and farm forestry.

Scope of Industrial Agroforestry

In India, the demand for wood ansd wood products are increasing at an alarming rate and it has been estimated that over 158 million m^3 of wood is required during the year 2020 against the actual supply of less than 50 million m^3. This has created over 50% of short fall and necessitated massive promotion of industrial agroforestry. The predominant industries in the country *viz.*, timber, pulp and paper, plywood and veneer, match splints and boxes, pencil packing, dendro power, biofuel and construction industries are requiring huge volume of wood and hence the industrial agroforestry has an excellent scope of application in India.

Policy directions

The National Forest Policy (1988) has directed all the wood based industries in the country to generate their own raw material needs without depending on forest department supply. It has also guided the industries to promote agroforestry plantations by establishing suitable linkages with farmers by facilitating quality seedling supply, technological support, support price for farm grown wood and also credit and insurance facility. This has stimulated several industries to initiate industrial agroforestry plantations in the country. Similarly, the National Agroforestry Policy (2014) has intensified the promotion of industrial agroforestry plantations by establishing suitable linkages with all stakeholders. It has also directed the government to liberalize the timber transit rules in order to attract more farmers towards industrial agroforestry. The policy also directed to extend credit and insurance facilities to agroforestry plantations besides establishing necessary market mechanism.

Major Wood based industries and the prioritized species

S.No.	Major industry	Species promoted and deployed
1.	Timber and construction industry	*Tectona grandis, Dalbergia latifolia, Gmelina arborea, Shorea robusta, Pterocarpus marsupium*
2.	Pulp and Paper	*Eucalyptus tereticornis, Eucalyptus camaldulensis, Casuarina equisetifolia, Casuarina junghuhniana, Acacia hybrids, Acacia auriculiformis, Acacia mangium, Leucaena leucocephala, Bamboos etc.*
3.	Plywood	*Populus deltoides, Eucalyptus, Melia dubia, Anthocephalus cadamba, Toona ciliata, Acrocarpus fraxinifolius etc.,*
4.	Matchwood	*Ailanthus excelsa, Populus deltoides, Albizia fulcataria*
5.	Dendro Biomass	*Prosopis, Leucaena leucocephala, Dalbergia sissoo, Acacias, Casuarina, Eucalyptus*
6.	Biofuels	*Jatropha curcas, Pongamia pinnata,Callophyllum inophyllum, Basia latifolia* etc.,

Successful industrial agroforestry plantations

Poplar based industrial agroforestry plantations in North India promoted by WIMCO and Eucalyptus by ITC in South India are successful evidences of industrial agroforestry plantations. The model developed in Tamil Nadu through Consortium of Industrial Agroforestry conceived by Tamil Nadu Agricultural University involving pulp and paper (TNPL and SPB), plywood (Century ply / Sharon/Ambiply), timber (Suresh timbers), Biofuel (Auromira Energy and Bannari Amman Group of Companies), Match (Vasan Matches and Ideal splints) have resulted in establishment of over 70000 ha of exclusive industrial agroforestry plantations. These developments have ensured availability of quality planting material, technology support, felling and transportation besides assuring support price for farm grown industrial wood trees. The introduction of tree insurance scheme in Tamil has attracted more farmers towards industrial agrofrestry.

Agroforestry in general and industrial agroforestry in particular is gaining significant attention across the world towards meeting the domestic and industrial wood requirement besides catering the needs of the risks associated with climate change. Though the agroforestry is an age old practice, still the achievement in this front is dismally modest for want of suitable science and technological advancements coupled with the associated supply chain process. To resolve these issues, several wood based industries are involved in promotion of industrial agroforestry through technological, organizational and marketing interventions which are at various stages of development and implementation.

8. Leaf fodder trees. (2005)

Trees and shrubs contribute substantial amount of leaf fodder especially during the lean period. Promotion of tree fodder based protein bank systems can help to meet the increasing demand for fodder requirement especially during dry and lean season. India heavily depends on green and dry fodder and it is estimated that India is in sort of 64% of feed which includes 61% green fodder and 21% of dry fodder during the year 2020 and this may accelerate further. Hence it is essential that if India wants to meet up the existing stage of fodder produce and wants to meet up the pre-requiste of the increasing rate of livelihood population which in turn will meet the requirements of growing human population, tremendous and serious efforts are needed. To make a proper balance, India has to fulfil the fodder deficit from all corners. Under such circumstances, the role of tree fodder in meeting the demand is very significant.

Advantages of tree fodder

- Provides nutritious feed and pods rich in proteins and minerals to livestock's
- Variety of products - fuel, timber, fibre & medicine
- Source of organic matter, increases soil N & soil structure
- Serves as fence/ hedge
- Serves as wind break
  - to prevent soil erosion and conserve soil moisture
- Provides shade for shade living plants
- Increase the yield and improve the quality of grasses
  - Acacia spp (velvel, karuvel), Agathi, Sithagathi, Subabul

Major fodder trees

Botanical name	Common name
A. nilotica	Karuvel, Black babul
A. leucophloea	Velvel, White babul
Albizzia lebbek	Vagai
Bauhina variegate	Orchid, Bauhina
Dalbergia sisoo	Sissoo
Erythrina indica	Mulmurungai
Glyciridia maculate	Glyricidia
Hardwickia binata	Acha
Leucaena leucocephala	Soundal, Subabul, Ipil ipil
Pitheclobium dulci	Kodukapuli
Melia dubia	Malaivembu
Prosophis juliflora	Semaikaruvel
Azadirachta indica	Neem

Melia azadirachta	Malaivembu
Entelobium saman	Rain tree
Anthocephalus cadamba	Kadam
Sesbania grandiflora	Agathi
Qurcus spp	Oak
Butea monosperma	Palash
Bassia latifolia	Mahua
Morus alba	White mulberry
Anogeissus pendula	Kardhai or Dhoy
Ziziphus mauritiana	Indian jujube, Ber

9. Describe agro-forestry systems under any two agro-climatic zones of India. (2006)

India has a variety of landscapes and climate conditions and this is reflected in the development of different soils and types of vegetation. There are 20 agroecological zones in India.

(i) Western Himalayan region

Treeson Agricultural field or on field bunds: *Grewia optiva* and other trees in Jammu & Kashmir, HP, Sikkim, Other north eastern hill region

Inter cropping on fruit orchards: Orange, other citrus and guava intercropped with cereals, tubers and rhizomes crop on Sikkim, Meghalaya.

Plantationcrops under shade of trees: Large cardamom, Coffee, etc under alder in Sikkim and Nagaland; Tea under leguminous trees in Assam and WB.

Silvipastoralpractices: Tree leaf fodder from Grewiaspp., *Celtis australis* and other species in Himalayas

(ii) Lower Gangeticplains:

Trees for soil reclamation (Sodic/saline/degraded soil): *Prosopsis chilensis, Acacia nilotica, Parkinsonia aculeate* and other for problematic soils.

Fodder trees in degradedgrazing lands: *Albizia lebbeck, Bahunia purpuria, D sissoo* etc. in grazing lands

Treeson boundaries of Agrifields: Populus spp, Tamarindus, *Ceiba pentandra*, Eucalyptus sppetc.

Fodder banksand wood lots: Fodder trees – *Azadirachta indica*, Melia, Albizia spp, *Syzygium cummini* etc.Wood lots - *Casuarina equisetifolia*, Eucalyptus, Bamboos, *Dalbergia sissoo*etc.

10. What are multipurpose tree species? Explain with examples. (2006)

Multipurpose tree species is defined as "Those trees and shrubs which are deliberately kept and managed for more thanone preferred use, product, or service, the retention or cultivation of thesetrees is usually economically but also sometimes ecologically motivated, in amultiple-output land-use system".

Importance of MPTs

- The trees should be suitable as raw materials for various wood based industries like pulpwood, paper, plywood, matchwood etc.
- Trees should be able to produce flowers and fruits for both animal and human consumption.
- The leaves should be palatable for human and animal consumption.
- The tree should act as a source for honey, silk and lac insects.
- The bark should contain tannin usable in leather industry.
- The trees should be able to produce gums, resins, dyes etc. The trees should give oils amenable for both domestic and human activity.
- The multipurpose trees species identified need not have all the properties mentioned above but should contain at least two or three characters which will facilitate its incorporation in agroforestry systems.
- In this system, various multipurpose tree species are scattered haphazardly or according to some systematic patterns on bunds, terraces or plot/field boundaries.
- The major components of this system are multipurpose trees and other fruit trees and common agricultural crops.
- The primary role of this system is production of various tree products and the protective function is fencing, social valuesand plot demarcation.
- Fast growing
- Nitrogen fixing ability

Examples of multipurpose trees:

Species	Utility
Leucaena leucocephala	*Fodder, fuel, pulpwood, nitrogen fixing ability*
Casuarina equisetifolia	*Fuel, pulpwood, poles and scaDolding, activated carbon, briquette, nitrogen fixing ability*
Azadirachta indica	*Oil, medicine, insecticide, timber, oil cake, fodder.*
Melia dubia	*Plywood, pulpwood, timber, fodder*
Acacia leucophloea	*Timber, fuel, fodder, nitrogen fixing ability, agriculture implements, tannin*
Prosopis juliflora	*Firewood, charcoal, activated carbon, fodder, nitrogen fixing ability*

11. *Write a brief note on India's experience with Joint Forest Management (JFM) in last fifteen years.* (2006)

The most significant aspect of the National Forest Policy of 1988 is that it calls for the involvement of the forest-dwellers, tribals and the neighbouring communities of the forest in the management of the forests. Joint Forest Management (JFM) is a management tool intended to protect the forests and manage the non-wood resources of forests in a sustainable manner with the active participation of forest-dependent communities ensuring their livelihood needs. Forest-dependent families are the members participating in the planning and implementation of various forestry and community-development programmes.

JFM originated in West Bengal at the Arabari Forest Range in West Midnapore, near Midnapore town in 1971. The major hardwood of Arabari is sal; a commercially-profitable forest-crop was grown with active participation of the local people. A few years later in 1977, JFM was employed in the State of Haryana to prevent soil erosion and deforestation. The program led to reforestation of many hills in the State. After the initial successes in West Bengal and Haryana, the JFM schemes received national importance in the legislation of 1988 and thrust in the Guidelines of 1990.

As on March 2011, there were 1,18,213 JFM Committees (JFMCs) across 29 States and Union Territories of Andaman and Nicobar Islands, which are managing 22.94 million ha of forests in the country. Madhya Pradesh has the largest number of JFMCs among all the States with 1,228 JFMCs and also the forest land being managed under JFM with 6.69 million ha. A total of 14.5 million families are involved all over the country, which includes 4.6 million, ST families and 2.5 million SC families. About one-third of the land with the forest department is being managed today in partnership with the people of the country.

In India, JFM has emerged as an important intervention in management of forest resources. In many parts of India, smallvillage groups have started to protect and reclaim degraded forest lands through collective action. The JFM Programme seeks to develop partnerships between local community institutions and State forest departments for sustainable management and joint benefit sharing of public forest lands. The primary objective of JFM is to ensure sustainable use of forests to meet local needs equitably while ensuring environmental sustainability.

12. Forests are key to tribal economy Justify. (2006)

Yes. It is true that the forests are key to the tribal economy as indicated below.

Role of forestry in tribal economy

(i) Direct Employment: Forests provide a wide spectrum of livelihoods for tribal communities in the form of direct employment, self employment and secondary employment. The direct employment is provided by the forest department and other departments (rural development, agriculture and co-operatives) in the form of managerial, technical, research, planning and executive jobs.

The other direct employments consisted of labour force for rural masses generated by these departments under regular forestry activities such as growth, development and maintenance of the forests, research and training, survey of forest resources, protection and conservation of forest resources, soil and water conservation, civil construction such as forest road, bridges and culvert, harvesting, collection, processing and marketing of NTFPs, preparation of nurseries, fencing, soil working, transplanting, planting, tending operations, watering, fertilizer and pesticide application, protection and management of plantation and infrastructure development. This gives much needed additional income to the tribals.

(ii) Forest-Based Industries: The self employment in forestry creates livelihoods for local people through the sale of fuel-wood and fodder, grazing, lopping and grass cuning, forest-based handicrafts and cottage industries, sericulture, lac cultivation, bee keeping, charcoal burning, leaf plate making, liquor making, rope making and basketry, medicines, collection, processing and marketing of NTFPs, cultivation of agricultural crops under agri-silvicultural practices, livestock rearing, social and farm forestry, and availing of rights and concessions. The application of local skills and village-level technology in wood-based and small-scale forest-based enterprises provides secondary employment and livelihood opportunities for tribal people; main amongst them are saw milling, rayon, pulp and paper, ply wood and panel products, wood seasoning and preservation, tanning, sports goods, match splints, veneers, wooden boxes, bamboo and cane products, agricultural implements, furniture, structural wooden items, musical instruments, bidi making, educational goods, wood carving, wooden utensils, etc. Some of the tribes have started small-scale and cottage industries such as bamboo cane and allied products, hand-made paper, tendu leaves, distillation of essential oil, katha manufacturing, charcoal making, starch making, tassar product, bee keeping, lac propagation, broom making and miscellaneous products.

(iii) Collection of NTFP: Forests provide significant social and economic benefits at all levels, especially in developing countries. Economics of people living in forest fringes has traditionally been dominated by subsistence agriculture.

NTFP such as fuel wood, medicinal plants, wild edible-vegetables, house building materials, fodder, fuel wood, mahua flowers (*Madhuca indica),* tendu leaves *(Diospyros melanoxylon)* wild mushrooms, tamarind *(Tamarindus indica)*soap nuts *(Sapindus emarginatus),* and wild brooms are an integral part of day-to-day livelihood activities, especially for tribal people. Forest products also can be used as raw materials for making a number of items for the home such as bidi, brooms, baskets, mats, rope, home-made toothbrushes and leaf plates. Some forest products such as charoli (*Buchanania lanzan),* lac (*Laccifer lacca),* dataoon as a teeth cleaning, leaf plates, and marula fruit (*Sclerocarya birrea)* are sold for a small cash income, which is sometimes used for buying clothing, oil or spices. These products may be used for subsistence or for sale, providing cash income.

More than 500 plants are being significantly used by the tribals as food, dyes, tannins, drugs, narcotic, drinks, housing instruments, weapons, fibres, medicine, etc. Thus, the NTFPs play a significant role in the livelihood of the forest dwellers and the communities living in the vicinity of the forest.

(iv) Forest Right Act: The STs and Other Traditional Forest Dwellers (Recognition of Forest Rights) Act, 2006, is a key piece of forest legislation passed in India on 18 December 2006. It has also been called the Forest Rights Act (FRA), the Tribal Rights Act, the Tribal Bill, and the Tribal Land Act The law concerns the rights of forest-dwelling communities to land and other resources, denied to them over decades as a result of the continuance of colonial forest laws in India.

(v) Ecotourism in Protected Areas: The country has a large treasure of natural beauty, and archaeological and architectural monuments. In addition, there are many hill and beach resorts, mountains, rivers, etc. The northern part of India is famous for its hills, plains, rivers, monuments, etc. - the Himalayas, Kashmir Valley, and the Ganges. There are colourful tribal lifestyles of the Northeastern States of Nagaland, Mizoram. Tripura and Manipur with their folk culture, which also is worth mentioning. In the Central Indian States of Odisha and Madhya Pradesh, tribal village life has resulted in a variety of artistically-executed handicrafts.

India has a huge ecotourism potential and this can be adjudged from this data that the country is having 634 nature-endowed areas, 103 national parks, and S31 wildlife sanctuaries (Ministry of Environment and Forest, 2015). Topography of homeland boasts of an abundant source of flora and fauna, and has numerous rare and endangered species in its surroundings. It is one of those rare countries, which has a wonderful blend of all the resources essential to make it as a golden star on the tourism map of the world.

Ecotourism, in simple terms, means management of tourism and conservation of nature in a way so as to maintain a fine balance between the requirements of tourism and ecology on one hand and needs of the local communities for jobs, new skills, income-generating employment and a better status for women on the other.

Tourism has long played an important role in the Indian economy as being the third most important industry sector of national economy. Within the tourism industry worldwide, ecotourism is one of the fastest growing sectors. The World Tourism Organization (WTO) has recently estimated that ecotourism is worth some $20 billion a year, and together with nature-based tourism, accounts for 20% of global international travel (WTO, 1998).

Positive impacts of tourism on indigenous populations in India in terms of economic growth and development (jobs and business opportunities); increased government expenditure on tourism infrastructure such as improvements of roads, water supply and sewage treatment, electrical facilities, education, communication that may also benefit residents and direct and indirect benefits due to tourism activities in tribal areas. Tribal people provide the services as guidance and interpretation, sightseeing, forest watcher, ecotourism enterprises such as sale of handicrafts, etc.

(vi) Joint Forest Management (JFM): The most significant aspect of the National Forest Policy of 1988 is that it calls for the involvement of the forest-dwellers, tribals and the neighbouring communities of the forest in the management of the forests. Joint Forest Management (JFM) is a management tool intended to protect the forests and manage the non-wood resources of forests in a sustainable manner with the active participation of forest-dependent communities ensuring their livelihood needs. Forest-dependent families are the members participating in the planning and implementation of various forestry and community-development programmes.

JFM originated in West Bengal at the Àrabari Forest Range in West Midnapore, near Midnapore town in 1971. The major hardwood of Arabari is sal; a commercially-profitable forest-crop was grown with active participation of the local people. A few years later in 1977, JFM was employed in the State of Haryana to prevent soil erosion and deforestation. The program led to reforestation of many hills in the State. After the initial successes in West Bengal and Haryana, the JFM schemes received national importance in the legislation of 1988 and thrust in the Guidelines of 1990.

As on March 2011, there were 1,18,213 JFM Committees (JFMCs) across 29 States and Union Territories of Andaman and Nicobar Islands, which are managing 22.94 million ha of forests in the country. Madhya Pradesh has the largest number of JFMCs among all the States with 1,228 JFMCs and also the forest land being managed under JFM with 6.69 million ha. A total of 14.5 million families are involved all over the country, which includes 4.6 million, ST families and 2.5 million SC families. About one-third of the land with the forest department is being managed today in partnership with the people of the country.

In India, JFM has emerged as an important intervention in management of forest resources. In many parts of India, small village groups have started to protect and reclaim degraded forest lands through collective action. The JFM Programme seeks to develop partnerships between local community institutions and State forest departments for sustainable management and joint benefit sharing of public forest lands. The primary objective of JFM is to ensure sustainable use of forests to meet local needs equitably while ensuring environmental sustainability.

(vii) Development of Agriculture: Though the people had changed their cultivation pattern from traveller's cultivation to settled cultivation, agriculture is still the most primitive occupation of the tribal people. The technology is transferred to them without affecting their sentiment. A number of appropriate technologies are generated by the government and have become sustainable under their farming system.

(viii) Tribal Development Schemes and Programmes: Government of India implemented different schemes through forest departments such as National Afforestation Programme (NAP), National Bamboo Mission (NBM), schemes for the welfare of STs living within the forest, Tribal Insurance Scheme (TIS), Forest Right Act 2006, Green India Mission (GIM), gas cylinder on subsidy, ecotourism, National Medicinal Plant Board Scheme, schemes of social forestry such as Kotwalia schemes, Malki plantation, Food grain, and Social security scheme.

13. Discuss di-erent agroforestry systems under di-erent agro ecological zones. (2007)

There are five agroecologial zones in India and they are

(i) Himalayan region
(ii) Indogangetic Plains
(iii) Arid and semi arid Region
(iv) Humid and Sub Humid Region
(v) Coastal and Iceland Region

(i) Himalayan region

AFS	Examples
Treeson Agricultural field or on field bunds	*Grewia optiva* and other trees in Jammu & Kashmir, HP, Sikkim, Other north eastern hill region
Inter cropping on fruit orchards	Orange, other citrus and guava intercropped with cereals, tubers and rhizomes crop on Sikkim, Meghalaya
Plantationcrops under shade of trees	Large cardamom, Coffee,etc under alder in Sikkim and Nagaland; Tea under leguminous trees in Assam and WB
Silvipastoralpractices	Tree leaf fodder from Grewia spp., *Celtis australis* and other species in Himalayas

(ii) Indogangetic Plains

AFS	Examples
Trees for soil reclamation (Sodic/ saline/degraded soil)	*Prosopsis chilensis, Acacia nilotica, Parkinsonia aculeata* and other for problematic soils.
Fodder trees in degradedgrazing lands	*Albizia lebbeck, Bauhnia purpurea, D sissoo* etc. in grazing lands
Treeson boundaries of Agrifields	Populus spp, Tamarindus, *Ceiba pentandra*, Eucalyptus spp etc.
Fodder banksand wood lots	Fodder trees – *Azadirachta indica*, Melia, Albizia spp, *Syzygium cummini* etc. Wood lots – *Casuarina equisetifolia*, Eucalyptus, Bamboos, *Dalbergia sissoo* etc

(iii) Arid and semi arid Region

AFS	Examples
MPTson Agricultural land	*Prosopis cineraria, Zizipus* spp, *Acacia senegal, Ailanthus excelsa* etc
Trees for reclamation of degraded soils	*Albizia lebbeck, Casuarina equuisetifolia, Acacia tortolis, neem* and other spp,
Windbreaks and shelterbelts	*Acacia nilotica, cassia spp, Azadirachta indica, Tamarix spp, Prosopis chilensis etc.*
Fodder banks	*Prosopis cineraria, Ziziphus mammularia, Salvadora persica etc*
Trees on rangelands(Silvopastoral system)	*Prosopis chilensis, Acacia senegal, Acacia senegal, Prosopis cineraria, Zizipus spp.* etc

(iv) Humid and Sub Humid Region

AFS	Examples
Homegardens/Homesteads	Homegardensin Kerala, Assam, WB etc., have a mixture of trees, shrubs, herbs etc. Coconut, Areca nut, Erythrina, Gliricidia, with black pepper, cocao, cassava, and other cash crops.
Multi typesystem or plantationcrops combination	Coconut, Arecanut, Erythrinaand other trees with coffee/ banana, pineapple papaya cacao/coffee and black pepper / betel vine/large cardamom.
Multipurposetrees in Agrifields	*Acacia ariculiformis, Aegle marmelos, Albizzia spp, Anogeisus latifollia, Anthocephalus chinensis, Artocarpus spp etc.*

(v) Coastal and Iceland Region

AFS	Examples
Plantation crop combination or multistoried	Coffee under *Erythrina lithosperma*, Cacao with coconut, Black pepper on Gliricidia spp, *Grevellia robusta,* Cardamum under *Toona ciliata,* Artocarpus etc
Trees with Aquaculture	Composite fish culture in ponds and MPTs in Home steads
Mangrovesplantations, Homesteads	*Acanthus ilicifolius, Avicenia oDcinalis, Azadirachta indica, Acacia Senegal* etc.
Shelterbeltsand windbreaks	*Casuarina equisetifolia, Azadirachta indica, Prosopis chilensis* etc.
Trees on boundaries of Agrifields	*Acacia auriculiformis, Aegle marmelos, Anogeisus latifolia, Calliandra calothyrus* etc

14. Write short notes on: (2011)

(i) Aqua forestry
(ii) Sacred groves
(iii) Water-use efficiency
(iv) Home garden

(i) Aqua forestry

Aqua forestry is defined as a multipurpose production system that allows production of fish in mangrove reforestations. Mangroves is a community of intertidal plants including all species of trees, shrubs, vines and herbs found on coasts, swamps, or border of swamps.

The aqua-forestry system comprises composite fish culture in farm ponds, and raising fodder, fuel, timber and fruit trees on waste lands along the bunds. Inland fish such as Catla, Rohu, Mrigal, common carp, silver carp and grass carp can be grown in the ponds for 6 to 10 months.

Aquaforestryis classified into three types

- **Aquaculture under natural forests:** Characteristics of these systems are variable water levels, as few trees tolerate permanent submergence under water. Aquaculture systems operating include fishing of unmanaged stocks, mainly with nets, lines and spears; also some stock-rearing in pens and cages. Examples of such systems are
  (i) Fisheries under flooded rainforest, *e.g.* in the Amazon in Brazil
  (ii) Mangrove forests and their hinterlands, e.g. West Africa
- **Aquaculture under plantations:** Characteristics of these systems are the addition of aquaculture production to an existing plantation system, often with notable nutrient exchanges between the tree and aquatic components, leading to a more efficient, diversified and profitable systems. Examples include:
  (i) Fisheries under plantation crops, e.g. coconut in India
  (ii) Flooded plantation forests, e.g. fisheries in the southern USA
  (iii) Ponds in plantation forestry, e.g. planting trees in ponds

- **Aquaculture in silvoarable systems:** Characteristics of these systems are the pre-existence of intensively managed agriculture as the principal component, with the inclusion of trees serving secondary productive and/or protective functions, and aquaculture as another peripheral activity leading to increased diversification, nutrient transfers and system efficiency. A livestock component may be present, but is less important, often being only the grazing of fallows, etc. The aquaculture component can be managed concurrently with the crop or between harvest and land preparation, and this temporal sub-classification is used below. Examples included:

  Fisheries in rice paddies, with trees as shelterbelts or hedges along field margins and / or irrigation channels. Eg: In South and South-East Asia.

(ii) Sacred groves

Sacred grove is defined as the "Segments of landscape containing vegetation, life forms and geographical features, delimited and protected by human societies under the belief that to keep them in a relatively undisturbed state is expression of an important relationship of humans with the divine or with nature".

Distribution of Sacred Groves

Sacred groves are distributed throughout the country particularly in the states of Maharastra, Kerala, Karnataka and Tamil Nadu. In North Eastern India most sacred groves are reported in Arunachala Pradesh, Mehalaya and Manipur. In India, it has been estimated that the number of sacred groves would vary between 100000 and 150000.

Classification of Sacred Groves

(a) **Traditional Sacred Groves:** It is the place where the village deity resides, who is represented by an elementary symbol

(b) **Temple Groves:** Here a grove is created around a temple and conserved.

Importance of Sacred Groves

- Sacred Groves provide shelter to thousands of species of other plants and animals.
- Sacred groves have great powers to heal body and spirit.
- Sacred Groves are important reservoirs of biodiversity.
- These are last refuse for endemic and endangered plant and animal species.
- These are storehouses of Medicinal plants valuable to village communities as well as modern pharmacopoeia.
- Sacred Groves contain relatives of crop species that can help to improve cultivated varieties.
- Sacred Groves help in keeping the water cycle in local areas.
- The scared grove improves soil stability, prevent top-soil erosion and provide irrigation for agriculture in drier climates.

(iii) Water-use e-ciency

Water use efficiencyis the ratio of water used in plant metabolism to water lost by the plant through transpiration. Two types of water-use efficiency are referred to most frequently:

- **Photosynthetic water-use e3ciency** (also called intrinsic or instantaneous water-use efficiency), which is defined as the ratio of the rate of carbon assimilation (photosynthesis) to the rate of transpiration, and
- **Water-use e3ciency of productivity** (also called integrated water-use efficiency), which is typically defined as the ratio of biomass produced to the rate of transpiration.

Increases in water-use efficiency are commonly cited as a response mechanism of plants to moderate to severe soil water deficits, and has been the focus of many programs that seek to increase crop tolerance of drought. However, there is some question as to the benefit of increased water-use efficiency of plants in agricultural systems, as the processes of increased yield production and decreased transpirational water loss are fundamentally opposed. If there existed a situation where water deficit induced lower transpirational rates without simultaneously decreasing photosynthetic rates and biomass production, then water-use efficiency would be both greatly improved and a desired trait in crop production.

(iv) Home gardens

Home gardens represent land use systems involving deliberate management of multipurpose trees and shrubs in intimate association with annual and perennial agricultural crops and livestock within the compounds of individual houses. The whole tree crop animal units are being intensively managed by family labour.Home gardens can also be called as Multitier system or Multitier cropping.

This practice finds expression in the states of Kerala and Tamil Nadu with humid tropical climates and where coconut is the main crop. Many species of trees, bushes, vegetables and other herbaceous plants are grown in dense and in random or spatial and temporal arrangements. Most home gardens also support a variety of animals. Fodder grass and legumes are also grown to meet the fodder requirement of cattle. In India, every homestead has around 0.20 to 0.50 ha land for personal production. Home gardens are highly productive, sustainable and very practicable. Food production is primary function of most home gardens.

Structure of Home Gardens: Home gardens are characterized by high species diversity and usually 3-4 vertical canopy strata. The layered configuration and compatible species admixture are the most conspicuous characteristics of all home gardens. Generally all home gardens consist of herbaceous layer near the ground, a tree layer at the upper levels and an intermediate layer. The lower layer can be partitioned in to two, the lowermost being at less than 1.0m in height, dominated by different vegetables and the second layer of 1.0 -3.0/m height comprising food crops such as banana, papaya and so on. The upper tree layer can also be divided into two, consisting of emergent, full grown timber and fruit trees occupying the upper most layers of 25m height and medium size trees of 10-20m occupying the next lower layer. The intermediate layer of 5-10m height is dominated by various fruit trees.

Choice of species:

(a) Woody species: *Anacardium occidentale, Artocarpus heterophyllus*, Citrus spp, *Psiduim guajava, Mangifera indica, Azadirachta indica, Cocus nucifera,*

(b) Herbaceous species: Bhendi, Onion, cabbage, Pumpkin, Sweet potato, Banana, Beans *etc.*

15. Di-erentiate between Agroforestry and silviculture (2012)

S.No.	Agroforestry	Silviculture
(i)	It is a sustainable land use system that combines agriculture, forestry and or animal husbandry and increase the productivity per unit area.	It is an art and science of raising tree species with predominant focus on single or double species approach.
(ii)	There are more than two components in agroforestry	Predominantly a singly species
(iii)	It helps to conserve all natural resources particularly soil and water erosion	It comparatively increases the soil and water erosion
(iv)	Productivity per unit area is high	Productivity per unit area is low
(v)	It provides multipurpose benefits like food, fuel, fodder and fiber.	It predominantly gives only one product
(vi)	Animal husbandry component is mandatory	Animal husbandry is not mandatory
(vii)	It helps to mitigate and adapt to climate change issues	Comparatively less
(viii)	It is a livelihood land use system	It is not a livelihood land use system
(ix)	It generates income and employment generation activities on a sustainable basis	It is not a sustainable income and employment generation activity

(i) Agroforestry

Agroforestry is any sustainable land-use system that maintains or increases total yields by combining food crops (annuals) with tree crops (perennials) and/or livestock on the same unit of land, either alternately or at the same time, using management practices that suit the social and cultural characteristics of the local people and the economic and ecological conditions of the area.

Agroforestry is a collective name for a land-use system and technology whereby woody perennials are deliberately used on the same land management unit as agricultural crops and/or animals in some form of spatial arrangement or temporal sequence. In an agroforestry system there are both ecological and economical interactions between the various components.

Importance of Agroforestry

(i) Efficient and multiple land use system
(ii) Provides fuel, food, fodder and fibre
(iii) Improves soil fertility and conserve soil

(iv) Serves multipurpose needs
(v) Improves water harvesting system
(vi) Enhances milk and milk production
(vii) Extends livelihood oppournites through continous income and employment generation activities
(viii) Mitigates and adapts to growing climate change associated risks.

(ii) Silviculture

'Silviculture is that branch of forestry which deals with the establishment, development, care and reproduction of stands of timber'.

Silviculture is defined as 'The art and science of cultivating forest crops'.

Importance of Silviculture

(i) Production of species of economic value
(ii) Production of larger volume per unit area
(iii) Production of quality timber
(iv) Reduction of rotation
(v) Raising forests in blank areas
(vi) Creation of manmade forests in place of natural forests
(vii) Introduction of exotics.

16. Write short notes on Agro ecological zone. (2012)

An "Agro-climatic zone" is a land unit in terms of major climates, suitable for a certain range of crops and cultivars. The planning aims at scientific management of regional resources to meet the food, fiber, fodder and fuel wood without adversely affecting the status of natural resources and environment. Crop yield is (FAO, 1983). Agro-climatic conditions mainly refer to soil types, rainfall, temperature and water availability which influences the type of vegetation. An agro-ecological zone is the land unit carved out of agro-climatic zone superimposed on landform which acts as modifier to climate and length of growing period.

Planning of Agro climatic zones of India

- Agro-climatic regions by the Planning Commission
- Agro-climatic zones under National Agricultural Research Project (NARP)
- Agro-ecological regions by the National Bureau of Soil Survey & Land Use Planning (NBSS & LUP)

Agro-climatic regions by the Planning Commission

The Planning Commission, as a result of the mid-term appraisal of the planning targets of the Seventh Plan, has divided the country into fifteen broad agro-climatic zones based on physiography, soils, geological formation, Climate, cropping patterns, and development of irrigation and mineral resources for broad agricultural planning and developing future strategies. Fourteen regions were in the main land and the remaining one in the islands of Bay of Bengal and the Arabian Sea. The main objective was to integrate plans of the agro-climatic regions with the state and national plans to enable policy development based on techno-agro-climatic considerations. In the agro-climatic regional planning, further sub-regionalization was possible based on agro-ecological parameters.

Agro-ecological regions by the National Bureau of Soil Survey & Land Use Planning

The National Bureau of Soil Survey & Land Use Planning (NBSS&LUP) came up with twenty agro-ecological zones based on the growing period as integrated criteria of effective rainfall, soil groups, delineated boundaries adjusted to district boundaries with a minimal number of regions. Subsequently, these twenty agro-ecological zones were sub- divided into 60 sub-zones.

(i) Western Himalayas
(ii) Western Plain, Kachchh, and part of Kathiwara Peninsula
(iii) Deccan Plateau
(iv) Northern Plain and Central Highlands including Aravallis
(v) Central Malwa Highlands, Gujarat Plains, and Kathiawar Peninsula
(vi) Deccan Plateau, hot semi-arid ecoregion
(vii) Deccan (Telengana) Plateau and Eastern Ghats
(viii) Eastern Ghats, Tamil Nadu Plateau and Deccan (Karnataka)
(ix) Northern Plain, hot sub-humid (dry) ecoregion
(x) Central Highlands (Malwas, Budelkhand, and Eastern Satpura)
(xi) Eastern Plateau (Chattisgarh), hot sub-humid ecoregion
(xii) Eastern (Chotanagpur) Plateau and Eastern Ghats
(xiii) Eastern Plain
(xiv) Western Himalayas
(xv) Bengal and Assam plains
(xvi) Eastern Himalayas
(xvii) North Eastern Hills (Purvanchal)
(xviii) Eastern Coastal Plain
(xix) Western Ghats and Coastal Plain
(xx) Island of Andaman Nicobar and Lakshadweep.

17. Comment on following statement - 'Pure stand of forests result incomplete utilization of the site'. (2010)

Yes it is true that the pure stand of forest results in complete utilization of site due to the following reasons and factors:

(i) Pure stands are stands created artificially and predominantly with one or two species. If it is one species then the site may not be utilized fully due to the reason that each and every portion of the site is occupied by pure stand. Because o pure stands are planted on a systematic plating pattern and with uniform spacing and density which results in incomplete utilization of land.

(ii) Pure stands of several species partial, confifers are reported to create adverse effect on soil. The pure stands may exhaust the soil and detoriate the quality.

(iii) There are pure stands are species specific and sometimes the identified species may not match with the locality. Under such circumstances, the site is not fully utilized.

(iv) The pure stands invites wide range of silvicultural operations part soil working between the stands. The areas not occupies by the trees will be exposed to all natural calamities and it cause serious loss. So pure stand not only fully utilizes the site but expose a site and various calamities.

(v) The pure stands are amenable for epidemic status of pest and diseases which weakens the productive capacity of the site.

(vi) Since pure stands are not fully distributed across, the site there are chances of annual and perennial weed growth. Hence the site is not fully utilized by the target trees rather it is utilized by the unwanted weeds.

These backdrops very clearly evidenced that pure stands are not fully utilizing the site and in turn they laed soil detoriation.

18. Comment on following statement - 'Planted forestry has high production potential but low conservation value'. (2010)

Yes it is true that the plantation forestry has high production value but no conservation value. The concept and the purpose of plantation forestry in the country has emerged as a source of rawmaterial to meet the domestic and industrial requirement. In India over 95 % of the forests are government owned forest and this forest are closed due to policy and legal compulsions.

Again the productivity of Indian forest is one of the lowest (0.5 to 0.6 m^3/ha/annum) compared to the global average (2.1 m^3/ha/annum). The forest area in the country is 24.16% against the mandate requirement of 33%. This low forest cover coupled with poor productivity of Indian forest has rushed in a total mismatch in demand and supply.

Considering this the plantation forest emerged at strong attraction to the traditional forest based production process. Hence the primary objective of plantation forestry was produced more that towards conservation of major plantation forestry species.

The growing demand of domestic and industrial requirement necessitated plantations like Teak, Eucalyptus, Popular, Casuarinas, Acacia, Sal, Silver Oak, Bamboos and etc. which are planted as large scale plantation to mandate to produce more wood and wood products and these plantations have low conservation value.

This is because the monocultures of plantations will exploit the site productive capacity at an alarming rate. This will reduce the fertility status of the soil due to rapid exploitation by monoculture plantations and do not support the new vegetation. More over these plantations are felled at a specific rotation period and hence they have high production potential and low conservation value. The planted forestry does not improve the soil and water conservation measures which in turn reduce the local biodiversity. Most of the planted forests are single species concept and hence species diversity may not be available. Under such circumstances it is clearly evident that planted forest has high production potential but low conservation value.

19. Write short notes on Management of urban forestry. (2010)

The following management practices are carried out in urban forestry management

(i) Survey and inventory
(ii) Providing tree guards for young saplings
(iii) Staking of saplings
(iv) Pruning the trees
(v) Irrigation
(vi) Fertilization
(vii) Proper pest and disease management practices

(i) Survey and inventory: The existing urban tree resources should be mapped using GIS and the urban tree resources demanding improvement to improve the urban dwelling have to be identified and approached through the following management technologies.

(ii) Providing tree guards for young saplings: The young saplings planted should be protected from all biotic damages by suitably providing individual tree guards.

(iii) Staking of saplings: Young trees should be able to support their own weight, but when they are planted or transplanted, they often need time to reestablish themselves. When staking a tree, the following points to be considered.

- Only stake the tree long enough for it to be able to stand on its own.
- Stakes should not be too tight – there should be room for tree to sway in wind.
- Stakes should not be too loose – the tree should not rub against stakes.
- Stakes should be buried at least 45 cm underground to provide ample support.
- Drive stakes into undisturbed ground before covering hole, to avoid damage to roots.

Larger trees should be guyed. To guy a tree, use three wires attached to three stakes. Thread a 30 cm piece of hose on each wire and have the hose-covered bit of wire loop around the tree at the lowest branch crotches. Make loops around the trunk very loose to avoid damaging the tree. Drive stakes into firm soil at least 45cm deep outside the perimeter of the planting hole. Anchor the loose ends of wires securely to stakes. Various fabric straps and rigid staking systems are available that may be easier to use than hose and wire.

Stakes and guys should be removed before wire girdles the trunk. Generally, remove supports after one growing season for a 4cm diameter tree seedling, two seasons for a 8 cm diameter tree seedling. For larger seedlings, guys may need to remain for three seasons or more. They should be inspected annually and adjusted to prevent trunk girdling.

(iv) Pruning the trees: Pruning of matured and unsized tree is very important in order to reduce the plant size and improve the growth and development. The pruning prevents from over growing of plants and allows the plant for the proper growth and development. It also helps to remove the undesirable growth on the plants particularly the climbers and the parasites. Well planned pruning helps to maintain shape, vigour and health of the plant by removing the over matured, diseased, moribund and dead branches. The pruning also helps for rejuvenation of young shoots and ensures youthful and natural growth habit. Pruning also helps to protect the buildings, adjoining structures and other properties nearby.

In urban areas pruning can be done in any time of the year without jeopardizing basic survival. The best time for the pruning is late summer and the beginning of early monsoon. For damaged and dead trees there is no time for pruning. As soon as it is found, the trees can be pruned.

(v) **Irrigation:** Newly planted trees to survive their first year, have to get the right amount of water. Over watering is just as harmful as under watering. On sandy soils, plants may require more water especially in summer while plants in poorly drained, clay soils need less frequent watering.

(vi) **Fertilization:** Tree fertilization is especially important in urban areas where soils are altered due to construction. These urban soils tend to be heavily compacted, poorly aerated, poorly drained, and low in organic matter. Even where soils have not been affected, fertilization may be needed as part of a maintenance programme to increase plant vigour or to improve root or top growth. Nutrients can be applied in organic and inorganic forms.

Slow release fertilizers and dry forms (granulated, pelletized, spikes, pulverized, encapsulated) are available now. Fertilizer should not be concentrated around stem or trunk of a tree, but should be applied over plant's root zone. For trees and shrubs, fertilizer should be applied over an area twice as large as the crown spread or drip line. Since most landscape plant roots grow in top foot of soil, surface or shallow application is recommended.

Fertilizer rates should be based on plant type, younger trees and shrubs generally receiving higher rates of N than mature plants. If a complete fertilizer is used, the ratio of N-P-K should be 3-1-2 or 3-1-1.

Evergreen trees need less (0.5-1.5 kg) while deciduous trees commonly need more (1.5-3 kg) of fertilizer per plant.

Application of organic manure at 2 to 5 kg per pit and neem cake at 100 g per pit is advisable for all tree crops.

(vii) **Proper pest and disease management practices:** Urban trees should be protected from all biotic damages particularly from pest and diseases. Any tree infested with pest and diseases should be controlled suitably spraying pesticide and fungicide. Heavily damaged trees should be removed as far as possible to avoid spread of diseases to the neighbouring trees.

(viii) **Burlapping technology**

- **Inventory and identification of trees:** The tall and elite trees intended for transplanting in urban area will have to be identified before transplanting. The crown to root ratio and the root pruning techniques will be carried out on case to case basis depending on the trees.
- **Newly transplanted trees:** During transplanting, bare root trees suffer root damage as well as a loss of roots. To compensate for this loss of roots, leaf area of the plant should

be reduced by 1/3 to reduce water loss due to transpiration and evaporation. Thinning should preserve natural form of newly planted trees. Remove branches and parts of branches by cutting at different lengths.

When the tree is approximately 2 years old, establish good branch spacing. Branches should arise alternately from main stem and be well spaced. This allows stronger limbs to develop. Select branches with the widest angles in the crotch. Wider the angle, stronger is the limb attachment.

- **Transplanting Techniques:** Transplanting of trees and shrubs can be categorized into two types.

  (i) Bare or naked root transplanting

  (ii) Balled and burlapped transplanting

  (i) Bare or naked root transplanting: Bare root transplanting is usually practiced for deciduous trees and shrubs. Bare root transplanting has to be done in early spring.

  (ii) Balled and burlapped transplanting: This method is usually preferred for evergreen trees and shrubs. The trees are moved or transplanted during the growing season. This method can be adopted throughout the year but very good success has been witnessed only when transplanting during the dormant season.

When a tree is transplanted, it usually passes through a period of shock. Nature provides a balanced root system and leaf crown. If the tree is from a reputable nursery, roots were probably pruned for several years to form a compact ball of roots that could be taken up with the tree. Such pruning avoids undue disturbance of this nature balance.

20. a. Write short notes on Canopy architecture (2009, 2012)

b. Outline the role of tree architecture in agroforestry (2015)

Definition and concept

The crowns of trees possess a striking forms and structures, which are difficult to interpret. This phenomenon is termed crown or canopy architecture.

- Plant architecture involves the manner in which the foliage is positioned in different microenvironments and determines the flexibility of a shoot system to take advantage of unfilled gaps in the canopy, to allocate and utilize assimilates, and to recover from herbivoury or mechanical damage.
- The architecture of a plant depends on the nature and on the relative arrangement of each of its parts; it is, at any given time, the expression of equilibrium between endogenous growth processes and exogenous constraints exerted by the environment.
- Tree crowns are highly variable. The general shape varies from relatively dense conoids for young and healthy trees of excurrent habits to wide open hemispheric shapes for older trees of deliquescent habits. Shapes and branching habits of tree trunks and crowns are far more variable in the tropics than in temperate regions.

Factors influence the forms and structures of tree crowns: The individual tree crown conditions is the result of a combination of many factors including

- Genetic traits,
- Growing site characteristics and
- Past and present external stresses (e.g., drought, insect outbreak, fire, etc.).

The results of these factors include different branching patterns, different shapes of boles, orientation of the leaves, branch angle, etc.Some other factors includes,

(a) **The Plant Design:** The shape of a plant is determined by the shape of the space that it fills, but most plants attain a characteristic shape when grown alone in the open due to an inherited developmental program. This developmental program usually implies the reiterative addition of a series of structurally equivalent subunits (branches, axes, shoots, leaves), which confers plants a modular nature. This developmental program is the result of plant evolution under some general biomechanical constraints. For instance, the shape of the crown of a tree is constrained by the fact that the cost of horizontal branches is greater than that of vertical branches.

(b) **Modular Nature of Plants:** In the crown of most vascular plants, it is easy to recognize a hierarchical series of subunits. The largest subunit is the branch, which is made up of modules. Module is a general term that refers to a shoot with its leaves and buds, and the term can be applied to either determinate (structures whose apical meristem dies or produces a terminal inflorescence) or indeterminate shoot axes.

(c) **Branching pattern:** The architecture of some trees consists only of a single vegetative axis during their whole life span, most display a more complex architecture consisting of several axes, one derived from another by a repetitive process known as branching.

Plants exhibit an extraordinary variety of branching patterns. The branching and consequent growth of trees and shrubs can be characterized by vertical or longitudinal, and horizontal or lateral symmetries. Vertical symmetry is characterized by growth of branches at the top (acrotony) or at the base (basitony), whereas lateral symmetry is characterized by branch growth at the upper or lower side of the lateral branch (epitony and hypotony, respectively). Logically, shrubs exhibit a basitonic branching, whereas trees are characterized by acrotonic branching. Analogously, while typical trees exhibit a hypotonic branching, most shrubs and small trees exhibit epitonic branching. The pyramidal shape of the crown of many conifers is due to the combination of basitonic branching (typically a shrub pattern) with a monopodial growth of the bole.

(d) **Apical Control and Apical Dominance Shaping Trees:** Shoot branching patterns can be greatly affected by apical control and apical dominance. Apical control is the inhibition of a lateral branch growth by shoots above it (distal shoots). It refers to the physiological condition governing the excurrent (single stem, pyramidal) or decurrent (branchy) pattern of growth. Species with strong apical dominance have weak apical control and this result in branchy trees whereas species with weak apical dominance have strong apical control and this result in columnar tree.

(e) **Plant Biomechanics: Coping with Gravity and Wind:** While plant architecture is an outcome of many selective pressures, the shapes of plant parts, their elasticity, and resistance to strain, are constrained by well-known mechanical principles. Because aerial plant parts face the obvious forces of gravity and wind, a fraction of the biomass must be devoted to support. As mechanical structures of similar shape become increasingly inefficient with increasing size, the fraction required to support plants increases rapidly with increasing plant size.

Role of tree architecture in Agroforestry

Tree architecture is one of the essential factors which decides the success or failure of agroforestry plantations. The tree architecture is based on the canopy length and breadth which ultimately has an enourmous effect on the underground vegetation or inter crops grown with tree crops. The trees with narrow architecture are able to provide more space and more light to the underground vegetation and helps in cogeneration of food and wood in the same land use system. At the same time, the influence of the narrow architecture in moderating the climate is low to average which may lead to excessive evapo transpiration and requires more water for enhancing agricultural productivity.

On the other hand, the dense architecture will have positive interaction in reducing the incoming solar radiation thereby influences the micro climate of the locality by reduction in temperature, increase in humidity and its effect on enhancing local precipitation. This micro climate moderation will help to utilize the natural resources rationally and enhances the productivity.

Hence tree architecture is a vital component in designing agroforestry model for cogeneration of wood and food in the same land use system.

21. a. Write short notes on Diagnosis and Design in agroforestry (2010)

b. What are D and D? Who can make use of D and D and how? (2012)

Definition

Agroforestry D & D is a family of procedures for the diagnosis of land management problems and potentials and the design of agroforestry solutions. The ICRAF has developed an approach to assist agroforestry researchers and development fieldworkers to plan and implement effective research and development projects.

Key Features of D & D

(i) **Flexibility:** D & D is a flexible discovery of procedure, which can be adopted to fit the needs and resources of different users

(ii) **Speed:** D & D has been designed with the option of a 'rapid appraisal' application at the planning stage of a project with In-depth follow-up during project Implementation.

(iii) **Repetition:** D & D is an open-ended learning process. Since initial designs can almost always be improved. The D & D process need not end until further Improvements are no longer necessary.

Basic logic of Agroforestry Diagnosis and Design

Basic question	Key factors to consider
Prediagnostic stage Which land-use system? How does the system work?	Distinctive combinations of resource technology and land-user objectives. Production objectives and strategy subsystems and components.
Diagnostic stage How well does the system work?	Problems in meeting objectives, causal factors, constraints, and intervention points.
Design and evaluation stage How can the system be improved?	Specifications for problem-solving or performance-enhancing interventions.
Planning stage How can the Improved technology be developed and disseminated?	D & D needs, extension needs.
Implementation stage How can the plan of action be adjusted to new information?	Feedback from research trials, independent farmer innovations etc.

Procedures of Agroforestry Diagnosis and Design

The procedures of Agroforestry D & D are usually done of two types:

(i) Macro D & D

(ii) Micro D& D

(i) Macro D & D: Macro D &D is a large scale analysis of an ecozone within a country (or) a group of countries.

Objectives

- To identify broad issues and problems constraining all the land use system in a given ecozone.
- To identify and prioritize areas for potential agroforestry inventions
- To identify research priorities and formulate research programme
- To identify needs, opportunities and mechanisms for inter – institutional collaboration for the technology development.

Steps in macro D &D: There are seven steps in macro D&D

(i) Identification of study ecozone

(ii) Delineation of land use systems within the ecozone

(iii) Description of LUS

(iv) Analysis of LUS constraints and its potentials

(v) Analysis of potential agroforestry technologies

(vi) Definition of agroforestry research needs and

(vii) Inter institutional coordination

(i) Micro D&D: Micro D &D focuses on one land use system (LUS) with in the larger ecozone that has special priority for agroforestry intervention.

Objectives

- To describe and analyze an LUS in order to identify its constraints and
- To design and evaluate agroforestry technologies to address the constraints

Steps in micro D&D

(i) Analysis of land use system and constraints
(ii) Design and evaluation of agroforestry technologies
(iii) Design and evaluation of research programs.

Since the resources are inevitably limited, a country institution (or) project will have to be selective in deciding which LUS should be subjected to micro D & D. The choice of LUS for micro D &D depends on criteria such as

(i) Political and economical importance of the system
(ii) Technical potentials for improvement of the LUS
(iii) Scientific expertise and other resources in the national collaborating institutions for carrying out research in LUS

Although it is not essential for a macro D&D exercise to precede a micro D&D exercise, the task of prioritizing LUS, defining research focus and defining areas for institutional collaboration will be much easier after macro D&D has been completed.

Diagnosis and design is a systematic and objective methodology used to initiate monitor and evaluate agroforestry research for development. It can be applied at the level of an archeological zone (or) at the level of a specific land use system. Methodological guidelines have been presented to explain how the D&D achieves its basic objectives, which are:

(i) Describing and analyzing existing land use systems
(ii) Diagnosing their constraints and casual factors
(iii) Designing appropriate agroforestry technologies
(iv) Designing appropriate research work and
(v) Identifying needs and opportunities for inter institutional collaboration.

The key methodological considerations for D&D implementation are:

(i) The composition of research team
(ii) The definition of research domains and recommendation domains
(iii) Data collection methods
(iv) Logical and operational aspects

Criteria of Good Agroforestry Design

A good agroforestry design should fulfill the following criteria:

(i) **Productivity:** There are many different ways to improve productivity with agroforestry *viz.,* increased output of tree products, improved yields of associated crops, reduction of cropping system inputs, increased labour efficiency, diversification of production, satisfaction of basic needs and other measures of economic efficiency or achievement of biological potential.

(ii) **Sustainability:** By seeking improvements in the sustainability of production systems, agroforestry can achieve its conservation goals while appealing directly to the motivation of low income farmers, who may not always be interested in conservation for its own sake

(iii) **Adoptability:** No matter how technically elegant or environmentally sound an agroforestry design may be, nothing practical is achieved unless it is adopted by its intended users. This means that the technology has to fit the social as well as environmental characteristics of the land-use system for which it is designed.

22. Write short notes on radiation absorption and energy balance in forests (2010)

The energy radiated by the sun reaches the earth in the form of electro magnetic waves of varying length ranging nearly from 290 m !: to 5300 m !:. The total energy received from the sun may be classified on percentage basis as under.

Ultraviolet about 1%

Visible spectrum about 39%

Infra-red about 60%

The maximum energy of solar radiation occurs in the green and yellow regions of the visible spectrum (400 m !: to 720 m!:). The incoming solar radiation was partly reflected and partly absorbs by the atmosphere. Fourty two percentage of incoming radiation was reflected and this is known as albedo. Forest trees generally require about 1 to 1.5% of full sunlight for their growth so that photosynthetic gain is more than the loss of due to respiration.

Energy balance in forests

Energy balance is the relationship between energy taken by the plant and energy released out to the environment. Energy can be transformed from one form to another form and the energy is constantly recycled through forest ecosystem. In the intial establishment the energy is taken by the plants and it is converted into plant biomas which in turn return to the soil through plant biomass. This plant biomas get decomposed through microbial organisms and released nutrient into the soil. That nutrients in turn is used by the plants for their growth and development thus the energy is balanced.

23. Discuss drawbacks of social forestry programs in India. (2010)

The social forestry programmes in India has witnessed several draw backs and the most debated drawbacks furnished below.

(i) Monoculture

(ii) Conversion of land use system

(iii) Proper supervision

(iv) People participation
(v) Land ownership
(vi) Lack of legal constitution
(vii) Long gestation
(viii) Lack of sustainability plan

(i) **Monoculture:** Monoculture is one of the objectionable issues in social forestry. It is said that monoculture has a negative impact on forest environment. The climatic conditions in natural forest which favour the growth of a variety of flora and fauna with in the shade of the leaf canopy are lacking in the case of monoculture. Monoculture also promotes breaking of trees by wind and creates favorable conditions for the spreading of diseases andpests. If trees having more leaves and spreading crown are selected and planted in mixed plantations, instead monoculture, it can create climatic conditions more or less identical to that of a natural forest. This would promote the growth of more flora and fauna, more deposit of humus and more yield of fuel and fodder and gradual natural regeneration of forest.

(ii) **Convertion of land use system:** Another serious criticism leveled against the social forestry programs in India is that instead of involving rural communities in raising and protecting useful species of trees on common and waste land, social forestry encourages farmers to convert land from food crop cultivation to farm forests or to replace useful trees by plantations.

(iii) **Proper supervision:** Social forestry programs need proper supervision and implementation. In almost all the states, there is a general complaint that the state forest department is not paying adequate attention to see that the treesplanted in social forestry programs attain maturity which lead to failure of a plantation. Once the scheme is over the people are also not involved in monitoring activities which lead to failure of a plantation.

(iv) **People participation:** The social forestry programme is aimed at forestry for the people, by the people and to the people. The original dwellers and the villagers near by forests should be actively involved in the social forestry programmes. However in many cases the involvement of people is very minimal which lead to failure of social forestry concept in many states. Similarly the coordination between people and the forest department was also not encouraging and results in lack of active participation of the people. There is no institutional mechanism created to ensure active participation of people which is believed to be one of the strong draw back of social forestry programme.

(v) **Land ownership:** Some of the social forestry programmes particularly the tank bed plantations established with *Acacia nilotica* experienced problem of ownership. The land belongs to Public Works Department and the community is involved in establishment and management. However at the time of felling conflicts arises between forest department, Public Works Dcpartment and the local people.

(vi) **Lack of legal constitution:** There is no legal frame work to control and manage the social forestry plantations due to involvement of too many stakeholders like forest department, community and other line departments.

(vii) **Long gestation:** The long gestation period of most of the trees promoted under social forestry programme without adequate linkages with supply chain detract the attraction of social forestry and leads to failure of a plantation.

(viii) **Lack of sustainability plan:** Once the project is over, the people are not involved in sustaining the project and there is no sustainability plan established in the frame work which resulted in failure of plantations and the concept of social forestry.

24. How can we make use of the traditional knowledge of the major tribes of India in forest conservation? (2010)

The indigenous and ethnic people of the world have learnt to live in most hostile environmental condition in this universe. The most interesting feature associated with these indigenous and ethnic has been found that, they live in localities which are immensely rich in biodiversity. These ethnic and indigenous people have played a vital role in conservation of environmental management and development process as they posse's traditional knowledge which has been useful in Eco-restoration. It has been noticed that these people know how to live with harmony in nature.

In India, 68 million people belonging to 227 ethnic groups and comprising of 573 tribal communities derived from six racial stocks namely - Negroid, Proto- Australoid, Mongoloid, Mediterranean, West Breachy and Nordic exists in different part of the country. These ethnic people mostly the indigenous tribals live close in the vicinity of forests and have managed and conserved the biodiversity of their localities since long time. These tribals take shelter from forest and utilize wild edible plants both raw and cooked. The flower and fruits are generally eaten raw whereas tubers, leaves and seeds are cooked. Tribals utilize forest produce, forest timber and fuel wood. These tribals are living in forest since ages and have developed a kind of affinity with forests.

India is a country with large ethnic society and has immense wealth due to which it is rich in biodiversity. There are 45,000 species of wild plant out of which 9,500 species are ethno botanically important species. Of these 7,500 species are in medicinal use for indigenous health practices. About 3,900 plant species are used by tribals as food (out of which 145 species comprise of root and tubers, 521 species of leafy vegetables, 101 species of bulbs and flowers, 647 species of fruits), 525 species are used for fiber, 400 species are used as fodder, 300 species are used in preparation and extraction of chemicals which are used as naturally occurring insecticides and pesticides, 300 species are used for extraction of gum, resins, dyes and perfume. In addition to these a number of plants are used as timber, building material and about 700 species are culturally important from moral, cultural, religious, aesthetic and social point of view of. Indian sub-continent is one of the twelve mega-centres of biodiversity representing two of the eighteen hotspots of biological diversity one occurring in Western Ghat and another in North- Eastern Himalaya (Zeven and Zhikovsky 3). Floristically 141 endemic genera belonging to over 47 families of higher plant occur in India In India 11.95% of the world's biodiversity has been conserved by ethnic people in many ways. Botanical survey of India has reported 46,214 plant species are found in India of global flora of these 17,500 represents flowering plants. Thirty seven of these are endemic and found in North -East of India.

- **Conservation of Edible plants:** Tribals follow environmental conservation rule in harvesting edible plants which establishes ecological prudence. Tubers of edible plants like those of

Dioscorea spp. are harvested by tribals when the leaves of the vine turn yellow and have physiologically matured. The wild tubers are dig carefully avoiding damage to associated species.

- **Conservation of Plants of Herbal Healers:** Many plant species are of great economic importance to tribals as rhizomes of such plants like Acorus calamus, stem bark of Bunchania lanzan, stem and leaves of Moringa oleifera, Achyrnthus aspera, Gynandropsis gynandra, Bombax ceiba are being used as antidote of snake - bite and scorpion sting. Paste is prepared from rhizome and applied on wounds. These plants are conserved for above purpose by tribals.
- **Conservation of plants used in Orthopedic treatment:** The root, stem and leaves of some plants are powdered and paste is prepared and applied by tribals on broken bone portions. The paste prepared from of stem and leaves of plants like Vanda tessala, Alternanthera sessiles and of roots of Cassia adnata, Sida cordata, Bauhina purpurea etc. are tied for healing of wound for 10 - 15 days on broken bones. These plants are conserved by tribal herbal healers in natural forests for orthopedic treatments.
- **Conservation of Medicinal Plants:** Primitive and indigenous people have been using several plants for combating disease from centuries and are found wide acceptance in traditional medicinal use. Plants like *Equisetum ramosissimum*, *Argemone maxicana* are dried, powdered and paste is applied on infested portion of skin and on wounds. Plants like *Bauhinia purpurea, Sida acuta, Jatropha curcus, Grewia hirsutum, Albizzia lebbeck, Capparis decidu*as are conserved and used in muscular pain, cure of fever, headcache, and body swelling.Decoction prepared from roots of *Curculigo orchiodes, Bombax ceiba*, to cure white discharge in urine of tribal women are also conserved by primitive tribes.
- **Conservation of Plants in shifting cultivation areas:** The shifting agriculture jhoom practice of cultivation of crops are practiced by ethnic societies in North -East region of India in states of Assam, Tripura, Mizoram etc.in Central India in states of U.P., Maharashtra, Orissa and Chharisgarh and in South India in states of Tamil Nadu, Andhra Pradesh, Karnataka and Kerala. In this practice a forest is cleared by felling of trees and plant bio-mass is burnt and the ashes collected which are source of essential plant nutrients are spread in the fields. After a year of cultivation, the land is abondened for several years for regeneration of fertility of soil. During this period the farmers are moved for cultivation in other land. The tribals do not perform complete felling of forest but they retain several useful species of horticultutral and agricultural importance such as *Mangifera indica* (Mango), *Citrus spps*. (Orange), *Musa spps*.(Banana), *Phyllanthus embilica* (Goose berry), *Zea mays* (Maize), *Saccharum spps*.(Sugarcane). Several useful plants like *Ardisia polycephala, Ardisia cripsa, Casearia glomerata, Meliosma ipñata, Rhus spps, Phoenisx spp*. etc. are colonized at abondened sites.
- **Sacred groves Conservation:** The ethnic people of India have played a vital role in preserving bio-diversity of several virgin forests and have conserved several flora and fauna in sacred groves of tribals, otherwise these flora and fauna might have been disappeared from natural eco -system. The sacred groves are the natural forests which are located in North - East, Central and Peninsular India. The interference of all kind of human activities are prohibited in sacred groves. In sacred grove of Maharashtra in Western ghat, the giant tree Mangifera indica

(Mango) which is covered by the twiner of Tinospora sinensis having hanging stem and looks like trunk of elephant. Due to magico-religious belief the tribals worship mango tree and have conserved these trees in natural eco-system.

25. Comment on the need of Research in Agroforestry in India. (2010)

In India, the need for wood and wood products both for domestic and industrial utility in increasing at an alarming rate. This increase is due to population explosion, industrialization, government policies on societal development which demanded huge volume of wood and wood products. It has been estimated that India demands over 87 million m^3 of softwoods and over 65 million m^3 of hardwoods during the year 2020. This huge wood requirement cannot be met from the existing natural forests due to policy and legal regulations. The forest area in the country is very low of 24.16% against the mandated requirement of 33%. This besides the productivity of Indian forests is very low of 0.5 to 0.7 m^3/ha/annum compared to the global average of over 2.1m^3/ha/annum. The low forest cover coupled with poor productivity and the associated policy and legal regulations restricted the supply of wood and wood products from the natural forests.

Under this circumstances, the growing demand of wood and wood products will have to be met from imports or growing them under agroforestry systems. The massive imports of wood and wood products will take significant volume of country's exchequer and hence the possible way is promote agroforestry in the agriculture landuse systems. Though agroforestry is an age old practice and many of the practices are still being followed in different agro-climatic regions of the country. The home gardens, silvi-pasture and the sporadic distribution of trees have been practiced since time immemorial. However the existing agro forestry system in the country may not meet the growing demands of wood and wood demand due to the absence of high yielding varieties, precision silvicultural technology, multifunctional agroforestry models and the agroforestry systems towards mitigating the growing climate change risks. Hence research on agroforestry is very essential towards meeting the growing demands. The following researchable issues will help to address the growing issue.

India is endowed with wide range of edapho climatic conditions and no single agroforestry system is amenable for wide range of agroclimatic zones. This necessitated intensive research in order to location specific agroforestry models amenable for multifunctional utility. Similarly the growing concerns on climate change and the associated causes and effects are critically viewed seriously across the globe due to its impact on agriculture, health and other landuse system. The intergovernmental panel on climate change has identified agroforestry as one of the robust landuse system which can mitigate and adapt to the growing climate related issues. However there is no unique model suitable to address this issues which demanded systematic research on agroforestry to resolve the issues.

26. What do you know about recent progress in agroforestry research and development in our country for sustainable development? (2012)

Though the agroforestry as a practice was very ancient in South Asia, the science is relatively new. The organized research in agroforestry was initiated with the establishment of the All India

Coordinated Research Project (AICRP) on agroforestry in April, 1983 by the Indian Council of Agricultural Research (ICAR). The council took lead in conducting systematic research in agroforestry initially through several coordinated projects and later by establishing National Research Center for Agroforestry [NRCAF] in 1988 at Jhansi. The center was established to cater basic, strategic and applied research needs in the field of agroforestry. The All India Coordinated Research Project (AICRP) on Agroforestry – a large agroforestry network operational since 1983 was transferred to NRCAF in 1997 by empowering Director, NRCAF as the Project Coordinator of this splendous program. At present, the AICRP is being operated at 25 State Agricultural Universities (SAUs), 11 ICAR institutes and one Indian Council of Forestry Research and Education (ICFRE) institute. In last three decades several agroforestry systems have been developed which have gone to farmers' fields and provided livelihood support to poor farmers.

Presently, the NRCAF has been upgraded as ICAR Central Agroforestry Research Institute and is recognized worldwide for its research and development capabilities, agroforestry data-base and information repository and natural resource management on watershed basis. The institute conducts basic and strategic research in agri-silviculture, agri-horticulture, silvopasture, tree improvement and human resource development. This besides the Tamil Nadu Agricultural University has established systematic research on industrial agroforestry through a consortia mode involving all stakeholders which are now started replicating across the country.

The major research developments and the associated sustainable development are furnished below.

(i) Industrial agroforestry research and development: This involves research on development of High Yielding Short Rotation clones and made significant impact on productivity improvement (over 40m^3/ha/annum) and reduced rotation (Less than 5 years). The development of varieties and clones in Eucalyptus, Casuarina, Populus, Melia, Subabul *etc.*, have made tremendous impact in development of organized industrial agroforestry plantations.

(ii) Poplar based agroforestry: Poplar based agroforestry (PBAF), in North India is a successful land use intervention for sustaining farming and generating benefits to the farming community in Indo-Gangetic plains in North India. This region is one of the highly productive belts for food grain production and was the torch bearer of green revolution. This revolution undoubtedly secured food for millions of Indias and we have been the food surplus Nation for decades now. This Poplar based agroforestry technology generates 8 million tonnes of fresh timber, 1.8 million tonnes of pulpwood, 3 million tonnes of firewood and generate over 100 million mandays of employment and contributes towards sustainable development.

(iii) Teak based agroforestry: Teak is one of the best timber tree and teak based agri-silvi hortipastural model involving teak, sorghum, groundnut, chilly, papaya, subabul and grass has been one of the successful model increased the productivity of both wood and food and practiced by 2-5% of the farmers of the Karnataka state.

(iv) Eucalyptus based agroforestry: Eucalyptus based agroforestry technology is well adapted in indo gangetic plains and also in Southern India. Eucalyptus has been incorporated in various agroforestry models and it has been promoted by several paper industries through agro and farm forestry. This has created 8-10 million ha of Eucalyptus plantations and is supplementing over 60% of the raw material demands of pulp and paper industries in the country.

(v) **Alder-Large Cardomon based agrisilviculture:** This system is very robust in North eastern states judicially incorporated Alder and Cardomon. This system is found to accelerate the nutrient cycling, increases the soil fertility, augments productivity, reduces soil erosion, conserves biodiversity, soil, water and severs as carbon sink besides improving the livelihood of the locality.

(vi) **Prosopis based agroforestry in arid and semi-arid:** The growing demand of fodder and fuel has identified Prosopis based agroforestry as one of the research output towards sustainable development. The *Prosopis cineraria* is grown at varied plant densities along with other annual crops. This system has ensured the availability of fodder to the cattle and fuel to the human system and contributes towards sustainable development.

(viii) **Climate resilient agroforestry:** Agroforestry systems can be better climate change mitigation option than ocean and other terrestrial ecosystem because of the secondary environmental benefits such as food security, secured land tenure, increased farm income, restoring and maintaining above and below ground biodiversity, maintaining watershed hydrology and soil conservation. Trees are believe to sequester more carbon and it is estimated that each and every tonne of wood produced is equal to 0.5 tonne of carbon sequester thereby help to balance the CO_2 and carbon-di-oxide ratio.

(viii) **Agroforestry as a means of adaptation:** The agroforestry options may provide a means for diversifying production systems and increasing the sustainability of small holder farming system. Diversifying production system judicially incorporating tree component may buffer against income risks associated with climate variability. Research into the contributions of agroforestry in buffering against climate variability is not well advanced but the research in intensified to prove towards its role in buffering capacity towards sustainable development.

27. How social forestry di-ers from other types of forestry? (2012)

(i) Social forestry has a definite principle of people participation from the inception upto the completion of the project. But in other regular forestry programme trees are planted and managed by government agencies whereas in social forestry peoples participation is mandatory from planting to utilization.

(ii) In social forestry the people participation is compulsory except in few cases where it is voluntary but in other forestry programmes people participation is not mandatory.

(iii) Social forestry is a forestry established by the people for the people and to the people. But other types of forsts are not mandate for this purpose.

(iv) The main objective of social forestsry is to satisfy and meet the fuel wood demands of the local community. Example: Agroforestry the purpose is to co-generate wood and food and not exclusively firewood alone. Similalry, the Urban Forestry concentrates on aesthetic and recreational benefit and not provides fuel wood. Similary the role of farm forestry is to enhance the farm income and not as prescribed in social forestry.

(v) Another major objective of social forestry is to provide fodder needs of the society. But other forestry programmes do not aim for meeting the fodder needs.

(vi) The main aim of social forestry is to provide adequate income and employment generation activites. But other forestry programs are not mandated to create income and employment generation activities.

(vii) Social forestry include water regulation, flood control etc. which may not be the objectivies of other types of forestry projects.

(viii) The social forestry also includes recreational, aesthetic and extensional value but other forestry type are very specific and do not involve comprehensively environmental protection.

Hence, in a holistic approach social forestry is different from other types of forestry in terms of objective, people participation, social and environmental protection.

28. Discus the need and scope of agroforestry for the benefit of people. (2015)

Need for Agroforestry

Agroforestry has great technical and economical potential to overcome the problems in the way of higher and sustained crop production and farm income. It can play a very effective role in the utilization of all the natural resources in the most effective manner for sustained crop production.

The different aspects in which agroforestry can help in enhancing the productivity of our lands to meet the demand of ever-growing human and livestock population and at the same time maintaining and improving quality of producing environment are as follows:

- To create an alternate food source in ecologically disadvantaged situations such as dry and arid lands.
- To utilize the available farm resources efficiently and judiciously.
- To maximize the per unit production of food, fodder, fuel, livestock and other forest products.
- To reduce pressure on protective and productive forests for meeting the local demands of fuelwood, fodder, building material and industrial timber.
- To maintain the ecological balance by mitigating green house effect and sequestering atmosphere carbon dioxide.
- To check the soil erosion, conserve soil moisture and increase the soil fertility.
- Alternate option for reclamation of degraded lands such as salt affected, mine spoils.

Scope of Agro-forestry In India

Agroforestry has tremendous scope in India due to the following reasons.

(i) Growing demand for wood and wood products: India is endowed with second populus country in the world and demands over 150 million m^3 of wood which included over 80 million m^3 of long rotation species and over 60 million m^3 of short rotation species. Indian forest's does not support provision for such a huge timber requirement and hence it has to be grown only through agroforestry and hence it has excellent scope in India.

(ii) Growing fodder demand: Similar to human population, India is robust in animal population which led to shortage of both green and dry fodder. The shortage is estimated as high as 61% in green fodder and 21% in dry fodder. The existing fodder resources are not amenable to meet the growing demands. The growing failures of monsoon, increasing temperature and water scarcity delimit the availability of annual fodders. Under such circumstances, agroforestry trees play a vital role in meeting the fodder needs particularly during lean season.

(iii) Combating climate change: Agroforestry has been identified as the potential land use system to mitigate and adapt to growing climate change and the associated environmental risks due to its carbon sequestration potential and other mitigation process. Hence agroforestry can play a vital role globally and has an excellent scope adapting to climate change.

(iv) Rehabilitation of degraded lands In India, over 175 million hectare has been categorized as degraded land due to increasing temperature, reduction in water availability, increasing flood, improper tillage practices etc., which resulted in soil and water loss. To resolve these issues, agroforestry as a land us system can conserve soil and water and improve the soil fertility. Hence agroforestry has an excellent scope in rehabilitation of degraded lands and other problem soils.

(v) Industrial Agroforestry: The National Forest Policy 1988 and National Agroforestry Policy 2014 have directed wood based industries in the country to generate their own raw material through agroforestry without depending on supply from natural forests. This has attracted several wood based industries towards agroforestry and hence it has excellent scope particularly in timber, pulp and paper, plywood, construction, sports goods, packing cases, match splints, dendro power and biofuel industries.

(vi) Soil and water conservation: Agroforestry has an excellent scope of application towards soil and water conservation. It has been estimated that over 175 million ha of land area in the country are degraded and it is believed that over 30 tonnes of soil per ha is eroded annually. For this purpose agroforestry can play a vital role in reducing the erosion, protecting the soil and improving the water harvesting system. Hence agroforestry has an excellent scope in all watershed projects towards sustainable soil and water conservation.

29. Elaborate upon the social objectives of agroforestry. (2015)

The social objectives of agroforestry involve the following issues.

(i) Employement generation: The main objective of agroforestry is to extend adequate employment generation activities in order to have sustainable livelihood system. The various agroforestry activities like land preparation, establishment of agroforestry model, nursery, livestock, fishing, apiculture, sericulture, medicinal plant cultivation etc. may provide necessary employment oppournities both through unskilled and semi skilled work.

(ii) Income generation: The society as a whole depends on income generation from various activities. Un and under employment are the major issues limit income generation. The agroforestry help to sustain income generation activities per unit area of the land. This income generation activity is useful to landless labourers, women selfhelp group and other stakeholders who are involved in the process of agroforestry establishment. The activities related to production, processing and consumption which includes felling and transportation will generate over 300 mandays of employment per hectare and hence continuously extend income generation.

(iii) Providing basic needs: One of the major social objectives of agroforestry is to provide the basic needs of the society particularly the food, fuel, fiber and fodder needs of the society. Till the recent past, fuel wood is the major issue for the society and agroforestry can very well help to meet it. Similarly the fodder shortage in the domestic cattles leads to reduced milk production and income generation particularly during dry season. The agroforestry system may help to resolve the issue by providing fodder during lean season. Similarly integration of annuals with perennials will extend cogeneration of food and wood in the same area thereby help to achieve the social objectives.

(iv) Enhancing agriculture productivity: The agroforestry system will help to enhance the soil nutrients by the addition of leaf litter and reduces the soil loss by arresting the erosion through wind and water. This help to increase the soil quality which will inturn help to improve the agriculture productivity and satisfy the social demands.

(v) Extending clean environment: The growing environmental issues particularly increasing temperature can cause global warming which may affect the health and wealth of the society. The increased concentration of green house gases may cause wide range of diseases like skin burn, asthma, bronchitis, cataract, stress etc. The practice of agroforestry system may help to reduce the green house gases and protect the human from all serious effects. Hence the agroforestry helps to contribute to the social objective of protecting the environment. It also helps to conserve soil and water which are essential features for the society.

30. Describe constitution of JFM Network by MOEF, GOI with its terms of reference (2014)

Objectives of JFM

The JFM programme in the country was reviewed critically by the Government of India from time to time in consultation with state governments, NGOs and other stakeholders to resolve the emerging issues. In 22 February 2000, the Government of India issued guidelines for strengthening the JFM programme.

- To provide a visible role to the local communities in planning, management and protection of forests and to give them a share in the benefits from these forests.
- To elicit active participation of villagers in (a) creation (b) management and (c) protection of plantations.
- To achieve ecological needs consonant with sustainable productivity of wood and other non timber forest resources.
- To wean away the land owning communities from shifting cultivation by adopting an alternative i. e. Tree farming.
- To productively utilize the degraded jhum-land thereby checking soil erosion.

- To conserve Biodiversity through people's action.
- To create and generate forest – based economy for the villagers.

Structure of JFM

Until recently, Forest Protection Committees were by and large not legal bodies as they were not recognized or registered under the Societies Registration Act or through related, enabling legislations. However, States such as Gujarat, Karnataka and Rajasthan had provided for registration of FPCs under the Cooperative Societies Act in the early stages itself. Structures and nomenclatures of JFM have varied from State to State. In general, one can delineate the following structure for JFM. The base of the JFM structure, which comprises village level institutions, is referred to as "forest protection committees". These bodies include a "General Body" and a "core", "Executive" or "Management Committee (MC)" elected by the General Body for discharging the assigned functions. The General Body comprises eligible members of the village. The eligibility criteria for membership vary from State to State.

In most states, adults are eligible for joining the General Body. The General Body elects the local community representatives in the Executive Committee. The Executive Committee has elected members ranging in numbers from 5 to 15. It also includes ex-officio, non-elected members drawn from the Forest Departments, local NGOs, village schools, village administrative and development officers and in some cases representatives of the Gram or Mandal Panchayats. The term of the Executive Committee is generally for two to five years in most of the States. The Executive Committee undertakes decision making, planning and implementation of management plans.

Powers of Forest Protection Committees

The functioning of FPCs in the post- 1990 phase indicate that there are widespread inter- State differences, when it comes to grant of execution of powers by Forest Protection Committees. Different States give different flexibilities to FPCs in the matter of administrative, executive and financial powers. Nevertheless, the standard parameters of power granted to forest protection committees can be broadly categorized as under

- Framing of rules governing management of forests.
- Administrative powers for convening meetings of General Body and Management Committee meetings.
- Voting rights in Management / Executive Committees.
- Punitive powers – i.e., powers for punishing and imposing fines on offenders.
- Rights to collect specific resources and sell them.
- Distribution of benefits arising from conservation and regeneration.
- Cancellation of membership of recalcitrant members.
- Financial powers – to incur expenses, maintain accounts.

The Joint Forest Management is initiated due to the policy directions of Government of India. As per the provisions of National Forest Policy (1988), the Government of India vide letter number 6.21/89 – PP dated: 1st June, 1990 outlined and conveyed to state governments a frame work for creating massive peoples movement through involvement of village committes for the protection, regeneration and development of degraded forest land.

This gave impetus to the participation of stakeholders in the management of forests situated in the vicinity of villages. The JFM programme in the country is structured on the broad frame work provided by the guidelines issued by the ministry. Based on these guidelines over 10.2 million hectare of forests have been brought under JFM programme through 36075 committes as on the year 2000.

31. Give the fundamental characteristics of tribal economy in India (2013)

The sources of subsistence and livelihood are varied so far the Indian tribals are concerned. Starting from the pure and simple parasitic habit of the nomadic hunters and food-gatherers who depend mostly on nature for the sources of subsistence to the settled agriculturists and the group of industrial laborers, we have the views of different economic set-up the Indian tribals. From this view point, we can classify the Indian tribals into six broad economic clusters:

(i) **Food-gatherers and hunters:** The nomadic primitive tribes like the Andamanese, Onge, Jarwa, Kadar, Kharia, Lodha, etc. are included in this category so far their subsistence economy is concerned. They usually live far away from rural-urban way of life and possess a simple type of social organization.

(ii) **Permanent Settled Cultivators:** Like the cultivators of the advanced societies, some of the tribals in India are taking resort to permanent settled agriculture. The Oraon, Munda, Gond, Bhumij, Ho, Santal, are efficient cultivators at present. They practice wet cultivation by transplanting method. Artificial irrigation and application of compost manure are not unknown to them. Rotation of crops is within the knowledge of these cultivators. They work in their owned fields as well as in the fields of others as share-croppers (Bhag-Chasi). The major bulk of the tribal population work as agricultural laborers. In search of jobs, these landless agricultural laborers participate in seasonal migration to the neighboring states. The social and religious organizations of these settled agriculturist tribals are much developed and highly complex. The traditional council of elders (panchayet) has to perform significant role in maintaining societal norms.

(iii) **Pastoral people:** The Bhotia of Almora and the Toda of the Nilgiri hills of South India live on pastoral economy. They do not practice agriculture, hunting, fishing, etc. They also live far away from the sophisticated modern world and possess a slightly developed but non-complicated social structure. Both of them practice polyandrous marriage system. They rear buffaloes and cows, the milk-products are being exchanged to procure the commodities of day-to-day use.

(iv) **Shifting Hill cultivators:** A section of the Gond, Naga, Kharia, Juang, Riang, Khasi, Garo, Savara practice such type of primitive cultivation by "slash and burn" method. It is known by different names in different areas. The Assam tribes call it Jhum, the Gond as Podu, etc. A hilly forested tract is selected for this purpose may have to be abandoned after three successive cultivating seasons as this soil likely to lose fertility. The plants, shrubs, and undergrowth's are cut down and left for drying for a month or so. Then, they set fire to them. The ashes serve as auto-manure to the soil. On the onset of monsoon, the soil is slightly loosened by a simple digging stick or hoe. Seeds of different Kharif crops, millet's like Bajra, Jowar, Kurthi, pulses, potato, tobacco, and sugar-cane are grown in this type of cultivation. It could support them partially but not wholly. They have to find out some other vocations as supporting source of subsistence.

(v) **Manual laboring group:** A substantial portion of the tribals has become landless laboring class owing to the economic hardship which they are facing at present. They earn their livelihood by selling their manual labor in different vocations.

(vi) **Craftsmen:** Some of the tribals are still retaining their traditional crafts along with the principle sources of subsistence. The Naga and the Khasi are experts in coloured hand-loom products and the Lohar are traditional black smiths. With the marginal profit in their traditional specialized crafts, these tribals are at present, taking resort to other types of jobs. Their mixed pattern of economy reflects back adversely upon their social system.

General Characterstics of Tribal Economy

(i) Forest based economy
(ii) Division of labour based economcy
(iii) Simple technology based economy
(iv) Absence of profits in economic dealings
(v) Cooperative unit based economy
(vi) Economy based on gift and ceremonial exchange
(vii) Periodical market economy
(viii) Economic backwardness

32. Elaborate upon the multipurpose tree species ideotype selection criteria (2013)

Ideotype is an ideal tree which is a base line for evaluating the multipurpose tree species that are actively available. It is a model in morphological and physiological terms of the ideal agroforestry species for a particular purpose. The ideotype is expected to perform greater quantities and higher qualities of yield in a definite environmental condition.

Category of ideotypes

Isolation ideotype: Characterized by wide branches, broad leaf coupled with aggressive growth attributes and grows best free standing specimens.

Competition ideotypes: This grows well in a community of their own kind with a tendency to dominate less aggressive varieties.

Crop ideotypes: It has an ability to share environmental resources in a community of their own kind

Associative ideotypes: It contributes to the fulfillment of agroforestry system's designed objectives maximizes environmental resource use in both space and time and have positive interaction with the crops.

Ideotypes for multipurpose tree species

Ideotypes for multipurpose tree species can be characterized based on its morphology, purpose and other attributes related to productive and service functions. No single species or individual tree can exhibit all the traits desirable for an agroforestry application. Some of the traits may be mutual exclusive like fruit and wood production for multipurpose trees ideotypes, the trees are needed for multiple objectives which demand precise and recent definition of desired ideotypes.

A	**Ideotype design needs**	
(i)	Production and service requirement	Fodder, fuel, food, pole, shade, timber, wind break
(ii)	General selection criteria	Vigour
(iii)	Ancelory information	Nitrogen fixer, chemical composition of leaves and pods
B	**Ideotype description**	
(i)	Stem	as straight as can be found in a population; multi-stemmed phenotypes acceptable, but long boles important
(ii)	Crown	fairly rounded, medium diameter (crown/bole ratio 25:1 or less) with many branches and positioned high up the stem; foliage medium to dense
(iii)	Roots	geotrophic angled rather than horizontally extending lateral roots
(iv)	Pods	large, averaging 10 cm long and 8 mm wide, in large quantities
(v)	Thorns	as few and as small as can be found
(vi)	Response to management	prolific regrowth after pollarding and individual branch pruning; reliable coppicing response
(vii)	Deciduousness	short period of dry-season leaflessness compared with population average

33. What entry point activities are recommended in joint forest management? (2013)

I. Creation of Self Help Groups (SHG):

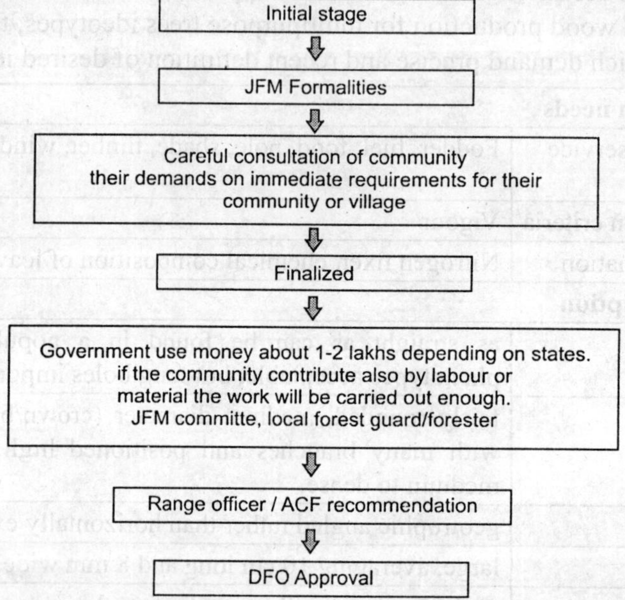

II. Developmental activities in Buffer zone

- The buffer zone includes patta lands and poromboke lands adjoining Reserve Forest.
- In order to enlist the cooperation of the villagers, their needs were fulfilled like construction of threshing floor, bore wall for drinking water etc. were taken up as part of buffer zone activities.
- The prioritized needs of the abuttant villagers were enumerated through Interaction with local village people and the following works were carried out in the target villages.
  - (i) Construction of threshing floors
  - (ii) Construction of bus shelter
  - (iii) Construction of cattle tub
  - (iv) Construction of community hall
  - (v) Construction of water tub
  - (vi) Providing retaining wall to the village pond
  - (vii) Supply of stoves for energy conservation
  - (viii) Providing Television set
  - (ix) Supply of seedlings - free of cost

(x) Conducting medical camps

(xi) Creating drinking water facilities

(xii) Creating forest produce processing centre

(xiii) Desilting and improvement of older village

III. Human resources development

- Participation was sought to be ensured by promoting human resources development activities in the JFM villages
- Human resource development was implemented through vocational training and skill development
- Training were organized on skill development on the following aspects,

  (i) Basket making

  (ii) Tree seedling and nursery raising

  (iii) Tailoring

  (iv) Mat weaving

  (v) Gem cutting

  (vi) Rexine Hand bag making

  (vii) Forest produce processing – Usil leaf powder making etc.

IV. Water harvesting activities

Water is the tangible product that can be obtained through a joint forest management programme. The activities of JFM on water harvesting measures included.

(i) Construction of series of check dams

(ii) Construction of percolation ponds

(iii) All this activities to facilitate the ground water recharge in the targeted villages.

34. Describe tree-crop allelopathy in agroforestry. (2013)

Allelopathy is the direct or indirect effect of one plant upon another through the production of chemical inhibitors that are released in to the environment, should also be recognized as another factor in analyzing mechanisms of plant interactions.The species interaction due to chemical influences is also designated as Allelochemistry, phytochemical ecology or Ecological biochemistry and Allelobiology.

The released allelo chemicals into the soil inhibit seed germination and establishment of certain agriculture crops and the allelopathic effect is very selective and vary with trees and crops. Among the various tree components the leaves and roots are the most potential sources of allelo chemicals which contain the toxic metabolites.

For example: the tannins present in Acacia species, Phyto toxins present in Eucalyptus and Casuarina are major inhibitor to the agriculture crops.

Allelochemicals secreted by di-erent Treespecies

S.No	Tree species	Allelochemicals
1	*Pinus ponderosa*	*CaDeic acid, Chlorogenic acid, Quercetin, Tannin*
2	*Acacia dealbata*	*Hydroxy benzoic acid, Protocatechuic acid, Vanillic acid, Gentisic acid, Coumaric acid, CaDeic acid, Ferulic acid, Sinapic acid*
3	*Leucaena luecocephala*	*Gallic acid, Protocathic acid, Hydroxy phenylactic acid, Hydroxybenzoic acid, Vanillic acid, CaDeic acid, Syringic acid, Ferulic acid*
4	*Eucalyptus camaldulensis*	*Cineole, phenols, Gallic acid, Ferulic acid, Coumaric acid, Chlorogenic acid, CaDeic acid, Sapthulenol*
5	*Delonix regia*	*Hydroxy benzoic acid, Dihydroxy benzoic acid, Gallic acid, Dihydroxycinnamic acid, Dinitrobenzoic acid, Azetidine,Carboxylic acid, Dihydroxy benzaldehyde,*
6	*Gliricidia sepium*	*Gallic acid, Protocatechuic acid, Hydroxy benzoic acid, Gentistic acid, Vanillic acid, Syringic acid, Cuomaric acid, ferulic acid and Sinapic acid, Resorcinol*
7	*Azadirachta indica*	*Myristic acid, Palmitic acid, Gallic acid, Stearic acid, Oleic acid, Linoleic acid acid, Benzoic acid*
8	*Eucalyptus globulus*	*CaDeic acid, Coumaric acid, Gallic acid, Gentisic acid, Hydroxybenzoic acid, Syringic acid and Vanillic acid, Catechol*
9	*E.citriodora*	*CaDeic acid, Coumaric acid, Gallic acid, Gentisic acid, Hydroxybenzoic acid, Syringic acid and Vanillic acid,Catechol*
10	*Acacia nilotica*	*Benzoic acid, Catechol, Salicilic acid, Gentisic acid, Protocatechic acid, CaDeic acid, Vanillic acid, Syringic acid, Hydroxy Cinnamic acid, Palmitic acid, Pyrogallol, Tannic acid, Oxalic acid*
11	*Gmelina arborea*	*Benzoic acid, Catechol, Salicilic acid, Gentisic acid, Protocatechic acid, CaDeic acid, Vanillic acid, Syringic acid, Hydroxy Cinnamic acid, Palmitic acid, Pyrogallol, Tannic acid, Thalic acid, Oxalic acid, Ferulic acid, Coumaric acid*
12	*Tectona grandis*	*Benzoic acid, Catechol, Coumaric acid, Gentisic acid,Vanillic acid, Syringic acid, Hydroxy Cinnamic acid, Palmitic acid, Pyrogallol, Thalic acid, Oxalic acid, Napthol, Phenol, Feruic acid*
13	*Simaruba glauca*	*Benzoic acid, Catechol, Coumaric acid, Vanillic acid, Cinnamic acid, Oxalic acid, Feruic acid, CaDeic acid, Salicylic acid, Resorcinol*
14	*Ailanthus excelsa*	*Benzoic acid, Catechol, Coumaric acid, Feruic acid, Salicylic acid, Protocatechu*

35. a. Describe the benefits and constraints of Agroforestry (2014)

b. Benefits of Agroforestry (2009)

Environmental benefits

- Agroforestry reduces the pressure on forest and enhanced better protection of ecological systems.
- It augments biological diversity by favouring environmental conditions.
- It improves microclimate of the area by lowering of soil temperature, reduction of evaporation and maintenance of soil moisture.
- It reduces run-off, nutrient leaching and soil erosion.
- It improves the soil fertility through addition of litter-fall.
- It enhances efficient recycling of nutrients with in soil at different depth.
- It improves soil structure by the addition of organic matter of soil.

Economic benefits

- Agroforestry yields increased outputs per unit area (more crop per unit area).
- Agroforestry system reduces the total crop failure.
- All these benefits resulted in increase of farm income

Social benefits

- Agroforestry improves the rural living standards by increasing their income in a sustained manner.
- It improvement the health and nutrition of the peoples by providing diverse, nutrient rich farm produce.
- Multiple outputs without the risks of total failure stabilize the farm communities.

Constraints in agroforestry

- Depression in crop yields due to interference effects caused by the tree
- Delayed liquidation of planting investments due to long gestation period
- Increased damage to crops due to birds which the tree attract
- Increased damage to crops due to pests for which the tree serve as alternate hosts
- Allelopathy

36. Can you di-erentiate between the terms mixed plantations in forestry and mixed cropping in agriculture? Justify the di-erences between them (2015)

S. No	Mixed planting in forestry	Mixed cropping in agriculture
(i)	Growing two or more trees simultaneously or sequentially without any distinct row arrangement	Involves planting two or more annual plants simultaneously in same field

(ii)	It is otherwise called as mixed plantation or multistory cropping	It is otherwise called as inter cropping or co cultivation or mixed farming
(iii)	The objective of mixed plantation is to utilize the vertical space more.	The objective is to utilize the horizontal space more.
(iv)	There exist a height gradient variation between the trees	No height gradient variation
(v)	It helps to conserve soil and water conservation very effectively	Comparatively low soil and water conservation
(vi)	Weed infestation is possible	No weed infestation
(vii)	Balanced nutrient composition	Balanced input and output of soil nutrients
	Eg: Casuarina + Melia + Teak Teak + mahogany + Terminalia	Eg: Green gram + black gram + onion Ground nut + onion + chillies

37. Provide scientific names of four potential NFTs each suitable for Tropical and Temperate conditions (2016)

Tropical nitrogen fixing trees

(i) ***Acacia albida***: The Acacia, native to Africa, are usually considered as highly valuable legume trees serving especially as soil improvers in agroforestry, and sources of fuel-wood, forage, and gum. Their nitrogen potential seems to be rather poor, but could possibly be improved by capitalizing on the great variability of the host plant. The subject has not yet been adequately srudied.

(ii) ***Leucaena leucocephala***: It is a fast growing nitrogen fixing tree with the capacity to fix 100 – 150 kg of nitrogen / ha/ annum. There are four types *viz.,* Hawaiian type, Solvador type, Peru type and Cunningham type. The tree is a fast growing and amenable for wide range of soil and climatic conditions. It regenerates profusely through natural and artificial means. The density of the plantation is 4500 / ha and is amenable for harvest in 3 years. The tree is an excellent fodder due to high protein content of around 20%. But has amino acid called mimosine which limits its use as fodder. It is a multi purpose tree and used as a fuel, fodder, pulp, scaffolding and also as a rehabilitator for degraded sites.

(iii) ***Albizia lebbeck***: Native to Bangladesh, Burma and Pakistan, this tree has been propagated in many tropical and subtropical regions (National Research Council 1980). Large perennial nodules of large size have been observed on adult trees in Senegal, but the potential nitrogen fixation of this species has not yet been evaluated.

(iv) ***Prosopis spp.***: The nitrogen-fixing characteristics of this genus, which is currently receiving the pride of place in semi-arid regions and saline soils, Measuring the nitrogen compartmentalization of the biomass, productivity of leaves,branches, trunk and reproductive tissues, a nitrogen-fixation rate of 23-36 kg N2/ha/yr with a 33 per cent stand'cover. From determinations of natural abundance l5N/I4N ratios, Prosopis fixed approximately 43-61

per cent of its nitrogen, this range of figures applying to six of the seven sites they studied in the Sonoran desert, California. Prosopis juliflora, a tree native to Central America and northern South America Has been introduced in many arid iones of the world; it grows fast, even on soils very low in nutrients, probably thanks to its good nitrogen- fixing potential. In some places, *P. juliflora* is an agressive invader and thus is considered a hindrance. Other species of Prosopis, namely *Prosopis cinerea* and *Prosopis tamarrigo* are widely utilized in arid countries: their nitrogen-fixing potential has never been evaluated. An experimental study of the nodulating ability of *Prosopis chilensis*, *P. ramarugo* and *P. alba* has been initiated recently. Among these species, *P. chilense* seems to respond best to inoculation. The effectiveness of the nitrogen fixation process appears to vary greatly with the different Prosopis populations.

Temperate nitrogen fixing trees

(i) ***Alnus nepalensis*:** *A. nepalensis* is a large deciduous alder with silver-gray bark that reaches up to 30 m in height and 60 cm in diameter. The leaves are alternate, simple, shallowly toothed, with prominent veins parallel to each other, 7–16 cm long and 5–10 cm broad. The flowers are catkins, with the male and female flowers separate but produced on the same tree. The male flowers are 10 to 25 cm (4 to 10 in) long and pendulous, while the female flowers are erect, 1 to 2 cm (0.4 to 0.8 in), with up to eight together in axillary racemes.Unusually for an alder, they are produced in the autumn, with the seeds maturing the following year.

(ii) **Betula *Spp:*** Birch species are generally small to medium-sized trees or shrubs, mostly of northern temperate and boreal climates. The simple leaves are alternate, singly or doubly serrate, feather-veined, petiolate and stipulate. They often appear in pairs, but these pairs are really borne on spur-like, two-leaved, lateral branchlets.The fruit is a small samara, although the wings may be obscure in some species. They differ from the alders (Alnus, other genus in the family) in that the female catkins are not woody and disintegrate at maturity, falling apart to release the seeds, unlike the woody, cone like female alder catkins.

(iii) **Elaeagnus spp:** *Elaeagnus angustifolia* is a small tree can reach a height of 5-10 m. The plant has shiny brownish red spines. Stems, leaves, flowers, and fruits are covered with silver-white scales. Leaves are inconspicuously veined, lance-shaped or linear lanceolate, 3-7 cm long and 1-1.3 cm wide, obtuse at the apex, and cuneate at the base. Appearing from May to June, the fragrant flowers are erect or nearly erect, and have bell-shaped calyx tubes and a conspicuous, glabrous, conical floral disc, which surround the base of the style. Fruits are pink, elliptic, 9-12 mm long and 6-10 mm wide, and mature in September.

(iv) **Elms:** Elms (*Ulmus carpinifolia*) come from North-Africa, Europe and south-western Asia. They are Ulmaceae plants and can reach up to 30 m tall. Their foliage is hemispherical, their branches are thin and pale brown, their flowers are small and red. Their bark is grey-brown with deep furrows; their leaves are oval with a pointed end and a slanting base.

38. Describe the traditional agroforestry system practiced in North-eastern India. (2009)

Traditional Agroforestry Systems of the NEH region

North Eastern India consisted of 8 states and occupies 5.6% of the total geographic area of the country. These regions are housed with more than 120 diverse ethinic groups who are maintaining their own cultural heritage. North Eastern region encompasses wide variation in altitude, climate, soil and rainfall. The cropping system, cropping pattern and agroforestry systems practiced vary from state to state over all about 70% of the land area is cultivated at one time or the other each year in the region.

(i) **Home garden:** The North Eastern people have been practicing agroforestry in variety of forms. The traditional home garden are best examples wherein 2 to 3 tier cropping systems are practice which provide the domestic and the livelihood needs of th people.

(ii) **Jhum cultivation:** The North Eastern experiences practice of shifting cultivation over several decades. This involves clearing a forest area and practicing agriculture with cerals, pulses and vegetables. This cultivation of agricultural crops may be continued for 2-3 years and once the productivity decreases the people shift to new area and again practice the same activity. This kind of shifting cultivation is a traditional age old practice carried out by the people in North Eastern India.

The harmful effect of shifting cultivation has been exposed to tribal communities and ethinic groups and they are directed towards practicing organized agroforestry practices like silviagriculture, silvihorticulture, home garden, plantation crops etc. through necessary awareness and capacity building programe which resulted in changes in the attitude in agroforestry practices.

39. Write the scientific names of ten Multipurpose Tree Species (MPT) only planted in social forestry. (2009)

Multipurpose Tree Species

Woody perennials that purposefully grown so as to provide more than one significant contribution of the production, and/or service functions (shelter, shade, land sustainability) of the land use system implement.

- *Acacia catechu*
- *Acacia nilotica*
- *Albizia lebbek*
- *Bamboos*
- *Butea monosperma*
- *Calliandra calothyrsus*
- *Cedrus deodara*
- *Celtis australis*

- *Diospyros melonxylon*
- *Gliricidia sepium*
- *Moringa oleifera*
- *Neolamarckia cadamba*
- *Robinia pseudoacacia*
- *Tamarindus indica*

40. Discuss various approaches required to motivate the members of a tribal community to introduce social forestry plantation in their farmland (2009)

The various approaches required to motivate the members of tribal community to introduce social forestry plantations in their farm lands are furnished below.

(i) Farm Forestry: Individual farmers are being encouraged to plant trees on their own farmland to meet the domestic needs of the family.

(ii) Community forestry: Community forestry is a village-level forestry activity, decided collectively and implemented on communal land, where local populations participate in the planning establishing, managing and harvesting of the forest crops. Hence, they receive a major proportion of the socio-economic and ecological benefits from the forest. Community forestry is a process of increasing the involvement of and reward for local people, of seeking balance between outside and community interests, and of increasing local responsibility. The animals are also allowed to graze in a particular plantation that is beyond damage, for maximizing the total output from the land.

(iii) Extension forestry: Planting of trees on the sides of roads, canals and railways, along with planting on wastelands is known as 'extension' forestry. Extension forestry helps in increasing the boundaries of forests. Under this project, there has been creation of forests on the village common lands, government wastelands and Panchayat lands. Schemes for afforesting degraded forests of the government, which are close to villages, are being carried out all over the country.

(iv) Agro-forestry: Agro-forestry means practice of agricultur and forestry on the same piece of land. Agro-forestry is the growing of trees as an integral part of the cropping or livestock system. The home gardens of South-east Asia are one of the most highly-developed forms of traditional agro-forestry. The ecological and economical benefits derived from a combination of trees and crops are manifold.

(v) Industrial Plantation: The promotion and establishment of industrial wood plantation in the form of pulpwood, plywood, fuelwood, timber and other non-wood forest plantation will help to encourage the tribal community to introduce the social forestry plantations. The industrial plantation will extend income and employment generation activities which will attract the tribal community towards social forestry planations.

41. Write general principles of benefits sharing under Joint Forest Management (2009)

Benefit sharing

Sharing the sustainable benefits from the degraded forests, Government lands and Community lands lying within the management unit with the members of Village Forest Committee are the most important component of Joint Forest Management. The guiding principle of this benefit sharing is equitable distribution among the members.

There are two major products in forestry and they are timber and non timber forest products. The JFM activities would concentrate on NTFP management and no alteration is permitted in the basic silvicultural prescription prescribed in the working plan but to promote regeneration, development and sustainable harvesting of NTFPs. These products are collected at free of cost or at some concessional rates as per the existing practice and the benefits are shared among the members of the JFM committees.

The benefit sharing mechanism will be different for good forest area. The JFM committees also be eligible for benefit sharing for timber, only if they have satisfactorily protected the good forest for a minimial period of atleast 10 years. The sharing percentage is kept limited to a maximum 20% of the revenue from the final harvest. The felling of trees and harvesting of timber will be as per the provisions of working plan. A certain percentage of revenue from final harvest should be ploughed back in the silviculture and management of forests.

The extent of good forests to be allowed for benefit sharing will depend upon the number of village households and should be restricted to a maximum limit of 100 ha and also limited to two kilometer from the village boundry.

In case of degraded forests the JFM committees can concentrate on areas upto five kilometers from the village boundry.

There is also a provision to get 25% share of the village community which should be deposited in the village development fund for meeting the conservation and development needs of the forests. The forest department is also share matching grant towards improving this activitiy.

The benefit sharing varies from state to state and some states are following 50% share of the benefits after ensuring the regeneration and development of a plantation after the harvest.

42. Briefly describe the merits and demerits of 'High Density Short Rotation' forestry. Enlist suitable species in this regard along with their productivity potential. (2010)

Definition

High Density Short Rotation forestry (SRF) is defined as the silvicultural practice in which high density, sustainable plantations of fast growing tree species produce wood biomass on agricultural land and fertile and also on degraded forest land.

Approaches for HDSR

- Production of feed stock for energy and wood based industries
- Alternate landuse system for the lands taken out of agricultural production
- Sequestration of carbon and mitigate climate change

Merits of HDSR

- The HDSR plantations are amenable for higher productivity and are capable of producing atleast 10 times of dry matter/ha/annum.
- This system utilizes optimally all natural resources using biological, physical, theoretical and practical knowledge in an ecologically acceptable way.
- The biomass produced in HDSR may replace the wood from natural forests in terms of timber, fodder, energy as well as wood for pulp and paper industries.
- It will also help to produce wide range of non-timber forest products
- HDSR has the capacity to sequester more and are amenable for climate adaptation and mitigation projects.

Demerits of HDSR

- Promotion of HDSR will lead to monoculture of species thereby the chances of biodiversity loss is expected.
- The promotion of fast growing and densely populated plantations will rapidly utilize the soil nutrients and the soil may become degraded due to over exploitation.
- Chances of invasion of epidemic status of pest and disease are more in case of single species based HDSR plantations.
- The HDSR plantations are specific to few industries and incase of over production there are chances of glut in the market and may affect directly the growers.
- In many cases the ecological impact of HDSR plantations are not known.
- Several HDSR plantations are established through exotic species like Eucalyptus, Subabul, and Casuarina have eliminated the indigenous species in the respective landuse system.

HDSR Species

Populus deltoides, Casuarina equisetifolia, Eucalyputs spp., Melia dubia, Anthocephalus cadamba, Leucaena leucocephala, Salix etc.

43. What is integrated land use management? Give a plan of integrated land use management for 10 hectare of land in tropics and sub-tropics parts of India. (2010)

Integrated Land Use System

Integrated land use system is a complete and sustainable landuse system judicially incorporates crop, livestock, forestry and fish systems which may function independently in certain farming system and their products are only additional benefits. Sometimes the output of one subsystem in the integrated farming system may become input for the other subsystem resulting in a greater efficiency of output of desired products from the land. There is a synergism in the IFS which involve working together of the different sub system which has greater total effect than the some of their individual effects.

Importance of Integrated Land Use System

- It helps to balance the economic, social and cultural opportunities in a specific area of land use system.
- It helps in maintaining the integrity of ecosystem and sustains it.
- It helps and coordinate all the interested parties to make decisions about how the land and resources can be used and managed in order to enjoy the economic benefit and to protect the ecosystem benefits

Plan for managing 10 hectare through Integrated Land Use System

(i) Integrated land use system with planation crops
(ii) Integrated land use system with coconut
(iii) Integrated farming system
(iv) Integrated agroforestry models

(i) Integrated land use system with planation crops: In the 10 ha of tropic and sub-tropic area integraed land use system with plantation crops is one of viable option. It includes establishment of modern commericial plantations of crops like rubber, coffee, tea, oil borne plantations which can be profitable if it is well managed the scope of integrative practices involved in plant associations in limited but still this model can be a viable option if it is integrated with other crops like pepper, ginger, garlic etc.

(ii) Integrated land use system with coconut: Coconut is one of the most widely grown tree crop in the tropic and can be integrated in land use system by judicially incorporated with other trees, annual and animal components. The coconut plantation are widely placed and in between them commericial short duration crops like banana, turmeric and ginger.

This model can have the following crop combination.

S.No.	Major crop	Tree component	Other Crops
1.	Coconut	Arecanut	Spicies and continents Ginger, turmeric, minor spicies
2.	Coconut	Cocoa	Root crops Cassava, sweet potato, yam
3.	Coconut	--	Cereals Rice, maize, finger millet

(iii) Integrated mixed farming: Mixed farming involve planting wide range of crops from annuals to perennials by judicially incorporating animal component. The following crop combinations will help to achieve greater economic returns.

S.No.	Tree component	Agriculture component	Grass component	Animal component
1.	*Tectona grandis*	Cow pea	Napier	Milching cows
2.	*Pterocarpus marsupium*	Green gram	Guinea	
3.	*Melia dubia*	Black gram	Stylus	

(iv) Integrated Agroforestry Model: The 10 acres of tropical areas can be established with integrated agroforestry models through the following approaches.

(a) Pulpwood based Integrated Farming System

Tree component	Animal component
Casuarina equisetifolia *Leucaena leucocephala* *Sesbania grandiflora*	Sheeps, Milch animals, Poultry, fish

(b) Silvipastoral model

Tree component	Grass	Animal
Acacia nilotica *Melia dubia* *Acrocarpus fraxinifolius* *Grewia robusta*	Napier, Guinea	Goat, Sheep, Milch animal, Poultry fish

(c) Plywood based Agroforestry model

Tree component	Other Crops
Melia dubia	Cereals, Pulses, Medicinal plants, grasses

(d) Timber based Agroforestry model

Tree component	Other Crops	Animal component
Tectona grandis	Cereals, Pulses, Fodder grasses	Milch animals

(e) Biofuel based Agroforestry model

Tree component	Other Crops	Animal component
Pungam	Cereals, Pulses, Fodder grasses	Milch animals Poultry fish

44. While selecting the species for agroforestry, the below-ground and above-ground interactions between the component species need to be considered. Discuss (2011)

Integration of trees in agroforestry system results in positive or negative interactions between trees and crops. Micro-climate amelioration and maintenance or improvement in soil productivity is the major positive interactions while competition for light, water and nutrients, and allelopathy are the major negative interactions in agroforestry systems. The balance between negative and positive interactions determines the overall effect of interactions in a given agroforestry system. Selection of suitable tree species for agroforestry is important in order to balance the above and below ground interaction to ensure sustained production. The above ground interaction is predominantly for light, space and the below ground interaction is predominantly for moisture, nutrients and the space. The management of trees both above and below ground is very significant in balancing the interaction.

Management stratergies for above ground interaction

Tree management/crown management

There are certain management techniques which are applied to trees and shrubs in agroforestry systems. Some of these techniques are similar to those used in the management of trees in forestry plantations, but others are different. The most important management techniques with regard to the part of the tree which is above ground are:

(a) Pruning
(b) Lopping
(c) Pollarding
(d) Coppicing
(e) Thinning

(a) Pruning: Removal of branches from the lower part of the tree crown is known as pruning or side pruning. While pruning a tree, branches are always cut near the stem.

The objectives of pruning in agroforestry are threefold:

- Reduction of shade for crops near the tree
- Improving the quality of the trunk, mainly for timber and poles
- Early harvest of branch wood for fuel or other use.

Pruning should be done at least up to the height the adjacent crops if trees are growing in fields. Such pruning facilitates farming operations and reduces competition. The best time for pruning is towards the end of the dry season when the work will not interfere with growing crops and when the workload in other agriculture tasks is not so heavy.

(b) Lopping: Lopping is distinguished from pruning in that branches are not cut from the base. Also lopping is not always done starting from the lower part of the tree but can be more haphazard. If any selection of branches is made, the main criterion is often a good green leafy biomass since the lopping is usually done to obtain branches for fodder.

(c) Pollarding: If all the branches and the top part of a tree are cut off this is known as pollarding. There can be several objectives with pollarding:

- Early harvest of wood, fodder or other biomass
- Production of wood or fodder that is out of the reach of livestock, hence there is no need for protection from browsing
- Reduction of shade for crops near by
- Regeneration of the tree crown to promote growth of the trunk for timber or poles

(d) Coppicing: Systematic coppicing is applied as the management technique in alley cropping, and it may be an option for trees on soil-conservation structures. In such a situation coppicing may be done annually, but in other situations, e.g. regeneration of *Eucalyptus* for pole production, it may be much less frequent. In that case, an interval of 6-8 years may be more suitable.

(e) **Thinning:** Trees established by direct seeding or that have been planted with little space between them will soon start to compete with each other. A dense stand initially promotes straight growth and small branches, but later the trees must be thinned otherwise they will grow too slender and eventually not reach the desired size. Thinning is particularly important for trees grown in woodlots, but applies also to other situations where trees are growing close to each other. Thinning can, for example, be done by removing every second tree or two out of every three trees. Thinning is also a way of obtaining some early harvest.

Management stratergies for below ground interaction

Root management and root pruning

Trees growing in cropland can have their shallow roots cut 0.3-0.6 m from the trunk when they reach a height of 2-3 m. This is applicable to species which would otherwise compete with crops. This also serves to minimize competition.

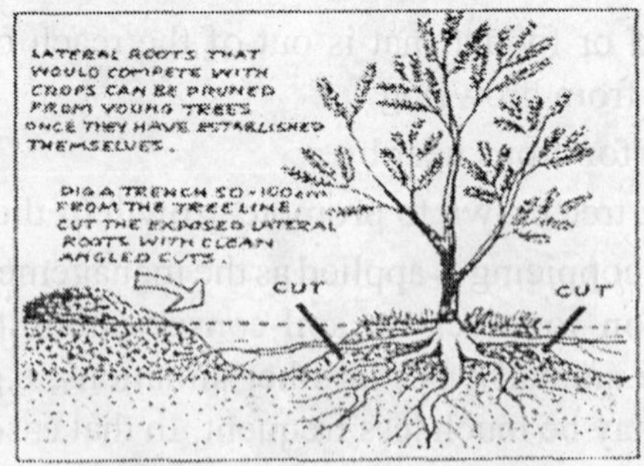

There is an inherent conflict in agroforestry between expected favorable effects of tree root systems, e.g. on soil fertility and nutrient cycling, and competition between tree and crop roots.

Root management attempts to optimize root functions and to stimulate facilitative and complementary interactions. It makes use of the plasticity of root systems to respond to environmental factors, including other root systems, with altered growth and physiology. Root management tools include species selection, spacing, nutrient distribution, and shoot pruning, among others.

Root distribution determines potential zones of root interactions in the soil, but are also a result of such interactions. Plants tend to avoid excessive root competition both at the root system level and at the single-root level by spatial segregation. As a consequence, associated plant species develop vertically stratified root systems under certain conditions, leading to complementarity in the use of soil resources.

Parameters of root competitiveness, such as root length density, mycorrhization and flexibility in response to water and nutrient patches in the soil, have to be considered for predicting the outcome of interspecific root interactions. The patterns of root activity around individual plants differ between species; knowing these may help to avoid excessive competition and unproductive nutrient losses in agroforestry systems through suitable spacing and fertilizer placement.

45. Write short notes on (2011)

(i) Tribal economy (ii) Chola Naickans (iii) Gujjars (iv) Gonds

(i) Tribal economy

The tribes and primitive societies have passed through several stages of economic development. There are four important stages which have been identified in the economic development of tribals viz.,

(a) Hunting and food gathering stage

(b) Pastoral and pre-agricultural stage

(c) Agricultural stage

(d) Post agricultural stage

The tribal economy type includes forest hunting types, hill cultivation type, plain agricultural type, simple artisan type, pastoral and cattle Herder's type, Folk Artist type, Agriculture and Non-agriculture types, Akilled white collar Job, traders type.

General characters of tribal economy

(i) Forest based economy

(ii) Division of labour

(iii) Simple technology

(iv) Absence of profit in economic dealings

(v) Tribal community acts as a cooperative unit

(vi) Gift and ceremonial exchange

(vii) Peirodical market

(viii) Economic backwardness

(ii) Chola naickans

- The Cholanaikkans are an ethnic group of India. They primarily inhabit the southern Kerala State, especially Silent Valley National Park, and are one of the last remaining hunter-gatherer tribes of the region.
- The Cholanaikkans speak the Cholanaikkan language, which belongs to the Dravidian family.
- The Chemmams are found widely scattered in the forest ranges. They subsist on food-gathering, hunting and minor forest produce collection.
- Their language is a mixture of Kannada, Tamil and Malayalam.
- They use dress and ornaments, household articles, tools and weapons in their daily life.
- They use rice as their staple food. They also use wild roots, tubers, seeds, fruits and meat.
- Their livelihood is totally dependent on the forest.
- The collection and selling of minor forest produce is the major source of income. There are still many customs, practices and taboos prevailing among the Cholanaikans.

(iii) Gujjars

- Gurjar or Gujjar are a pastoral agricultural ethnic group with populations in India (Distributed in Rajathan, Uttrakhand, Uttar Pradesh, Madhya Pradesh, Himachala Pradesh, Jammu and Kashmir etc.) and Pakistan and a small number in northeastern Afghanistan.Alternative spellings include Gurjara, Gurjar, Gojar and G jjar.
- Although they are able to speak the language of the country where they live, Gurjars have their own language, known as Gujari.
- They variously follow Hinduism, Islam, and Sikhism.The Gurjars are classified as Other Backward Class (OBC) in some states in India; however, Gurjars in Jammu and Kashmir and parts of Himachal Pradesh are categorized as a Scheduled Tribe.
- Hindu Gurjars were assimilated into various varnas in the medieval period.

(iv) Gonds

- The Gondi or Gond people are Adivasi people of central India, spread over the states of Madhya Pradesh, eastern Maharashtra, Chhattisgarh, Uttar Pradesh, Telangana, Andhra Pradesh, Bihar and Western Odisha.
- The Gondsis also known as the Raj Gond. The term was widely used in 1950s, but has now become almost obsolete, probably because of the political eclipse of the Gond Rajas.
- The Gondi language is closely related to the Telugu, belonging to the Dravidian family of languages. About half of Gonds speak Gondi languages while the rest speak Indo-Aryan languages including Hindi.
- According to the 1971 census, their population was 5.1 million. By the 1991 census this had increased to 9.3 millionand by 2001 census this was nearly 11 million.
- For the past few decades they have been at the receiving end of the NaxaliteMaoist insurgency in the central part of India.
- Gondi people are made to act as a militiary against Naxalites by the Government of Chhattisgarh through Salwa Judum.

46. Explain the various components of a hydrological model for an agroforestry system. (2011)

Hydrological modelling

Hydrological model is a simplified representation of real world system. The best model is the one which give results close to reality with the use of least parameters and model complexity. Modelsare mainly used for predicting system behaviour and understanding various hydrological processes. A model consists of various parameters that define the characteristics of the model. A runoff model can be defined as a set of equations that helps in the estimation of runoff as a function of various parameters used for describing watershed characteristics. The two important inputs required for all models are rainfall data and drainage area. Along with these, water shed characteristics like soil properties, vegetation cover, watershed topography, soil moisture content, characteristics of ground water aquifer are also considered. Hydrological models are now a day considered as animportant and necessary tool for water and environment resource management.

Hydrologic models are simplified, conceptual representations of a part of the hydrologic cycle. They are primarily used for hydrologic prediction and for understanding hydrologic processes. Two major types of hydrologic models can be distinguished:

Stochastic Models

These models are black box systems, based on data and using mathematical and statistical concept to link a certain input (for instance rainfall) to the model output (for instance runoff). Commonly used techniques are regression, transfer functions, neural networks and system identification. These models are known as stochastic hydrology models.

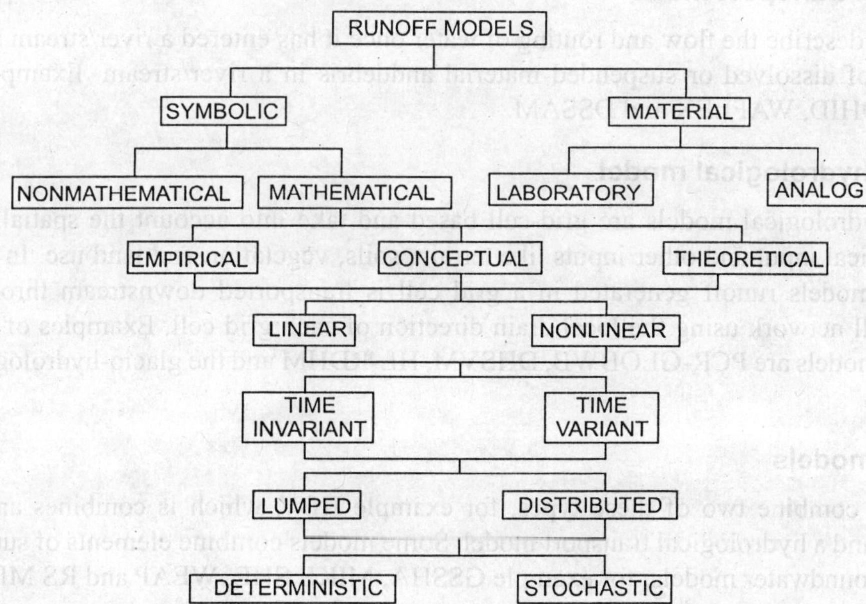

Various Components of hydrological model for Agroforestry

Process-Based Models

These models try to represent the physical processes observed in the real world. Typically, such models contain representations of surface runoff, subsurface flow, evapotranspiration, and channel flow, but they can be far more complicated. These models are known as deterministic hydrology models. Deterministic hydrology models can be subdivided into single-event models and continuous simulation models.

Groundwater models

Groundwater models are computer models of groundwater flow systems, and are used by hydrogeologists. Groundwater models are employed to simulate and predict aquifer conditions.

Surface water modelling

Runo- model (empirical)

These models use an empirical method to convert rainfall volume into runoff volume. An example is the Curve Number method. The runoff curve number (also called a curve number or simply CN) is an empirical parameter used in hydrology for predicting direct runoff or infiltration from rainfall excess. It can be used with a unit hydrograph to derive the runoff rate from the direct runoff by convolution.

Runo- model (reservoir)

These models describe rainfall-runoff relations using the concept of a (non) linear reservoir.

Hydrological transport model

These models describe the flow and routing of water once it has entered a river/stream system and the transport of dissolved or suspended material anddebris in a river/stream. Examples include MIKE 11, MOHID, WAFLEX and DSSAM.

Distributed hydrological model

Distributed hydrological models are grid-cell based and take into account the spatial variability of meteorological input and other inputs like terrain, soils, vegetation and land use. In distributed hydrological models runoff generated in a grid cell is transported downstream through a grid cell to grid cell network using the local drain direction of each grid cell. Examples of distributed hydrological models are PCR-GLOBWB, DHSVM, HL-RDHM and the glacio-hydrological SPHY model.

Composite models

Many models combine two of these types, for example HBV, which is combines an empirical runoff model and a hydrological transport model. Some models combine elements of surface water models and groundwater models, for example GSSHA, MIKE SHE, WEAP and RS MINERVE.

Impacts of Hydrological Modelling on Agroforestry

An overall assessment of whether AF systems use rainfall more effectively than a combination of woodlots and open fields can only be made when all components of the water balance are understood and compared. Furthermore, as it is not possible to make direct measurements of all possible tree and crop combinations, it is necessary to combine direct measurements with modeling.

Complementary and Competitive Interactions for Resources

AF is known to increase overall system productivity and enhance soil structure, nutrient recycling, and sustainability, while producing timber, firewood, fruit, fodder, and green manure in addition to annual food crops (Young, 1997). However, the potential benefits depend on complex spatial and temporal interactions between the biological, physical, hydrological, and climatic components of the system; these include the tree and crop species involved, soil type, depth and fertility, quantity and distribution of rainfall, solar radiation, and air temperature and atmospheric humidity. A major aspect affecting the overall outcome of integrating trees in farming systems is the increasing competitiveness of trees as they establish and grow larger. Tree roots extend to much greater lateral distances and depths than those of crops, thereby exploiting a much greater soil volume. There is therefore the potential for both complementarity and competition depending on the distribution and activity of tree and crop roots in the soil profile. Rooting depth is potentially unlimited in deep soils lacking impenetrable hardpans or stony horizons and the roots of some tree species may reach depths of 20–30 m in search of deepwater reserves.

Brief description of few models

SWAT model (Soil and Water Assessment Tool) Development of SWAT model is an ongoing process and it is the successor of "the Simulator for Water Resources in Rural Basins" model (SWRRB). SWAT model is a complex physically based model and was designed to test and forecast the water and sediment circulation and agriculture production with chemicals in ungauged basins. It is efficient in performing long term simulations. The model breaks the entire catchment in to sub catchments which are further divided in to hydrologic response units (HRU), land use, vegetation and soil characteristics. Daily rainfall data, maximum and minimum air temperature, solar radiation, relative air humidity and wind speed are the inputs used by this model and is able to describe water and sediment circulation, vegetation growth and nutrients circulation. Based on amount of precipitation and mean daily air temperature rate of snowfall can be determined. Penman Monteith, Priestly- Taylor and Hargreaves methods are used for the estimation of evapotranspiration. In order to obtain accurate forecasting of water, nutrient and sediment circulation, it is necessary to simulate hydrologic cycle which integrates overall water circulation in the catchment area and hence the model uses the following water balance equation in the catchment.

$$SW_t = SW_o + \sum_{i=1}^{t} (Rv - Qs - Wseepage - Et - Qgw)$$

Where SW_t is the humidity of soil

SW_O is base humidity

Rv is rainfall volume in mm water

Qs is the surface runoff

Wseepage is seepage of water from soil to underlying layers

ET is evapotranspiration

Qgw is ground water runoff and t is time in days)

$$SW_1 = 1\ SW_0$$

MIKE SHE model (Systeme Hydrologique European)

It is a physically based model and hence it requires extensive physical parameters and was developed in 1990. The model accounts various processes of hydrological cycle such as precipitation, evapotranspiration, interception, river flow, saturated ground water flow, unsaturated ground water flow etc. It can simulate surface and ground water movement, their interactions, sediment, nutrient and pesticide transport in the model area and various water quality problems and can be applied for large watersheds.

47. Management challenges in urban forestry are unique as compared to other social forestry programmes. Discuss. (2011)

The urban forestry management poses serious challenges compared to social forestry programme due to following reasons

(i) **Components of urban forestry:** The urban forestry included wide range of components from peoples, industries, educational institutions, hospitals, common property resources, parks, botanical garden and etc. And hence their management their complex system compared to social forestry.

(ii) **Species diversity:** Compared to social forestry, urban forestry has wide range of species diversity and it has been estimated that an average distribution of over 200 species in one particular urban area as seconds single or two species management in social forestry. This species diversity in urban forestry pose series challenges towards managing them for aesthetic and recreational benefits

(iii) **Crown size and architecture:** The species richness and diversity in urban forestry sector exhibits wide range of crown size and architecture. And hence their management in terms of pruning and thinning pose series problem in each and every species with wide range of canopy architecture is a major challenge towards canopy management.

(iv) **Root management:** The planted trees in urban area exhibited wide range of root pattern from deep to shallow. The species planted near the residential complexes pose series problem of uprooting even in a small wind or cyclone this is probably due to shallow rooting pattern as the tree roots are confirmed only top layer due to the availability of water from domestic sewages. Hence root management is a series challenges compared to social forestry programme.

(v) **Societal issues:** In case of social forestry, the forest has been establish for the purpose to benefit the people but in case of urban forestry the ultimate benefit is to the society in the form of recreation and aesthetic value. The diverse societal section and their management will be major challenges in an urban forestry sector.

(vi) Biotic management: The planted trees and plants in urban areas need significant protection towards this establishment and management. Since the urban trees is planted in open areas like roads, highways, residential complexes, hospitals, common property resources etc. their protection from domestic animals and some cases in human interference, does challenges arise in social forestry such in urban forestry is very unique.

Considering all these factors into account, the management of urban forestry is very very challenging task compared to social forestry program which is very easy to implement and management.

48. Discuss the characteristics which are shared by the diverse tribal groups all over India. (2011)

(i) The tribal groups live in a definite territory which is well defined and occupy a definte territorial identity and serve as a homeland for tribal community.

Example: The area occupied by Gond tibes is referred as Gondwana.

(ii) Tribals also speak a common language or a dialect but most tribal languages do not have a script. The following are commong languages shared by Indian tribes.

S.No.	Tribal Name	Language
1.	Gond tribe	Gondi language
2.	Korku	Korku
3.	Gavit	Bhilli
4.	Kholan	Kolami
5.	Madia	Gondi
6.	Warli	Warli

(iii) **Blood relationship:** The tribal communities are believed to be originated from a common ancesters which may be real or mythical. They are close blood relatives because of tge endogamy that exists among tribal community.

(iv) **Endogamy:** It is a common characteristic feature of tribals in India. Endogamy means the members of trib marry with in a group which may be caste, class, rule, ethinic group, religious group or even a village. The rule of endogamy stipulates that the partner must be from the same group and the marriage outside the group is strictly prohibited.

(v) **Common culture:** It is one of the characteristic feature by tribe shared among them. The cultural similarity is an important factor to regulate the tribal organization. The tribal community is rich in their culture, folk tales, folk song and folk stories.

(vi) **Common religion:** It is believed that the tribal have a common religious and / or inspired by a mantras which is common to all tribal groups.

(vii) **Common name:** Commonis the characteristic feature of tribal community and has greater significance from sociological point of view which binds the members of tribe with unite felling.

(viii) **Common occupation:** Most tribes have started their livelihood from forestry and agricultural activities which are common across the country. The occupational activities include hill cultivation, cattle rearing, hunting, agriculture, folk arts, artisans, handicrafts and collection of minor forest products.

(ix) **Political and Social organization:** Every tribal community is known to be strong social and political event unit. These units are directed to maintain peace, harmony, unity and integrity in a tribal population. This unity is weted with an individual called Mukhia who is responsible for solving any dispute in a tribal society. Mukhiya has a control ove the decision and all members have fear, faith and devotion towards mukhiya.

(x) **Tapoos, totum and Myths:** Tribes believe in tapoos, which are certain forbidden things. They also believe in totum which are certain objects having mythical relationship with the members of the tribes. Tribals also believe in certain myths which may be religious or magical origin.

(xi) **Others:** The tribals have also show the common charactertistics of primitive and simple way of life, shy of stronger, love for drink and dance, nomadic life, murder and crime, illetracy ignorance, etc.

49. Important Indian Tribes. (2011)

The tribal population of India is found all over the country. There are over a hundred tribes and sub-tribes in our country. As the country is marching towards self-reliance and poverty alleviation, it becomes essential for us to have in-depth knowledge of the various tribes; so that they can be merged in the national mainstream. The process of tribal development is very complex and is a function of time.

Major Tribal Groups in India

(i) Gonds	(ii) Santhals	(iii) Bhils
(iv) Kolis	(v) Manipuris	(vi) Nagas
(viii) Garos	(viii) Khasis	(ix) Mikirs
(x) Jantirs	(xi) Ladakhis	(xiii) Other Himalayan Tribes

(i) Gonds: The Gonds of Madhya Pradesh are one of the largest tribes in India. They have a rich social, cultural and religious life. Found widely distributed in remote areas of Madhya Pradesh and parts of Bihar, Orissa and Rajasthan. Agriculture forms the main occupation of the people. Principal crops are rice, maize, jowar, wheat, millets etc., besides; they also raise vegetables and fruits. Certain groups rear sheep, goats, cows and buffaloes and earn their living by selling their products. They also indulge in fishing, honey collection-and collection of minor forest produce.

The Gonds have a rich cultural life. They play a wide variety of musical instruments such as jhuia, tabla, drums etc. Group dancing and singing is also very common. Main festivals are Holi and Dushera. During holi, brides are selected by the youth. In certain areas, the Gonds have youth dormitories which are separate for boys and girls. Here they are taught finishing, hunting, knowledge of agriculture and other economic activities. They are also imparted sex education during youth.

(ii) Santhals: Santhals inhabit parts of Bihar, Orissa, Madhya Pradesh Bengal and Assam. Originally they belong to the santhal paragons of Bihar and later migrated to other areas. They are dark complexioned strong in physical built and healthy in appearance. Scantily dressed lion cloth for males and dhoti for females. They are a simple and peace loving people. In the past, they were a hunting tribe and bow-arrow was sacramental to them. Hunting is still given such a level of priority that when young santhal boy attains a certain age, he is ceremoniously initiated for hunting .The placenta of a newly born child is also detached by means of an arrow and the same is placed at the side of the baby for five days with a belief that child will be protected from evil spirits.

(iii) Bhils: Bhils inhabit the western hills of India mainly the Aravalli and Salpura ranges. Their population is found in the states of Rajasthan, Madhya Pradesh, Maharashtra and Gujarat. The district of Bhilwara has been so named as it contains a large population of Bhills.

Each tribe has hereditary head men, who till the turn of this century, had the power of life and death. He is still treated with a considerable degree of respect. Usually, the headman has a number of persons to aid him, who together form the village council. Matters of social and religious interest are settled by this body. The tribal's residing in the hills has been described as active and hardy with high cheek bones, wide nostrils and almost Negroid features. Those of the plains are of a good built, tall stature and handsome features, very often depicting an Aryan origin.

(iv) Kolis: Kolis inhabit the western part of India. They resemble the bhills so far as ethnology is concerned. However, all along the Western hills, the characters, institutions, religion and modes of life of the Kolis differ in a remarkable way from that of other tribes. The Kolis and Bhils occupy a region which has been exposed to the onslaught of invaders. At present, the Kolis are found in parts of Rajasthan, Gujarat, Maharashtra and Madhya Pradesh.

(v) Manipuris: Manipuris is a mountainous state comprising of about 92% of the total land area in the form of hills. On basis of physiographic, the state may be divided into two broad divisions-hill tract and valley tract. These two divisions are inhabited by the Tribal's and Maties respectively. For the purpose of this work, both communities have been discussed under the broad heading - Manipuris.Maties are Hindus and dominate all sections of the state. They comprise of seven different tribes / clans.

- Khuman tribe
- Lamang tribe
- Moirang tribe
- Meitie tribe
- Ninghouja tribe
- Khaba-Naganea tribe
- Chenglei tribe

(vi) Nagas: Nagaland is a separate hill state in northeastern India. It is largely a hilly tract with plains around Dimapur valley bed and along the foot hills. Except the Kachcharis who live around

Dimapur, all the other residents of Nagaland are known as Nagas. As a matter of fact, Nagas area group of tribes comprised mainly of the Anganis, Chakhasangs. Kukis, Rerigmas, Sejflas, Aos, Lothas, Changs, Sanglams, Phorus, Konvaks, and Zithaaas etc.

(vii) Garos: Garos live in the Garo hill tract of Meghalaya state. Legend prevalent in the garo people is that originally their ancestors inhabited a province in Tibet named Toura, Excavations show that civilization existed in this area during the Neolithic times. Goswammi (1973) writes,these implements are typologically similar to the stone implements of Neolithic age of Burma, Thailand, Java and some other southeast Asian countries. In these countries the Neolithic cultures survived up to 2000 year B.C. From linguistic evidence it can be said that the garos are the people of other Tibeto-Burman speaking groups of East India, such as the Naga, Mikir, Boro, the Dimasa etc., and as such they can be supposed to have a common original home in the remote "*prehistoric past*'.

(viii) Mikirs: Mikirs, locally called *orlengs* inhabit the Mikir hills of Northeastern India. They are well built, tall and usually of a light, yellowish brown complexion. The hair is tied in a knot behind the head. Mikir males wear dhoti, sleeveless jacket and headgear. Females adorn stripped petticoats and an upper garment, which tightly laces the body. Main occupation is shifting cultivation. There are numbers of allied sources of employment e.g. weaving, carving, fanning and labourers

Rice is the staple food and fond of meat and fish but not beef. Not much fond of dances and music. A sole dance is performed at funerals. Due to lack of interest in music the main musical instrument comprise of only flute, drum, one stringed fiddle and fife. Little dancing takes place during the harvesting season.

(ix) Jantias: Jantias inhabit the Jantia hills of Meghalaya. They are fair complexioned and short satured. The jantias society, more so, its backward segment, is still in the clutches of priests and money lenders. In some areas, priests own about a fourth of the land of cultivators. For cultivators any part of the land, a peasant has to pay tax to the priests. When the crop is ready for harvesting, it cannot be reaped, unless inspected and permitted by priests.

(x) Ladakhis and lahulis: Ladakh and Lahul-Spiti are the trans-Himalayan area of Jammu & Kashmir (J.K) and Himachal Pradesh (H.P.) states respectively. As a matter of fact, these are the only two extensive areas in the country which lie across the main Himalayan range and the conditions prevailing in Tibet. Ladakhis-Lahulis are short in stature, with high cheek bones, from which the face rapidly narrows downwards and the chin is small and retreating. One most characteristic feature is that the out comers of the eyes are drawn out and the upper eyelids appear to be overhung by a fold of the skin above them facial hair is lacking and the face tends to become highly wrinkled, as soon as old age begins to set in. This may probably be due to the extreme cold conditions under which they live. Eyes are usually brown coloured the nose depressed at the bridge. Mouth is large and expressionless, lips projecting but not thick. Hair may be cut close to the front and at the sides of the head; it may be collected behind to form a small pig tail.

50. Compare nutrient cycling in a natural forest, an agroforestry system and an agricultural field. Discuss how it helps to sustain soil fertility. (2011)

Nutrient cycling in natural forest

A nutrient cycle in natural forest is the movement and exchange of organic and inorganic matter back into the production of living matter. The process is regulated by food web pathways that decompose matter into mineral nutrients. Nutrient cycles occur within ecosystems. Ecosystems are interconnected systems where matter and energy flows and is exchanged as organisms feed, digest, and migrate about. Minerals and nutrients accumulate in varied densities and uneven configurations across the planet. Ecosystems recycle locally, converting mineral nutrients into the production of biomass, and on a larger scale they participate in a global system of inputs and outputs where matter is exchanged and transported through a larger system of biogeochemical cycles.

Nutrient cycling in agroforestry system

In agroforestry systems, trees can provide nitrogen inputs by 2 processes biological nitrogen fixation in the case of nitrogen-fixing tree species(NFTS),and deep nutrient capture.Although the magnitude of biological nitrogen fixation is methodologically difficult to quantify, the overall recorded rates of fixation for trees both leguminous as well as non-leguminous range from 20 to 300 kg N/ ha / yr. The ad vocation of inclusion of NFTS in agroforestry systems has led to the belief that nitrogen fixed in the root nodules may be used by the companion crop though direct evidences to support it are still lacking. NFTS can also supply considerable nitrogen tocrops through litterfail.On the other hand, there is ample evidence that trees such as cherry and mandarin, where nitrogen-fixation does not occur, accumulate as much or more N in their leaves as doNFTS,presumably because of their greater root volume and ability to capture nutrients from sub surfacesoil. However, these non NFTS are only cycling nutrients and not adding them to the system.

Deep nutrient capture the second process through which trees can provide nitrogen inputs is the uptake of nutrients by tree roots at depths where crop roots are not present. This is an additional nutrient inputin agroforestry systemsbecause such nutrients are leached and lost for crop use, butthey become an input on being transferred to the soil through tree litter decomposition.

Nutrient cycling in agriculture

In agriculture the land area is more open which indicated that the turn over within the system is relatively low compared to forestry and agroforestry. Simultaneously the loses the particularly the soil and nutrient loss are very heavy compared to other two systems. This indicate that the agriculture system requires more inputs for sustaining the fertility and the sources of nutrient addition in agriculture system is very low.

51. What are the fundamental bases of classification of agroforestry systems? (2013)

Agroforestry systems classified based on the following four criteria.

A. Structural Basis

B. Functional basis

C. Socio economic Basis

D. Ecological basis

A. Structural Basis

(a) Nature of Components

(a) Arrangements of Components

(a) Nature of Components

(i) Agricultural systems

(ii) Silvopastoral systems

(iii) Agrosilvopastoral systems

(iv) Other systems

(i) Agricultural systems

- Improved fallow species in shifting cultivation
- The taungia system
- Multispecies tree gardens
- Alley cropping
- Multipurpose trees and shrubs on farmlands
- Crop combinations with plantation crops
- Agroforestry fuel wood plantations
- Shelter belts
- Wind breaks
- Soil conservation hedges

(ii) Silvopastoral systems

- Protein bank
- Live fence of fodder trees and hedges
- Trees and shrubs on pastures

(iii) Agrosivopastoral systems

- Home gardens
- woody perennials for browse, mulch, green manure, soil conservation

(iv) Other systems

- Apiculture with trees
- Aquaforestry
- Multipurpose wood lots

(b) Arrangement of components

(i) Spatial arrangement

(ii) Temporal arrangement

B. Functional Basis

(a) Productive functions

(b) Protective functions

(a) Productive functions
- Food
- Fodder
- Fuel wood
- Cloths
- Shelter
- NTFPs

(b) Protective functions
- Wind breaks
- Shelterbelts
- Soil conservation
- Soil improvement
- Shade

C. Socio Economic Classification

(a) Commercial systems
(b) Intermediate systems
(c) Subsistence systems

D. Ecological Classification

(a) Humid / sub humid
(b) Semiarid / arid
(c) Highlands

52. Discuss why land-use system is often more appreciated in agroforestry than in pure agriculture (2015)

Agroforestry is often more appreciated for the following reasons

(i) E3cient land use system: Compared to pure agriculture agroforestry system play a vital role in efficient utilization of land by integration of tree component, crop component and animal component. This ensures judicial utilization of land for generation of wood, food, milk and meat thereby enhances the over all productivity and profitability. This is a reason agroforestry is more appreciated than pure agriculture.

(ii) Improved soil health: Agriculture degrades the soil health and it is estimated that over 50% of the land use system in the country are degraded due to faulty agricultural practices. But agroforestry system helps to improve the soil by addition of organic matter, reduction of soil loss through erosion and augmentation of micro climate in the locality. This is particularly true in case of marginal and sub marginal lands wherein the increased production of organic matter help to maintain the soil fertility by augmenting the soil physical, chemical and biological properties. Root development absorb many nutrient from deep soil and help to tap the nutrient, litter fall and soil physical, chemical properties.

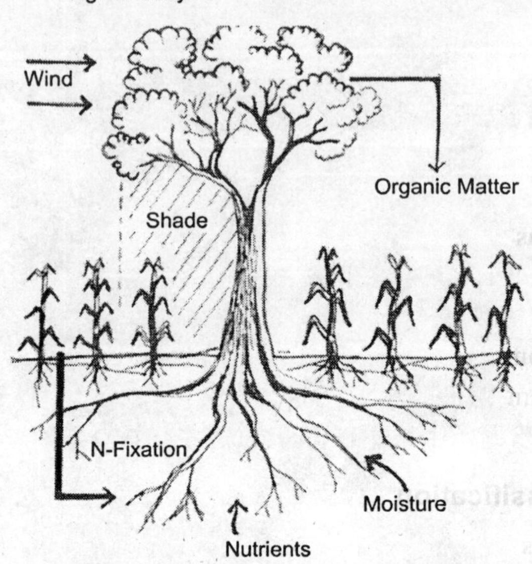

(iii) Microclimate amelioration: The agroforestry based landuse system is more appreciable than pure agriculture for the reason that the tree based agroforestry component help to moderate the micro climate of the locality. This results in reduction in temperature of the locality and the soil, increase in humidity which results in increased precipitation. Hence agroforestry as a whole reduces the evaporation from the soil and water requirement thereby help to sustain the productivity. It is very well believed that the agroforestry model increases the local precipitation and helps in agriculture. Due to this micro climate moderation agroforestry is more appropriated than pure agriculture.

(iv) Suppression of pest and diseases: The pure agriculture system is housed with wide range of pest and diseases and in extreme cases there are chance of pest outbreak. But the agroforestry systems due to the nature of its crop diversity helpts to avoid or reduces the pest and disease incidences and protect the agriculture crop from total damage.

(v) Income and employment generation: Compared to pure agriculture farming which are seasonal and monsoonic dependents the agroforestry models are sustainable land use system which generates continuous income and employment generation activities due to varied sub system integrated within the model. This ensures continuous employment and income generation activities and that is the reason agroforestry is more appreciated than pure agriculture.

(vi) Climate change mitigation: Compared to pure agriculture which contributes to climate change, agroforestry is recognized as a potential land use system by IPCC which indicated that the agroforestry has an excellent potential of mitigating the climate change effects and adapting to the changing environment. This helps in reducing the carbon-di-oxide level and other pollution hazards thereby protect the agriculture, human population and ecosystem. Hence agroforestry is more appreciated than pure agriculture.

53. Adaptation of Agroforestry practices to increase the cattle (2015)

Agroforestry is the sustainable land use system and judicially incorporate tree and crop components along with animal components. In order to increase the productivity and profitability per unit area of the land, incorporation of cattle is mandatory component in all agroforestry practices and help to increase the cattle population.

(a) **Dryland Agroforestry Model:** Almost 25% of the people inhabiting dryland areas are adopted towards rearing cattle. This is because the people owning drylands are traditionally practicing agroforestry. This dryland agroforestry has incorporated fodder trees like Albizzia, Neem, Subabul, Hardwickkia, Glyricidia which are acting as excellent fodder resources to the cattle. The dryland regions experience extreme temperature variation and hence availability of annual crop based fodder is the major issues. Under such circumstances, the trees can cut as excellent fodder resources especially during lean season and help to increase the cattle population.

(b) **Silvipastoral Model:** The silvipastoral systems involves growing of ideal combination of grasses, legumes and trees for producing nutritious fodder and forage, fuel wood and timber and optimizing land productivity. This model is other called as Animal Agroforestry which includes livestock as the major component. This silvipastoral model provides in-situ based two tier fodder system one from the annuals and other from perennials.

This model increases the carrying capacity of the land and hence increases the cattle population. This model ensures sustainable fodder availablit because during rainy season the underground grasses act as a fodder resource and during dry season the trees act as a fodder resource. Hence this agroforestry adaptation practices increases the cattle population.

(c) **Protein Banks:** It is a form of silvipastoral practices including two plant species spatially separated along with the animal component. In protein banks, the trees are planted in and around the form and the foliage is cut and carried over to a place where livestock can consume. Alternatively, the cattles are allowed to enter into these systems at regular intervals and act as a complement to the feed stock. This model ensures quality fodder due to rich protein source simultaneously providing the huge volume of leaf fodder.

Many leguminous trees like Subabul, Prosopis, Acacia, Glyricidia and Non leguminous crops like Melia, Thespesia, Kadam, Mulberry etc., are planted as protein bank as maintained as hedges which ensured sustained availability of quality tree fodder. This model attracts more towards animal husbandry and increases cattle populations.

(d) **Live fence of fodder trees and hedges:** Various fodder trees are planted as live fences to protect the field from animals and other biotic influences. Trees with thorns and high fodder values are raised as live fences in dry regions. These lie fences provide excellent feed stock to the animals and hence increases the cattle population.

(e) **Trees and shrubs on pasture:** In this model trees and shrubs are planted in a systematic pattern to supplement fodder production. This enhances availability of quality fodder particularly during lean season and increases the cattle populations.

Hence various agroforestry practices particularly dryland agroforestry and silvipastoral models incorporated trees and grasses as fodder crop which provide the quality fodder to the cattle populations. Due to the availability of quality fodder as a sustainable basis the practice of animal husbandry is gaining significant attraction which lead to increases in cattle population.

54. Define Agroforestry. Describe an agroforestry model suitable to rehabilitate degraded hills. Discuss the gains of agroforestry system (2009)

Agroforestry

Agroforestry is a land use system that integrates trees, crops and animals in a way that is scientifically sound, ecologically desirable, practically feasible and socially acceptable to farmers

Rehabilitation of degraded hills

For establishment of agroforestry model is degraded hill it is very essential to identify the causes of degradation and factors of locality in order to intiate the rehabilitation process.

(i) **The causes of degradation in hillocks:** The possible causes of degradation in a hillock are deforestation, overgrazing, soil erosion, wind erosion, loss of top soil and the associated natural calamities.

(ii) **Rehabilitation measures:** The following rehabilitation measures are very essential before intiating the agroforestry process.

(i) **Land preparation:** It involves ploughing to influence intake of water, obstruction to surface flow of water and soil, deep ploughing to reduce erosion due to cloddy soil structure.

(ii) **Contour bunding:** Contour bunding is a traditional process and found effective physical measure for soil and water conservation. It consists of narrow based trapezoidal bunds on contour to impound run of water behind them such that it can gradually in filtrate in to the soil for use by the plants. This method of contour bunding is recommended for areas with low rainfall of less than 600 mm and the slope up to 6%.

(iii) **Graded bunding**: This is another method of soil and water conservation practiced in areas of more than 600 mm rainfall. This method will help to remove the excess water and avoid water stagnation.

(iv) **Bench terracing**: It is a combination of ridge and channel belt across the slope. It consists of transforming relatively steep land into a serious of level strips or platforms across the slope to reduce the slope length and the associated erosion. The field is made into serious of benches by excavating soil from upper part and filling them in lower part of the terrace. This method is practiced for the lands with the slope ranged between 16 and 33%.

(v) **Contour trenching:** This is one of the popular methods of soil and water conservation wherein the trenches are made along the contour. This method is practiced in hill slopes. The trenches made will break the slope length; reduce the velocity of runoff leading to retarded scouring and carrying capacity. This method is practiced with the slopes exceeding 20%. The trench size is usually 16 x 45 cm and the distance between the trenches varied between 10 x 30 m.

B. Biological measures: The biological measures include agricultural and silvicultural practices viz., contour cultivation, strip cropping, mulching, cover cropping, silvipasture, agroforestry, wind break, shelter belts etc.

(i) **Choice of crop:** Generally, pulse crop and leguminous tree crops are found effective towards soil and water conservation. These crops are considered erosion resistant crop and provide better cover and protection to soil by way of minimizing the impact of rain drop and obstructing to runoff.

(ii) **Contour cultivation**: The contour made in the slopes is cultivated by sowing the crop across the slope. The crops like cereals, pulses and grasses can be cultivated in the contour which will ensure utilization of the land coupled with preventing soil erosion.

(iii) **Strip cropping:** The erosion resistant crop and erosion permeable crops are grown in alternate strips on contour with the objective of breaking long strips to prevent soil loss and run off. The resistant crops reduce the velocity of water movement and filter the sediment from the run off to deposit them in the field.

(iv) **Cover cropping:** Any crop planted to control erosion is termed as cover crop. The pulses with drought resistant are generally grown for this purpose. In some arid areas drought tolerant grasses are grown.

(v) **Crop rotation:** Crop rotation means alternate growing of crops one after the other. The Legumes followed by grasses are preferred crop rotation which will improve the soil besides controlling the erosion.

(vi) **Mulching:** Mulching of crop residues on or near the soil surface reduce evaporation besides conserving the soil and water. The residues on surface observe wind force and reduce the wind erosion.

(vii) **Forestry practices:** The forestry component particularly leguminous trees like Acacia, Leucaena, Dalbergia etc. play a major role in arresting soil erosion and conserving moisture through interception and through fall. The desirable characters of trees are drought resistance, leguminous, deep root system and will have multipurpose utility.

(viii) **Windbreaks:** Windbreak is a single row plantation and act as a barrier for protection from winds. The effectiveness of wind break depends on its density, height and length. The spacing of wind breaks depends on the density and angle of wind break to the direction of hazardous wind. Eg. *Casuarina*, *Eucalyptus* etc. The length of the wind break is about 24 times the height of the tree. The distance between the wind break is 10 times the height of the tree.

(ix) **Shelter belts:** Shelter belt is a longer barrier than a wind break consisting of combination of shrubs and trees intended for protecting field crops and conserving soil. The shelter belt includes the following design.

(a) **Central rows:** *Acacia Arabica, Albizzia lebbek, Azadirachta indica, Dalbergia sissoo, Eucalyptus rostrata, Prosopis spicigera, Tamarindus indica.*

(b) **Flank rows:** *Acacia Senegal, Anacardium occidentale, Casuarina equisetifolia, Cassia siamea, Prosopis juliflora.*

(c) **Outer rows:** *Acacia jacquemontii, Agave spp., Cassia auriculata, Gliricidia maculata.*

Gains of agroforestry systems

Appropriate agroforestry systems with a combination of woody components and herbaceous crops essentially contribute to both productivity and sustainability of farming system on marginal and sub marginal lands by increased production of organic matter and maintaining soil fertility, reduced erosion and create micro climate favorable for associated crops and animals. The agroforestry systems can extend the following gains.

(i) Agroforestry systems play an important role in protecting the land from the process of erosion and prevent the ultimate deterioration of land, reducing the soil loss and maintaining the soil productivity on a sustainable basis.

(ii) Agroforestry systems keeps the land almost continuously covered thus improves the infiltration rate of soil , reduce soil erosion and build up the fertility status and reduces the runoff and conserves water.

(iii) The agroforestry systems play a vital role in increasing the soil fertility due to accumulation of high organic matter. Growing of nitrogen fixing trees will help to improve the soil nitrogen status.

(iv) The agroforestry system helps to improves soil physical properties and increases the porosity and activates the soil faunal population.

(v) The agroforestry systems play a crucial role in microclimate amelioration through temperature reduction, increasing humidity and ultimately increased precipitation.

(vi) The agroforestry systems decreases the incidence of pest and disease due to diversity in plant species and pest outbreak

(vii) The agroforestry system is accepted as the potential land use system and generates steady income and employment generation to the subsistence farmers.

(viii) The agroforestry practices help to provide adequate employment and income generation activity contributing towards socio economic development.

(ix) The agroforestry land use system has a profound utility in mitigating and adapting to growing climate change related issues and help to ameliorate the environmental issues.

55. How can NGOs, schools, banks and industry help to carry out an a-orestation programme? (2010)

A-orestation programme

Afforestation is the renewal of forest is an area where it has not seen the vegetation during the last 50 years. In these areas, afforestation is possible only through participatory approach involving the local communities, village level institution, educational institutes and the wood based industries. Hence it is possible to intiate and establish afforestation programme using the above indicated stakeholders.

(i) **A3orestation through Non Governmental Organizations:** Non Governmental Organizations (NGOs) are the village level institutions which can play a vital role in afforestation activities. Most of the NGO's in the country have mandated tree planting and afforestation in their agenda activities. Hence, the local NGO's can be involved in afforestation projects particularly to identify factors of locality coupled with other issued towards successful establishment of forests. The NGO's can develop nurseries and mass multiply the tree seedling required for afforestation activities. The agencies involved afforestation programme can involve NGO's towards identifying factors involving local community, developing nursery, pitting and plantation development activities. On a long NGO's can act as a local institution even to protect the afforested area. Hence NGO's can help in all afforestation programme.

(ii) **Schools:** The school particularly school children can plya a vital role in all afforestation activities. Each and every school can be linked with one afforestation sites and they can implement a process called "one student one tree" by judicially integrating students in the afforestation program. This process may start from sixth standard onwards. Each and every sixth standard student may be assigned with one tree and they may be asked to develop tree upto their tenth standard. Once the student can completed tenth standard, trees can be evaluated and the students who have grown successful tree are to be evaluated. This kind of encouragement helps the students to actively involving the afforestation activities.

(iii) **Banks:** Banks are financial institution which involves both public and private sector undertakings and they can play a crucial role in afforestation projects. Now the National Bank for Agriculture and Rural Development (NABARD) has been identified as National Implementing Entity for climate change projects which included both afforestation and reforestation projects. Hence NABARD can direct all the banking sector undertakings to fund afforestation. Projects through GCF or through any other mechanism towards successful establishment afforestation projects.

The afforestation projects have been identified as one of potential land use system to combat climate change associated risk and trees are potential mitigators and adopters. Hence banking sector can play a vital role in all afforestation projects.

Industries: Industires in general and wood based industries in particular can help in completing the afforestation projects. The wood based industries can take afforestation programme on a commericial scale which can act as a sustainable source of raw material besides protecting the soil, environment and the other natural resources in the habitat. The non wood based industries can help in afforestation programme htorugh their (CSB) projects. All the CSR projects should involved 2% of the total revenue and the same can be used in afforestation projects. Hence industries can support the aforstation project financially and also play a vital rolc in implementing the afforestation projects.

56. Name di-erent tribes of India state wise and their specific characteristics. How can we make use of their traditional knowledge in forest conservation (flora fauna) (2014)

- The share of the Scheduled Tribe population in urban areas is a meager 2.4%.Madhya Pradesh, Maharastra, Orissa, Gujarat, Rajasthan, Jharkhand, Chhattisgarh, Andhra Pradesh, West Bengal, and Karnataka are the states having larger number of scheduled tribes. These states account for 83.2% of the total Scheduled Tribe population of the country.
- Assam, Meghalaya, Nagaland, Jammu & Kashmir, Tripura, Mizoram, Bihar, Manipur, Arunachal Pradesh, and Tamil Nadu, account for another 15.3% of the total scheduled tribe population. The share of the remaining states / union teretories is negligible.
- The scheduled Tribes in India form the largest proportion of the total population in Lakshadweep and Mizoram followed by Nagaland and Meghalaya.
- Madhya Pradesh has the largest number of scheduled tribes followed by Bihar. Bastar district of Madhya Pradesh consists of largest number of scheduled tribes.
- There are no Scheduled Tribes in Punjab, Delhi, Chandigarh, Pondicherry, Haryana.

Statewise Distribution of Tribes in India

States	Tribes
Andhra Pradesh	Andh, Sadhu Andh, Bhagata, Bhil, Chenchus (Chenchawar), Gadabas, Gond, Goundu, Jatapus, Kammara, Kattunayakan, Kolawar, Kolam, Konda, Manna Dhora, Pardhan, Rona, Savaras, Dabba Yerukula, Nakkala, Dhulia, Thoti, Sugalis.
Arunachal Pradesh	Apatanis, Abor, Dafla, Galong, Momba, Sherdukpen, Singpho.
Assam	Chakma, Chutiya, Dimasa, Hajong, Garos, Khasis, Gangte.
Bihar	Asur, Baiga, Birhor, Birjia, Chero, Gond, Parhaiya, Santhals, Savar.
Chhattisgarh	Agariya, Bhaina, Bhattra, Biar, Khond, Mawasi, Nagasia
Goa	Dhodia, Dubia, Naikda, Siddi,Varli.
Gujarat	Barda, Bamcha, Bhil, Charan, Dhodia, Gamta, Paradhi, Patelia.
Himachal Pradesh	Gaddis, Gujjars, Khas, Lamba, Lahaulas, Pangwala, Swangla.
Jammu and Kashmir	Bakarwal, Balti, Beda, Gaddi, Garra, Mon, Purigpa, Sippi.
Jharkhand	Birhors, Bhumij, Gonds, Kharia, Mundas, Santhals, Savar.
Karnataka	Adiyan, Barda, Gond, Bhil, Iruliga, Koraga, Patelia, Yerava
Kerala	Adiyan, Arandan, Eravallan, Kurumbas, Malai arayan, Moplahs, Uralis
Madhya Pradesh	Baigas, Bhils, Bharia, Birhors, Gonds,Katkari, kharia, Khond, Kol, Murias.
Maharashtra	Bhaina, Bhunjia, Dhodia, Katkari, Khond, Rathawa, Warlis.
Manipur	Aimol, Angami, Chiru, Kuki, Maram, Monsang, Paite, Purum, Thadou
Meghalaya	Chakma, Garos, Hajong, Jaintias Khasis, Lakher, Pawai, Raba

Mizoram	Chakma, Dimasa, Khasi, Kuki, Lakher, Pawai, Raba, Synteng
Nagaland	Angami, Garo, Kachari, Kuki, Mikir, Nagas, Sema
Odisha	Gadaba, Ghara, Kharia, Khond, Matya, Oraons, Rajuar, Santhals
Rajasthan	Bhils, Damaria, Dhanka, Meenas(Minas), Patelia, Sahariya
Sikkim	Bhutia, Khas, Lepchas
Tamil Nadu	Adiyan, Aranadan, Eravallan, Irular, Kadar, Kanikar, Kotas, Todas
Telangana	Chenchus.
Tripura	Bhil, Bhutia, Chaimal, Chakma, Halam, Khasia, Lushai, Mizel, Namte
Uttarakhand	Bhotias, Buksa, Jannsari, Khas, Raji, Tharu
Uttar Pradesh	Bhotia, Buksa, Jaunsari, Kol, Raji, Tharu
West Bengal	Asur, Khond, Hajong, Ho, Parhaiya, Rabha, Santhals, Savar.
Andaman and Nicobar	Oraons, Onges, Sentinelese, Shompens
Little Andaman	Jarawa
North-East	Abhors, Chang, Galaong, Mishimi, Singpho, Wancho.

57. Why are agroforestry systems becoming popular in Himalayan tarai regions, western plains and southern parts of India? (2014)

Agroforestry systems

Agroforestry systems are very popular in Himalayan tarai region, western plains and southern parts of India due to their acceptance, adaptability and sustainability towards meeting the demands of wood and wood products, besides providing sustainably livelihood options.

The specific reasons for region wise are furnished

(a) **Himalayan Tarai Region:** Agroforestry systems are accepted as one of the sustainable land use management systems for provisioning functions adopted by mountain communities. The Agroforestry systems conserve the soil by improving the fertility levels, erosion and providing quality water for local consumption, besides providing fodder, fuel and timber for domestic utility. The integration of cash crop particularly the cardamom based Agroforestry systems gives good economical returns and promotes poverty alleviation and contributes to the human well being and good health. E.g: cardamom based Agroforestry systems with trees.

(b) **Western Plains:** The western plains particularly in the states of Haryana and Punjab, the Agroforestry systems are very popular due to the introduction of popular based Agroforestry systems. Popular is one of the fast growing species amenable for varied Agroforestry systems and compatible to grow along with wheat, rice, mustard. It is a fast growing and short rotation tree crop amenable for various industrial utility. The tree is harvested within 6 years which has given good economic returns in addition to agriculture.

The high yielding short rotation with assured marketing from plywood and poplar based Agroforestry systems is very popular which has provided increased revenue on a sustainable basis. The government initiatives on liberalization felling and transport of farm grown trees also helped in popularization of Agroforestry systems in western India.

(c) **Southern India:** The Agroforestry systems is very popular in southern India due to the promotion of Industrial Agroforestry in southern India and mushrooming of Industrial Agroforestry particularly timber, pulp and paper, plywood, Biomass, packing cases, Pencil, Value addition industries have demanded cultivation of more trees on farm lands. This has promoted timber based agroforestry, pulpwood based agroforestry, plywood based agroforestry and etc., and augmented the productivity and profitability. The establishment of contract farming models by some of the wood based industries have made Agroforestry more popular in Southern India, the species like Teak, Casuarinas, Eucalyptus, Bamboos, Melia, Ailanthus and etc., have found profound utility in Agroforestry systems. The availability of High Yielding Short Rotation Varieties, Precision Silviculture Technology, Profitability Agroforestry Models, Assures Markets and Trade etc., have made many people attracted towards Agroforestry systems. The liberalization of transit rules also helps the popularization of Agroforestry systems in southern India.

58. Discuss the role of agroforestry in nutrient cycling and soil conservation. How is saline soil reclaimed? (2016)

Agroforestry and soil conservation

The agroforestry system plays an important role in protecting the land from erosion. This system is considered as long lasting and recurring measure with a view to prevent ultimate deterioration of land, decreases soil loss and maintaining the soil productivity on a sustainable basis. Tree species and agroforestry systems prevent the soil from blown away by reducing the wind velocity at the ground surface. This is achieved by acting as a barrier or as a cover. This barrier function is to control by checking the runoff water and suspended sediments. The cover function involves reducing rain drop and run off by increasing soil cover through accumulation of large volume of leaf litter. Hence the combined intervention of agronomy, forestry and engineering will reduce the soil erosion and helps land remaining productive on sustainable basis.

Nutrient cycling in agroforestry systems

In agroforestry systems, trees can provide nitrogen inputs by 2 processes biological nitrogen fixation in the case of nitrogen-fixing tree species(NFTS),and deep nutrient capture.Although the magnitude of biological nitrogen fixation is methodologically difficult to quantify, the overall recorded rates of fixation for trees both leguminous as well as non-leguminous range from 20 to 300 kg N/ ha /yr. The advocation and inclusion of NFTS in agroforestry systems has led to the belief that nitrogen fixed in the root nodules may be used by the companion crop though direct evidences to support it are still lacking. NFTS can also supply considerable nitrogen tocrops through litterfail.On the other hand, there is ample evidence that trees such as cherry and mandarin, where nitrogen-fixation does not occur, accumulate as much or more N in their leaves as do NFTS, presumably because of their greater root volume and ability to capture nutrients from sub surfacesoil. However, these non NFTS are only cycling nutrients and not adding them to the system.

Deep nutrient capture the second process through which trees can provide nitrogen inputs is the uptake of nutrients by tree roots at depths where crop roots are not present. This is an additional nutrient inputin agroforestry systemsbecause such nutrients are leached and lost for crop use, but they become an input on being transferred to the soil through tree litter decomposition.

Saline soils

Saline soils are defined as soils having a conductivity of the saturation extract greater than 4 dS m^{-1} and an exchangeable sodium percentage less than 15. The pH is usually less than 8.5. Formerly these soils were called white alkali soils because of surface crust of white salts.

Saline Soil Reclamation

All saline soils can be reclaimed easily with good quality water if available. Since the salts in this soil are soluble in natureusing quality water they can be solubilised and leached off from the field. In the absence of good quality water in becomes necessary to manage saline soils for better growth of plants.

Management of saline soils

- Growing crops that are tolerant tosoil salinity
  Eg: Cotton, Ragi,Barley, Sugar beat,
- Mulching the soil prevents evaporation which reduces accumulation of salts due to capillary rise of water at surface of soils.
- Providing drainage in water logged areas also helps to reduce salt accumulation.
- Addition of extra dose of nitrogen to the tune of 20 – 25% of recommended level will compensate the low availability of N in these soils.
- Addition of organic manures like FYM, helps in reducing the ill effect of salinity due to release of organic acids produced during decomposition.
- Green manuring and green leaf manuring also counteracts the effects of salinity.
- Proportional mixing of good quality water with saline water and then using for irrigation reduces effect of salinity. Alternate furrow irrigation system favours growth of plant than flooding. Drip and sparkler irrigation systems aims at reduced use of water which is favorable for growth of plant since salt accumulation also reduced with low usage of water.

59. Write an account on coastal a-orestation? (2005)

The term coast is used to indicate the zone of contact between land and the sea. Coastal areas suffer from air laden with salt, torrential rain, howling gales and much higher light levels than inland, as well as heightened temperatures.

Reclamation of salt affected soils requires additional agricultural inputs such as amendments, fertilizers, water and infra-structure for farm operations and drainage. Requirement of extra resources on recurring basis has been a major bottle neck for the rehabilitation of salty soils on an extensive scale. Therefore, enhancement in the pace of the reclamation of salty soils calls for adoption of strategies which permit horizontally extensive use of salt affected soils with phased-out use of agricultural inputs. This requirement is very well met by setting up afforestation programs, especially on 'wasted' village community and other governmental lands. In other situations, agroforestry or farm forestry can be practiced for speedy rehabilitation of salty soils. Besides providing fuel, fodder and timber, afforestation will also lead to bioamelioration of salt affected soils. Afforestation of salt affected soils is not only prudent from ecological and environmental considerations but also offers a unique opportunity for relieving pressure on traditionally cultivated lands.

Practices of A-orestation in coastal areas

Creation of favorable root environment for proper establishment of tree saplings on salt affected soils is very important towards successful coastal afforestation. In addition the brackish waters demands for the adoption of special package of practices. The following are the approaches for afforestation of coastal areas.

- Identification of the nature of salt problem
- Assessment of availability and quality of irrigation water and selection of suitable tree species
- Choice of pitting and planting methods for alkali and saline soils
- Desiging suitable soil and water management
- Physical and social fencing during initial years

Both alkali and the saline soils differ in their physical and chemical characteristics and thus nature of constraints for plant growth. Therefore, diagnosis of the nature and magnitude of the problem is must. This can be ascertained by sending the soil and water samples to a nearby soil testing laboratory. Once the causative factors are known, specific instructions for undertaking afforestation programs on such soils may be followed.

Selection of Tree species

Choice of proper tree species depends upon the local agro-climate, land capablity, purpose of planting, tolerance to salinity / alkalinity and water logging drought stresses. In general, plantations for fuelwood are rated better for salty soils than the timber wood tree species. In addition to inter and intragenic variations in salt tolerance of forest tree species, tolerances also vary with the growth stages. It may also be pointed out that besides salt tolerance, socio-economic conditions and the ameliorative role of trees should also be given due consideration· in selecting tree species for afforestation programs. Numerous tree species have been evaluated for their tolerance to saline and alkali conditions. Amongst the principal species of arid and semi-arid regions, some species have been rated better than the others. A short list of consistently better performing species which could be recommended for growing under saline and alkali conditions of soils or irrigation waters, has been presented in the following table.

Suitable tree species for coastal region based on soil pH

Soil pH_2	Firewood/timber/fruit species down to 1.2m
> 10.0	*Acacia nilotica*
	Butea monosperma
	Casuarina equisetifolia
	Prosopis cineraria
9.0-10.0	*Aegle marmelos*
	Albizzia lebbeck
	Carissa carandua
	Cassia siamea
	Eucalyptus tereticornis
	Feronia limonia
	Pongamia pinnata
	Psidium guajava
	Sesbania sesban
	Tamarix articulate
	Terminalia arjuna
	Zizyphus maurtiana
	Zizyphus jujuba
8.6- 9.0	*Azardirachta indica*
	Da lbergia sissoo
	Emblica oDcianaJis
	Grevillia robusta
	Hardwickea binnata
	Kajellea pinnata
	Morus alba
	Mangifera indica
	Pyris communis
	Populus delteoides
	Tectona grandis
	Syzium cumuni
	Vitis vinifera

60. Write a short notes on agro ecological zones? (2009)

Agriculture is highly dependent on soils and climate. The ever-increasing need for food to support the growing population in the country demands a systematic appraisal of its soil and climate resources in order to prepare effective land-use plans. India has a variety of landscapcs and climate conditions and this is reflected in the development of different soils and types of vegetation. Based on 50 years of climate data and an up-to-date soil database, the country has been divided into 20 agro-ecological zones (AEZs). Each AEZ is as uniform as possible in terms of physiography, climate, length of growing period and soil type for macrolevel land-use planning and effective transfer of technology.

Important Features of Agro-ecological Zones of India

AEZ No.	Agro-ecological region	Geographical area (million ha)	Gross cropped area (million ha)	Physiography	Precipitation (mm)
1.	Cold arid ecoregion with shallow skeletal soils	15.2 (4.7%)	0.07	Western Himalayas	<150
2.	Hot arid ecoregion with desert and saline soils	31.9 (9.7%)	20.85	Western Plain & Kachchha Peninsula	< 300
3.	Hot arid ecoregion with red and black soils	4.9 (1.9%)	4.18	Deccan Plateau	400–500
4.	Hot semi-arid ecoregion with alluvium-derived soils	32.2 (9.8%)	30.05	Northern Plain & Central Highlands including parts of Gujarat Plains	500–800
5.	Hot semi-arid ecoregion with medium and deep black soils	17.6 (5.4%)	11.04	Central (Malwa) Highlands, Gujarat Plains & Kathiawar Peninsula	500–1000
6.	Hot semi-arid ecoregion with shallow and medium (dominant) black soils	31.0 (9.5%)	25.02	Deccan Plateau	600–1000
7.	Hot semi-arid ecoregion with red and black soils	16.5 (5.2%)	6.19	Deccan (Telangana) Plateau & Eastern Ghats	600–1000
8.	Hot semi-arid ecoregion with red loamy soils	19.1 (5.8%)	6.96	Eastern Ghats (Tamil Nadu uplands) & Deccan Plateau (Karnataka)	600–1000
9.	Hot subhumid (dry) ecoregion with alluvium-derived soils	12.1 (3.7%)	11.62	Northern Plain	1000–1200
10.	Hot subhumid ecoregion with red and black soils	22.3 (5.8%)	14.55	Central Highlands (Malwa & Bundelkhand)	1000–1500
11.	Hot subhumid ecoregion with red and yellow soils	11.1 (4.3%)	6.47	Eastern Plateau (Chhattisgarh Region)	1200–1600

12.	Hot subhumid ecoregion with red and lateritic soils	26.8 (8.2%)	12.09	Eastern (Chhota Nagpur) Plateau and Eastern Ghats	1000–1600
13.	Hot subhumid (moist) ecoregion with alluvium-derived soils	11.1 (3.4%)	10.95	Eastern Plains	1400–1600
14.	Warm subhumid to humid with inclusion of perhumid ecoregion with brown forest and podzolic soils	18.2 (5.6%)	3.20	Western Himalayas	1600–2000
15.	Hot subhumid (moist) to humid (inclusion of perhumid) ecoregions with alluvial-derived soils	12.1 3.7%)	8.99	Bengal Basin and Assam Plain	1400–2000
16.	Warm perhumid ecoregion with brown and red hill soils	9.6 (2.9%)	1.37	Eastern Himalayas	2000–4000
17.	Warm perhumid ecoregion with red and lateritic soils	10.6 (3.3%)	1.56	North-Eastern Hills	1600–2600
18.	Hot subhumid to semi-arid ecoregion with coastal alluvium-derived soils	8.5 (2.6%)	6.12	Eastern Coastal Plains	900–1600
19.	Hot humid perhumid ecoregion with red, lateritic and alluvium-derived soils	11.1 (3.6%)	5.70	Western Ghats and Coastal Plains	2000–3200
20.	Hot humid / perhumid island ecoregion with red loamy and sandy soils	0.8 (0.3%)	0.05	Islands of Andaman & Nicobar and Lakshadweep	1600–3000

61. Di-erentiate between Shelter belts and wind breaks (4 mark) (2010)

Shelter Belt

These are belts / blocks consisting of several rows of trees established at right the purposes are:

(a) To deflect air currents

(b) To reduce the velocity of winds.

(c) To providc gcncral protection to the leeward areas against the effects of wind erosion.

(d) To protect the leeward areas from desiccating effects of hot wind.

(e) To provide fuel, fodder timber etc.

Characteristics of Shelter Belts

(a) **Shape and Composition:** Shelterbelts have a spiracle, triangular shape by raising tall trees in the centre.

(b) **Density and Width:** Shelterbelts up to 50m width are considered ideal under Indian conditions. A certain degree of penetration by wind is planned.

(c) **Orientation:** Orientation depends on the direction and velocity of the winds.

(d) **Length:** The minimum length of a shelterbelt should be about 25 times its height.

Dodonaia viscosa, *Cassia* sp., *Clerodendron*, *Acacia arabica*, *Dalbergia, Eucalyptus*, *Parkinsonia*, *Prosopis* etc.

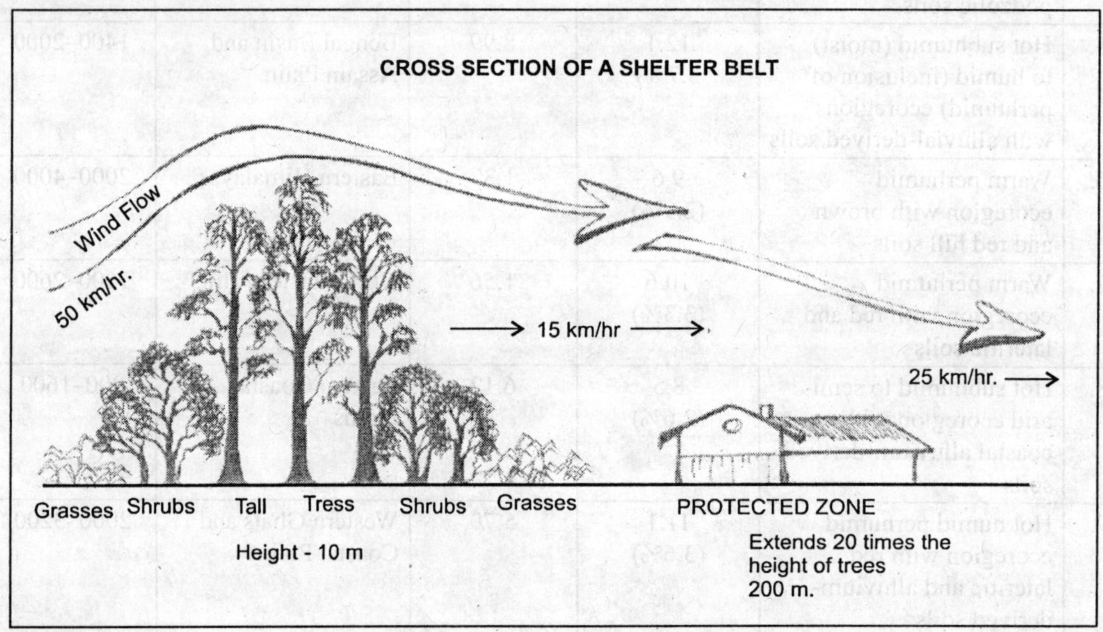

Wind - Break

Wind-breaks are strips of trees and or shrubs planted to protect fields, homes cannels or other areas from wind and blowing soil or sand.

Purpose of Raising Wind Breaks

(i) To protect field crops / livestock from cold / hot wind.
(ii) To prevent soil erosion.
(iii) To reduce evaporation from farmlands.
(iv) To improve the microclimate.
(v) For fencing and boundary demarcation.
(vi) For productive role-fuel, fodder, etc

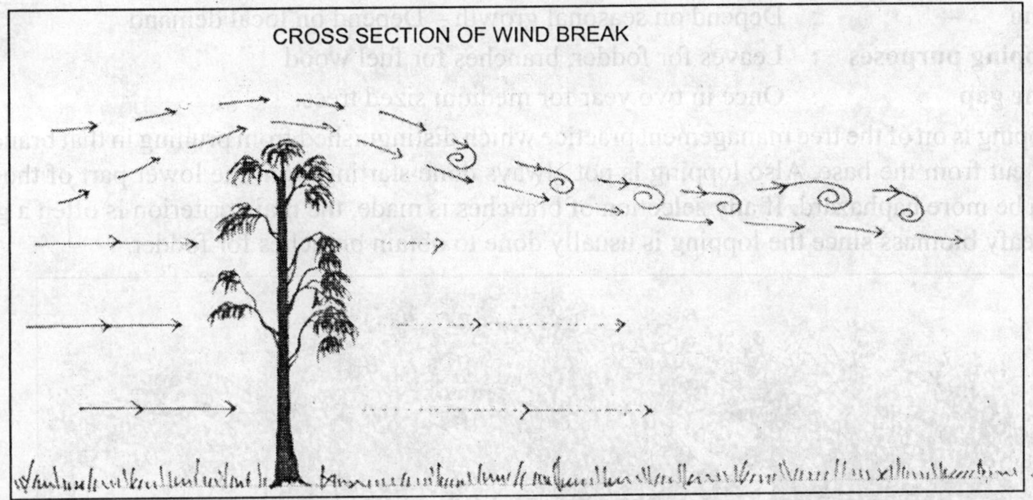

Main Features of Wind Break

(a) **Permeability:** The primary purpose of raising windbreaks is to filter and break up the force of the wind. Permeable windbreaks, which allow some wind to pass through, are the most suitable. The desired permeability can be obtained by carefully selecting trees and shrub species. Species such as Eucalyptus and Casuarina will form dense mind break but most native species are more permeable.

(b) **Orientation:** For best results, wind-breaks should be-raised at right angles to the direction of wind. N - S direction is good compromise. They should give better shading of adjacent crops and pastures.

62. Write short notes Lopping Management (2009), (2010)

Lopping is defined as the cuttings of branches or stems of trees between the branch unions.

Species amenable for loppping

- *Quercus species in Himalaya*
- *Butea monosperma*
- *Azadirachta indica*
- *Prosopis cineraria*
- *Acacia nilotica*
- *Inga dulce*
- *Anogeissus pendula*

Lopping is based on species capacity to produce foliage depend on climate. Care should be taken to avoild excessive lopping.

Implement used : Sharpe edged – Avoid damage of any part of trees

Time	:	Depend on seasonal growth – Depend on local demand
Lopping purposes	:	Leaves for fodder, branches for fuel wood
Time gap	:	Once in two year for medium sized trees.

Lopping is on of the tree management practice which distinguished from pruning in that branches are not cut from the base. Also lopping is not always done starting from the lower part of the tree but can be more haphazard. If any selection of branches is made, the main criterion is often a good green leafy biomass since the lopping is usually done to obtain branches for fodder.

Lopping

Demerits of Lopping

- Harm the growth due to lower – Poor seed production – Low regeneration
- It changes crown to root ration – Affects – Creating of food and energy other than fodder and fuel wood.
- Reduce the growth due to lower photosynthesis
- Invite pest and diseases
- Regeneration from lopped part is weak and unstable likely to break by wind
- Doing it for ornamental purposes. Use to remove wild unstable parts.
- Reduce the value of the tree in product such as timber and recreation point

63. Trace the history of Joint Forest Management in India. Narrate any one success story with details (2016)

Origin and Evolution of JFM

Though the initial set of policies and laws on forestry date back to the colonial period and the immediate post-independence period, one notices a paradigm shift in India's forest policy and legislations in the 1980s, with the passage of the Forest (Conservation) Act, 1980. This Act highlighted the primacy of conservation of forests over the previous emphasis on utilizing 'forests' for meeting the requirements of agriculture and industry.

The first forest policy of India was enacted during the British period in 1894. This policy was centrally influenced by the Voelcker Commission Report, which described forests to be a biomass provider for the agricultural sector.

After independence, the Government of India enacted a new forest policy in the year 1952 which largely subscribed to the philosophy of the 1894 policy, nevertheless highlighted the functional classification of forests. The 1952 policy classified forests into Protection forests, National forests, Village forests and Tree lands. The policy recommended weaning of tribes from shifting cultivation practices and controlling of grazing and other activities in forest areas.

The first one was a community-based forest conservation movement initiated in Arabari in Midnapur District of West Bengal State during 1971-72, while the second one was a grassroots movement in the Sukhomajri village in Haryana in the 1980s, to rejuvenate forests and agricultural systems in the village. In Arabari, the movement was triggered by a Silviculturalist of the State Forest Department, who by offering incentives to local communities induced them to protect and regenerate degraded Sal forests. The local communities, which were organized into 'Forest Protection Committees' (FPCs), successfully protected the degraded forests from illegal felling, overgrazing, fire, and encroachment. In Sukhomajri, Haryana State, the movement started from amongst the people. Construction of earthen dams stabilized agricultural output in the village. Forests in and around Sukhomajri village regenerated as a result. The regenerated forests in turn provided valuable biomass, including bhabar grass to local communities.

These instances awakened the policy makers at the Central Government to the need to go beyond the legalistic "Forest Conservation Act 1980". Institutional measures for arresting the alarming trend of forest depletion in the country were actively considered. The National Forest Policy (1988) was enunciated by the Government of India with a strong focus on conservation, environmental stability and ecological balance through association of tribal's and local communities in protection, regeneration and development of forests. In pursuance of the National Forest Policy of 1988, the Government of India issued a 'circular' in June 1990 for involvement of village communities and village associations (VAs) in the regeneration of degraded forest lands. This marked the birth of the Joint Forest Management (JFM) movement in India.

On 1st June 1990, Government of India, issued guidelines to the state governments for involving local communities in the protection and development of the degraded forests. Based on this, the JFM programme has now been widely adopted in twenty two states of the country.

Success stories of JFM in India

(a) Arabari experience: The Arabari forest range of Midnapur Forest Division, West Bengal has experienced complete degradation of natural forests particularly the sal forests and the associated natural regeneration. The local people have exploited the area which leads to loss of genetic resources. Hence the participatory approach of forest management has started its inception in Arabari by the then DFO Dr. A.K.Baneerji. He has earmarked 1272 ha area of degraded forests for this approach and successfully established this area with plantations through the participation of local dwellers. Under this participatory approach the villages were provided with employment generation activities coupled with exclusive rights to collect all non wood forest products at free of cost. The villages were allowed to cultivate paddy, fodder, grasses, maize and groundnut. They have inturn protected the forests by establishing forest protection committees. This has shown excellent protection and the forests have been regenerated which formed the basis for implementation of JFM in the country.

(b) **Behroonguda experience:** Behroonguda is one of 77,000 hamlets and villages in Andhra Pradesh. The 97 families who belong to the Gond and Naikpod tribes formed themselves into a forest protection group in 1990, but the state government officially recognized their efforts in 1993. In 1998, five years after the recognition of the committee - referred to as Vana Samarakshana Samithi, or VSS - Behroonguda residents began to derive usufruct benefits from the forest. The costs they have borne and the benefits they have derived from protecting 500 hectares of degraded forest allotted to them as part of the joint forest management (JFM) in Andhra Pradesh. The VSS is widely regarded as being successful.

An analytical method is developed and applied to two case studies of communities managing a mixed teak forest system and a sal coppice forest system in Madhya Pradesh the analytical results show that-

(a) Economic returns to JFM are considerable in both forest systems.

(b) There is an increase in revenues from the forest to the communities but a theoretical loss to forest departments.

(c) Income flows into the communities increase significantly, partially offset in the sal coppice system by declining income from collection of non-timber forest products (NTFPs).

(d) JFM benefits are not always distributed equally, resulting in collectors of firewood and some NTFPs losing, though overall gains are sufficient to compensate losers.

Forest Soils, Soil Conservation and Watershed Management

1. Write short notes on carbon sequestration (2007, 2011)

Carbon sequestration is the process of capture and sequer storage of carbon from the atmosphere into long lived pools such as terrestrial, ocean by photosynthetic pathway of plants and microorganisms.

"Carbon sequestration is the process involved in carbon capture and long-term storage of atmospheric carbon dioxide". Carbon sequestration involves long-term storage of carbon dioxide or other forms of carbon to mitigate or defer global warming.

Carbon sequestration describes long-term storage of carbon dioxide or other forms of carbon to either mitigate or defer global warming and avoid dangerous climate change. It has been proposed as a way to slow the atmospheric and marine accumulation of greenhouse gases, which are released by burning fossil fuels. Carbon dioxide is naturally captured from the atmosphere through biological, chemical or physical processes. Some artificial sequestration techniques exploit these natural processes, while some use entirely artificial processes. Carbon dioxide may be captured as a pure by-product in processes related to petroleum refining or from flue gases from power generation.

Kinds of carbon sequestration

(i) Ocean Sequestration

(ii) Geologic Sequestration

(iii) Terrestrial Sequestration

(iv) Forests as carbon sequestration

(i) Ocean Sequestration: Carbon stored in oceans through direct injection or fertilization.

(ii) Geologic Sequestration: Natural pore spaces in geologic formations serve as reservoirs for long-term carbon dioxide storage."The process of removing carbon from the atmosphere and depositing it in a reservoir" When carried out deliberately, this may also be referred to as carbon dioxide removal, which is a form of geo-engineering.

(iii) Terrestrial Sequestration: A large amount of carbon is stored in soils and vegetation, which are our natural carbon sinks. Increasing carbon fixation through photosynthesis, slowing down or reducing decomposition of organic matter, and changing land use practices can enhance carbon uptake in these natural sinks.

(iv) Forests as carbon sequestration: Forests in general and trees in particular are excellent sequester of carbon. Trees absorb carbon from the atmosphere and utilize them through photosynthesis and produce biomass which utilizes the plant growth and development. Due to this process the concentration of carbon-di-oxide in atmosphere is very well balanced and reduces the risks of global warming. Due to this sequestration ability, the IPCC has identified agroforestry as one of the potential landuse system which has sequest huge volume of carbon and balance concentration of atmosphere.

2. Di-erentiate between Photosynthetic e-ciency and nutrient use e-ciency. (2010)

Photosynthetic e-ciency

The photosynthetic efficiency is the fraction of light energy converted into chemical energy during photosynthesis in plants and algae. Photosynthesis can be described by the simplified chemical reaction

$$6H_2O + 6CO_2 + \text{energy} \longrightarrow C_6H_{12}O_6 + 6O_2$$

Where $C_6H_{12}O_6$ is glucose (which is subsequently transformed into other sugars, cellulose and lignin). The value of the photosynthetic efficiency is dependent on how light energy is defined – it depends on whether we count only the light that is absorbed and on what kind of light is used. It takes eight (or perhaps 10 or more) photons to utilize one molecule of CO_2.

Nutrient use e-ciency

Nutrient use efficiency (NUE) may be defined as yield per unit input. In agriculture this is usually related to the input of fertilizer, whereas in scientific literature the NUE is often expressed as fresh weight or product yield per content of nutrient. Improvement of NUE is an essential pre-requisite for expansion of crop production into marginal lands with low nutrient availability. The nutrients most commonly limiting plant growth are N, P, K and S. NUE depends on the ability to efficiently take up the nutrient from the soil, but also on transport, storage, mobilization, usage within the plant and even on the environment.

3. Write short notes on carbon-nitrogen ratio in forests. (2009)

Carbon-Nitrogen Ratio

Carbon-Nitrogen ratio (C/N ratio) is defined as 'the ratio of the weight of organic carbon to the weight of total nitrogen in a soil, or organic material.' The ratio varies with soil typo and within the same soil type with climate.

Importance

Importance of C/N ratio - Carbon nitrogen ratio is a measure of nitrate status of soil. The lower the ratio, the greater is the amount of nitrogen converted into ammonia and finally into nitrates to be available to plants. The conversion goes on satisfactorily till the ratio is 17 : 1 or less but when ratio exceeds 33: 1, nitrogen gets completely immobilized. Similarly high C/N ratio interferes with nutrient availability for which 10 : 1 ratio is considered suitable.

4. Write short notes on Forest leaf litter (2009)

Forest leaf litter is a dead plant material, such as leaves, bark, needles, twigs, and cladodes; that have fallen to the ground. This detritus or dead organic material and its constituent nutrients are added to the top layer of soil, commonly known as the litter layer or O horizon ("O" for "organic").

Litter is an instrumental factor in ecosystem dynamics, it is indicative of ecological productivity and useful in predicting regional nutrient cycling and soil fertility.

Litter fall is characterized as fresh, undecomposed and easily recognizable plant debris. This can be anything from leaves, cones, needles, twigs, bark, seeds/nuts, logs, or reproductive organs (e.g. the stamen of flowering plants). Items larger than 2 cm diameter are referred to as coarse litter, while anything smaller is referred to as fine litter or litter. The type of litter fall is most directly affected by ecosystem type. For example, leaf tissues account for about 70 percent of litter fall in forests, but woody litter tends to increase with forest age. In grasslands, there is very little aboveground perennial tissue so the annual litter fall is very low and quite nearly equal to the net primary production.

The forest leaf litter is very important and protects the forest floor from stroking of both rainfall directly and rains coming from the crown closure. The litter within the forest area can vertically infiltrate precipitation and protects the soil and improves the ground water level.

The forest leaf litter can infiltrate most intensive rainfalls and it is estimated that a litter consists of decomposed mull can infiltrate 150 mm per hour intensity very easily.

The forest leaf litter reduces stream flow velocity, postpones time of stream flow to be maximum and decreases stream sedimentation. The forest leaf litter with 6.5 cm thickness has delayed 32 minutes infiltration time of a rainfall with 23 mm per ha intensity.

The forest leaf litter on decomposition helps to increase the soil fertility status in terms of the essential nutrients. This in turn will help to improve the productivity of a plantations.

5. Write short notes on Nutrient cycling in natural forests. (2010)

Nutrient cycling in natural forest ecosystem

Nutrient cycling is the continuous process of transfer of minerals in and out among various nutrient pools. Nutrient cycling is very significant to maintain the soil fertility. The following are the major nutrient cycle prevalent in natural forest.

(i) Carbon cycle
(ii) Nitrogen cycle
(iii) Oxygen cycle
(iv) Phosphorus cycle

Nutrients are elements or compounds that are essential for the growth and survival of plants. Plants require large amounts of nutrients such as nitrogen (N), phosphorus (P), carbon (C), hydrogen (H), oxygen (O), potassium (K), calcium (Ca) and magnesium (Mg), but only small amounts of others such as boron (B), manganese (Mn), iron (Fe), copper (Cu), zinc (Zn) and chlorine (Cl) (micronutrients).

Forest nutrient cycling is defined as the exchange of elements between the living and nonliving components of an ecosystem. The processes of the forest nutrient cycle include:

- Nutrient uptake and
- Storage in vegetation perennial tissues,
- Litter production,
- Litter decomposition,

Nutrient transformations by soil fauna and flora, nutrient inputs from the atmosphere and the weathering of primary minerals, and nutrient export from the soil by leaching and gaseous transfer.

Features of the major nutrient cycles

Element	Uptake by the trees	Tree uptake	Limiting situation
Carbon	Atmosphere	Atmosphere	Atmospheric concentration may limit growth
Oxygen	Atmosphere	Atmosphere	Waterlogged soils
Hydrogen	Atmosphere	Atmosphere	Extremely acidic and alkaline conditions
Nitrogen	Soluble NO_3 and NH_4; N_2 for nitrogen fixing species	Soil organic matter; atmospheric N_2 for nitrogen fixing species	Most temperate forests, many boreal forests and some tropical forests
Phosphorus	Soluble phosphorus	Soil organic matter; adsorbed phosphate and mineral phosphorus	Old soils high in iron and aluminum, common in subtropical and tropical environment
Potassium calcium	Soluble Resoluble Ca^{2+}/ soluble Mg"+	Soil organic matter; exchange complex and minerals	Miscellaneous situations and some old soils
Magnesium			

Gersmehl's Nutrient Cycles

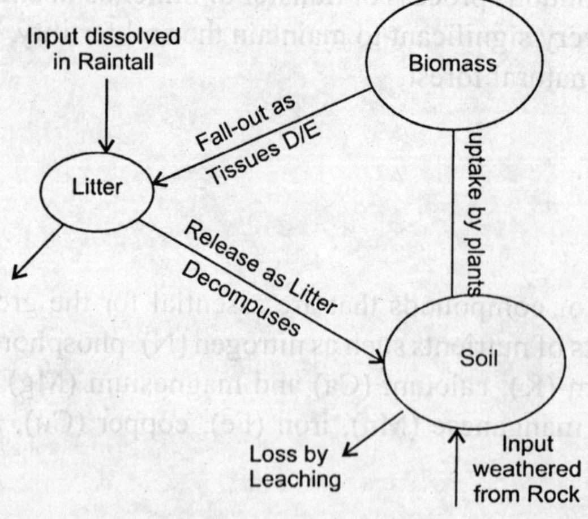

Nutrient cycling in forest ecosystems is controlled primarily by three key factors: climate, site, abiotic properties (topography, parent material), and biotic communities.

6. Among the causes of "soil erosions some are manageable". Comment. (2010)

Yes some of the causes of soil erosion are manageable. Soil erosion is the detachment and transportation of soil from one place to another through the action of wind, water, ice, tidal waves or by beating action of rain drops. The process of soil erosion involves erosivity, erodability, detachability and transportability. There are several factors which causes soil erosions and they are

(i) Excessive deforestation
(ii) Over grazing
(iii) Improper agricultural practices
(iv) Human activities such as mining, constructions, roads and etc.
(v) Slope of the land

Among these factors some of the factors are easily manageable and they are

(i) Control of excessive deforestation: There is always the possibility for the control of deforestation through legal measures and also through creating awareness. India has implemented several acts particularly the Indian Forest Act, 1927 and Forest Conservation Act 1980 which has classified forests into reserve and protected forests and do not permit deforestation and illegal felling. Hence deforestation can be easily managed by enforcing the laws indicated above.

Alternatively awareness programme has to be created among the people, the causes and effect of deforestation. Generally deforestation happen mainly to meet the domestic wood needs and also for livelihood options. Hence the forest department takes the necessary action to promote agroforestry models in order to meet the domestic timber demands. Similarly alternate livelihood option may also be identified to provide income generation activities which will restrict deforestation activities.

(ii) Control of over grazing: Grazing is the major cause for soil erosion because the grazing animals affect the regeneration which leads to soil degradation and erosion. Over grazing can easily be managed by establishing fodder bank and grazing lands by planting the improved high yielding varieties. The practices of rotational grazing, controlled grazing and differed grazing may help to control the grazing pattern thereby soil erosion can be controlled.

(iii) Implementing advance agricultural practices: The unscientific and improper agriculture practices may play a major role in soil erosion. This can be easily managed by introducing **scientific** and modern agricultural practices. The planting of cover crop, mulching, wind breaks, shelter belts, crop rotation, strip cropping and etc. will help to minimize the loss due to erosion.

(iv) Human activities: The human activities like mining, construction, roads and other development projects can be controlled through suitable policy and legal regulations.

(v) Slope of the land: Though slope of the land is not manageable but soil erosion can be controlled through suitable physical and biological measures. The physical measures like formation of contour bunds, development of bench terraces coupled with cultivation will reduce the soil erosion.

7. Di-erentiate between erosivity and erodibility. (2010)

Erosivity

Erosivity is a term that is used to describe the potential for soil to be washed off from disturbed, de-vegetated areas and move with surface waters during storms. It may also be defined as the potential ability of rain to cause the erosion.

It depends upon the physical characteristics of rainfall, which include raindrop size, drop size distribution, kinetic energy, terminal velocity *etc.* For a given soil condition, the potential of two storms can be compared quantitatively, regarding soil erosion to be caused by them. The power of overland runoff flow to erode soil material is partly a property of the rainfall, and partly of the soil surface. Rainfall erosivity is highly related to soil loss. Increased rain erosivity indicates greater erosive capacity of the overland water flow. Soil erosion by running water occurs where the intensity and duration of rainstorms exceeds the capacity of the soil to infiltrate the rainfall. The potential for erosion is based on many factors which include soil type, slope, and the energy or force of precipitation expected during the period of surface disturbance.

Erodibility

Soil erodibility is an estimate of the ability of soils to resist erosion based on the physical characteristics of each soil. Generally, soils with faster infiltration rates, higher levels of organic matter and improved soil structure have a greater resistance to erosion. Sand, sandy loam and loamy textured soils tend to be less erodible than silt, very fine sand, and certain clay textured soils. On the basis of erodibility, a soil can be compared quantitatively with other soils for a given rainfall condition. The soil erodibility depends on mechanical composition of soil, such as sand, silt, and clay, presented by the ratio as:

$$\text{Erodibility, } E = \frac{\%\,\text{sand} + \%\,\text{silt}}{\%\,\text{clay}}$$

The range of particle diameter of clay, sand and silt is:

Clay	< 0.002 mm
Silt	0.002 – 0.006 mm
Sand	0.06 - 2.0 mm

8. Mention ten species (Scientific name) of trees tolerant to salinity (2010)

- *Acacia nilotica*
- *Albizia lebbeck*
- *Azadirachta indica*
- *Calophyllum inophyllum*

- *Casuarina equisetifolia*
- *Erythrina indica*
- *Eucalyptus tereticornis*
- *Madhuca latifolia*
- *Lannea coromandelica*
- *Thespesia populnea*

9. How shelterbelt and wind breaks are helpful in sand dune stabilization and desert control? (2012)

Sand dunes are artificial land use system created due to shifting sands. The shifting sands from the desert in the nearby area and provide resistance to flow of sands from that locality. This had led to development of sand dues which has a very gentle slope of 5-10° with the horizontal on windward side and leeward side with stand a steep slope of 30°. The height of dunes depends on the size of particles, moisture content and wind velocity.

The movement of sand dunes occurs during dry season and seriously affects the houses, constructions, buildings, roads, railways, canals and water courses unless these sand dunes are stabilized the shifting sands create a problem.

Stabilization of sand dunes

Sand dunes can be stabilize using trees particularly through shelterbelt and wind breaks.

(i) Shelterbelt: Shelter belts are belt of rows of trees established at right angles to the prevailing wind direction. Shelter belt deflects the air currents and thus by reduces the wind velocity and erosion.

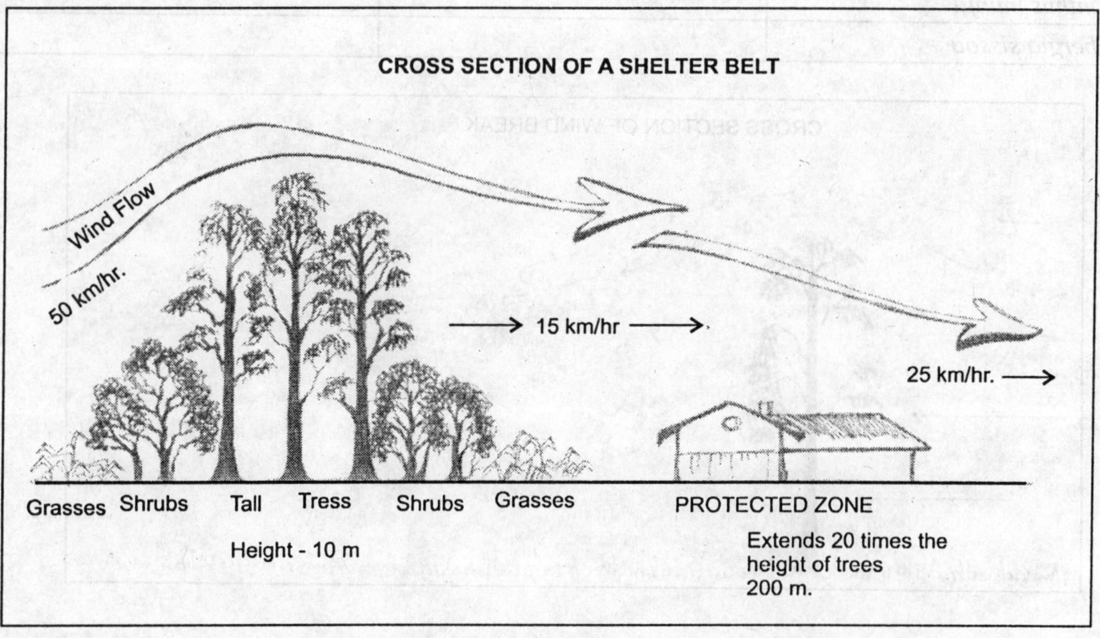

This shelter belts arrests the shifting sands moving from one place to other since the shelter belt are 3-5 rows in structure. They play efficient and effective barrier role in stabilizing the sand dunes and to control the shifting of sands. The size and shape of shelter belt can be designed based on the intensity of the wind and the associated degradation process. The following species can be potential species for establishment of shelter belts. In this the shelterbelt can be designed the tall tree components in central region flanked by shrubby bushes followed by short herbs and grasses.

Species suitable for shelterbelt establishment

Grasses: *Saccharum spontaneum, S. munja, Panicum antidotale* and *Cenchrus* spp.

Shrubs: *Calotropis procera, Clerodendron phlomoides, Cassia auriculata, Dodonaia viscosa.*

Tree species: *Acacia arabica, Acacia leucopholea, Dalbergia sissoo, Lannea coromandelica, Eucalyptus spp., Tamarix articulata, Parkinsonia aculeata, Prosopis juliflora, Prosopis spicigera, Casuarina equisetifolia.*

(ii) Wind break: Windbreak refers to the strip of trees and / or shrubs planted in order o protect fields, homes, canals or other areas from wind and blowing soil. Tall plants of the following species may be used as wind breaks for sand dune management.

Species suitable for wind break establishment

Trees	Bushes	Grasses
Prosopis cineraria	*Crotalaria burhia*	*Cenchurs ciliaris*
Prosopis chilensis	*Calligonum polygonoides*	*Lasiurus sindicus*
Acacia tortilis	*Aerva javanica*	*Panicum turgidum*
Acacia radiana	*Ziziphus spp.*	
Ziziphus mauritiana		
Erianthus munja		
Dalbergia sissoo		

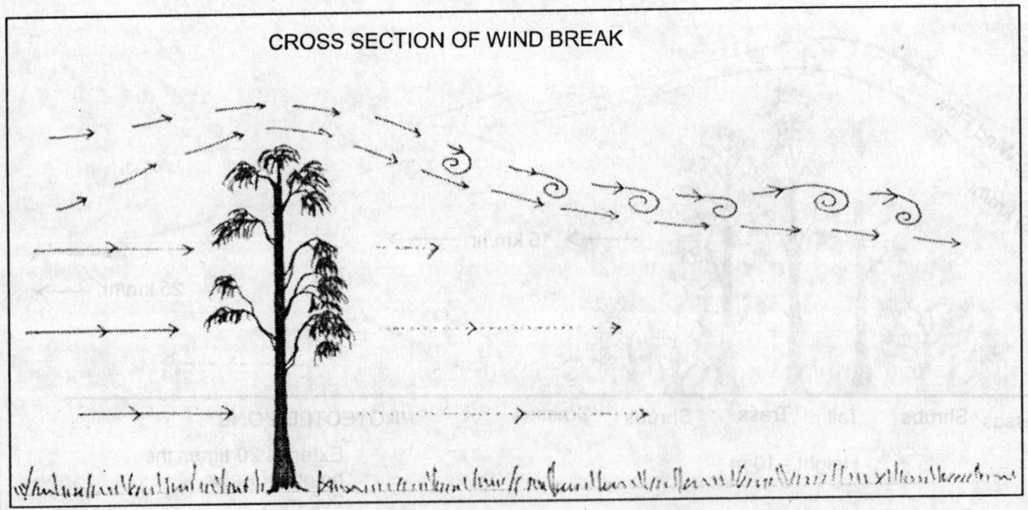

10. Discuss the role of forests in interception, surface runo-and infiltration of rainfall maintaining soil fertility. (2012)

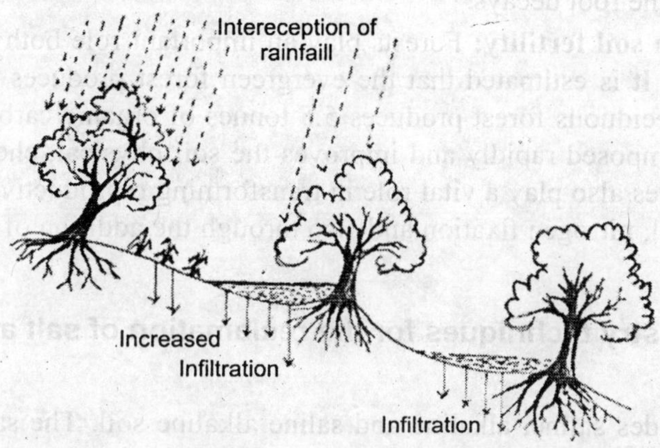

(a) Role of forests in interception: Forests filter and regulate the flow of water, in large part duc to their leafy canopy that intercepts rainfall, slowing its fall to the ground and the forest floor, which acts like an enormous sponge, typically absorbing up to 18 inches of precipitation (depending on soil composition) before gradually releasing it to natural channels and recharging ground water.

Average interception of rainfall by a forest canopy ranges from 10-40% depending on species, time of year, and precipitation rates per storm event. In urban and suburban settings a single deciduous tree can intercept from 500 to 760 gallons per year; and a mature evergreen can intercept more than 4,000 gallons per year. Even young, small trees help. In a recent Forest Service study a single small tree (callery pear) that was only 9 years old, was able to intercept 58 gallons of storm water from a ½ inch rain event (67% of the rain that fell within the canopy).

(b) Role of forests in surface runo3: Trees and forests improve stream quality and watershed health primarily by decreasing the amount of runoff and pollutants that reaches our local waters. Trees and forests reduce runoff by capturing and storing rainfall in the canopy and releasing water into the atmosphere through evapotranspiration. In addition, tree roots and leaf litter create soil conditions that promote the infiltration of rainwater into the soil. This helps to replenish our groundwater supply and maintain during dry periods.

The presence of trees also helps to slow down and temporarily store runoff, which further promotes infiltration, and decreases flooding and erosion downstream. Trees and forests reduce pollutants by taking up nutrients and other pollutants from soils and water through their roots, and by transforming pollutants into less harmful substances. In general, trees are most effective at reducing runoff from smaller and more frequent storms.

(c) Role of forests in infiltration of rainfall: Living and decaying roots create a network of well-connected channels in the soil called macropores. Flow through these macro pores can be up to several hundred times faster than flow through the soil matrix. In addition, organic matter from leaf litter and tree roots improves soil structure, which can increase infiltration rates. Soil structure is improved as soil particles are cemented together by humus, by organic glues created by fungi and bacteria decomposing organic matter, and by polymers and sugars excreted from roots. Areas with trees generally have greater infiltration rates than lawn or pasture without trees because trees create more stable macro pores. Woody plant roots have lining on their roots that further increase stability of the macro pores after the root decays.

(d) Role of forests in soil fertility: Forests play an important role both in the formation and conservation of the soil. It is estimated that the evergreen forest produces 20 tonnes of organic matter every year and deciduous forest produces 5.5 tonnes of organic carbon every year. These organic carbon get decomposed rapidly and improves the soil physical, chemical and biological properties. Forest and trees also play a vital role in transforming reproductive soil into productive soil through this litter fall, nitrogen fixation and also through the addition of nutrients through the root exudation.

11. Give suitable forestry techniques for the reclamation of salt a-ected soils. (2012)

Salt affect soils includes saline, alkaline and saline alkaline soil. The saline soils contain an excess of sodium salts having EC more than 4 dsm^{-1} and white encrustation is seen on the surface.

The alkaline soils are complex soils which contains appreciable quantities of exchangeable sodium. Alkaline soil may or may not contain excess salts.

Reclamation of salt a-ect soils through forestry practices

Before starting any forestry practices the salt affected soils should be treated with chemical amendments like addition of lime for saline soils and addition of gypsum for alkaline soils.

Another major reclamation process is leaching the excess water with good water. After this the following biological treatment helps to salt affected soils.

(i) Green manuring: The growing of *Sesbania aculeata* has proved to be most useful reclamation of salt affected soils. It enhances the reclaiming effect of gypsum and enriches soil with organic matter and nitrogen besides improving soil permeability. Similarly growing "dhaincha" will also improve the soil physical and chemical properties.

(ii) Reclamation through silvipasture: This model incorporates simultaneous growing trees and grasses in order to reclaim the soil and to provide the forage needs of domestic animals. The tree component increases soil permeability and augments the fertility status by addition of organic matter in the form of litter fall and stem fall. It also mitigates the carbon-di-oxide effect and improves the micro climatic of the region. The tree species like *Acacia nilotica, Acacia leucophloea, Acacia planifrons*, *Azadirachta indica*, *Casuarina equisetifolia, Prosopis juliflora, Leucaena leucocephala* are highly amenable for salt affected soil. The grass component includes kernel grass (*Diplachne fusca*) and *Cenchrus ciliaris* which improves the soil physical and chemical properties. This silvipasture model able to reduce pH and salt concentration on the soil, which result in conversion of unproductive salt soils into productive soils.

(iii) Reclamation through mixed wood lot: The salt affected soils can very well reclaimed through the establishment of block plantation of leguminous trees like, *Acacia spp, Albizzia, Subabul, Casuarina, Prosopis juliflora, Sesbania* etc. These leguminous trees improves their soil fertility by fixing atmospheric nitrogen through their micro symbiont present in the root nodules and adds to the soil. The trees add leaf litter to the soil which on decomposition helps to improve the soil physical, chemical and biological properties. This litter fall coupled with addition of nitrogen reclaims the soil and convert them into productive soil.

(iv) **Woodlots of Nitrogen fixing trees:** Nitrogen fixing trees like acacias, Albizia, Casuarinas, Pterocarpus etc., can be planted as woodlots. These trees are able to fix atmospheric nitrogen and ameliorate the soil fertility. The leaf litter of these nitrogen fixing trees will alter the soil physical, chemical and biological properties. Thereby help to reclaim the soil.

12. Explain the terms (i) Catchment (ii) Watershed and (iii) Micro-watershed. (2009)

(i) **Catchment:** A catchment is an area of land where water collects when it rains, often bounded by hills. As the water flows over the landscape it finds its way into streams and down into the soil, eventually feeding the river. Some of this water stays underground and continues to slowly feed the river in times of low rainfall. Every inch of land on the Earth forms part of a catchment.

(ii) **Watershed:** A watershed is an area of land that catches rain and snow and drains or seeps into a marsh, stream, river, lake or groundwater. Homes, farms, cottages, forests, small towns, big cities and more can make up watersheds. Some cross municipal, provincial and even international borders. They come in all shapes and sizes and can vary from millions of acres, like the land that drains into the Great lakes, to a few acres that drain into a pond.

(iii) **Micro-watershed:** The watershed is a natural hydrological entity that covers a specific aerial expanse of land surface from which the rainfall runoff flows to a defined drain, channel, stream or river at any particular point. It is a general phenomenon governed by topography of the terrain. Based on the size, the hydrological unit is termed as water resource region, basin, catchment, sub-catchment, watershed, sub-watershed and micro-watershed respectively. The smallest hydrologic unit in the hierarchal system is termed as Micro watershed having size of 500-1000 ha.

13. List the Pioneers flora of sand dunes. (2012)

Recommended trees, shrubs and grass species in sand dunes areas based on rainfall

(i) Rainfall zones of 15 cm to 30 cm.

Trees: *Prosopis juliflora* (Vilayathi babool), *Acacia tortilis* (Israeli babool), *Acacia Senegal* (Kumat).

Shrubs: *Calligonum polygonoides* (Phog) and *Zizyphus sps*. (Bordif Calotropis) species.

Grasses: *Lasiurus sindicus* (Sewan).

(ii) Ranfall zone of 30 cm to 40 cm.

Trees: *Acacia tortilis*, *Acacia Senegal*, *Prosopis juliflora*, *P. cineraria* (Khejri), *Tecomelta undulata* (Rohida), *Dichrostachys glomerata* and *Casuarina equisetifolia*.

Shrubs: *Zizyphus spp.*,*Calligonum polygonoides*, *Ricinus communis* (Arandi) *Cordia gharaf*.

Grasses: *Lasiurus sindicus, Cenchrus ciliaris (Dhaman), Cenchrus setigerus (Kala Dhaman), Saccharum munja (Moonj).*

(iii) Rainfall zone of 40 cm and above:

Trees: *Prosopis juliflora, Prosopis cineraria, Acacia tortilis, Acacia Senegal, Acacia nilotica sub-spsindica* (Kikar). *Tecomella undulata, Parkinsonia aculeata, Albizzia lebbek (Siris), Dalbergia sissoo* (Shisham), *Dichrostachys glomerata, Colophospermum mopane.*

Shrubs: *Ricinus communus, Zizyphus spp., Cassia auriculata etc.*

Grasses: *Cenchrus ciliaris, Cenchrus setigerus, Panicumanti dotale* (Gramna), *Saccharum munja* etc.

The success of sand dune stabilization depends on choosing the right species, planting sturdy stocks, planting in time and deep timely replacement of casualties and after care. No grazing should be allowed within the area. Cropping and other hiotic interferences should be stopped.

14. Justify the statement "Forests substantially check soil erosion and control run o-". (2016)

Yes, the sustainability of forest will help to control soil erosion and runoff.

Sound farming and forestry practices can reduce runoff. The retention of plantation residues like leaf litter, branch fall etc., will help to reduce flooding. Forest plays a vital role in shaking of soil erosion and controlling runoff.

The litter fall from the trees is organic matter which acts as a cementing agent in the formation of aggregates of soil particles. In general higher the organic matter content larger the size of aggregate and less will be the erodability of soil. Plant roots also provide biological cohesion and help to establish the soil.

The root of the trees helps to bind the soil which prevents the soil erosion through less erodability. Plant roots also provide biological cohesion and help to stabilize the soil.

Control run o-

The forest cover retards the surface velocity of runoff and reduces the transportation capacity. The root system of the plant acts as the binding agent on the soil and reduce the detachability on the soil and controls the runoff.

The decaying of roots, leaves etc. increases the porosity of soil and capacity to absorb the water also increases.

Addition of humus increases the biological activity which is favourable to the growth of microflroa and fauna. This micro flora and fauna population creates favourable condition for absorption and holding of more moisture inside the soil. The tree canopy intercepts the rainfall and reduces the velocity which in turn reduces the runoff.

A well distributed tree canopy not only slows down the water movement but also tend to prevent the concentration of water. These effect greatly reduce the erosion capacity of runoff water.

15. Explain types of rocks based on formation and minerals based on chemical composition (2016)

Rocks

Rocks are aggregate of minerals. They are the basic units, which constitute the earth's crust.

Petrology

It is that branch of geology that deals with the study of rocks.

Primary rocks

It is the first to have formed on the earth's crust. Eg: Igneous rocks.

Secondary rocks

These are formed due to the weathering and deposition or metamorphism of existing primary or secondary rocks viz. sedimentary and metamorphic rocks.

Definition: A rock may be defined as a mass of mineral water. It is composed of one or more minerals. Some rocks are hard and compact e.g.: granite and basalt, some are loose and feebly aggregated e.g.: sandstone and loose sand. Each rock possesses certain characters like structure, colour, specific gravity, cleavage, fracture and mineralogical make up, which help to distinguish one from the other.

There are three main categories of rocks based on their origin or mode of formation.

1. Igneous rocks: Formed by action of heat. These are the first to be formed when molten magma cooled and consolidated into solid rock. The other rocks were formed subsequently. Igneous rocks constitutes 95% of the earth's crust. They are crystalline and known as **Crystalline rocks**. Also called as massive rocks as they occur in masses. Igneous rocks mainly consist of primary minerals of which quartz, feldspar, amphiboles, pyroxenes and micas are the most common. Igneous rocks are formed in different ways depending on the manner and conditions under which molten mass consolidated. They are

(a) Extrusive: Formed when the molten mass was poured out on the surface of the earth where it consolidated on cooling. Among this includes the rocks of volcanic origins, which consolidated at the surface as lavo-flows or were forced between rocks near to surface. Rhyolite, obsidian, trachyte, dolerite and basalt – e.g. of extrusive rocks if volcanic origin.

(b) Intrusive rocks: Sometimes the molten mass cooled and consolidated beneath the surface. Those that consolidated at much greater depth than others are Plutonic rocks. e.g. Granite, syenite, dolerite and gabbio.

Some intrusive rocks consolidated in vertical cracks and formed wall like masses called as dykes. Others consolidated into horizontal cracks called as sills. In some cases the molten material is forced in irregular and narrow cracks or fissures. This is called as vein.

Many igneous rocks, especially those of volcanic origin contain water under great pressure even in molten condition. As the molten magma comes to the surface, the pressure is released and water is turned into steam. As steam cannot escape, it forms a globular cavity known as vesicle and the rock container; this is called as vesicular rock. There are minerals embedded in these vesicles and such rocks are amygdaloidal rock. The minerals embedded in this are quartz, calcite, zeolite, glauconite, etc.

Based on chemical composition igneous rocks are divided into four classes.

(a) Those that contain large amount of basic components – Basic rocks.

(b) Those that contain more of acidic components – acidic rocks.

(c) Those intermediate between the two – sub acid or

(d) Sub-basic rocks according to the silica acid content.

(i) Acidic rocks	-	>65 % silica. eg. granite, rhyolite, obsidian, pitchstone
(ii) Sub-acidic rocks	-	60.65 % silica eg. syenite and trachyte
(iii) Sub-basic rocks	-	55.-60 % silica. eg. diorite and andesite
(iv) Basic rocks	-	45-55 % silica. eg. gabbro, dolerite, bassalt
(v) Ultra basic rocks	-	< 45 % silica

Sedimentary rocks: Formed through the agency of water. It is also called as aqueous rocks formed from sediments brought by water. The sediment may contain various types of substances and sizes of particles. The particles are cemented by silica, iron oxide or lime to give a consolidated form. The rocks are mostly deposited in layers or strata – so called as **Stratified rocks**. Sometimes they are formed by cooling, evaporation or by direct chemical precipitation. Anyway they are of secondary origin.

Sedimentary rocks are divided into six groups as follows:

(i) Arenaceous: Formed of the deposits of coarse grained particles. They are composed of siliceous material derived from the disintegration of older rocks. The fragmental material so derived is deposited in beds of varying thickness through the agency of water. Depending upon the nature of cementing material present, some arenaceous rocks are hard and refractory, but most are loose and fall away very easily. E.g. Sandstone, grit, conglomerate and breccia.

(ii) Argillaceous rocks: Consist of small sized particles known as clay. They are composed of hydrated silica of alumina in admixture with sand, various other silicates and calcareous matter. When clay is deposited mainly of silicate of alumina, it is known as kaolin or China clay. E.g. clay, mudstone, shale and fuller's earth.

(iii) Calcareous rocks: Consists of carbonate of lime or lime and magnesia. They may be of sedimentary origin or formed by chemical precipitation or by organic agency. When, they are of organic agency, they are composed mainly of debris from plant and animal life. They are formed either by growth and decay of organisms in situ or by the transport and subsequent accumulation of their remains. The rocks so formed are found in layers, which vary considerably in depth of thickness. When formed by chemical precipitation, the calcareous material is deposited in the form of layers/sheets from waters containing calcium carbonate in solution. The important calcareous rocks of aqueous origin are limestone, chalk, magnesian, ferruginous limestones, dolomite, marks of various varieties and coral.

(iv) **Carbonaceous rocks:** Formed from decomposing vegetation under anaerobic conditions. When plants undergo decomposition under restricted air supply, is greater portion of the carbonaceous matter is retained and the material is slowly converted into coal. E.g. peat, lignite, coal, anthracite.

(v) **Siliceous rocks:** Siliceous rocks of organic origin formed from parts of minute plants and animals like diatoms, radiolarian etc. Some are soft and friable and crumble to powder very easily. Others like flint and chert are hard and compact.

(vi) **Precipitated salts:** Consist mainly of deposits formed as rock masses either by cooling, evaporation or by chemical precipitation. Water charged with acid or alkaline material, acting under pressure as it does under subterranean regions, dissolves various mineral substances from rocks with which it comes in contact. The salts thus formed deposit as rocks and such rocks vary in composition. They are

(i) Oxides – Eg: Hematite, limonite, bauxite and quartz.
(ii) Carbonates – Eg: Stalactite, stalagmite, magnetite and limestone.
(iii) Sulphates – Eg: Gypsum and anhydrite
(iv) Phosphates – Eg: Phosphorite
(v) Chlorides – Eg: Rock salt.

(vii) **Metamorphic rocks:** The igneous and sedimentary rocks after they were first formed sometimes undergo a change. When the change is considerable, the rock is said to have undergone metamorphosis and the new rock is known as a metamorphic rock. The metamorphism is brought about by the action of water, heat or pressure or by the combined action of any one of these or all. The change brought about by water is hydro-metamorphism.

The change brought about by heat is thermo-metamorphism. The change brought about by pressure is dynamo- metamorphism. The changes that are brought about are both physical and chemical in character. In some cases the metamorphism is so pronounced that the new rock looks quite different from the original.

The action of water tends to remove some material or introduce new materials. By the introduction of a cementing material like silica, lime or iron oxide, loose sand may be turned into sandstone or sandstone into a quartzite. By the removal of certain constituents by percolating waters, basalt or granite may be converted into a laterite.

The action of heat hardens the rock and develops new crystals in it. Crystalline marble is produced this way from amorphous limestone by the action of heat and pressure. Due to pressure, the crystals of the original rock get pressed or flattened and the new rock is foliated. When foliation is slight, the layers are inseparable and it is called as gneiss. It foliation is complete with distinct and separable layers it is called as schist.

16. a. Describe the role of micro-organisms in ameliorating soil. (2009)

b. Write a note on 'role of microorganism and Rhizobium in amelioration of forest soils'. (2013)

Every gram of a typical healthy soil is home to several thousand different species of bacteria. One square metre of soil can contain about 10 million nematodes and 45 000 microarthropods (springtails and mites). It has more species in it than 1 $km^2$of rainforest. In addition to bacteria, soil is home to microscopic fungi, algae, cyanobacteria, actinomycetes, protozoa and nematodes, and macroscopic earthworms, insects and the occasional wombat.

All of these organisms can be divided between autotrophs ("self-feeders"), such as plants, algae and cyanobacteria, and heterotrophs ("different-feeders"), such as fungi and bacteria, which decompose organic matter.

(a) Role of microorganism in soil

(i) **Soil microbes break down organic matter:** Microorganisms play an important role in the decomposition of organic matter. Different types of microbes are specialized to different types of organic matter, between them covering just about everything.

(ii) **Soil microbes recycle nutrients:** Soil microbes play a crucial role in returning nutrients to their mineral forms, which plants can take up again. This process is known as mineralization.

(iii) **Soil microbes create humus:** When the soil microbes have broken down all they can, what's left is called humus, a dark brown jelly-like substance that can remain unchanged in the soil for potentially millennia. Humus helps the soil retain moisture, and encourages the formation of soil structure. Humus molecules are covered in negatively charged sites that bind to positively charged ions (cations) of plant nutrients, thus forming an important component of a soil's cation exchange capacity. Humus is also suspected of suppressing plant diseases.

(iv) **Soil microbes create soil structure:** Some soil microbes secrete polysaccharides, gums and glycol proteins, which glue soil minerals together, forming the basis for soil structure. Fungal hyphae and plant roots further bind soil aggregates together. Soil structure is essential to good plant growth.

(v) **Soil microbes fix nitrogen:** Agriculture depends heavily on the ability of certain microbes (mainly bacteria) to convert atmospheric nitrogen (N_2 gas) to ammonia (NH_3). Some live freely in the soil, while others live in association with plant roots the classic example is Rhizobium bacteria in the roots of legumes. The process of conversion is known as nitrogen fixation.

Biological nitrogen fixation contributes about 60% of the nitrogen fixed on Earth. In contrast, manufactured fertilizers contribute 25%. As the cost of energy continues to raise the cost of manufactured nitrogen fertilizers will also raise, so biological nitrogen fixation is likely to have ever increasing importance in food production.

(vi) **Soil organisms promote plant growth:** Some soil microbes produce a variety of substances that promote plant growth, including auxins, gibberellins and antibiotics.

(vii) **Soil microbes control pests and diseases:** The best known example of the use of soil microbes in pest control is the commercial production of the soil bacterium Bacillus thuringiensis (Bt) to control caterpillar pests of crops. Some strains of Bt are used to control beetles and flies as well. Several strains of the fungal genus Trichoderma have been developed as biocontrol agents against fungal diseases of plants, mainly root diseases. Various other genera of fungi are used for the control of insect pests.

(b) Role of Rhizobium: "Rhizobia are soil bacteria that fix nitrogen once they found a base inside root nodules of legumes". In other terms, they can be defined as legume root-nodule bacteria. Fixation of nitrogen cannot be done independently. That is rhizobia require a plant host. It represents a vital source of N input in agricultural soils including those in arid regions. That is they convert dinitrogen into ammonia. Ammonia is toxic in nature. Hence they are rapidly absorbed into organic compounds.

Nitrogen fixation helps in increasing productivity of low-N soils and soil fertility. The various behavioral factors such as drought stress, nutrient deficiency, salt stress, fertilizers, pesticides of nitrogen-fixing systems.

17. Explain the various applications of geo-textiles for soil conservation. (2011)

"Geotextiles are permeable fabrics which, when used in association with soil, have the ability to separate, filter, reinforce, protect or drain. Typically made from polypropylene or polyester, geotextile fabrics come in three basic forms: woven, needle punched, or heat bonded.

Geotextile composites have been introduced and products such as geogrids and meshes have been developed. Overall, these materials are referred to as geosynthetics and each configuration geonets, geosynthetic clay liners, geogrids, geotextile tubes, and other scan yield benefits in geotechnical and environmental engineering design

Geotextiles and related products have many applications and currently support many civil engineering applications including roads, airfields, railroads, embankments, retaining structures, reservoirs, canals, dams, bank protection, coastal engineering and construction site silt fences or geotube. Usually geotextiles are placed at the tension surface to strengthen the soil. Geotextiles are also used for sand dune armoring to protect upland coastal property from storm surge, wave action and flooding. A large sand-filled container (SFC) within the dune system prevents storm erosion from proceeding beyond the SFC. Using a sloped unit rather than a single tube eliminates damaging scour.

Erosion control manuals comment on the effectiveness of sloped, stepped shapes in mitigating shoreline erosion damage from storms. Geotextile sand-filled units provide a "soft" armoring solution for upland property protection. Geotextiles are used as matting to stabilize flow in stream channels and swales. Geotextiles can improve soil strength at a lower cost than conventional soil nailing. In addition, geotextiles allow planting on steep slopes, further securing the slope. In building demolition, geotextile fabrics in combination with steel wire fencing can contain explosive debris.

Coir (coconut fiber) geotextiles are popular for erosion control, slope stabilization and bioengineering, due to the fabric's substantial mechanical strength. Coir geotextiles last approximately 3 to 5 years depending on the fabric weight. The product degrades into humus, enriching the soil.

18. Discuss the factors to be considered for e-cient recycling of harvested water. (2011)

Recycling of harvested water

Water harvesting is the process of collecting water from various forms of runoff from different sources. Normally the water harvesting is then both in-situ and ex-situ methods depending on the locality and socio economic status of the farmers.

The water is stored in farm ponds, percolation ponds and silt detergent tanks. The water stored in farm pond is directly used for irrigation. Then the water stored in other structure is recharging in the ground water and used for supplementary irrigation.

E-cient water recycling

For efficient recycling of water the following factors are very important.

(i) **Configuration of land:** Appropriate land configuration viz., broad fit, furrows, gradual models, strips, inter rows and inter plot, water harvesting system are very essential not only to store the water but also for efficient recycling of stored water.

(ii) **Inter and intra water harvesting and recycling:** Similarly the inter and intra row dimensions are very important for harvesting are very efficient recycling of stored water. For this purpose the furrow are about 30-40 cm diameter alternated by ridges of 60 -70 cm are formed at right angles to the field slopes. These practices reduce the runoff and at the same time water is concentrated in furrows causing better availability.

The crops are planted in furrows in lighter soil and sites of ridges in heavy soils in order to efficient utilization of harvested water.

The method of water harvesting on inter row basis also helps in improving the crop stand.

19. Explain the di-erent processes of soil erosion. Briefly describe them giving examples as to how the vegetation including trees can help in conserving soil and water. (2012)

Soil erosion

Soil erosion is a detachment and transportation of soil materials from one place to another through the action of wind, water and tidal waves and beating action of rain drops.

Process of erosion

Soil erosion process involves four factors viz., erosivity, erodability, detachability and transportability.

(i) **Erosivity:** It is the potential ability of rainfall to cause the erosion. It depends on the intensity of rainfall, amount and distribution on the size of the rain drop.

(ii) **Erodability:** It refers to the vulnerability or susceptibility of the soil to get eroded. This erodability depends on soil structure, texture and management practices.

(iii) **Detachability:** It refers to the separation of individual soil particles from the soil system through the action of erosive agents.

(iv) **Transportation:** It refers to the movement of soil particles from one place to another through the action of transporting agents of wind and water.

Role of vegetation in soil and water conservation

The biological measures include agricultural and silvicultural practices viz., contour cultivation, strip cropping, mulching, cover cropping, silvipasture, agroforestry, wind break, shelter belts etc.

(i) **Choice of crop:** Generally, pulse crop and leguminous tree crops are found effective towards soil and water conservation. These crops are considered erosion resistant crop and provide better cover and protection to soil by way of minimizing the impact of rain drop and obstructing to runoff.

(ii) **Contour cultivation:** The contour made in the slopes is cultivated by sowing the crop across the slope. The crops like cereals, pulses and grasses can be cultivated in the contour which will ensure utilization of the land coupled with preventing soil erosion.

(iii) **Strip cropping:** The erosion resistant crop and erosion permeable crops are grown in alternate strips on contour with the objective of breaking long strips to prevent soil loss and run off. The resistant crops reduce the velocity of water movement and filter the sediment from the run off to deposit them in the field.

(iv) **Cover cropping:** Any crop planted to control erosion is termed as cover crop. The pulses with drought resistant are generally grown for this purpose. In some arid areas drought tolerant grasses are grown.

(v) **Crop rotation:** Crop rotation means alternate growing of crops one after the other. The Legumes followed by grasses are preferred crop rotation which will improve the soil besides controlling the erosion.

(vi) **Mulching:** Mulching of crop residues on or near the soil surface reduce evaporation besides conserving the soil and water. The residues on surface observe wind force and reduce the wind erosion.

(vii) **Forestry practices:** The forestry component particularly leguminous trees like *Acacia*, *Leucaena*, *Dalbergia* etc. plays a major role in arresting soil erosion and conserving moisture through interception and through fall. The desirable characters of trees are drought resistance, leguminous, deep root system and will have multipurpose utility.

(viii) **Windbreaks**: Windbreak is a single row plantation and act as a barrier for protection from winds. The effectiveness of wind break depends on its density, height and length. The spacing of wind breaks depends on the density and angle of wind break to the direction of hazardous wind. Eg. *Casuarina*, *Eucalyptus* etc. The length of the wind break is about 24 times the height of the tree. The distance between the wind break is 10 times the height of the tree.

(ix) **Shelter belts**: Shelter belt is a longer barrier than a wind break consisting of combination of shrubs and trees intended for protecting field crops and conserving soil. The shelter belt includes the following design.

- **Central rows** – *Acacia Arabica, Albizzia lebbek, Azadirachta indica, Dalbergia sissoo, Eucalyptus rostrata, Prosopis spicigera, Tamarindus indica.*

- **Flank rows** – *Acacia Senegal, Anacardium occidentale, Casuarina equisetifolia, Cassia siamea, Prosopis juliflora.*
- **Outer rows** – *Acacia jacquemontii, Agave spp., Cassia auriculata, Gliricidia maculata.*

20. What are the sources and nature of elements considered as essential plant nutrients. (2013)

Sixteen chemical elements are recognized as being essential for the growth of all plants. Five others, silicon, sodium, cobalt, vanadium, and nickel, have been recognized as necessary for the growth of some plant species. Although certain essential elements can exist in nature in a number of ionic forms, plants can use only specific ones.

(i) Nitrogen: Other than carbon, hydrogen, and oxygen, nitrogen is the nutrient required by plants in the greatest quantity. The nitrogen concentration of plants ranges from about 0.5 to 5% on a dry weight basis. Since most plants have a rather high nitrogen requirement and most soils can't supply sufficient nitrogen to meet this demand, nitrogen normally must be supplemented from organic or inorganic fertilizer sources.

The ultimate source of all nitrogen in soils is the atmosphere, which is approximately 78% N_2. Although this quantity represents an almost unlimited supply, nitrogen as N_2 is not directly available for uptake by most plants. Recall that legumes in symbiosis with particular species of the bacterial genus, Rhizobium, transform gaseous N_2 to plant available form, with capacities to fix N_2 ranging from about 40 to greater than 300 lbs N/acre/year. Nonsymbiotic fixation of N_2 by free living soil microorganisms also occurs but quantities are usually less than 10 lbs N/acre/year. Lightning discharge in thunderstorms can oxidize atmospheric N_2 to plant available nitrate, but quantities are generally less than 20 lbs N/acre/year. N_2 can also be transformed to plant available forms through the fertilizer manufacturing process. The chemical triple bond that exists between the two N atoms in N_2 is very strong. All the above processes that convert N_2 to plant useable forms, therefore, are highly energy intensive.

Nitrogen is an extremely reactive element with many of its possible transformations being depicted in the

(ii) Phosphorus: Phosphorus in soil organic matter accounts for about 20 to 65% of the total phosphorus found in soils. Therefore, phosphorus mineralization from soil organic matter is an important source of available phosphorus for plant growth. Phosphorous ranks second to nitrogen as a limiting nutrient for plant growth. Although plant available forms of this element are anionic, phosphorus is immobile in soils with appreciable colloid content because it tends to be tightly bound to these tiny particles. Phosphorus may also form water insoluble compounds such as insoluble calcium phosphates in alkaline soils and insoluble iron and aluminum phosphates in acid soils. The concentration of phosphorus in soil solution is normally much less than one part per million (ppm), even in fertilized soils, and often is only hundredths of a ppm in unfertilized soils.

(iii) Potassium Potassium is required by plants in amounts second only to nitrogen. Unlike nitrogen and phosphorus, potassium is not organically combined in soil organic matter. Different potassium-containing minerals, such as micas and feldspars, therefore, are the principal sources of potassium in soils. Clay-sized micas weather more rapidly to release potassium than feldspars because of their much greater surface area. Soils that contain considerable micaceous clays may be able to supply a crop's entire potassium requirement without fertilization. Acid, weathered soils are those most likely to be deficient in available potassium.

(iv) **Calcium and Magnesium:** Calcium is the predominant exchangeable cation in soils, even in the majority of acid soils, followed by magnesium. This occurs because of the large number of minerals in soils that contain calcium and/or magnesium. Actual plant deficiencies of these elements are infrequent because problems associated with soil acidity, such as aluminum toxicity, become limiting first.

(v) **Sulfur:** Approximately 85% of total soil sulfur is found in soil organic matter. Microbial mineralization of the soil organic fraction is an important source of available sulfur for plant growth. Reactions of sulfur in soils are very similar to those of nitrogen. Sulfides, such as pyrite and other reduced forms of sulfur, are commonly unearthed in metal and coal mining. Upon exposure to oxygen, sulfuric acid which is produced through chemical and biological oxidation can result in soil acidification and acid mine drainage. Soils may also receive sulfur through atmospheric deposition. Soils near large metropolitan areas may receive greater than 150 lbs S/acre/year from the combustion of fossil fuels. Volcanic eruptions can also emit large quantities of sulfur gases. Soils most commonly deficient in available sulfur are sandy, leached soils that are low in organic matter.

(vi) **Micronutrients:** Iron, zinc, manganese, copper, chlorine, boron, and molybdenum are classified as micronutrients. Micronutrients are plant essential elements that are required by plants in much smaller amounts than the other essential nutrients. Generally, less than 1 lb/acre of each micronutrient will be present in the aboveground portion of crops. This small quantity contrasts with the 200 lb/acre or more of nitrogen.

The total quantity of many micronutrients in soils doesn't necessarily relate to plant availability. Most soils, for example, will contain from 20,000 to 200,000 lbs total iron/acre to a depth of six inches, but may not be able to supply a crop with sufficient available iron for uptake of 1 lb/acre. Iron deficiency is usually associated with highly alkaline soils because iron solubility roughly decreases 1000-fold for each one unit increase in soil pH.

Zinc, manganese, and copper availabilities are also decreased by alkalinity and by high organic matter concentrations (> 10%). These three elements form very stable bonds with soil organic matter which decrease their availability.

Plant micronutrient deficiencies are becoming more widespread because of greater quantities required by higher yields and decreasing micronutrient impurities in fertilizers.

21. What is watershed management? Explain in objectives and role in rural development. (2013)

Watershed

A watershed, also called a drainage basin or catchment area, is defined as an area in which all water owing into it goes to a common outlet. People and livestock are the integral part of watershed and their activities affect the productive status of watersheds and vice versa. Watershed is not simply the hydrological unit but also socio-political-ecological entity which plays crucial role in determining food, social, and economical security and provides life support services to rural people

Watershed management is the study of the relevant characteristics of a watershed aimed at the sustainable distribution of its resources and the process of creating and implementing plans, programs, and projects to sustain and enhance watershed functions that affect the plant, animal, and human communities within the watershed boundary. Features of a watershed that agencies seek to manage include water supply, water quality, drainage, storm water runoff, water rights, and the overall planning and utilization of watersheds. Landowners, land use agencies, storm water management experts, environmental specialists, water use surveyors and communities all play an integral part in watershed management.

Objectives of watershed management

The different objectives of watershed management programmers are:

(i) To control damaging runoff and degradation and thereby conservation of soil and water.

(ii) To manage and utilize the runoff water for useful purpose.

(iii) To protect, conserve and improve the land of watershed for more efficient and sustained production.

(iv) To protect and enhance the water resource originating in the watershed.

(v) To check soil erosion and to reduce the effect of sediment yield on the watershed.

(vi) To rehabilitate the deteriorating lands.

(vii) To moderate the floods peaks at downstream areas.

(viii) To increase infiltration of rainwater.

(ix) To improve and increase the production of timbers, fodder and wild life resource.

(x) To enhance the ground water recharge, wherever applicable.

Role in Rural Development

(i) **Women Promotion/ Gender Integration:** For gender mainstreaming and empowerment of women, women are involved at every stage of the project. The VWC (Village Watershed Committee) has at least 30% representation by women. Further, there is a special provision of Women's Development Fund under the programme by earmarking 5% of project funds for forming SHGs and promoting savings and lending activities, and for taking up 'Social Development' and 'Income generating activities'. A trained Lady Social Worker is employed to address women related issues. The activities which have been taken up by women out of this fund so far include:

- Drinking water schemes
- Drainage repair
- Soak pits
- Kitchen gardens
- Community halls

- Flour mill on group basis
- Dairy, poultry, stall fed goat rearing

(ii) Participation in community development programmes.

- Drinking water can be taken up as entry point activity. It can be augmenting the existing source or creating a new source or attending to fluoride problem etc.
- Soil moisture conservation measures like soil bunding, field bunding, raising agro forestry species on the bunds, rock fill dams; check dams can be taken up by women user groups.
- Raising plantations in the common lands and avenue plantation can be taken up.
- Horticulture, Vermiculture, green manuring.
- On-farm crop demonstrations of improved agronomic practices involving women farmers.Raising green fodder for cattle and teaching scientific feeding practices.
- Adopting smokeless Stove.
- Taking up various livelihood activities like tailoring, basket making, mat weaving, dairy, poultry etc.
- **Literacy:** Enrolling the women in adult literacy programmes and encouraging them to send their children to schools and seeing that drops-out are made to rejoin in the school.

(iii) Health and Sanitation: Utilizing the services of a health organizer in bringing awareness among women regarding nutrition/malnutrition, communicable diseases, family planning and personal hygiene. The women have to be encouraged to go for individual sanitary latrines.

- Women SHGs can play a vital role in attending to the above issues.
- Communication of information Information regarding the latest technologies in Agriculture, Agro Horticulture, Animal Husbandry etc., should be communicated to the women SHGs.

(iv) Training Programmes: Skill enhancement programmes and other training programmes on natural resources management, livelihood aspects and leadership aspects should be imparted to women.

The activities undertaken in these programmes include soil and moisture conservation measures like construction of check dams, water harvesting structures, desilting of village ponds, treatment of drainage lines/ gullies, land levelling, bunding of farms, treatment of problem soils, agro-forestry, agri-horticulture, silvi-pasture, organic farming, use of bio-fertilizers, value addition and marketing of produce through farmers groups, training & Capacity Building of stakeholders.

(v) Land Treatment

(a) Soil and Moisture Conservation: Land leveling, graded binding, contour bonding, vegetative bonding, contour cultivation, drainage line treatment, gullies stabilization, gully plugging, percolation tanks and farm ponds.

(b) **A3orestation:** Tree plantation in degraded forests, Panchayat lands, community lands, private lands etc.

(c) **Pasture Development:** In village community lands, pasture lands with suitable grass and fodder species.

(vi) Cropping Pattern

- Introduction of suitable crops, improved crop varieties, inter-cropping, contour cultivation and crop management practices;
- Live stock development fodder cultivation, milch cattle distribution, establishment of milk co-operatives
- Integration of other activities such as sheep rearing, fisheries, piggery, poultry, bee-keeping etc.

(vii) Employment Generation Activities

- Creating more employment through land based and productive activities
- Raising backyard nurseries;
- Wage earning through community assets creation such as community buildings, village roads etc.
- Cottage industries based on bamboo, wood craft, cane craft, etc.

22. Explain the survival strategies of the following group of plants (10 mark) (2014)

(i) Halophytes
(ii) Phreatophytes
(iii) Xerophytes
(iv) Succulent

(i) Halophytes: The plants, which grow in saline habitats, are called Halophytes. There is high concentration of salts like sodium chloride, $MgSO_4$ etc. in these habitats. As the habitat is physiologically dry due to salts, the halophytes show characters similar to Xerophytes.

Adaptations

- The leaves are reduced to small scaly structures (Casuarinas) or spines (Opuntia)
- The leaves are shed when water is scarce
- Water storage structures develop in the leaves
- They have a thick cuticle and a multiple layered epidermis
- They have sunken stomata
- They have long roots, which go in search of water
- The stem becomes green and takes over the function of photosynthesis

(ii) Phreatophyte: A phreatophyte is a deep-rooted plant that obtains a significant portion of the water that it needs from the phreatic zone (zone of saturation) or the capillary fringe above the phreatic zone.

- Phreatophytes are plants that are supplied with surface water and often have their roots constantly in touch with moisture.

- A phreatophyte is one that absorbs its water from a constant source on the ground.
- Phreatophytes are plants that grow extremely long roots, called tap roots, that allow them to obtain water from deep in the ground.
- Mesquites (Prosopis sp.) have the longest tap root of any desert species; it may reach down 25 m.
- The creosote bush (*Larrea tridentata*) is one of the most successful of all desert species. because it uses a combination of several
- It has adaptations like, deep tap root, a shallow root system, and wax-coated leaves that close their pores during the day to avoid loss of water.

(iii) Xerophytes: A xerophyte (xero meaning dry, phyte meaning plant) is a plant which is able to survive in an environment with little available water or moisture, usually in environments where potential evapotranspiration exceeds precipitation for all or part of the growing season

- The root system is very well developed with root hairs and root caps. e.g. *Calotropis*.
- The roots are fasciculated as in Asparagus.
- Stems are stunted, woody, dry, hard, ridged, and covered with thick bark, may be underground, e.g. *Saccharum*. In *Opuntia* phylloclade is covered with spines.
- Stem is covered with thick coating of wax and silica in *Equisetum* or dense hairs as in *Calotropis*.
- Stems may be modified into a thorn e.g. *Ulex* or cladodes e.g. *Asparagus*.
- Leaves are very much reduced, small scale-like, appearing only for a brief period (Caducous) sometimes modified into spines or scales as in *Casuarina, Ruscus* and *Asparagus*.
- Lamina may be narrow or needle like as in *Pinus* or divided into many leaflets as in *Acacia* or succulents as in *Aloe*.
- In *Euphorbia* and *Zizyphus jujuba* stipules become modified into spines.
- Xerophytes like *Calotropis* have hairy covering on the leaves and stems to check transpiration.

(iv) Succulent

Adaptations

- Crassulacean acid metabolism (CAM) to minimize water loss
- Absent, reduced, or cylindrical-to-spherical leaves
- Reduction in the number of stomata
- Stems as the main site of photosynthesis, rather than leaves
- Compact, reduced, cushion-like, columnar, or spherical growth form
- Ribs enabling rapid increases in plant volume and decreasing surface area exposed to the sun
- Waxy, hairy, or spiny outer surface to create a humid micro-habitat around the plant, which reduces air movement near the surface of the plant, and thereby reduces water loss and creates shade

- Roots very near the surface of the soil, so they are able to take up moisture from very small showers or even from heavy dew
- Ability to remain plump and full of water even with high internal temperatures (Eg. 52 °C or 126 °F)
- Very impervious outer cuticle

23. Describe di-erent textural classes of soil and the say they a-ect plant growth. (2015)

Soil texture

- Soil texture is defined as the relative proportion of sand, silt and clay and it is concern with the size of mineral particles. The classification of soil texture is based on size of mineral particles is furnished below.

Classification System	Size of the particle
Sand	2.0-0.05 mm
Silt	0.05-0.002 mm
Clay	< 0.002 mm

- The soil texture is considered as a permanent feature and it is the basic properties of the soil. If the soil contains more than 85% sand the soil is called sandy soil. When it contains more than 80% silt is called silty soil and it contains more than 40% clay then it is called clay soil.

Soil texture and plant growth

- The three fundamental group soils to texture classes are sandy, loamy and clay. Loamy soils are best suited for agriculture production because it contains mixture of sand, silt and clay particles. Due to this structured mix of soil loamy soils are able to retain more water and nutrients than sandy soil and exhibit better drainage, aeration and tillage properties which influences the plant growth.
- If the soil texture is sandy means the porosity will be maximum and the water holding capacity is very minimal and this is not support the plant growth and development.
- Similarly if the soil texture contains only clay means, it imparts cohesion and stickiness and fine texture to the soil. This induces resistance to the passage of implements and makes tillage operation very difficult. Due to this character clay soils are considered heavy soils and sandy soils considered light soils and in both the cases the only few and specific plant can be grown.
- The soil texture influences soil physical and chemical properties like water holding capacity, nutrition repression, nutrient fixation, nutrient availability, drainage, strength, compressability and thermal regime which influences growth and development of plants. Hence soil texture is the significant physical property which influences plant growth.

24. a. Saline and alkaline soil reclamation by a-orestation. (2006)

b. Reclamation of saline and alkaline soil. (2009)

c. What are saline and sodic soils? (2010)

d. Why saline-alkali soil is considered problematic? Can you suggest any procedure to make it suitable for plant growth? (2015)

e. Write distinguishing features of saline and alkaline soil (2016)

Saline soils

Saline soils are defined as soils having a conductivity of the saturation extract greater than 4 dS m^{-1} and an exchangeable sodium percentage less than 15. The pH is usually less than 8.5. Formerly these soils were called white alkali soils because of surface crust of white salts.

Alkali or sodic soil

Alkali or sodic soil is defined as a soil having a conductivity of the saturation extract less than 4 dS m^{-1} and an exchangeable sodium percentage greater than 15. The pH is usually between 8.5 - 10.0. Most alkali soils, particularly in the arid and semi-arid regions, contain $CaCO_3$ in the profile in some form and constant hydrolysis of $CaCO_3$ sustains the release of OH^- ions in soil solution. The OH^- ions so released result in the maintenance of higher pH in calcareous alkali soils than that in non - calcareous alkali soils.

The saline and alkaline soil are considered problematic because, excess soil salinity causes poor and spotty stands of crops, uneven and stunted growth and poor yields, the extent depending on the degree of salinity. The primary effect of excess salinity is that it renders less water available to plants although some is still present in the root zone. This is because the osmotic pressure of the soil solution increases as the salt concentration increases. Apart from the osmotic effect of salts in the soil solution, excessive concentration and absorption of individual ions may prove toxic to the plants and/or may retard the absorption of other essential plant nutrients

Reclamation of Saline and Alkali Soils

- Adequate drainage.
- Availability of sufficient water to meet crop use and also leach the salt below the root zone in the soil.
- Better than average soil management to include perfect land leveling, good bunding for irrigation and advanced agronomic practices.
- Protection and reclamation to be taken in large blocks.
- Irrigation water should be of good quality.

I. Saline Soil Reclamation and Management

Saline soils in which the soluble salts contain appreciable amounts of calcium and magnesium do not develop into alkali soils by the action of leaching water. The reclamation is comparatively easy in such soils. The main problem is to leach the salts downward below thc root zone and out of contact with subsequent irrigation water.

Following methods may be used for removal of salts

(A) Mechanical Methods

(i) Flooding and leaching down of the soluble salts: The leaching can be done by first ponding the water on the land and lowering it to stand there for a week. Most of the soluble salts would leach down below the root zone. After a week, standing water (dissolved with soluble salts) is allowed to escape. Such, 2 to 3 treatments are given to reclaim highly saline soils. Sometimes gypsum is also added to flood water when the soluble salts are low in calcium to check development of alkalinity.

(ii) Scrapping of the surface soil: When the soluble salts accumulate on the soil surface, scrapping helps to remove salts. This is a temporary cure and salinity again develops on such lands.

(B) Cultural Methods (Crop, Soil and Water Management)

(i) Providing proper drainage: If the soil is not free draining, artificial, drains are opened or tile drains laid underground to help wash out the salts.

(ii) Use of salt free irrigation water: Salt free good quality of irrigation water should be used.

(iii) Proper use of irrigation water: It is known that as the amount of water in the soil decreases the concentration of salts in the soil solution increases, thus, moisture should be kept at optimum field capacity.

(iv) Planting or sowing of seeds in the furrow: The salt concentration even in smaller amounts is most harmful to the germinating seedlings. Water generally evaporates from the highest surface by capillarity and hence, these points have maximum salt concentrations. If the seeds or seedlings are planted inside the furrows, they escape the zone of maximum salt concentrations and thus, can germinate and develop properly during their early growth stage.

(v) Use of Acidic Fertilizer: In saline soil, acidic nature of fertilizers (e.g., Ammonium sulphate) should be used.

(vi) Use of organic manures: The organic manures have very high water-holding capacity. When sufficient amount of these manures are added the water-holding capacity of soil increases and as a result the conductivity of the soil solution decreases.

(vii) Ploughing and leveling of the land: Ploughing and leveling of the land increases the infiltration and percolation rate. Therefore, salts leach down to the lower levels.

(viii) Retardation of water evaporation from soil surface: Water may be conserved in the soil retarding the water evaporation. Thus, salts may remain in the lower level with the water.

(ix) Growing of salt tolerant crops

(a) High salt tolerant crops : Para grass, barley, sugar beet, etc.

(b) Moderately salt tolerant crops : Wheat, rice, sorghum, maize, flax etc.

(c) Low salt tolerant crops : Beans, radish, white clover etc.

(d) Sensitive crops : Tomato, potato, onion, carrot etc.

Tree species suitable for saline soils

Acacia aneura, Acacia holosericea, Acacia salicina, Annona squamosa, Balanites aegyptica, Butea mososperma, Capparis decidua, Cassia siamea, Casuarina equisetifolia, Colophospermum mopane, Dalbergia sissoo, Ficus bengalensis, Moringa oleifera, Phoenix dactylifera, Pongamia pinnata, Prosopis juliflora, Salvadora persica, Tamarindus indica, Tamarix aphylla, Tecomella undulata, Ziziphus mauritiana and *Ziziphus nummularia.*

Tree species suitable for Alkaline soils

Acacia aneura, Acacia nilotica, Acacia salicina, Acacia tortilis, Albizia lebbek, Balanites aegyptica, Butea monosperma, Capparis deciduas, Casuarina equisetifolia, Colophospermum mopane, Dalbergia latifolia, Emblica oDcinalis Eucalyptus camaldulensis, Ficus bengalensis, Leucaena leucocephala, Madhuca latifolia, Pithecelobium dulce, Pongamia pinnata, Prosopis cineraria, Prosopis julifolra, Salvadora persica, Tamarindus indica and *Tamarix aphyll.*

25. Why are carbon recycling important/ what are its influences on Climate? Discuss your points for or against. (2015)

Carbon is one of the major element required in large quantity by the plants and it is available in three forms. Carbon is available as gaseous form in the atmosphere and as dissolved carbon in ozone and also it is stored as sediment in fossil fuels and other organic matter as solid form.

Carbon recycling is important because it influences all the biological growth and development simultaneously building up resources like fossil fuels for human use and sustenance. Carbon occurs in the ocean as a largest pool with about 39000 billion tonnes which is equivalent to 93% of the earth's total carbon content. All the carbon is bound chemically in carbon-di-oxide. Plants absorb this carbon-di-oxide through photosynthesis. This carbon is very essential for photosynthesis of green plants and synthesis of fossil fuels through organic carbon and it is also involved in respiration process. Hence carbon is very essential and important factor in biological sustainability.

Carbon cycle: The carbon in the atmosphere in the form of carbon-di-oxide enters into the biological cycle through photosynthetic process wherein the carbon along with hydrogen and oxygen are utilized by the plants and convert them into organic compounds. The carbon-di-oxide assimilated by the plants is stored in woody tissues and act as a reservoir. The carbon is released from the plants during respiration. The death of plants and animals leads to continuous sedimentation and decomposition which forms large reservoirs of fossil fuels and on burning them, they are once again released into the atmosphere. Hence carbon cycle is very essential for biological cycling and geological wealth.

Negative influence of carbon on climate

Carbon in the form of carbon-di-oxide is one of the serious green house gases which are trapping the ultra violet rays and keep the globe warm resulting in increase of earth's temperature which is now pronounced as global warming.

- This global warming has serious implication in changing the climatic pattern which lead to erratic temperature fluctuations and variability in monsoon pattern thereby it affects the agriculture.

- The increase in carbon-di-oxide and the associated temperature increase lead to melting of glaciers and changes in the arctic ecosystem.
- Due to melting of glaciers there are chances of increase in sea level and may lead to sinking of various islands and may affect the society.
- The increase in carbon content and the associated increase in temperature can cause serious human diseases like pressure, skin burn, stress, cataract etc.
- The increase in global warming may lead to changes in climatic conditions and may lead to migration of biodiversity and changes in composition of biodiversity.
- The soil ecosystem may find significant changes due to alteration in soil temperature, soil structure, texture and biota which may lead to conversion of productive soil into unsuitable soil.
- There are chances of increasing aridity and may lead to decline in agriculture productivity.

Positive influence of carbon

- Carbon is one of the essential element which are involved in the biological process of plants. The plants during photosynthesis require carbon in the form of carbon-di-oxide which are utilized to produce energy in the form of carbohydrates; hence it is very essential.
- Respiration is a biological process and during this process the carbon has to be released outside in order to complete the successful respiration process. Hence it is very essential to complete the biological process.
- The carbon balance in the atmosphere is very significant in order to protect the earth from severe cold and freezing temperature. The carbon-di-oxide is essential to keep the earth warm in order to balance the agriculture productivity and activate the biological process of plants and animals. Hence carbon-di-oxide is essential in required quantity.
- Carbon-di-oxide is essential in the form of organic source particularly it is able to generate wide range of petroleum products and thus contributing towards energy requirement of the society.
- Carbon is also available in the form of sediments particularly in the form of limestone which act as a mineral source and find profound utility. Hence carbon is essential.

26. Good watershed management must consider the social, economic and environmental sustainability and institutional factors. Comment (2015)

Yes, the watershed management should consider the social, economic and environmental sustainability.

Watershed development aims to balance the conservation, regeneration and use by humans of land and water resources within a watershed. Common benefits from successful watershed development projects include improved agricultural yields and increased access to drinking water. The overall attributes of the watershed development approach, by and large, are three fold, viz. promoting economic development of the rural area, employment generation, and restoring ecological balance. However, the multiple objectives include:

- **Social factors:** It includes alleviation of poverty, awareness generation, improving skills of the local community, capacity building activities, women's participation in decision-making process, empowerment of the community.
- **Economic factors:** It draws attention for increase in cropping intensity through inter and sequence cropping, maximizing farm income through agricultural related activities such as dairy, poultry, sheep and goat farming, improved and sustained livelihood status of the watershed community with special emphasis on the poor and women.
- **Environmental factors:** Protecting vegetative cover throughout the year, to create ecological balance in the watershed area, protecting fertile top soil, utilizing the land based on its capabilities, in situ conservation of rain water, increasing ground water recharge.
- **Institutional factors:** It includes formation of watershed committees and self-help groups, establishing sustainable community organization.
- **Equity factors:** To develop equitable distribution of the benefits of land and water resources development and the consequent biomass production, involvement of village communities in participatory planning, implementation, social and environmental arrangement, maintenance of assets and to operate in a more socially inclusive manner.

Watershed and society

The success of any watershed development program required volunteer participation of the society or community as a whole. The watershed program should ensure the regular income and employment activities to the local communities. The activities such as nursery development, plantation establishment, collection and harvesting of wood and etc., will help to have sustain employment and will ensure sustainability of watershed project.

Watershed and Economic sustainability

All the watershed programmes are designed in order to generate economic sustainability for the local community and society. The watershed development programme should involve all economic improvement activities like sustained and profitable agriculture, integrated livestock production system, sericulture production technology, fish production, biogas production, poultry production, agroforestry production, dryland technology coupled with efficient water use management system. These integrated farming system coupled with efficient use of natural resources will extend economic sustainability on a continuous basis. The integrated farming system judicially incorporates agriculture, horticulture, animal husbandry, fisheries etc. which will generate adequate income and employment generation activities and extend scope for economic sustainability.

Watershed and Environment

The increasing human populations and the human activities have compelled sustainable management of watershed. The watershed management judicially incorporates sustainable management of environmental issues. This includes encouraging and increasing the harmony between the society and the environment. Hence the watershed is planned in such a way in order to protect the environment as a whole which lead to increased soil and water conservation, reduced soil loss, increase in components of vegetation and biodiversity and ultimately it influences the local climate of the region. An ideal water shed resolve all the issues related to soil and water conservation besides augmenting and ameliorating the soil and climate. Hence the watershed development programmes designed to ensure environmental sustainability.

27. Integrated Watershed Management Approaches. (2006)

Watershed

A watershed also called a drainage basin or catchment area, is defined as an area in which all water owing into it goes to a common outlet. People and livestock are the integral part of watershed and their activities affect the productive status of watersheds and vice versa. Watershed is not simply the hydrological unit but also socio-political-ecological entity which plays crucial role in determining food, social, and economical security and provides life support services to rural people.

Watershed Management

Watershed management is the study of the relevant characteristics of a watershed aimed at the sustainable distribution of its resources and the process of creating and implementing plans, programs, and projects to sustain and enhance watershed functions that affect the plant, animal, and human communities within the watershed boundary. Features of a watershed that agencies seek to manage include water supply, water quality, drainage, storm water runoff, water rights, and the overall planning and utilization of watersheds. Landowners, land use agencies, storm water management experts, environmental specialists, water use surveyors and communities all play an integral part in watershed management.

- **Integrated watershed management:** This approach suggest the integration of technologies within the natural boundaries of a drainage area for optimum development of land, water and plant resources to meet the basic needs of people and animals in a sustainable manner. This approach aims
  - (i) To improve the standard of living of common people by increasing his earning capacity or by offering all facilities required for optimum production.
  - (ii) In order to achieve its objective, integrated watershed management suggests to adopt land and water conservation practices,
  - (iii) Water harvesting in ponds and recharging of groundwater for increasing water resources potential
  - (iv) Stress on crop diversification,
  - (v) Use of improved variety of seeds
  - (vi) Integrated nutrient management
  - (vii) Integrated pest management practices.
- **Consortium Approach:** Consortium approach emphasizes on collective action and community participation including of primary stakeholders, government and non-government organizations, and other institutions. Watershed management requires multidisciplinary

skills and competencies. Easy access and timely advice to farmers are important drivers for the observed impressive impacts in the watershed. These lead to enhance awareness of the farmers and their ability to consult with the right people when problems arise. It requires multidisciplinary proficiency in field of engineering, agronomy, forestry, horticulture, animal husbandry, entomology, social science, economics and marketing. It is not always possible to get all the required support and skills-set in one organization. Thus, consortium approach brings together the expertise of different areas to expand the effectiveness of the various watershed initiatives and interventions.

Recommendations for Practioners

- Select watershed sites where dire need exists in terms of improving soil and water conservation, enhancing productivity and improving livelihoods.
- Adopt holistic and participatory consortium approach from the beginning ie, from selection of watershed.

28. Describe techniques for eco-restoration of water-logged waste-lands. (2009)

Water logging is a problem occurs due to stagnation of water for most of the year due to improper and poor drainage. Water logging is a common feature in coastal lands, canal irrigation area etc. and it is defined as an area where the water table is increased to an extent that the soil pores in the root system in the crop saturates resulting in restriction of normal circulation of air, decreased oxygen and increased carbon dioxide.

Techniques for the eco-restoration of water logging soil

Before eco-restoration the Water logging area have to be treated properly in order to enhance the drainage and reduce the levels of stagnation. The following are the measures used in water logged waste lands.

(i) **Ditching:** In moist land, ditching is commonly used to lower the water table by draining the excess water. In some countries tractors are used to turn over the subsoil for raising dykes on one or both sides of the ditches, which are then planted. The depth and width of the ditches mainly depend on the water table and its fluctuation; the higher the water table with greater fluctuation; the deeper and wider the ditches. The water table and its fluctuations are generally controlled below 50 cm from the ground surface.

(ii) **Terracing:** In waterlogged lands the water table is closed to or even above the ground surface. Just ditching is not enough to drain the excess water. The soil removed from the water logged and immerged lands are used to raise spaces (terraces) on one side and to create ponds on the other. The size of the terrace and pond varies greatly in accordance with the water situation and agroforestry practice. In general, water and land are such proportions as 0.5:1.5, 1.0:1.0 and 1.5:1.5.In immerged land, terraces should be higher and ponds should be bigger and deeper. For planting trees with a deep root system the terraces should be much larger and higher. If fishery is considered as a part of the agroforestry system, bigger ponds with deeper depths are needed.

(iii) Establishment of drainage canals: Wherever water logging is a problem, canals have to be dug up to drain the excess water and divert the inflow water

Restoration process

- **Mound planting:** In Water logged area, mound planting can be practiced which will help to establish the plantation very easily. Once established the tree stabilizes the soil and increased the infiltration rate through their action.
- **Ridge cum trench planting:** In Water logged area, based on the species, the trenches are made based on the Espacement plan for each species. The soil excavated from the trench is made as ridges along the trenches and planting of trees are encouraged along the ridges in order to avoid the damage due to excess water.

Restoration of Water logged area through trees

Eucalyptus robusta, Syzygium cumini, Salix spp., Populous nigra, Terminalia arjuna, Casuarinas equisetifolia and *Acacia nilotica.*

Restoration through agroforestry

A combination of agriculture crop like jute, sugarcane with palatable grass like Napier and guinea grass along with tree component is highly effective in eco-restoration of water logged areas.

29. Explain di-erent kinds of soil erosion their causes and management of soil eroded areas. (2006)

Soil Erosion

The soil forming processes and the erosion processes of running water and wind go on simultaneously. However, generally, there is a balance between these two processes. Sometimes, such a balance is disturbed by natural or manmade factors, leading to a greater rate of removal of soil. With increasing population, the pressure on the land increases and forests are removed for human settlement, for cultivation, for grazing animals and for various other needs. The two most important agents of soil erosion are wind and water. Wind erosion is significant in arid and semi-arid regions. Water erosion is significant in regions with heavy rainfall and steep slopes. Water erosion which is more serious and occurs extensively in different parts of India, takes place mainly in the form of sheet and gully erosion.

There are four kinds of soil erosion which can be arranged in an order of

(i) Splash erosion

(ii) Sheet erosion

(iii) Rill erosion

(iv) Gully erosion

(i) Splash Erosion: Raindrop splash erosion is a result of the energy of falling raindrops causing detachment of soil particles and down-slope movement of sediment. Maintenance of ground cover, such as in reduced or no-till operations, use of cover crops, and enhancement of the stability of soil aggregates can be important in reducing detachment of soil particles. The effect of manure application in enhancing soil aggregation also leads to reducing soil erodibility due to the raindrop splash effect.

(ii) Sheet Erosion: Sheet erosion, although less noticeable than other types of erosion, typically is the main erosive force. Sheet erosion is less noticeable, as it does not leave obvious cuts in the soil surface as with gully erosion. Sheet erosion is the removal of a relatively uniform, although thin, layer of soil from the land surface by unchanneled runoff, or sheet flow. Sheet erosion takes place on level lands after a heavy shower and the soil removal is not easily noticeable. But it is harmful since it removes the finer and more fertile top soil.

(iii) Rill Erosion: Rill erosion is the process by which numerous small channels–less than three inches in depth–are formed. This type of erosion results from concentration of overland water flow associated with sheet erosion. Rill erosion can be especially serious on recently cultivated land. Rill erosion is best minimized by minimizing sheet flow, such as by maintaining crop residues and utilizing cover crops. Physical barriers, such as terraces, and vegetative barriers can be effective in stopping or reducing rill erosion.

(iv) Gully Erosion: Gully erosion refers to the cutting of narrow channels called gullies. The gullies can be caused by small channels of approximately 3 to 12 inches deep. Gullies may be one to several feet deep. Gully erosion cuts deep and removes the surface soil as well as deeper soil that may still have substantial amounts of total nutrients but less compared to the surface soil. Gully erosion needs to be prevented, as it is difficult to check once started. Gully erosion is common on steep slopes. Gullies deepen with rainfall, cut the agricultural lands into small fragments and make them unfit for cultivation. A region with a large number of deep gullies or ravines is called badland topography.

Causes of Soil Erosion

Land use: Humans play a major role in soil erosion through their use and abuse of natural resources, for example deforestation, grazing, arable land use, faulty farming systems, high crop intensity, housing construction, mining etc.

Climate: The two most important climatic factors having a direct effect on erosion are precipitation and wind velocity. Other climatic factors have an indirect effect on soil erosion, such as water balance, evapo-transpiration, temperature and relative humidity. Indirect factors affect the erosivity of rainfall by altering the soil moisture regime and the proportion of rainfall that may become surface runoff. For erosion control it is necessary to investigate physical characteristics of rainfall, including the amount, distribution, intensity, energy load, seasonality and variability of rainfall and the formation and course of surface runoff.

Soil: The susceptibility of a soil to erosion is influenced by its physical, hydrological, chemical and mineralogical properties as well as its soil profile characteristics. Important soil physical and hydrological properties that affect the resistance of a soil to erosion include texture, structure, water retention and transmission properties.

Hydrology: Infiltration, surface detention, overland flow velocity, and subsurface water flow are important soil erosion components of the hydrological cycle. The different types of flow and their velocities may be turbulent or laminar, steady or unsteady, uniform or non-uniform and influence the extent of erosion

Landforms: Slope gradient, slope length and shape of slope are the important variables of landform that affect erosion processes for all types of soil erosion, e.g., splash, sheet, rill, and gully erosion.

Management of soil eroded areas

The soil eroded areas can be managed with the following strategies.

Soil and water conservation can be effectively done through physical and biological approaches. The physical approach included land preparation, contour forming and other associated mechanical structures. The biological approach includes choice of species, contour cultivation, mulching, strip and cover cropping, shelter belts and wind breaks etc.

A. Physical or mechanical measures

(i) **Land preparation:** It involves ploughing to influence intake of water, obstruction to surface flow of water and soil, deep ploughing to reduce erosion due to cloddy soil structure.

(ii) **Contour bunding:** Contour bunding is a traditional process and found effective physical measure for soil and water conservation. It consists of narrow based trapezoidal bunds on contour to impound run of water behind them such that it can gradually in filtrate in to the soil for use by the plants. This method of contour bunding is recommended for areas with low rainfall of less than 600 mm and the slope up to 6%.

(iii) **Graded bunding:** This is another method of soil and water conservation practiced in areas of more than 600 mm rainfall. This method will help to remove the excess water and avoid water stagnation.

(iv) **Bench terracing:** It is a combination of ridge and channel belt across the slope. It consists of transforming relatively steep land into a serious of level strips or platforms across the slope to reduce the slope length and the associated erosion. The field is made into serious of benches by excavating soil from upper part and filling them in lower part of the terrace. This method is practiced for the lands with the slope ranged between 16 and 33%.

Contour Bonding

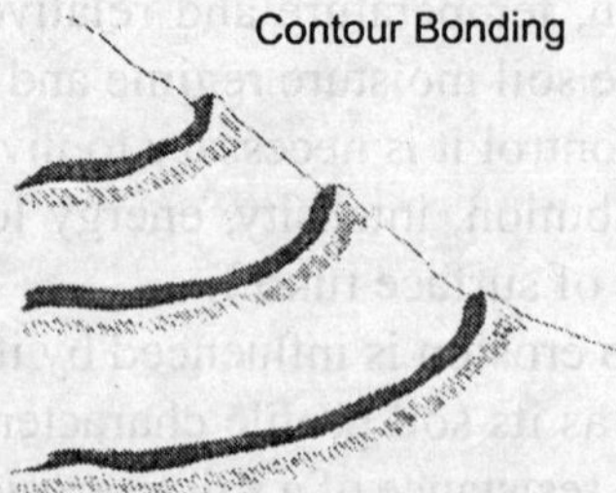

(v) **Contour trenching:** This is one of the popular methods of soil and water conservation wherein the trenches are made along the contour. This method is practiced in hill slopes. The trenches made will break the slope length; reduce the velocity of runoff leading to retarded scouring and carrying capacity. This method is practiced with the slopes exceeding 20%. The trench size is usually 16 x 45 cm and the distance between the trenches varied between 10 x 30 m.

B. **Biological measures:** The biological measures include agricultural and silvicultural practices viz., contour cultivation, strip cropping, mulching, cover cropping, silvipasture, agroforestry, wind break, shelter belts etc.

(i) **Choice of crop:** Generally, pulse crop and leguminous tree crops are found effective towards soil and water conservation. These crops are considered erosion resistant crop and provide better cover and protection to soil by way of minimizing the impact of rain drop and obstructing to runoff.

(ii) **Contour cultivation:** The contour made in the slopes is cultivated by sowing the crop across the slope. The crops like cereals, pulses and grasses can be cultivated in the contour which will ensure utilization of the land coupled with preventing soil erosion.

(iii) **Strip cropping:** The erosion resistant crop and erosion permeable crops are grown in alternate strips on contour with the objective of breaking long strips to prevent soil loss and run off. The resistant crops reduce the velocity of water movement and filter the sediment from the run off to deposit them in the field.

(iv) **Cover cropping:** Any crop planted to control erosion is termed as cover crop. The pulses with drought resistant are generally grown for this purpose. In some arid areas drought tolerant grasses are grown.

(v) **Crop rotation:** Crop rotation means alternate growing of crops one after the other. The Legumes followed by grasses are preferred crop rotation which will improve the soil besides controlling the erosion.

(vi) **Mulching**: Mulching of crop residues on or near the soil surface reduce evaporation besides conserving the soil and water. The residues on surface observe wind force and reduce the wind erosion.

(vii) **Forestry practices:** The forestry component particularly leguminous trees like *Acacia*, *Leucaena*, *Dalbergia* etc. plays a major role in arresting soil erosion and conserving moisture through interception and through fall. The desirable characters of trees are drought resistance, leguminous, deep root system and will have multipurpose utility.

(viii) **Windbreaks**: Windbreak is a single row plantation and act as a barrier for protection from winds. The effectiveness of wind break depends on its density, height and length. The spacing of wind breaks depends on the density and angle of wind break to the direction of hazardous wind. Eg. *Casuarina*, *Eucalyptus* etc. The length of the wind break is about 24 times the height of the tree. The distance between the wind break is 10 times the height of the tree.

(ix) **Shelter belts:** Shelter belt is a longer barrier than a wind break consisting of combination of shrubs and trees intended for protecting field crops and conserving soil. The shelter belt includes the following design

- **Central rows:** *Acacia arabica, Albizzia lebbek, Azadirachta indica, Dalbergia sissoo, Eucalyptus rostrata, Prosopis spicigera, Tamarindus indica.*

- **Flank rows:** *Acacia senegal, Anacardium occidentale, Casuarina equisetifolia, Cassia siamea, Prosopis juliflora.*
- **Outer rows:** Acacia jacquemontii, Agave spp., Cassia auriculata, Gliricidia maculata.

30. Suggest suitable trees for di-erent soil type? (2005)

Sandy soils: The following tree species are suitable for sandy soils of a watershed. *A. leucophloea* (Israeli kikar), *A. modesta* (Plhulani), *A. nilotica, A. tortilis, , Ailanthus excelsa, Albizia lebbeck* (siris), *Azadirachta indica* (Neem), *Casuarina equisetifolia* (Jortor), *Dalbergia sissoo* (Shisham), *Eucalyptus hybrid, Melia azedaraoh* (Dhrek), *Parkinsonia, Prosopis cineraria* (Jand), *Salvadora oleoides, Tecomella undulata, Zizyphus mauritiana* (ber).

Water-logged soils: The following tree species are suitable for water-stagnated or water logged areas of a watershed. *Anthocephalus chinensis* (Kadam), *Barringtonia acutangula, Bombax ceiba (Semul), Dendrocalamus* spp. *(Bamboos), Eucalyptus robusta, Eugenia cuspidata, Pongamia pinnata, Salix* spp. *(Willows), Syzyquium cumini (Jamun), Terminalia arjua (Arjun)* and *Trewia nudiflora.*

Shallow, rocky soils: The following tree species are suitable for shallow and rocky soils of a watershed. *Acacia holosericea, Acaica leucophloea, Acacia salicina, Acacia senegal, Anacardium occidentale, Annona squamosa, Azadirachta inidca, Balanites aegyptica, Borassus flabellifer, Boswellia serrata, Butea monosperma, Capparis decidua, Cassia siamea, Celtis australis, Cordia obliqua, Dendrocalamus strictus, Dichrostachys cineraria, Eucalyptus tereticornis, Grewia tenax, Hardwickia binata, Holoptelia integrifolia, Madhuca latifolia, Parkinsonia aculeata, Pithecellobium dulce, Pongamia pinnata, Prosopis juliflora, Rhus mysorensis, Wrightia tinctoria* and *Zizyphus nummularia.*

Sandy, duny soils: The following tree species are suitable for sandy and duny soils of a watershed. *Acacia albida, Acacia aneura, Acacia ferruginea, Acacia holosericea, Acacia leucophloea, Acacia salicina, Acacia senegal, Acacia tortilis, Balanities aegyptica, Borassus flabellifer, Boswellia serrata, Butea monosperma, Capparis decidua, Cassia siamea, Casuarina equisetifolia, Colophospermum mopane, Dalbergia sissoo, Dichrostachys cineraria, Hardwickia binnata, Moringa oleifera, Morus alba, Phoenix dactylifera, Prosopis cineraria, Prosopis juliflora, Salvadora oleoides, Tamarix aphylla, Tecomella undulate* and *Zizyphus nummularia.*

Saline soils: The following tree species are suitable for saline soils of a watershed. *Acacia aneura, Acacia holosericea, Acacia salicina, Annona squamosa, Balanites aegyptica, Butea mososperma, Capparis decidua, Cassia siamea, Casuarina equisetifolia, Colophospermum mopane, Dalbergia sissoo, Ficus bengalensis, Moringa oleifera, Phoenix dactylifera, Pongamia pinnata, Prosopis juliflora, Salvadora persica, Tamarindus indica, Tamarix aphylla, Tecomella undulata, Ziziphus mauritiana* and *Ziziphus nummularia.*

Alkaline soils: The following tree species are suitable for alkaline soils of a watershed. *Acacia aneura, Acacia nilotica, Acacia salicina, Acacia tortilis, Albizia lebbek, Balanites aegyptica, Butea monosperma, Capparis deciduas, Casuarina equisetifolia, Colophospermum mopane, Dalbergia latifolia, Emblica oDcinalis, Eucalyptus camaldulensis, Ficus bengalensis, Leucaena leucocephala, Madhuca latifolia, Pithecelobium dulce, Pongamia pinnata, Prosopis cineraria,*

Prosopis julifolra, Salvadora persica, Tamarindus indica, Tamarix aphylla, Tecomella undulata, Terminalia arjuna, Zizyphus mauritiana and *Zizyphus nummularia.*

31. Why is a lot of emphasis laid on research relating to soil conservation? Discuss (2015)

Lot of emphasis has been oriented towards research relating to soil conservation because soil and water are the basic resources of the country and they must be conserve as carefully as possible. It is estimated that about 2 billion hectares of land is degraded globally and annually 5-7 million hectares of productive land are getting lost. In India, out of 328 million hectares of geographic areas 68 million hectares are critically degraded while another 107 million hectare are severely eroded.

In the estimated 175 million hectare of degraded cultivated plan, nearly all of them are subjected to serious soil erosion hazards. The arid areas are subjected to wind erosion and the semi arid areas are subjected to sheet and gully erosion. The ravines of Yamuna and Chambal are continuously moving to fertile gangetic plain.

The demand for plant and animal products is increasing at an alarming rate due to increase in population and the associated increase in standard of living. The basic resources needed to meet the demand are soil and water without which the growing of plants is not possible. Hence increasing research emphasis has been given to conservation of soil.

For growth and reproduction the plants require 16 different elements. Similarly the animals and human beings need 14 of the same in addition to 6 more elements. The plants absorb elements which they do not need, including the 6 needed exclusively by the animals and people. Many species and individuals of animal kingdom live exclusively on plants. But these plants cannot grow without soil on a commercial scale and hence emphasis has been given to soil conservation research as the soil adds all the required nutrients to the plants.

Due to the above facts and figures it is evidenced that soil is lost heavily which results in loss of top soil, loss of fertility, silting of reservoirs, reduction and ground water potential and productive land into unproductive land. To resolve these issues and to augment the productivity of agriculture increasing emphasis has been given for soil conservation research.

32. Discuss the impact of forests on the climate of locality. How do forests a-ect soil and water conservation? (2007)

The forests play a vital role on the climate of the locality.

(i) **Impact on temperature:** The presence of trees and the forests rapidly reduces the local and micro climate by reducing the temperature due to their filtering mechanism. The tree canopy obstructs the incoming radiation thereby help to reduce the temperature both in the atmosphere and also in the soil. The effect of this temperature reduction is witnessed around the forested areas as well. Dense ever green forests are likely to have more influence on air temperature than deciduous forests.

(ii) **Impact on wind:** The forests reduces the wind velocity thereby help to protect the other vegetation due to the action of heavy wind and storm. The forests and the trees due to their broader canopy and strong base of the stem help to reduce the wind velocity. Dense forests with heavy foliage with the greatest effect and it has the capacity to reduce 20 – 60% of the wind velocity.

(iii) **Impact on precipitation:** It is believed forests increases precipitation. It plays a very big role in local condensation and increases the precipitation. Evapotranspiration from forests is more rapid and a large amount of moisture is available for precipitation. It is also proved that forest tend to enrich the moisture laiden winds which helps in condensation of clouds and causes precipitation.

(iv) **Impact on humidity:** Forests absorbs large quantity of water from the soil and releases large volume of water vapour to the atmosphere through evapo-transpiration process. Hence the relative humidity inside the forests is always higher than the open conditions. However the effect on humidity due to forests vary due to locality, forest type, density, foliage etc.

(v) **Impact on global warming:** Forests significantly reduces the increasing global temperature due to its moderating effect. On a long term basis forest absorbs the green house gases and indirectly it reduces the impact of global warming.

(vi) **Impact on climate change:** Forests are the potential land use system that has been identified to mitigate and adapt to growing climate change risks and issues. The forests due to their action of utilizing carbon-di-oxide in their photosynthesis process and release of oxygen and helps to balance the carbon-di-oxide and oxygen content. Hence forests and trees have a great impact on the climate of the locality.

Forest and soil conservation

Forests play a vital role in conserving the soil. Soil is one of the most important natural resource which are degraded due to natural calamities. In India, out of 328 million hectare nearly 175 million hectare are degraded which results in loss of nutrients and conversion of productive land into unproductive soil. Under such circumstances, trees improve the soil fertility and conserve the soil.

The tree species contain large quantity of living biomass and 20 – 25% of living biomass is added continuously to the soil through dead and decayed leaves and roots. These plant biomass gets decomposed and add nutrition to the soil. This also improves soil structure and texture thereby help to improve the porosity and water holding capacity. This in turn helps to control the soil erosion and conserve the soil.

Some of the tree species has the great potential of nitrogen fixation through the action of micro symbionts present on the root nodules. These nodulating bacteria fixes atmospheric nitrogen and adds to the soil thereby enriches the soil fertility.

Tree roots are acting as an excellent soil binder which helps to protect the soil from being eroded. The roots of tree species releases wide range of chemical substances which increases the populations of micro flora and fauna and improves soil fertility and reduces soil erosion.

Trees acting as a wind break and they are obstructing heavy speed of wind and reduces the wind erosion. Hence forests and trees play a vital role in soil conservation.

Forest and water conservation

Forests play a vital role in water conservation through their action of interception, litter fall, through fall, stem fall and reducing the intensity of water erosion. The heavy and thickly foliated trees intercept the heavy precipitation and reduce the velocity of rain. This reduction in velocity will help to reduce the erosion. Tree by their action of litter fall through and stem fall receives the water and directs them to the soil and improve the ground water through infiltration and percolation. The tree root acts as a binder and reduces the running water and helps to direct them into the soil. The tree by their action of soil protection helps to improve the porosity and water holding capacity which results in increased soil and water conservation.

33. Discuss strategies and plan adopted for the treatment of catchment areas. (2011)

Catchment areas are reservoirs to conserve and store water on a larger area which will be placed for the agriculture production in and around the catchment localities. However the sequential process of erosion, entrainment, transportation, deposition and compaction of sediment lead to siltation and catchment area capacity to store water is reduced which in turn will affect the agriculture. The removal of top fertile soil from catchment advisory affects the agriculture production. Hence the catchment area needs to be treated judicially with appropriate strategies and plans. The following are the strategies to reduce erosion, conserve water and enhance the vegetation ecosystem.

Treatment of catchment area

A. Physical or mechanical measures

(i) **Land preparation:** It involves ploughing to influence intake of water, obstruction to surface flow of water and soil, deep ploughing to reduce erosion due to cloddy soil structure.

(ii) **Contour bunding:** Contour bunding is a traditional process and found effective physical measure for soil and water conservation. It consists of narrow based trapezoidal bunds on contour to impound run of water behind them such that it can gradually in filtrate in to the soil for use by the plants. This method of contour bunding is recommended for areas with low rainfall of less than 600 mm and the slope up to 6%.

(iii) **Graded bunding:** This is another method of soil and water conservation practiced in areas of more than 600 mm rainfall. This method will help to remove the excess water and avoid water stagnation.

(iv) **Bench terracing:** It is a combination of ridge and channel belt across the slope. It consists of transforming relatively steep land into a serious of level strips or platforms across the slope to reduce the slope length and the associated erosion. The field is made into serious of benches by excavating soil from upper part and filling them in lower part of the terrace. This method is practiced for the lands with the slope ranged between 16 and 33%.

(v) **Contour trenching:** This is one of the popular methods of soil and water conservation wherein the trenches are made along the contour. This method is practiced in hill slopes. The trenches made will break the slope length; reduce the velocity of runoff leading to retarded scouring and carrying capacity. This method is practiced with the slopes exceeding 20%. The trench size is usually 16 x 45 cm and the distance between the trenches varied between 10 x 30 m.

B. Biological measures: The biological measures include agricultural and silvicultural practices viz., contour cultivation, strip cropping, mulching, cover cropping, silvipasture, agroforestry, wind break, shelter belts etc.

(i) Choice of crop: Generally, pulse crop and leguminous tree crops are found effective towards soil and water conservation. These crops are considered erosion resistant crop and provide better cover and protection to soil by way of minimizing the impact of rain drop and obstructing to runoff.

(ii) Contour cultivation: The contour made in the slopes is cultivated by sowing the crop across the slope. The crops like cereals, pulses and grasses can be cultivated in the contour which will ensure utilization of the land coupled with preventing soil erosion.

(iii) Strip cropping: The erosion resistant crop and erosion permeable crops are grown in alternate strips on contour with the objective of breaking long strips to prevent soil loss and run off. The resistant crops reduce the velocity of water movement and filter the sediment from the run off to deposit them in the field.

(iv) Cover cropping: Any crop planted to control erosion is termed as cover crop. The pulses with drought resistant are generally grown for this purpose. In some arid areas drought tolerant grasses are grown.

(v) Crop rotation: Crop rotation means alternate growing of crops one after the other. The Legumes followed by grasses are preferred crop rotation which will improve the soil besides controlling the erosion.

(vi) Mulching: Mulching of crop residues on or near the soil surface reduce evaporation besides conserving the soil and water. The residues on surface observe wind force and reduce the wind erosion.

(vii) Forestry practices: The forestry component particularly leguminous trees like *Acacia*, *Leucaena*, *Dalbergia* etc. play a major role in arresting soil erosion and conserving moisture through interception and through fall. The desirable characters of trees are drought resistance, leguminous, deep root system and will have multipurpose utility.

(viii) Windbreaks: Windbreak is a single row plantation and act as a barrier for protection from winds. The effectiveness of wind break depends on its density, height and length. The spacing of wind breaks depends on the density and angle of wind break to the direction of hazardous wind. Eg. *Casuarina*, *Eucalyptus* etc. The length of the wind break is about 24 times the height of the tree. The distance between the wind break is 10 times the height of the tree.

(ix) Shelter belts: Shelter belt is a longer barrier than a wind break consisting of combination of shrubs and trees intended for protecting field crops and conserving soil. The shelter belt includes the following design.

- **Central rows** – Acacia Arabica, *Albizzia lebbek, Azadirachta indica*, Dalbergia sissoo, Eucalyptus spp., Prosopis spp., Tamarindus indica.
- **Flank rows** – *Acacia Senegal, Anacardium occidentale*, Casuarina equisetifolia, Cassia siamea, Prosopis juliflora.
- **Outer rows** – *Acacia spp., Agave spp., Cassia auriculata, Gliricidia maculata.*

34. Discuss the factor that inhibits tree growth in alkali and saline soils. How is alkali soil managed for tree species plantation. Write scientific names of 4 tree species suitable for plantation in alkali soils. (2011)

Factors inhibiting growth

Alkaline conditions limit the availability of nutrients such as iron, phosphorous and zinc. Symptoms of saline soil can be seen in plant leaves, including yellow stripes on the upper leaves caused by zinc and iron deficiency or dark green or purple coloration of stems and lower leaves caused by phosphorus deficiency. Drought symptoms such as wilting or yellowing of foliage are also common in saline soils. A powdery substance can also be noticed on the soil, as the salts are dispersed to the soil surface.

Management of alkaline soil

Alkaline soils can be managed by the following activities.

(i) Improving the drainage facility of the site will help to reduce the alkalinity and make the soil amenable for cultivation.

(ii) Leach the salts by irrigation with quality water to reduce the alkalinity and to convert them into productive use.

(iii) The alkaline soils can be managed with good soil management practices by applying adequate organic matter in the form of farmyard manure or other residues.

(iv) Encouraging cultivation of leguminous crop and plough back them into the soil will help to reduce the alkalinity and improve the soil quality.

(v) Application of gypsum to the tune of 4-5 tonnes / hectare will help to reduce the alkalinity.

Species for Alkaline soil

- *Prosopis juliflora*
- *Azadiracta indica*
- *Lannea coromandalica*
- *Thespesia populenea*
- *Calophyllum inophyllum*
- *Basia latifolia*

35. What are the specific problems of the following wastelands? Suggest atleast 3 species for planting in each of them (2012)

((i) ***Hot desert***

(ii) ***Saline and alkaline soils***

(iii) ***Wet lands***

(i) **Hot desert:** Hot deserts are the areas where the seasons are generally warm throughout the year and very hot in the summer. The winters usually bring little rainfall. Temperatures exhibit daily extremes because the atmosphere contains little humidity to block the Sun's rays. Desert surfaces receive a little more than twice the solar radiation received by humid regions and lose almost twice as much heat at night. Many mean annual temperatures range from 20-25° C. The extreme maximum ranges from 43.5-49° C. Minimum temperatures sometimes drop to -18° C.

Rainfall is usually very low and/or concentrated in short bursts between long rainless periods. Evaporation rates regularly exceed rainfall rates. Sometimes rain starts falling and evaporates before reaching the ground. Rainfall is lowest on the Atacama Desert of Chile, where it averages less than 1.5 cm. Some years are even rainless. Inland Sahara also receives less than 1.5 cm a year. Rainfall in American deserts is higher – almost 28 cm a year.

Soils are course-textured, shallow, rocky or gravely with good drainage and have no subsurface water. They are coarse because there is less chemical weathering. The finer dust and sand particles are blown elsewhere, leaving heavier pieces behind.

(a) **Problems in hot desert**

- Increased temperature
- Higher wind
- Poor soil fertility
- Shifting sand dunes

(b) **Species for a3orestation in hot desert**

- *Prosopis juliflora* or *P. cineraria*
- *Acacia senagal* or *Acacia nilotica* or *A. tortilis*
- *Tecomella undulata*
- *Salvadora oleoidis*

(ii) **Saline and alkaline soils**

(a) **Problems**

- High proportion of bases parent material Eg. Basalt, Sand stone.
- Low rainfall
- High evaporation
- Poor drainage
- Poor quality irrigation waters
- High water table
- Sea water intrusion

- Base forming fertilizers

(b) Species for a3orestation in saline and alkaline soils.

- *Prosopis juliflora*
- *Azadiracta indica*
- *Albizia lebbeck*
- *Pongamia pinnata*
- *Lannea coromandalica*

(iii) Wet lands

(a) Problems

- Water logging due to stagnant or impeded water
- Presence of swampy areas
- Occurrence of marshy areas
- Presence of saline muddy soil
- Saline and alkanity
- Poor drainage
- Excessive salt
- Poor nutrient availability
- Physiological drought

(b) Species for a3orestation in wet lands

- *Syzygium cumini*
- *Terminalia arjuna*
- *Acacia nilotica*
- *Eucalyptus saligna*

36. Explain the importance of soil and air temperature on the growth of forest trees (2014)

Soil temperature influences on trees in the following ways:

(i) Soil temperature has a profound influence on absorption of soil moisture which increases markedly with the rise in temperature upto a certain limit. When soil temperature rises above 35ºC, there is a decrease in absorption as the permeability of plasma membrane is adversely affected. On the other hand, if there is a fall in soil temperature below 27ºC, water absorption is greatly reduced till at 0ºC it becomes in significant. Cold soils are, therefore, physiologically dry.

(ii) Soil temperature also affects cambial activity, particularly in temperate climate. Cambial activity in trees starts earlier in warmer soils than in colder soils. Thus growth starts earlier in warmer soils than in colder soils.

In short, increase in temperature creates conditions in which trees and other vegetation grow well. This means higher temperature is indicate of multiplicity of vegetation while low temperature that of limited vegetation.

Air temperature influences on trees in the following ways:

(i) The solar radiation directly as well as through its influence on air temperature, provides heat to the plant body and helps in satisfactory initiation and continuation of various physiological activities, *e.g.,* transpiration, photosynthesis and respiration. High temperature increases transpiration while low temperature decreases it. Though photosynthesis takes place under a wide range of temperatures varying with species and locality, increases in temperature upto 25°C increases photosynthesis, after which it deceases sharply. The rate of respiration increases as temperature rises from 0°C to 40°C but it decreases in temperature lower than 0°C and higher than 40°C.Thus temperature exerts a greater influence on the vital physiological activities of trees.

(ii) Air temperature increases microbiological activity on soil surface resulting in decomposition of organic matter and release of nutrients to be available to trees.

(iii) Air temperature affects activities of enzymes, which are practically stopped at temperatures above 50°C or below 1°C.

(iv) Air temperature increases cambial activity in the shoot portion.

(v) Through its effect on the vital physiological activities and cambial activity, air temperature affects growth of trees.

(vi) Temperature is essential for germination of seeds.

37. Define watershed? Describe tree based models for reclamation of degraded hills? (2016)

Watershed

Watershed is the geographical unit with the common drainage outlet.

Reclamation of degraded hills

The degradation of hill is mainly due to devoid of vegetation, accelerated soil erosion and associated degradation process. Hence the reclamation of degraded hills can be approached through physical and biological measures.

(a) Physical measures: This involves treatment of the hillock land on a watershed basis to reduce the erosion and improve the water holding capacity judicially utilizing the available natural resources.

(i) Establishment of contour bunds: The reclamation process should start with establishment of contour bunds across the slope of the land on a contour, breaking the long slope into a series of small ones as indicated in the figure.

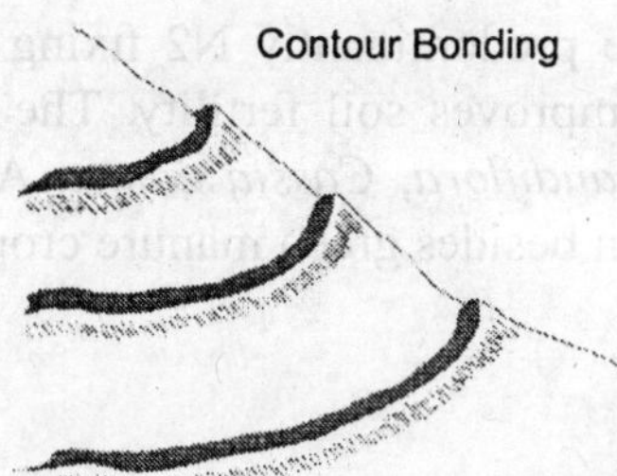

Each contour bund will act as a barrier to the run of and reduces its velocity and allows more percolation of more water into the soil and reduces the soil erosion. The size of contour and the spacing between the contours is based on the slope, intensity of rainfall, soil type and depth and the intensity of degradation.

(ii) **Establishment of bench terraces:** If the hillock slope is more than 10 – 25 %, then the establishment of bench terrace is a recommended practice.

b. Biological measure

(i) **Contour cultivation:** Within the contour bund areas the land should be leveled and it can be amenable for tillage practices wherever possible the cover crops and strip crops can be practiced in order to avoid the rapid evaporation process.

(ii) **Establishment of block plantations:** Once the hillock is ready with the contour formation, suitable tree species can be planted within and between the contours as a block plantation. Between two contours block plantation is encouraged and on the contour species amenable for wind breaks can be planted. The following species are recommended for block plantations in the hillock areas.

Hardwickia binnata	*Tamarindus indica*
Albizia lebbeck	*Ziziphus maurutiana*
Pterocarpus santalinus	*Ailanthus excelsa*
Azadirachta indica	*Acacia tortilis*

(iii) **Establishment of wind breaks:** The hillocks can be planted with wind breaks on the contour line in order to conserve soil and water besides reducing the wind velocity. The species like Eucalyptus, Casuarina, Subabul are recommended for planting as wind breaks.

(iv) **Soil conservation hedges:** Along the contour planting of Glyricidia, Agave, Cassia, Subabul etc. will help to arrest the soil and water erosion besides improving the soil fertility through nitrogen fixation. These crops due to their inherent lopping amenability can act as a potential fodder species besides acting as a soil stabilizers.

(v) **Cover crop:** Any crop planted to control erosion is termed as cover crop. The pulses with drought resistant are generally grown for this purpose. In some arid areas drought tolerant grasses are grown.

(vi) **Alley cropping:** It is otherwise called as hedge-row intercropping. In this method, perennial tree crops are grown as hedges and agricultural crops are grown in alleys.

The hedge species are predominantly N2 fixing and leaves are collected and used for mulching which improves soil fertility. The common tree crops are Leuceana, Gliricidia, *Sesbania grandiflora*, *Cassia siamea*. Agricultural crops like cereals, pulses and oil seeds are grown besides green manure crops.

Chapter

7

Environmental Conservation and Biodiversity

1. How industrial pollution can be controlled with suitable tree species? (2005)

Air pollution due to industries and increased traffic is becoming an increasingly dominant problem, especially in the developing countries. The levels of one or more air pollutants are higher than the permissible limits set by Central Pollution Control Board in many industrialized / urbanized pockets of India. Control of air pollution is possibly by two ways, namely control at source by implementing industrial ecology and by restricting the spread of air pollution through development of greenbelt around industries and plantations in urban areas.

Selection of species

Careful attention could be devoted to the selection of plant species. Selection of the appropriate plant species, which are tolerant to the particular pollutants of that area, can be done on the basis of climatological conditions and ecological status of that region.

The ideal characters of trees control industrial pollution are furnished.

(i) Fast growth rate for quick development of canopy

(ii) Strong branches, thick and durable canopy which can withstand storm

(iii) Large leaf size for greater retention of pollutants

(iv) Dense foliage for better trapping of pollutants

(v) Preferably perennial and evergreen species

(vi) The species should be Indigenous

(vii) Resistant to specific air pollutants, diseases and insects

(viii) Able to maintain the ecological and hydrological balance of the region

(ix) Leaves with hairy, resinous, scaly, and coarse surfaces could capture more particles than smooth leaf

(x) Species must have the ability to tolerate the unique features of urban soil.

Role of trees in controlling industrial pollution

The tree species by nature of its carbon sequestration potential absorbs more carbon-di-oxide and release more oxygen thereby help to balance carbon-di-oxide and oxygen in the atmosphere. It is estimated that each and every tonne of biomass produced by the tree sequest 0.5 ton of carbon-di-oxide.

Reduction in noise pollution

Dense vegetation with heavy foliage will help to control the noise pollution the sound observing potential of the trees , the structure, density and type of vegetation, the dense foliage based vegetation reduces noises to the greater extent. The dense belt of 50 m width of trees is able to reduce the noise level 20 – 30 decibel.

Choice of species for various industrial pollutants

The major industrial pollutant are SO_2, NO_2 suspended particles, noise pollution etc., and the trees suitable for each category of pollution are furnished.

S.No.	Pollution type	Trees suitable
1.	Dust pollution	All broad leaved and coarse textured species Teak, Ficus, Mango, *Peltoforum, Terminalia*
2.	Noise pollution	Alstonia scholaris, Acer negundo, Erythrina spp., Butea monosperma, Alnus indica, Grevellia, *Juniperus chinensis, Terminalia arjuna.*
3.	Lead pollution	Cassis siamea, Zizyphus mauritiana
4.	SO_2	Albizzia lebbeck, Ailanthus, Alstonia, Neem, Ficus, Polyalthia, Terminalia arjuna, Quercus palustries
5.	Ozone	Acer campestral, A.platanoides, A. negundo, Quercus rubra
6.	Peroxy Acetyl Nitrite (PAN)	Acer platanoides, Juniperus, Quercus spp., Eunonymus europalaus
7.	Oxides of nitrogen	Quercus rubra, Robinia pseudocacia, Sambucus nigra, Taxus buccata, Ulnus Montana, Pinus spp.
8	CO_2	All types of trees

2. a. How forests and tree plantations help in environmental conservation? (2006)

b. Explain the role of forests in environmental conservation. (2009)

Forests in general and trees in particular play a significant role in protection and conservation of environment. The forest influences each and every aspect of environment which includes physical, economic, social and cultural value. The influence of forest on physical environment has been debated widely and it is proven that forest and trees have positive role in augmenting and ameliorating the environment.

(i) **Forest and climate:** Forest play a vital role in moderating the climate of the locality and more precisely it has significant influence on the microclimate. Trees with dense foliage are capable of reduce the temperature within the forest area to the tune of 1-2°C compared to open ground. Similarly, the forest increase the temperature during the winter season.

Hence forest and trees play a vital role in moderating the local climate.

(ii) **Forest and rainfall:** The most stressed and inconclusive is a debate on the influence of forest and rainfall of the region. Various observations and experiments have led to different conclusion. However it is clear that, forests and trees may or may not influence rainfall on a region or a continental scale but they do affect the incidence and distribution of rainfall locally. Forests modify the microclimate by reducing the temperature, raising humidities and bring conditions that favours increased rainfall in a high land forest it has been estimated that the total precipitation level in increase more 1-2%.

(iii) **Forest and wind:** Forests considerably reduces wind velocity and protect agriculture and other land use system on the leeward side from erosion and the desiccating effects of wind. The reduction in wind velocity varied between 20 and 60% on the protection may be offered to a distance of 40 times the height of the tree.

(iv) **Forests and water:** Forests influence the water quality and quantity by canopy interception and evaporation from foliage. It also intercepts the water through litter and finally increasing infiltration in the ground as a result of mixing of organic matter into the soil and to the creation of channel through the plant roof. The canopy and leaf litter interception reduced the runoff velocity and direct them to infiltrate and percolate into the soil. Forest not only reduces the impact on soil of rainfall of heavy intensity but also make soil less erodible. The organic matter acts as a cementing agent in the formation of aggregate from the soil particle. As a general rule higher the organic matter, the larger the size of aggregates and lesser will be the erodability of the soil.

Forests also influence water quality because water flowing from forest is excluded from human activities and is of good quality. It contains less sediments, less dissolved nutrients, lower in temperature and has more dissolved oxygen.

Hence, forests play a vital role in water conservation by augmenting its quantity and quality.

(v) **Forests and soil:** Forests play an important role both in the formation and conservation of the soil. It is estimated that the evergreen forest produces 20 tonnes of organic matter every year and deciduous forest produces 5.5 tonnes of organic carbon every year. These organic carbon get decomposed rapidly and improves the soil physical, chemical and biological properties. Forest and trees also play a vital role in transforming reproductive soil into productive soil through this litter fall, nitrogen fixation and also through the addition of nutrients through the root exudation.

(vi) **Forest and air:** Forests are the one of the important sources in ameliorating the atmosphere by assimilating the gaseous and particulate pollutants. The increasing concentration of carbon-di-oxide in atmosphere has led to increased in temperature of earth forest act as a sink in reducing the carbon-di-oxide and increasing concentration of oxygen and hence they are regarded as excellent sequesters of carbon.

Forest not only act as a sink for carbon-di-oxide but also reduces the concentration of other gaseous and particulate pollutants. This scavenging ability of tree is remarkable and tree leaves can filter out dust, shoot and smoke to the tune of 120 !:g/sq.m of the leaf surface. The large surface of leaf absorbs pollutants through numerous stomatal opening and help in conservation of environment.

(vii) **Forest and Biodiversity:** Forests play a vital role in biodiversity conservation. It is estimated that India has 45,000 species of plants with 15,000 flowering plants, 4100 species of vertebrates and over 30000 species invertebrates. Indian forests is housed with four hotspots distributed in Himalayan and western ghats region and hence play a vital role in floral and faunal conservation.

3. a. Write a note on Environmental Impact Assessment (EIA). (2006)

b. Describe environmental impact assessment tool in forestry. (2007)

c. What are the objectives of carrying out EIA. Discuss sequentially, the di-erent phases of an EIA study. (2016)

Environment Impact Assessment (EIA) It is an exercise or a process to be carried out before any project or developmental activity is undertaken to ensure that these developments will not in any way harm the environment on a short term or a long term basis. The EIA will also identify and predict the impact of a project on environment and human health.

Features of EIA

- The EIA will identify the possible and negative impact of the project on the environment.
- The EIA provides plan for implementation and will help to resolve the negative impacts and lead to minimal environmental degradation.
- The EIA will establish monitoring mechanism to assess the impacts of the project or developmental activity on the environment and the community living around the environment.
- The EIA will help to promote developmental process that is sustainable and optimizes resource' use as well as management opportunities.

Types of EIA

The Environmental Impact Assessment is primarily depend on the type of activity and is classified into four types. They are-

- **Strategic environmental assessment:** This assessment involves systematic analysis of the developmental policies, plans, programmes and strategic actions on the environment.
- **Regional EIA:** This assessment involves the environmental impacts of regional development plans and programme and this assessment is confined at the sub country level. It is also referred as economic, environmental development plan.
- **Sectoral EIA:** This assessment involves addressing the specific environmental problems that may be encountered in planning and implementing sectoral developmental projects.
- **Project Level EIA:** This assessment is at project level and it is relatively simple and draw a meaningful conclusion. In India, project level environmental impact assessment is more dominant and approached on a case to case basis.

Components of EIA

The Environmental Impact Assessment takes into consideration the following components for impact assessment and also for future recommendations.

- **Land environment:** The EIA should take into consideration all the land ecosystem particularly the factors of locality, drainage, landscape and existing land use pattern.
- **Biological environment:** It includes survey of all flora and fauna clearly delineating their status before the implementation of the project. Assessment of potential damage to terrestrial and aquatic flora and fauna due to industrialization may also be taken into consideration. The mitigation measures to prevent and protection of damages to all biological organisms are to be furnished.
- **Atmosphere:** It determines the impact zone and should indicate the monitoring network schedule. Monitoring of the existing status of ambient air quality, monitoring the site specific meteorological data, identification, quantification and evaluation of potential emissions into the atmosphere are to be considered. The EIA should also include the mitigation measures to resolve the issues.
- **Socio-economic environment:** The EIA should incorporate the demographic and the associated socio-economic data pertaining to the project area. The demographic data should include all prominent endemic diseases and morbidity rates among the population within the project zone. The projection of expected changes in the socio-economic and health due to project activities may also be indicated along with mitigation measures if any.
- **Risk assessment:** All the risk associated with the project activity and development should be incorporated. The analysis of failures and the accidents in the form of fire, explosion, hazardous chemical release and other emergency risk are to be planned and incorporated along with disaster or risk management plan.

Management plan: The EIA should include the management plan clearly delineating the mitigation measures including prevention and control of each environmental component, rehabilitation and resettlement plan.

Baseline data for EIA: The environmental impact assessment of any project should include all the essential base line data in order to have proper monitoring system and to resolve the issues. The base line data should include the following.

- **Climatic data:** The climatic data of the project site particularly the rainfall (pattern, mean, mode, seasonality), temperature (max, min), wind (direction, speed), relative humidity (max, min) etc.
- **Geological data:** The geological data pertaining to soil, rock, structure, texture, nutrient status, mineral resources, oil resources etc., should be included.
- **Topographic data:** The slope, orientation, longitude, latitude, altitude, special land forms etc. should be incorporated in the base line data.
- **Water resource data:** The availability of water, status of ground water (water table, local aquifer, specific yield), water quality, surface water (rivers, lakes, ponds, gullies), water discharge, waste water accumulation etc. should be properly recorded.
- **Air data:** Thc ambient air quality, air composition, air climatic conditions etc., should be documented and reported.
- **Biological data:** All biological data incorporating all flora and fauna, their type, density exploitation, current status etc. should be included in the base line data.

- **Land use pattern:** The existing land use pattern in the form of agriculture, horticulture, forestry, industries, urban settlements etc., need to be reported in the base line.
- **Forest:** The existence of natural forests (dry deciduous, moist deciduous, evergreen, temperate forest etc.) and plantation forests, presence of animals, movement of animals, existence of rare, endangered and threatened species and the presence of wild flora and fauna should be documented.

Developmental data: The availability of roads, communication facilities, industries, dams, settlements etc., should be included in the base line data.

EIA Phases

Screening to determine which projects or developments require a full or partial impact assessment study;

Scoping to identify which potential impacts are relevant to assess (based on legislative requirements, international conventions, expert knowledge and public involvement), to identify alternative solutions that avoid, mitigate or compensate adverse impacts on biodiversity (including the option of not proceeding with the development, finding alternative designs or sites which avoid the impacts, incorporating safeguards in the design of the project, or providing compensation for adverse impacts), and finally to derive terms of reference for the impact assessment;

Assessment and evaluation of impacts and development of alternatives, to predict and identify the likely environmental impacts of a proposed project or development, including the detailed elaboration of alternatives;

Reporting the Environmental Impact Statement (EIS) or EIA report, including an environmental management plan (EMP), and a non-technical summary for the general audience.

Review of the Environmental Impact Statement (EIS), based on the terms of reference (scoping) and public (including authority) participation.

Decision-making on whether to approve the project or not, and under what conditions;

Monitoring, compliance, enforcement and environmental auditing. Monitor whether the predicted impacts and proposed mitigation measures occur as defined in the EMP. Verify the compliance of proponent with the EMP, to ensure that unpredicted impacts or failed mitigation measures are identified and addressed in a timely fashion.

4. What are the reasons for loss of Biodiversity? How is the loss of Biodiversity overcome? (2006)

Biological diversity (Biodiversity)

It refers to the variety and variability among living organisms and the ecosystem complexes in which they occur. Diversity can be defined as the number of different items and their relative frequency.

Causes for loss of biodiversity

Biodiversity in India, particularly forests, grasslands, mountains, desserts and marine ecosystem experience intensive pressure due to population explosion coupled with the associated science and technology development resulting in threat to biodiversity. The general causative factors for the loss of biodiversity are

I. Developmental process

- Population
- Industrial Development
- Dams and irrigation projects
- Mining
- Roads and transportation
- Communication

II. Alternate land use system

- Expansion of agriculture and allied activities
- Fisheries and allied activities
- Grazing
- Horticulture
- Monoculture of trees
- Expansion of forest fringe villages
- New settlements
- Expansion of area under plantation crops

III. Over exploitation and utilization

- Food, fruit, fibre and fuel gathering
- Over exploitation of timber and non timber forest products
- Hunting and poaching of animals
- Unregulated collection of medicinal plants and orchids
- Collection made by scientific and educational institutions

IV. Manmade disaster

- Human induced fires
- Hunting
- Grazing
- Deforestation

V. Management of natural resources

- Over exploitation of resources
- Introduction of exotics
- Pest and diseases
- Increased competition among bioresources

VI. Management of human resources

- Change in human lifestyle
- Behavioural changes
- Lack of awareness
- Ignorance
- Erosion of indigenous knowledge
- Tourism development

VII. Political and policy issues

- Lack of policy guidelines
- Change in legal status
- Lack of interdepartmental coordination
- Military and defence activities
- Political activities and pressure

Biodiversity conservation

Conservation of biological diversity means maintenance of ecological stability and productivity towards sustained development and stable national economy.

Types of conservation There are two types of conservation. They are

(i) ***In-situ* conservation:** It means conservation in the natural, original place or position. It includes conservation of plants and animals in their native ecosystem where they naturally occur. This is commonly practiced for wild flora and fauna and not to the domesticated plants and animals. This type of conservation is aimed at protection and preservation of land races, wild relatives, wild forms etc. Eg: National Parks, Sanctuaries, National monument, Cultural landscape, Biosphere reserve etc.

(a) **National Park:** It is an area dedicated to conserve the scenery, natural objects and the wildlife therein. In National parks, all private rights are nonexistent and all forestry operations and other alternatives such as grazing of domesticated animals are prohibited. However the general public can enter into the National Parks for observation and study purpose. The tourism is also allowed in certain part provided that it is not disturbing the animals. In India, 103 National Parks are established which serve the purpose of *in-situ* conservation.

(b) **Wildlife Sanctuary:** It is an area dedicated to protect the wildlife but it considers the conservation of species only and also the boundary of it is not limited by state legislation. In sanctuary, hunting or capturing of any species of birds and mammals is prohibited except by or under the control of the highest authority in the forest department responsible for the management of sanctuary. Private ownership, forestry and other practices are allowed to the extent that these activities do not adversely affect the wildlife. In India, there are 531 wild life sanctuaries.

(c) **Biosphere Reserve:** It is an area created to conserve the biological diversity and genetic integrity of plants, animals and microorganisms in their totality as part of the natural ecosystem to ensure perpetuation and unhindered evolution of the living organism.

(ii) *Ex-situ* conservation: It refers to conservation of components of biological diversity outside their natural habitats. This includes conservation of both domesticated and wild materials. In *ex-situ* conservation, the seeds or vegetative plant materials like plant cells, tissue, organs etc., are preserved under appropriate conditions for long storage in gene bank. e.g.: Zoological parks, Botanical garden, arboretum, tissue culture, clonal complex, seed germplasm, cryopreservation etc.

(a) Zoological parks: It is a park, building, cage, enclosure or premise in which animals are kept for public exhibition. Zoos play a vital role in awareness creation among common people about the need for conservation of animals and nature. In some zoos breeding of animals of rare and endangered category are done for rehabilitation and conservation.

(b) Botanical gardens: They are called plant zoos of the world. They have been established to conserve the rare and threatened plant species. There are about 1600 botanical gardens across the world which ensures conservation of global genetic resources. Most botanical gardens serve as seed bank for several species.

(c) Field Gene Bank: Different species and the families in each species are assembled and maintained in the form of field gene bank for immediate utility and for future improvement programme. These field gene banks may act as seed production areas and seed orchards towards supply of quality seeds.

5. a. List di-erent types of air pollutants. What are their e-ects on growth and productivity of forests? (2006)

b. Growth and productivity of forest plants are a-ected by di-erent types of air pollutants. What are those? Suggest suitable remedies. (2015)

Air pollutants

Any undesirable change in the physical, chemical and biological properties of air that adversely affects the life is called is air pollution. The factors responsible for air pollution are called air pollutants. The air pollutants are of the following types

(i) Dust: The dust is of two types which may be natural (includes seeds, spores, pollen grains, algae, fungi, bacteria, virus etc.) and industrial which includes mechanical disintegration of handling, crushing, grinding, detonation, decrepitation of organic and inorganic materials.

(ii) Smoke: It consists of soot, fly ash and other solid and liquid particles which are less than 1 micron in diameter.

(iii) Fumes: It is small particles formed due to condensation, sublimation and chemical reaction and consists of metals or metallic oxides.

(iv) Smog: It is a combination of smog and fog.

(v) Gases and vapours: This includes sulphur oxides, nitrogen oxides, carbon mono oxide, hydrogen sulphide, chlorine, fluorine, aldehydes, hydro carbons, inorganic and organic acids, radioactive dust and gaseous.

(vi) Particulate pollutant: It includes dust and particulate substances

E-ects of Air pollutants on Forests / Trees

(i) The air pollutants affect the forests through creating injuries to the plants.
(ii) It affects the flowering pattern and fruiting behaviour many forest tree species.
(iii) The pollutants also inhibit the rate of growth of trees by influencing the morphological and physiological activities of the forest vegetation.
(iv) The air pollutants adversely affect the germination, root and shoot development and cause injury to the plants.
(v) Some of the air pollutants cause structural change in the leaves particularly the sulphur-di-oxide which causes interveinal chlorosis. Similarly carbon monoxide causes leaf curling and reduction in leaf size.
(vi) Some of the air pollutants like nitrogen oxides causes defoliation and directly inhibit the growth and development of trees.
(vii) The particulate matter prevents gas and vapours exchange and affects the physiological functions of the trees.
(viii) Some of the air pollutants like sulphur and nitrous oxides alter the soil chemistry through acid rain and changes the soil ecosystem. This affects the growth and development of tree species coupled with natural regeneration of several tree species.

Remedial measures to control air pollution

The following are the major remedies.

(i) Source correction method
(ii) Pollution control equipments
(iii) Forest and vegetation

(i) **Source correction method:** To control pollution one of the major remedy is their control in the production system itself. The industries are the major source of pollution and the degree of pollution can be reduced by controlling their formation and emissions at the source level deploying recent technologies by replacing the age old technologies

(a) **Raw material substitution:** It is always possible to reduce the degree of pollution by substituting suitable raw material which may be high in quality and significantly reduced the pollution formation low sulphur fuel can be used instead of high sulphur fuel for oil and petroleum gas industries, the recent refinery technology can be used.

(b) **Process modification:** One of the major drawback in the industries emitting pollution level is high. These age old technology has be replaced with newer technology which substantially reduce the emission load. Similarly the existing machinery may be modified or replaced with environmental free. It is also very important to maintain the equipment in order to reduce pollution.

(ii) **Deployment of PCE:** Due to the advancement in science and technology, new products and equipments for controlling pollution have been developed which are environmentally safe. The equipments like gravitational setting chamber, cyclone separator, fabric filter, electrostatic precipitation, wet collectors, ventury scrubber, cyclone scrubber etc. can very well help to control the pollution.

(iii) Pollutant di3usion in Air: It is one of the alternate method to control pollution by reducing its concentration. The dilution of contaminants in the atmosphere is achieved with the use of tall stacks which penetrate the upper atmosphere layer and disperse the contaminants. This will drastically reduce ground level pollution.

(iv) Forests as a sink for air pollutants: Forests in general and trees in particular are acting as an excellent sink for gaseous and particulate pollutants. Trees by their large surface of leaves absorb pollutants through their stomatal openings. The pollutants after entering through this stomata get modified utilized by the pits for their growth and development. Some trees have hairy leaves which trap the atmospheric pollutants and utilize them for their growth and development.

(v) Agro-forestry as a carbon sequester: Forests in general and agroforestry in particular is one of the potential land use system which can sequester the carbon during the photosynthetic process. Among various pollutants, carbon-di-oxide is very notorious which trap the incoming radiation and leads to global warming. The practice of agroforestry can help in absorbing more carbon-di-oxide through photosynthetic process and release oxygen. The absorbed carbon-di-oxide is stored by the tree as a wood, it is referred carbon sequestration. It is estimated 1 ha of a well manage agroforestry plantation is able to sequester 50 tonnes of carbon. Hence, by practicing agroforestry and farm forestry the carbon-di-oxide and oxygen balance can be maintained and pollution can be controlled.

(vi) Zoning of industries: It is yet another process in controlling air pollution through efficient pre-planning strategies in terms of identifying the suitable location for the industry to reduce the effect of pollution. Through this approach, industries can be located in a cluster approach similar to SIPCOT in Tamil Nadu and the location should be away from the residential area. In this approach, pollution load can be monitored by establishing pollution monitoring system and suitable strategies can be evolved to reduce the concentration.

6. Discuss briefly the role of forests in control of air, water and noise pollution. (2006)

A. Air Pollution: Air pollution defined as a wide range of inorganic gases, organic compounds, inorganic metallic substances and root particles, is discharged into the atmosphere by motor vehicle, factories, power plants, have furnaces and waste incineration plants.

Following are the major air pollutants:

- Gases
- Carbon Monoxide. CO
- Inorganic Metallic Substance

Role of forests in control of Air Pollution

(a) Control of gaseous pollutants: The gaseous pollutants include oxides of carbon, sulphur, nitrogen, hydro carbons, methane etc. The trees reduce gaseous and chemical pollutants from the atmosphere through the process of precipitation, chemical reaction, dry deposition and absorption.

The water soluble pollutants are absorbed by plants along with water and other elements. Since sulphu-di-oxide, hydrogen fluoride, nitrogen-di-oxide and ozone are readily soluble in water, they are easily absorbed and taken by the plants.

(b) Control of particulate pollutants: Forests in general with a huge crown volume and multi-layer structure are efficient filters for dust and other particulate substances. The dust filtering capacity of forests is large and various from 50 tonnes to 150 tonnes per hectare depending on the forest type, density, composition and structure.

The hairy leaves of several tree species form a suitable surface through which the dust particles are observed. The air carrying the dust particles when passes through the forests, the air speed is reduced and particles touching the leaves are other parts of the tree get deposited on them. The canopy surface of the forests is very rough which favours turbulence in the air and supports sedimentation of dust particles.

The forests are able to filter dust from the air to such an extent that the weight of dust may exceed the weight of twigs and foliage several times.

The filtered dust reaches the soil through the action of rainfall and litterfall which gets decomposed during the process of development and help to reduce the particulate pollutants.

B. Water Pollution: Water pollution is the presence of any foreign substance, organic, inorganic or biological substance in water which tends to degrade its quality so as to impair the usefulness of water.

(a) Role of forests in control of water Pollution: Trees and forests are not as effective for control of water pollution as in the case of air pollution. But the role of forests in controlling water pollution by sand and silt is significantly high. The streams and rivers flowing in well wooded areas carry clean water due to the filtering action of the trees and forests.

The forest also helps in improving the water quality by reducing the salt and other mineral contents in the soil. This is due to the fact that the trees and vegetation absorb large quantity of salt and other desirable and undesirable minerals with water during the process of growth and development.

The forests also play a role in controlling water pollution indirectly. Several trees are reported to tolerate highly polluted water and can grow at a reasonably faster rate if irrigated by the effluents flowing out of the industries. The industrial pollutants are consumed and utilized by the trees and only little waste is available to cause pollution in the river. Hence trees and forests play a vital role in controlling water pollution.

C. Noise Pollution: The unpleasant or disagreeable loudness sound or sound without value that causes discomfort to the listener. It may be an excessive, offensive, persistent or startling sound which makes the uncomfortable situation.

Causes of Noise Pollution

Today traffic is the major source of human communication, which ultimately leads to noise pollution.

(a) Role of forests in control of water Pollution:

- Tree species did not differ greatly in their ability to reduce noise levels, provided that deciduous trees are in full leaf. Proper highway design, Well design planting in such situation can aid tremendously in sound control. It is concluded that plants play a vital role in noise abetment. They can become more effective in playing their role of noise control if planted in massive form. Thus trees, shrubs and all other vegetation aid in noise pollution reduction if used properly.
- The tree is reduces the noise pollution by absorbing, suppressing or reflecting the sound waves.
- The heavy crown of forest reflects the sound waves.
- The intensity of waves is also reduced by absorption.
- The cylindrical needles of several tree species distribute the sonic waves and their intensity is reduced.
- The branches and leaves of trees produce constant murmuring sound which will have soothing effect and help to reduce the high pitched noise.

7. Write short notes on nitrate pollution. (2011)

The nitrate is the source of nitrogen and is a basic nutrient that is essential for growth and development of plants. If nitrate concentration is higher, it will affect the plant growth and development.

Nitrate pollution caused due to mushrooming of industries will increase the concentration of nitrate in the atmosphere. Suppose if these exist nitrate concentration reaches the earth surface it will alter the soil quality.

The excessive nitrate concentration in the surface of water bodies leads to extensive growth of microscopic plants and degrades the water quality.

The nitrate pollution released from fertilizer industries can enter into the ground water and changes the composition and quality of ground water and this water may not be suitable for domestic and agricultural utility.

Sometimes the ground water is used for drinking purposes if the ground water contains excessive nitrate concentration, it will create serious and immediate health through human being and new born babies under the age of 3 months. The nitrate irons react with blood hemoglobin and reduces the bloods ability to carry oxygen and causes the diseases. This disease is termed as blue baby or methemoglobinemia.

8. Write short notes damage caused by forest fire on soil. (2009)

- Forest fire has conservable damage on soil. Any fire of the reasonable intensity burns down everything on soil. The forest fire makes a soil barren and the barren soil exposed to all sort of natural calamities and affects the soil fertility.

- Due to barrenness after burning of leaf litters and humus layers the soil is exposed to wind and soil erosion which leads to top fertile soil. It is known that to develop 1cm soil layer it takes approximately 600 years but simple fire followed by splash erosion will damage entire soil accumulation.
- The burning due to fire has direct implication on the soil structure and texture which have serious impact on soil porosity, water holding capacity and volume thereby converts quality soil into problem soils.
- The fire has direct implication on the nitrogen resources on the soil which leads to loss of soil nitrogen.
- The soil after burning due to forest fire become compact and impervious to water and other nutrients which make them unsuitable for regular cultivation.
- If the soil is clay in nature, then the effect due to forest is very heavy because the clay soil get extremely hardened and show the symptoms of crack.
- The fire also alters and changes the soil biology particularly the microflora, microfauna and nematode population is very sensitive to increase temperature.
- The alteration of soil biota will change the soil decomposition process and affect the soil quality.

9. a. Biodiversity of tropical regions and temperate region. (5) (2010)

b. Tropical regions are rich in floral biodiversity as compared to temperate region? Do you agree. (2010)

Yes. I agree that tropical regions are rich in floral diversity compared to temperate regions because tropical regions are characterized by wide range of species due to favourable climatic conditions. The tropical regions occur near the equator within the area bounded by latitudes 23.5 degrees north and 23.5 degree south. One of the major characteristics that decide the floral diversity is their distinct seasonality; winter is absent and only two seasons viz., rainy and dry seasons are present.

Another important factor which decides the floral diversity is the day length. In tropical regions, the day light is 12 hours with little variations. The day light factor has contributed significantly the greater diversity pattern of tropical region.

Third important factor is the temperature of the tropical region which as an average ranged between 20 and 25 degrees with little variability throughout the year. The average temperature of the three warmest and three coldest month do not differ more than five degrees. This factor is most attributable to increased floral diversity.

The fourth important factor which leads to increased floral diversity is the precipitation which is evenly distributed throughout the year with an annual rainfall increasing over 2000 mm.

The soils in the tropical region are characterized with wide range of nutritional quality and the soils are mostly acidic in nature which leads to improved soil fertility and results in greater diversity.

Due to these factors, the floral biodiversity is very rich and one square kilometer may contain as many as 100 different tree species. The following four tropical region are characterized as regions of excess biodiversity.

(i) **Tropical rain forest:** This tropical region supports the greatest diversity of living organism on earth and it is estimated that it covers 7% of the earth surface and housed with 50% of all life on the planets land mass. Around 40 – 75% of all biotic species are indigenous to rain forests which includes two third of the flowering plants. Because of this diversity the tropical rain forests are pronounced as world largest pharmacy.

(ii) **Coral reef:** It occupies 1% of the ocean floor and extends millions of people with food, medicine, protection from storm and revenue from fishing and tourism due to their increased diversity.

(iii) **Tropical oceanic region:** This region has high temperature and high precipitation and due to this factor the biodiversity is excess in this region.

(iv) **Wetland:** The coastal and inland wet land houses wide range of coastal biodiversity and supports millions of life due to their richness.

10. Discuss Rio conference. (2010)

The United Nations sponsored Earth Summit was held in Rio de Janerio in 1992 from 3rd and 14th June. In this summit, leaders of 100 nations, over 10,000 delegates from over 160 countries, are nearly 9000 journalists participated. The focus of the conference was the Agenda 21, the problems of 21st Century and the treaties, biodiversity and climatic change. The summit ended with the declaration of 27 principles which came to be known as Agenda 21 (the Agenda for 21st Century). Maurice F. Strong, a Canadian and Secretary General of the UNCED, summarized in his conference address.

The Five Earth Summit Agreements

Five key agreements that emerged from the Earth Summit conference are summarized below.

(i) **Climate change framework:** This legally binding agreement is a first-ever attempt to evaluate and address global warming on an International scale. This was signed in 1992 by 154 nations including Canada and USA.

(ii) **Biological Diversity:** This legally binding agreement is the first international attempt to protect the Earth's biodiversity. It provides more equitable rights among nations in biotechnology and gnetic wealth of tropical ecosystems in particular.

Out of 161 signatories, the United States, Vietnam, Singapore and Kribati (a pacific Island nation) refused to sign the original treaty.

(iii) **Management, Conservation and Sustainable Development of all Types of Forests:** This non-binding agreement guides world forestry practices toward a more sustainable future of forest yields and diversity.

(iv) **The Earth Charter:** This is a non-binding statement of 27 environmental and economic principles. They establish an ethical basis for a sustainable human-earth relationship.

An important emphasis is inclusion of environmental costs in economic assessments. Improvisation and utilization of air, soil, water and ecosystems sometimes is mistaken for progress. The environment is not an inexhaustible mine of resources to be tapped indefinitely. In terms of natural capital – air, water, timber, fisheries, petroleum – Earth is indeed a finite physical system.

(v) **Agenda 21 (Sustainable Development):** This non-binding action programme is an 800 page guide for all nations into 21st Century. The idea of 'sustainable development' as proposed in a business proposal of 'sustainable growth' is examined in Agenda 21.

11. Tree species for dust and pollution control. (2010)

Forest in general and trees in particular act as a sink not only for CO_2 but also for various gaseous and particulate pollutants. Among the particulate pollutants dust is the one of the serious pollutants which causes human diseases and also affect the growth and development of plants.

Trees have remarkable scavenging ability and can filter out dust soot and smoke. Tree leaves can reduce the particulate matter 120 mg/sq.m of leaf surface over a distance of 200m. It has been estimated that Indian trees have the ability to collect or scavenge 1.44 to 5.35 grams of dusts / sq.m on their leaf surface area. Among various types of trees the evergreen tree species with simple leaves having rough and hairy surface are a potential dust collector than deciduous tree. The layer surface of a tree absorbs dust pollutants through their numeral stomatal apertures. The dust collecting efficiency of few tree species are indicated below.

S.No.	Species	Dust collecting e3ciency
1.	*Tectona grandis*	5.35 g/sq.m
2.	*Terminalia arjuna*	4.49 g/sq.m
3.	*Madhuca indica*	4.05 g/sq.m
4.	*Azadirachta indica*	2.92 g/sq.m
5.	*T. indica*	2.08 g/sq.m

12. List five each of the important resident and migratory birds noticed in Indian forests. (5) (2010)

List of Birds of India

S.No	Common Name	Scientific Name	Migratory Status
1	Rufous-faced Warbler	*Abroscopus albogularis*	local resident
2	Black-faced Warbler	*Abroscopus schisticeps*	local resident
3	Yellow-bellied Warbler	*Abroscopus superciliaris*	local resident
4	Narcondam Hornbill	*Aceros narcondami*	local resident
5	Rufous-necked Hornbill	*Aceros nipalensis*	local resident

Migratory birds

S.No	Common Name	*Scientific Name*	Migratory Status
1.	Imperial Eagle	*Aquila heliaca*	widespread winter visitor
2.	Steppe Eagle	*Aquila nipalensis*	widespread winter visitor
3.	Lesser Spotted Eagle	*Aquila pomarina*	widespread resident
4.	Spotted Owlet	*Athene brama*	widespread resident
5.	Common Pochard	*Aythya ferina*	widespread winter visitor

13. Do forests influence the rainfall? If so, how? (2012)

Yes, forests influence the rainfall.

Precipitation is formed when the atmosphere draws moisture from oceans as water vapor, which condenses and falls as rain, hail or snow. Globally, about two-thirds of this precipitation returns to the atmosphere as water vapor, and the majority of that falls again on land. Winds travelling through forests typically produce more than twice as much rain as those that blow over open land, leading to predictions.

Trees influence cloud formation by emitting carbon-based chemicals called volatile organic compounds (VOCs) into the atmosphere.

Some of those compounds are deposited on tiny airborne particles such as dust, bacteria, pollen and fungal spores. As the particles grow with the deposition of VOCs, they promote condensation and gather the resulting moisture, hastening cloud formation and lead to occurrence of rainfall.

14. What are the main reasons for decline of the forest cover in our country (2012)

Reasons for declining forest cover in India

Since independence, forests are witnessed serious exploitation both for domestic utilization and for other developmental projects which had led to decline in forest cover. The following are the major causative factors for decline of forest.

(i) **Population explosion and over exploitation:** The growing population and the associated need for forest and forest products are one of the major reasons for decline in forest area. Since independence the population has been growing at an alarming rate from 300 m to 1200 m population. This population increase has demanded wood for construction, shelter and other forest products which resulted in decline of forest cover. It has been estimated that forest decline in the country was 1.5 m ha/annum till the recent past.

(ii) **Developmental projects:** After independence the G.O.I has established several river valley projects, dams, roads and communication, railway development which have take heavy toll of Indian forests. Examples of Narmada river valley projects, konkan railway project, coal minings etc. are the major cause for declining forest cover.

(iii) **Increase in wood demand:** Traditionally forests have been looked for its production potential and it has provided wide range of goods and services both for domestic and industrial utility. The 1952 National Forest Policy has also made a provision for providing raw material resources for the wood based industries. All these issues have led to decline forest cover.

(iv) **Diversion forest for agriculture:** The people in general and the tribal people in particular heavily rely on the forest land resources for their livelihood. The political compulsions have led to diversion of forest land for agriculture purpose which acted as a major cause for declining forest cover.

(v) **Shifting cultivation:** It is one of age old practice of agroforestry practiced predominantly in North Eastern states and other forest types of North and South India. The ethinic groups dwells within the forest have used shifting cultivation process as a source of subsistence and livelihood farming. Through this approach the local dwellers clear the forest area and practice agriculture to meet their food requirement. The practice will be continued for 2.3 years and based on the productivity the area will be shifted to new area. This type of shifting cultivation across the country is one of the major cause for declining forests in the country.

(vi) **Over grazing:** Similar to human population in India is also witnessed higher livestock population the livestock population growth has not been commensurated with fodder developing programme which resulted in shortage of both green and dry fodder. This has compelled the graziers to extensively deploy the livestock in the forested area which is one of the major reasons for decline of forest cover.

(vii) **Forest Fire:** The fire both natural and intentional is one of the major cause for declining forest cover. The occurrence of natural fire is western ghats and eastern Himalayas have eroded several species. Similarly the smugglers and poachers intensively set fire to collect animal products, charcoal production, honey collect and other forest products collection have led to serious damage and cause the declining of forest cover.

(viii) **Extreme climatic conditions:** The decline in forest cover is also attributable due to extreme climatic factors. The increase in temperature due to climate change has created more acidity and create decline in forest covers. Similarly heavy precipitate intensive snow and frost also act as a major causative factor for declining the forest cover.

These are the major causative factors which play a critical role in declining forest cover which resulted in forest cover of around 21% compared to the projected cover of 33%.

15. What is the role of di-erent techniques involved in connection with conservation and multiplication of threatened species (2012)

Threatened species are any species which are vulnerable to endangerment in the near future.

Conservation of threatened species

The threatened species can be conserved by

(i) ***In-situ* conservation:** The threatened species need to be identified and conserved at the site of its original habitat. The factors responsible for its threatened needs to be identified and resolved through suitable silvicultural approaches. The habitat management and the individual tree management are very essential in order to conserve the species. Similarly protecting the species from wild animals and other extreme climatic conditions are also very significant towards in-situ conservation.

(ii) ***Ex-situ* conservation:** The threatened species may be identified and rehabilitated in other locations in order to save them from extinction. For this purpose the seeds or cuttings can be collected and multiplied through progeny or vegetative propagation method. Once they are successful they can be conserved in the form of field gene bank, incorporated in the botanical garden or arboretum or established in the form of clonal garden for mass multiplication and rehabilitation.

Techniques of multiplication of threatened species

The threatened species can be multiplied through the following approaches towards conservation and rehabilitation.

(i) Tissue Culture: Any vegetative plant part can be collected from the threatened species and mass multiplied under invitro conditions. The explants collected have to be added in shoot multiplication medium followed by rooting medium and hardening. This will ensure development of mother plants which could be further utilized for mass multiplication and rehabilitation in their original ecosystem.

Example: *Juglans regia* and *Santalum album*

(ii) Vegetative multiplication: The cuttings of threatened species can be collected and on suitable treatment with auxins and cytokinins they can be rooted under controlled environmental conditions. The rooted plants can hardened and planted in the mother garden for secondary multiplication process.

Example: *Acacia ferruginea*

(iii) Seed propagation: If the threatened species is able to produce seeds which can be used for propagation duly following necessary seed treatment methods.

Example: *Adinandra parviflora*

(iv) Miniclonal multiplication: Due to the advancement in technology there is always a possibility to mass multiply the threatened species through mini clonal technology. In this process apical shoot cuttings can be collected from the threatened species rooted either in tissue culture or in controlled chamber and established in the form of clonal mother garden. From this garden more number of apical shoots can be multiplied and it will facilitate rapid multiplication towards conservation and rehabilitation

(v) Layering: There is also possibility to induce air layers in case of highly recalcitrant nature. The successful layered plants can mass multiplied through secondary clonal multiplication process.

16. a. What do you understand by the term 'hot spot' in relation to floral biodiversity? Explain methods of ex-situ and in-situ conservation. (2011)

b. Write short notes on Ex-situ and in-situ conservation. (2012)

Floral biodiversity hotspots in India

A biodiversity hotspot is a biogeographic region with significant levels of biodiversity that is under threat from humans.

To qualify as a biodiversity hotspot, a region must meet two strict criteria: it must contain at least 0.5% or 1,500 species of vascular plants as endemics, and it has to have lost at least 70% of its primary vegetation.

In-situ conservation

It means conservation in the natural, original place or position. It includes conservation of plants and animals in their native ecosystem where they naturally occur. This is commonly practised for wild flora and fauna and not to the domesticated plants and animals. This type of conservation is aimed at protection and preservation of land races, wild relatives, wild forms etc. Eg: National Parks, Sanctuaries, National monument, Cultural landscape, Biosphere reserve etc.

(a) **National Park:** It is an area dedicated to conserve the scenery, natural objects and the wildlife therein. In National parks, all private rights are nonexistent and all forestry operations and other alternatives such as grazing of domesticated animals are prohibited. However the general public can enter into the National Parks for observation and study purpose. The tourism is also allowed in certain part provided that it is not disturbing the animals. In India, 103 National Parks are established which serve the purpose of *in-situ* conservation.

(b) **Wildlife Sanctuary:** It is an area dedicated to protect the wildlife but it considers the conservation of species only and also the boundary of it is not limited by state legislation. In sanctuary, hunting or capturing of any species of birds and mammals is prohibited except by or under the control of the highest authority in the forest department responsible for the management of sanctuary. Private ownership, forestry and other practices are allowed to the extent that these activities do not adversely affect the wildlife. In India, there are 531 wild life sanctuaries.

(c) **Biosphere Reserve:** It is an area created to conserve the biological diversity and genetic integrity of plants, animals and microorganisms in their totality as part of the natural ecosystem to ensure perpetuation and unhindered evolution of the living organism.

Ex-Situ Conservation

Ex-situ conservation is the preservation of components of biological diversity outside their natural habitats. This involves conservation of genetic resources, as well as wild and cultivated or species, and draws on a diverse body of techniques and facilities. Such strategies include establishment of botanical gardens, zoos, conservation strands and gene, pollen seed, seedling, tissue culture and DNA banks.

(i) **Seed gene bank:** These are cold storages where seeds are kept under controlled temperature and humidity for storage and this is easiest way to store the germ plasma of plants at low temperature. Seeds preserved under controlled conditions (minus temperature) remain viable for long durations of time.

(ii) **Gene bank:** Genetic variability also is preserved by gene bank under normal growing conditions. These are cold storages where germplasm are kept under controlled temperature and humidity for storage; this is an important way of preserving the genetic resources.

(iii) **Cryopreservation:** This is the newest application of technology for preservation of biotic parts. This type of conservation is done at very low temperature (196°C) in liquid nitrogen. The metabolic activities of the organisms are suspended under low temperature, which are later used for research purposes.

(iv) **Tissue culture bank:** Cryopreservation of disease free meristems is very helpful. Long term culture of excised roots and shoots are maintained. The meristem culture is very popular in plant propagation as it's a virus and disease free method of multiplication.

(v) **Long term captive breeding:** The method involves capture, maintenance and captive breeding on long term basis of individuals of the endangered species which have lost their habitat permanently or certain highly unfavorable conditions are present in their habitat.

(vi) **Botanical gardens:** A botanical garden is a place where flowers, fruits and vegetables are grown. The botanical gardens provide beauty and calm environment. Most of them have started keeping exotic plants for educational and research purposes.

(vii) **Animal Translocation:** Animal translocation is also one of the *ex-situ* conservation method which is in practice for most endangered and endemic animals.

Translocation is carried in following cases:

(i) When a species on which an animal is dependent becomes rare.

(ii) When a species is endemic or restricted to a particular area.

(iii) Due to habit destruction and unfavorable environment conditions.

(iv) Increase in population in an area.

(viii) **Zoological Gardens:** In zoos wild animals are maintained in captivity and conservation of wild animals (rare, endangered species). The oldest zoo, the Schonbrumm zoo which exists today also, was established in Vienna in 1759.

In India, the 1st zoo came into existence at Barrackpore in 1800. In world there are about 800 zoos. Such zoos have about 3000 species of vertebrates. Some zoos have undertaken captive breeding programmes.

Advantages of *Ex-situ* preservation

(i) It is useful for declining population of species.

(ii) Endangered animals on the verge of extinction are successfully breeded.

(iii) Threatened species are breeded in captivity and then released in the natural habitats.

(iv) Ex-situ centres offer the possibilities of observing wild animals, which is otherwise not possible.

(v) It is extremely useful for conducting research and scientific work on different species.

17. What is the penalty prescribe in Section 15 of the Environment (Protection) Act, 1986 for contravention of the provisions of the Environment Act rules and orders? (2015)

Penalty for contravention of the provisions of the Act and the rules, orders and directions: Whoever fails to comply with or contravenes any of the provisions of this Act, or the rules made or orders or directions issued there under, shall, in respect of each such failure or contravention, be punishable with imprisonment for a term which may extend to five years or with fine which may extend to one lakh rupees, or with both, and in case the failure or contravention continues, with additional fine which may extend to five thousand rupees for every day during which such failure or contravention continues after the conviction for the first such failure or contravention.

- If the failure or contravention referred to in sub-section (1) continues beyond a period of one year after the date of conviction, the offender shall be punishable with imprisonment for a term which may extend to seven years.

18. "Configuration of the land surface has an impact on local climatic conditions and wind movement, which in turn have a bearing on forest." Comment. (2013)

Yes the configuration of land surface influence the climatic condition and wind movement, which inturn have benefit the growth and development of forest. The configuration influences vegetation through its effect on temperature, wind movement. For example: In high hills and valleys, due to proximity of high hills on either side sunlight reaches the valley late in the morning and disappears early in the afternoon. The shade of nearby hills makes the valley colder in winter.

In summer, though the shade prevents the rise in temperature, early in the morning, yet the heat radiated by the nearby hills makes the valley extremely hot and it continues even in night.

The valleys due to proximity of hills, have for greater variation in diurnal and seasonal temperatures than the places on the hill proximity of hills also results in collection of cold air in the land lacked valleys and the vegetation in the valley is affected due to pool frost.

Configuration and wind movement

Configuration of land surface also affects the wind movement particularly in the valleys. The wind movement is a characteristic feature of hill which experiences strong noisy wind up and down the valley which is pronounced "Dadu" in North India. The configuration of ground after concentrates the monsoon wind which leads to heavy precipitation.

Due to the temperature and wind effect, the hills experience high humidity and reduced temperature which is far different from the plain area with the same mean annual temperature. This had lead many foresters to describe montane climate as a class and it decides the forest type and class.

Hence, the configuration of land surface has an impact on local climate which ultimately has an impact on the local forest type.

19. What is 'carbon trading'? How can India reap the benefit out of it? (2009)

Carbon trading an approach used to control carbon-di-oxide (CO_2) pollution by providing economic incentives for achieving emission reductions. It is also called as Carbon Emission Trading. It is an exchange of carbon credit between nations designated to reduce C emission. The concept of carbon trading came into existence as a result of increasing awareness towards pollution control. It was formalized in Kyoto protocol as an agreement made to set target for reduction of emission of green house gaseous from 1990 levels by about 5% in the period between 2008 and 2012.

Carbon Markets

There are two types of carbon markets viz.,

(i) Compliance or Regulatory market: It generates and trade green house gas emission reduction known as Certified Emission Reductions (CERs), Emission Reduction Units (ERUs), Assigned Amount Units (AAUs) and Removal Units (RMUs) which are regulated and directly initiated under the Kyoto protocol.

(ii) Voluntary market: It generates and trade green house gas emission reductions that are not regulated or directly initiated by the Kyoto protocol and is pronounced as Verified Emission Reductions (VERs)

Two important steps

- Set a limit of Carbon-di-Oxide emission by inter organization
- Transfer of credit – the buyer pays to pollute, the seller is rewarded for emission reduction.

All company are issued permits to hold credits in order to emit an equivalent amount of carbon-di-oxide. So credit cannot exceed the limit of that level. If company needs to increase the credit / limit, they have to buy credit from lower polluter.

Kyoto protocol established trading mechanism. It is a new form of commodity in the form of emission. India is also a manufacturer of energy, metal products, mining chemicals, transport, agriculture etc. and contributes to emission

Carbon trading opportunities in India

Lowering greenhouse gas emissions is not only good for environment, but it also may benefit an entity bottom line. India has a huge potential for carbon financing, which is an emerging business opportunity for Indian companies. In recent years a small, albeit fragmented, market for carbon has started to emerge, and the institutional superstructure for clean development mechanism (CDM) is being set up both internationally and nationally.

The carbon market is expected to grow to $100 billion in the near future, and Indian firms want to reap some of the benefits. The CDM under the Kyoto Protocol allows richer countries to trade their emission-reduction targets with developing countries by buying carbon credits earned by the latter for projects reducing emission of greenhouse gases.

Recent estimates predict that uncontrolled carbon emissions could cost the global economy more than $200 billion annually by 2030 unless the pollution levels are controlled. Environmental group Greenpeace has said that shifting to renewable energy and reducing carbon emissions could save Southeast Asia $80 billion annually.

As far as India is concerned, emission trading holds enormous potential with it already being one of the leading generators of certified emission reductions (CERs) through CDM. The potential for generation and sales of CERs should be harnessed on a sustained basis. India has a great scope of earning and trading carbon credits through its diversified and potential sources of developing technologies.

Proposals relevant to Indian context to reap the carbon benefits

(i) **A3orestation and reforestation projects:** India is robust country towards afforestation projects and all these projects can be linked through CDM approach in order to benefit the carbon sequestration potential of forests towards carbon trading mechanism.

(ii) **Agroforestry projects:** Agroforestry has been identified as the potential land use system which can mitigate and adapt to growing climate change associated risks. The IPCC has identified agroforestry land use system has a greater potential for carbon sequestration compared to other system and hence it can play a vital role in reaping the carbon benefits.

(iii) **Bioenergy project:** India has already an installed capacity of over 12000 MW of biomass power project established in the form of dendro biomass project, cogeneration project, gasification project etc. This biomass project is a carbon neutral project and hence carbon benefits through Certified Emission Reductions can be reaped through this project.

(iv) **Other renewable energy project:** India is one of the robust country for the solar and wind based energy generation and hence it is contributing towards emission reduction target for each and every unit of energy generated. This can be a potential and a viable option in India.

(v) **Promotion of Biofuel:** India has mandated blending 20% of biofuel with the fossil fuel petroleum products and hence mandated to reduce the emission levels through the reduction and usage of fossil fuels. This estimated reduced utility can be reaped in for carbon benefits.

(vi) **Clean India Programme:** India has mandated to make comfortable environment for each and every citizen in the country and the aim is to reduce, recycle and reuse the waste generated. This can play a vital role emission reduction target and can play significantly in carbon finance.

Hence this trade facilitates countries that have higher carbon emission to purchase (CC) the right to release more carbon-di-oxide into the atmosphere from the countries that have lower carbon emission.

20. Discuss the direct use value of biodiversity. (2011)

Direct values

The direct value include food resources like grains, vegetables, fruits which are obtained from plant resources and meat, fish, egg, milk and milk products from animal resources. These also include other values like medicine, fuel, timber, fiber, wool, wax, resin, rubber, silk and decorative items. The direct values are of two types (i) Consumptive use value and (ii) Productive use value.

(i) Consumptive use value: These are the direct use values where the biodiversity products can be harvested and consumed directly. Example: Food, fuel and drugs. These goods are consumed locally and do no figure in national and international market.

(a) **Food:**

(i) **Plants:** The most fundamental value of biological resources particularly plants is providing food. Basically three crops i.e. wheat, maize and rice constitute more than two third of the food requirement all over the world.

(ii) **Fish:** Through the development of aquaculture, techniques, fish and fish products have become the largest source of protein in the world.

(b) **Fuel:** Since ages forests have provided wood which is used as a fuel. Moreover fossil fuels like coal, petroleum, natural gas are also product of biodiversity which are directly consumed by humans.

(c) **Drugs and medicines:** The traditional medical practice like ayurveda utilizes plants or their extracts directly. In allopathy, the pharmaceutical industry is much more dependent on natural products. Many drugs are derived from plants like

(i) Quinine: The famous anti malaria drug is obtained from cinchona tree.

(ii) Penicillin: A famous antibiotic is derived from pencillium, a fungus.

(iii) Tetracycline: It is obtained from bacterium.

(iv) Recently vinblastin and vincristine, two anti cancer drugs have been obtained from catharanthus plant which has anti cancer alkaloids.

(ii) Productive use values: These are the direct use values where the product is commercially sold in national and international market. Many industries are dependent upon these values. Example- Textile, leather, silk, paper and pulp industry etc. The productive values include timber, pulp and paper, plywood, match splints, sports goods, furnitures, agriculture implements etc. which have direct utility in terms of revenue generation to the forest department and to the country.

Similarly wide range of non timber forest products like gums, resins, tannins, dyes, essential oil, bheedi leaves, edible fruits, honey, lac, silk etc. are also productive values which generate adequate income and employment generation to the ethnic groups living within the forests and the local villages living around the forests.

21. Name the biodiversity hot spots in India. Discuss the major threats to biodiversity. (2011)

Biodiversity Hotspots in India

(i) **Himalaya:** Includes the entire Indian Himalayan region (and that falling in Pakistan, Tibet, Nepal, Bhutan, China and Myanmar)

(ii) **Indo-Burma:** Includes entire North-eastern India, except Assam and Andaman group of Islands (and Myanmar, Thailand, Vietnam, Laos, Cambodia and southern China)

(iii) **Sundalands:** Includes Nicobar group of Islands (and Indonesia, Malaysia, Singapore, Brunei, Philippines)

(iv) **Western Ghats and Sri Lanka:** Includes entire Western Ghats (and Sri Lanka)

Major threats to biodiversity

Some of the main threats to biodiversity are:

(i) Human Activities and Loss of Habitat

(ii) Deforestation

(iii) Desertification

(iv) Marine Environment
(v) Poaching and
(vi) Climate Change

(i) Human Activities and Loss of Habitat: Human activities are causing a loss of biological diversity among animals and plants globally estimated at 50 to 100 times the average rate of species loss in the absence of human activities. Two most popular species in rich biomes are tropical forests and coral reefs.

Tropical forests are under threat largely from conversion to other land-uses, while coral reefs are experiencing increasing levels of over exploitation and pollution. If current rate of loss of tropical forests continues for the next 30 years (about 1 percent per year), the projected number of species that the remaining forests could support would be reduced by 5 to 10 percent relative to the forest in the absence of human disturbance.

The rate of decline would represent 1000 to 10,000 times the expected rate of extinction without deforestation by humans. Some studies suggest that, globally, as many as one half of all mammal and bird species may become extinct within 200 to 300 years.

Biodiversity loss can result from a number of activities, including:

(a) Habitat conversion and destruction
(b) Over-exploitation of species
(c) Disconnected patches of original vegetation
(d) Air and water pollution

Over the decades, the human-induced climate change may increasingly become major factor in reducing biological / biodiversity. These pressures on biodiversity are, to a large extent, driven by economic development and related demands including the increasing demand for biological resources.

Activities that reduce biodiversity, jeopardize economic development and human health through losses of useful materials, genetic stocks, and the services of intact ecosystems. Material losses include food, wood, and medicines, as well as resources important for recreation and tourism. Losing genetic diversity, like losing species diversity, makes it even more likely that further environmental disturbance will result in serious reductions in goods and services that ecosystems can provide.

Decreased biodiversity also interferes with essential ecological services such as pollination, maintenance of soil fertility, flood controls, water purification, assimilation of wastes and the cycling of carbon and other nutrients.

(ii) Deforestation: Forest ecosystems contain as much as 80 percent of the world's terrestrial biodiversity and provide wood fiber and biomass energy as well as critical components of the global cycles of water, energy and nutrient. Forest ecosystems are being cleared and degraded in many parts of the world.

Current projections suggest that demand for wood will roughly double over the next 50 years, which will make increasing use of sustainable forest practices more difficult. In addition to threats to biodiversity and potential shortages in the supply of forest products, the degradation of forests represents an enormous potential source of green house gas emissions.

Forest ecosystems contain about three times the amount of carbon currently present in the atmosphere and about one-third of this carbon is stored above ground in trees and other vegetation and two-third is stored in the soil.

When forests are cleared or burned, much of this carbon is released into the atmosphere. According to the current estimates, the tropical deforestation and burning accounts for about one quarter of carbon emissions into the atmosphere from human activities.

(iii) Desertification: Desertification and deforestation are the main causes of biodiversity loss. Both processes are decisively influenced by the extension of agriculture. The direct cost of deforestation is reflected in the loss of valuable plants and animal species. Desertification process is the result of poor land management which can be aggravated by climatic variations. Converting wild lands to agriculture often involves ploughing the soils which leads in temperate regions to an average decline in soil organic matter between 25 and 40 per cent over twenty five years.

Decreasing soil organic matter is always a clear indication of soil degradation, and often is accompanied by reductions in water infiltration, fertility, and ability to retain fertilizers. Ploughing also exposes the soil to wind and water erosion resulting in large-scale pollution of freshwater resources.

(iv) Marine Environment: Oceans play a vital role in the global environment. Covering 70 per cent of the earth's surface, they influence global climate, food production and economic activities. Despite these roles, coastal and marine environment are being rapidly degraded in many parts of the globe.

In coastal areas, where human activities are concentrated, pollution, over-exploitation of resources, development of critical habitats such as wetlands, and mangroves, and water-flow from poor land-use practices have led to drastic reductions in near shore fisheries production and aquatic biodiversity.

(v) Poaching: Large mammals such as the tiger, rhinoceros and the elephant once faced the distinct possibility of complete extinction due to rampant hunting and poaching. However, efforts by conservationists since the 1970s have helped stabilize and grow these populations. Still, the trade of tiger hides, elephant tusks, tiger teeth, rhinoceros horn remains profitable and rampant.

(vi) Climate Change: As climate warms, species will migrate towards higher latitudes and altitudes in both hemisphere. The increase in the amount of CO_2 in the air affects the physiological functioning of plant and species composition. Moreover, aquatic ecosystems, particularly coral reefs, mangrove swamps, and coastal wetlands, are vulnerable to changes in climate.

In principle, coral reefs, the most biologically diverse marine systems, are potentially vulnerable to changes in both sea level and ocean temperature. While most coral systems should be able to grow at a sufficient pace to survive a 15 to 95 centimeter sea-level rise over the next century, a sustained increase of several degrees centigrade would threaten the long-term viability of many of these systems.

22. What are the e-ects of particulate air pollutants on the regeneration of a forest ecosystem? Discuss. (2011)

Particulate pollution is a pollution caused due to accumulation of solid material from industries.

Types of Particulate pollution

The particulate pollutants include aerosols, mist, dust, smoke, fume, blume, fog and smog which are emitted from wide range of factories and industries and causes particulate pollution. These particulate pollutants have major impact on plant and human system.

E-ect of particulate pollutants on regeneration

- The particulate pollutants like aerosols which are suspended particles in the air create a problem in regeneration. The accumulation at aerosols on the forest floor will inhibit the regeneration process and affect the growth and development of already regenerated plants.
- The aerosols in the form of mist which are liquid droplets and if these droplets are dropped on the forest floor inhibits the seed germination, affect growth and development and play a vital role towards infection due to diseases.
- The dust particulate pollutants which consist of solid particles that are blow to the air and their accumulation on the leaves closes the stomata and arrest the transpiration process and the moment of water and nutrients. This in turn affects the growth and development of plants.
- The smoke particulate pollutants which consist of solid particulate or mixed at solid and liquid particles used by time and other activities enhances the concentration of other pollutants and affect the photosynthetic and respiration process of the plant system.
- The particulate pollutants in the form of fog and smog and the deposition of the forest, inhibit regeneration, growth and development of plants.
- The particulate pollutants besides its effect on plants also create problem to the human being and cause cancer and lung disease.

23. Discuss in detail the protective role of National Forests in India. (2012)

Forests in general and trees in particular play a significant role in protection and conservation of environment. The forest influences each and every aspect of environment which includes physical, economic, social and cultural value. The influence of forest on physical environment has been debated widely and it is proven that forest and trees have positive role in augmenting and ameliorating the environment.

(i) **Role of Forest on climate:** The National forest play a vital role in moderating the climate of the locality and more precisely it has significant influence on the microclimate. Trees with dense foliage are capable of reduce the temperature within the forest area to the tune of 1-2°C compared to open ground. Similarly, the forests increase the temperature during the winter season.

Hence forest and trees play a vital role in moderating the local climate.

(ii) Role of Forest on rainfall: The most stressed and inconclusive is a debate on the influence of forest and rainfall of the region. Various observations and experiments have led to different conclusion. However it is clear that, forests and trees may or may not influence rainfall on a region or a continental scale but they do affect the incidence and distribution of rainfall locally. Forests modify the microclimate by reducing the temperature, raising humidities and bring conditions that favours increased rainfall in a high land forest it has been estimated that the total precipitation level in increase more 1-2%.

(iii) Role of Forest on wind: The National Forests considerably reduce wind velocity and protect agriculture and other land use system on the leeward side from erosion and the desiccating effects of wind. The reduction in wind velocity varied between 20 and 60% on the protection may be offered to a distance of 40 times the height of the tree.

(iv) Role of Forests on water: The National forests influences the water quality and quantity by canopy interception and evaporation from foliage. It also intercepts the water through litter and finally increasing infiltration in the ground as a result of mixing of organic matter into the soil and to the creation of channel through the plant roof. The canopy and leaf litter interception reduced the runoff velocity and direct them to infiltrate and percolate into the soil. Forest not only reduces the impact on soil of rainfall of heavy intensity but also make soil less erodible. The organic matter acts as a cementing agent in the formation of aggregate from the soil particle. As a general rule higher the organic matter, the larger the size of aggregates and lesser will be the erodability of the soil.

Forests also influence water quality because water flowing from forest is excluded from human activities and is of good quality. It contains less sediments, less dissolved nutrients, lower in temperature and has more dissolved oxygen.

Hence, forests play a vital role in water conservation by augmenting its quantity and quality.

(v) Role of Forests on soil conservation: The National forests play an important role both in the formation and conservation of the soil. It is estimated that the evergreen forest produces 20 tonnes of organic matter every year and deciduous forest produces 5.5 tonnes of organic carbon every year. These organic carbon get decomposed rapidly and improves the soil physical, chemical and biological properties. Forest and trees also play a vital role in transforming reproductive soil into productive soil through this litter fall, nitrogen fixation and also through the addition of nutrients through the root exudation.

(vi) Role of Forest on air quality: The National forests are the one of the important sources in ameliorating the atmosphere by assimilating the gaseous and particulate pollutants. The increasing concentration of carbon-di-oxide in atmosphere has led to increased in temperature of earth forest act as a sink in reducing the carbon-di-oxide and increasing concentration of oxygen and hence they are regarded as excellent sequesters of carbon.

Forest not only acts as a sink for carbon-di-oxide but also reduces the concentration of other gaseous and particulate pollutants. This scavenging ability of tree is remarkable and tree leaves can filter out dust, shoot and smoke to the tune of 120 !:g/sq.m of the leaf surface. The large surface of leaf absorbs pollutants through numerous stomatal opening and help in conservation of environment.

(vii) **Role of Forest on Biodiversity Conservation:** The National forests play a vital role in biodiversity conservation. It is estimated that India has 45,000 species of plants with 15,000 flowering plants, 4100 species of vertebrates and over 30,000 species invertebrates. Indian forests is housed with four hotspots distributed in Himalayan and western ghats region and hence play a vital role in floral and faunal conservation.

(viii) **Role of forests on recreational benefits:** The National forests play a vital role in providing recreational and aesthetic benefits to the society. A variety of trees, shrubs, animals and birds attract large number of people towards them. The national parks and sanctuaries which are rich in flora and fauna extends recreational aesthetic benefits.

(ix) **Role of forests on climate change:** The National forests play a vital role in combating pollution and mitigating the effects of global climate change. The carbon sequestration potential of national forests is very significant which balances the carbon in the atmosphere and reduces the impact of global warming and the other associated risks.

24. Explain the following: (2013)

(i) Criteria Pollutants

(ii) Pollutant Standards Index

((iii) Severance Tax

(iv) Smog

(i) **Criteria Pollutants:** Air pollution contributes to a wide variety of adverse health effects. EPA has established national ambient air quality standards (NAAQS) for six of the most common air pollutants – carbon monoxide, lead, ground-level ozone, particulate matter, nitrogen dioxide, and sulfur dioxide – known as "criteria" air pollutants (or simply "criteria pollutants"). The presence of these pollutants in ambient air is generally due to numerous diverse and widespread sources of emissions.

Criteria pollutants (or) Criteria air contaminants (CAC) are a set of air pollutants that cause smog, acid rain, and other health hazards. CACs are typically emitted from many sources in industry, mining, transportation, electricity generation and agriculture. In most cases they are the products of the combustion of fossil fuels or industrial processes.

The Clean Air Act (CAA) of 1970 identified six common air pollutants of concern, called *criteria pollutants*. The criteria pollutants are carbon monoxide, lead, nitrogen dioxide, ozone, particulate matter, and sulfur dioxide. Criteria pollutants are the only air pollutants with national air quality standards that define allowable concentrations of these substances in ambient air. Exposure to these substances can cause health effects, environmental effects, and property damage. Health effects include heart or lung disease, respiratory damage, or premature death. Environmental effects include smog, acid rain, radiation, and ozone depletion.

(ii) **Pollutant Standards Index:** The PSI is based on six pollutants particulate matter (PM), fine particulate matter (PM), sulphur dioxide (SO_2), carbon monoxide (CO), ozone (O_3) and nitrogen dioxide (NO_2). For each pollutant, a sub-index is calculated from a segmented linear function that transforms ambient concentrations onto a scale extending from 0 through 500.

(iii) Severance Tax: Severance taxes are incurred when non-renewable natural resources are extracted (or severed) within a taxing jurisdiction. Resources that typically incur severance taxes when extracted are oil, natural gas, coal, uranium, and timber. Some jurisdictions use other terms like gross production tax.

Severance taxes are used in jurisdictions where most resource extraction occurs on privately owned land and/or where sub-surface minerals are privately owned (for example, the United States). Where the resources are publicly owned to begin with (for example, in most Commonwealth and European Union countries), it is not a tax but rather a resource royalty that is paid. In the case of the forestry industry, this royalty is called "stumpage".

(iv) Smog: Smog is basically derived from the merging of two words; smoke and fog. Smog is also used to describe the type of fog which has smoke or soot in it. Smog is a yellowish or blackish fog formed mainly by a mixture of pollutants in the atmosphere which consists of fine particles and ground level ozone. Smog which occurs mainly because of air pollution, can also be defined as a mixture of various gases with dust and water vapor. Smog also refers to hazy air that makes breathing difficult.

25. Describe the anthropogenic causes of forest destruction. How can these be checked? (2013)

The anthropogenic or human induced causes are the major sources of forest degradation and they are furnished below

- **Logging:** The increased logging operation to meet the timber demands is one of the most dangerous anthropogenic factors which cause destruction during the period between 1950s and 1980s. During this period, it was estimated that the deforestation rate in the country was nearly 1.5 million ha / annum which has now come down due to the policy and legal implications.
- **Deforestation and over exploitation:** The growing population and the associated need for forest and forest products are one of the major reasons for decline in forest area. Since independence the population has been growing at an alarming rate from 300 m to 1200 m population. This population increase has demanded wood for construction, shelter and other forest products which resulted in decline of forest cover. It has been estimated that forest decline in the country was 1.5 m ha/annum till the recent past.
- **Developmental projects:** After independence the Government of India has established several river valley projects, dams, roads and communication, railway development which have take heavy toll of Indian forests. Examples of Narmada river valley projects, konkan railway project, coal minings etc. are the major cause for declining forest cover.
- **Increase in wood demand:** Traditionally forests have been looked for its production potential and it has provided wide range of goods and services both for domestic and industrial utility. The 1952 National Forest Policy has also made a provision for providing raw material resources for the wood based industries. All these issues have led to decline forest cover.

- **Diversion forest for agriculture:** The people in general and the tribal people in particular heavily rely on the forest land resources for their livelihood. The political compulsions have led to diversion of forest land for agriculture purpose which acted as a major cause for declining forest cover.
- **Shifting cultivation:** It is one of age old practice of agroforestry practiced predominantly in North Eastern states and other forest types of North and South India. The ethinic groups dwells within the forest have used shifting cultivation process as a source of subsistence and livelihood farming. Through this approach the local dwellers clear the forest area and practice agriculture to meet their food requirement. The practice will be continued for 2.3 years and based on the productivity the area will be shifted to new area. This type of shifting cultivation across the country is one of the major cause for declining forests in the country.
- **Over grazing:** Similar to human population, India is also witnessed higher livestock population the livestock population growth has not been commensuated with fodder developing programme which resulted in shortage of both green and dry fodder. This has compelled the graziers to extensively deploy the livestock in the forested area which is one of the major reason for decline of forest cover.
- **Forest Fire:** The fire both natural and intentional is one of the major cause for declining forest cover. The occurrence of natural fire is western ghats and eastern Himalayas have eroded several species. Similarly the smugglers and poachers intensively set fire to collect animal products, charcoal production, honey collect and other forest products collection have led to serious damage and cause the declining of forest cover.
- **Extreme climatic conditions:** The decline in forest cover is also attributable due to extreme climatic factors. The increase in temperature due to climate change has created more acidity and create decline in forest covers. Similarly heavy precipitate intensive snow and frost also act as a major causative factor for declining the forest cover.

These are the major causative factors which play a critical role in declining forest cover which resulted in forest cover of around 21% compared to the projected cover of 33%.

These causative factors can be checked by the following alternate ways.

(a) Strict enforcement of existing laws towards arresting deforestation and over exploitation

(b) Promotion of agroforestry and farm forestry to meet the basic demands of the people in terms of food, fuel, fodder and fiber.

(c) Regular patrolling to monitor illegal felling, forest fire etc.

(d) Utilization of remote sensing and GIS technology to identify the forest fire occurrence and resolving them through rapid action.

(e) Providing alternate livelihood system for the local tribals to avoid the practice of shifting cultivation.

(f) Educating the people towards the issues of deforestation.

(g) Creating awareness among the people regarding the importance forests and its role in production and protection aspects.

(h) Controlled grazing and practicing rotational or periodic grazing in order to encourage more regeneration.

26. Briefly describe the national legislation of biodiversity. Name the activities for which states can seek funds from the Central Government (2013)

Important National legislations

- Wildlife (Protection) Act, 1972.
- Indian Forest Act, 1927 and Forest (conservation) Act, 1980.
- Protection of plant varieties and Farmers Right Act, 2001.
- The Scheduled Tribes and other Traditional Forest dwellers (Recognization of Forest Rights) Act, 2006.
- The Geographical Indications of Goods Act, 1999.
- Biological Diversity Act, 2002.
- The Indian Patents Act, 1970 and Amendment, 2005.
- The National Green Tribunal Act, 2010.

The states can seek fund from the central government for the following activities.

- Afforestation activities.
- Waste land development project.
- Medicinal plants establishment and conservation .
- Conservation of rare and endangered species.
- Restoration of degraded forest.
- Conservation of biodiversity.
- Tribal development projects.
- Conservation projects (project elephant, project tiger, etc).
- Soil and water conservation
- Watershed development
- Climate change mitigation and adaptation
- Promotion of agroforestry through agroforestry mission
- Bamboo promotion through bamboo mission
- Promotion of urban forestry through green India mission
- **Green India mission**
- Education, Research and Development projects
- Development of Infrastructure to protect biodiversity
- Fund for management for protected areas (National Park, Wild life sanctuaries, Tiger Reserve, etc.)
- Eco-development projects
- Forest Development Agency (FDA)

27. What do you understand by the term greenhouse gases? Explain how these gases disturb the ecological balance of nature and suggest suitable remedies. (2014)

Greenhouse gases

Greenhouse gases are a group of compounds that are able to trap heat (longwave radiation) in the atmosphere, keeping the Earth's surface warmer than it would be if they were not present. These gases are the fundamental cause of the greenhouse effect. Increase in the amount of greenhouse gases in the atmosphere enhances the greenhouse effect which is creating global warming and consequently climate change.

Impact on Earth

Greenhouse gases allow sunlight (shortwave radiation) to pass through the atmosphere freely, where it is then partially absorbed by the surface of the Earth. But some of this energy bounces back out towards space as heat. Of the heat emitted back to space, some is intercepted and absorbed by greenhouse gases in the atmosphere. This is because these compounds are made of three or more atoms. This molecular structure allows them to absorb some of the escaping heat and then re-emit it towards the Earth which increases global temperatures.

The ability of these gases to trap heat is what causes the greenhouse effect. So the more greenhouse gases you have in the atmosphere, the more heat stays on Earth. This process, which is very similar to the way a greenhouse works, is why the gases that can produce this effect are collectively known as greenhouse gases.

There are 2 ways that a greenhouse gas (often abbreviated GHG) can enter our atmosphere. One of them is through human activities. The main human sources of GHG emissions are: fossil fuel use, deforestation, intensive livestock farming, use of synthetic fertilizers and industrial processes. The other is through natural processes like animal and plant respiration.

Suitable remedies

To reduce the future impacts of climate change, we obviously need to reduce the amount of greenhouse gases we put into the atmosphere – in particular, we need to reduce the amount of carbon.

- Develop less polluting energy generation, transport and other industry
- New agriculture methods
- Re-forestation
- Use less animal products, especially in food
- Carbon sequestration (storage of carbon rather than allow it to enter the atmosphere

The following is a list of 10 steps to reduce greenhouse gas emissions:

(i) **Reduce, Reuse, Recycle:** Buying products with minimal packaging will help to reduce waste. By recycling half of your household waste, you can save 2,400 pounds of carbon dioxide annually.

(ii) **Use Less Heat and Air Conditioning:** Adding insulation to your walls and installing weather stripping or caulking around doors and windows can lower your heating costs more than 25 percent, by reducing the amount of energy you need to heat and cool your home. Turn down the heat while you're sleeping at night or away during the day, and keep temperatures moderate at all times. Install a programmable thermostat because setting it just 2 degrees lower in winter and higher in summer could save about 2,000 pounds of carbon dioxide each year.

(iii) **Replace Your Light Bulbs:** Wherever practical, replace regular light bulbs with compact florescent light (CFL) bulbs. Replacing just one 60-watt incandescent light bulb with a CFL will save you $30 over the life of the bulb. CFLs also last 10 times longer than incandescent bulbs, use two-thirds less energy, and give off 70 percent less heat. If every Canadian family replaced one regular light bulb with a CFL, it would eliminate 90 billion pounds of greenhouse gases, the same as taking 7.5 million cars off the road.

(iv) **Drive Less and Drive Smart:** Less driving means fewer emissions. Besides saving gasoline, walking and biking are great forms of exercise. Explore the York Region Transit system and check out options for carpooling to work or school. When you do drive, make sure your car is running efficiently. For example, keeping your tires properly inflated can improve your gas mileage by more than 3 percent. Every gallon of gas you save not only helps your budget, it also keeps 20 pounds of carbon dioxide out of the atmosphere.

(v) **Buy Energy-E3cient Products:** Home appliances now come in a range of energy-efficient models, and compact florescent bulbs are designed to provide more natural-looking light while using far less energy than standard light bulbs.

(vi) **Use Less Hot Water:** Set your water heater at 120 degrees to save energy, and wrap it in an insulating blanket if it is more than 15 years old. Buy low-flow showerheads to save hot water and about 350 pounds of carbon dioxide yearly. Wash your clothes in warm or cold water to reduce your use of hot water and the energy required to produce it. That change alone can save at least 500 pounds of carbon dioxide annually in most households.

(vii) **Use the "O3" Switch:** Save electricity and reduce global warming by turning off lights when you leave a room, and using only as much light as you need. And remember to turn off your television, stereo and computer when you're not using them. It's also a good idea to turn off the water when you're not using it. While brushing your teeth, shampooing the dog or washing your car, turn off the water until you actually need it for rinsing.

(viii) **Plant a Tree:** If you have the means to plant a tree, start digging. Trees absorb carbon dioxide and give off oxygen. A single tree will absorb approximately one ton of carbon dioxide during its lifetime.

(ix) **Get a Report Card from Your Utility Company:** Many utility companies provide free home energy audits to help consumers identify areas in their homes that may not be energy efficient. In addition, many utility companies offer rebate programs to help pay for the cost of energy-efficient upgrades.

(x) **Encourage Others to Conserve:** Share information about recycling and energy conservation with your friends, neighbors and co-workers, and take opportunities to encourage public officials to establish programs and policies that are good for the environment.

28. Discuss the role of National Parks in conservation of biodiversity. (2014)

National parks protect the best of our natural heritage: stunning landscapes, extraordinary wildlife and majestic forests. Together with other protected areas they form the basis of our economic and social wellbeing, attract millions of visitors annually, and help to protect unique wildlife by acting as a refuge for threatened species. Although their primary purpose is the protection of biodiversity, National Parks also deliver other invaluable economic, social, cultural and health benefits to people. Future generations deserve the right to see these natural values intact and protected as we do today.

National parks help to take care of places with natural and historical value. Unlike humans, who are extremely good at surviving in all sorts of conditions, some plants and animals find it very difficult, or impossible, to survive in areas where their habitat has been disturbed or changed. National parks preserve habitats for a wide range of native plants and wildlife. Parks maintain biodiversity and protect endangered species. They provide people with opportunities to learn about natural flora and fauna as well as to explore and admire the beauty of diverse environments.

National Parks (NPs) are mainly designed to protect the remaining "wilderness" of a given country and have primarily focused on the conservation of extraordinary areas or emblematic species. National parks have many roles among which "preserving nature" has become a matter of considerable social, political, economical and scientific concern. One of the major problems concerning National Parks is how to preserve their landscapes and biodiversity. While the diversity of plants and animals can be experimentally assessed, their protection involves the maintenance of their ecosystems and periodic monitoring. Any change in an environment can certainly have some effect on the plants and animals living there and so, the consequences of changes at a variety of scales is hard to predict although variations often lead to a reduction or homogenization of animal and plant diversity. In the long run, the positive role played by National Parks for nature conservation and tourism will be maintained if we ensure that social, economic and environmental goals are closely aligned.

Role of National Parks in biodiversity conservation

- National Park protects ecological integrity
- National Park as a protected area exclude exploration of all biological resources thereby help to conserve biodiversity
- The National Park provides opportunity for education and research which indirectly help to conserve the biodiversity.
- The National Park extends the recreational and aesthetic benefits which help to conserve the biodiversity.
- The National Park extends participation of local people through eco-development committee which indirectly help to conserve the biological resources.
- The National Park provides legal protection to the in-situ conservation of flora and fauna.
- The National Parks are declared under Wild Life Protection Act 1972 and classified the animals and plants into various schedules as per their status and any damage to these resources would result in punishment to the offenders for the period upto 7 years.

- The National Parks protects the habitats and facilitates natural regeneration through continued succession which helps to conserve the biodiversity.
- The National Parks protect the native vegetation from alien weeds, pests and diseases.

29. Write the chemistry of Ozonosphere and list the adverse e-ects of ozone layer depletion. (2015)

The stratosphere is in a constant cycle with oxygen molecules and their interaction with ultraviolet rays. This process is considered a cycle because of its constant conversion between different molecules of oxygen. The ozone layer is created when ultraviolet rays react with oxygen molecules (O_2) to create ozone (O_3) and atomic oxygen (O). This process is called the Chapman cycle.

- An oxygen molecules is photolyzed by solar radiation, creating two oxygen radicals:

$$h!; + O_2 \rightarrow 2O$$

- Oxygen radicals then react with molecular oxygen to produce ozone:

$$O_2 + O \rightarrow O_3$$

- Ozone then reacts with an additional oxygen radical to form molecular oxygen:

$$O_3 + O \rightarrow 2O_2$$

- Ozone can also be recycled into molecular oxygen by reacting with a photon:

$$O_3 + h!; \rightarrow O_2 + O$$

It is important to keep in mind that ozone is constantly being created and destroyed by the Chapman cycle and that these reactions are natural processes, which have been taking place for millions of years. Because of this, the thickness the ozone layer at any particular time can vary greatly. It is also important to know that O_2 is constantly being introduced into the atmosphere through photosynthesis, so the ozone layer has the capability of regenerating itself.

REGIONS OF THE ATMOSPHERE

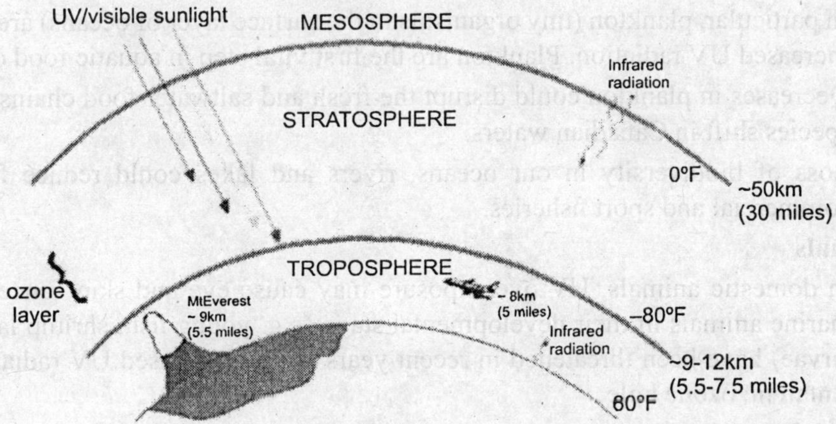

Adverse effects of ozone layer depletion are

(i) Harm to human health
(ii) Adverse impacts on agriculture, forestry and natural ecosystems
(iii) Damage to marine life
(iv) Animals
(v) Materials

(i) Harm to human health

- More skin cancers, sunburns and premature aging of the skin.
- More cataracts, blindness and other eye diseases: UV radiation can damage several parts of the eye, including the lens, cornea, retina and conjunctiva.
- Cataracts (a clouding of the lens) are the major cause of blindness in the world. A sustained 10% thinning of the ozone layer is expected to result in almost two million new cases of cataracts per year, globally (Environment Canada, 1993).
- Weakening of the human immune system (immune suppression). Early findings suggest that too much UV radiation can suppress the human immune system, which may play a role in the development of skin cancer.

(ii) Adverse impacts on agriculture, forestry and natural ecosystems

- Several of the world's major crop species are particularly vulnerable to increased UV, resulting in reduced growth, photosynthesis and flowering. These species include wheat, rice, barley, oats, corn, soybeans, peas, tomatoes, cucumbers, cauliflower, broccoli and carrots.
- The effect of ozone depletion on the Canadian agricultural sector could be significant.
- Only a few commercially important trees have been tested for UV (UV-B) sensitivity, but early results suggest that plant growth, especially in seedlings, is harmed by more intense UV radiation.

(iii) Damage to marine life

- In particular, plankton (tiny organisms in the surface layer of oceans) are threatened by increased UV radiation. Plankton are the first vital step in aquatic food chains.
- Decreases in plankton could disrupt the fresh and saltwater food chains, and lead to a species shift in Canadian waters.
- Loss of biodiversity in our oceans, rivers and lakes could reduce fish yields for commercial and sport fisheries.

(iv) Animals

- In domestic animals, UV overexposure may cause eye and skin cancers. Species of marine animals in their developmental stage (e.g. young fish, shrimp larvae and crab larvae) have been threatened in recent years by the increased UV radiation under the Antarctic ozone hole.

(v) Materials

- Wood, plastic, rubber, fabrics and many construction materials are degraded by UV radiation.

30. a. What is 'acid rain'? Explain. (2005)

b. Briefly explain the process of acid rain formation and its adverse e-ects on buildings and aquatic bodies. (2015)

Acid rain

Acid rain is a rain or any other form of precipitation that is unusually acidic, which means it possesses elevated levels of hydrogen ions (low pH).

Forms of Acid Rain

(a) **Wet Deposition:** When the wind blows the acidic chemicals in the air to the areas where the weather is wet, the acids fall to the ground in the form of rain, sleet, fog, snow or mist. It removes acid from the atmosphere and deposit them on the earth's surface.

(b) **Dry Deposition:** The wind blows the acidic chemicals in the air to the areas where the weather is dry, the acidic pollutants slip into dust or smoke and fall to the ground as dry particles. These stick to the ground and other surfaces such as cars, houses, trees and buildings. Almost 50% of the acidic pollutants in the atmosphere fall back through dry deposition.

Chemical Equations for the Formation

Formation of Nitrogen Dioxide: $N_2(g) + 2O_2(g) \rightarrow 2NO_2(g)$

Both sulfur dioxide and nitrogen dioxide are acidic oxides and react with water to form acids.

Sulfur dioxide reacts with water to form sulfurous acid.

$$SO_2(g) + H_2O(l) \rightarrow H_2SO_3(aq)$$

Substances in the upper atmosphere then catalyse the reaction between sulfurous acid and oxygen to form sulfuric acid.

$$2H_2SO_3(aq) + O_2(g) \rightarrow 2H_2SO_4(aq)$$

Similarly, nitrogen dioxide reacts with water to form a mixture of nitric acid and nitrous acid.

$$2NO_2(g) + H_2O(l) \rightarrow HNO_3(aq) + HNO_2(aq)$$

Substances in the atmosphere then catalyse the reaction between nitrous acid and oxygen causing the formation of more nitric acid.

$$2HNO_2(aq) + O_2(g) \rightarrow 2HNO_3(aq)$$

Both sulfuric acid and nitric acid are soluble in water and are the major acids present in acid rain. As this forms and falls onto the Earth's surface, these strong acids are also brought to the surface causing harmful effects on the built and the natural environment.

E-ects of Acid Rain

Acid rain causes acidification of lakes and streams and contributes to damage of trees at high elevations (for example, red spruce trees above 2,000 feet) and many sensitive forest soils. In addition, acid rain accelerates the decay of building materials and paints, including irreplaceable buildings, statues, and sculptures that are part of a nation's cultural heritage. Prior to falling to the Earth, SO_2 and NO_x gases and their particulate matter derivatives, sulfates and nitrates, contribute to visibility degradation and harm human health.

(a) **E3ects on materials and buildings:** Acid rain and the dry deposition of acidic particles contribute to the corrosion of metals (such as bronze) and the deterioration of paint and stone (such as marble and limestone). These effects seriously reduce the value to society of buildings, bridges, cultural objects (such as statues, monuments, and tombstones), and cars. Dry deposition of acidic compounds can also dirty buildings and other structures, leading to increased maintenance costs.

(b) **E3ects on water bodies:** The ecological effects of acid rain are most clearly seen in the aquatic or water environments such as streams, lakes, and marshes. Acid rain flows into streams, lakes, and marshes after falling on forests, fields, buildings, and roads. Acid rain also falls directly on aquatic habitats. Most lakes and streams have a pH between 6 and 8, although some lakes are naturally acidic even without the effects of acid rain. Acid rain primarily affects sensitive bodies of water, which are located in watersheds whose soils have a limited ability to neutralize acidic compounds (called "buffering capacity"). Lakes and streams become acidic (pH value goes down) when the water itself and its surrounding soil cannot buffer the acid rain enough to neutralize it. In areas where buffering capacity is low, acid rain also releases aluminum from soils into lakes and streams. Many lakes and streams examined in developed countries suffer from chronic acidity, a condition in which water has a constant low pH level. Streams flowing over soil with low buffering capacity are as susceptible to damage from acid rain as lakes. The acidification problem grows in magnitude if "episodic acidification" is taken into account. Episodic acidification refers to brief periods during which pH levels decrease due to runoff from melting snow or heavy downpours. Lakes and streams in many areas are sensitive to episodic acidification. A lot more lakes and streams become temporarily acidic during storms and spring snowmelt. Acid rain causes a cascade of effects that harm or kill individual fish, reduce fish population numbers, completely eliminate fish species from a water body, and decrease biodiversity of water bodies.

(c) **E3ects on forest floors:** A spring shower in the forest washes leaves and falls through the trees to the forest floor below. Some trickles over the ground and runs into a stream, river, or lake, and some of the water soaks into the soil. That soil may neutralize some or all of the acidity of the acid rainwater. This ability is called buffering capacity, and without it, soils become more acidic. Differences in soil buffering capacity are an important reason why some areas that receive acid rain show a lot of damage, while other areas that receive about the same amount of acid rain do not appear to be harmed at all. The ability of forest soils to resist, or buffer, acidity depends on the thickness and composition of the soil, as well as the type of bedrock beneath the forest floor.

(d) **E3ects on plants and trees:** Acid rain does not usually kill plants and trees directly. Instead, it is more likely to weaken them by damaging their leaves, limiting the nutrients available to them, or exposing them to toxic substances slowly released from the soil.

Ways to reduce the harmful e-ects of acid deposition

- Clean up smoke stacks and exhaust pipes:
- Almost all of the electricity that powers modern life comes from burning fossil fuels like coal, natural gas, and oil and here use alternative energy sources
- Restore a damaged environment: Take action as individuals:
- Turn off lights, computers, and other electric appliances when they are not being used.
- Use energy efficient appliances: lighting, air conditioners, heaters, refrigerators, washing machines, etc.
- Insulate the houses as best as possible so as to reduce energy losses.
- Use public transportation rather than personal cars, or even walk or bicycle whenever doable.
- Buy vehicles with low NO_x emissions, and maintain all vehicles well so that they release lesser emissions.

31. What is arsenic pollution? Discuss the strategies to mitigate it? (2015)

Arsenic contamination of groundwater is a form of groundwater pollution which is often due to naturally occurring high concentrations of arsenic in deeper levels of groundwater.

Mitigation strategies for arsenic pollution

- Awareness raising in rural areas only to use those shallow hand-pump tube wells confirmed as arsenic safe (and marked by green labelling) or sand-filtered water from ponds for drinking-water purposes
- Provision of arsenic-removal devices at household or community level (mostly through the administration and supervision of NGO's) in rural areas where no shallow hand-pump tube wells were found to be arsenic-safe
- Drilling of deep hand-pump tube wells in the coastal zone (where the shallow arsenic- safe ground water is mainly unfit for drinking due to high salinity levels)
- Construction of piped water-supply schemes (for urban and rural multi-village centers) from arsenic safe sources (deep groundwater, river-bank infiltration or treated surface water)
- By the usage of the following filters the arsenic pollutant load can be reduced.
  - Clay Pitcher Arsenic Removal Filter, the
  - Gravity flow arsenic removal cartridge filter
  - Arsenic removal cartridge filter

32. How are "Environment", "Environment Pollutant" and "Hazardous substance" narrated in Environment (Protection) Act, 1986? (2016)

The environment, environment pollutant and hazardous substance are narrated in EPA 1986 are as follows.

"Environment" includes water, air and land and the inter- relationship which exists among and between water, air and land, and human beings, other living creatures, plants, micro-organism and property;

Environmental pollutant" means any solid, liquid or gaseous substance present in such concentration as may be, or tend to be, injurious to environment

Hazardous substance" means any substance or preparation which, by reason of its chemical or physico-chemical properties or handling, is liable to cause harm to human beings, other living creatures, plant, micro-organism, property or the environment

33. What is global warming? Discuss how it disturbs the ecological balance of nature, and suggest suitable remedies. (2016)

Global Warming is the increase of earth's average surface temperature due to the effect of greenhouse gases, such as carbon dioxide emissions from burning fossil fuels or from deforestation, which trap heat that would otherwise escape from Earth. This is a type of greenhouse effect.

Global warming is an average increase in earth's temperature which in turn causes changes in climate variables. The global warming is caused due to green house gases which trap the solar energy from the sun and causes rise in temperature. The gases responsible for this effect are carbon-di-oxide, nitrous oxide, sulphur oxide, methane, CFC's etc. and hence they are pronounced as greenhouse gases.

Environmental Issues

There are certain issues and problems that are related to ecological imbalance. These are problems that have evolved because of the disruption of ecological equilibrium. Probably, there are three major problems which effects of imbalances in the ecosystem:

(i) **Global problems** – these are problems that affect different nations and can only be resolve through solidarity of affected nation. Some global problems are:
- Global warming or Greenhouse effect
- Acid Rain
- Pollution (Air and Marine Pollution)
- Depletion of ozone layer in the atmosphere
- Radioactive fallout because of nuclear war

(ii) **National problems** – these are problems that affect a country and can only be resolved within the country. These national environmental issues are:
- Pollution (air, water and soil)
- Degradation of natural resources such as soil erosion, deforestation, depletion of wildlife, shortage of energy, degradation of marine ecosystems and depletion of mineral resources
- Alteration and inconsistent land use like the conversion of agricultural land into industrial estates, conversion of mangrove swamps into fishponds and salt beds.

(iii) **Community problems** – these are problems that affect in a particular localities or communities and can only be resolved in that exact level.
- Broken and not flowing drainage
- Stench damping site (Pollution)
- Widespread of epidemic in localities

E-ects on ecological balance

Global warming leads to increase in temperature which may results in melting of ice glaciers

The melting of ice glaciers in the arctic region may increase the sea level. It has been estimated that there is a possibility of sea level rise of 9-88 cm at end of the century. Such a rise will submerge many islands and lead to changes in ecological balance of nature.

The global warming leads to warmer countries with more humid which may increase the rainfall and may alter the ecological balance.

Similarly due to global warming some parts of the world may become dry and aridity is a major issue this will alter the climatic, edaphic and biological system and hence the ecological balance is disturb in such regions. The global warming may cause changes in weather patterns which may be extreme in both ends (Heavy cold and increased temperature) which may alter the ecological balance in the habitat.

The global warming may leads to epidemic status of pest and diseases which may be a threat to crop plants and trees. This will alter the ecosystem balance. The animals and plants will find difficult to adjust the altered environment due to global warming. This may lead to plant and animal migration and hence the ecological balance is under threat.

The global warming also cause serious problem in human health particularly due to stress and irritation which may affect the human beings.

There are chances of epidemic spread of disease like Malaria, Dengue fever, Yellow fever and encephalitis which may be fatal to humans.

It is also predicted that some of forest type may disappear due to global warming which is a major threat to ecological balance.

Remedial measures

(a) **Management of Natural Resources:** Efforts must be concerted in order to maintain sustainable manner to use the natural resources for maintenance of ecological balance. The expansion of civilization has put on burden over the ecosystem. Fossil fuels, minerals and different natural resources are at alarming rate. Therefore human beings have to shift their usage to other alternatives.

(b) **Controlling Population:** The problem of over population has to be controlled despite of cultural, emotional or religious sensitivity. For example as excessive fishes in an aquarium will foul the water, over population on the planet is quite capable enough to destroy ecological balance. Therefore the government of every state has to involve themselves to control population to reduce stress on ecosystem.

(c) **Establishment of Forestry and Agroforestry Plantation:** To keep the carbon-di-oxide out of atmosphere and to balance the carbon-di-oxide concentration, forestry and agroforestry can play a vital role For this purpose, promotion of agroforestry and farm forestry will help to sequester more carbon-di-oxide and can balance the carbon-di-oxide and oxygen ratio in the atmosphere. It is estimated that one tonne of wood produced is equivalent to 0.5 tonne of carbon sequestered. Hence promotion and development of agroforestry plantations in all the degraded lands of the country will help to sequester more carbon and may reduce the chances of global warming.

(d) **Reduction of Pollutant by the Industries:** To reduce the production of green house gases, the industries releasing pollutants have to replace the old technology with recent advance technology in order reduces gaseous, particulate and water pollutants.

(e) **Recycle, reuse and reduce:** One of the ways to reduce global warming is to use the natural resources in a judicious approach through the concept of reduced use, reuse and recycling of resources which will limit the pollutant release and ultimately global warming will be reduced.

34. Write a short note on ozone layer depletion? (2007)

Ozone is an allotrope of oxygen consisting of three atoms of oxygen bound together in a non linear shape which is symbolized as O_3. The ozone layer exists in the atmosphere between 20 km and 50 km. The ozone layer is degraded due to the ozone depleting substances like chloro fluro carbon, halon, carbon tetrachloride, methyl bromide, hydrobromo fluro carbon etc., and causes hole in the ozone layer. The significance of ozone layer is it protects the earth from the action of ultra violet rays. But due to the activities of Ozone Depleting Substances (ODS) the layer gets depleted and causes serious consequences on agriculture, human beings, food production and marine ecosystem.

Chemical reaction for ozone depletion

$$CFCl_3 + \text{UV light} \rightarrow CFCl_2 + Cl$$
$$Cl + O_3 \rightarrow ClO + O_2$$
$$ClO + O \rightarrow Cl + O_2$$

The free chlorine atom is then free to attack another ozone molecules

$$Cl + O_3 \rightarrow ClO + O_2$$
$$ClO + O \rightarrow Cl + O_2$$

And again

$$Cl + O_3 \rightarrow ClO + O_2$$
$$ClO + O \rightarrow Cl + O_2$$

And again for thousands of times

The cholor fluro carbon reacts with ultra violet rays and releases chlorine. This free chlorine reacts with ozone and converts them into oxygen and chlorine oxide. This chlorine oxide again reacts with oxygen and once again chlorine gets freed. This free chlorine again reacts with ozone and splits into chlorine oxide and oxygen. This process continues for thousands of time which results in ozone hole formation.

Mechanism of ozone layer depletion

The ozone depleting substances have long life time from 20 – 120 years and they do not break in the lower layers of atmosphere. The CFCs do not dissolve in rain and hence the rain cannot wash them. They are also heavier than air. Despite their high molecular weight, these gases reach atmosphere due to the constant movement of air current and also due to the fact that they have longer life time.

These ozone depleting substances have either chlorine or bromine which are essential to break the ozone. The ultra violet rays break the CFCs or CCL_4 into chlorine and bromines from hallon or methyl bromide. This free chlorine or bromine reacts with O_3 and break it up.

Besides bromine and chloride the sulphur particles also do their job of ozone depletion.

E-ects of Ozone layer depletion

(i) The depletion of ozone layer leads to excessive penetration of ultra violet radiation which causes several human diseases like skin cancers, eye related problems and sun burns.

(ii) It also affects the human immunity system and make the body more vulnerable to diseases.

(iii) The depletion of ozone layer and the associated penetration of ultra violet rays is known to cause serious changes in species composition and affects ecosystem and biodiversity.

(iv) The ozone layer depletion and the associated penetration of ultra violet rays may alter the agriculture productivity. This may lead to increased or decreased productivity.

(v) Due to ozone depletion the flowering and phonological characters of several plants may be influenced and chances of some plants becomes sterile is also possible.

(vi) The nitrifying bacteria is infected which leads to changes in nitrogen cycle particularly in water bodies.

(vii) The depletion of ozone layer alters marine sulphur cycle.

(viii) The depletion of ozone layer leads to photo degradation of material substances and bringing down their life.

(ix) Due to changes in climate variables because of ozone depletion, the animals and the plants may migrate and there are chances of genetic erosion.

Tree Improvement and Seed Technology

1. Selection and breeding for resistance to diseases and insects. (2009)

Selection and breeding for pest and disease resistance is in practice for most tropical trees which are commercially grown in the country. The fundamental principles and methods adopted are furnished.

Disease development process

The disease in plants develops in the following sequence;

(i) **Contact:** The first infection process occurs through contact. The causal agent (i.e., pathogen) contacts with the host plants and initiates the infection process.

(ii) **Infection:** After physical contact, the pathogen infects the plant. Infection is easier through the wounds in plant parts.

(iii) **Establishment:** After initial infection, pathogen derives its nourishments from the host and establishes itself better inside the host.

(iv) **Development:** After establishment, pathogen develops well and causes injury.

Factors a-ecting disease development

(i) **Host:** The resistance plants are not vulnerable to pathogen because of its adaptability. The host plants are classified as four classes based on the susceptibility to pathogens viz., i) Susceptible host, ii) Immune host and iii) Tolerant hot.

(ii) **Pathogen:** The virulent pathogen infects the host strongly whereas light pathogenic strains cause less impacts on host.

(iii) **Environment:** Certain pathogens need specific environmental conditions for its development. Dry rot is more common in dry condition rather than irrigated conditions. Similarly, damping off develops well under waterlogged conditions.

Types of disease resistance

(i) **Vertical disease resistance:** It is determined by major genes or oligogenes. This is specific to pathotype as well as race specific. This develops resistance against particular strain alone in plants.

(ii) **Horizontal disease resistance:** This is controlled by poly genes. This resistance is pathotypc non-spccific and produce partial or general resistance to plants.

Table: Di3erences between vertical and horizontal resistance

Features	Vertical resistance	Horizontal resistance
Pathogen specificity	Specific	Non specific
Nature of gene action	Oligogene	Polygene
Response to pathogen	Hyper sensitive response	Resistance response
Phenotypic expression	Qualitative	Quantitative
Selection and evaluation	Easier	Difficult
Suitable for		
Heat	Annual not perennials	Both
Pathogen	Immobile, e.g. Soil borne pathogen	All (Mobile and air borne)
Efficiency	Highly efficient against specific races	Variable but operate against all races

Mechanism of disease resistance

(a) **Mechanical resistance:** This resistance is arises due to the anatomical features of the plant.

(b) **Hypersensitivity:** This refers the immune response of plants for pathogen infection. Here after infection, the surrounding host cell dies and leads to the death of pathogen or prevents its spore formation.

(c) **Nutritional mechanism:** In this resistance, the resistance host does not fulfil the nutritional requirements of pathogen and hence pathogen cannot able to survive inside the host plants.

Genetics of disease resistance

(i) **Oligogenic resistance:** In this mechanism the resistance is offered by the one or few dominant genes against the particular pathogen. This gives specific resistance for the particular pathogen.

(ii) **Polygenic resistance:** In this, resistance is offered by the many genes for the general protection. This is the non-specific resistance and hence plants are resistance to number of pathogen.

(iii) **Cytoplasmic resistance:** In this mechanism, genes presents in the cytoplasm offer resistance for disease resistance.

Sources of disease resistance

(i) Some known variety
(ii) Germplasm collection
(iii) Mutation
(iv) Somoclonal variants
(v) Related plant species
(vi) Unrelated plant species

Methods of breeding for disease resistance

(i) **Selection:** Selection is the process of favouring selected plants due to its advantages. It is quickest and cheapest method for getting desired improvement towards disease resistance. Normally selection programme has to be carried in the heavily infested population.

(ii) **Introduction:** This refers to the introduction of resistance species where it is not present before. This is also quickest way of obtaining better improvements.

(iii) **Mutation:** In this, mutant variant is used in the breeding program for obtaining a disease resistance.

(iv) **Hybridization:** This is the most common method of obtaining disease resistance mechanism. Here the resistance gene may be obtained from wild or some other varieties. These genes are transferred into the susceptible host through crossing methods. Back cross method is suitable for oligogeneic traits whereas pedigree method is useful for polygenic resistance.

(v) **Somoclonal variation:** Various cells are cultured in medium and treated with pathogens. The disease free cells are screened from the callus cultures and used for development of disease resistance seedling.

(vi) **Genetic engineering:** It is done for developing disease resistance in plants. But this is costlier method and requires more technical knowledge.

Limitations of breeding for disease resistance

- Development of resistance to disease is difficult when disease is controlled by one pair of dominant and recessive genes.
- The genetic complexity of many host parasite systems creates problems for resistance development.

Advantages of disease resistance over pest resistance

- The pathogens are less mobile and usually do not exhibit the wild fluctuations in population number.
- Artificial inoculation of diseases is easier than forced feeding of an insect.
- Chances of success are high in breeding for disease resistance than insects.

Breeding for insect resistance

(A) Mechanism of insect resistance

(i) **Non preference:** It refers to the plant characteristics that lead insects away from a particular host. This includes of both plant and insect. This is due to the unattractiveness or unsuitable condition of the host plant for colonization, oviposition or both for the insect. This is also referred as antixenosis or avoidance.

(ii) **Antibiosis:** It refers to the adverse effect offered by the resistant plant on the development and reproduction of the insect pest. It involves morphological, physiological and biochemical features in developing resistance mechanism.

(iii) **Tolerance:** Tolerance refers to the ability of the host plant to withstand the pathogen attack without much damage to the economic yield.

(iv) **Avoidance:** It refers to the escape of the plant from the infection but it is not a true resistance mechanism.

(B) Nature of insect resistance

(i) **Morphological resistance:** This refers the different morphological conditions, which avoids the insect attack. It includes thick bark, hairiness of leaf (leaf beetles), colour of foliage (red cabbage less preferred by butterflies than green varieties) and thicker leaves (against jassids).

(ii) **Physiological resistance:**

- High osmotic concentration of cell sap shows resistance against sap sucking pests.
- Gummy exudates show resistance against aphids.
- Presence of resins and alkaloids in cell sap gives resistance against certain beetles.
- Amount of carbohydrate reserves in the host controls bark beetles.

(iii) **Biochemical resistance:**

- High silica content in leaf offers resistant to borer attack.
- Concentration of phenolic compounds in leaves gives resistance to some insects.
- High saponin content induces general insect resistance.

Genetics of insect resistance

(i) Olygogenic resistance

(ii) Polygenic resistance

(iii) Cytoplasmic resistance

Methods of breeding for insect resistance

(i) **Introduction:** The resistant genotypes / provenances / seed sources / clones / progenies can be introduced from other centers of origin for adoption. This will ensure resistant towards the targeted pests.

(ii) **Selection:** It is one of the most acceptable breeding method for insect resistance. Normally in the heavily infested population, selections can be made for pest free and resistance. Once the trees are identified the resistant genotypes can be mass multiplied through clonal technology and deployed as resistant clones.

(iii) **Hybridization:** The resistant variety can be crossed with commercial susceptible variety and the resultant hybrid can be used as a resistant variety. In case, the F1 is also susceptible, repeated back crossing can be facilitated till the resistance is obtained.

(iv) **Genetic engineering:** One of the most recent and robust technology is developing genetically engineered plants. The success evidenced in cotton bole worm resistance through genetic engineering can be deployed for tree species as well in order to develop resistant variety.

Limitations in breeding for insect resistance

- It leads to susceptibility for another pest.
- Sometimes it may decreases the quality of produce.
- Many times genes available from wild varieties are difficult to transfer.
- Sometimes, disease resistance associated with linkage drag.
- Screening for pest resistance is difficult and more critical. It needs helps of both breeder and entomologies.
- It is the long term program and needs adequate financing.

2. Write short notes on strategies for conventional tree improvement programme. (2010)

The tree improvement refers to the control of parentage combined with other forest management activities for the improvement of yields from a forest both in quantity and quality aspects. Tree improvement is effective when it combined with silvicultural practices.

Tree improvement strategy

The tree improvement strategy depends on the objective of management coupled with availability of capital investment. This besides the strategy depends on the existing genetic resources, availability of adequate variation both within and between the genetic resources and their utilization strategies. So in a holistic perspective the tree improvement strategy aims at improving the tree genetic resources either for utilization or for conservation. But all these strategies depends on the objectives of any tree improvement programme which are furnished below.

Objectives of tree improvement

- Improvement of the trees for better yield and quality
- Change the inheritance pattern of a given characters
- Improve the variability towards the desired direction
- Development of varieties for biotic and abiotic tolerance
- Attaining the desired characters such as straight bole, dense wood, etc
- Development of trees for difficult sites
- Development of high yielding short rotation varieties
- Improvement of wood quality for better timber, paper, pulp and other uses.

Strategies and methods for improvement

To satisfy the above objectives and strategies the following methods of tree improvement programme will help to achieve the targets.

(i) Introduction: Introduction is the fundamental strategy of improving any species which involve introduction of one tree species from the place of their origin to such areas where they were never grown earlier.

Example: Ecualyptus, Casuarina, Simarouba

(ii) Selection: The tree improvement programme involves primarily selecting the superior genetic stocks by exploiting the existing variability.

- **Evaluation:** The selected genetic stocks are evaluated through progeny / provenances / seed sources / clonal tests and genetically proven materials are screened for further evaluation.
- **Amplified test:** Once the genetically superior resources are identified they must be amplified through a larger experiments in order assess both wood yield and wood quality parameters.
- **Deployment:** Once the identified genetic stocks is amenable for higher productivity and wood quality then the genetic resources can be deployed operational tree improvement programme. These are the general strategies followed in any tree improvement programme.

  Example: Teak, Casuarina, Red sander, Sandal, *Dalbergia sissoo*, Albizzia

(iii) Hybridization: In case the existing variability is not sufficient and they are not able to meet the objectives envisaged, then the genetic resources have to undergo systematic hybrid development programme involving inferior and superior varieties. The resultant F1 hybrids are evaluated and the superiority can be fixed through clonal approaches.

Example

- Eurograndis hybrids between *Eucalyptus urophylla* and *Eucalyptus grandis.*
- Casuarina hybrids between *Casuarina equisetifolia* and *Casuarina junghuhniana.*
- Jatropha hybrids *Jatropha curcas* and *Jatropha integerrima.*

(iv) Mutation: Some species require sudden genetic changes like development of thornless variety, superior oil quality, increased wood quality etc. which can be achieved through mutations both through physical and chemical mutagens.

Example

- Development of thornless prosopis through gamma radiation
- Development of thornless in *Caesalpinia sappan* through gamma radiation
- Changing oil quality in *Jatropha curcas* through gamma radiation

(v) Tissue culture: The superior genetic stocks can be mass multiplied through tissue culture approaches for deploying them in operational tree improvement programme coupled the conservation of genetic resources.

Example: Teak, Sandal, Eucalyptus, Casuarina and Bamboos

(vi) Genetic engineering: This gene transfer technology has eliminated the obstacle exists in the traditional tree improvement programme which facilitated transfer of genes across biological organisms.

Example: GMO in *Allocasuarina verticillata*, GM poplars

3. Discuss selection as a method of tree improvement (2010)

Selection is one of the traditional genetic tool employed in the tree improvement programme. This selection depends on the tree species, objective of management and number of traits involved. Accordingly selection methods are different for single trait and multiple trait selection which are furnished below.

I. Selection methods for single traits

(a) **Mass selection:** It involves choosing individuals on the basis of their phenotype. It does not consider any information about the performance of ancestors, siblings, offspring or other relatives. It is the only type of selection that can be used in natural stands or in plantation where tree parentage is unknown.

(b) **Family selection:** It involves the choice of entire families on the basis of their average phenotypic values. Family selection works best with traits having low heritability where individual phenotypes are not a good reflection of genotypes. The family averages are based upon large number of individuals and hence environmental variance is reduced. This leads to increased rates of inbreeding and reduces the total genetic base of the population.

(c) **Sib selection:** In this, individuals are chosen on the basis of the performance of their siblings and not on their own performance. This is applicable to destructive sampling process.

(d) **Progeny testing:** It involves selection of parents based on the performance of their progeny. It is very precise as it estimates breeding values. It discards genetically poor parents and gives better results.

(e) **Within family selection:** In this selection, individuals are chosen on the basis of their deviation from the family mean. The family values per se are given no weight when selections[1] are made.

(f) **Family pins within family selection:** This is the two stage method involving selection in families followed by selection of individuals within families, it consists of choosing the best families along with the best individuals in them.

II. Selection methods for several traits

(a) **Tandem selection:** In this selection, one desire trait is selected and breeding proceeded to get a desired improvement. After getting desired improvement in the trait concerned, selection and breeding efforts are concentrated for the next traits. E.g. Disease resistance and cold hardiness.

(b) **Independent culling:** It is a method of multi-trait selection. In this minimum values for each trait of interest is set first. Individuals falls below this standard is removed and individuals falling above this value is retained for breeding program.

(c) **Selection index:** It is a form of multi-trait selection that combines information on all traits of interest into a single index. Selection index combines family plus individual information for all traits into one index.

(d) **Recurrent selection:** It is devised in relation to heterosis breeding. The idea was to ensure the isolation of superior inbreeds. It increases the frequency of desired alleles in the progenies as it is repeatedly crossed with any one of the parent from which desired characters are transferred in to the progeny. It may be of simple recurrent selection or reciprocal recurrent selection. This is suitable for development of varieties for specific disease or insect resistance.

4. Selection intensity a-ects genetic gain. (2010)

Selection is based upon the principle that the average genetic value of selected individuals will be better than the average value of individuals in the population as a whole. The improvement that can potentially be made from selection for a characteristic is a function by the heritability of the trait and the variation for the trait that exits in the population.

Selection di3erential: The average phenotypic value of the selected individuals, expressed as a deviation from the population mean.

The greatest opportunity to improve gain from selection is to increase the selection differential. The selection differential is dependent upon 2 factors.

(i) The Proportion of individuals in the population that are selected i.e. the intensity with which selection is done.

(ii) Phenotypic standard deviation

$G = ih^2 \sigma_p$

i = Intensity of selection ;

h^2 = heritability

σ_p = Phenotypic standard deviation

This formula indicates that both selection intensity and phenotypic variation influence gains that can be made.

If the phenotypic standard deviation and the intended selection intensity is known, a response of selection can be predicted before selections are ever made.

Alternatively, selection intensity are varied to determine how many individuals must be chosen to obtain a certain desired gain. Selection intensity is related to the proportion of individuals in the population that are selected. If this proportion is known, it can be calculated directly from the value.

5. Discuss the use of tree improvement in natural forest and stand improvement. (2012)

The principles of tree improvement can be used profitability both in regeneration of natural stands and also in certain forest management activities which can be applied in already established stands though the genetic gain through the use of genetic tools will be less compared to plantations. But still the genetic tools can be applied very quickly for genetic manipulation.

Characteristics of natural forest

The natural forests have the following characteristics which are unique for the application of tree improvement principles.

- Stands are truly uneven aged and representation of all age classes may be present.
- The desired species may occur sporadically in a scattered form and there won't be any comparison trees.
- The species may be a vigorous sprouter and trees growing near the candidate tree can be on a common root system and have same genotype.
- The stand contains mixture of species.

Use of tree improvement

A. Selection for provenances / seed sources / progenies /clones: The natural forests are acting as centers of origin for many species and will act as a base population for a tree improvement programme.

Hence the tree improvement tools will help to identify the superior genetic stocks which could be useful for productivity (or) can be a resource for conservation.

B. Genetic gain and seed production: The best genetic gains can be obtained by adopting tree improvement tools in natural forest. Traditionally the forest trees are selected based on phenotypic superiority which leads to dysgenic selection and only the inferior and emaciated population. Hence the tree improvement programme should ensure leaving the best individuals for seed production following regeneration either naturally or artificially. This type of improvement programme can be practiced in shelter wood system.

C. Breeding for biotic resistance: Particularly pest and disease resistance is gaining significant attraction due to their epidemic nature. For this purpose natural forests acts as source for disease resistance plants.

Example: Southern pine is improved through the selection of rust free population from heavy infested population and hence useful in developing disease resistance varieties. The genetic approach has considerable potential in naturally regeneration for stands. When it is applied to the establishment of new stand and followed by silviculture treatment the best genetic gain can be obtained through proper thinning.

The tropical hardwood in natural forest can be very well managed using breeding tools selection, domestication and further improvement. Similarly tree improvement play a vital role in conservation of endemic, endangered threatened and vulnerable species of the natural population. The tree improvement principles will help to convert any natural stand into a seed production area by judicious application of cleaning and thinning operations.

6. Write short notes on seed orchards. (2012)

Seed orchards (SO): A seed orchard is a plantation of selected clones or progenies which is isolated or managed to avoid or reduce pollination from outside sources and managed to produce frequent, abundant and easily harvested crops of seed.

The genetic improvement obtained in seed orchard is usually greater than in seed production areas and depends primarily on the intensity of selection and the heritability of the traits on which selections were based.

While planning the establishment of seed orchard, to meet the seed requirement of individual species, information is required on a number of items such as (i) expected yield of seeds per tree, (ii) the age at which the species normally reaches the reproductive stage, (iii) flowering and fruiting patterns of the species, (iv) breeding systems of the species and (v) the most appropriate vegetative propagation method for the species.

Seedling seed orchards and clonal seed orchards are the two types of seed orchards. From the individual selected trees, clones are collected for establishment of clonal seed orchards where as for establishment of seedling seed orchards from individual selected trees, seeds are collected (Table).

Table: Types of Seed Orchards

S. No.	Seedling Seed Orchard (SSO)	Clonal Seed Orchard (CSO)
1	Established from seedlings i.e., by seed propagation from selected parents. Genetical identity may be lost through segregation.	Established by using grafts, cuttings, air layer plants, tissue culture plantlets or other methods of vegetative propagation. Genetical identity of the material is maintained.
2	Preferred if a genetic test can be conducted fulfilling both the production (economic trait ie.,wood/oil yield & seed yeild) function at one time. It is important if vegetative propagation is difficult	Generally preferred when vegetative propagation is possible and when seed material as well as vegetative propagules are expected from the seed orchards.
3	Selfing cannot be ruled out as the pollination may take place within the family	Selfing is reduced by avoiding planting of same ramet near by
4	Broader genetic base than CSO but less selection differential	Narrower genetic base than SSO but higher selection differential.
5	An outstanding genotype will appear only once	An outstanding genotype will appear many times.
6	Testing is on family level; 100 or more parental trees are used. Height hampers control of pollination studies.	Testing is on individual level, 20-40 clones are used. Flowering at lower height aids control of cross pollination studies.
7	First flowering and fruiting usually occur later than CSO. Applicable for species with early flowering.	Generally first flowering occurs earlier than in SSO. Applicable for species with late physiological flowering.
8	Harvest of fruits/seeds is generally more difficult than from CSO	Harvest of fruits/seeds easy due to low crown branching.
9	**Advantages** (i) Larger genetic gains because two cycles of selection is combined in one operation (ii) Avoids problems with vegetative propagation (iii) Cost is lower	**Advantages** (i) Seed production begins within few years after orchard establishment (ii) Orchard can be established at best site for seed production

10	**Disadvantages** (i) Progeny testing and seed production have different requirements. Eg., Genetic tests are planted on lands that are typical of those that are to be reforested with improved trees. These areas may not be suitable for the seed orchards. (ii) For seed production well developed crown is required. So, wider spacing is adopting but wider spacing impairs progeny test.	**Disadvantages** (i) Only one cycle of selection in one operation (ii) Difficulties in vegetative propagation (iii) Cost is higher

Principle steps in seed orchard establishment

(i) Plus trees are intensively selected form natural stands or unimproved plantations.

(ii) Clones and progenies of these selected trees are established together in a seed orchard. This is first generation seed orchard where selection of plus trees is done only on the basis of phenotypic characters. Under seed certification scheme, the seeds collected form this seed orchard comes under "Reproductive Material collected from Untested Seed Orchards".

(iii) Breeding value of each tree is evaluated by progeny testing.

(iv) Genetically undesirable individuals are rouged from the orchard on the basis of progeny test results. Remaining trees are allowed to cross-pollinate for commercial production of genetically improved seeds.

(v) Best individuals within the progeny trials are selected based on the genotypic selection i.e., half-sib and full–sib progeny trials. They are established together in second generation or advanced generation seed orchards. Under Seed Certification Scheme the seeds collected from this orchard is called "Tested Reproductive Material".

Examples of seed orchard in India

S.No.	Species	State
1.	Teak Both SSO and CSO	Kerala Karnataka Andhra Pradesh Madhya Pradesh Maharashtra Manipur Tripura Uttar Pradesh Tamil Nadu

2.	*Dalbergia sissoo* SSO	Bihar Maharashtra
3.	Eucalyptus Both SSO and CSO	Tamil Nadu Uttar Pradesh
4.	*Bombax ceiba* SSO	Arunachal Pradesh Uttar Pradesh
5.	*Santalum album* Both SSO and CSO	Karnataka Tamil Nadu
6.	*Gmelina arborea* SSO	Tripura Arunachal Pradesh
7.	Casuarina Both SSO and CSO	Tamil Nadu
8.	*Pinus kesiya* SSO	Manipur

7. Distinction between selection intensity and heritability (2012)

The selection differential that the breeder uses is dependent upon two factors. One is the proportion of individuals in the population that are selected; that is, the intensity with which selection is done. The other factor is the phenotypic standard deviation, which, as we have seen, is a description of the variation in the population and is expressed in the same units as the population mean. Many breeders prefer to express response to selection or gain by the formula.

$$G = ih^2\sigma_p$$

Where

i = Intensity of selection ; h^2 = heritability

σ_p = Phenotypic standard deviation

This formula indicates that both selection intensity and phenotypic variation influence gains that can be made.

A comparison of the two formulas for selection response shows that

$$G = h^2 S = ih^2 \sigma_p$$

Therefore, $S = i\sigma_p$ and $i = S/\sigma_p$

The selection intensity, or i, measures how many standard deviations the mean of the individuals that were selected exceeds the mean of the base population. For example, a selection differential (S) of 10 indicates that the mean of the selected population is 10 units better than the mean of the whole population. If the phenotypic standard deviation (σ_p) is equal to 5, then $i = s/\sigma_p 2$, and the mean of the selected population is two phenotypic standard deviations better than the mean of the whole population. Because of the characteristics of the normal distribution, the equation is a convenient way to calculate genetic gain. If the breeder knows the phenotypic standard deviation and the intended selection intensity, a response to selection can be predicted before selections are ever made.

Alternatively, selection intensity can be varied to determine how many individuals must be chosen to obtain a certain desired gain.

Selection intensity is related to the proportion of individuals in the population that are selected. If this proportion is known, it can be calculated directly from that value. Calculation of selection intensity involves a more in-depth knowledge of statistics. When the population from which selections are to be made consists of only a few individuals, selection intensities will be lower for any given proportion saved.

8. What are the unique requirements of tree improvement in agroforestry? (2013)

Agroforestry is an age old idea of having trees in form land. Now a days trees are growing as an important money yielding enterprise equal to agriculture. Hence the farmers are looking for some improved genetic material for their farm land. Hence the requirement of tree improvement to satisfy the demands of agroforestry is entirely different from the conventional tree improvement strategies. The following are the requirements of tree improvement in agroforestry.

(i) **Development of high yielding varieties:** The agroforestry deals with the farmers and the economic returns depend on the productivity of plantations. The farmers will always compare the economic differences between the crops and they will choose the crop with higher volume and economic returns. Under such circumstances the tree improvement activity should aim for high yielding varieties and the successful evidences are already available.

Example: Eucalyptus, Casuarina and Poplar clones

(ii) **Development of short rotation variety:** The farmers are always looking for early yield and returns and hence the traditional long rotation tree crop needs to be replaced with short duration variety. Hence the tree improvement strategies have to be oriented in such a way in order to produce high yielding and short rotation varieties.

(iii) **Multi-purpose tree species:** The agroforestry system is aimed at satisfying the demands of domestic utility besides enhancing the productivity of the landuse system. Under such circumstances, one of the unique requirements of tree improvement for agroforestry is development of multi-purpose tree species amenable for food, fuel, fodder and fiber utility.

(iv) **Fast growth with apical dominance:** One of the unique requirements of tree improvement for agroforestry is development of fast growing trees with apical dominance character. This will ensure dominance of apical growth and suppress the horizontal growth thereby reduces the competition with the agricultural crops.

(v) **Nitrogen fixing trees:** The tree improvement for agroforestry should aim for identifying and introducing nitrogen fixing trees with a potential to fix 100-150 kg of nitrogen / ha/ annum which will ensure soil health management.

(vi) **Trees for industrial agroforestry:** The tree improvement strategy for agroforestry should inventorize high yielding short rotation varieties amenable for varied industrial utility like timber, pulp and paper, plywood, match splints, bio-power, and biofuel utility. The improvement in these varieties will help the farmers to adapt them in their agro and farm forestry system.

Industry	Tree improvement requirement
Paper and pulp	• HYSR variety • High cellulose and low lignin
Plywood	• HYSR • Straight bole with high density
Pesticide	• Higher seed yield • High Azadiractin content
Timber	• Straight bole • Fast growth • High density • Early harvest
Match	• High fissility • High cellulose low lignin
Dendro power	• High calorific value • Fast growth • Coppicing ability

(vii) Improving trees for climate change: The growing concern on the agriculture is impact of climate change on agriculture productivity due to vagaries of monsoon failure, temperature fluctuations and natural calamities. Trees in general and agroforestry in particular are identified as potential land use system which can mitigate and adopt growing climate change risks. Hence tree improvement should be aim to develop varieties and trees amenable for varied climate associated risk and to mitigate and adapt to the climate change.

9. Explain the possible causes of genetic variation in forest trees (2014)

Causes of genetic variation: The genetic variation in a species arises because of the four causes *viz.,* 1) mutation, 2) gene migration, 3) selection and 4) genetic drift.

(i) Mutation: Mutation is the main sources of variation which designates the sudden heritable change in the genetic constitution of an organism. The mutation occurs as a random event at very low frequency. Most of the mutations produce deleterious effect on plants. But these can be suitably used to obtain a genetic improvement in trees.

(ii) Gene migration: This refers the migration of alleles from one population to another. Gene flow take place at species level through a process called introgression. But in introgressive breeding, the new progeny may not be well adapted to competent with the other species. Hence, introgressive progeny is backcrossed with any one of its parents to get the desired characters for its survivability. Gene flow with recombination produces great variation at genome level.

(iii) Selection: Selection is the process of maintaining the desired individuals for the human benefits. As it facilitates encouragement of particular genotype, it reduces the variability in a natural stand.

(iv) **Genetic drift:** It is a complex mechanism that operates the chance of fluctuations in allele frequencies within a population. It depicts how the population obtained is deviate by chance from the parent populations. Thus genetic drift causes reducing variation by fixing or loosing alleles. Genetic drift occurs in forestry due to natural catastrophes.

10. Write on the provenance e-ect on seed quality of silvicultural plant species. (2009)

Provenance e3ect on seed quality: Provenance is the original geographic region in which the species naturally exists. Provenance also marks the natural origin of the species. The effect of provenance on seed quality is scientifically proved and few reasons for improved quality of seed is given below.

(i) **Provenance e3ect on physical and physiological quality of seed:** The climate and edaphic factors of the provenance is highly positive which makes the trees to flower robustly and positive fruit setting. The seeds are filled equally and attained physiological maturity uniformly. This surely improves the seedling parameters such as germination percent, seedling vigour, dry matter production etc. Hence the provenance has direct effect on physical and physiological quality of seeds.

(ii) **Provenance e3ect on the genetic quality of seeds, genetic worth of seeds:** The provenances also extend significant genetic quality and leads to improved genetic resources. The provenance resources marks robust number of inter breeding trees and the process of natural selection. During the process of evolution, it naturally removes the inferior genotypes from the population. Only the elite genotypes thrive and continuously interact with the environment to attain phenotypic superiority.

Through sexual reproduction new allele combinations are created and these individuals show improved silvicultural characters. Thus the provenance is the place where tree improvement takes place naturally and this ensures genetic worthiness of the seeds collected from provenance.

Some examples of provenance e3ect on seed quality of plant species

- *Tectona grandis* – Nilambur teak for its wood quality quality
- *Santalum album* – South Eastern Ghats
- *Pterocarpus santalinus* – Andhra Pradesh (Chittor at shesachalem district)

11. a. Discuss the possibilities of biotechnological interventions in tree improvement programmes. (2010)

b. Discuss the importance of tissue culture techniques as a tool in tree improvement. (2011)

Biotechnological and tissue culture interventions: The advancements made in science and technology particularly in the biotechnology and tissue culture has excellent scope and application in tree improvement programme. The traditional tree improvement programme is delimited due to slow growing nature of trees coupled with long rotation. This besides, the existing genetic resources and the associated variability needs to be exploited rapidly towards meeting the growing demand. The rapid deforestation and the increasing demand for wood and wood products have compelled the incorporation of new technologies towards improving the productivity and profitability. Under such circumstances, the biotechnology and the tissue culture tools will help to accelerate the tree breeding and improvement activities through the following approaches.

(i) **Mass multiplication of elite genetic resources:** Conventional methods of asexual propagation (Vegetative propagation) like grafting, budding, layering etc. are often too slow or fail completely in several tree species. Micro vegetative propagation using tissue culture allows much greater control and manipulation of the development of tissues within the culture tube than conventional methods. In normal cuttings, each cuttings can result in only one plant, whereas by micro propagation thousands of plants can produce from a single piece of plant tissue explants. Not only is the rate multiplication increased, but the mean generation time is also decreased because the process can continue all round the year under controlled laboratory conditions.

Methods of tissue culture are now available for rapid multiplying "elite" teak and eucalyptus trees, growing in the forests of chandrapur and Tamil Nadu respectively. There are also other reports where tissue culture methods have been developed in India for forests and fruit trees viz., apple, citrus, *Dalbergia sissoo*, *D.latifolia*, *Albizzia lebbeck*, tamarind, sandal, rubber etc.

(ii) **Production of disease free plants:** Virus infections are major problem with vegetatively propagated species. Conventional methods of eliminations of virus from stock by heat treatment are useful only with some varieties. The growing tip (meristem) is usually univaded by viruses and plants obtained by the culture of these meristem tips remains virus free. Morel amd Martin (1952) made and first attempt to demonstrate of dahlia plants. This technique has been exploited to produce virus free plants of fruit trees like citrus, apples, etc. In general, meristem culture results in the production of completely disease free root stocks and other plants. The plants produced have been found to be healthier, more vigorous and to produce higher yields.

(iii) **Embryo rescue technology:** In traditional plant breeding, hybrid embryos of many interspecific crosses fail to grow to maturity mainly due to the degeneration of the endosperm or an abortion of the embryos at an early stage. Culture of interspecific hybrid embryos has now found wide utilization in the fruit trees. It has been successfully used for peach, plum, pear and apple cultivars. Another application of embryo culture is to overcome seed dormancy which with many trees takes several years for germination under natural conditions.

(iv) **Development of haploids:** Haploid plants being gametophytic in origin possess only half the normal numbers used to produce homozygous lines which are invaluable for any breeding programmes and also various other genetic manipulation. After the first successful report on regeneration of haploid plants from pollen grains of the cultured anthers of datura this techniques has been demonstrated in a large number of herbaceous species. However, the techniques of culturing anther and pollen have found only limited success when applied to forest and fruit species. Haploid callus has been obtained from cultural anthers of forest tree species of Pinus, Vitis, Citrus, etc.

(v) Early flower induction: Trees unlike agricultural crops, take years to attain sexual maturity and to flower. Thus, tree breeders have to wait upto twenty years or even more. This problem is further aggrevated in some trees such as bamboo, which may flower once in forty years. Thus the early induction of flowering by the application of growth regulators in vivo or their use in vitro cultures would help to reduce the breeding cycle.

(vi) Production of Somatic hybrids: Somatic hybridization through protoplast fusion opens an avenue for synthesizing characteristics which were not possible hitherto. Even in wide crosses through embryo cultures, the degree of variability is low. Somatic hybridization is an alternative to sex in order to combine the entire genomes from incompatible parents and is expected to results in hybrids.

(vii) Development of GMOs: Genetic transformation using agro bacterium mediated transfer is another important technique in biotechnology in which a gene or group of genes encoding for a specific trait can be isolated and cloned. Successful results are already available in gene transfers in Eucalyptus, Poplars, Beutla, Casuarina etc. for biotic and abiotic resistance coupled with qualitative traits.

(viii) Cryopreservation for ex-situ conservation: The rapid rate of diminishing of the genetic resources has caused concern of great magnitude for the conservation of important and elite germplasm. For this, Cryopreservation techniques have been evolved which involved freeze preservation of cells, tissues and organs as a meaningful tool for long term conservation (-196°C), establishment of gene banks and international exchange of germplasm.

(ix) Clonal repositiories and germplasm banks: Meristem cultures and in vitro plantlets can be packed in card board and foam boxed transported to international destinations. This distributions and the exchange of desirable germplasm can also be effectively carried out by the transfer of frozen cells and tissues in portable liquid nitrogen cylinders and can be transported by air. Hence it is highly desirable to set up "germplasm banks" and clonal repositories of the rare, elite and other desirable genetic stocks of trees. Such banks should be responsible for the storage, maintenance and the exchange of germplasm of trees both at national and international levels.

(x) Production of secondary metabolites: The biotechnology help to add value in the traditional tree improvement programme through the production of secondary metabolites. Several tree species are known to contain wide range of alkaloids but their exploitation through the traditional technology is the time consuming process coupled with low recovery. The biotechnology has resolved the issue through cell suspension cultures and generation of secondary metabolites. The biotechnology facilitated production of taxol, an anti cancer alkaloid from *Texus buccata*, quinine – anti malarian alkaloid, vincristine and vimblastine from *Vinca rosea* are some of the examples of biotechnological interventions in metabolite production using forestry tree species. Similarly wide range of compounds have been identified for antibacterial, antifungal and anti viral properties through biotechnology mediated cell suspension cultures.

Types of cultures

Plant tissue cultures can be divided into five classes based primarily on the type of material used on the medium.

(a) **Meristem culture:** This term is often used loosely to refer to very small shoots apices dissected from terminal or lateral buds. Strictly speaking, it refers to the microscopic apical dome with only the smallest leaf primordial evident, usually less than 2 mm across. The advantages of using shoot meristem is that they are most likely to be free of internal pathogens.

(b) **Callus culture:** Sometimes explants produce callus rather than new shoot growth particularly where high levels of hormones are applied. In other cases, callus may be induced intentionally because of its potential for mass production of new plantlets. The limiting factors are the difficulty in inducing the initiation of new shoot apices, especially in woody species.

(c) **Cell suspension culture:** This is essential a product of callus culture i.e. callus usually refers to a mass of undifferentiated cells. Once these are separated in liquid culture, it becomes a cell suspension. This culture may be used to produce a product directly from these cells without regenerating new plants. These cells may be genetically engineered to increase the synthesis of different secondary metabolites.

(d) **Protoplast culture:** This is a next step beyond cell suspension culture where the cell walls of suspended cells are removed using enzymes to digest the cellulose to leave the isolated protoplast. With the cell wall removed, it is possible to insert or remove foreign materials including the basic genetic materials DNA and RNA or to fuse together cells from entirely different species.

(e) **Organ culture:** The culture of embryos, anther, shoots, roots or other organs on a medium is called organ culture.

12. Describe the various approaches for obtaining genetically superior seed, giving advantages of each (2011)

Approaches for obtaining genetically superior seeds

(a) **Provenance collection:** It is the original geographical location to which plants are native and within which their genetic characteristics have been developed through natural selection. Hence provenance yield large amount of quality planting materials.

Advantages: Very cheap method to get genetically superior seeds.

(b) **Seed source collection:** Progeny of individuals whose provenance is known.

Advantages: Foresters avoid shipping seeds indiscriminately for long distances from provenance. Ample seeds can be collected from local seed sources.

(c) **Collection of seeds from seed production area:** SPA refers to phenotypically superior stand made up of vigorously growing healthy trees, upgrading by thinning to remove poor phenotypes which is treated and managed to cause abundant seed production.

Advantages: First source of superior planting material. Interim source until progeny tested seeds can be obtained from the seed orchard.

(d) Collection form seed orchard: Seed orchard is a plantation of selected clones (or) progenies which is isolated, managed to avoid pollination from inferior sources and managed to produce frequent, abundant and easily harvested crops of seed.

Advantages: The genetic gain is maximum compared to all other approaches.

(e) Progeny collection: The identified superior progenies after evaluation through progenies test can be advanced to seedling seed orchard by eliminating the inferior populations and retaining only the superior genetic resources. This will ensure supply of quality seeds.

Advantages: Higher genetic gain

(f) Clonal collection: The genetic resources tested through clonal evaluation can be advanced to clonal seed orchard by eliminating the inferior clones and retaining only the superior clones. This will ensure production of genetically superior seeds.

Advantages: Maximum genetic gain

13. Describe the operational use of vegetative propagation in tree improvement. (2013)

Operational use of vegetative propagation: The vegetative propagation reproduces clones which contain all the genetic information of the parent tree. The term clone is used to mean a genetically uniform plant material derived from a single individual and propagated exclusively by vegetative means.

Before going for a clonal propagation, the first step is selection of genetically superior tree or elite or mother tree. After selecting the trees, the clonal materials are collected and used for propagation. The cultural requirements of all the individual members of a clone are the same. Environment can modify the growth and development of individual members of a clone. Clones are usually less broadly adopted, even though each member of a clone has the same genotypes, the individual genotype can possess a considerable ability for adaptation to differing adverse environments. It is possible to select clones that possess greater adoptability than that possessed by the seedling. A forest tree needs great adaptability mainly to survive and to reproduce.

Production and Maintenance of a clone: The production and maintenance of a clone have the following requirements.

(a) Selection of a Clone: The first step in selection of a clone is correct identification, proper labelling and should be genetically true to type. A dominant tree with good apical dominance, free from pests and diseases and free from genetical disorders with a history of superior phenotypic performance is generally selected.

(b) How many clones should be selected: There is no absolute answer, although a good ultimate can be made based on experience and knowledge for the variation of the species. The question again relates to reduced risk, that is how one can achieve the greatest improvement by selecting only the best clones while staying within acceptable limits of the dangers of destructive losses in the forest. Usually biologists are cautions and they sacrifice too much gain to be certain that they take little risk, whereas financially oriented people often emphasize gain to the point of real dangers from destruction by pests and diseases or adverse environments.

The important question relating to how many clones are necessary, to assure reasonable safety and to make optimal gain is the major issue. It all depends on rotation age, intensity of forest management, genetic variability of the species and clones involved, the likely risks and the acceptable loss levels. It is also recommended that 7 to 30 clones should be used, which would be a safe and reasonable number.

Generally, the more severe the stress a given species or provenance may face, the greater the number of clones, that should be used. There are special conditions such as, severe insect or disease areas or very severe environment wherein, very few clones may be justified.

(c) **Maintenance of a Clone:** Once a disease free, true to type tree has been selected, it must be maintained and multiplied under strict sanitary conditions to prevent reinfection. Measures should also be taken for immediate detection of any reinfection or any change in the original stock. Propagation stocks can be maintained pathogen free by

- Maintaining proper isolation distance
- Observing strict sanitary measures during propagation
- Periodic inspection and testing to ensure that, it does not become infected again.

(d) **Multiplication of a Clone:** Clones of proven superior quality are assembled in an isolated area and properly labelled. Clones are then multiplied in a large scale vegetatively. Once the planted ramets are well established, large scale plantations are raised from clones obtained from Clonal Multiplication Area.

(e) **Testing of a Clone:** Testing of clones is absolutely necessary for systematic screening of clones for productivity attributes and their adoptability to adverse sufficient number of ramets from each clone. This should be designed statistically to avoid experimental errors. The general principles of genetic testing in seedling forestry are applicable to clonal forestry and issues such as mating designs, statistical consideration of plots and interpretation of estimated parameters have been described by many workers. However, the inclusion of a vegetative propagation step requires additional testing to determine the clonal fidelity. There are also benefits of using clonally replicated trees in genetic tests even for programmes using seedling based deployment such as obtaining additional genetic information and increased precision of estimates.

The primary reason for clonal testing is to identify suitable clones for deployment. Clonal testing generally involves several candidate clones evaluated at several different locations with respect to traits of interests. The clones may be vegetatively propagated from breed, selected trees or crosses of such trees. Testing of multiple test sites provides information on adoptability of clones in varying environmental conditions known as genotype x environment interactions. Normally, testing of a large number of clones will result in larger genetic gain than testing of a small number. The testing is constrained by limitations in operation and resources. Thus there is a need to decide the optimal number of clones, test sites and number of ramets per clone and usually "six or less" ramets per clone which is much less than 10-20 trees per family for family ranking.

14. Discuss the mechanism of drought resistance, drought tolerance and drought avoidance in plants. (2011)

Drought is a metrological event which implies the absence of rainfall for a period of time, long enough to cause moisture depletion in soil and water deficit with a decrease of water potential in plants. Under drought conditions, water stress develops in the plant, as the demand exceeds supply of water. This may occur either due to atmospheric or soil conditions and is reflected in a gradient of water potentials developed between the soil root interface, leaf and the transpiring organ. Moisture stress is likely to develop to different rate in different plant organs along with the gradient. Drought like many other environmental stresses has adverse effect on plant growth and yield.

Drought resistance mechanism

As the water stress increases the older leaves senescence to various degrees and reduces the leaf area of plants. Loss of leaf function and premature onset of senescence of older leaves may occur. The production of ethylene may serve to regulate leaf performance throughout its life span as well as to determine the onset of natural senescence and mediate drought induced senescence. The plants may develop such mechanism in order to resist the drought. However drought resistance is very difficult situation for the plants and hence they develop either tolerant or avoidance mechanism.

Drought tolerance mechanism

It is the property of given variety to show a relatively small yield reduction upon exposure to drought. The drought tolerance mechanism is grouped into two categories

(i) **Drought escape:** The ability of the plants to escape periods of drought, in particular the most sensitive periods of its development are flower initiation and grind filling and is termed as drought escape. Early maturity is one of the most important attribute of drought escape.

(ii) **Dehydration avoidance:** The ability to endure or withstand a dry period by maintaining a favourable internal water balance under drought conditions is termed as dehydration avoidance. The dehydration avoidance mechanisms are results of various physiological, biochemical and metabolic process of plants that are involved in growth and yield not being internally exposed to stress and thereby they are protected from water stress. The associated high water use efficiency with plants that avoid drought are described into two types *viz.,* water savers and water spenders.

Drought avoidance mechanism

- Stomata control with the drought-hardy seedlings transpiring rapidly when water was available but conserving water under stress. (A similar pattern has been reported for some species of *Eucalyptus*.)
- Root morphology with the drought-hardy trees having deeper, more fibrous root systems..
- Needle morphology in which the drought-hardy trees have smaller, deeper stomatal pits.
- Number of stomata per unit needle length.

Dehydration tolerance

The ability of plant cells to withstand water stress and continue metabolic functions is termed as dehydration tolerance. Cellular dehydration causes significant disorders in membrane structure, composition and function. Maintenance of membrane integrity and function under a given level of dehydration stress can be taken as a measure of tolerance. Dehydration tolerance of a genotype means that a significantly lower level of changes is induced in it than those in another genotype when both of them are subjected to the same level of dehydration.

15. Comment upon the relationship of general combining ability and breeding value in forest tree improvement programme. (2015)

General Combining Ability (GCA) is the average performance of the progeny of an individual when it is mated to a number of other individuals in the population.

Although general combining abilities may be expressed in absolute units, it is usually more convenient and meaningful to express them and deviations from the overall mean. Thus, a parent with GCA of 0 has an average general combining ability. A positive GCA indicates a parent that produces above-average progeny, whereas a parent with a negative GCA produces progeny that perform below average for the population.

For example,

Female parents	Male Parents 1	2	3	4	Progeny means
5	9	17	12	14	13
6	10	16	12	10	12
7	11	20	10	15	14
8	14	15	6	17	12
Progeny means	11	17	10	14	Test mean 13

The GCA of male Parent 2 would be calculated as

$$GCA_2 = \text{mean of parent 2} - \text{test mean}$$
$$= 17 - 13 = +4$$

Parent 2 therefore has a general ability (GCA_2) for volume of +4 units. Other general combining abilities would be calculated in the same way. For example, general combining ability for Parent 4 is +1 volume units, whereas GCA of Parent 3 is -3 units.

The breeding value of an individual is defined as twice its general combining ability. The difference between breeding value and general combining ability is largely conceptual in nature. Breeding values are meaningful because the parent in question contributes only half of the genes to his or her offspring, the other half coming from other members of the population. The breeding value of Parent 2 would be calculated as

$$2\,(GCA_2) = 2(4) = 8$$

16. Discuss the statement "Even in large experiments with many families, hertabilities are not estimated without error". (2011)

Yes the statement of "Even in large experiments with many families, hertabilities are not estimated without error" is true because heritability estimates can give only a general idea of the strength of inheritance.

Heritability itself is a variable ratio and is subject to changes. Whenever heritability is used in determining the amount of genetic gain are breeding strategies, the values should be qualified with a statement that the confidence level is less than 100%.

Heritability is a ratio between genetic and phenotypic variances: thus, it is not a fixed value for a given characteristic of a given species. Estimates of heritability are not estimated without error; therefore the, ratios obtained are only a relative indication of genetic control and should not be interpreted or invariant values.

The heritability estimates for a group of trees grown in a green house would not be appropriate for the same trees growing in a field environment. Height in the green house may not be influenced by exactly the same genes as height in the field. Even if the traits were the same, the estimates of heritability obtained in the green house will usually been higher than those from the field, due to less environmental variation in the green house.

Hence because of the influence of environment on the heritability ratio, the h^2 estimates for a given characteristic in a species in one geographic area probably will not be the same as those found in another region. The heritability values of a given character in a population often change with age when the environment changes and when the genetic control of the characters changes as a tree mature.

Hence heritability is a variable factor and its estimation may create error even in larger population with larger families.

17. Comment on the statement "Controlled pollination helps to test the genetic purity of seeds". (2009)

Yes the controlled pollination helps to test the genetic purity of seeds. Controlled pollination is done by bagging female flower in sealed bags and applying pollen from a known superior male parent so the genetic quality of both parents are known for the controlled cross.

Controlled pollination are specially designed and managed to produce seed with greater genetic trait and more genetic uniformity than spontaneously pollinated plants and also eliminate pollen contamination problems.

Genetic purity assessment of the seeds refers to the contribution of alleles from the parents involved in pollination. In a seed orchard, several pollen and seed parents are participating and the presence of any allele, which is not from the participating parents and is of foreign origin, indicates the loss genetic purity. Identification of such contamination is difficult as many parents need to be genotyped. Loss in genetic purity may lead to huge loss in terms of productivity.

However in a controlled pollinated system the assessment of genetic purity is much easy as the pollen and seed parents are known. In the purity tests presence/absence of parental alleles can be spotted faster.

18. How can magnitude and type of variability be manipulated to obtain good gains in some tree characteristics? (2015)

Variability is a very essential factor which helps to improve the trees for the desired quality and product. Generally the variations in trees are caused by environment and the genetics. Environmental variation is understood by foresters and its management is possible through silvicultural activities. Sometimes it is very difficult to control environmental variations which demand developmental of varieties and breeds suitable to that locality.

Environmental variability is complex but if its magnitude and type are known, it is very well be used on operational forestry programme. This magnitude and type of genetic variations can be manipulated to obtain good gains the genetic variation is broadly incorporated two components viz., additive and non additive components.

Genetic variance = Additive variance + Non additive variance

The additive variance is due to the cumulative effects of alleles at all gene loci influencing a trait.

Non additive genetic variance can be divided into two types. Dominance variance is due to interaction of specific alleles at a gene locus, whereas epistasis variance is due to interactions among gene loci. For now, it is sufficient to know that the additive portion is the one of value in population improvement programmes. Non additive variation can be exploited only by use of other, more specialized production programs that involve making specific crosses or using vegetative propagation for the commercial production of planting stock. In most tree improvement programs, the non additive types of genetic variability have generally been given little attention, because the additive portion of genetic variance is easier to utilize.

Most characters of economic importance in tree species are under additive genetic control which can be successfully used in simple selection system. The characters like wood specific gravity, bole straightness and other quality characters are strongly under genetic control.

The pest and disease resistance involve both additive and non additive variance depending on the infecting pest and diseases. But good gains are possible through selection programme than the use of additive portion of the genetic variation.

Under such circumstances, the breeders can identify the existing magnitude and type of variations and incorporate them in the breeding programme in order to achieve the short term objectives. However on a long term basis, the magnitude and type of variations can be manipulated through long term breeding programme in order to obtain the desired tree characteristics which include both quantitative and qualitative traits.

19. How is knowledge of forest genetics essentially needed for the management of forest plantations? Explain. (2013)

The knowledge on forest genetics is very essential for the management of plantations because of the following.

(i) **Objective of forest plantations:** The knowledge of forest genetics are very essential towards fixing of objective of forest plantation establishment. For example the objective is to produce quality seeds then the knowledge on forest genetics is very essential to tree with superior genetic quality. If the objective is to produce wood, then the knowledge on forest genetics is very significance towards establishing plantation of improved genetic quality. Suppose if the objective is to developing the breeding orchard, the knowledge on forest genetics on the intended species is very essential.

(ii) **Knowledge on genetic origin:** The origin of species and the associated land races are very significant in successful development of any plantations. The provenances, seed sources, progenies and clones are basic breeding populations which are deployed for identifying superior genetic stocks towards successful plantation establishment. It is of recent origin that the tree parentage is important and that changes improvements in tree growth and quality can be brought about through breeding and parental control. The knowledge on parentage is very important in forest trees which can be manipulated in forest management.

(iii) **Knowledge on variability:** Without the knowledge on sufficient genetic variability associated with qualitative and quantitative traits, the improvement and productivity of plantation is not possible. Therefore the knowledge on genetics through the existing variability is a critical factor for tree improvement programme which lead to the development of improvement plantations. The most successful plantation development programmes in the whole world are predominantly confined only to a few identified species like Eucalyptus, Casuarina, Populus and Acacia's. In all these species the plantation development programme has been started by exploiting the existing genetic variability which leads to the development of plus trees and these trees are deployed in forest plantation development. The estimates of genotypic coefficient of variation and phenotypic coefficient of variation will help to identify the characters which are strongly under genetic variability and the knowledge on this aspect will help the breeder to establish successful plantations.

(iv) **Heritability and Genetic gain:** The knowledge on heritability is very essential for successful transfer of heritable traits in all plantation development programme. The most selection programmes in forestry are following the characters which are strongly under genetic control. The height which is based on the site quality is not under genetic control. Hence selection based on height of the trees may not be replicated in actual plantations. Hence the knowledge on heritability, genetic gain and other associated genetic parameters are very essential in order establish successful plantations.

Without the knowledge on the genetics and the associated tree improvement, intensive forestry programme is generally of marginal value and there must be a union between these two if maximum gains are to be achieved. This combination of genetics and silviculture has shown tremendous yield improvement in Eucalyptus, Poplars, Casuarina etc., and contributed significantly towards productivity improvement. Trees with higher genetic value will have great potential across different environments. Under such circumstances, the knowledge on genetics is very significant for the management of successful forest plantations.

20. Write short notes on Quick viability test for forest seeds. (2009)

Before initiating germination or supplying seeds, the seeds have to be tested for their viability which is grouped into physical and biochemical tests.

(i) **Physical viability tests:** There are three methods of physical viability tests predominately used and they are floating, colour and cutting tests.

- **Floating test:** In this method, it is assumed that empty seeds will float on the surface of water which can be removed. However, there is every possibility that seed may float because of oven drying or presence of wing or light weight. The minute seeds like that of poplar, *Eucalyptus* cannot be tested with this method.

  Eg: Acacia, *Delonix regia* and *Subabul.*
- **Colour test:** Different species exhibit specific colour of the seed coat when the seed had ripened. This change in colour can act as a test for viability. Generally green colour is changed to yellowish-green and ultimately to grayish-brown or brownish-yellow thus signifying move towards maturity and hence more viability.
- **Cutting Test:** The seeds are cut and examined to see the viability. If the embryo is milky or viscous, seed in unviable. Also if seed is empty, decayed, attacked by insect etc it is treated as unviable. Only if embryo has turned firm it is declared as viable. This test is suitable for large sized seeds

  Eg: *Quercus, Fagus, Fraxinus* etc.

(ii) **Biochemical viability test:** Viability of seed is tested through the following biochemical methods. In this method, the living cells are able to convert certain chemicals during respiration and thus produce distinct colour in cells of seeds.

- **Tetrazolium chloride test:** In this method, 2, 3, 5 Triphenyl Tetrazolium Chloride (TTC) is employed for determining the viability of the seeds. Here the seeds are soaked in water for overnight and they are divided into 2 parts exposing the embryo (for large seeds), the small sized seeds are directly put into the solution of TTC (Concentration varies from 0.25 to 1%) and kept in dark place. The seeds are soaked in the prepared solution for 1 to 18 hrs at a temperature of around 30° C. The viable seeds will show red colour and nonviable seed will be colourless.

  Eg: Neem, Teak, *Albizia* and *Melia.*
- **Excised embryo test:** In this method the embryos are excised and placed on the moistened filter paper and kept in the seed germination room with controlled environmental conditions. The embryos are examined daily and up to 2 weeks. The embryos which produce normal seedlings are treated as viable and others are treated as unviable seeds.

  Eg: Neem, Acacia, Albizia, Cassia, *Wrightia tinctoria* and *Subabul.*
- **Hyrdogen peroxide method:** In this method, the seed coat is cut at the tip of radical end and seed is put into 1 per cent H_2O_2 (Hydrogen peroxide) solution and kept at room temperature for a week. At the end of week the seeds which showed radical having size

more than 1 mm are counted as viable. This method is relatively easier and less time consuming.

Eg: Sandal, Red sanders

- **X-ray test:** In this method, the use of the principle of semi-permeability of living cells is practiced. The seeds are soaked in solution of $Bacl_2$ (Barium chloride) or Sodium iodide for 1 to 2 hours. The dead tissues are penetrated by the solution but living tissue is not penetrated due to their semi-permeability. The seeds are then washed with water to remove the solution and dried at 70 degrees Celsius temperature. These seeds are exposed to x-rays. An embryo is treated as viable if it doesn't have any impregnation and for endosperm 25 per cent impregnation is allowed.

  Eg: Teak, *Pterocarpus santalinus, Pinus sylvestris* and *Picea abies.*

21. Di-erentiate between Exogenous dormancy and Endogenous dormancy (2015)

Exogenous dormancy: It is otherwise called as seed coat dormancy which is due to hard or impermeable seed coat or pericarp. This dormancy is further classified into physical, chemical and mechanical dormancy.

(i) Physical dormancy: This dormancy is predominantly due to impermeable seed coat by the presence of thick walled, cutinized palisade like macroscleroid cells layer.

Eg: *Acacia, Prosopis* and *Albizia* spp.

(ii) Chemical dormancy: This is due to the presence of chemical inhibitors present on the seed coat.

Eg: *Betula* Spp. and *Ailanthus* spp.

(iii) Mechanical dormancy: This dormancy is due to the presence of hard woody fruit wall especially endocarp and sometimes mesocarp. The endocarp acts as a mechanical barrier to a germinating embryo.

Eg: Species of *Terminalia* and *Pterocarpus*

Endogenous Dormancy

It is otherwise called as embryo dormancy or internal dormancy which is caused due to immature embryo development or due to the presence of chemical inhibitors. This dormancy is further classified into morphological, epicotyl and physiological dormancy.

(i) Morphological Dormancy: In this dormancy the fruit looks like matured but the embryo may not be fully grown to get differentiated into radicle and cotyledon. This limits the germination until the differentiation and the growth occurs.

Eg: *Fraxinus*

(ii) Epicotyl dormancy: Certain species expresses dormancy during germination especially when the radicle growth proceeds normally but the embryo and embryonic axis remain dormant at root temperature. The hypocotyl or root of these seeds may elongate and develop till the stored food is consumed and after that it dries and dies.

(iii) Physiological dormancy: This dormancy occurs due to physiological factors due to oxidation and also due to the presence of inhibitors. The seed coat may restrict transfer of oxygen to the embryos or due to the presence of inhibitors like phenols which may fix oxygen by oxidation and limits the availability of oxygen to embryo and leads to problem in germination. This physiological dormancy may be non deep, intermediate and also deep physiological dormancies.

22. Write short notes on Exotics in Indian Forestry. (2007)

Definition

- An Exotic is a plant that is not native.
- Exotic is strictly defined as "An introduction from a foreign country".
- Exotic is one which is grown outside the limits of its natural range.

Advantages of exotics

- In tropics and subtropics, the exotic coniferous timbers have a great need to their superior wood quality because the indigenous forests frequently do not contain the required quality wood.
- The productivity of indigenous forests are very low of 0.5 m^3/ha/annum but the exotic plantations have the production potential of 25-30 m^3/ha/annum and hence they are preferred predominantly for productive industrial wood plantations.
- Exotics have wider adaptability and multipurpose utility compared to the indigenous species which are attracted by several forest based industries.
- Most of the exotics are fast growers which resulting in lowering of the rotation period and increased economic returns. Some of the exotics like *Casuarina* and *Eucalyptus* are harvested in less than 5 years wherein no indigenous species are harvested within this period.
- The tropical and indigenous hard wood forests are so variable and are difficult to manage silviculturally than the exotics. On the other hand, exotics are easy to manage both silviculturally and ecologically due to their short rotation, fast growing and multiple utility. Hence exotic are preferred in place of native species.
- Most foresters generally prefer to use species that are more uniform and whose products are known and accepted. Hence the exotics are widely accepted as a species with uniformity and easy to manage.
- Marketing potential and manufacturing technology favors the use of exotic woods such as Pines, Poplars, *Eucalyptus* etc., compared to indigenous species.
- Availability of quality seed and planting stock is a key to success in any plantation forestry programme. The seeds of the indigenous species are sometimes hard to obtain due to poor understanding of reproductive biology. On the other hand, several exotic species have been developed into improved genetic stock resulting in better quality and more uniform trees from which seeds are easily available.

Problems of the exotics species

- Immediate failure of the introduced plantations is one of the major problems of the exotics.
- There are instances of delayed failure of exotics. Initially the exotics species exhibit good survival and growth but over periods they do not develop into useful forests.
- The exotics may have a continued substandard performance resulting in low production and poor quality.
- The exotics have initial survival and growth but later on attacked by pest and diseases resulting in complete failure of the plantations.
- Growth of the exotics may be unsatisfactory due to shortage or absence of mycorrhizae.
- Adverse weather conditions like drought, cold and violent weather are the causes for failure of exotics. Extremes of weather result in loss in growth and quality and sometimes lead to the death of the plants.
- The newly established exotics may be initially pest free but later on it is infected with pest and diseases resulting in complete failure of the plantations.
- In certain cases exotics also one of the causal factor for poisoning of the soil through allelopathy effect due to the secretion of toxic substances.
- Few exotics develop abnormality in its new environment resulting in loss of apical dominance, crooked development and with multiple tops.
- Exotics also can change the quality and amount of water supply that drains from the forests.
- Some of the exotics like eucalyptus and proposes have been debated towards their role in depletion in water table and is a major thread.

Some Successful exotics in India.

- *Casuarina, Eucalyptus, Poplars*, Silver Oak, Wattles, Tropical pines, *Simarouba* etc.

23. Explain Orthodox and Recalcitrant seeds (2014)

Orthodox seeds: Orthodox seeds are the seeds which desiccate naturally on the mother plant. These seeds can be dried down to a low moisture content of around 5% and successfully stored at low or sub-freezing temperature for longer periods.

Orthodox seeds are having longer viability and in most cases it is having protected embryos. But some of the seeds are not only having resistant seed coat but a hard bony endosperm.

Orthodox seeds can be classified into two categories based on the seed coat.

(i) **Orthodox seed with hard seed coats:** In this case, the seed coat is impermeable to water and do not swell even after keeping in a moist medium for longer period.
Eg: *Acacia, Prosopis, Albizzia, Cassia* and *Bauhinia*

(ii) **Orthodox seeds without hard seed coats:** In these types of seeds, the seed coat is permeable to water.
Eg: *Pinus, Eucalyptus, Pongamia pinnata, Pterocarpus marsupium* and *Lagerstroemia parviflora*

Recalcitrant seeds

Recalcitrant seeds cannot be dried to moisture contents below 30 percent without injury and are unable to tolerate freezing. At optimal conditions, these seeds can live only for short durations because their high moisture content encourages microbial contamination and results in rapid seed deterioration.

Eg: *Quercus, Syzygium cumini, Bassia latifolia, Swietenia mahogani*

Since the recalcitrant seeds loss their viability rapidly due to drying of moisture, these seeds have to be sown immediately after collection. For example Mahua seeds collected and sown after 45 days resulted in less than 20% germination compared to over 65% germination in case of fresh seeds. Similarly, the seeds of *Lannea coramandalica* are also loss their viability within one week which necessitated immediate sowing after collection. Otherwise the failure of germination leads to loss of one year period for seedling production and the associated plantation development.

24. Discuss the necessity and procedure for in-situ conservation of forest genetic resources. (2009)

Necessity of *In-situ* Conservation

- *In-situ* conservation is an important component of the conservation and management of genetic resources. It supplements the *ex-situ* conservation efforts of local, national, and international collections and provides some important advantages.
- *In-situ* conservation sites preserve potentially important and useful genes, many of which may be unrecognized today. Their existence enables the selective and adaptive processes that give rise to new genetic traits to continue in response to environmental stresses.
- *In-situ* conservation areas can be sources of genetic traits not already captured in *Ex-situ* collections.
- *In-situ* reserves can also provide living laboratories for studying the genetic diversity of the wild species that are the progenitors of modern crops.
- *In-situ* conservation is very essential for the species which are rare, vulnerable and threatened.

Procedure for *In-situ* conservation

In-situ conservation can be established based on the need and the conservation status of particular species. The following process and procedures will help to establish *in-situ* conservation for forest genetic resources.

(i) The information related to the area to be converted into *in-situ* conservation area may be identified.

(ii) Development of objectives and specific targets for the *in-situ* conservation

(iii) Once area is identified, the next step is to identify the target species, its status and setup of priorities for its conservation.

(iv) Once the species is identified, establish basic conservation methods like active, passive, dynamic and static.

(v) This can be followed by identification and selection of populations within the target area.

(vi) The management guidelines are to be prepared and utilized during conservation period.

(vii) A specific monitoring system need to be establishment towards periodically monitoring the status and the impact of conservation on building up of the population.

Methods of *In-situ* Conservation

- National Parks
- Biosphere reserve
- Wild life Sanctuaries
- Tiger reserve
- Protected areas

25. Write a note on "Pollen Dilution zones in seed orchard" Discuss various factors a-ecting their e-ectiveness. What are the options available in Lieu of pollen dilution zone? (2013)

Pollen dilution zones: Pollen dilution zone is zone surrounding a seed orchard or seed production area in order to protect the seed orchard from contamination of outside pollen. There are, three modes of establishing pollen dilution zone, they are:

(i) Open type with grass or low vegetation

(ii) Trees of different species, which do not pollinate with that species

(iii) Large stands should be surrounded by a zone of good phenotype.

Pollen dilution zone in seed orchard: In seed orchards, the areas cleared between the seed stand and the possible contaminating sources is termed as pollen dilution zone. The pollen dilution zone varies with species and wind exposure. For example for wind pollinated species, the pollen dilution zone is usually between 200 to 300 m. In case of insect pollinated species, the width of pollen dilution zone depends on the behaviour of the pollinator.

Establishment of Pollen Dilution Zone: Cleared areas are established between the seed stand and possible contaminating sources. The pollen dilution zone is a zone where trees of the same species and species that can hybridize with the species are removed. The wideness depends on the species and pollination pattern. The zones also serve as fire-breaks. For wind pollinated species a short vegetation of the pollen dilution zone is desired, usually 200–400 m, depending on species and wind exposure. If there is a prevailing wind during the pollination period, the width of the pollen dilution zon`e may be diminished from the other sides. For insect-pollinated species the width and the vegetation of the pollen dilution zone depends on the behaviour of the pollinator. Information should be sought on individual species.

Pollen Dilution Zone

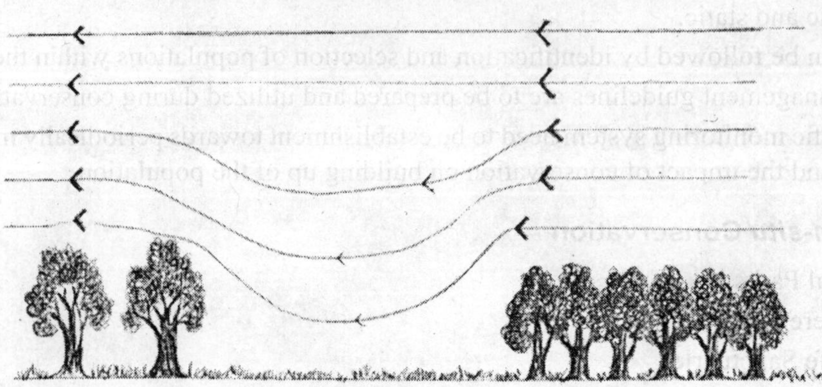

Factors a-ects the e-ectiveness of Pollen Dilution Zone

(i) Species
(ii) Wind exposure
(iii) Width of pollen dilution zone
(iv) Vegetation of pollen dilution zone

Options available in lieu of pollen dilution zone

The options available for establishment of pollen dilution zone

(a) Establishment of wind break
(b) Development of shelterbelts
(c) Establishment of linear plantations
(d) Increasing the width of the pollen dilution zone
(e) Establishing plantations with sterility

26. What is genetic drift? State and discuss Hardy – Weinberg law with its significance. (2013)

Genetic drift: Genetic drift is defined as change in gene frequency of the Mendellian population. Genetic drift is concerned with gene frequency of a reproducing small population. This theory was developed by Sewall Wright in 1930 that is why genetic drift is otherwise called as Sewall Wright effect.

In a small population not all the alleles which are representative of the species may be present. Thus the process of inheritance is in violation of Hardy – Weinberg law.

In such a small population, a chance, even may increase the frequency of the character that has little adaptive value. Thus the genetic drift may remain a significant factor in the origin of new species on islands and other isolated populations.

Due to loss of alleles with low frequency, amount of genetic variation may get reduced in small population.

Further continuous mating within such population may cause decrease in the proportion of heterozygotes and increase in the number of homozygous. This decrease in heterozygosity and increase in homozygosity is termed as Genetic Drift.

However a small population as the whole may develop characters different from that found in main population. Such deviation may even lead to speciation or formation of new species.

When a small group of individual due to genetic drift become founders of new population and is termed as founder principles. The new population often has genotypic frequencies different from the parent populations. Sometimes genotypic frequencies in small population may get changed due to natural calamities. When the population regains its original sites, the members of the small population may have diversed genetically from the parental population. Hence interbreeding between the member of small and large population may not be possible and the small population might have evolved into a new species. This type of genetic drift is named as "Bottle Neck Effect".

Hardy – Weinberg Law: The law states that there is no change in gene and genotype frequency in the Mendelian population if the selection, mutation, migration and genetic drift is absent.

The Hardy – Weinberg Law states that the relative frequencies of various kinds of genes in a large and randomly mating sexual populations tends to remain constant from generation to generation in the absences of mutation, selection and migration.

Significance of Hardy – Weinberg Law

- This law is foundation for population genetics and modern evolutionary theory.
- As per the law of the normal Mendelian genetic frequency are maintained only under certain conditions.
- If the conditions are not followed, the gene frequency will change deviations and cause variations.
- Such variation due to change of gene frequencies will act as a source for future evolution.

27. (a) Write short notes on **(2009)**

(i) Provenance testing of tree species

(ii) Seed production area

(iii) Heritability and genetic gain

(iv) Progeny testing

(b) What is progeny testing? Discuss various objectives of progeny testing in tree improvement. **(2007)**

(c) What do you understand by the term provenance trial? Explain the stepwise procedure for this trial followed in a forest species **(2010)**

(d) Discuss the role of provenance test in tree improvement (2014)

(i) Provenance testing of tree species: The term provenance indicates the original geographic occurrence. So testing the populations or families collected from the original geographic occurrence areas are termed as provenance tests. The provenance tests include range wide provenance phase, restricted provenance phase and the provenance proving phase.

a. Range wide provenance phase

Objective: To determine the extent and pattern of variations between provenances of promising species with wide natural variation.

Provenance tests:

Number of provenances : 10 - 30

Period of evaluation : Quarter to Half rotation (0.25 – 0.5 x rotation age)

b. Restricted provenance phase

Objective: To test the suitability of provenances to site under tests

Provenance tests:

Number of provenances : 3 - 5

Period of evaluation : Half rotation (0.5 x rotation age)

c. Provenance proving phase

Objective: To test the genetic worthiness of the identified provenance towards the objectives envisaged (either productivity or resistance)

Provenance tests:

Number of provenances : 1 or 2

Period of evaluation : Half rotation (0.5 x rotation age)

Spacing : 3 m × 3m

Number of trees : 25 (5 × 5 sq. plot)

Simultaneously the general provenance tests can also been carried out through the following standard evaluation tests.

Number of provenances : 15 - 30

Number of replication : 3-10

Spacing : 3m × 3m

Number of trees / plot : 9 or 16 or 25 or 36 or 49

(Depending on the resources available)

Evaluation period : Half to full rotation

Example: Teak provenances test carried out at FCRI, Mettupalayam

Breeding Objective : To identify the best provenance suitable for semi-arid locality for higher productivity and wood quality

Number of provenances : 30 (28 Indian provenances and 1 each from Syyaburry and Chittangoong)

Plot size : 81 m^2

Number of trees / plot : 9 trees / plot

Spacing : 3 m × 3m

Evaluation period : Half rotation (20 years)

(ii) **Seed production area:** The seed production area or seed stand is defined as a crop of vigorously growing, middle aged to mature trees of good quality which is managed exclusively for seed collection. The seed stands or seed source refers to the population of trees possessing sufficient uniformity in composition, constitution and arrangement to be distinguishable from the adjacent population.

Seed production areas have three attributes that are vitally important. These are as follows:

(i) Seed collected will have better genetic qualities than seed from commercial collections, especially in adaptability, bole and crown characteristics, and pest resistance.

(ii) When seed production areas are established in natural stands (and in some plantations), the geographic origins of the parent trees are known, thus yielding seed from a suitable source. This is not true for many exotic plantations, but selection of the best individuals in an exotic plantation will result in the development of a land race.

(iii) Seed production areas are reliable sources of well-adapted seed at modest cost. Assurance of seed supply is becoming of ever-increasing importance, especially in rapidly expanding plantation programs that are using species suitable to tropical or subtropical conditions.

Specifications for a Seed Production Area

The best natural stands or plantations that are near full stocking are used for the development of seed production areas. There are no specific age limitations, other than the fact that the stand must be old enough to produce seed, and the individual trees must have sufficient crown surface areas so that they can produce large seed crops.

For collection efficiency and to assure ac equate cross-pollination, it is usually- best if 50 seed trees of acceptable phenotype per acre (125/ha) can be retained, although the optimum number of "leave" trees will depend on tree size and selection intensity. It is sometimes impossible to obtain 50 good trees per acre. However, an area with 20 to 30 seed trees per acre (50 to 75/ha) after thinning is considered to be stocked well enough for operational use. Economic and biological considerations generally dictate that if less than 10 to 12 trees per acre (25/ha) cannot be retained after roguing has been completed, the potential seed production area should not be used. A very real danger for some species on certain sites is "windthrow" that is caused by opening up the stand too much.

Objectives of seed production areas

(a) To produce and collect improved as well as good quality seeds.

(b) To improve the germinative capacity of the seeds.

(c) To make seed collection easier and well organized.

Seeds production area may be selected either in plantations or in natural forests. In both cases the selected area should not only be free from disease and deformities but the crop should be middle aged, healthy, vigorous, clear and having of high quality.

Selection of Trees for a Seed Production Area

Desired attributes of the trees left in a seed production area are similar to, but less rigorous than, the qualifications required for a select tree to be used in an intensive tree improvement program. Both because of their growth and seed-producing potential, only trees in the dominant and codominant crown classes are considered for retention. The seed production area tree must have a high level of vigor, be straight; have desirable limb form and pruning, and be free from insects and diseases. No tree below the desired standard should be left, regardless of spacing. Trees exhibiting the potential for good cone production are given preference, although the evidence of past seed production is not essential if the trees have been growing in a tightly closed stand. Excellent seed crops are often produced after heavy thinning by trees that exhibited little seed production prior to thinning. However, it is necessary that stands be thinned at a young enough age so that the trees retained will have crowns that are large enough and healthy enough to produce large seed crops.

General guidelines for selection of individuals from evenaged natural forests

- Selection is done in outstanding stands or from plus trees.
- Selection is not concentrated in good site quality alone.
- Selection should not be done in young or very old plantations.
- Selection of stands such as logged or thinned should be avoided.
- Heavy flowering is not given for much emphasis.
- Comparison or check tree method is followed wherever it is possible.
- Preferably one tree is selected for one small area.

Criteria for identification of potential seed production areas

- The origin of seeds from which the stand has been established should be known.
- The are a should contain a large number of good phenotypes.
- The trees should be mature for seed production but not too old.
- The area should provide sufficient quantity of seeds.
- The area should be easily accessible.
- The area should not be exposed to major natural disasters.
- It should be protected from major noxious/destructive animals.
- It should be free from pests and diseases.
- The trees should have proven to produce flowers and seeds in the area.
- The area should not have been exposed to logging.
- It should be verified whether necessary operations viz., rouging, thinning and possible establishment of pollen dilution zone are practically possible.

Establishment and treatment of seed stands

In a natural forest seed stands are established in the following ways;

- Delineate the area of superior stands. This should have a minimum area of 4 ha.
- Borders or surrounding area is cleared and demarcated with poles.
- Trees to be retained and removed are selected.
- Inferior trees are removed.
- 'C' grade thinning is carried out for this stand.
- Rouging and thinning should be done carefully in order to avoid or minimize damage to the remaining seed trees.
- Removal of debris and cut material is done immediately to avoid insect, disease, and fire hazard.
- Establishment of pollen dilution zone (PDZ).

Maintenance of seed stand

- Seed stand should be maintained by judicious thinning to reduce crop competition.
- Thinning is done before flowering sets in so that good quality seeds are obtained.
- Inferior phenotypes are removed in order to enhance quality seeds.
- The trees should be protected against fire, disease, grazing and human interference.
- Fertilization is done to improve the growth of plus trees retained.
- Flower induction treatment is given for better flowering and seed production.
- Pests like squirrels and animal damage (monkeys etc.,) should be controlled.
- Competing grasses, plants and climbers should be eliminated.
- Regular tending operations are necessary.
- The provision for fertilization and irrigation should be made.
- Grazing should be totally banned.

Advantages of SPA

- The seed stands can be managed especially for seed production.
- It enhances the higher seed production from the individual tree.
- It minimizes the pollination from inferior trees.
- Individual trees in the seed stand are likely to be pollinated by good phenotype.
- The seed collection is cheaper.
- Seed is produced in a shorter time.

Disadvantages of SPA

- Selection differential is less.
- The management of seed stands implies considerable management cost.
- Individual trees in a seed stand may create a chance for inbreeding.
- Little genetic gain is obtained except from provenance selection.
- Risk of wind thrown, fire etc, is increased due to increased spacing.

(iii) Heritability and genetic gain: Heritability is one of the most important and most used factor in tree improvement programme. Heritability is the ratio of additive genetic variance to phenotypic variance.

$$h^2 = \frac{\sigma^2_A}{\sigma^2_P}$$

Where h^2 is the heritability

s_A is the additive genetic variance

σ_P is the phenotypic genetic variance

Types of Heritability: There are two types of heritability depending on the type of genetic variance utilized for tree improvement.

(a) Narrow sense heritability: It is the ratio of additive genetic variance to the total phenotypic variance or it is the proportion of total variance that is attributable to the additive effects of genes. It is described as

$$h^2 = \frac{\sigma^2_A}{\sigma^2_A + \sigma^2_{A/A} + \sigma^2_E}$$

(b) Broad sense heritability: It is the ratio of total genetic variance (additive + non additive) to the total phenotypic variance.

$$H^2 = \frac{\sigma^2_A + s^2_{NA}}{\sigma^2_A + \sigma^2_{NA} + \sigma^2_E}$$

Applications of heritability

- Heritability helps to select individual with high heritability.
- It helps to predict the genetic gain predicted through tree improvement.

Genetic Gain: Genetic gain is the product of heritability and selection differential which is described as

Genetic gain = $h^2 \times$ Sd

Where h^2 is the heritability

Sd is the selection differential

Application: When individuals are selected based on their phenotypic values only, without information on relatives and the response to the selection is estimated through genetic gain.

(iv) Progeny testing: It is a test to verify the genetic worthiness of a parent through the performance of their offsprings or progenies.

Objectives of the progeny tests

- To test the genetic worthiness of a parent
- To identify the best and suitable progenies for varied objectives
- To establish seedling seed orchard
- To establish seed production area
- To conserve the genetic resources of the identified species

Method of progeny test

Number of progenies	–	15 to any number
Design	–	RBD (Single tree plot
Method	–	Single tree plot (if number of progenies are more) Line plot (if area is restricted) Block plot (Ideal method)
Spacing	–	3 m × 3m but variable due to species and objective
Period	–	0.5 to 1 × rotation age

Example : Melia progeny tests at FCRI, Mettupalayam

Number of progenies	–	20
Design	–	RBD
Method	–	Block plot (9 tree plot at 3 × 3)
Spacing	–	3 m × 3m
Period	–	10 years

28. a. What is gene conservation? Discuss in brief the di-erent methods of *ex-situ* conservation of forestry species? (2007)

b. Short notes on Breeding Arboretum (2012)

Gene conservation relates to activities directed at saving gene pools for use to prevent loss of genes, gene complexes, and genotypes and, in extreme instances, to prevent extinction of whole taxonomic categories of trees.

Gene conservation involve conservation of all possible genes or gene complexes within species or race and those who espouse saving those genes or gene complexes that will be most helpful in a long-term breeding program.

Methods of Conservation

***In-situ* Conservation**

In-situ conservation is conservation at the original habitat. *In-situ* conservation is applied by setting aside and preserving stands of the desired species or complexes to prevent further losses, usually from the activities of humans. One of its major benefits is the conservation of ecosystems. There is no other effective method to achieve conservation of ecosystems than through the "*in-situ* approach."

A major need for *in-situ* conservation of genetic resources, which is either not understood generally or is ignored, is to manage the preserved stand in order to maintain the desired genetic composition.

One common problem with in situ conservation is that often the wrong forest populations are saved. Instead of conserving the truly endangered populations, stands are often chosen that are growing at high elevations or on other unique poor-site types, because they are less expensive, easier to obtain, or are the desired size. In so doing, the objective of conserving the populations occurring on more typical sites is defeated. Great care must be used in the selection of stands for in situ conservation.

Eg: Biosphere reserve, National parks, Sanctuaries, Preservation plots etc.

Ex-situ Conservation

It is the conservation outside its natural habitat.

Because of the difficulties and costs of in situ gene conservation, it is likely that ex situ methods, especially those related to packaging the desired genes, preserving by vegetative propagation, or seed and pollen storage followed by multiplying by crossing, will be the most widely used methods of gene conservation in forest trees in the future.

Eg: Zoological parks, Botanical garden, arboretum, tissue culture, clonal complex, seed germplasm, cryopreservation etc.

Methods of _ex-situ_ conservation

(i) **Conservation through vegetative propagation:** The most common of *ex-situ* conservation is through reproduction by the conventional vegetative propagation techniques of grafting, rooting cuttings, or air layering. Foresters have a special opportunity, in that given genotypes (or gene complexes) can essentially be conserved "forever" through vegetative means. Rather than savings hundreds or thousands of acres of trees, a few vegetative propagules of the desired trees can be established, maintained, and crossed when and as desired. Although the objective is gene conservation for special uses, seed orchards effectively maintain and increase the frequency of desired genetic characteristics.

(ii) **Zoological parks:** It is a park, building, cage, enclosure or premise in which animals are kept for public exhibition. Zoos play a vital role in awareness creation among common people about the need for conservation of animals and nature. In some zoos breeding of animals of rare and endangered category are done for rehabilitation and conservation.

(iii) **Botanical gardens:** They are called plant zoos of the world. They have been established to conserve the rare and threatened plant species. There are about 1600 botanical gardens across the world which ensures conservation of global genetic resources. Most botanical gardens serve as seed bank for several species.

(iv) **Field Gene Bank:** Different species and the families in each species are assembled and maintained in the form of field gene bank for immediate utility and for future improvement programme. These field gene banks may act as seed production areas and seed orchards towards supply of quality seeds.

(v) **Arboretum:** Arboretum is an important *ex-situ* conservation method involving deliberate collection and management of woody perennials for the purpose of display to public, education and research. The trees in arboretum are valuable genetic resources thus serving the purpose of field gene bank. Most of the tree seeds are recalcitrant and chances of losing their viability is very high. Such species can be conserved only through field gene bank (or) arboretum.

The elite genotypes in arboretum mate among themself and produce elite progenies. The trees can be kept under evaluation for longer number of years.

Advantage: The genotype interacts with the environment and expresses a phenotypic character. Appearance of new combination of alleles takes place which may be highly adaptable to future environmental conditions.

Disadvantage: Occurrence of natural disaster may leads to death of genotypes. Loss of genotype due to pest and disease infestation has also been recorded in some arboretum.

29. Discuss the regression selection method for plus tree selection in uneven aged stand (2011)

The regression selection is employed for grading of uneven aged or mixed species type stands. A regression selection is built by sampling a number of trees for a desired character, such as volume growth on a given site, and then plotting them against age. A reliable regression curve for height or volume can be made with about 50 trees, if there is a reasonable age-class distribution. Once the curve has been developed, the regression is used as follows,

- A candidate tree is chosen, based on the judgement of the selector and measured for the characteristics desired, such as height or volume.
- The trait is plotted on the regression graph using the proper age and site. If the candidate tree falls above the curve, it has the desired growth for its age whereas some may fall on the line i.e. average, therefore, its use depends on other characteristics and some fall below the curve, therefore it has inferior growth for its age and should not be used.
- The tree which falls above the regression line are selected and deployed in tree improvement programme. This method is very unique in case of selecting trees in uneven aged natural forests.

Regression Selection Method

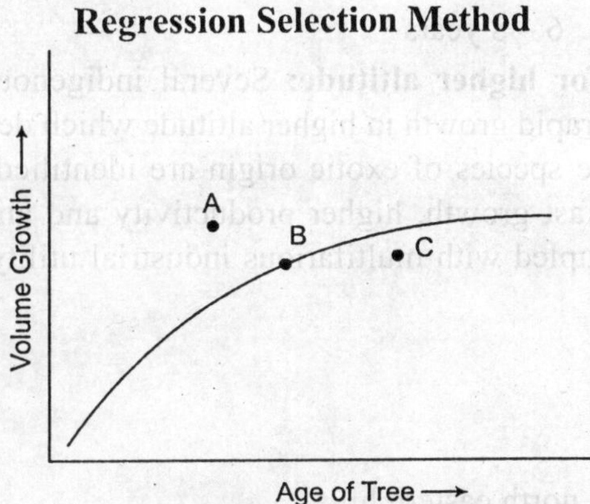

30. Suggest suitable steps to select exotics species or provenances for plantations (2015)

The following steps are essential towards selection of exotics species or provenances.

(i) Objective of selection: The fundamental aspect of selecting any exotic or provenance depends on the objective of selection which may be for timber production, fuel wood production, pulp and paper utility and other domestic and industrial utility.

The classical example is *Prosopis juliflora* which was introduced from Central American countries mainly to meet the fuel wood requirement of India. Similarly wide range of Eucalyptus and Casuarina provenances has been introduced from Australia, Indonesia and Thailand to meet the pulpwood raw material requirement.

(ii) **Species for biotic and abiotic tolerance:** Another major choice of species selection towards exotic or provenances is for biotic and abiotic tolerance. Several species are selected towards resistance against pest and diseases. For example several Eucalyptus provenances have been introduced for resistant against Eucalyptus gall wasp. Similarly several casuarinas provenances have been introduced for resistant against wilt disease.

(iii) **Selection based on higher productivity:** Most of the exotic are selected based on their rapid productivity. The productivity of Indian forests is one of the lowest of $0.5m^3$ / ha/ annum which is not sufficient to meet the growing wood and wood products demand. Hence several exotics viz., Poplars, Casuarina, Eucalyptus, Acacia, Subabul etc., are introduced due to their higher productivity of over $25m^3$/ha/annum.

(iv) **Selection of exotics due to short duration:** The indigenous species in India is predominantly a slow grower and takes several years to reach harvestable maturity. Hence selection of exotics or specific provenances in that exotics is the major factor which will be a fast grower and amenable for harvest in short duration.

Example:

Poplar	:	4 – 6 years
Eucalyptus	:	3 – 6 Years
Casuarina	:	3 – 5 Years
Silver oak	:	6 – 8 years

(iv) **Selection of exotics for higher altitude:** Several indigenous species may not be fast growing amenable for rapid growth in higher altitude which detract their attraction. Hence many fast growing tree species of exotic origin are identified and introduced for higher altitude. Due to their fast growth, higher productivity and amenability towards extreme climatic conditions coupled with multifarious industrial utility these exotics are selected and deployed.

Example:

Wattles in Nilgiris,

Blue gum in Nilgiris

Silver oak in south and north eastern hills

(v) **Selection of exotics for multifarious industrial utility:** The demand for wood and wood products is increasing and there is no concomitant plantation development programme which necessitated selection and introduction of exotics to meet the industrial raw material requirement. The following species are introduced exclusively for industrial utility.

Casuarina	–	Pulp and Paper
Eucalyptus	–	Pulp and Plywood

Wattles	–	Pulp and Tannins
Silver oak	–	Timber
Acacia auriculiformis	–	Timber

31. Comment on the statement that "Vegetative propagation is a dead end of Tree Breeding". (2011)

The statement that vegetative propagation is a dead end of Tree Breeding has two different thought and approaches.

(i) Yes. It is a dead end because once the tree is identified for vegetative propagation then the traditional technology will never allow for further improvement programme. The vegetative propagation of the identified tree act as a last stage of improvement programme and only the same vegetative propagule can be mass multiplied and there is no possibility of further breeding in the traditional approach. That is the reason it is said to be a dead end.

(ii) No. It is not a dead end and rather vegetative propagation is a beginning of a tree improvement due to advancements made in controlled breeding approaches. Under this concept, vegetative propagation particularly grafting / layering / budding have been used as a method for multiplication of identified elite trees. After multiplication these plants are planted in controlled breeding chamber or controlled breeding orchard.

Grafted plant based controlled breeding: In this method the identified trees are grafted on the root stock and the successful grafts are planted in bigger plastic containers or cement pots. These plants are applied with paclobutrazol as soil application which induces early flowering. The grafted plants are produced for both the sexes (male and female) separately. This early flowering helps to induce hybrids and these hybrids are beginning of tree improvement programme. Hence the statement of vegetative propagation is a dead end of tree breeding is not correct.

This type of vegetative propagation based controlled breeding has been successfully reported in Eucalyptus in Brazil which lead to the development of the world famous Urograndis hybrid which has the production potential of over 40m^3/ha/annum compared to less than 10m^3/ha/annum of the traditional breeding method.

In India also similar successful results in Casuarina (between the vegetative propagules of *Casuarina equisetifolia* and *Casuarina junghuhniana*)and several species of Eucalyptus have been evidenced. Hence vegetative propagation is a beginning of a tree breeding programme and is not the dead end.

32. As a community of interbreeding individuals, what parameters would need to be known to describe a population of forest trees? (2015)

In a population with interbreeding individuals the Hardy Weinberg law is applicable which states that in the absence of selection, mutation, migration and random drift. The allele frequencies remain unchanged. It should be remembered that the properties of population with respect to any determine characters depend on gene frequencies.

(i) **Random drift:** Random changes of gene frequencies are called gene drit. If the gene frequency in any one small population followed. It may be seemed to change in an erratic manner in generation to generation with no tendency to revert to its original value.

(ii) **Description of population in the interbreeding process, di3erentiation between the sub population:** The occurrence of random drift is different sub population leads to genetic differentiation between the sub population. The inhabitantsof a large area constitute a single large population because mating takes place more often between inhabitants of the same region. Therefore, the natural population are divided into local or sub populations and they differ their genetic frequencies if the number of individuals in the group is small.

(iii) **Uniformity with in sub population:** Genetic variations within each population become progressively reduced and the individuals become more like genotype. This genetic uniformity is the reason for widespread use of inbred strains.

(iv) **Increased homozygosity:** The frequency of homozygosity increased at the expense of heterozygosity, the general tendency for deleterious alleles to be recessive is the genetic basis for the loss of fertility and viability that almost and always results from inbreeding.

33. Exotics have potential, do you agree or disagree. Justify your response. (2014)

Exotics: An Exotic is a plant that is not native. Exotic is strictly defined as "An introduction from a foreign country". Exotic is one which is grown outside the limits of its natural range.

I agree because

- In tropics and subtropics, the exotic coniferous timbers have a great need to their superior wood quality because the indigenous forests frequently do not contain the required quality wood.
- The productivity of indigenous forests are very low of 0.5 m^3/ha/annum but the exotic plantations have the production potential of 25-30 m^3/ha/annum and hence they are preferred predominantly for productive industrial wood plantations.
- Exotics have wider adaptability and multipurpose utility compared to the indigenous species which are attracted by several forest based industries.
- Most of the exotics are fast growers which resulting in lowering of the rotation period and increased economic returns. Some of the exotics like Casuarina and Eucalyptus are harvested in less than 5 years wherein no indigenous species are harvested within this period.
- The tropical and indigenous hard wood forests are so variable and are difficult to manage silviculturally than the exotics. On the other hand, exotics are easy to manage both silviculturally and ecologically due to their short rotation, fast growing and multiple utility. Hence exotic are preferred in place of native species
- Most foresters generally prefer to use species that are more uniform and whose products are known and accepted. Hence the exotics are widely accepted as a species with uniformity and easy to manage.
- Marketing potential and manufacturing technology favours the use of exotic woods such as Pines, Poplars and Eucalyptus etc., compared to indigenous species.

- Availability of quality seed and planting stock is a key to success in any plantation forestry programme. The seeds of the indigenous species are sometimes hard to obtain due to poor understanding of reproductive biology. On the other hand, several exotic species have been developed into improved genetic stock resulting in better quality and more uniform trees from which seeds are easily available.

I Disagree because

- Immediate failure of the introduced plantations is one of the major problem of the exotics.
- There are instances of delayed failure of exotics. Initially the exotics species exhibit good survival and growth but over periods they do not develop into useful forests.
- The exotics may have a continued substandard performance resulting in low production and poor quality.
- The exotics have initial survival and growth but later on attacked by pest and diseases resulting in complete failure of the plantations.
- Growth of the exotics may be unsatisfactory due to shortage or absence of mycorrhizae.
- Adverse weather conditions like drought, cold and violent weather are the causes for failure of exotics. Extremes of weather result in loss in growth and quality and sometimes lead to the death of the plants.
- The newly established exotics may be initially pest free but later on it is infected with pest and diseases resulting in complete failure of the plantations.
- In certain cases exotics also one of the causal factor for poisoning of the soil through allelopathy effect due to the secretion of toxic substances.
- Few exotics develop abnormality in its new environment resulting in loss of apical dominance, crooked development and with multiple tops.
- Exotics also can change the quality and amount of water supply that drains from the forests.

34. Discuss the reasons for widespread use of exotics for plantation and specific advantages of exotics over native species? (2016)

Exotics: Exotics are used extensively for plantation development because of their easy adaptation, fast growing nature compared to the indigenous species. Another major widespread utility of exotic is its higher productivity of over 20m^3/ha/annum. Above all, the utility of exotics is widespread due to its multifarious utility and satisfies the demand from wide range of wood based industries. Due to these reasons, the utility of exotics is widespread.

Reasons for widespread use of exotics:

- In tropics and subtropics, the exotic coniferous timbers have a great need to their superior wood quality because the indigenous forests frequently do not contain the required quality wood.
- The productivity of native species is very low of 0.5 m3/ha/annum but the exotic plantations have the production potential of 25-30 m3/ha/annum and hence they are preferred predominantly for productive industrial wood plantations.

- Exotics have wider adaptability and multipurpose utility compared to the indigenous species which are attracted by several forest based industries.
- Most of the exotics are fast growers which resulting in lowering of the rotation period and increased economic returns. Some of the exotics like Casuarina and Eucalyptus are harvested in less than 5 years wherein no indigenous species are harvested within this period.
- The tropical and indigenous hard wood forests are so variable and are difficult to manage silviculturally than the exotics. On the other hand, exotics are easy to manage both silviculturally and ecologically due to their short rotation, fast growing and multiple utility. Hence exotic are preferred in place of native species
- Most foresters generally prefer to use species that are more uniform and whose products are known and accepted. Hence the exotics are widely accepted as a species with uniformity and easy to manage.
- Marketing potential and manufacturing technology favours the use of exotic woods such as Pines, Poplars, Eucalyptus etc., compared to indigenous species.
- Availability of quality seed and planting stock is a key to success in any plantation forestry programme. The seeds of the indigenous species are sometimes hard to obtain due to poor understanding of reproductive biology. On the other hand, several exotic species have been developed into improved genetic stock resulting in better quality and more uniform trees from which seeds are easily available.

Successful exotics species in India: Eucalyptus, Poplar, Subabul, Casuarina, Silver oak, Phyllodinous Acacia, Wattles, Pines etc.

Forest Management, Management Systems and Forest Working Plan

1. What is working circle? How is it decided in working plan exercise? (2010)

Working circle (W.c.) is a forest area (forming the whole or part of a Working Plan area) organized with a particular object and subject to one and the same silvicultural system and the same set of working plan prescriptions. The unit of forest management is a Working Plan, usually covering an area of Forest Division.

The Working Plan is a written Scheme of management aiming at continuity of policy and action and controlling the treatment of a forest. Since the entire Working Plan area is usually large and heterogeneous in site conditions and crop composition, different silvicultural treatments may have to be given (i.e., different Silvicultural Systems prescribed) in different parts of the Working Plan area and different working rules, called prescriptions, drawn up for different parts.

Working circle prescription as per National Working Plan Code

The national working plan code prescribed by Government of India has given provision for creation of working circle under the part II programme with the subdivision 1.3. As per this direction, the mapping and listing of working circles (Including overlapping ones), their area and distribution giving justification for their constitution amongst various options for sustainable management of forests are to be carried out by the working plan officer. There will be a separate chapter for each working circle including the overlapping ones. The last chapter identified with the working circle is presumed as N (last in row) which has numerical values like 15 and 16.

The working plan code briefs the general constitution of working circle which states that mapping and summarization of working circle area statement by ranges, blocks and compartments included in the circle indicating gross area and showing forest type and density classes as adopted by Forest Survey of India. Normally the working circles are marked on GIS maps in the scale of 1:50000.

2. Describe the procedure for allotting di-erent types of periodic blocks in a forest (2010)

A Periodic Block (P.B.) is defined as: the part or parts of forest set aside to be regenerated, or otherwise treated, during a specified period. The regeneration block is called "Floating" or "Single" when it is the only P.B. allotted at each Working Plan revision. When all P.Bs. are allotted and retain their territorial identity at Working Plan revision they are termed as "Fixed" or "Permanent".

Periodic Blocks: Felling series in Uniform System

Under the regular shelterwood (or Uniform) System of natural regeneration, age-classes take the place of age-gradations, and periodic blocks (P.Bs) take the place of annual coupes, each containing one age-class instead of one age-gradation.

Arrangement of Period Blocks.

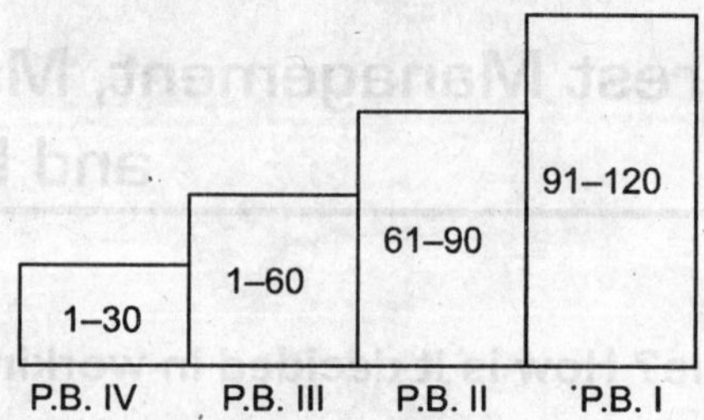

Example:

Species	:	Blue pine (*Pinus wallichiana*)
Silviculture system	:	Uniform Shelterwood System
Rotation	:	120 years
Regeneration period	:	30 years

The allotment and arrangements of Periodic block is done by the following formula.

$$PB = \frac{\text{Rotation}}{\text{Regeneration period}}$$

In case of blue pine it is $\frac{120}{30} = 4$ blocks.

It is assume that each and every periodic block is approximately even aged and it depends on the regeneration period. The regeneration period varies from species to species, location and the associated factors of locality. In practice the regeneration status in the periodic block is not uniform and may spread over a period of 1-30 years. This would result in continued regeneration with varying age gradations. But for all practical utility the trees within the periodic block are considered even aged.

Hence the blue pine rotated under 120 years rotation has been allotted with four different periodic blocks and the age classes are 1-30, 31-60, 61-90 and 91-120 for the periodic blocks IV, III, II and I respectively as indicated in the figure.

3. What is intermediate yield? How does it di-er from final yield? (2010)

The intermediate yields are normally obtained during the process of thinning. Thinning is the process of felling made in an immature stands for the purpose of improving growth and form of the trees that remain without permanently breaking the canopy.

The thinning involves reduction in number of trees per ha and it decrease with increase the age of trees. The decrease is faster in earlier period and slower when the tree reaches maturity. The thinning helps to improve the hygiene coupled with getting the desired crop and intermediate yield.

Intermediate: Intermediate yield is the yield which includes all materials from thinning or operations preceding the main felling in a regular forest.

Final yield: Final yield is the yield that is obtained from the main fellings in a regular forest.

S.No.	Intermediate yield	Final yield
i.	Yield from thinning or any intermediate felling	Yield obtained at the final harvest
ii.	The yield obtained from thinning yield used predominantly for firewood and other secondary utilization	Yield obtained and final yield is used for primary utility like timber and timber products
iii.	The economic value of intermediate yield is low.	Economic value is high
iv.	In long rotation crop, intermediate yield will give returns at regular intervals.	The economic return is only at rotation period.

4. What are the peculiarities you have observed in forest management practices? Suggest ways to overcome them. (2014)

Forest in general and forest management in particular is very peculiar and attracts different practices compared to other landuse system. The peculiarities that are observed in forest management are.

(i) **Long rotation of the species:** One of the major peculiarity of the tree is its long gestation period. Traditionally forests are considered as a base for hunting, recreation and aesthetic purposes. However due to population growth, urbanization, industrialization forests have been considered as the potential source of raw material for several wood based industries. However long rotation is a peculiar feature which is a problem faced by several industries. Now forests are considered as an essential entity both for productivity and to extend ecosystem services.

(ii) **Multiple functions of the forest:** Forests by virtue of its natural has multiple functions which is very peculiar and entirely different from other sectors of landuse system. Forest do produce timber, fiber, fruit, fodder, gums, tannins, fuel, medicinal value etc. Besides, the forests play a vital role in protecting the environment like reducing the temperature, increasing the humidity and its role in soil and water conservation. Due to these multiple functions, forests have been considered as a peculiar ecosystem.

(iii) **Long term investment:** Unlike agriculture and other industrial investments forests is not a short term investment till the recent past. Once invested it takes several years to return it which ranged between 10 years and 60 years. Due to this long term investment forests have been considered as a both expenditure as well as investment entity.

(iv) **Identify of the product and manufacturing plant:** Forests itself is considered as a product of human utility and can act as a raw material for many wood based industries. The wood from trees are directly used in construction purposes, simultaneously the wood is value added into various products. Hence it is very peculiar as a source of direct product development and as a source of raw materials to different industries.

(v) **Flexibility nature:** Forests in general and trees in particular has a flexible role to meet the various productive and service demands of the society. If the demand is for meeting the fuel wood requirement of the society it can be harvested in shortest possible time and if it is for industrial utility it can be maintained for medium rotation and for aesthetic benefits it can be a long rotation. So based on utility and object of management the forests can be a flexible entity.

5. Describe the general principle of yield regulation in Uneven – aged forest crop (2014)

Yield regulation based on increment: The yield regulation in uneven aged forest is based on the principle of determination through stock, increment and the combined effect of growing stock and increment.

The general principles involved in yield regulation on uneven aged forest involve determination of area, volume and combination of both the parameters. The principle factors responsible for yield regulation are growing stock, increment, objective of management and the silvicultural system

In general principles methods involved in yield regulation of uneven aged forest

(i) Yield regulation based growing stock

(ii) Yield regulation based on increment

(iii) Yield regulation based growing stock and increment

(i) Yield regulation based growing stock: In this method, the growing Stock is primary factor which decides the yield regulation. The formula utilized for yield regulation in regular forest has been modified and deployed for yield regulation in uneven aged forest.

$$Y = \frac{V_o}{R} + \frac{V_r}{½R}$$

Where,

V_o = Volume of growing stock 7.5 feet girth

V_r = volume of remained of the enumerated stock

R = Rotation

There are few modification proposed in this formula but their utility is very limited.

(ii) Yield regulation based on increment: It is based on principle that, the yield depends on the continuous accumulation of increment throughout the life of the tree. The following formula indicates the yield regulation based on increment.

$$Y = \frac{V + a - V_n}{n}$$

Where,

Y - Annual yield,

V - Present volume

V_n - Volume 'n' years ago

a - Volume removed during the previous 'n' years.

There are two methods prescribed for yield regulation of uneven aged forest based on increment and they are,

(i) Swiss method
(ii) Bioley's check method

(i) Swiss method:

$$\text{Annual yield} = \frac{Y+Z}{c.c}$$

$$c.c. = \frac{Y+Z}{X_i + Y_i + Z_i}$$

Where,

X = Volume of the diameter classes below the diameter limit.

Y = Volume of the diameter classes above the diameter limit and upto the limit up to which the C.A.I is considered satisfactory.

Z = volume of the diameter classes above the satisfactory diameter limit

I = The C.A.I of various dia classes

$c.c$ = The period during which the volume of the oldest dia. Classes should be removed, or the cutting cycle.

(ii) Bioley's check method:

Annual yield = $I + V_n - V_a$

Where, I = Calculated Increment
V_a = Actual growing stock
V_n = Desired growing stock (or normal stocking)

(iii) Yield regulation based growing stock and increment: This method takes into consideration both the volume of growing stock and increment. Several formulae have been developed for yield regulation which are furnished below.

1. Hufnagl's Diameter class method: This method was proposed by Hufnagl based on growing stock and increment and proposed that all diameters above exploitable diameter are to be removed during the period and the formula is furnished.

Variation I:

$$AY = \frac{V + V_{xi} + \frac{r}{4}}{\frac{r}{2}}$$

Where, AY = Annual Yield
r = Rotation in years
V = Volume of half rotation age
i = CAI

Variation II:

$$AY = \frac{\frac{V + a \times i \times r}{4}}{\frac{r}{2}}$$

Where, AY = Annual Yield

V = Volume of the crop of half the rotation age and above it

i = Final MAI (CIA may also be taken)

r = Period of rotation

a = Area occupied by the crop

2. Brandis diameter class method

Yield = Average number of trees reaching exploitable size annually + a fraction of surplus number of trees of over the exploitable girth/size.

3. Volume unit method

$$Y = \frac{V + I/2}{P}$$

Where, AY = Annual yield

I = Increment

P = Period taken for one class to pass on to the next

4. *Smythie's* safeguarding formula

$$Y = \left(\frac{x}{1 + \frac{x}{2}} \times 100 \right)$$

Where $x = \frac{f}{t}(\text{II} - z\,\%\text{II})$

f = Felling cycle

t = Time taken by a II class, tree to pass into I class

z = percentage of II class trees that do not pass into I class in 't' years due to mortality removal of thinnings etc.

6. What is increment percent? Discuss the relationship between CAI and MAI of a forest stand. (2005,2006,2007,2012, 2016)

- **Increment percent:** Increment percent is the average annual growth in volume or basal area over a specified period expressed as a percent of the volume or basal area either at beginning or more usually half way through the period.

- **Current Increment Percent:** It is the relation of the increment during a given year to the volume at the beginning of the year.
- **Periodic Increment Percent:** It is the relation of the increment during a given period to a basic volume which may be taken as the mean or average volume for the period, or the volume at the beginning of the period.
- **Mean Annual Increment Percent:** It is the percent ratio which the M.A.I for a given age bears to the total volume at the age.

Current and Mean Annual Increments (CAI & MAI)

Relation between CAI and MAI: In any forest crop / individual tree, the CAI is small in the early /initial stages (seedlings and saplings). It increases slowly at first, then more rapidly to the maximum after which it begins to decline and finally ceases with the mortality or felling of the crop / individual tree.

The sum of the CAIs laid on for the entire period give the total volume which when divided by the age gives the MAI (average of the past production). MAI coincides with CAI only twice during the life of an individual tree / forest crop, first at the end of the first year and secondly when it culminates and equals the CAI.

In the initial stages, the rising CAI raises the MAI also but the curve of the MAI is not as steep as that of CAI. The reason for the smoothness of the MAI curve is due to the fact that increasing growth each year is spread over all the previous years. Subsequently, CAI curve reaches its maximum and culminates followed by a decline. However, the successive MAI values for each year after culmination of CAI continue to rise since the amount of growth added to the crop during these years is still greater than the average or mean.

When the CAI finally decreases to an amount equal to the average mean for the entire period, the curve of MAI has reached its maximum. It is advisable to fix the rotation of tree species at this particular point when the MAI culminates

CAI and MAI Curve

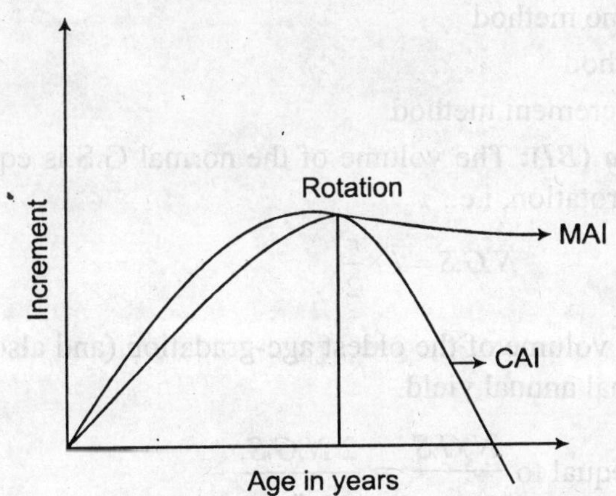

The conclusions from the above descriptions, tabulated data and trends of the MAI and CAI curves may be summed up as:

- To begin with the MAI keeps below the CAI
- While the CAI is more than MAI, the MAI is rising
- The CAI attains the maximum before the MAI
- The CAI is falling when the MAI is still rising

When the CAI is equal to the MAI (the two curves intersect) the MAI is stationary, and has attained its maximum - the culmination point. The MAI curve is flat at this point and the MAI is about the same for some years before and some years after it. If the crop is felled at this age, at which the MAI is maximum and maximum average volume per unit area per annum will be obtained. This age is the rotation for maximum volume production.

7. How yield is regulated? Describe the Von Mantel's formula for yield regulation in forests. (2014)

Yield regulation: Yield regulation is defined as the determination of the yield and prescribed means of realizing it. It means the fixing in advance usually for a short period (the working plan period) the amount of timber or other produce which may be reviewed from the forests annually or periodically.

Objectives of regulating yield

(i) To cut each crop or tree at maturity
(ii) To obtain maximum yield of a desired produce
(iii) To cut approximately the same quantity of material annually or periodically
(iv) To limit the area to be felled to that which can be regenerated

Basis of Yield regulation

(a) By area method
(b) By volume method
(c) By area and volume method
(d) By increment method
(e) By volume and increment method

***Von Mantel's Formula* (*BI*)**: The volume of the normal G.S.is equal to the m.a.i. of the F.S. multiplied by the half the rotation, i.e.,

$$N.G.S = I \times \frac{r}{2}$$

Where I represents the volume of the oldest age-gradation (and also the total annual increment of the F.S) and is the normal annual yield.

$$I \text{ is, therefore, equal to } \frac{N.G.S}{\frac{r}{2}} = \frac{2\,N.G.S.}{r}$$

Von Mantel's simple formula derived from the above relation is :

$$\text{Annual Yield} = \frac{\text{Actual Volume of } G.S.}{\frac{1}{2}\text{Rotation}} = \frac{2\ G.S.}{r}$$

According to Von Mantel, the annual yield from any forest must bear the same proportion to the actual growing stock as normal increment bears to the normal growing stock.

$$\frac{\text{Actual Yield}}{\text{Actual G.S.}} = \frac{\text{Normal Yield}}{\text{Normal G.S.}} : \text{ or } \frac{Y_a}{V_a} = \frac{Y_n}{V_n}$$

Von Mantel's formula may be written as:

$$I \times r = 2\text{ G.S., or } Ir = \text{G.S.} + \left(I \times \frac{r}{2}\right)$$

Therefore, yield during the rotation is equal to twice the G.S. The existing G.S. supplies the yield for half the rotation; yield for the other half of the rotation coming from the increment.

Advantages of Von Mantel's Formula

- It is easy to apply; requires determination of G.S. and rotation only. It is handy.
- It regulates the yield according to actual G.S. Over-stocked forests will be heavily felled and under-stocked forests lightly felled.
- It is useful as a preliminary step in yield regulation, specially in irregular forests.
- It gives conservative yield and helps in enrichment of the G.S.

Disadvantages of Von Mantel's Formula

- This formula does not take into consideration the difference between actual and normal growing stock.
- This formula is inelastic because the surplus or the deficient in the growing stock is spread over half the rotation.
- It involves complete enumeration of the growing stock which is a tedious process.
- It does not take into consideration the age class distribution and the rate of growth.
- The formula does not take into consideration the thinning yields.

8. Progressive yield concept di-ers from sustained yield. Under the present situation, which could you suggest and why? (2015)

Sustained Yield

(a) The material that a forest can yield annually (or periodically) in perpetuity.

(b) As applied to policy, method or plan of management (Sustained Yield Management), it implies continuous production with the aim of achieving at the earliest practical time at the highest practical level an approximate balance between net growth and harvest by annual or somewhat longer periods.

Progressive Yield: The concept of Sustained yield has now been replaced by that of Progressive Yield, originally advocated by a German forester, Hartig. This takes into account both the gradual evolution of the economy as well as the progress in the silvicultural techniques, as a result of experience and research, which is considered as an important ingredient of scientific management.

The concept of Progressive Yield envisages raising the productivity of soil and of the crop, by silvicultural treatments, judicious tending, enrichment of the forest by changing the crop composition and by replacement of the original inferior forest by valuable forest species. It also stipulates avoidance of loss of increment by effective protection and tending and adoption of quick and efficient regeneration techniques. The new concept signifies a dynamic outlook in Man to build up and maximise productive enterprises through technological efficiency. It connotes a synthesis of ecological and economic considerations, and a long range view of forest products requirement for economic self sufficiency.

Current scenario in India

In the present scenario, the requirement of wood and wood products in the country is growing at an alarming rate. It has been estimated that country demands 158 million $m^3$of wood during 2020 which includes both hard and soft woods. The concept of sustained yield may be very good if the requirement is constant over periods. But in India as indicated above the demand is growing rapidly year by year and hence the concept of progressive yield is more opt than sustained yield.

The progressive yield is achievable due to the deployment of high yielding varieties which will ensure productivity improvement and amenable for harvest and early stages which will meet the growing requirements.

Similarly, the introduction and practice of precision silvicultural practices taking into consideration all factors of locality will give significant volume increment. The management of irrigation and fertilization coupled with intensive tending operations will help to achieve increased and progressive yield which can meet the current demand.

For example, two decades before the Eucalyptus plantations were rotated with 10 year rotation cycle and were able to produce less than 10 m^3/ha/annum of wood annually through a sustained yield concept but today same Eucalyptus plantations are managed in 3-5 years cycle with the productivity of over 25 m^3/ha/annum through the deployment of high yielding short rotation clones coupled with intensive silvicultural management practices.

Hence in the present scenario the concept of sustained yield has to be replaced with progressive yield with improved genetic resources and precision silvicultural practices in order to improve the productivity per unit area and to satisfy the growing demands of wood and wood products.

9. Show by your interpretation, either graphically or theoretically, how the forest yield depends upon the growing stock. (2015)

The forest yield which may be timber and timber products depends primarily on the growing stock. The growing stock with normal series of age gradation, normal growing stock coupled with normal increment will help to achieve the required yield in required time. The normal growing stock managed under sustained yield is depicted below.

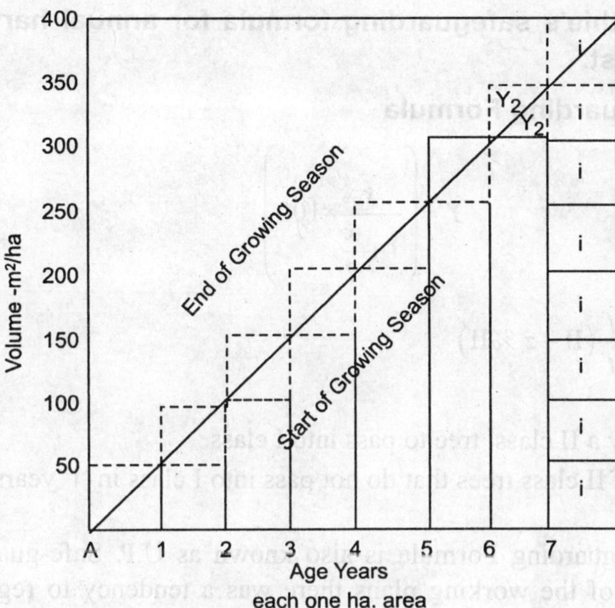

Species: Eucalyptus clonal plantation

Rotation: 8 Years

The above graph represented Eucalyptus in 8 years rotation. This mature wood would represent the increment on the whole forest and the difficulty of removing the annual increment from each unit area, say one hectare, is overcome by removing accumulated production of 8 hectare on 1/8th part of the total area, as illustrated in fig. Such a series of trees or crops of all ages, from seedling to maturity, so as to enable the removal of the oldest is known as a complete series of age-gradations.

As the forest in the above example has equal area of every age in it, an equal area will be available for felling at maturity. The establishment of such a series of age-gradations, as illustrated above is one form of the crop necessary for Sustained Yield Management (or Sustention / Sustenance) and for maintaining it in perpetuity.

Relationship Between G.S. and Yield

A complete picture on the NGS at the end of the first rotation is given in fig. The 8 year old crop will be felled at the beginning of the next rotation and will be immediately regenerated by planting or coppice. Similarly, every plantation will reach maturity and, in its turn, felled to give annual yield. Thus the G.S. removed during a rotation period of r years will be

I (annual coupe/yield) × *r* (rotation) = $I \times r$,

As represented by rectangle ABCD, which is twice the triangle ABC, then N.G.S. triangle. Thus, during a rotation the yield is twice the existing G.S. – the other half coming from the increment put on by the G.S. during the interval (rotation); in other words, during a rotation, half the yield is provided by the existing G.S and the other half by increment. Only half the increment occurring during the rotation is used in this way. The other half goes to form the G.S. of the next rotation. The volume equal to the increment alone only is felled during the rotation (i x r).

10. Discuss Smythie's safeguarding formula for annual harvest of timber from a selection forest. (2011)

Smythie's Safe-guarding Formula

$$Y = \left(\frac{x}{\mathrm{I} + \frac{x}{2}} \times 100 \right)$$

Where $x = \frac{f}{t}(\mathrm{II} - z\,\%\mathrm{II})$

f = Felling cycle

t = Time taken by a II class, tree to pass into I class

z = percentage of II class trees that do not pass into I class in 't' years due to mortality removal of thinnings etc.

Smythies's safe-guarding Formula is also known as U.P. Safe-guarding formula. Smythie's found that in most of the working plans there was a tendency to regard everything above the selection limit as surplus stock; this tendency had been responsible for decrease in the yield from selection trees in many areas. He, therefore, considered it essential to maintain sustained yield, mostly from Selection Class of trees, which he termed as Selection Yield. This formula stipulates:

(i) Sustained yield of Selection trees and not total yield of large and small timber.

(ii) No method of yield calculation could by itself safe-guard the Selection Yield, including methods of area yield.

(iii) Area yield with a check on Selection Trees can be applied to all type of forests, and eliminates most of the objections of Indian Management Systems.

The formula is based on the following assumptions:

(i) That exploitable (Selection) trees already exist and are being recruited everywhere (and in any particular compartments only) and that fellings pass regularly through the forests in consecutive annual coupes in a felling cycle of 10, 15, 20, 25 or 30 years.

(ii) The basic object of management is to ensure, as far as possible, sustained or increasing yield of exploitable trees above the exploitation limit at every felling cycle.

(iii) The middle-aged and younger dia. Classes are well represented and the present rate of recruitment into the two highest dia. Classes (I and II) will not decrease for quite a few decades.

(iv) Sustained yield of Selection trees is adequately safe-guarded if, in any area, the number that may be felled is limited to the number that will pass up from the below by the time felling come round again.

Limitations and Disadvantages of the Method:

(i) This formula is based on the enumerated values of I, II and class III trees. However, for these class trees the time taken to move from I class to other class and the percentage of the trees moving from one class to other class cannot be accurately measured.

(ii) The yield bound to vary due to the removal of class I trees which depend on the silvicultural management and the associated regeneration.

(iii) The yield in this system may fall below the expected because the yield in second felling cycle depends on the behaviour of class III trees.

(iv) The formula does not have any relation to the increment of the forest.

(v) It is the merely modification of brandis formulae and the yield is based on the stocking of class II trees and any surplus growing stock is not taken into consideration.

11. What are the main objects of preparation of Divisional Working Plans for the forest Plantations. (2007)

The working plan preparation for establishing a forest plantation is very crucial to the divisional working officer because the normal working plan pertaining to the division is different from the one planned for forest plantations. The difference is mainly due to objective and the associated management of plantations.

Hence, a proper working plan must determine and devise, according to time and locality, the entire management of the forest plantations so that the objectives, for which it is maintained, may be realised as full and as economically as possible; A working plan will deal both with long-term planning, at national or State level to ensure continuity of purpose and policy, as well as short term planning for each forest – action plan for about ten years indicating the volume of wood to be generated in a specific period and time.

Broadly a Working Plan involves

(i) Survey and assessment of the past results, present facts, resources and constraints.

(ii) Analysis of facts, with respect to prospects leading to conclusions on policy, objects of managements – for a long term plan.

(iii) A plan for future action in a definite period in conformity with the long-term-plan, i.e., short-term-plan.

(iv) Provision for control of prescriptions, for maintenance of records and collection of new facts by research and new techniques (e.g., aerial photography, mechanized logging, sophisticated inventory methods etc.)

The Working plan should be prepared within the frame work of national policy, five year development plans and other consideration of national and regional level planning. For efficient management of forests, it has to be flexible and not rigid, dynamic and not static.

Objectives of working plan

(i) To meet the long term as well as short term objectives by establishing suitable plan of work.

(ii) To meet the requirements of goods and services with respect to ecosystem needs.

(iii) To produce timber and other products to satisfy the demands of wood based industries.

(iv) To augment the sources of non timber forest products for use by the local dwellers.

(v) To establish a written document for long term vision towards managing and protecting the forests.

(vi) To enrich the soil and water resource of the working plan area through judicial management approaches.

Importance of working plan code

The working plans are prepared as per the directions described under the national working plan code. The national working plan code was adopted in the year 2004 which gives detailed procedures for preparation and writing up of working plan. Recently in the year 2014 the National Working Plan Code 2014 was enunciated for sustainable management of forest and biodiversity in the country.

This working plan code has given necessary inputs for the working plan officer towards preparation of maps, thematic layers, satellite imageries, compartment history, preparation of preliminary working plan report etc. It also extend necessary directions to the working plan officer towards drafting working plan, submission, amendment and other related issues. The national working plan code has issued the writings of working plan indicating the contents of the plan, numbering, paging, the standard designs, symbols and colours to be used along with various working schemes. So the working plan officer should ensure that the guidelines indicated in the national working plan code are followed for the preparation of working plan.

12. a. Describe a stock map. Discuss the scheme of recording crop composition (including colouring pattern used) and crop density. (2013)

b. Stock Mapping (2011)

A Stock Map is defined as map showing distribution of different forest or stand types, which have a bearing on Management, with information about their composition, age-classes, etc. The process of preparation of stock map is known as stock mapping.

These stock maps will include the following parameters.

(i) Forest type or crop composition

(ii) Crop density

(iii) Site quality

(iv) Distribution of age classes

(v) Regeneration status

(i) **Forest type or crop composition:** This will indicate the composition of the crop like pure or mixed, dry or moist forests. For the species composition the following colours are used.

Example:

Sal forest	–	Blue
Teak forest	–	Burnt sienna
Mixed forest	–	Green
Cultivation	–	Uncoloured
Barren land	–	Yellow

(ii) Crop density: The crop density is estimated based on ocular process and depicted on the map. The stock map contains the following stock density classes.

Closed forest	–	1.0
Dense forest	–	0.8 to 1.0
Fairly dense forest	–	0.6 to 0.8
Thinly dense forest	–	0.4 to 0.6
Open forest	–	0.2 to 0.4
Sparse vegetation	–	less than 2

13. Basic Factors (Attributes / Characteristics) of Normality. (2010)

The basic factors of normality are

(i) A normal series of age-gradations or age-classes.

(ii) A normal increment.

(iii) A normal growing stock.

(i) Normal series of age-gradations or classes: It means the presence in the forest, in appropriate quantity, trees of all ages from one-year-old to rotation age. When the trees of each age occur on separate areas, they constitute a series of age-gradations. When trees falling within certain age limits occur mixed together on the same area, they form an age-class. In very irregular forests there may neither be age-gradations nor age-classes; in such cases is the sign of normality the proper distribution of trees of all ages.

(ii) Normal Increment: It is the best or maximum increment attainable by a given species and for a given rotation, per unit area on a given site. An abnormal increment may be caused by faulty formation, faulty treatment injurious external influences and also unequal distribution of age-classes.

(iii) Normal Growing Stock: It is the volume of stands in a forest with normal age-classes and a normal increment. In practice, this is taken to be the volume indicated in Yield Tables for each age-class. It must, however, be remembered that a normal forest represents an ideal condition rarely, if ever, attained in practice; while the Yield Tables are the averages of actual Sample Plots, which were varying degrees short of the normal.

It will be seen that the conditions determining the normality of a forest are, in fact, only two, viz., normal age-classes and normal increment; normal growing stock follows as a matter of course, if these two conditions are satisfied. On the other hand, the presence of a normal growing stock does not necessarily imply a normal forest. A natural sal forest of low density may consist entirely of mature and over-mature trees and may carry a volume of timber equal to or even greater than that indicated in the Yield Table but such a forest is definitely abnormal because of the absence of younger age-classes, and cannot produce sustained yield.

The easiest way to visualize the conception of a Normal Forest is to consider is as a series of even-aged plantations of equal area, each of one age-gradation, worked under Clear-felling or coppice system. A plantation of one hectare was planted every year for ten years.

These are, therefore ten age-gradations of equal size, constituting a normal series of age-gradations, of which one is assumed to be ripe for harvesting every year. At the end of tenth year, the plantation planted first of all is cut and regenerated by Coppice, natural seeding, sowing or planting. At the end of next year, this regenerated coupe becomes one-year old age-gradation; the series is complete again and the oldest plantation is now ten-year-old and due for felling.

14. How is De Liocurt's principle utilized to ensure normality concept in selection forest? (2010)

De Liocourt's law: De. Liocourt discovered that in a fully stocked selection forest, the number of stems falls off from one diameter class to the next in geometrical progression, which means that the percentage reduction in the stem number from one diameter class to the next is constant. This is referred to as *De Liocourt's Law*.

If the quotient of the series is known and the number of stems in any class is given, the whole series can be worked out and this should give the proportionate distribution in an Ideal Selection forest or its balanced composition. This series is represented by the geometrical progression:

$$a, aq^{-1}, aq^{-2}, aq^{-3} \ldots\ldots\ldots\ldots\ldots aq^{-(n-1)}$$

where a = number of stems in the lowest dia class.

q = Co-efficient of reduction in the number of stems; the quotient.

By itself, it is a very important discovery, but it does not readily lead us to the ideal selection forest. Firstly, it is not known what the ideal number in known class should be and, secondly, what the quotient should be. Later researches found that the normal stem number distribution for different site qualities varies. It was still in geometrical progression but the quotient was different for different qualities. For example, this quotient q for silver fir Q.I is reported as 1.30, 1.35 for Q.II, and 1.40 for Q.III and 1.50 for Q.IV.

(i) The number of stems in successive diameter classes of true selection forest of tropical wet ever-green, southern moist deciduous or hill sal type follows De Liocourt's law, which was hitherto known to hold well for temperate forest.

(ii) A set of 'empirical yield Tables' for selection *Sal* forest in one locality, and 'empirical stand Tables' for true selection sal forest for various qualities have been prepared. It is felt that there is urgent need for a study of the rate of growth and the composition in a selection forest and for the preparation of stand tables for different types of forest.

15. What are the di-erent regeneration categories that are observed and recorded during sal regeneration survey? (2010)

In regeneration surveys, it should suffice to record the occurrence of the most promising categories of plants (regeneration) only. In case of sal, it should suffice to concentrate on poles, saplings or other established plants, or on woody shoots on way to establishment, and on large whippy saplings – establishment height being taken as 2–3 metres (2.5 m in Dehra Dun) depending on local adverse factors.

Method of recording is as follows:

S.No.	Symbol	Category of Regeneration	Remarks
1.	*e*	Woody shoots of estt. Ht. or over. (Weightage Factor-W.F=5)	*e* symbol indicates presence of atleast one such plant sufficient to stock the quadrat.
2.	*w*	Woody shoot, less than estt. Ht., healthy and vigorous to become estd.	In the absence of (1), symbol *w* indicates at least one such plant to stock the quadrat completely.
3.	*u*+	Non-woody (whippy) old, unestd. Regeneration(W.F.=2)	In the absence of (1) and (2), symbol *u*+ indicates more than one such plant.
4.	*u*	-do- -do –do-(W.F.=1)	-do- -do- one such plant only
5.	*r*	Recruitment (W.F.=0.1)	In the absence of 1, 2, 3 and 4, *r* may be recorded.
6.	*o*	No seedlings: blank	-

16. a. Define rotation and describe the various types of rotation prescribed (2013,2014)

b. Briefly discuss the relative importance of physical and silvicultural rotations in respect of existing forest resources of India (2010)

c. Distinguish between silvicultural rotation and rotation of the maximum volume production (2009)

Rotation: The period which a forest crop takes between its formation and final felling is known as Rotation or Production Period

Types of rotation:

(i) Physical Rotation
(ii) Silvicultural Rotation
(iii) Technical Rotation
(iv) Rotation of Maximum Volume Production
(v) Rotation of Highest Income
(vi) Financial Rotation

(i) Physical Rotation: It is the rotation which coincides with the natural lease of life of a species on a given site. The natural life-span (longevity) of trees varies greatly with species and the site factors. This rotation is applicable only in case of protection and amenity forests, park lands, and in some case roadside avenues. It is very variable, fairly long and also indefinite. Another interpretation of Physical Rotation is the age upto which the trees remain sound, or producc viable seed in high forests and, in coppice crops, can put forth reliable coppice shoots. This rotation is not of any relevance to economic forestry.

Example: Sandal harvest is based on physical rotation.

Trees in roadside,

Trees in parks and towns

(ii) Silvicultural Rotation: It is the rotation through which a species retains satisfactory vigour of growth and reproduction on a given site. It can neither be lower than the age at which trees start producing fertile seed in sufficient quantity, nor beyond the age when they stop doing so. It is also necessary that soil conditions remain satisfactory for germination and establishment of seed. This rotation is now precisely managed through precision silvicultural technologies in order to achieve maximum growth in reduced rotation period.

(iii) Technical Rotation: It is the rotation under which a species yields the maximum material of a specified size or suitability for economic conversion or for special use. It aims at producing the maximum material of specific dimension / quality for specific purposes, such as railway sleepers, saw-logs, mine-props, transmission poles, match-wood, paper-wood, etc. Since one and the same tree any yield different assortments of material, and the trees in a crop may attain given size at different times, the technical exploitable age offers no reliable fixed point for fixing the rotation. It does, however, allow for fixation of limits within which a tree or stand is adopted for the production of assortments in greater number or better quality. Technical rotation is adopted, particularly, by industrial firms which own forests / plantations for the purpose of supplying raw material for their plants.

Eg: Timber industry – 30 inches and above girth
Plywood industry – 18 inches and above girth
Paper industry – 2 to 8 inches

(iv) Rotation of Maximum Volume Production: It is the rotation that yields the maximum annual quantity of material; i.e., the age at which the Mean Annual Increment (M.A.I) culminates. The M.A.I referred to is that of the stand and not that of individual trees.

The quantity (usually the volume of wood above a minimum thickness) referred to naturally includes material from all thinnings, as well as the final volume felled at the end of rotation. The length of this rotation will coincide with the year when the average rate of growth, or volume increment per unit area, reaches the maximum, i.e., the age indicated by the point of intersection of C.A.I. (Current Annual Increment) and the M.A.I. curves.

(v) Rotation of Highest Income/Revenue or Forest Rental: It is the rotation which yields the highest average annual gross or net revenue irrespective of the capital value of the forest. It is calculated without interest and irrespective of the times when the items of income or expenditure occur. This rotation is important from the overall national point of view. With Forestry in the public sector, attainment of highest gross revenue is more important than that of net income because larger expenditure and investment generates several social benefits, and indirect advantages to the trade and industry. The private owner of a forest estate is interested in maximum net revenue (gross income minus expenditure, both discounted to date) by keeping the rotation period as short as possible.

The average net annual revenue or rental obtained from a stand of trees is expressed by the formula:-

$$\text{Mean annual net revenue per unit area} = \frac{Y_r + \sum T_r - C - \sum e}{R}$$

Where, *Yr* = Value of final felling (final yield) per unit area,

Tr = Value of all thinnings during rotation period R, per unit area,

C = Cost of formation of stand, per unit area. e = Annual cost of administration / maintenance,

R = Rotation (years).

(vi) Financial or Economic Rotation: It is rotation which yields the highest net return on the invested capital. It differs from the Rotation of Highest Net Income in that all items of revenue and expenditure are calculated with compound interest at an assumed rate, usually the rate at which the Govt. it able to borrow money. It is also defined as:

- The rotation which gives the highest discounted profit, usually at its commencement."
- The rotation which is most profitable."
- The rotation which gives the highest net return on capital value, i.e., under which the *Soil Expectation Value* (S_e) calculated with a given rate of interest is the maximum.

17. Describe the normal forest concept for commercial plantations (2006,2009)

The normal forest is the forest with normal series of age gradations, normal increment and with normal growing stock.

The commercial plantations are established for the purpose of commercial value with the determined productivity and period. So the concept of normal forests is very well applicable to commercial plantations.

A commercial plantation with 5 year rotation cycle will have 5 years of age gradations with one year interval. The increment of each year may be 10 m^3/ha. This indicates that the established plantation has normal series of 1-5 years age gradation, it has normal increment 10 m^3/ha/annum. Hence the commercial plantation can be harvested at the rate of 10 m^3/ha/annum from 6th year onwards and continued upto 10 years.

If the species selected for the commercial plantation has a coppicing ability then next rotation follows from 11-15 and 16-20 years with the similar age gradation, increment and the growing stock volume.

Hence the concept of normality and normal forests is very well applicable to all commercial plantations which can be managed and harvested annually or periodically (periodic blocks in case of timber species with long rotation). The commercial plantations with normality concept is depicted in the following figures a and b.

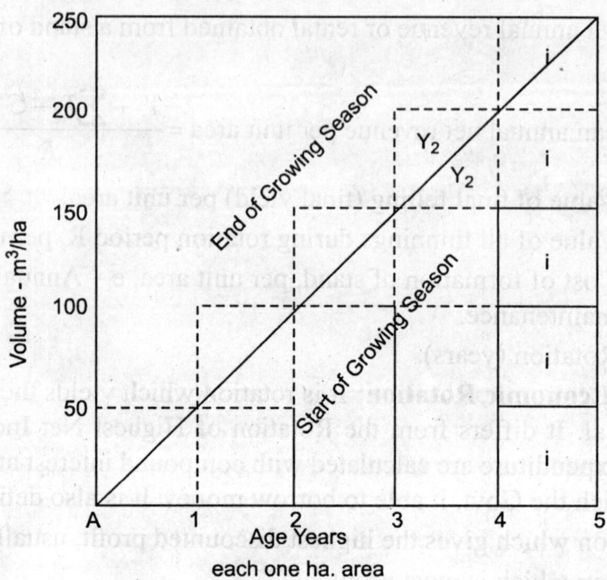

Fig a. Commercial pulpwood plantation with normality concept – Eucalyptus

Arrangement of Period Blocks.

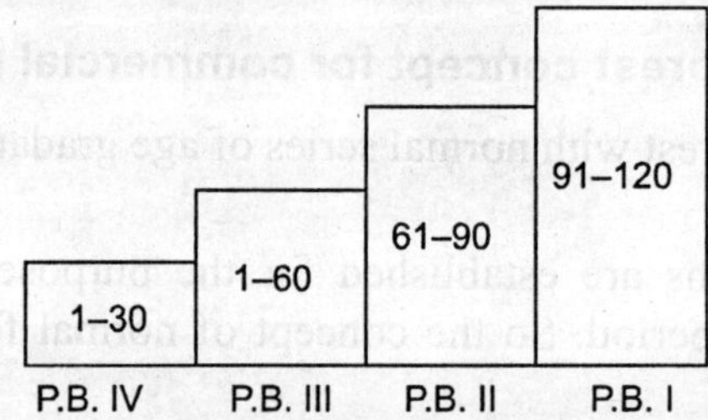

Fig. b. Commercial Teak plantations with normality concept

18. How is the soil expectation value helpful for deciding financial rotation? (2010)

Soil Expectation Value (S_e)

It is the value that a piece of land is expected to provide a continued net income with the acceptable rate of interest and this expressed by the following formulae.

Expressed by formula, $S_e = \dfrac{X}{0.0\,p}$

Where,

X = Rupees

p = Rate of interest, percent and

This soil expectation value is useful in fixing financial rotation because the identified land due to its superior site quality is able to provide continuous yield which is calculated in terms of financial value on an annual basis. This financial rotation though based on overall financial returns but it is obtained from the tree where it grows in a piece of land in which the soil is expected to give revenue in terms of rupees annually.

19. Explain the components of compartment description. (2010)

Compartment Description: The compartment is a basic unit of forests. Traditionally the compartment history files are maintained in India to record the condition of forests, stocking position, subsequent removals and regeneration operations that were carried out. During the working plan preparation, the working plan officer and his team visit each and every compartment of a forest and based on that the compartment is described. For compartment description, the National Working Plan Code (2014) of Government of India has indicated the following guidelines and format.

Title of description	Explanatory notes to write the description
Area (in Hectares)	Total as also under each species or type, blank, unworkable, etc.
Situation	As accurate as possible with the help of GPS.
Boundaries	N, S, E, W (mention boundary pillars, ridges, spurs, streams, etc.)
Legal status	Legal status of the forests may be as reserve forest, protected forest, un-classed forest, national park, private forest, private land with tree owned by government and undetermined, etc.
Land Use	The land use type of the plot may be categorised as closed forests, dense forests, open forests, scrub, bamboo brakes, shifting cultivation, young plantations of forestry species, tree in line, forest roads, grass lands, and barren, agricultural land without trees in surround, agricultural land with trees in surround, non-forestry plantations, habitation, and water bodies.
General topography	The topography of the area may be determined from the toposheets and the same may be confirmed by field observations. It may be categorized as flat, gently rolling, hilly and very hilly.
Altitude	In meters. In case of hilly areas give lower and upper limits.
Aspect	Main aspect and variations if any; The direction of the slope may be recorded as northern, north-eastern, eastern, south-eastern, southern, south-western, western, north- western and no aspect.
Gradient/slope	Precipitous, very steep, steep, moderate, gentle, etc.
Configuration	Rugged, undulating, flat, etc.
Rock and geology	Mention main underlying rocks including alluvial deposits
Soil	Texture, depth, permeability, drainage, surface compaction, humus, etc.
Soil erosion	Heavy/ moderate/ mild/ no erosion
Crop composition	Mention of major tree species predominant in the area and extent of their representation in terms of percentage may be made.
Regeneration status	Regeneration status may be observed and recorded as adequate/ moderate/ poor/absent for major tree species.
Injury/ damage to crop, if any	Insect attack, fungal infestation, leaf defoliator, top drying, girdling, scarring, lopping, damage by natural calamities/ wildlife/fire may be observed assessed and recorded.

Grazing incidence	Depending upon the pressure of grazing exerted on the forests by livestock the incidences of grazing may be categorized as heavy/ moderate/ light/ none.
Presence of bamboos	If yes, brief description of bamboo density, bamboo quality, bamboo regeneration and bamboo description may be made.
Presence of grasses	Ground cover on an area of about 2 ha around the plot (grid) centre may be intensively observed to classify the area for grasses as very dense, dense, moderate, scanty and absent.
Presence of weeds	Ground cover on an area of about 2 ha around the plot (grid) centre may be carefully observed to classify the area for presence of weeds as very dense, dense, moderate, scanty and absent.
Plantation status	If a plantation is in existence in a plot (grid) area fully or partially, the details of the plantation with regard to its area, year of plantation, species, spacement, general growth conditions, average crop diameter, any specific events or happenings related to plantations may be observed, assessed and recorded.
Water bodies capital	Name, type, extent, seasonality of the water body may be explored and recorded. Potability i.e., safe enough for drinking has to be ascertained.
Drivers of degradation	Biotic: Brief account of degradation sources may be given and on the basis of various factors of degradation like grazing, browsing, fire, lopping, girdling, illicit fellings, mining, encroachment, etc. the intensity of degradation may be categorized as heavily degraded, moderately degraded, mildly degraded and not degraded. Natural calamities: Degradation due to calamities such as landslides, avalanches, floods, frost, cyclones, droughts may be categorized as heavily degraded, moderately degraded, mildly degraded and no calamities.
Faunal sighting, if any	The flagship species including mammals, birds, reptiles, amphibians, plants, etc. which may be very significant to the area may be identified. Suitable habitats and micro-habitats for such key faunal species may be identified and recorded so as to prescribe appropriate measures needed to conserve and improve.
Faunal traces of flagship species	The entire 2 ha area around the centre of plot (grid) may be scanned for any faunal traces of flagship species. If found the details of the same may be observed and recorded.

20. Kinds of Abnormality (2011)

Abnormality in forests may occur due to the following factors:

- **Over-stocked:** A forest past the age of maturity, or having excessive distribution of older age-classes, will have more volume per unit area that the normal.
- **Under stocked:** This condition may be caused due to preponderance of younger age-classes, poor density or extension in rotation. In such cases conservative treatment, i.e.,

felling less than the permissible amount in order to build up a proper reserve of growing stock is indicated. The accumulation of growing stock to remedy a deficit is often secured by adopting a rotation longer than that necessary, or by cutting less then the possibility.

- **Normal growing stock but abnormal distribution of age-classes or age-gradations:** i.e., they may be disproportionate, some of them may be even missing. This is the worst form of abnormality because the entire forest may be practically of one age-class. The conversion of such a forest to normality is a very difficult problem, and involves either an interruption in sustained yield or sacrifice of material by decay or unsoundness.
- **The increment may be sub-normal:** This condition may be caused by defective density, fire disease, etc., or it may be due to preponderance of over mature trees.
- **Normal increment volume in an abnormal forest:** Annual increment of a forest may be the correct volume for normal increment, but unless it is laid on to trees of the right size-class in the right proportion it is not a normal increment.

21. Describe the scope and objectives of forest management. What is the impact of restrictions laid down by Hon'ble Supreme Court on green felling in forests? (2013)

Scope of Forest Management: Management of Forests, as that of any other enterprise, involves a process of making and implementing policy decisions to achieve the objectives of the owner. These decisions involve, in turn, a plan of action. Planning is the responsibility of the States and the Centre in case of State-owned forests, broad principles for which are embodied in the National and the State Forest Policies. Detailed plans are prepared by Forest Managers at the professional level, and executed by the technician level staff. Forest Manager has to constantly manage the growing stock to achieve given objects of management; in this process he has to decide: 'how much, when where and how to cut.

Scope of Forest Management is very extensive; it encompasses broadly, the following main activities.

Control of growing stock: The objective of forest management is to control the growing stock in terms of its structure and composition which involve the activities from site preparation upto protection of the resources. The following are the major scope of forest management activity under this category.

- Site adaptation
- Choice of species
- Manipulation of stands
- Harvesting the produce
- Regeneration
- Protection.

Distribution and marketing of produce: The forest management has an excellent scope of harvesting, transportation, marketing and sale of forest produce towards generating revenue through the forest management and hence it has an excellent scope in the following activities.

- Transportation and communication
- Logging Plan
- Marketing data
- Sale of produce
- Revenue

Administration of Forest Property: Forest management has significant scope towards administering the forest property. The administration includes men and material management and hence the forest management has excellent scope on the following activities.

- Forest organization
- Management of Personnel
- Monitoring and control or works
- Labour management and welfare
- Financial control and economy efficiency
- Fulfillment of social obligations
- Record for present and future reference.

Objectives of forest management

- Maintaining and, as far as possible, raising the productive capacity of the soil of the forest stands consistent with the maximum site potential.
- Promoting the protective effect of the forest, against soil erosion, avalanches, floods and protection of the physical factors such as natural scenery, local flora and fauna.
- Execution of silvicultural operations and regulation of fellings in such a way so as to bring the forest to a condition of as near normality as possible; in simple words, attainment of a normal forest is one of the principal objects.
- Satisfaction or rights of the right holders in respect of timber, firewood, grazing, etc., in particular, and to meet the bonafide requirements of the local population in general.

Subject to the above Silvicultural, Conservational and Social considerations, providing the maximum possible volume of valuable timber for constructional and industrial purposes, and other forest produce for meeting the market demands and securing the highest possible financial results.

Impact of Honourable Supreme Court restriction: The intervention of Honourable High Court on felling has made significant impact towards restriction of felling and diversion of forest areas. The following are the major impact

(i) **Name of the case:** T. Godhavarman case Vs Union of India 1995 – Conversion is taking place by the provision of Forest Conservation Act 1980 and time to time guidelines are issued many times from the Honourable Supreme Court with return to Godhavarman Case.

(ii) The felling in reserve forest area has totally been banned which resulted in restricted logging operations, reduction in illegal felling operations and the associated damages to ecosystem has drastically been reduced.

(iii) The wood based industries deriving raw material from the natural forests have started establishing agro and farm forestry based industrial wood plantations which has created self reliance in raw material security and protected the natural forests.

(iv) The Honourable Supreme Court intervention has made significant improvement in forest area from 23% to 24.16% (2015 survey) which has made tremendous improvement forest area increase and the associated natural resource improvement.

(v) The restriction on felling and conversion operations have improved the populations of wild animals and ensures sustainable food chain management in their respective habitates.

(vi) The loss of soil, water and the associated natural resources has been drastically reduced due to ban on felling operations.

(vii) Honourable supreme clear cut directions made the necessary of approved working plan for felling in any forest area.

22. Define growing stock. Explain the estimation of growing stock and density. (2014)

Growing stock: It is defined as the "sum (by numbers or volume) of all the trees growing in the forest, or a specified part of it".

Estimation of growing stock: The growing stock can be estimated through the following methods.

(i) Total enumeration: This is rarely practised over large forest areas as the time consumed is more and it is also labour intensive. In case of inaccessible forests spread over hundreds of acres, this method is practically impossible and hence avoided. However, this method is practised even today when the tree species under consideration is valuable and when the distribution of the species is also to a limited extent.

(ii) Partial enumeration /sample enumeration: This is normally practised in forestry because of its advantages and its acceptability on statistical grounds. The results of such enumerations are reasonably accurate. This is mostly followed when the area to be covered is large and when the heterogeneity in site conditions is observed to be minimum / negligible.

(iii) Sample plot method: This method is applied to selected representative areas of the crop. Of late, aerial photographs are being widely used for inventorizing large areas. Sample plot method is widely used in the estimation of growing stock in various regions of the country. The importance of determination of the actual growing stock and idea about the composition, age gradations and age classes is of paramount importance especially when working plans are prepared / revised. Such estimations will indicate if the growing stock in a forest is at its optimum or whether there is a need for any suitable silvicultural treatment to make the crop attain its optimum increment. Frequent / periodical enumerations aid in indicating the direction in which the growing stock is progressing and help to arrive at management decisions to rectify the shortcomings if any.

23. How do you visualize the concept of "Normal Forest' in Indian context? Describe the e-ect of silvicultural systems of Normality. (2016)

A Normal Forest is a forest on a given site with given objective of management will have an ideal growing stock, an ideal distribution of age classes of the component crop and shall add ideal increment in tune with the growth and development of the plantations.

This normal forest will yield either annually or periodically without endangering future yield and without affecting the site quality. Normal forest with the above characteristics is considered essential for sustained yield and is identified as good scientific management.

E-ect of Silvicultural System on Normality

The Silvicultural system has significant effect on the normality. In general, the silviculture systems are grouped into different categories based on regeneration system as indicated below.

(i) Even-aged system - e.g., Clear felling, Coppice and Uniform.

(ii) Uneven-aged system - e.g., True or Single Tree selection and group selection.

The basic difference between the two groups is that in an even-aged system any small unit of area, even half an hectare, contains trees of only one age, although the forest as a whole contains all ages upto maturity. In the uneven-aged forest, however, each small unit or area contains trees of all ages; in the single tree selection system the unit of area is very small, say only half an hectare or even less, but in group selection system it will be larger perhaps two hectares or even more.

In very intensive even-aged forestry, in order to take full advantage of differences in site individual stands of one age may also be small, perhaps 0.2 or even 0.1 hectare. The forest then approaches and begins to merge into an uneven-aged one.

The basic difference between the two groups is whether separate ages and areas can be distinguished, and delineated on the ground and on the map or not. In even aged forestry they can be distinguished so that management can be based on area and age. If they cannot be distinguished, as in true selection forests, then management must be based on individual trees and their size and not on age and area.

Therefore, there must be two main types of Normal Forest, viz.,

- The Normal Even-aged Forest and
- The Normal Uneven-aged Forest.

24. Explain the concept of rotation and its application in regular and irregular forests. (2016)

Rotation is the period that elapses between formation of a crop and final felling. It is otherwise termed as production period. The concept of rotation varies with species, objective of management and the final product expected from the species of interest. For example, the species grown for site protection may not be suitable for timber production. Hence the concept of rotation is very significant in deciding the species, the objectives and the final product expected from the species.

Concept of rotation in regular and irregular crops: The term Rotation, strictly speaking, is correctly applicable to regular crops only. In Clear-felling System and plantations, rotation is a definite period of interval between the year of formation and final felling. In these, and regular forests in general, entire crops of trees of a sizeable area are felled at a time or during a comparatively short period (regeneration period in Regular Shelterwood System) when ready for felling. There is, more or less, a clear production period which can be planned in advance to give timber which satisfies the object of management.

In case of irregular forests, the rotation is fixed for the whole working circle as a unit, as the average length of time between the establishment of crops and their harvesting. Though this system facilitates better planning and organization of work, it does not take into consideration that:

- Rate of growth will vary with site variation, even for the same species.
- It involves sacrifice of immature trees / crops, as some will not have reached exploitable size.
- Accidents, such as fire, disease, and wind-throw may happen, necessitation gelling earlier than planned.
- If profit is the main object of management, some difficulties may arise, as the degree of profit is affected by rotation, rate of growth, size / price gradients of timber and the cost of growing it. To obtain desired profits, stands will have to be finally at various times depending on their rate of growth.

It is evident from the above statement that irrespective of the object of management size of timber or profit rotation is associated with final felling age or removal age and, even with a planned rotation, the removal age may vary with rate of growth. On better sites rotation will be shorter; damage / mortality due to accidents may necessitate earlier removal of some parts thereby increasing the tendency to break up uniformity of original crops into smaller segments, until they start approaching the structure of uneven aged forest.

In uneven aged (irregular) selection forests, trees are selected individually on their merit for felling, depending on:

- Qualities of size, vigour and suitability for markets.
- Adjustment of proportion of different sizes.
- Silvicultural principles; e.g., removal of inferior stems in favour of better ones.

Such a system clearly has greater flexibility, and enables forester to adopt felling to suit different rates of growth caused by variation in site or species. Moreover, forest is a perpetual entity and never suffers complete clearance of trees on any part of the area, except periodical thinning. Therefore, size being the criterion for felling, age is known, and there being no final harvesting, there is no rotation as defined above.

However, one could say that its rotation period is equal to that of the average age of the exploitable size trees removed the exploitable age, at which they attain the size required to fulfill the objects of management.

Therefore, in selection / irregular forests, concept of Rotation assumes, by and large, only an academic importance for accounting purposes. In these forests, there is no definite correlation between age and diameter; the latter depending on site and available light conditions and even individual characters also. The trees of all ages are mixed together and the crop as a whole, on any unit area, does not reach the age of final felling at a time. Consequently, the term Rotation or Production Period, is not correctly applicable to the age at which individual trees reach the age of maturity and are remove. Maturity in selection forests is related to size, and exploitable size is fixed for removal of individual trees. Correct term expressive of maturity in selection forests is, as started above, exploitable age, or size. Size should therefore, be used as a standard of exploitability, and not age.

25. How enumeration of the growing stock is done? Describe in brief the various methods of enumeration of preparing a working plan. (2014)

Growing stock enumeration

Diameter Classes: All enumerations should, ordinarily, be in 10 cm dia. classes, or 30 cm girth classes, except when the exploitable dia. is so low that 5 cm classes, at least for the lower part of the range, are desirable. If the exploitable dia. falls within a dia. class, it may be necessary to sub-divide the class; e.g., if the exploitable dia is 35 cm, it may be necessary to enumerate 30-40 cm dia. class in two parts, viz., 30-35 cm and 35-40 cm.

Species to be enumerated: The species to be enumerated and recorded will be given in the first Preliminary Report. In view of the increasing importance of miscellaneous species for major and minor forest products, it would be advisable to enumerate and collect statistics of as many important, and potentially important, species as possible. These may be lumped together in the final abstracts of the W.P., but should be shown separately in the Compartment Histories. It may, however, be kept in mind that it is not possible to enumerate a large variety of species accurately; and enumeration of more than six species simultaneously adds considerably to the cost of enumeration.

Classification: Trees should be classified into sound or unsound, fit or unfit, saleable or un-saleable, according to the local practice in vogue. In case of Chir forest where twist is prevalent, it is usual (in U.P.) to note upto 7° left-handed and upto 15° right- handed twist only, as beyond these limits the trees are unfit for sawing.

System of Enumeration: The Working Plan Conservator will decide the system of enumeration, in consultation with the Silviculturist, for each Working Circle, or part thereof, the factors on which the choice of the system will depend are: the extent of the area, topography, variation in species composition, density of valuable species, intensity of management and productivity of the forest.

- Choice between total and partial enumeration: based on factors of intensity of management, cost, time, labour, standard of accuracy required.
- In case of partial enumerations: the intensity of sampling methods thereof (viz., stratified random or systematic sampling by whole compartments, sub-compartments or by topographical sampling units *or* systematic sampling by strips in plains forests).
- Stock maps would be necessary for preliminary planning and selection of sampling units.
- Stratification, i.e., the division of the area into a convenient pattern of blocks from each of which a certain number of samples is taken.
- Enumeration staff and labour.
- Instructions to the Enumerators.
- Check Enumerations.
- Equipment required.
- Norms of output.
- Point Sampling—Horizontal and Vertical forms. Use of Wedge prisms and Relaskops.
- Lay out of grid for sampling.

Methods of Enumeration for preparation working plan

There are two types of Enumeration

(i) Total or complete enumeration: It is the enumeration of desired species about the specified diameter limit is carried out over the entire area of forest unit under consideration. Being expensive and time consuming it is done in comparatively smaller areas valuable and intensively managed forests where maximum possible accuracy is desired in the estimate. Total enumeration is also done in case of miscellaneous forests where the number of species is very large or the species of economic importance form a smaller portion and are confined only to favourable localities.

(ii) Partial or sample enumeration: In partial enumeration, enumeration is to done only in a representative portion of the whole forests. For example a forest may be of 1000 ha out of which only 100 ha have been selected for enumeration and estimate of the whole population of 1000 ha is made, it is called partial or sample enumeration. Therefore sample enumeration is the enumeration of a representative part of a whole population. The ratio of the sample to the whole population is called sampling fraction or intensity of the sampling which is expressed as the percentage. Eg. 5%, 10% and 20%. A sampling unit need not to be compact area but may consist of several areas. The record of each sampling unit is kept separately for working out sampling errors other statistical parameters.

26. How would you view the Indian initiatives for sustainable forest management? Discuss. (2013)

The forestry sector in India is among the first few in the world to be managed on the lines of modern scientific management. Establishment of modern and scientific forest management system and process, which started from the middle of the eighteenth century, coincided with the Industrial Revolution in the west. The forests were considered and used as important resources during the pre-independence period, as the demand of raw materials produced from forests increased. They were extensively exploited to expand the railway network, for example, in the shape of wooden sleepers in that period. Forestry was thus production oriented at that time. However, the basic change in perception that forest as a resource cannot be exploited unhindered as brought about by the National Forest Policy of 1952. This tried to shift focus to production forestry and also had an eye on meeting objectives regarding the maintenance of ecological balance and meeting the needs of various other stakeholders in the best way possible.

Initiatives to integrate elements of SFM in current Indian Forestry

- The criteria and indicators approach was developed with the idea of a set of specific forestry conditions evolved through various international processes among participating countries.
- In this initial phase, it was thought to develop a frame work for SFM for the Indian context also to establish a benchmark for sustainability according to the prevailing global policy framework.

- In 1999, a workshop on "Development of National Level Criteria and Indicators for the Sustainable Management of Dry Forests in Asia" (also called the Bhopal India Process) was held at the Indian Institute of Forest Management (IIFM), Bhopal, with support from the Food and Agriculture Organization of the United Nations and the United Nations Environment Programme in collaboration with the ITTO, the US Department of Agriculture's Forest Service, and the IIFM, Bhopal.
- The outcome and approach is now widely referred to as the "Dry Forest in Asia Process" in which 10 Asian countries jointly developed a regionally applicable set of national-level criteria and indicators relevant for dry forests in the region (FAO 2000).
- The Asia regional initiative was later endorsed by the "National Task Force on Sustainable Forest Management", appointed by the Ministry of Environment and Forests (MoEF), Government of India, and these have led to a national draft set of criteria and indicators.
- Thus the Indian initiative of criteria and indicators approach for SFM has been spearheaded by the IIFM, in collaboration with the ITTO and the MoEF (IIFM 2000).
- The IIFM, through a series of national technical workshops and consultation meetings, has also been a pioneer in sensitizing communities, forest managers, NGOs, and researchers and emphasizing the need for developing a set of criteria and indicators at the national, state, and forest management unit levels.
- Over a period of time, eight criteria and 51 indicators specific to Indian forestry conditions were evolved after a consultative process involving a large gamut of stakeholders. The Bhopal India Process has helped evolve the national draft standards after a lot of deliberations and field-testing over the years. But Indian forestry has still to go a long way to achieve sustainability.

27. What is the yield regulation in forest management? How can the yield regulation by Judeich method be used? (2013)

Yield regulation: Yield regulation in forest management is applied to the determination of the yield and the prescribed means of realizing it. In any forest management activity the yield regulation is concentrated on the following.

- To cut each crop at optimum maturity
- To obtain maximum yield of the desired produce
- To cut the same quantity of yield annually or periodically
- To limit the area to be felled over so as to carry out the regeneration operations

Judeich Stand Selection Method: This method was practiced by Judeich in Germany, long before the F.P.B method was developed in France. The principle in both the cases is same, i.e., treatment of each crop according to its needs. It was originated by Cotta in 1841 and perfected by Judeich. The basis of Judeich's method is the fact that no method of yield regulation over long periods can be accurate, if it is not frequently revised. He, therefore, regulates yield for a period of, say, 10 years at a time and at the end of that period the Working Plan is revised.

The essence of this method is the careful selection of mature stands (maturity being technical, financial or silvicultural, depending on the object of management) and the allotment of, a more or less, proportionate area of such stands to be felled and regenerated during the Working Plan (W.P) period.

The procedure is as follows:

(i) The forest is divided into a number of compartments of suitable size.

(ii) The compartments are inspected and described with reference to the treatment they require during the next Working Plan Period, say 10 years.

(iii) The rotation and the Working Plan period are fixed.

(iv) The compartments are selected for felling and regeneration during the W.P. Period on the following principles.

(v) Stands which must be felled to meet the necessary silvicultural or protective necessities.

(vi) Mature and over-mature stands; maturity being related to objects of management.

(vii) Stands of doubtful maturity, whose felling may be desirable for reasons of convenience; e.g., onc occurring between two mature stands.

(viii) The size of the mean annual coupe and the area to be felled during the W.P. period is determined

(ix) If the total area of the compartments selected for felling under 4 (i), (ii) and (iii) is within, say 10% of the mean periodic coupe as calculated in (5) above, no further adjustment is necessary. It is larger than this, then some of the areas brought in under 4 (iii) may be omitted. If it is smaller than the calculated minimum area, then some more compartments, even though not quite mature, will have to be included.

28. How is yield regulated in forest which is worked under clear felling system. (2011)

Yield regulation in clear-felling system: In clear felling system, yield is regulated by the following method:

(i) Yield regulation by area

(ii) Yield regulation by reduced area

(a) Annual Coupes by Area: This is the oldest and simplest form of regulating yield from a forest. It was initially applied to the coppice crops worked on short rotations, upto twenty years. Subsequently, with the introduction of Coppice-with-Standards System, the same method of area control continued to be applied, wherein a certain number of standards was reserved for two, three or four coppice rotations.

In India, this method is widely adopted in plantations, particularly those coppice for fuel. It is adopted in Coppice-with-Standards and Coppice-with-Reserves Systems. It regulates the Final Yield i.e., fellings in areas to be finally felled at rotation age and regenerated; it does not take into account Intermediate Yield from cleanings and thinning in the younger crops.

In its simplest form, the area of the forest or the Felling Series is divided into a number of annual coupes equal to the number of years in the rotation. Usually, the sequence of annual coupes is laid down in the Working Plan and shown on the map and sometimes demarcated on the ground as well, but there should be no objection to felling a coupe, nearly mature but damaged by fire, in advance of its prescribed date and making suitable adjustments in the felling order.

This method of yield regulation will ensure equal sustained yield in the second and subsequent rotations though it may not in the first. Annual coupes so formed and equal in the area on the ground are known as equi-extensive coupes.

(b) **Annual Coupes by Reduced Area:** Since the density and site quality may vary from coupe to coupe, reduction factors should be applied for equalising annual yields and areas allotted to each coupe made equi-productive, rather than equi-extensive. Reduction for density is seldom made, as this variation may be only a temporary feature and full stocking may be expected after regeneration operations, and this factor will not affect future yields. However, yield variation due to site quality may be a permanent feature of the locality, and is taken into account by determining and applying suitable 'Reduction factors' before formation of equi-productive coupes.

Advantages of Yield Regulation by Area:

(i) It is easy to apply

(ii) Leads to absolute regularity of age-gradations

Disadvantages

(i) It is very rigid, every change of rotation will necessitate re-division of the coupes.

(ii) Prescribes fellings without consideration of the crop conditions

Limitations

(i) In unmanaged forests, it is not possible to lay out equi-productive coupes.

(ii) In mixed natural forests, regulation by area is only a crude method; it was given up in Europe in early 19th Century in favour of regulation by area and volume.

29.a. What is normal growing stock? Explain the determination of NGS in clear felling system with graphical illustration and numerical examples. (2016)

b. How are yield table data used for assessment of normal growing stock. (2011)

c. Explain the concept of normal growing stock in clear felling system. (2009)

The concept of normal growing stock in clear felling system is very easy and applicable in plantation forestry sector. The normal growing stock with normal series of age gradation and normal increment will have uniform productivity and hence amenable for harvest in the specified rotation. This concept is applicable to pulp wood plantations in India. In the clear felling system the normal growing stock is determined by the following methods.

NGS in Clear Felling System

(a) **Based on final MAI:** The simplest example may be taken of a firewood Eucalyptus plantation of ten hectares, one hectare of which has been planted annually for ten years - the proposed rotation period. There are, therefore, ten age-gradations of equal/equi productive areas, consisting of a normal series of age-gradations, of which one 8 year old is felled and allowed to coppice to form future crop supplemented with planting, if an when necessary. So, at the end of next year, this regenerated area becomes one-year old age-gradation and the series is complete again, and the oldest plantation now 8 year old is harvested and regenerated.

Example: The volume of 8-year old gradation on one hectare area has grown 400 m³, so the final MAI at this rotation age 400/10 = 40m³, and this is shown as the volume on one-year old hectare as *i.*

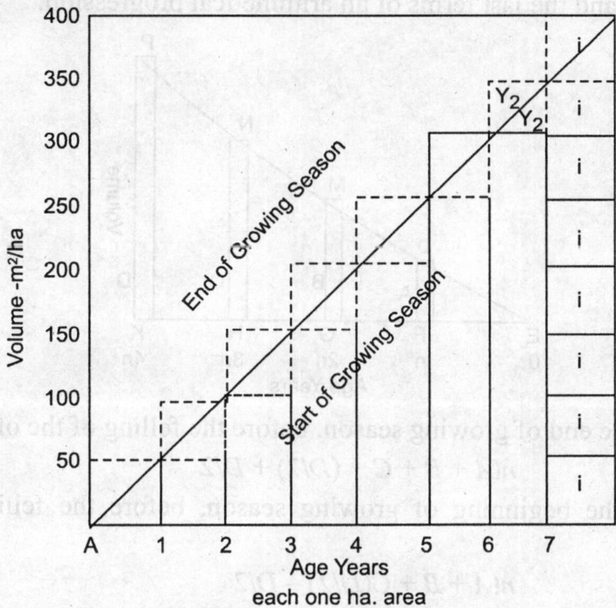

The volume r x i standing on the oldest age-gradation at the end of *r* years (in this case, 8 × 50 = 400 m³) is also the sum of the MAIs of all the r (8) age-gradations, and may be termed as I to represent the increment of the whole series. Therefore, by felling the rotation-age ha each year, the normal increment for the rotation of this normal series is being harvested. Careful study and interpretation of this diagram brings out that the volume of the oldest or rotation (r) age-gradation is equal to $r \times i = I$ (or)

- Final MAI of anyone age-gradation *i* multiplied by rotation *r* (or)
- The sum of CAIs, i, of all the *r* age-gradations
- The total MAI of the series

(b) Calculation of N.G.S. Form yield table: If a yield table gives data for intervals of one year, that is, it gives the volume of each age-gradation, the G.S volume can be readily calculated by adding up the volumes of successive years. But the yield tables usually give data for intervals of five or ten years. In such cases, the volume of N.G.S. can be accurately determine by plotting the yield table data on a graph paper, drawing a smooth curve and computing the area below the curve either by a plani-meter, area-square or by counting the squares. This is however, a cumbersome procedure.

If we assume that during each Yield Table interval, say five years, the rate of increment is uniform (although it may vary from interval to interval), the trend of the growth curve from one measurement to the next is smooth either rising or falling, and the increment put on by the crop changes in equal quantities annually, or in an arithmetical progressing. Though this assumption is not quite correct, but as the period between two successive measurements is usually mathematically by regarding the two successive entire in the yield table as the first and the last terms of an arithmetical progression.

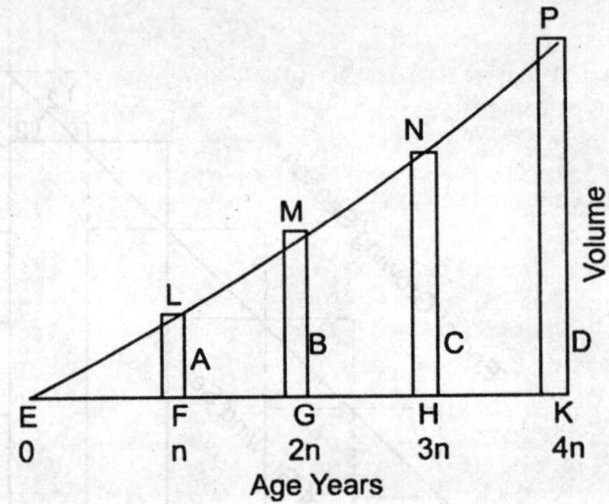

Calculation at the end of growing season, before the felling of the oldest age gradation,

$$n(A + B + C + (D/2) + D/2$$

Calculation at the beginning of growing season, before the felling of the oldest age gradation,

$$n(A + B + C(D/2) - D/2$$

Calculation at the mid-season of the growing season

$$n(A + B + C + (D/2)$$

Where, n – yield table interval

A, B, C & D–Volumes given in the yield table at ages n, $2n$, $3n$ and $4n$ years respectively

30. a. Maps are an integral part of forest management. Give your perceptions about the kind of maps you consider important for managing a forest circle (2015)

b. Explain various types of maps prepared by Working Plan O-cer (2011)

Yes. Maps are an integral part of forest management which facilitates the forest officials to describe the forest division, determine the composition, forest type, soil type etc. in order to efficiently manage the forests. Hence maps are very essential for the forest managers in order to implement forest management activities for efficient administration of biotic resources. The following seven maps I consider very important for managing a forest circle.

(i) **Management Maps:** These are generally on 1: 50,000 scale (corresponding to the old l'= l mile) showing new Working Circles; Felling Series. Periodic Blocks arid other details of management. Each Working Circle is coloured with a distinct colour wash; P.B.I, areas, if any, are usually given a deeper shade of the appropriate colour.

(ii) Stock Maps: These are generally prepared on 1: 15,000 (old 4'=1 mile) scale. These maps give, ordinarily, the distribution of forest types and main forest species, non forest areas, blanks, etc. If (he- management is intensive, age-classes, quality-classes and density are suitably differentiated and super-imposed by hatchings, etc. If old stock maps are available, these are checked and brought upto-date.

(iii) Regeneration Survey Map: The regeneration map is very significant for managing a forest circle. Normally the regeneration maps are prepared with 1: 5000 or 1: 4000 scale.

(iv) Working Plan Map: These are prepared on 1: 60,000 (corresponding to old 1"=1 mile) to 1 : 30,000 scale (old 2=1 mile); generally 1: 50,000. These show all features which are reasonably permanent, such as physiographic features, territorial boundaries including pillars and their numbers, names of blocks, compartments, also showing Forest Rest Houses (F.R.H,), roads, railways, and main-features of the country around the reserved forests.

(v) Enumeration Map: In case of partial enumerations, an enumeration map, on 1:50,000 scale, showing location of plots, strips, topographic units, compartments and sub-compartments actually enumerated. Tracings are also made on 1: 15000 scale maps and deposited for future reference.

(vi) Forest Type Map: This is prepared on 1: 50000 scale showing forest types, according to the prescribe colour guide (as per State W.P. Code). The W.P.O. will prepare one copy in consultation with the State Silviculturist and get it approved by the C.F, (W P,), Copies of the same are prepared and distributed as per prescribed practice.

To prepare such maps, it will, in most cases, be sufficient to distinguish climax types (Champion and Seth's classification) by flat wash of distinct colour; serai and edaphic types could be shown by black patterns super-imposed as follows (draft suggestions for the All-India W.P. Code) :

Primary Serai	–	Black horizontal lines
Secondary Scral	–	Black Vertical lines
Edaphic	–	Black dots
Degradation	–	Black cross-hatching

For each main group, the following colour scheme may beused. Sub-groups, sub-types, etc., may be distinguished by the super-imposed pattern given above.

S.No.	Group Designation		Colour
1.	Tropical Wet Evergreen	–	Violet
2.	Tropical Scmi-evcrgreen	–	Purple
3.	Tropical Moist-dcciduous	–	Cobalt blue
4	Littoral and Swamp forests	–	Prussian blue
5.	Tropical Dry Deciduous	–	Yellow
6.	Tropical Thorn	–	Red

7.	Tropical Dry Evergreen	–	Pink
8	Sub-tropical Broad-leaved Hill forests	–	Vermillion
9.	Sub-tropical Pine	–	Orange
10.	Sub-tropical Dry Evergreen	–	Burnt umber
11.	Northern Wet Temperate	–	Olive Green
12.	Himalayan Moist Temperate	–	Green
13.	Himalayan Dry Temperate	–	Yellow
14.	Sub-alpine	–	Burnt sienna
15 & 16	Alpine	–	Sepia

(vii) Soil Map: Soil map is considered to be one of the significant which classifies the actual soil type within the working circle. The type of soil, soil series, presence of rocks and minerals etc., will be depicted in the soil map. This soil map will help the district forest administration to plan for establishment and regulation of growing stock towards sustained yield and sustainable forest management.

31. Introduction of Join Forest Management in various states in India was found positive in biodiversity conservation. Discuss in detail. (2007, 2014)

Joint Forest Management (JFM) is the partnerships in forest movement involving both the state forest departments and local communities. The policies and objectives of Joint Forest Movement are detailed in the Indian comprehensive National Forest Policy of 1988 and the Joint Forest Movement Guidelines of 1990 of the Government of India. Joint Forest Management (JFM) programme in the present form can be traced to the Arabari experiment initiated by foresters in the state of West Bengal. As on March, 2011 there are 1,18,213 JFM committees across 29 States and UT of Andaman & Nicobar Islands which are managing 22.94 million ha of forests in the country and play a vital role in conservation of biodiversity.

Role of JFM in restoration of degraded forests – Arabari experience: The Arabari forest range of Midnapur Forest Division, West Bengal has experienced complete degradation of natural forests particularly the sal forests and the associated natural regeneration. The local people have exploited the area which lead to loss of genetic resources. Hence the participatory approach of forest management has started its inception in Arabari by the then DFO Dr. A.K.Baneerji. He has earmarked 1272 ha area of degraded forests for this approach and successfully established this area with plantations through the participation of local dwellers. Under this participatory approach the villages were provided with employment generation activities coupled with exclusive rights to collect all non wood forest products at free of cost. The villages were allowed to cultivate paddy, fodder, grasses, maize and groundnut. They have in turn protected the forests by establishing forest protection committees. This has shown excellent protection and the forests have been regenerated which formed the basis for implementation of JFM in the country.

JFM and biodiversity conservation:

- In the context of India, Joint Forest Management (JFM) programme is more than a decade old and has been able to bring forth the important understanding regarding incentives for the sustainable management of biodiversity.
- Hence, the case of JFM in India has been considered here as a mechanism to understand non monetary positive incentives for the conservation and sustainable management of biodiversity.
- It laid out the broad guidelines for an institutional arrangement, involving the local people, to jointly protect and manage forest resources.
- The co-management approach by recognizing the dependence of the local communities on the forests and involving them for the sustainable management of the natural resources has brought out that the non monetary incentive mechanisms such as policy level changes, local empowerment, awareness programmes, etc. plays a crucial role in successful implementation of sustainable forest management practice all over the country.
- Some of the non monetary positive incentives for the conservation and sustainable use of biodiversity being addressed under the JFM programmes like,
- **Institutionalisation of peoples' involvement -** Ministry of Environment and Forests (MoEF) issued detailed guidelines on June 1, 1990, which clearly laid down the procedure for people's involvement in forest conservation and management through appropriate village level institution and under a proper scheme
- **Empowerment at local level-** As a result of empowerment, people have started protecting the forests by their own initiative.
- **Awareness of the need to manage environmental resources**
- **Accrual of benefits -** With the participation of JFM committees, the degraded forests have been rehabilitated and biodiversity has been improved. The quality and quantity of forest produce has also been increased in many areas
- **Learning experience for governance** – The JFM programme has been a learning experience for governance in natural resource management. Decentralization of implementation and involvement of local communities have not only reduced the burden of the forest department but also have been able to provide a learning experience throughout the process of implementation for the foresters and the local communities.
- **Adaptive approach** -The adaptive approach has helped communities and forest department to develop better understanding of the requirements for natural resource management.

  All these JFM activities implemented in various states were found profound increase in biodiversity conservation.

32. Why is a balance between production, social and environmental objective necessary in sustainable forest management plan. (2011)

Sustainable forest management is defined as the stewardship and use of forests and forest lands in a way, and at a rate, that maintains their biodiversity, productivity, regeneration capacity, vitality and their potential to fulfill, now and in the future, relevant ecological, economic and social functions, at local, national, and global levels, and that does not cause damage to other ecosystems.

Sustainable forest management is the management of forests according to the principles of sustainable development. Sustainable forest management has to keep the balance between three main pillars: ecological, economic and socio-cultural.

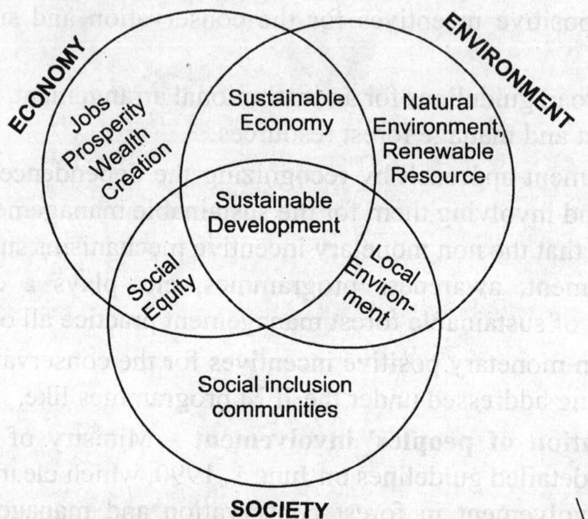

Forest managers must assess and integrate a wide array of sometimes conflicting factors – commercial and non-commercial values, environmental considerations, community needs, even global impact – to produce sound forest plans.

Environmental Sustainability: In a truly sustainable environment, an ecosystem would maintain populations, biodiversity, and overall functionality over an extended period of time. Ideally, decisions that are made should promote equilibrium within our natural systems and seek to encourage positive growth. Unnecessary disturbances to the environment should be avoided whenever possible. If there is a disturbance, it should be mitigated to the maximum practicable extent. When decisions are made, one part of the discussion should always be the environmental impacts of the proposed outcome or result.

There are several items that are directly related to environmental sustainability. One of the concepts that is of the utmost importance is the proper management of our natural resources. Using the Z-squared approach to sustainability, we can minimize our impacts to the environment. In some cases we can even promote habitat restoration and preservation as means to negotiate a successful solution to a problem. Economic Sustainability is a balancing act. Profitability and cost of a decision must be balanced with the environmental and social impacts of its results.

Economic Sustainability: Similar to environmental sustainability, economic sustainability involves creating economic value out of whatever project or decision you are undertaking. Economic sustainability means that decisions are made in the most equitable and fiscally sound way possible while considering the other aspects of sustainability. In most cases, projects and decisions must be made with the long term benefits in mind (rather than just the short term benefits). Keep in mind that when only the economic aspects of something are considered, it may not necessarily promote true sustainability.

For many people in the business world, economic sustainability or growth their main focal point. On the large scale, this narrow-minded approach to management of a business can ultimately lead to unsatisfactory results. However, when good business practices are combined with the social and environmental aspects of sustainability, you can still have a positive result that is for the greater good of humanity.

There are several key ideas that make up economic sustainability. For example, governments should look to promoting "smart growth" through no-nonsense land use planning and subsidies or tax breaks for green development. Strong financial support for universities, education programs, and research & development is an important part of economic sustainability as well. In addition to this, an emphasis should also be placed on other areas such as reducing unnecessary spending and cutting red tape.

Social Sustainability: Social sustainability is based on the concept that a decision or project promotes the betterment of society. In general, future generations should have the same or greater quality of life benefits as the current generation do. This concept also encompasses many things such as human rights, environmental law, and public involvement & participation. Failing to put emphasis on the social part of decision or action can result in the slow collapse of the spheres of sustainability.

Balance between production, social and environmental objectives: Forests provide a wide range of benefits at local and national levels. Log production is usually the main objective and revenue earned for governments and companies is the major driving force in tropical forest harvesting. Revenue earned from log harvesting will usually be the main funding source for long-term sustainable tropical forest management.

Many communities depend heavily upon non-wood forest products for subsistence and as a basis for local trade, for example, canes, medicinal and food plants, gums, resins and wildlife. Tropical forests are an essential source of energy for many communities, directly through burning of wood for cooking and heating and indirectly to protect watersheds as source of water for hydroelectricity generation. They have an important role in protecting physical and biological environments at local and provincial levels.

Tropical forests are dwelling places for many millions of people and are increasingly of value for recreation and tourism, notably "eco-tourism". They are important havens for wildlife and are the habitats of many endangered species of plants and animals. It is essential in tropical forest management planning to achieve a balance between the long-term between wood production, social and environmental management objectives.

The overlapping-circles model of sustainability indicated in the figure acknowledges the intersection of economic, environmental, and social factors.

33. Di-erentiate Working Plan and Annual Plan of Operations. Suggest the changes needed in the working plan preparation for more e-ective application in forest management. (2016)

Working Plan: Working plan is an "A written scheme of management aiming at continuity of policy and action and controlling the treatment of a forest".

Components of a good working plan

- Be simple and brief, exact and definite, for the subordinate staff to follow the prescriptions clearly.
- Based on the thorough knowledge of the tract to be dealt with.
- Aim at achieving realistic and not idealistic objectives; workable in practice.
- Avoid limiting the discretion of the executive staff to an irrational degree, particularly in silvicultural matters.
- Recognize practical difficulties in respect of marketing and transport problems, limitations of the technical abilities of the staff; co-ordination of administrative resources in men, money and equipment.
- Provide a work of reference, containing all relevant information and a foundation on which all subsequent management plans can be built up.

Objectives and scope of working plan

- Survey and assessment of the past results, present facts, resources and constraints.
- Analysis of facts with respect to the forest policy and objects of managements.
- A plan for future action in a definite period in conformity with the long-term plan, i.e., a short-term plan.

Provision of control and maintenance of records and collection of new facts by research and new techniques.

Annual plan of operations: An annual plan of operation is "a tabular statement showing the order and extent of all works to be carried out during the year. The plan is based on the prescriptions of the working plan where it exists, otherwise on general silvicultural principles as modified by financial provisions

"Cultural and other forestry operations are correlated with funds – annual budget estimates" for revenue and expenditure. The district forest officer sanctions the annual plan of operations of the Ranges, keeping in view the distribution of work to range staff, supplies to various consumer industries, etc.

Changes needed in the working plan preparation: The existing working plan is very elaborate and need changes in terms of workable and practically applicable model. Hence the following suggestions may be thought of for establishing sound working plan.

(i) The working plans are normally written by the forest officers following the traditional working plan written several years before. Currently there are advancements in science and technology and hence the working plan should involve the scientists from the disciplines of forestry, soil science, botany, ecology, remote sensing etc. Hence a scientific team may be planned for preparation of working plan.

(ii) The divisions have already the past history of vegetation present within that. This past history has to be seriously reviewed and the areas may be segregated into several working divisions taking into consideration the demands of the society and the conservation needs of the forests.

(iii) The working plan should always consider the productive division, protective division, wildlife, recreation and tourism and other protected area networks and accordingly the divisional working plan should take care of all the working divisions equally.

(iv) The use of GIS and remote sensing technology for digitizing the work plan division may be thought of.

(v) The working plan should also include the sustainability plan in all aspects of objectives envisaged.

(vi) The directions / guidelines of National working plan code there be strictly followed while preparing the working plan.

34. Describe the characteristics and structure of even-aged stands. (2014)

An even-aged stand of trees is one in which there are only small differences in age among the individual trees. even-aged management - A forest management method in which all trees in an area are harvested at one time, called clear cutting or in several cuttings over a short time to produce stands that are all the same age or nearly so. This management method is commonly applied to shade-intolerant conifers and hardwoods.

Characteristics of Even Aged Stand

- Although even-aged stands often occur naturally, they rarely exist in a pure form. Natural even-aged forests occur after a major disturbance initiates the processes involved in stand regeneration.
- Often, these stands maintain their even-aged structure until the next major disturbance occurs. In managed even-aged stands, the regeneration period is not greater than 20% of the length of a standard managed rotation.
- This time period will be reflected in the age difference between the oldest and youngest trees.
- Even-aged stands generally have one age class, although two age classes can be found in some two-layered natural or managed stands.
- For managed even-aged stands, regeneration cuttings are concentrated near the end of the rotation because even-aged management is easier to plan, conduct, and regulate.
- Large-scale natural disturbances, such as wildfire, have naturally favoured even-aged stands of seral species with a niche closely associated with these disturbance types.

Structure of Even-Aged Stand

- Pure even-aged stands generally have a nearly bell-shaped diameter distribution.
- This means that most trees are in the average diameter class.
- However, diameter distributions should be viewed cautiously because diameter can be a poor criterion for age.
- These stands generally have a well-developed canopy with a regular top at a uniform height.

- The smallest trees in natural even-aged stands are generally spindly, with vigour suppressed by the over-storey.

35. How do variations in density and quality of a forest influence annual yield estimation. (2010)

Variation in Density: Density of a stand which is not normally / fully stocked is expressed by decimal. E.g.: 0.6. It is usually assessed by comparison of actual basal area with that of yield Table basal area, per hectare. The area occupied by the crop is multiplied by its density of stocking to get the equivalent area in terms of crop with normal density (1.0).

Eg: If the crop has the density of 0.7 at present may reach 0.9 during the process of growth and development. The stocking factor is not a permanent factor and it varies from species to species, age to age and rotation to rotation. Under such circumstances, there is no need for application of reduction factor for density because in the first rotation annual yields may be unequal and with prospects of normal equalized yield in the future yields. Hence density play a vital role in annual yield estimation.

Variation for quality: Variation in growing stock due to yields capacities of different site qualities is, more or less, a permanent factor of the locality and has to be taken into account. For all important species, Yield tables are usually available, separate for different site qualities.

In order to determine the areas under different qualities which will produce the same yield, a reduction or modification of the area will have to be made on the basis of their production capacity, i.e., M.A.I. Reduction may be made in either of the three ways

(i) Reduction to the most prevalent quality
(ii) Reduction to the mean quality of the whole area
(iii) Reduction to any suitable quality – whether present in the area or not.

36. a. What are the main contribution of Dr. D. Brandis in the Indian Forestry? Enumerate the various stages of working plan. What is the role of silviculture system in the working plan. (2013)

b. Role of Silvicultural system in the working plans. (2013)

a. (i) Contributions of Dr. Dietrich Brandis in Indian Forestry: Sir Dietrich Brandis was born in Bonn, Germany and worked as a lecturer of Botany. He joined the British service in 1856 as the superintendent of the teak forests in eastern Burma. He has been deputed to India by the then British rulers and he has made significant contribution towards scientific management of Indian Forests besides establishing institutions for forestry professionals and instrumental in designing a law for forest protection. The major contributions of Sir Dietrich Brandis to Indian forest are furnished below

Contribution to forestry management in India

- In 1864, the British felt the need of an organized Forest Department for the complete administration of the forests in India. They invited Dietrich Brandis for advice and to organize the administration of the forests in India.

- He was the First Inspector General of Forests of India and is also commonly known as the **Father of Tropical Forestry**.
- Brandis formulated new forest rules and helped establish research and training institutions.
- He set up the Indian Forest Service in 1864 and helped formulate the Indian Forest Act of 1865.
- To help spread the study of forestry, Brandis founded the Imperial Forest Research Institute at Dehra Dun in 1906. He introduced and started the practice of 'scientific forestry' or commercial forestry in India.
- Brandis documented the sacred groves in Rajputana and Kans (woodlands) of Mysore, the Garo and Khasia hills which he visited in 1879, the Devarakadus of Coorg in 1868, and the hill ranges of the Salem district in the Madras Presidency in 1882, the Swami Shola on the Yelagiris, the sacred grove at Pudur on the Javadis and several sacred forests on the Shevaroys.
- He was among the earliest in India to formally link forest protection with local peoples.
- He also took an interest in the forest flora of northwest and central India and Indian trees. Even after retirement Brandis continued to work on Indian forestry and at the age of 75 he started his principal botanical work, Indian Trees, dealing with 4400 species. It was first published in 1906 and re-issued several times afterwards, the last time in 1971.
- He was posted at Balaghat in M.P. as a principal of forester training institute for a long time in his service period.

a. (ii) Preparation of working plans

Preliminary working plan report: The principal object of the W.P. Preliminary Report is to examine the result of the past working, to decide how far the prescriptions of the previous W.P. should be followed, and where and in what direction modifications are necessary. It is, therefore, an important document for the preparation and revision of a W.P.

First preliminary report: The Conservator of Forest (Working plan) asks the D.F.O. (Territorial), through the C.F.(T), at least six months in advance, for a preliminary reconnaissance report on the operation of the current W.P. This report briefly reviews for each working circle

- The results of management
- The success or failure of past prescriptions
- Suggestions for any changes that may be necessary

This is followed by a discussion between the C.F. (W.P.) and the C.F. (T) and, if necessary, the C.C.F. is consulted to arrive at the greatest measure of agreement. In case of a new W.P., the first P.R. will contain a short description of the forest concerned, facts relating to their management, working and reproduction, the future treatment recommended along with reasons therefore, and also a sketch map showing, roughly, the proposed Working Circles. In the following season, the C.F. (W.P.), inspects the forests and prepares the first P.R. This is submitted to the C.F. for comments, discussed and got approved by the C.C.F.

The preliminary W.P. report forms the basis of the W.P.O.'s immediate work. In some States, such reports are prepared by the W.P.O. himself, after discussion with the territorial D.F.O. and the C.F., and submitted by him through the C.F. (W.P.) to C.C.F. for approval. In respect of the changes in management, generally only tentative proposals are made pending collection of further information by the W.P.O. during the course of the first year's field work. This report also indicates the information to be collected, the maps to be prepared, the type and intensity of enumerations to be carried out, and so on. The skeleton of the first PWR is gives here under:-

PART I: Summary of facts on which proposals are based

Chapter 1: Deals with area under working plan

(i) Name of Situation
(ii) Geographical Location
(iii) Description of Land
(iv) Boundaries
(v) Distribution of Areas
(vi) Legal Position

Chapter 2: Deals with the locality factors of the area under working plan

(i) Climatic factors
(ii) Geology, Rock and Soil
(iii) Drainage
(iv) Rights, Concessions and Privileges
(v) Encroachments
(vi) Injuries to which crop is liable
- Natural Calamity
- Grazing
- Insects and Pests

Chapter 3: Deals with the forest resources

(i) Composition and Condition of the crop
(ii) Status of Natural regeneration
(iii) Resource Assessment

Chapter 4: Deals with utilization of forest produce

(i) Agricultural Customs and Wants of the people
(ii) Main markets and marketable products
(iii) Lines of Export
(iv) Methods of exploitation and their costs
(v) Past and Current Prices

Chapter 5: Deals with establishment & labor

(i) Staff
(ii) Labor

Chapter 6: Deals with result of previous management

(i) General History
(ii) Past Systems of Management
(iii) Past Yield
(iv) Past Revenue and Expenditure

Chapter 7: Deals with growth & yield statistics

(i) Statistics of Growth

(ii) Volume and Yield

PART II: Future management

Chapter 1: Discussion and Formation of Working Circle

(i) General Objects of Management

(ii) Methods of treatment to be adopted

Chapter 2: Working Circles

(i) Formation of Working Circles

(ii) Period of working Plan

(iii) Abstract of Working Circles

Chapter 3: Working Plan for Development/Block Forest Working Circle

Chapter 4: Working Plan for Road Working Circle

Chapter 5: Working Plan for Canal Working Circle

Chapter 6: Working plan for Rail Working Circle

Chapter 7: Working Plan for Bands & Drains Working Circle

Chapter 8: Wild Life Management

Chapter 9: Joint Forest Management

Chapter 10: Miscellaneous Regulations.

Chapter 11: Maintenance of Records for Control

Chapter 12: Financial Estimates and Expenditure

Second preliminary report

- The second preliminary report is prepared by the W.P.O., after the first season's field work. In this report, Part II of the W.P. is written up as complete as possible, Chapter by Chapter, using the standard paragraph headings in their proper order and making it fully self-contained.
- The main outline is that of the first P.R., with such additions or alterations as considered necessary. This report is then submitted to the Conservator of Forest (W.P.), who forwards it to the Conservator of Forest (T) and the chief conservator of forest.
- After considering the comments of the C.F. (T) and discussing it with him, the C.C.F. indicates such alterations as he considers necessary, and finally sanctions it.
- The Director Forestry Research, F.R.I, and Colleges, Dehradun has suggested the constitution of a Working Plan Committee, comprising the C.C.F., C.F. (W.P.), C.F. (Research) and the C.F.(T), W.P.O., D.F.O. (T) and the Forest Utilization Officer to discuss the second P R. in detail, before finalizing and adopting the report.
- Once the report is sanctioned by the C.C.F., no major alterations are ordinarily made.

b. Role of silviculture system in the working plan: The silviculture system is the method of silviculture procedure worked out in accordance with accepted sets of silviculture principles by which the crop constituting forest area, tended harvested and replaced by new crop of distinctive origin. It is the planned silvicultural treatment specific to a tree crop throughout its life which leads desired growth and development.

Similarly, working plan is the written scheme of management aiming at continuity of policy and action and controlling the treatment of forest.

Hence, silviculture system and the working plan have integrated role because silviculture system is a program involve procedure of treatment and working plan is a legal document to implement the set of silvicultural treatments.

The working plans are prepared for each and every division with various objectives of producing timber, producing Non-Timber forest products and protecting the forest resources. For all production oriented purpose the working plans are depending pure on silviculture system.

India is the robust country and the forest in the country has been classified into five major and 16 sub-types. Each and every forest type in the country needs to be managed as the working plan prescription. Working plan prescriptions suitably incorporate set of silvicultural treatments as envisaged silviculture system specific to certain forest type without which the forest department cannot manage the natural forest and manmade plantations.

Hence, silviculture system and working plans are inseparable in order to manage the Indian forest to achieve the objectives envisaged.

As a general rule, the working plans incorporate the operational guidelines and management prescriptions for working circle and these operational guidelines and management prescriptions are obtained only from silviculture system. The working plan has two components. One is past history of the plantations and the second one is the new treatment proposed to be imposed. These new treatments proposed to be implemented in the working circle are based on specific silviculture system amenable to either to the forest type or to the species concentrate.

37. Justify the statement. Control of stand density of desired species helps in production of maximum volume. (2009)

Yes. It is correct that the control density of desired species helps in production of maximum volume because stand density is a measure of the relative completeness of the tree stocking expressed as a decimal coefficient, taking normal number of trees, basal area or volume.

Stand density is a quantitative measurement of a stand in terms of basal area, number of trees and volume per acre and hence control of stand density helps to produce maximum volume. In India, the stand density is equated with crop density and is normally pronounced as over stocked, full stocked and under stocked in order to describe the stand density if it exceeds, equals or less than one.

The prediction of growth and yield of any stand depends on its density. The volume production in general depends on the stand density. The production of volume and yield is based on the stand density. For any given stand the current volume is estimated based on the current stand density and the future yield is predicted based on the same density. Hence control of stand density will decide the volume production. The normal stand may become over stocked or under stocked due to the existing natural resources which may result in over volume production or under volume production. Hence controlling and managing stand density is very essential in order to achieve the desired growth in terms of height and diameter coupled with clean bole without any pest and diseases will ensure maximum volume production.

38. How annual yield is calculated in a forest management under Uniform System of Management. (2005)

In the uniform system, the crop is regular and is divided into different periodic blocks which extend over a period of time. In this periodic block system, the old and matured crop is removed and the young crop is regenerated and hence this system always includes both old and young trees.

For removing part of the growing stock, Fischer has conceived the following formula for estimation of yield under normal growing stock in the uniform silvicultural system.

$$N.G.S = (V + V_1) \times \frac{P}{2} \times D$$

Where, V - is the initial GS;

V_1 - is the G.S. at the end of regeneration period;

P - is the regeneration period and

D - is the crown density

39. How is allotment of areas made in permanent periodic blocks? (2012)

In permanent periodic block method, equal areas of felling are allotted to individual parts of the rotation. These are termed as periods. In this method, the following is the procedures for allotment of areas.

(i) The rotation is divided into number of periods.
For example, if the crop is 80 years rotation, it can be divided into four periods of 0-20, 21-40, 41-60 and 61-80.

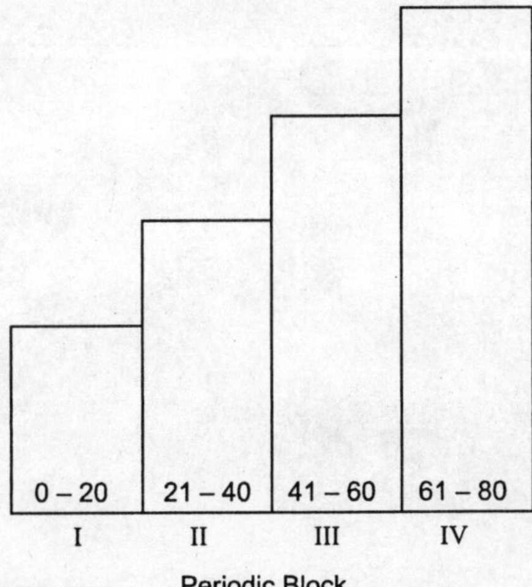

Periodic Block

(ii) Each period is allotted to one or more compartment on the basis of age of the crop. The area of each such block remains the same.

(iii) The area of each periodic block will be

$$\text{Area} = \frac{A}{N}$$

Where A is the area of felling series and
N is number of periods.

(iv) Suppose if there is any difference in the areas of different periods, compartments are shifted from one period to other till it becomes equal.

(v) Another consideration in compartment shifting is that areas which have less vigorous growth, inspect of their age, may be felled first, so that they are replaced by better crop. The old crops which are vigorous may be felled during the latter period.

(vi) Now due to advancements in forestry technologies, the concept of shifting equal areas is changing into equi-productive areas.

(vii) In case if the allotment is made on the basis of volume, the rotation is divided into number of periods and the forests for which yield is to be prescribed are divided into these periods using volume as an equalizing factor.

(viii) In case if the permanent allotment is made on the basis of area and volume, the allotments are made based on age classes and the equalization is based on either volume or area.

Chapter

10

Forest Mensuration and Remote Sensing

1. Write on the importance of stem analysis in forest mensuration. How does the stem analysis help in knowing the volume of the stand? (2009)

Stem analysis is the analysis of a complete stem by measuring annual rings on a number of cross-sections at different heights in order to determine its past rate of growth.

Importance of stem analysis

(i) Stem analysis helps to determine age-diameter, age-height, age-volume relations throughout the life of the tree-analysed and thus assess the average rate of diameter, height and volume increment.

(ii) The effect of external factors e.g., lopping, defoliation etc. on the rates of growth of trees, can also be determined.

(iii) The data of the stem analysis can also be used for preparing local volume tables by correlating diameter volume relations.

Determination of Volume

Stem analysis helps in assessment of volume through the following approach. The trees selected for stem analysis are felled and the following measurements are taken.

(i) Height of the tree

(ii) Crown width

(iii) DBH

(iv) Total Height

(v) Diameter over bark (20 cm)

The trees are cut into sections of the following category

(i) Lowest section with the length of 2.74 m which ensure the midpoint at DBH

(ii) Other sections are cut into lengths of 3m so that the mid points will be 4.27, 7.24 and 10.24, etc.

(iii) The final section vary depending on the condition.

(iv) In all sections the mid diameter is recorded.

(v) These felled sections are cross cut at mid points using power chain saw for counting of annual rings.

Counting of annual rings

(i) The average radius for each section is determined based on two diameters. For this purpose the two radii are marked and analyzed from pith outwards.

(ii) The first ring is counted in the basal section of the tree at DBH (ie. 1.37 m)

(iii) Counting of rings is done by decades from pith to circumference by separately inserting a pin for each decade.

(iv) The rings in remaining sections are counted and correlated with rings at DBH by comparing the incomplete rings in successive sections which will give diameter of wood Under Bark (U.B).

(v) Based on the ring counting, the age is determined and the difference between number of rings at breast height and any other section gives age to that section.

Computation of volume

The data recorded are used for calculating the volume of the trees and to the stand.

For calculating volume, the diameter and the basal area is very important and hence for each section diameter can be determined from the D.U.B / height curve.

This diameter is used for calculating basal area.

The basal area is then multiplied with the length of the log and volume is estimated. The same methodology is followed for all the sections except the top most one. The addition of all sections volume gives the volume of the trees.

Advantages

As compared to other methods for collection of data for increment of trees, stem analysis offers the following advantages:

(i) The data are collected from standing trees carefully selected and so it is more reliable. This is more so because the entire tree is analysed and data are not collected only from the lower part of the stem as in stump analysis or in increment boring.

(ii) It gives complete information about the growth of trees in respect of diameter, height and volume and so it is self-contained.

Recent Advances in Stem analysis

The conventional method of stem analysis is both time consuming and laborious, Pluth has developed a computer programme for complete stem analysis. This has, however, yet to be tested on Indian trees.

2. Write the procedure to measure the height of tree in the following situations where

(i) Observer's eye is above the top of the tree,

(ii) Observer's eye is below the base of the tree. (2009)

(i) The tree is above the observer and the lean away from him: Let AB be a tree leaning away from the observer. Let α and β be the angles to the top and base of the tree from the observer's eye E. Let BC be a vertical line drawn from the foot of the tree B meeting AE at C and the horizontal line from eye at D. Let ∠CBA, which denotes the lean of the tree, beθ.

In the triangle ECD, ∠ECD = 90° — K because∠EDC is 90° and ∠CED is K.

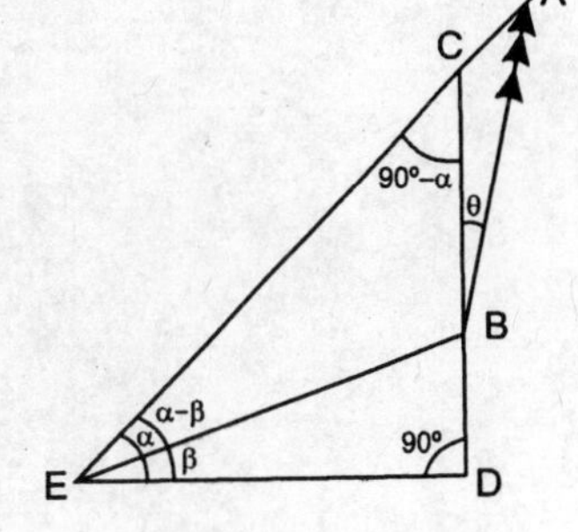

In triangle ACB,

∠CAB = exterior ∠ECB – interior ∠CBA

$$= 90° - K - \theta = 90° - (K + \theta)$$

In the triangle AEB,

$$\frac{AB}{\sin \angle AEB} = \frac{EB}{\sin \angle EAB}$$

$$\text{Therefore} \quad AB = \frac{EB \sin \angle AEB}{\sin \angle EAB}$$

$$= \frac{EB \sin(\alpha - \beta)}{\sin\left[90° - (\alpha + \theta)\right]} = \frac{EB \sin(\alpha - \beta)}{\cos\left[(\alpha + \theta)\right]}$$

(ii) The tree is below the observer and the lean away from him: Let AB be a tree leaning away from the observer. Let K and !2 be the angles to the top and base of the tree from the observer's eye E. Let BC be a vertical line drawn from the foot of the tree B meeting AE at C and the horizontal line drawn from observer's eye at D. Let ∠CBA which denotes the lean of the tree, be θ.

In triangle ECD,

∠ECD = 90° – K =∠ACB being opposite angle.

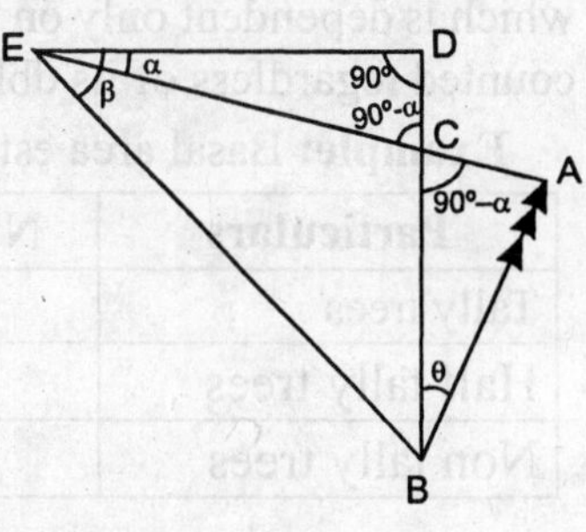

In the triangle ACB,

$$CAB = 180° - [\angle ACB + \angle ABC]$$

$$= 180° - [90° - K + \theta]$$

$$= 90° + (K - \theta)$$

$$\frac{AB}{\sin \angle AEB} = \frac{EB}{\sin \angle EAB}$$

$$\text{Therefore} \quad AB = \frac{EB \sin \angle AEB}{\sin \angle EAB}$$

$$= \frac{EB \sin(\beta - \alpha)}{\sin\left[90° + (\alpha - \theta)\right]} = \frac{EB \sin(\beta - \alpha)}{\cos[(\alpha - \theta)]}$$

3. Discuss the concept use of Horizontal Point Sampling when wedge prism is used. (2009)

Concept of Horizontal Point Sampling: In horizontal point sampling, a series of sampling points are selected randomly or systematically distributed over the entire area to be inventoried. Trees around this point are viewed through any angle-gauge at breast-height and all trees forming an angle bigger than the critical angle of instrument are counted.

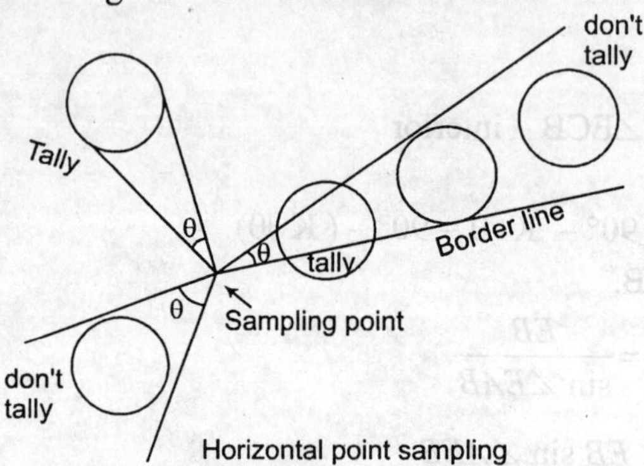

Fig. Point sampling for diameter or basal area

The circles represent the cross sections of trees at bh and the lines indicate the critical angle of the instrument projected from the sampling point.

It may be noted in the above figure that even though all the trees are of the same basal area, some are counted in the tally while others are not because being far away from the sampling point they did not form an angle bigger than the critical angle of the instrument. On the other hand, at the same distance from the sampling point a bigger tree is counted while a smaller tree is not. Thus the inclusion of trees in tally for a given angle depends upon the sizes of trees and their distances from the observer or sampling point. The number of trees counted when multiplied by a constant factor, which is dependent only on the size of the angle, gives the basal area per ha. It means that each tree counted regardless of its dbh represents the same basal areas per ha.

Example: Basal area estimation in teak

Particulars	Number of trees	Basal Area Factor	Basal Area
Tally trees	10	0.5	5.0
Half tally trees	2	0.5	1.0
Non tally trees	5	0.5	2.5

4. Discuss the features of electromagnetic spectrum mention bandwidth spectral range in remote sensing (2009)

All recording devices require energy to produce images. The most common and important source of energy used to produce image is the sun. The solar energy is received in the form of electromagnetic waves travelling at the speed of light (i.e., 299000 km per second).The wave motion

is described by wavelength, frequency and velocity. The wavelength (w) of the electromagnetic waves is the distance from one wave crest or peak to as shown in figure indicated below.

The frequency (f) is the number of peaks passing through a fixed point in space per unit of time. The velocity (c) is constant at the speed of light, the three are related as follows:

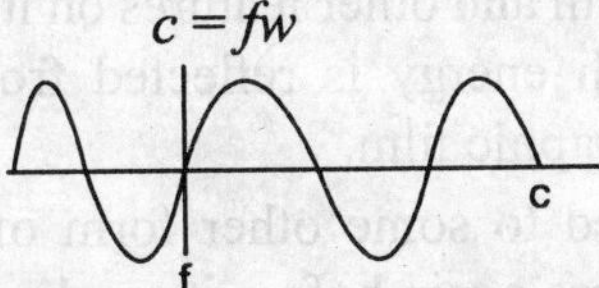

The usual unit of measurement of wave length is micrometer (!:m) i.e., one thousandth part of a millimeter. Thus it is numerically equal to 10^6 m.

The electromagnetic radiation occurs as a continuum of wave length and frequencies from short wave length and high frequency cosmic waves to long wave lengths and low frequency radio waves and is known as electromagnetic spectrum as shown in the following fig.

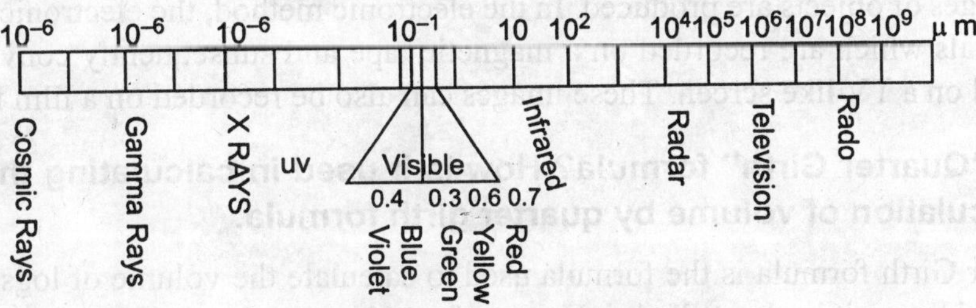

Fig. Electromagnetic Spectrum

The following nomenclature refers to the wave length spectrum.

S.No.	**Name of the region of electromagnetic Spectrum**	**Wave Length arrange**
1.	Cosmic rays	$<10^7$!:m
2.	Gamma rays	10^7 !:m to 10^4 !:m
3.	X rays	10^4 !:m to 10^2 !:m
4.	Ultraviolet rays	10^2 !:m to 0.4 !:m
5.	Visible part of the spectrum	0.4 !:m to 0.76 !:m
6.	Infra-red part of the spectrum	0.76 !:m to 10^2 !:m
7.	Micro wave	10^2 !:m to 10^6 !:m
8.	Radio wave	$>10^6$!:m

The energy radiating from sun has to pass through atmosphere before it reaches the earth and other features on it. While passing, part of the energy is either absorbed or scattered due do dust particles, water vapour and gases present in the atmosphere. Scattering takes place when energy interacts with tiny dust particles resulting in scattered blue wavelength seen in the form of blue sky. The presence of gas molecules results in absorption of electromagnetic waves in certain bands. The bands of the spectrum in which there in little or no absorption are known as atmospheric windows.

In the bands in which there are no atmospheric windows, the energy is filtered out. For instance, the harmful cosmic, gamma and x-rays are filtered out by the ozone layer of the atmosphere and so they never reach the earth. The atmospheric windows exist in wavelengths 0.4 to1.5!:m, 3 to 5.5!:m, 8 to 14 !:m and also in wavelengths larger than10^{8}!:m. After passing through the atmospheric windows, the energy reaches the earth and other features on it. Here it may be transmitted, reflected or absorbed. The manner in which energy is reflected from an object helps in detection and appearance of the object on photographic film.

The absorbed rays get converted to some other form of energy such as heat. The reflected energy passes through the atmosphere again before it reaches the sensor. As a result of its passage through the atmosphere twice and its interaction with it as well as die earth and other objects on it. Electromagnetic energy undergoes a number of changes in wavelength magnitude, polarization and direction, *etc.* The science of remote sensing deals with the technique of detecting these changes photographically or electronically.

In the photographic method, a film is used to detect and record energy variation and when film is developed images of objects are produced. In the electronic method, the electronic sensor produces electronic signals which are recorded on a magnetic tape and subsequently converted into image form displayed on a TV-like screen. These images can also be recorded on a film for record.

5. What is "Quarter Girth" formula? How is it used in calculating the volume of a log? Calculation of volume by quarter girth formula. (2009, 2011)

The Quarter Girth formula is the formula used to calculate the volume of logs. It is defined by Mr. Hoppus and hence it is also called as Hoppus formula.

In India and Britain, volumes of logs are often calculated by the quarter girth formula:

$$\text{Volume of log} = \left(\frac{g}{4}\right)^2 \times l$$

Where g is the girth of the log at the middle and

l is the length of the log.

Derivation of the formula

$$\text{Volume} = \pi r^2 l$$

But $\quad g = 2\pi r \quad$ and $\quad$ so $r = \dfrac{g}{2\pi}$

Volume $= \pi\left(\dfrac{g}{4}\right)^2 \times l$, Substituting the value of r in terms of g

$$= \pi \frac{g^2}{4^2} l$$

$$= \frac{g^2}{4} \times l$$

π is actually equal to 3.1416 but it has been roughly taken to be equal to 4.

$$\text{Therefore volume} = \frac{g^2}{4\,x 4} \times l$$

$$= \left(\frac{g}{4}\right)^2 \times l$$

The formula has made approximation for the value p which is originally 3.14 and moderated to the whole number 4. Hence the volume derived from quarter girth formula is lower than the true value as indicated below.

$$\text{Full circular volume } V_1 = \frac{g^2}{4} \times l$$

$$\text{Quarter girth volume } V = \left(\frac{g}{4}\right)^2 \times l \quad \text{or} \quad \frac{g^2}{16} \times l$$

$$\text{Thus, } \frac{V}{V_1} = \frac{4\pi}{16} = 0.785 \text{ or } 78.5\%$$

The quarter girth formula estimates only 78.5% of the tree volume of the log due to the moderation.

6. Write the salient di-erences between local volume table and general volume table. (2009)

S. No	Local volume tables	General volume tables
(i)	Local volume table is derived from general volume table	General volume table can be prepared only from field data and observations
(ii)	Prepared from measurements of trees growing in restricted locality	Prepared based on the average volume of trees growing over a large geographical area.
(iii)	Based on one independent variable viz., d.b.h. (o.b.).	Based on two independent variables viz., d.b.h. and total tree height
(iv)	Applicable to restricted region or locality as coupe or a compartment	Applicable to large area
(v)	Requires measurement of only one variable	Requires measurement of two independent variable
(vi)	Gives volume for a given diameter or height (any one independent variable) in a particular region.	Gives volume of trees by diameter classes and in each diameter class by height classes.
(vii)	Helps in estimating the volume of trees before felling with the objective of regulating yield	Helps in estimation of volume to regulate yield and comparison of increment put forth by various localities.

7. Commentonthecomparativesignificanceofcaliperandtapeford.b.hmeasurement. (2010)

S. No.	Caliper	tape
(i)	Calipers directly give the diameter of tree stems (2*r*)	tapes give only the circumference of the stem (2! *r*)
(ii)	The calipers touch only three points in the stem	The tape negotiates entire circumference of the tree stem
(iii)	Large instrument inconvenient to carry	Handy and carried easily
(iv)	Require two measurements in case of eccentric stems	Single measurement is enough
(v)	Digital calipers are now available for measurements related to nursery evaluation.	Tapes are not applicable for nursery studies
(vi)	The errors are both positive and negative and therefore the chances are that they may neutralize it give more accurate results.	Only positive errors occurs in tape

8. What are the pictorial elements used for interpretation of aerial photographs. (2010)

The first thing to do in photo interpretation is the identification of objects. For forest mensuration, trees are the most important objects and therefore identification of trees species is an essential preliminary. This is done on basis of following pictorial elements.

(i) Tone: Tone is defined as each distinguishable shade variation from white to black in crowns of trees caused by differential reflection of light. The standardized tone scale has 10 shades or variations from white (1) to black (10). The objects which reflect considerable sunlight back towards the camera are lighter in tone in the photograph; while those which absorb light to a high degree have darker one. Thus, tone variations are very important pictorial element in identifying species or groups of species as they attract the attention of the interpreter immediately. A classical example of the use of tone is provided by *Dipterocarpus alatus* trees in the forests of Thailand because their small, glossy leaves reflect light so efficiently that the crowns, of the species appear as light conspicuous spots in otherwise dark environment of other crowns.

(ii) Size: The size of the crowns of mature trees as well as their heights often give a good indication for identification of certain species particularly when tone, shape and other pictorial elements may not differentiate them from other species. For example, *Poeciloneuron indicum* in Karnataka forests can be easily identified by their size which makes them conspicuous by their towering height over the rest of the trees of the area.

(iii) Shape of the crown: Though the trees of different species have distinctive shapes when viewed horizontally and so can be identified easily on that basis from the ground, the problem is not so simple when viewed from air particularly in a vertical photograph. Therefore the interpreter should know what the shape of the crown of a particular species looks like when viewed from air at a particular time of the year. Figure shows the difference in shapes of crowns of chir pine and deodar. Oblique photographs are, however, more useful for the purpose as they give the shape of the crown as if viewed from a side. Though preparation of key of shapes of crowns of various tree species, as has been done in some western countries, is an useful aid in identification, size of the crown of the species should be such that its shape can be easily seen on the photograph of the usual scale adopted. In order to be identifiable, the crown size on the photograph should be at least 0.5 mm and this means that a tree should have a crown width of atleast 20m so that it can be identified on a photograph of a scale of 1:40,000.

(iv) Shadow: Shadows of the trees falling on ground help in identification of species as they give an indication of the shape of the crown. But this depends upon the time of photography and the direction of flight. Wind, however, affects the shadows and makes identification difficult. While shadow is advantageous in identifying the species, it has disadvantages too, as long shadows obscure ground details.

(v) Texture: Texture is defined as the frequency of change and arrangement of tones in an image. It describes the coarseness or smoothness of images. It is produced by aggregation of unit features that cannot be seen individually e.g., leaves and their shadows. Smooth and velvety texture is often associated with young trees and coarse texture with older trees. Therefore texture is useful in interpreting stand structure. Texture is also affected by scale. As the scale increases, texture becomes coarser. On the other hand as scale is reduced, it becomes finer.

(vi) Pattern: Pattern relates to spatial arrangement of objects. The regularity and characteristic arrangement of objects gives them a pattern which is helpful in identifying them and is therefore a useful aid in photo interpretation. For example, the characteristic linear arrangement of plants in a photograph, helps in identifying it as a plantation in an otherwise irregularly growing trees of the natural forest.

The characteristics described above pertain to the objects of interpretation which in case of forest management are the trees. In addition to these characteristics,, the following characteristics of the environment are also helpful in identifying species.

(vii) Location: Location relates to the situation of objects with reference to site factors such as climate, topography, soil, etc. As different species are found in different places under the influence of locality factors, location is helpful in identifying species. For example, broad- leaved species are generally confined to valleys and along nalas in hills while the ridges and slopes are covered by coniferous species. Thus, by seeing nalas and valleys in a photograph of hilly region, it can be concluded that the trees growing along them are broad - leaved species. Similarly, altitudinal zonation of species and effect of aspect on their distribution, help in identification of species. In plains, only certain species are found in particular places and so can be identified there. For example, khair and shisham can be identified in riverain sands and in river islands.

(viii) Association: Plant associations also help in identification. Many tree species look exactly similar in aerial photograph and so their identification becomes difficult but if they belong to different plant associations or if they occur on different topographical sites, in different plant associations or topographical range of the tree species, this helps in their identification. For example, broad-leaved species growing in coniferous forests cannot be identified as sal even though tones may be similar. Similarly, species growing with khair in riverain areas will be identified as shisham, even though it may not be very clear.

9. Explain the principle and use of Abney's level. (2010)

Principle: Abney's level instrument is based on the properties of trigonometric principles.

Let ABC be a right-angled triangle. The trigonometrical ratios of the ∠ACB are defined as follow:

$$\frac{AB}{AC} \text{ i.e., } \frac{\text{Perpendicular}}{\text{Hypotenuse}} \text{ is called sine}$$

(abb. sin) of ∠ACB

$$\frac{BC}{AC} \text{ i.e ., } \frac{\text{Base}}{\text{Hypotenuse}} \text{ is called Cosine}$$

(abb. cos) of ∠ACB

$$\frac{AB}{BC} \text{ i.e ., } \frac{\text{Perpendicular}}{\text{Base}} \text{ is called tangent}$$

(abb. tan) of ∠ACB

If AB is assumed to be a tree and C the position of observer, then AB can be found out from tangent ratio as $BC \times \angle ACB$ where BC is the horizontal distance of the observer from the tree and $\angle ACB$ can be measured by any angle measuring instrument. This is known as the tangent method.

Trigonometry also tells us that in any triangle sines of angles are proportional to the opposite sides. Thus in the triangle ABC in fig.

$$\frac{\sin ACB}{AB} = \frac{\sin CAB}{BC}$$

$$= \frac{\sin ACB}{AC}$$

The knowledge of this relationship can also be used in calculation of heights of trees and is known as the sine method.

Thus, trigonometrical ratios can be used for calculation of heights of trees by one of the following methods:

(i) Tangent method (ii) Sine method

Use of Abeney's Level

In order to use the instrument, the observer stands away from the tree at a place from where the top of the tree and the base are visible. Then the top of the tree is sighted through the pin hole of the eye piece after pulling it out. This makes the instrument inclined and the bubbled is not seen in the mirror. Therefore while sighting the top, the screw is rotated to bring the spirit level in a horizontal position. As the spirit level approaches horizontal position, the bubble appears on the edge of the mirror. The spirit level is continued to be moved slowly to the position when the bubble image is bisected by the line of horizontal wire on the mirror and in the other half, the tree top is seen touching the horizontal wire. At this position, the index arm reads the angle of elevation to the top of the tree on the circular arc. Similarly the angle of depression to the base can also be read and the height of the tree determined by any of the formulae mentioned earlier.

Abney's Level

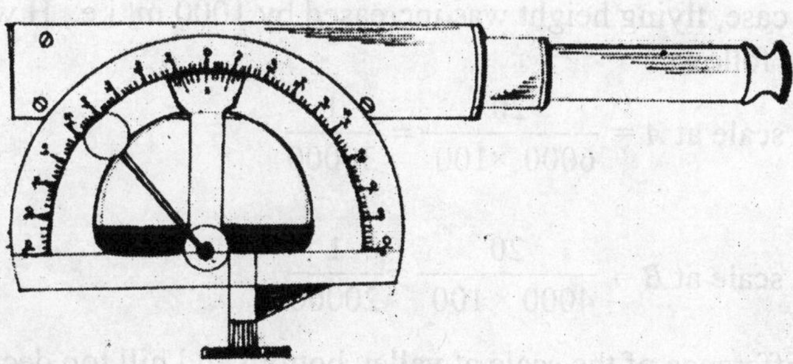

Advantages of the Abney's level

- It gives accurate angles of elevation or depression with the vernier with a least count of 10 minutes.
- Readings can be taken after sighting the tree without disturbing the index arm. The instrument is small and light and can be used even in hills without difficulty.

Disadvantages

- In the absence of a stand, shaking of the hand makes the sighting of the top or bottom of the tree a little difficult and time-consuming but this drawback can be removed by using a forked branch as a stand to give support to the instrument.

- The spirit level has to be adjusted by moving the head of the screw while simultaneously looking to the top or bottom of the tree. This is quite tiresome.

10. How does the fying height influence the scale of aerial photographs in hilly areas? (2010)

The scale of an aerial photograph is the ratio of a distance on a photograph to its corresponding distance on the ground. It is either expressed as a ratio such as 1:20,000 or as a representative fraction such as 1/20,000.This fraction equals the ratio of focal length (f) of the camera to the flying height (H) above the ground level; Therefore, scale or RF of an aerial photo- graph is equal to f/H both being in the same units. As already stated, one of the principal differences between an aerial photograph and map is that in vertical photograph taken over variable terrain, there are indefinite number of different scales because of change in flying height (H) with respect to the ground.

For the same flying height, if the altitude of land decreases, the photographic scale decreases. Conversely, if the altitude of land increases, photoscale is increased as H is reduced due to the land coming closer to the camera as illustrated below

In fig

$$\text{Photo scale at } A = \frac{20}{5000 \times 100} = \frac{1}{25000}$$

$$\text{Photo scale at } B = \frac{20}{3000 \times 100} = \frac{1}{15000}$$

But if in the above case, flying height was increased by 1000 m, i.e., H was 6000 m, the scales at A and B would be as follows:

$$\text{Photo scale at } A = \frac{20}{6000 \times 100} = \frac{1}{30000}$$

$$\text{Photo scale at } B = \frac{20}{4000 \times 100} = \frac{1}{20000}$$

Thus the relative difference of the scale at valley bottom and hill top decreases with increasing flying height. Therefore, in mountainous country any measurement on the photograph requires the calculation of scale of the particular locality on the photograph.

This, in turn, requires additional information about the altitude of the locality. The flying height and focal length can, however, be ascertained from records on the film itself. If neither flying height nor focal length is known, the scale can be worked out by selecting a distance possibly Bear the principal point1 which can be easily recognized both on the ground as well as on the photograph and measuring it both on photo as well as on ground. For example, if the ground distance is 100 m and the distance on the photograph is 2 cm, the RF = 2/100 × 100 = 1/5000.

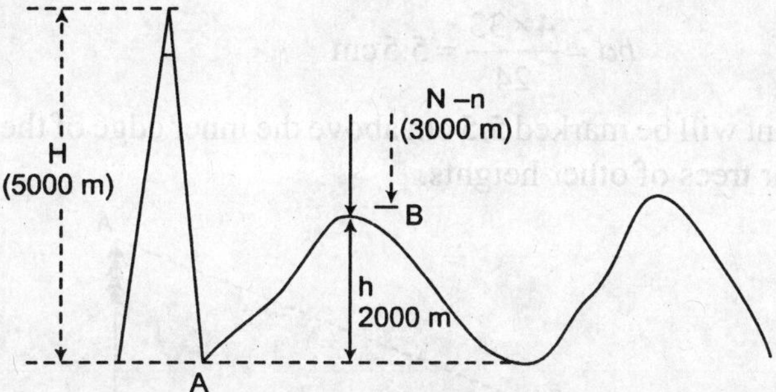

Fig. Flying height of Ariel photograph

The larger the scale, the larger are the images of the photograph and consequently the interpretation task becomes easier. But this increases the variations in the scale which among other things affect the accuracy of measurement of the tree height and crown diameter.

11. Write principle of Christen hypsometer and its use (2010)

Principle of the instrument

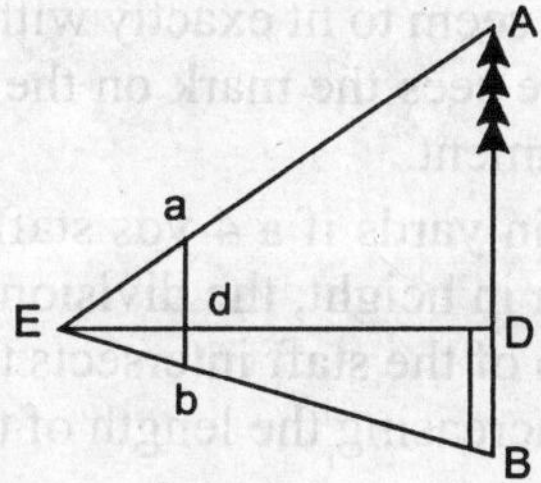

The instrument is based on principles of similar triangles. In Fig., let *AB* be a tree and *ab* a linear object. Let DB be a pole of known length.

In similar triangles *ABE* and *abE*

$$AB : ab : BE : bE$$

But in similar triangles *BDE* and *bdE*

$$BE : bE : BD : bd$$

Therefore $AB : ab : : BD : bd$

or $$AB = \frac{BD \times ab}{bd}$$

if *BD* and *ab* are made constant, *AB* varies inversely as *bd*. It is on this principle that the graduations of Christen's hypsometer are based. For graduations *bd* for different values of AB, the formula $\frac{BD \times ab}{bd}$ is used.

Example: Find the length *bd* for a tree of 24 m height when *BD* (staff) is *4m* and *ab* (distance between flanges of Christen's hypsometer) is 33 m.

$$bd = \frac{4 \times 33}{24} = 5.5\,\text{cm}$$

This means that *d* point will be marked 5.5 cm above the inner edge of the lower flange. Similarly, graduations are made for trees of other heights.

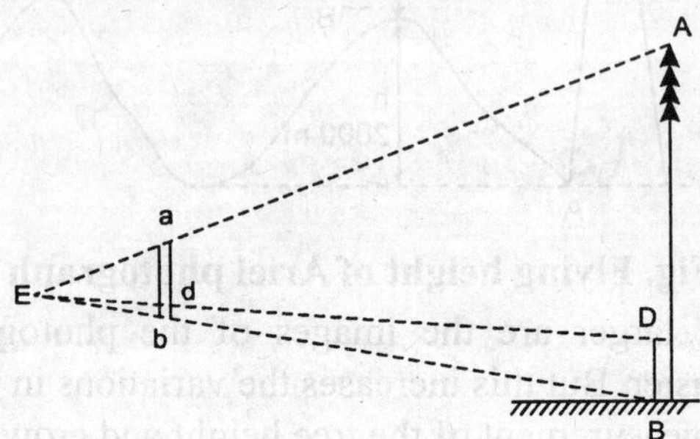

In order to use the instrument, the observer stands at a suitable place from where both the base and top of the tree can be seen between the flanges after placing a staff vertically against the tree. He holds the instrument vertically suspended from a thread and moves the hand backward or forward till the top and base of the tree seem to fit exactly within inner edges of the upper and lower flanges respectively. In this position he sees the mark on the instrument where the line of sight to the top of the staff intersects the instrument.

The mark reads the height of tree in yards if a 4 yds staff is used or in metres if a 4 m staff is used. When the tree is more than 30 m in height, the divisions are rather close and mark where the line of the sight from the eye to the top of the staff intersects the instrument is difficult to read. This difficulty can be overcome either by increasing the length of the staff or the instrument, though this will make them unwieldy.

Description

It consists of a strip of metal, thin wood or even card board about 2.5 cm wide. It has two flanges or protruding edges one at the top and other at the bottom. Each flange has a hole in it, the upper one to suspend the instrument by some thread passing through it and the lower one to suspend a weight from it to prevent it from swinging. It is used with and graduated for a staff of known length. The distance between the inner edges of the flanges varies with instruments used indifferent countries. The one generally used in India has this distance as 33 cm (13 inches) and is graduated to read yards when used with a staff of 3.6 m (12 ft). In some instruments the words '4 units' are written. This means that the staff could be of any linear measure but it should have 4 unit lengths. For example, if a 4 yds or 12 ft staff is used, the instrument will give height in yards. But if a 4 m staff is used it will give the height in meters. The graduations are wider apart above and closer below. Theoretically, all heights beginning with the height of the staff used, to infinity can be measured but the divisions become too close below 30 m (100 feet) mark. Therefore heights exceeding 30 m cannot be read accurately.

Figure. Christen's Hypsometer

Uses of Christen's Hypsometer

(i) It is light, easily made and easily transported.

(ii) The heights of trees can be read directly.

(iii) As the distance between the observer and the tree is not required, the instrument can be used from any point from where the top and base of the tree can be seen.

(iv) It is quicker to use and so it is useful in conditions where speed is required.

12. Describe reference point of diameter / girth measurement on a standing tree. (2011)

Standard Rules for diameter at breast height measurement

The following are the standard rules governing breast height measurements in a sample plot:

(i) The breast height should be marked at 1.37 m (4 ft 6 in) from the base and above the ground level and it can be marked by using a stick and by using the paint the position of breast height should be marked with a cross mark.

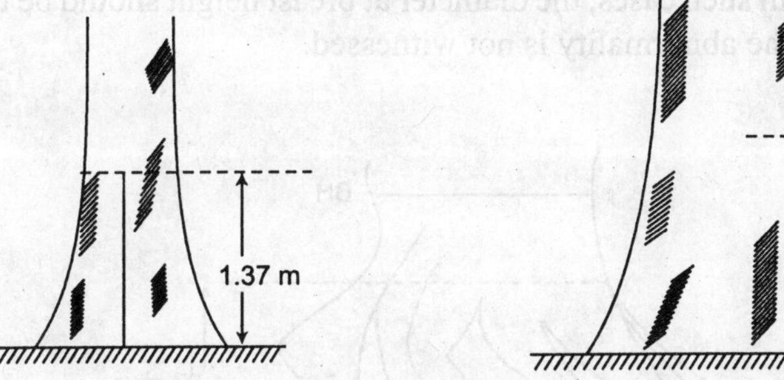

(ii) In case of trees growing on slopping ground the diameter at breast height should be measured on the uphill side only.

(iii) Suppose when the tree is leaning, the dbh should be measured along the tree stem and particularly on the leaning side, when it grows on flat ground. Suppose the leaning occurs in sloping area the dbh should be measured on the uphill side.

(iv) Suppose when the tree trunk is abnormal at 1.37 m, dbh should be measured either above or below the point of abnormality based on judgment by the observer and the measurements can be made.

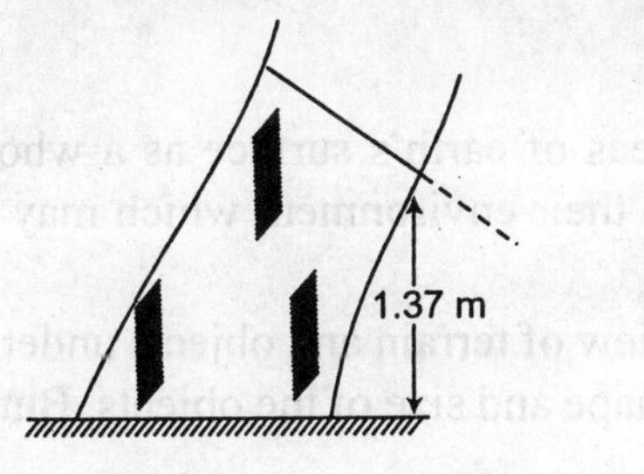

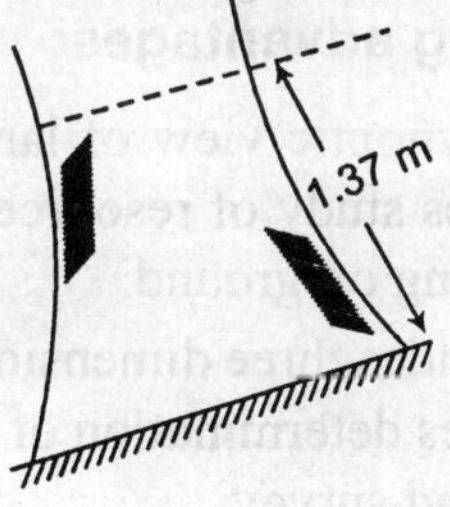

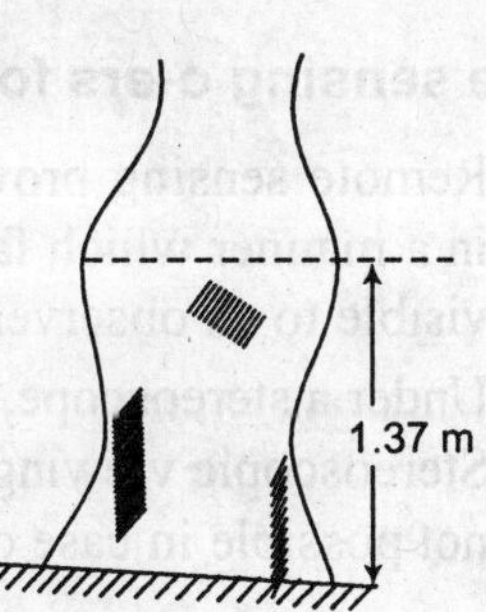

(v) During the growth and development of trees forking or multiple branching occurs at several stages. Suppose when the forking occurs at 1.37 meters, based on the observers judgment the tree can be measured either singly or considered as double trees depending on the place of measurement.

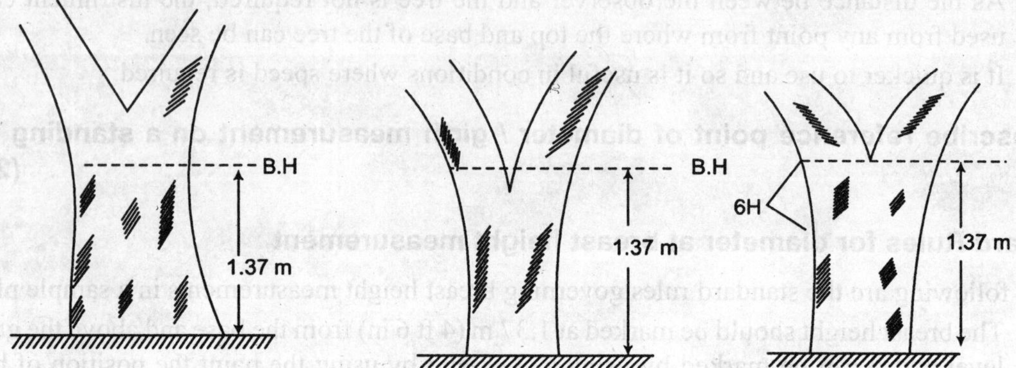

(vi) Many trees like Banyan and Terminalias exhibit buttress formation from the base and upto breast height level. In such cases, the diameter at breast height should be taken at the lowest point above which the abnormality is not witnessed.

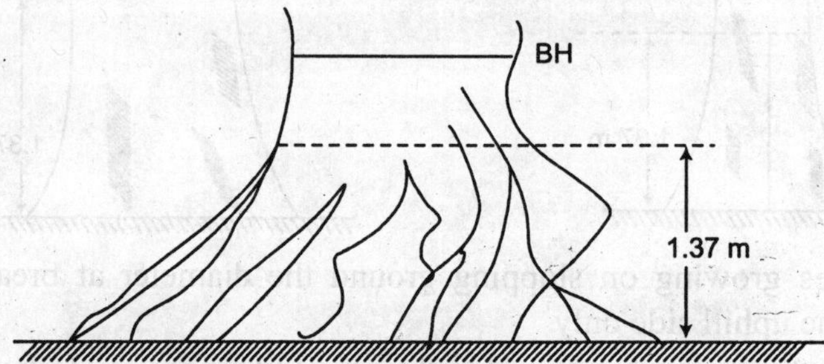

(vii) Before measurement the secondary growth on the stem of the trees like creepers, climbers, lichens, moss and lose bark must be removed.

(viii) Diameter at breast height should be measured in cm and to the nearest multiple of 2mm.

13. How is remote sensing advantageous as compared to ground surveys? (2011)

Remote sensing o-ers following advantages:

(i) Remote sensing provides synoptic view of large areas of earth's surface as a whole and in a manner which facilitates study of resources and their environment which may not be visible to an observer working on ground.

(ii) Under a stereoscope, it provides three dimensional view of terrain and objects under study. Stereoscopic viewing enables determination of the shape and size of the objects. But this is not possible in case of ground survey.

(iii) Characteristics of objects not visible to the human eye can be transformed into image form and studied. For instance, diseases and insect attack on trees or stand can be detected by remote sensing much earlier than they can be seen by human eye.

(iv) Remote sensing provides a permanent and unbiased representation of the earth, its resources and their environment to be studied intensively at leisure and in circumstances more favourable than those present during direct observation in field.

(v) It provides information about the resources and facilitates its analysis and interpretation at a much faster rate than that of the conventional methods.

(vi) It is an easier method to obtain information about resources situated in inaccessible areas. However ground survey is difficult in inaccessible area.

(vii) The remote sensing is useful in detecting the early fire occurrence which otherwise is a tedious process through ground survey. Remote sensing monitors the forest fire on a day today basis and direct the forest administration through online warning system which is very useful in monitoring forest resources. This cannot be done in the traditional ground survey.

(viii) The remote sensing plays a vital role in identification and management of watershed which is otherwise very difficult in ground survey. The remote sensing tools helps in delineation of watershed area, identification of natural resources and the associated management issues.

It may, however, be noted carefully that the imageries obtained in remote sensing only complement, improve or reduce field work rather than replace conventional methods.

14. Explain principles of height measuring instruments. (2011)

The height measuring instruments are designed based on the following two principles viz.,

(i) Principles of similar triangle: Two triangles are said to be similar, when the corresponding angles are equal and the corresponding sides are proportional. The fact that two triangles are similar is established by one of the following conditions:

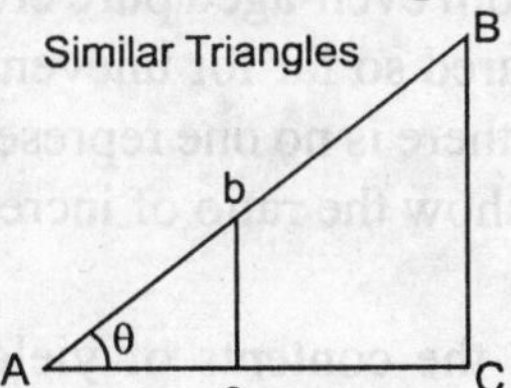

(a) Each angle of a triangle is equal to its corresponding angle of the other triangle

(b) Each side of a triangle is proportional to the corresponding side of the other triangle and

(c) One angle of triangle is equal to one angle of the other and the corresponding sides which subtend the equal angles are proportional

$$A'B' : AB :: B'C : BC$$

or $$A'B \times BC = AB \times BC$$

Therefore $$A'B' = \frac{AB \times B'C}{BC}$$

(ii) Principles of trigonometry: In a right angled triangle, the trigonometrical ratios or functions of angles other than the right angle can be defined in terms of the sides of the triangle.

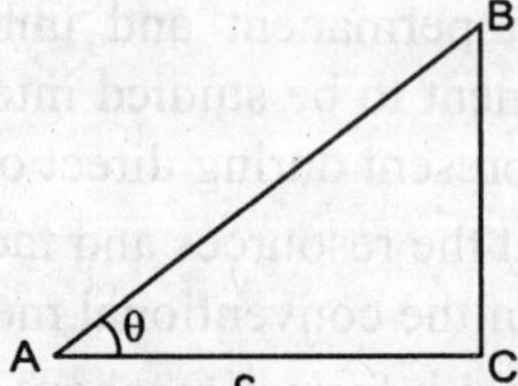

Let *ABC* be a right angled triangle. The trignometrical ratios of theÐ*ACB* are defined as follow:

$$\sin = \frac{\text{Perpendicular}}{\text{Hypotenuse}} = \frac{AB}{AC}$$

$$\cos = \frac{\text{Opposite}}{\text{Hypotenuse}} = \frac{BC}{AC}$$

$$\tan = \frac{\text{Perpendicular}}{\text{Hypotenuse}} = \frac{AB}{BC}$$

15. What is yield table? How contents of yield table are utilized in forestry? (2011)

Yield table is a tabular statement which summarizes on per unit area basis all the essential data relating to the development of a fully-stocked and regularly thinned even-aged crop at periodic intervals covering the greater part of its useful life.

As stated in the above definition, yield table is not applicable to uneven- aged forest because in its present form it has been compiled from even-aged pure crops and is therefore applicable to them alone. Yield tables have not been prepared so far for uneven aged crops because of the difficulties involved and the main difficulty is that there is no one representative average age. Some tables have been prepared for such crops but they show the ratio of increment (current or mean annual) to help in deciding the policy of management.

Contents of Yield Table: Though the contents of yield table can vary depending upon the circumstances of data collection or the objects with which it has been prepared, most of the yield tables in India give the following information by site qualities per unit area basis at interval of 5 or 10 years for the greater part of the useful life of the crop:

Main Crop

(i) Average diameter

(ii) Average height

(iii) Total basal area

(iv) Number of trees

(v) Stem timber form factor
(vi) Total (stem and branch) small wood form factor
(vii) Standing volume stem timber
(viii) Standing volume total small wood
(ix) Total standing volume, *i.e.,* stem and small wood

Thinning

(x) Volume stem timber
(xi) Volume total small wood
(xii) Total volume

Final Yield

(xiii) Volume stem timber
(xiv) Volume total small wood
(xv) Total volume

Accumulated Yield of Thinning

(xvi) Volume stem timber'
(xvii) Volume total small wood
(xviii) Total volume

Total Yield

(xix) Volume stem timber
(xx) Volume total small wood
(xxi) Total volume

Mean annual increment

(xxii) Volume stem timber
(xxiii) Volume total small wood
(xxiv) Total Volume

Current Annual Increment

(xxv) Volume stem timber
(xxvi) Total volume

The above contents of the yield table can be divided into two classes:

Primary data comprises of volume of main crop, thinning at that age, final yield, accumulated yield of thinnings up to that age and the accumulated yield of thinning up to the previous age), total yield.

Secondary data includes crop averages for diameter and height, number of stems for unit area, crop basal area, form factor MAI and CAI. In addition to the above information, yield tables give the following additional information:

(i) Top height by site quality and age

(ii) Spacement table
  (a) by age and site qualities
  (b) by crop diameter and site qualities
(iii) Number of trees per unit area by crop diameter and site qualities
(iv) Normal growing stock of stem timber volume for 40 hectares (100 acres) under different rotations, in case more than one rotation is in vogue
(v) Stand table is table showing the distribution of stems by diameter classes for each of a series of crop diameters. It is a common practice in India to give stand table along with yield table.

Utility of Yield Table content: The contents of yield table are used in determining yield and the associated growth attributes which are indicated below.

- Determination of site quality or fractional site quality
- Estimation of total yield or growing stock
- Determination of increment of stand
- Determination of rotation
- Preparation of stock-map by site qualities
- As a guide to silvicultural thinning

16. Describe various formulae for calculation of volume of logs (2012)

Calculation of volume of logs: Due to biological nature of trees felled logs are not real cylindrical but look like a cylindrical form and the base always look like a frustum of a neloid. The middle portion look like a frustum of a parabloid and top portion look like a frustum of cone. Hence various scientists have designed several formulae taking into considerations the farmers felled logs and the formulae are furnished below.

(i) Quarter Girth Formula: The Quarter Girth formula is the formula used to calculate the volume of logs. It is defined by Mr. Hoppus and hence it is also called as Hoppus formula. In India and Britain, volumes of logs are often calculated by the quarter girth formula:

$$\text{Volume of log} = \left(\frac{g}{4}\right)^2 \times l$$

Where, g is the girth of the log at the middle and
l is the length of the log.

(ii) Smalian's formula: If the form of wood log is Paraboloid, smalian's formula is used for calculation of volume of log. Hence the formula is,

$$\text{Volume of log} = \frac{S_1 + S_2}{2} \times l$$

Where, S_1 is the sectional area at the thick end in square units
S_2 is the sectional area at the thin end in square units and
l is the length of the log or height of the solid in linear units

(iii) Huber's formula

Volume of log = $S_m \times l$

Where, S_m is the sectional area at the middle in square units

l is the length of the log or height of the solid in linear units

(iv) Newston's formula: If the log is in Neiloid shape, Newton's formula is used

$$\text{Volume of log} = \left(\frac{S_1 + 4S_m + S_2}{6}\right) \times l$$

Where, S_1 is the sectional area at the thick end in square units

S_m is the sectional area at the middle in square units

S_2 is the sectional area at the thin end in square units and

l is the length of the log or height of the solid in linear units

17. Describe compound interest formula for calculation of diameter increment percent. (2012)

Compound interest formula: Diameter increment percent and volume increment percent was determined by compound interest formula. In these formulae, d is the initial diameter and D is the diameter after n years and p is the rate of incremental percent.

Diameter increment percent

$$D = d\left(1 + \frac{p}{100}\right)^n$$

or
$$\frac{D}{ed} = \left(1 + \frac{p}{100}\right)^n$$

or
$$1 + \frac{p}{100} = n\sqrt{\frac{D}{d}}$$

or
$$\frac{p}{100} = \left[\left(\frac{D}{d}\right)^{1/n} - 1\right]$$

or
$$p = 100\left[\left(\frac{D}{d}\right)^{1/n} - 1\right]$$

Volume increment percent: Compound interest formula is also used for calculating the volume increment percent. In this formula v is the initial volume, V is the volume after n years and p is the rate of volume increment percent.

$$p = 100\left[\left(\frac{V}{v}\right)^{1/n} - 1\right]$$

Schneider's formula: Schneider's formula is simpler which can be used any time without knowledge of initial diameter if the species shows annual rings and pressler's increment borer is available.

$$p = \frac{400}{nD}$$

Where D is the present diameter (U.B.) at the point of boring,

and n is the number of rings in the outermost centimetre of the radius.

This formula may be evolved from the Pressler's formula as under, assuming that D and d are separated by one year's growth.

Since n is the number of rings in the outermost centimetre then the increase in diameter per year will be $2/n$. Thus d will be $D - 2/n$.

Then $p = \dfrac{D^2 - \left(D - \frac{2}{n}\right)^2}{D^2 + \left(D - \frac{2}{n}\right)^2} \times 200$ as the period of increment has been assumed to be one year.

$$= \frac{\left(D + D - \frac{2}{n}\right)\left(D - D + \frac{2}{n}\right)}{D^2 + D^2 - \frac{4D}{n} + \frac{4}{n^2}} \times 200$$

$$= \frac{2\left(D - \frac{1}{n}\right) \times \frac{2}{n}}{2\left(D - \frac{2D}{n} + \frac{2D}{n^2}\right)} \times 200$$

$$= \frac{\frac{4}{n}\left(D - \frac{1}{n}\right)}{2\left[\left(D - \frac{1}{n}\right)^2 + \frac{4}{n^2}\right]} \times 200$$

$\dfrac{\frac{4}{n}\left(D - \frac{1}{n}\right)}{2\left(D - \frac{1}{n}\right)^2} \times 200$ = Ignoring $1/n^2$ which is very small as compared to $(D-1/n)^2$

As $1/n$ is very small as compared to D, it can be ignored, and therefore $p = \dfrac{400}{nD}$

18. Discuss the Metzer's theory of stem form (2012)

Metzger's Theory of tree form: Several theories have been put forward to explain variations in taper from tree to tree and in the same tree as well but the most plausible theory is that of Metzger, a German forester. According to him the tree stem should be considered as cantilever beam of uniform size against the bending force of the wind. In other words, the stem of the tree is built up in such a way as, with the minimum of material, to offer uniformly the greatest resistance to the stresses to which it is subjected.

The wind pressure acts on the crown and is conveyed to the lower parts of the stem in an increasing measure with the increasing length of the bole. Thus the greatest pressure is exerted at the base and there is a danger of the tree snapping at that place.

To counteract this tendency, the tree reinforces itself towards the base. The limited growth material is so distributed along the tree stem that it affords a uniform resistance all along its length to that pressure. As the pressure in the upper part of the tree is less, due to smaller length of the level in that portion it is allocated lesser growth material than the lower part where the pressure gets increased with the increased length of the bole.

The pressure of wind on crowns keeps on changing as the tree is growing in open or crowded portion. Trees growing in complete isolation have larger crowns and so the pressure exerted on them is the greatest.

If such a tree is to exist, it should allocate most of the expense of height growth. Allocation of more growth material for height will have double disadvantage in such cases. While the pressure will be increased with the increased arm of the lever, there will not be enough material at the base to counteract it. That is why trees growing in complete isolation or exposed situations have short but rapidly tapering boles while the trees growing in dense crops, which are therefore subjected to lesser wind pressure have long and nearly cylindrical boles.

This can be proved mathematically as follows:

Let a force *P* be applied to a cantilever beam at its free end, l the distance of a given cross section from the point of application of this force (i.e., the leverage), *d* the diameter of the beam at this point and *S* kg/cm² the bending stress, then by the rule of mechanics,

$$S = \frac{P + l}{d^3} \times \frac{32}{\pi}$$

As the force *P* in case of trees consists of components W = wind pressure per unit area and F = crown area, it will be $P = W \times F$. Then,

$$S = \frac{W \times F \times l}{d^3} \times \frac{32}{\pi}$$

or

$$d^3 = \frac{32 \times W \times F \times l}{\pi \times S}$$

For a given tree, W, F, S can be considered as constant; therefore $d^3 = kl$, where k is a constant. Thus, the diameter raised to the third power increases proportionately with lengthening of the level or with the increasing distance from the central point of application of wind force, which can be assumed to be at the centre of gravity of the crown. According to this logic, the tree stem must have the shape of a cubic paraboloid.

19. Di-erentiate between graphical method and regression equation method for preparation of general volume tables. (2013)

S.No.	Graphical Method	Regression Equation Method
(i)	The relationships between volume, dbh, height and form etc are expressed in graphical form.	The relationships between volume as dependent variable and dbh, height and form as independent variable are expressed mathematically in the form of regression equation.
(ii)	Graphical method is tedious and time consuming	Regression equation is easy to prepare and use
(iii)	Principle sources of error in graphical method is (a) Failure to select proper trees with sufficient number (b) In this method, only two variable can be dealt at a time. (c) Drawing of curves will always introduce the personal factor and consequent bias.	In this method choosing the correct regression model is difficult.

Preparation of general volume table: The volume table is based on the average volume of trees growing over a large geographical area. Generally based on two variable i.e. dbh and tree height. This table shows volume of trees by diameter classes and in each diameter class by height classes. It has limited direct application. It can be application to prepare local volume table.

A. Graphical method

B. Regression equation method or method of least squares

A. Graphical method

1. Selection of trees

- Trees with typical height and developed need to be selected
- Evenly distribution of selected trees over the entire area
- Defected trees should not be selected
- Select smaller number of trees rather much larger number of improper selection –best result
- Care: not to select trees above the average
- Number of trees depends on:
- The grouping adopted
- Precision required

- Deviation of individual trees volume from the mean
- Regression method requires lesser number than Graphical method
- Special fellings are rarely carried out for preparation of volume tables

2. **Measurements**
   - Measurement techniques vary with the kind of general volume table
   - For example, standard volume tables require measuring standard timber and small wood while commercial volume table require only commercial timber volume.
   - These standard timber, standard small wood, etc of a felled tree are measured by dividing the total length into sections of 3 m length, odd lengths being included in the last section, which should not exceed 1.5 m measuring the under bark diameter at the middle of each timber section and over bark diameter at the middle of each small wood section, computing their volumes separately and adding them to give the volume of the tree.
   - The data of individual trees is then summarized and finally grouped into height and diameter classes in a table called table of basic averages.

3. **Fitting the curve**

Step-1:

- Average diameter is plotted against corresponding average volume separately,
- Smooth curves will be drawn
- Based on smooth curve, read volume and tabulated

Step-1a:

- Average heights is plotted against corresponding average diameter class
- Smooth curves will be drawn
- Based on smooth curve, read average heights and tabulated

Step-2:

- Corresponding Average volume tabulated from step1 against average heights tabulated from step1a will be plotted
- Smooth curves again drawn
- Volumes against average height class will be read and tabulated

Step-3:

- Volumes tabulated from step2 again plotted against the middle of diameter classes for each height class.
- Smooth curve drawn
- Final volumes are read from these curves against middle of diameter classes and tabulated

B. Regression equation method: In regression method, the basic data viz., height and diameter at breast height remains the same as followed in graphical method. However the relationship between volume as dependent variable and diameter at breast height, height and form as independent variable are given mathematical expression through a regression equation. A plethora of workers have developed several regression equation which are furnished below.

Some of them are given as below:

(i) $V = a + bD^2 + cD^2H + dH^2 + eDH^2$ (Meyor modified)

(ii) $V = a + bD^2 + cH + dD^2H$ (Austrian)

(iii) $V = a + bD^2H$ (Combined variable)

(iv) $V = aD^2H$ (Constant Form Factor)

(v) $\log V = \log a + b \log D + c \log H$ (Schumacher)

(vi) $\log V = a + b \log D + (3-b) \log H$ (Dwight)

(vii) $\log V = \log a + b \log (D^2H)$ (Logarithmic)

Where

a, b, c, d, e are regression constant and coefficients,

D = Diameter at breast ht,

H = Tree height (m),

V = Volume (m^3)

20. Di-erentiate between random sampling and non random sampling. Describe di-erent methods of non random sampling that are used in forest inventories. (2013)

S.No.	Random Sampling	Non random sampling
1	It is a sampling method wherein all the samples have equal chances of being selected	There is no equal chances for all the samples but they are selected based on certain conditions and criteria's
2	It is predominantly used in all forestry enumerations purposes	It is not predominantly used but find only in selective utility
3	It can easily cover larger area	It has less application in larger area
4	It is fairly accurate and easy process	It is a tedious process and need pre occupied ideas
5	It is less time consuming process	It is a time consuming process

(A) Random Sampling: Random sampling is that method of sampling in which sampling units comprising a sample are selected in such a manner that all possible units of the same size have equal chance of being chosen.

In this method, sampling units are selected by using a table of random numbers, by unbiased drawing of cards from a well-shuffled pack or by drawing lots.

The following variations of this sampling method are used in forest inventories.

(i) Unrestricted or simple random sampling

(ii) Stratified random sampling

(iii) Multi-stage sampling

(iv) Multiphase sampling

(v) Sampling with varying probability

(vi) List sampling

(B) Non-Random Sampling: Non-random sampling is that method of sampling in which samples are selected according to the subjective judgement of the observer on the basis of certain rules or guidelines indicating what sample should be chosen.

As the selection is based on judgement of the observer, it is likely to be biased. The following variations of this method are used in forest inventories:

(a) Selective sampling: Selective sampling consists of choosing samples according to the subjective judgement of the observer. Selective sampling may give good approximations of population parameters if it is properly used by a person with intensive knowledge of the population.

(b) Systematic sampling: Systematic sampling is that method of non-random sampling in which sampling units are selected according to a predetermined pattern without recourse to random selection. Most commonly, the pattern envisages regular spacing of units. For example, sampling units may be located 80 metres apart in rows which may be 200 metres apart. The selection of the first unit may be either random or according to a fixed arbitrary rule and this will result in distinguishing two varieties of systematic sampling. When the first unit is selected at random and other according to a fixed pattern, the sampling is referred to as systematic sampling with a random start.

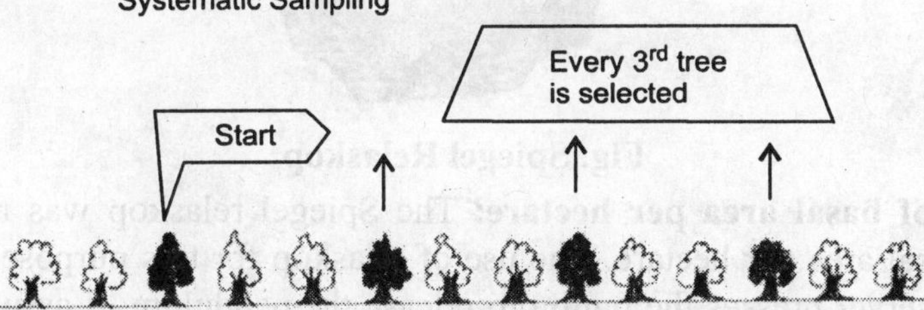

This method of sampling offers certain advantages as compared to random sampling and is therefore, frequently used in forest inventories in-spite of its shortcoming in estimating the variances of the result. The advantages are even greater in tropical forests where environmental conditions hamper field work. Among these advantages are the simplicity and ease in following it, ease of location of sampling units and consequent reduction in cost in reaching it, greater certainty of objectivity in the selection of sample and more uniform distribution of the samples.

(c) Sequential sampling: Sequential sampling is a method of sampling whose characteristic feature is that the number of observations in the sample is not determined in advance but sampling units are taken successively from a population. Each sample includes all the sampling units of the former sample. The confidence interval at each stage will go on decreasing at each stage of sampling. Sampling is stopped when the desired precision is reached. This type of sampling is usually adopted to test a hypothesis. The hypothesis is tested each time a new observation is made to determine whether the hypothesis should be accepted or rejected. When the decision is made and the hypothesis is accepted or rejected, then the sampling procedure is completed and no further sampling unit is selected beyond the number necessary to make a decision.

21. "Spiegel Relaskop is an instrument of great use in forestry"; Justify with reasons. How is the basal area per hectare determined by this instrument? Explain. (2014)

Yes, Spiegel Relaskop is an instrument used to measure both height and diameter simultaneously. It helps to assess the volume by measuring basal area and height of the tree. Since it gives volume with simple measurements, it is of great use in forestry

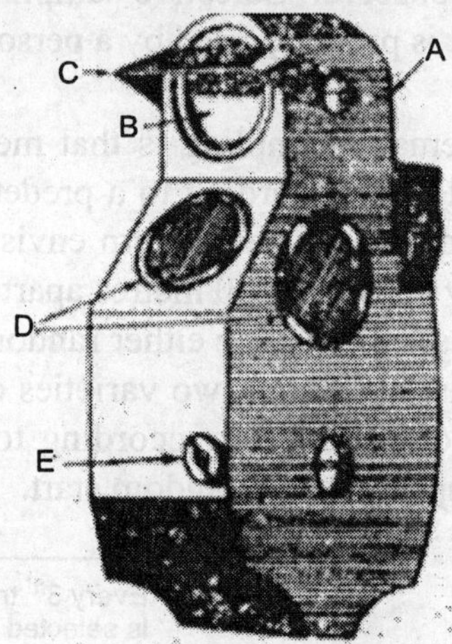

Fig. Spiegel Relaskop

Determination of basal area per hectare: The Spiegel relaskop was really developed for determination of basal area per hectare. The use of relaskop for this purpose is simple. Standing at one point, the observer presses the button to release the pendulum. It assumes again a vertical position resulting in an automatic adjustment of the variable band width since the graduations have been reduced by the factor of the cosine of the slope angle. The measuring bands representing different critical angles or basal area factors are magnified through a lens, projected on a mirror and appear in the lower half of the field of vision. The width of band with the peep-hole produces the angle. The observer then makes a sweep of the surrounding trees and compares the breast- height diameter of each tree with the bad. The tree that appears bigger than the chosen scale (band width) at breast-height is a tally tree. Trees appearing smaller than the chosen scale are non-tally trees. The basal area is obtained by multiplying the number of trees by the basal area factor of the band used

22. Which of the instruments has Wheeler's pentaprism replaced and why? (2015)

It has replaced the following instruments.

- Calliper
- Finnish parabolic calliper
- Optical forks

- Optical calliper
- Short base range finder

Reasons for replacing the instruments: The optical calliper employs two parallel lines of sight, tangent to the tree at the point of diameter determination. Two lines of sight can be considered equivalent to the arms of a calliper. The distance between the two lines of sight, d, is equivalent to the diameter, regardless of the range from the tree.

The instrument have been devised using this principle is called Wheeler Penta Prism Calliper. It is a precision optical-prism device to measure diameters of trees accurately at any point on the tree stem from any clear, convenient, sighting position. No base line distance is required. It gives diameters correct to ± 2 mm. It uses two penta-prisms to maintain parallelism in the lines of sight. Of these left hand prism is Fixed while the right hand prism can be moved along the scale. The reflection of the right side of stem is brought into coincidence with the left by moving the right hand prism.

Description of Wheeler Penta Prism Calliper: In order to use the instrument, it is held about 8 to 10 cm from the eye. The observer then looks through viewing slot. Through the upper part of the slot, he sees the left edge of the tree and through the lower part, the right edge of the tree reflected in the stationary prism. The observer then slides the movable prism with his right hand until the right edge of the tree reflection is brought into direct vertical alignment with the left edge, mid-way between the two vertical guide lines. The diameter is then read on the scale as indicated by the pointer.

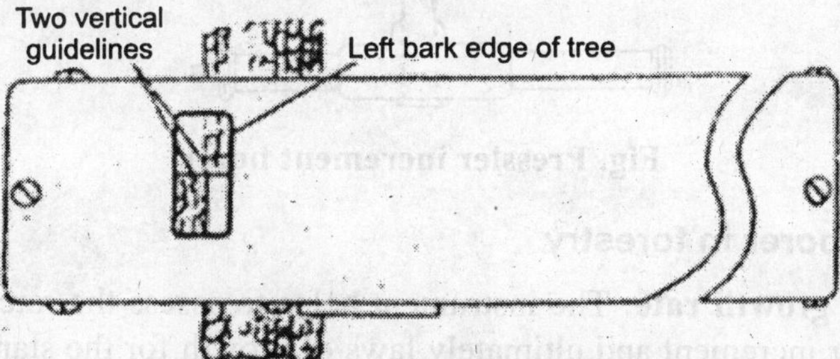

Fig. Wheeler Penta Prism

22. a. How important are the increment borers in forest health and growth analysis? Does Pressler's borer have any adverse e-ects on the tree sampled? Elaborate on the issue. (2015)

b. What is an increment borer? Describe its role in forestry (2014)

Increment borer: Increment borer is a precision instrument specially designed to extract a thin cylinder of wood from a tree, shrub, log or pole for the purpose increment analysis.

Importance of increment borers

- Increment borer is used to determine the age of the trees by counting annual rings from the wood.

- This will facilitate to determine age of the trees without felling.
- In natural forests for most of the trees age is not known. In order to assess the age without felling the tree it is only increment borer will help to resolve the issue.
- The increment borer also help to analyse the diameter increment over years thereby help to assess the incremental growth pattern.
- It is also useful to assess the rate of growth of tree, and the period in which the tree is growing rapidly and the age at which it ceases the growth.
- It helps to identify the maximum increment pattern and to fix the rotation based on diameter increment.

Description of the instrument: It is available in a variety of sizes ranging in length from 4 inches to 40 inches. Although the increment borer is essentially a very simple instrument, it is relatively expensive. Proper operation and maintenance can prolong the life of the increment borer and minimize problems. This manual provides tips on increment borer selection, operation, and maintenance to minimize problems in' the field and maximize the utility of the borer. Experience in a specific field application is needed to develop the best and most efficient technique.

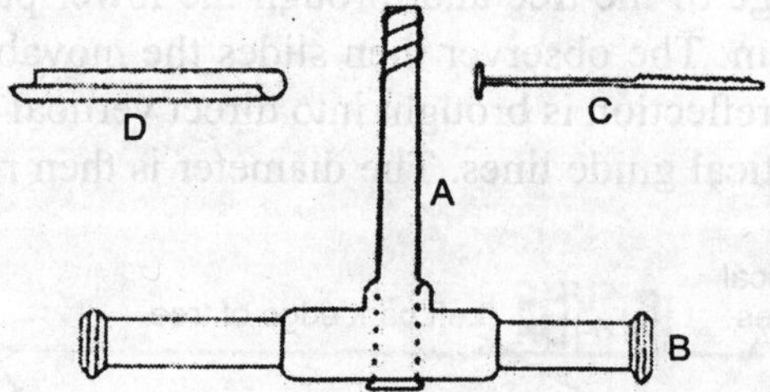

Fig. Pressler increment borer

Role of increment borer in forestry

Determination of growth rate: The instrument helps to assess the rate of growth of trees in terms of age, diameter increment and ultimately laws of growth for the standing trees. Otherwise the trees have to be felled for stump and stem analysis in order to assess the growth increment. Hence it is a handy and easy instrument to assess the growth rate of trees.

Determination age: Determination of age is a critical factor in many widely grown tree species particularly in natural forests. For several species growth attributes are available but for want of age determination the growth and volume tables are not prepared. Under such circumstances, Presslers increment borer is an ideal instrument to assess age of the trees based on annual growth rings without destructing the tree.

Determination tree soundness: The increment borer is very essential and useful to assess the soundness of the tree. In many even-aged and uneven aged plantations fully matured trees experienced heart rot, borer incidences and affects the quality of the wood. This cannot be visualized from external appearances. Under such circumstances, the increment borer will help to assess the soundness and integrity of standing timbers.

Penetration of chemicals towards wood treatment: Many matured trees experience the symptom of yellowing of leaves, leaf shedding, dye back, gummosis and other stem borers and diseases in soft and heart wood. To control all these pests and disease problem particularly for mature trees stem injections are preferred. The holes developed due to Pressler's increment borer will be one of the viable option to treat the pest and diseases through suitable wood preservative chemicals.

Specific gravity studies of wood: The increment borer is also used to assess the density and specific gravity of wood through the core sample taken from the standing trees and hence it is very useful.

Adverse a3ect of Pressler's borer: Yes, Pressler's borer has an adverse affect on the tree which has been bored for ring analysis. After boring the whole will act as a entry point for pest and diseases. Several heart rot borers like Hoplocerambyx enters through the whole, breeds and housed inside the whole. This in turn creates tunnels and the entire timber will be unecomical to use and hence cause huge economic loss.

Eg. Sal borer

Similarly after the whole formation, some of the wood will secrete resins substances, which will be attracted by several bacteria and fungi. Once it is infected, the fungi and bacteria multiplies rapidly and causes heart rot diseases. This will create discolouration and straining of wood, loss of cellulosic materials and creation of holes and finally death of the tree.

Eg. Heart rot disease

Sometimes the creation of boreholes on the tree will create stress and hence the entire physiological metabolism of the trees will be changed. There are incidences of sudden flowering, rapid leaf loss, rapid leaf drying etc. have been noticed due to hole creation. This sudden physiological changes alters the vegetative and reproductive biology of tree species and hence the growth of tree is affected.

Eg. Sudden flowering in *Tamarindus indica*

24. Discuss the significance of stump analysis in forest mensuration. How does it help in understanding the past growth in trees? (2016)

Stump analysis: Analysis of cross section of the stump by measuring annual rings in order to estimate the height of tree and its past rate of growth in terms of diameter and basal area.

Significance of stump analysis

(i) The stump analysis helps to determine the progress of diameter throughout the life of the tree.

(ii) It helps to establish age-diameter relationship in order to fix the rotation age corresponding to a desired exploitable diameter in the absence of yield table.

(iii) It helps to assess the influence of the external factors on the diameter increment.

(iv) It helps to estimate the age of the trees in the absence of basic records.

(v) It helps to assess the basal area of plantation.

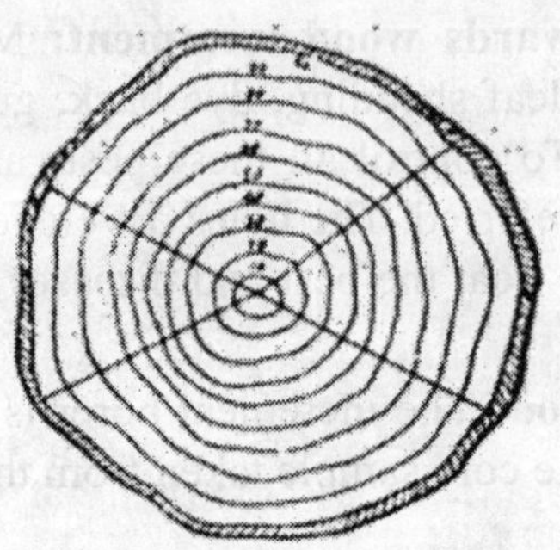

Fig. Stump Cross Section

Stump analysis and past growth of stands: The stump analysis helps to determine the past growth of stumps for this purpose the trees of ideal size at rotation age are selected and felled. Before felling the dia breast height is measured.

The tree is felled at stump height and under bark girth of the stump is measured.

The average radius under bark is calculated for each stump. In each stump the 4 radii drawn angularly and are marked on the stump as indicated in the figure.

The growth rings are counted from pith outwards and a pin is inserted at each decade and radii are measured in centimetre and millimetre.

(i) The radii at successive decades are totalled and halfed to get average diameter at each decade.

(ii) Similarly trees of different diameter classes and their under bark diameter at 15 cm, 45 cm and 75 cm from ground level and DBH over bark are measured, added and averaged.

(iii) The 2-3 m height grown seedlings in the locality are cut at 10, 20, 30 cm for ring countings to determine age at different heights.

The data collected from step 1, step 2 and step 3 are used to draw following stump analysis curve.

(i) **Seedling height curve**: This curve is drawn from data collected from step 3 and age corresponding to average stump height is read.

(ii) **DBH against age**: Curve is drawn which gives DBH over bark corresponding to convenient stump diameter under bark.

Stump diameter Verses number of rings can be plotted as a curve which lead to diameter of a curve. Stump DBH over bark against total age hence the stump analysis helps to determine the past growth of stand particularly the diameter growth at various stages of growth and development. The radial and diameter growth of at every decade indicate the active period of diameter increment and its relationship with corresponding climatic factors and precipitation.

If due to certain natural calamities of the diameter growth is affected that can also be absorbed through the stump analysis and the reason for reduced growth at a particular decade can also decided and in holistic approach the stump analysis helps to study the laws of growth about trees and identify the reasons prevailing besides helping us to fix the rotation age with optimal diameter growth.

Advantages: As compared to other methods, stump analysis offers the following advantages:

(i) Data can be collected from stumps of felled trees as long as the wood remains sound. Thus it can be carried out with or without special fellings.

(ii) Data can be collected any time with minimum of manual labour.

(iii) The data can be multiplied to any desired extent with no objection other than the time taken in measuring.

(iv) Field work is simple and easily learnt.

(v) Each stump provides data for the whole life of the tree.

25. Define Photogrammetry. Discuss in detail its application in forest management. (2016)

Photogrammetry is the art and science of obtaining reliable measurements and maps from Ariel photographs.

Applications of Photogrammetry in Forest Management: The forest management involved identification of species composition, assessment of growing stock and regulating the yield on a sustainable basis. This forest management requires wide range of measurement of trees in terms of height, diameter, volume, crown characters, stem characters, growing stock estimation etc. which are the essential managements that help in forest management. The photogrammetry is useful in the following aspects towards forest management.

- Determination of growing stock
- It helps to protect forests through the damages due to biotic and abiotic factors
- The conservation of wildlife and the associated management process also possible through the application of photogrammetry.
- The areas prone to soil erosion and the associated demand for watershed development can be easily managed through the application of ariel photographs and the associated photogrammetry.
- Photogrammetry also helps in forest monitoring and assessment in order to effectively manage the forests.
- Photogrammetry will helps to develop suitable planning models towards conservation and management of forests.
- Photogrammetry is very fast and accurate towards growing stock assessment and the associated forest management planning process.
- Forest inventory and enumeration is very easy and accurate due to photogrammetry.

26. What is stand density? How spacing is used to control stand density? Discuss. (2014)

Stand Density: Stand density is a quantitative measurement of a stand in terms of square feel ofbasal area, number of trees or volume per acre. It reflects the crowding of stems within the area.

Spacing and Stand Density

The spacing of trees play a significant role in deciding the density of the plantations. For example the following are the spacing and density of plantations for Casuarina.

Spacing	Density
1 m × 1m	4000 / acre
2 m × 1 m	2000 / acre
3 m × 1 m	1111 / acre
4 m × 1 m	1000 / acre
2 m × 2 m	1000 / acre
3 m × 3 m	444 / acre
4 m × 4 m	250 / acre
5 m × 5m	160 / acre

The above table indicates the stocking density of species which is decided by the spacing. Stocking is a relative term used to describe the adequacy of a given stand density in meeting the management objective. Thus a stand with 4000 trees per acre may be termed as over stocked, 2000 trees per acre may as moderately stocked and 160 per acre as under stocked. Hence spacing play a vital role in stand density. If it is over stocked the final material will be used as a raw material for paper industry, if it is understocked it may be used as a raw material for poles and scaffolding materials. Hence spacing is a significant factor which decides density of a stand and ultimately the management objective.

27. Explain Frequency distribution and Binomial & Poisson distribution. (2006)

(i) Frequency distribution: Frequency graphs for a series of data show certain common characteristics. The varieties are concentrated close to the mean and generally the frequency decreases towards the extreme ends. The graphs approach a smooth curve as the sample size increases. Theoretical distributions closely resembling the observed distribution have been worked out mathematically.

(ii) Binomial distribution: Essentially the variables in a binomial distribution are discrete and have only two characteristics or attributes. For a population of size *N* where *A* is the number of variables having attribute, the proportion of items in the population having this attribute is $P = \frac{A}{N}$ and the proportion not having this attribute is $1 - p = q$.

Hence, $p + q = 1$

If *p* be the chance of success of an event or the probability if its occurrence (say heads of a coin) and *n* the number of coins thrown, the probability of throwing *n, n-1, n-2,* etc. heads is given by the binomial expansion of the expression.

$$(p+q)^n = p^n + np^{n-1}\ q + \frac{n(n-1)}{1 \times 2} p^{n-2} q^2 + \ldots\ldots\ldots + q^n$$

When $p = q$, the binomial distribution is symmetrical and become similar to a normal distribution.

(iii) Poisson distribution: The poisson series is the limit to which binomial expansion tends when p is made exceedingly small and n exceedingly large. Like the binomial this is the distribution of a discrete variable. Thus, if m is the expected mean frequency of occurrence of some event in the population of sub-sample and if the frequency follows the poisson distribution then the solvent will occur 0, 1, 2 times in sub-samples with relative frequency equal to the successive terms of the series $1/e^m / e^m, m^2/2e^m, m^3(2\times3)e^m, m^4(2\times3\times4)e^m$ ....
Where e is the base of Napierian logarithm (= 2.71828)

Poisson series is of considerable value in biological science in respect of blood counts and research on bacterial growth.

28. Di-erentiate between Hojer's formula and Behre's formula for tree form (2013)

(i) Hojer's formula: The following general formulae have been suggested to determine diameter quotient (i.e., the ratio of the diameter of a stem at any given height to its breast – height diameter) for each form class.

$$\frac{d}{d.b.h} = C\log\frac{c+l}{c}$$

Where d is the diameter at any point on the stem, C and c are constants for each form class, and l is the distance from the top of the tree to the point at which d is measured, its value being given as a percentage of the length of the tree between b.h. and the top.

(ii) Behre's formula:

$$\frac{d}{d.b.h} = \frac{l}{a+bl}$$

Where a and b are constants for each form class, such that $a+b = 1$ and d and l have the same meaning as given for Hojer's formula.

These formulae have been used for preparation of taper tables in western countries. The senior author tried all the three formulae viz., Metzger's, Hojer's and Behre's for a number of Indian trees. These formulae were converted into linear forms for fitting linear regressions.

29. a. Sample plots laid down in di-erent forest species: (2015)

b. Write down the methods of laying out sample plots for periodic recording of growth data. (2010)

Sample plots are representative areas identified for enumeration. In tree species, sample plots are used to study the growth and development in order to prepare volume and yield tables. The sample plots are of two types viz., temporary and permanent plot. Temporary sample plots are laid for the purpose of immediate enumeration including growth attributes, survival, diversity pattern etc. The permanent sample plots are laid out for determination of growth attributes like height, diameter, volume, age, biomass etc to prepare yield tables.

Sample plots in di3erent tree species

(i) **Selection of plots:** Irrespective of the species the plots are generally laid out in evenaged, fully stocked and pure crops. In case if the trees are sporadically distributed as in natural forests then sample plots can be laid out with uniform stocking population.

(ii) **Number of plots:** It is always better to have more number of sample plots in order to increase the accuracy and reduce the error. However it has been estimated that for each and every thinning regime approximately 600 sample plots have to be laid out.

(iii) **Shape of the plot:** The sample plots are laid out in various shapes like circular, rectangular and square plots and it depends on the purpose and locality. For all practical purposes rectangular or square plots are preferred.

(iv) **Plot size:** Normally the plot size may vary due to objective and it is indicated that an ideal plot size of 0.1 to 0.2 ha is considered sufficient for enumeration.

(v) **Layout of plot:** Once the area is identified the plot should be laid out by establishing suitable surround, demarcation of the plot boundary, marking and numbering of trees and preparation of plot chart.

Bamboo Enumeration using sample plots: Two stage stratified random sampling method is generally used. In the first stage, sampling is carried out to estimate the number of clumps per unit area; second stage sub-sampling is carried out to estimate the number of culms of various ages per clump. In case of sampling by topographical units (say 20%), the total number of bamboo clumps is enumerated in the sampling unit concerned; this constitutes the first stage. Simultaneously, as the second stage, every 20th clump is also sub-sampled (5%) for recording the number of culms. In this way, ultimate sampling intensity for culms is 1%. In case of strip sampling, first stage consists in enumerating all the clumps in the strip concerned (40 m, i.e., 2 chain wide); the second stage comprises detailed sub-sampling all the culms in 40 mx20 m rectangle formed along the strip at 200 m intervals. Strip sampling being 10% in intensity, the ultimate intensity of culm sampling is 1%.

Sal enumeration through sample plots: The survey is usually combined with partial enumeration by strips; sub-sampling can be done, 2 to 5% of aggregate area. Regeneration is recorded in quadrats, 2 m × 2 m formed on either side of the cruise lines. If 10% strip sampling is being done, in strips 20 m wide (200 m apart), and 2mx2m quadrats are laid on either of cruise line, regeneration survey will be 20% of strip, or 2% of the total forest area. Cruise lines should not be too far apart, not more than 200 m to get a fairly complete picture of disposition of regeneration, and to make it possible to prepare an accurate stock map. When regeneration survey data comes to be plotted, each surveyed strip is representative of a rectangle 20 m × 200 m in case of 2% sampling, and 20 m × 80 m in case of 5% sampling.

30. Di-erentiate sample plots and preservation plots. Discuss the role in management. (2011)

Sample plots are representative areas identified for enumeration. In tree species, sample plots are used to study the growth and development in order to prepare volume and yield tables. The sample plots are of two types viz., temporary and permanent plot. Temporary sample plots are laid for the purpose of immediate enumeration including growth attributes, survival, diversity pattern etc. The permanent sample plots are laid out for determination of growth attributes like height, diameter, volume, age, biomass etc to prepare yield tables.

Sample plots in di3erent tree species:

(i) **Selection of plots:** Irrespective of the species the plots are generally laid out in evenaged, fully stocked and pure crops. In case if the trees are sporadically distributed as in natural forests then sample plots can be laid out with uniform stocking population.

(ii) **Number of plots:** It is always better to have more number of sample plots in order to increase the accuracy and reduce the error. However it has been estimated that for each and every thinning regime approximately 600 sample plots have to be laid out.

(iii) **Shape of the plot:** The sample plots are laid out in various shapes like circular, rectangular and square plots and it depends on the purpose and locality. For all practical purposes rectangular or square plots are preferred.

(iv) **Plot size:** Normally the plot size may vary due to objective and it is indicated that an ideal plot size of 0.1 to 0.2 ha is considered sufficient for enumeration.

(v) **Layout of plot:** Once the area is identified the plot should be laid out by establishing suitable surround, demarcation of the plot boundary, marking and numbering of trees and preparation of plot chart.

Preservation plot: Preservation plots are ecological models of existing forest type and are important for studies of ecological succession recording diversity in plant population with time. Intense biotic interference in forests has wiped out number of plant species and many of them are listed as endangered and rare ones.

31. Estimation of Crown volume depends on which factors and what are di-erent geometrical shape of crown? Write down the various formulae for measurement of crown volume. (2012)

Factors contributing to Crown Volume

(i) Crown spread

(ii) Crown thickness

(iii) Crown shape

Crown Volume Calculation

(i) Crown volume calculation using this formula is:

Crown volume = $[(\text{radius}_{\text{minor axis}})(\text{radius}_{\text{major axis}})]^{0.5}$

(ii) Crown volume = Volume equivalent cylinder

= (!)($h(r)^2$)

= (!) (thickness of crown) [(crown shape ratio)(average maximum crown spread)] ^{2}/4,

where average maximum crown spread = 2 average maximum radius

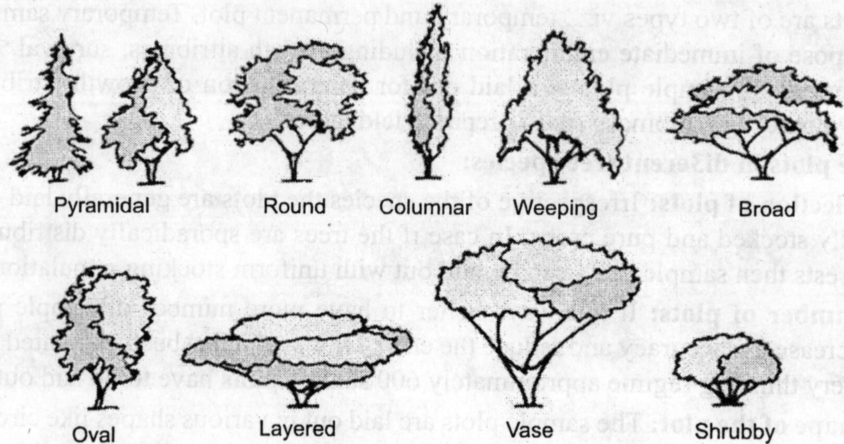

Di3erent geometrical shape of Crown

The constants can be rearranged to derive a crown form (CF) factor:

CF = [(!)(crown shape ratio) ^{2}]/4

The overall volume equation can then be rewritten as follows:

Crown volume = (CF) × (crown thickness) × (average maximum crown spread)^2

32. What is the importance of estimating form factor of a tree? Write common formula(e) used to estimate the form factor. (2009)

Form factor is the ration between the volume of the tree or its part of the volume of cylinder having the same length and the cross section as the tree. It is an important factor which helps to determine the growth and development of trees. It is also used as correction factor while calculating volume of the tree considering tree as a full cylinder.

Importance of Form Factor

(i) To estimate the volume of standing trees: Form factors may be compiled into tabular form giving average form factor of trees of different dimensions by diameter and height classes. These tables can be used to estimate the volume of standing trees by measuring their diameter and height. As these tables are prepared from data of large number of trees, their application to individual trees is not likely to give satisfactory results. They can, however, be used for estimation of volume of a group of trees. Their use is also limited to trees growing under same conditions as the samples on which the tables are based.

(ii) To study laws of growth: Form factor along with form point and form quotient (to be described later) give an insight into the laws of growth, particularly the stem form, of trees.

Normal (or True) Form Factor: In this form factor basal area is measured at a constant proportion of the total height of the tree, e.g., 1/10th, 1/20th, etc., of the total height and the volume refers to the whole tree above ground level.

This form factor has several disadvantages, viz., (i) the height of the tree has to be determined before the point of measurement can be fixed and (ii) the point of measurement may be very inconvenient in case of very tall as well as very short trees.

Absolute form factor and normal form factor are no longer used. Unless otherwise stated, form factor in India implies artificial form factor with basal area calculated at 1.37 m above ground level. Expressed as an equation, this form factor may be described as:

$$\text{Form factor} = V/sh$$

Where, F is the form factor

V is the tree volume in cubic units

S is the basal area of breast height in area units and

h is the height of the tree in linear units

33. In stem analysis, diameters of 30th ring (as computed on BH section) at di-erent height sections was found as follows: (2013)

Ht. of Section (m)	01.37	04.24	07.24	10.24	13.24	14.74
Diam. (Cm)	29.5	25.2	21.0	16.00	10.40	06.60

The diameter curve of 30th ring cuts the height axis at 18.24 m. (The average seedling takes 10 years to reach 1.37 m). Calculate the MAI at 40 years of age of the tree

Mean tree volume computation for decade is calculated by following method

Mean tree volume computation: Diameter of 30th ring is computed as BH section. This is computed by considering age at breast height as 0. Correction to age to breast height should be done in order to calculate Mean Tree Volume (MAI)

	Height of section in metres						
Diameter	1.37	4.24	7.24	10.24	13.24	14.74	
	29.5	25.2	21.0	16.00	10.40	06.60	
Basal area (m²)	0.068	0.049	0.034	0.02	0.008	0.003	
Volume (m³)	0.093	0.211	0.25	0.204	0.112	0.050	0.92

The average seedling takes 10 years to reach breast height (1.37 m). Then the age of tree is

$$\text{Age} = 30 + 10$$

$$\text{Age} = 40$$

In 40 years of age, Mean tree volume computation for decade is 0.92m³.

$$\therefore \text{MAI for 40 years of age} = \frac{0.92m^3}{10} = 0.093\text{m}^3$$

Mean tree volume computation for decades

34. If the angle of elevation to the tip of the tree is 30° and 45° respectively, measured from two sides of a ravine, and width of the ravine at the top is 15 m and height of the eye of observer from the ground is 1.5 m, find the height of the tree. (2012)

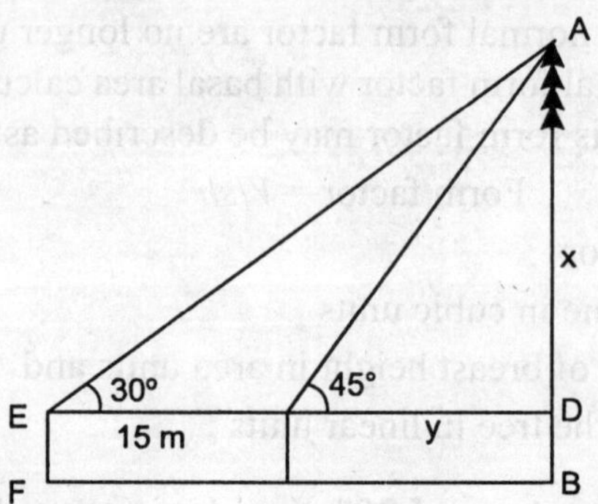

Height of the tree AB = AD + BD

$$\tan 30° = \frac{x}{15+y}$$

$$\frac{1}{\sqrt{3}} = \frac{x}{15+y}$$

$$15 + y = \sqrt{3}\,x$$

$$15 + x = \sqrt{3}\,x \qquad \left(x = y; \tan 45° \angle GDB = \frac{x}{y}, 1 = \frac{x}{y},\quad x = y\right)$$

$$x\sqrt{3} - x = 15$$

$$x(0.732) = 15$$

$$x = \frac{15}{0.732} = 20.5$$

$$AB = AD + BD = EF + x = 1.5 + 20.5$$

$$AB = 22m$$

Height of the tree is 22 m

35. A tree with elliptical c/s when measured at BH by a calipers gives two values as 71 cm and 65 cm. Girth of the tree at BH by tape is 1.82 m. Calculate the basal area by three di-erent methods, listing the methods clearly, and discuss which method is the correct one. (2012)

1. Basal area = r^2
2. Basal area = $\left(\frac{g}{4}\right)^2$

With 2 radii in case of eccentric tre stem

3. Basal area = $\left(\frac{a+b}{2}\right)^2$

Calipers = 71 cm & 65 cm

$D_1 = 71$

$D_2 = 65$

Girth – Tape = 1.82 m

4. Basal area $= \left(\frac{g}{4}\right)^2 = 0.207 \text{ m}^2$

5. Basal area $= \left(\frac{a+b}{2}\right)^2$

$D_1 = 71 \qquad D_2 = 65$

$r_1 = 35.5 \text{ cm} \qquad r_2 = 32.5 \text{ cm}$

$$r = \frac{35.5 + 32.5}{2}$$

$$r = 34 \text{ cm}$$

$$\text{Basal area} = \left(\frac{35.5 + 32.5}{2}\right)^2 = 0.362 \text{ m}^2$$

6. Basal area $= \frac{d^2}{4}$

$$G = 2\pi r = 1.82 \text{ m}$$

$$d = \frac{1.82}{\pi} = 0.579$$

$$\text{Basal area} = \left(\frac{0.579^2}{4}\right) = 0.263\text{m}^2$$

Section of the tree is seldom a circle; it approaches sometimes to an ellipse and therefore calculation of a sectional area requires the measurement of 2 diameters.

36. Calculate the values of i) Bark thickness; ii) Log volume OB; iii) Log volume UB; iv) Volume of bark, and v) Bark percentage, for a log with measurements (2012)

DBH OB = 130 cm; DBH UB = 124 cm and Length = 4.8 m

(i) Bark thickness

$$g' = g - 2\pi t$$

$$\text{Bark thickness} = 408.2 - 389.36/6.28$$

$$= 3 \text{ cm}$$

Log volume OB

$$\text{Volume} = \left(\frac{gob}{4}\right)^2 \times l$$

$$= 4.9939 \text{ m}^3$$

(ii) Log volume UB

$$\text{Volume} = \left(\frac{gub}{4}\right)^2 \times l$$

$$= 4.548 \text{ m}^3$$

(iii) Volume of Bark

Volume of bark = Volume OB –Volume UB

= 0.45 m^3

(iv) Bark percentage

Bark % = Volume of bark / volume OB * 100

= 9.02%

37. What is stand table? Give brief description for preparation of stand table? (2014)

Stand table: Stand table is a table showing the distribution of stems by diameter classes for each of the series of crop diameters. In India, it is usually appended at the end of yield table of the species and the distribution is given in the following forms:

- As percentage of trees over given diameter limits in crops of various diameters
- As number of trees per hectare (or acre) over given diameter limits, in crops of different qualities for successive 5 years age interval
- As percentage of the total number of trees by 10 cm (4 inches) diameter classes in crops of various diameters
- As percentage of the total number of trees by 10 cm (4 inches) diameter classes for various partial crop diameters excluding trees below a certain limit
- As number of trees by 10 cm (4 inches) diameter classes for 40 ha (100 acres) of normal forest under different rotations before the annual fellings

Objectives of stand table

- Stand table is very useful for forest management and it is today generally accepted that yield table is not complete without it.
- The average diameter of the main crop stems and the stems removed in successive thinnings can be ascertained from the yield but the range of diameters comprised in these averages cannot be known unless stand table is also prepared.
- The calculation of the final values of forests, preparation of money yield table and application of commercial volume tables made for local standards of conversion, to yield table crop data require knowledge of trees per unit area under different diameter classes and this is given by stand table.

Application and uses of stand table: Being statistical averages, stand table may not be applicable correctly to individual cases unless they are normally stocked. In normally stocked crops, however, they give a clear idea of the distribution of different diameter classes and can be put to several uses, some of which are given below:

- Determination of the financial value of crops and
- Preparation of money yield table

Stand volume table: Every forester is confronted with the problem of quick assessment of stand volume. This can be easily done if stand volume table is prepared. Chaturvedi has found out that a linear relationship exists between volume/ha of a stand and the product of crop height and basal area/ha as shown below:

$$V = a + b\,BA \times HC$$

Where, V - Volume in m^3/ha

BA - Basal area in m^2/ha

HC - Crop height in m

a and b are regression constant and coefficient respectively.

38. Di-erentiate between Geostationery and sun-synchronous satellite and satellite imagery in remote sensing (2014)

Geostationary: Geostationary satellites are always located directly above the equator with a zero angle of inclination. Geostationary orbit, therefore, is really just a special type of equatorial orbit. When a satellite is in geostationary orbit, its instruments are looking at a certain part of the Earth. That part of the Earth is called a footprint. A footprint can be pretty big. For example, the footprint for most Canadian communications satellites is almost the whole of Canada.

Height of Geostationary Satellites:

- The height of the geostationary orbit is 35786 kilometers above earth
- In Geostationary Orbit, the satellite moves with an orbital speed of 11068 km per hours.
- A minimum of three satellites are needed to cover the entire earth Super synchronous orbit is a disposal / storage orbit above GSO.
- From earth, they would seem drifting in westerly direction.

Sun-synchronous: A sun-synchronous orbit is an orbit around the Earth, where the movement of the satellite always looks the same when viewed from the Sun. A satellite in a sun-synchronous orbit still orbits the Earth, but does so in such a way that over the course of the day, its distance to the Sun will change in a consistent pattern no matter the time of year.

Notice that its position relative to the Sun varies across each orbit of the Earth, but the way that the orbit appears from the point of view of the Sun is the same no matter where the Earth is in its yearly orbit.

This kind of orbit also has many applications. The big advantage of sun-synchronous orbits is that the Earth will be illuminated in a similar way year round. For this reason, it's often used for weather and spy satellites, since they create images of the Earth. They're also used in ocean or atmospheric remote-sensing satellites, since they require sunlight to take their data.

Parameters	Geostationary satellites	Sun synchronous satellites
Altitude of orbit	36000 km	700 – 900 km
Orbital direction	Along equator	Pole to Pole
Orbital speed	East to west Equal to earth's rotation	North to south Less than earth's rotational speed
Global coverage	Not possible	Is possible at frequent interval
Continuous coverage of particular region	Possible	Not possible
Affected by atmospheric disturbance	No	Yes
View of extra terrestrial body	Possible	Not possible
Example	INSAT, GOES	Landsat, SPOT, IRS

Parameters	Satellite imagery	Remote sensing
Nature	It is an product / image taken from satellite	It is a science which obtain information about an object without being in contact
Terminology and scope	Narrow	Wider

39. Thermal remote sensing has specific application in forest management? Describe. (2016)

Thermal remote sensing: The earth surface features emit radiation in thermal infrared part of the electromagnetic spectrum which lies in the wavelength range of 3 to 14 μm. Because of atmospheric absorption, the thermal infrared images can be obtained in the wavelength regions of 3 to 5.5 μm and 8 to 14 μm only.

The interaction of thermal infrared radiation with vegetation, soil and water: The vegetation absorbs electromagnetic radiation in visible part of spectrum and emits at thermal infrared wavelengths. The emittance in thermal infrared part of spectrum is a function of radiant temperature of emitting surface. In case of forests, the temperature of canopy is modified by transpiration which depends on (a) characteristic of vegetation e g. leaf area, leaf structure and root to shoot ratio, (b) environmental factors e.g. atmospheric humidity, temperature, availability of light, wind speed, soil moisture etc. The environmental factors affect the rate of transpiration more than characteristics of vegetation. The increase in atmospheric humidity will decrease the rate of transpiration while the increase in temperature, availability of light, wind speed and soil moisture will increase it.

Due to the above reasons, the diurnal range of temperature variation in case of vegetation is small compared to surrounding air for majority of the day.

Due to large surface area in case of vegetation canopy, the radiation rate is very high compared to surrounding soil. This as a result further reduces canopy temperature both during day as well as night. The thermal emissivity from vegetation canopy also increases with its moisture content.

The use of Trad to estimate T kin in case of vegetation canopy has met with only partial success due to structural and dynamic nature of Vegetation. Trad is affected by the soil temperature, angle of sensor, gaps in vegetation canopy and moisture content in Vegetation.

For the proper interpretation of infrared imagery, the time of acquisition of data should also be known.

Characteristics of thermal infrared images: The scanner images contain a number of geometric distortions that affect the interpretation process. These distortions are discussed in the following sub-sections.

Scanner distortion: The thermal imageries show serious distortion in the direction perpendicular to flight path. The scanner mirror rotates at a constant angular rate while the imagery is recorded at a linear rate and hence the image details towards the edges get compressed. Distortion free pictures can be obtained by first recording the picture digitally on a magnetic tape and obtaining photographic prints after processing the data.

Spatial Resolution: Spatial resolution of thermal imagery depends on the flight altitude and IFOV and is given by where H is flight altitude and P the IFOV. Hence the thermal scanners sense radiated energy from ground resolution cells of varying sizes

Use of thermal imagery in forestry: In forestry applications, thermal infrared imageries can be used to locate ground fires in dense forest. Fire damage can be assessed under smoke conditions also. More research work is required to be done to estimate the potential of thermal imageries in forestry applications especially in vegetation classification and mapping.

40. Sampling has very important role in forest inventory? Discuss with a help of kinds of sampling. (2016)

Sampling has an important role in forest inventory:

(i) Sampling is very important in forest in order to determine volume of trees growing in the forest with a view to determine the yield. It is essential not only for timber species but also sampling is gaining significance for the estimation of non use values like recreation, tourism etc.

(ii) The sampling is very essential to determine the growth attributes for sales proceeding and also to design the management issues.

(iii) Sampling is important for research and development issues in order to identify, characterize and assess the volume for productive and protective utilization.

(iv) Sampling is essential to assess the current stocking position and to forecast the future requirement which help to plan forestry program on sustainable basis.

(v) Sampling also gaining significance for assessing the wildlife and avifaunal population to develop suitable conservation strategies.

(vi) Sampling is very significant towards assessing biodiversity status of natural forest or wildlife habitat or ecosystem.

Factors that determine the enumeration:

(i) Extent of the area to be covered,

(ii) Variation in composition and density,

(iii) The intensity of management and consequent degree of accuracy required and

(iv) The resources of labor, time and funds available.

Kinds of sampling: There are two main kinds of sampling used in forest inventories:

(a) Random sampling

(b) Non-random-sampling

Random sampling method: Random sampling is that method of sampling in which sampling units comprising a sample are selected in such a manner that all possible units of the same size have equal chance of being chosen. In this method, sampling units are selected by using a table of random numbers, by unbiased drawing of card from a well shuffled pack or by drawing lots.

Types of random sampling:

(i) Simple random sampling: The simple random sampling is one in which sampling units comprising the sample are selected by some strictly random process form the whole population or area without dividing it into homogenous blocks.

(ii) Stratified random sampling: It is one of the method of sampling in which the population is first divided into subpopulations of different strata and then sampling units are selected from each of them in proportion to their size.

(iii) Multi-stage sampling: It is a sampling in which samples are selected in different stages. In each and every stage the sample unit is reduced and the sampling is done.

(iv) Multiphase sampling: This method involves sampling in different phases. For example aerial photography sampling can be done at a larger area. Then ground truthing can be done in an identified area. Since sampling is done in two different phases it is called multiphase sampling.

(v) Sampling with varying probability: In this method, the probability of selection is not equal as described in earlier method and will vary.

(vi) List Sampling: This is another form of sampling with varying probability. This method consists of making a list of sampling units along with their measure of size in any order

(vii) Non-Random Sampling method: Non-random sampling is that method of sampling in which samples are selected according to the subjective judgement of the observer on the basis of certain rules or guidelines indicating what sample should be chosen. There are three non random sampling methods viz., selective, systematic and sequential

Selective sampling

- Selective sampling consists of choosing samples according to the subjective judgement of the observer.
- For quick and rough estimates, this sampling is often employed.

Systematic sampling

- Systematic sampling is that method of non-random sampling in which sampling units are selected according to a predetermined pattern without recourse to random selection.

- Most commonly, the pattern envisages regular spacing of units. The advantages are even greater in tropical forests where environmental conditions hamper field work.
- Among these advantages are the simplicity and ease in following it, ease of location of sampling units and consequent reduction in cost in reaching it, greater certainty of objectivity in the selection of sample and more uniform distribution of the sample.

Sequential sampling: Sequential sampling is a method of sampling whose characteristic feature is that the number of observations in the sample is not determined in advance but sampling units are taken successively from a population.

- Each sample includes all the sampling units of the former sample.
- The confidence interval at each stage will go on decreasing at each stage of sampling is stopped when the desired precision is reached.

This type of sampling is usually adopted to test a hypothesis.

41. a. Recent developments in GIS and digital image processing make forest cover assessment and mapping more easier and accurate. Explain your views. (2016)

b. What are the advantages of GIS over traditional forest management. (2006)

c. What is GIS? Describe the role of GIS for natural resource management. (2005)

Yes, it is true that the recent development in GIS and digital imagery make forest cover assessment and mapping very easy. Due to this advancement, the FSI under MoEF is carrying out assessment of India's forest cover once in every two years and bring out the status of natural forest and trees outside forest. For the assessment of forest cover, the satellite data ofLANDSAT having 80 m resolution was visually interpreted on 1:1 million scales. The minimum mappable area was 400 ha. The data of Indian remote sensing satellite RESOURCESAT - 1(IRS P6 sensor LISS – III) having resolution of 23.5 m and has been digitally interpreted on 1: 50000 scale with a minimum mappable area of 1 ha.

Digital image processing: It refers to the manipulation and interpretation of digital images with the aid of a computer. Here the digital image is fed into a computer one fixed at a time. The possible form are manipulated and given below.

Processing of raw satellite images data to correct geometric distortions, radiometric calibration to estimate noise.

Image enhancement: Produces applied to data in order to effective display the data for visual interaction. This involves techniques for increasing distinction features in a scene.

Image classification: Quantitative technique for automating the identification of features in a scene. Involves application statistical based decision rules to identify the fixed in an image if the rules are based on spectral radiations. It is called as spectral pattern recognition.

Data merging and GIS integration: These procedures are used to combine different data for a given geographic area. Frequently the remotely sensed data is combined with other source of information in GIS environment.

Biophysical modeling: Relates digital data to biophysical features and phenomena measured on the ground. Eg. Estimation of crop yield, pollution, concentration, watt depth likewise remotely sensed data are used for environmental modeling in GIS.

Geographic Information System: Geographic Information System (GIS) is defined as an information system that is used to input, store, retrieve, manipulate, analyze and output geographically referenced data or geospatial data, in order to support decision making for planning and management of land use, natural resources, environment, transportation, urban facilities, and other administrative records.

GIS is a computer based information system which is used for presenting and analysis of the geographic patterns present in earth surface of events that fully place on it. Simply it is a computer insisted information system for acquisition storage analysis display of geographic data in desired manner. GIS is designed to work with data referred by geographical coordinates.

Now days the importance of spatial dimension is recognized because of organizing of analysis capability. GIS is widely exploited in field of education, research, forest engineering and planning of other uses. It has ability to divergent source of data both spatial on non-spatial.

Basic functions of GIS

Functions	Sub-Functions
Data Acquisition and preprocessing	Digitizing Editing Topology building Projection transformation Format conversion Attribute assignment etc.
Data base management and retrieval	Data Archival Hierarchical modeling Network modeling Relational modeling Attribute query Object-oriented database etc.
Spatial measurement and Analysis	Measurement operations Buffering Overlay operations Connectivity operations etc.
Graphic output and visualization	Scale Transformation Generalization Topographic map Statistical map 3D Birds eye view etc. 1

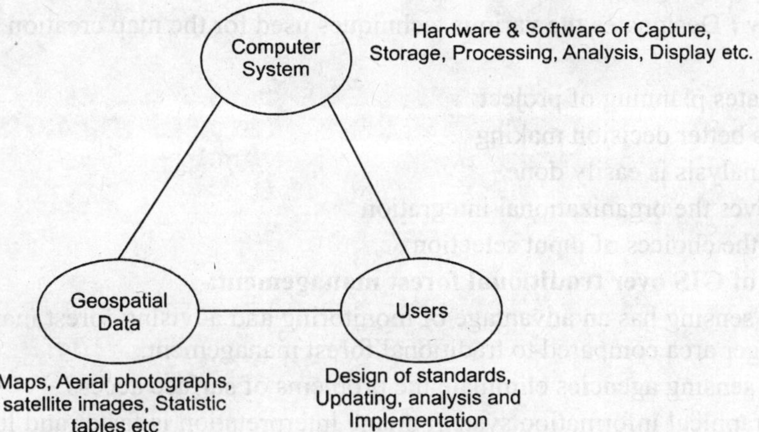

Key Components of GIS:

- **Hardware:** GIS refers the physical component of which the GIS software will run. It include component of plotter computer hardware such as CPU, Monitor, Printer, Scanner etc. It has both input and output device.
- **Software:** Provides the functions and tools that is needed to share analysis and display geographic information.
- **DATA:** Geographic feature can be controlled on purchased from specific categories. A GIS will integrate spatial data with other data resources. GIS stores vector and raster type of data that are found on map.
- **Vector data:** Represents boundaries on course of features such as a series of points
- **Raster data:** Represents features geographically.

4 Line Wave: Refers to the GIS experts and technical specialist who design and maintain system for daily work.

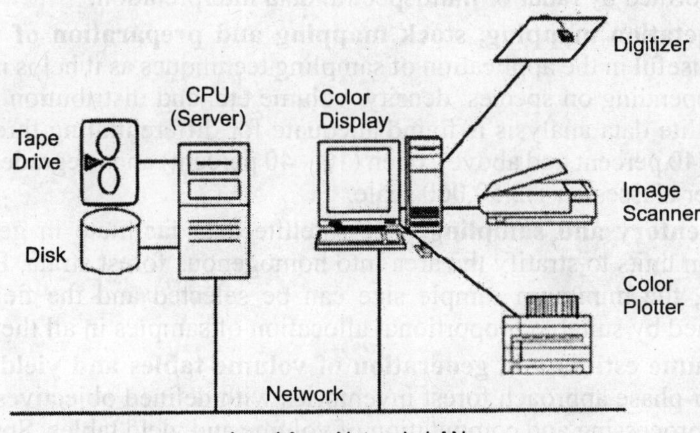

Components of GIS Hardward System

Source:http://www.westminster.edu/sta3/athrock/GIS/GIS.pdf.

Methodology : Designates the various techniques used for the map creation and future usage.

Advantages:

- It facilitates planning of project
- It makes better decision making
- Visual analysis is easily done
- It improves the organizational integration
- It helps the choices of input selection

Advantages of GIS over traditional forest management:

- Remote sensing has an advantage of monitoring and advising forest management strategy for a larger area compared to traditional forest management.
- Remote sensing agencies eliminate the problems of surface access
- In geographical information system image interpretation is faster and less expensive than traditional surveys involved in forest management
- GIS techniques are faster than traditional management system
- GIS techniques are applicable even in inaccessible areas wherein it is difficult to practice the traditional management system.
- Compared to the traditional forest management system the application of GIS based management is more accurate and timely.

GIS for Natural resource management

(i) **Reconnaissance mapping and forest cover monitoring:** Objectives to be met by National Forest or environmental agencies include forest cover updating, depletion monitoring, measuring bio physical properties of forest stands, forest cover type discrimination, agroforestry mapping.

(ii) **Species identification, forest typing and biodiversity assessment:** GIS provide a means of quickly identifying and delineating various forest types, a task that would be difficult and time consuming using ground surveys. Large scale species identification can be performed with multispectral, hyperspectral or air photodata while small scale cover type delineation can be performed by radar or multispectral data interpretation.

(iii) **Forest vegetation mapping, stock mapping and preparation of working plan:** It is extremely useful in the application of sampling techniques as it helps in correct delineation of strata depending on species, density, volume etc. and distribution of sampling units in them. Satellite data analysis is found adequate for differentiating three levels of stocking i.e, closed (40 percent and above), open (10 – 40 percent) and (degraded (below 10 percent) crown cover classes on 1:2,50,000 scale.

(iv) **Forest inventory and sampling:** The satellite data facilitate in generation of primary stratification units to stratify the area into homogenous forest strata. Based on the priority knowledge, the minimum sample size can be selected and the field inventory can be accomplished by suitable proportional allocation of samples in all the strata.

(v) **Forest volume estimation, generation of volume tables and yield table:** The satellite based multi-phase approach forest inventories with defined objectives provide ample data for further processing and computation of volume and yield tables. Spatial forest modeling using GIS can be substantial enhance the planning, harvesting strategies, location of forest stand, timber inventory data, ecologically sensitive area, key attributes of the terrain and other important factors could be mapped and include for spatial analysis necessary in harvesting plan preparation.

42. How is the remote sensing used in monitoring forest cover (2007)

Remote sensing: Remote sensing is a science and art of obtaining information about an object, area or phenomenon through the analysis of data acquired by a device that is not in contact with a object, area or phenomenon under investigation

Applications of remote sensing

Forest cover monitoring: The remote sensing helps to monitor and assess the forest cover. The Government of India since 1987 uses remote sensing technology to assess the forest cover monitoring system. The satellite data of LANDSAT having 80 m resolution was visually interpreted on 1:1 million scales. The minimum mappable area was 400 ha. The data of Indian remote sensing satellite RESOURCESAT - 1(IRS P6 sensor LISS – III) having resolution of 23.5 m and has been digitally interpreted on 1: 50000 scale with a minimum mappable area of 1 ha.

This has excluded the gaps / blanks in forest area with more than 1 ha distribution. Similarly the isolated forest patches of more than one hectare is captured through this satellite and included in the forest cover monitoring system. Since these maps are in digital form it is very easy and fast to generate maps on any desired scale with desired area for different stakeholders like central and state forest department.

Since 2003 onwards the satellite data (IRS – P6 LIS III) is used which is found to be superior in terms of radiometry and shadow compared to the earlier version. The digital interpretation has also made it possible to include one more canopy density class in forest cover mapping system.

FSI and Remote Sensing: The Forest Survey of India of Ministry of Environment and Forests, Government of India uses remote sensing tool to monitor India's Forest.The Forest Survey of India also updates the status of India's Forest once in two years using the remote sensing data. Similarly the Forest Survey of India use as remote sensing tool to detect and monitor the occurrence of fire across the country on a day today basis which also to monitor the forest cover.

43. To compute volume of standing trees which parameters are required to be measured. Write instruments used and brief procedure of their handling. (2007)

The volume of standing trees can be measured by using the parameters of height and diameter at breast height. The height can be measured using the instruments like Altimeter, Hypsometer, Abney's level etc. These instruments are working based on the principal of similar triangle and trigonometry. The diameter at breast height can be measured using tree calipers or girth can be measured using tailors tape and the diameter can be derived.

Procedure for handling altimeter for height measurement: Altimeter is the most common instrument used in measurement of height of trees. It has a gravity control, damped, pivoted pointer and a series of rotable scales on a hexagonal bark in a metal and a pistal shaped case. The top of the case is slightly rounded at the rare end of which has an eye piece with a pin hole and sight van in the front. It has a hallow tube of 1.5 cm long and 2 cm dia with two prongs protruding from the opposite side inside the tube. The prongs are in a horizontal line with a slight gap between the tool. Below the rounded top the instrument has a hexagonal bark which can be rotated by a knob on the front face of the instrument in line with the bar inside. The bar has a separate scale which are 15, 20, 25, 30 and the percent. The scale of unit 15, 20, 25 and 30 are the unit of horizontal distance from the tree to the observer's eye.

In order to use instrument, first the basal distance has to be fixed based on which the scale can be fixed by adjusting the knob. Once the distance is fixed, the pointer should be released by depressing the button and the top of the tree should be cited.

Through the pinhole of the eye piece, the sighting prongs should be aligned on the target site (tree top). After sighting the target in correct alignment, the trigger is forced to lock the pointer. Then the height of the top above high level can be read from the scale and noted. Similarly without changing the position, second observation is then directed towards the base. The pointer is again released and same procedure is followed to release the pointer and sight the base. Then the reading can be noted and the readings of top and bottom are added to obtain the height of the tree.

Height of the tree = Top reading + Bottom reading

Procedure for handling tree caliper for diameter measurement: Tree calipers are predominantly used to measure diameter of the standing tree directly. The caliper consists of a graduated rule and two arms. Among the arms, one is fixed at right angles to one end of the rule and the other is moveable. The length of the rule and the arms depends on the size of the trees to be measured. Each arm should be atleast half the length of the rule.

In order to use the caliper, handles of two arms of calipers are held in two hands. The moveable arm is tilted inwards so that it can move freely and help to insert the tree towards diameter measurement. The two arms are separated enough in order to receive the tree. When the tree touches the graduated rule, the moveable arm is shifted inwards in order to touch the tree. In this case, the fixed arm touches the left side of the tree and the moveable arm touches the right side of the tree trunk. The caliper should be placed at breast height level of 1.37 before doing the exercise. After fixing two arms in a correct position, the diameter of the tree can be read directly from the graduated rule.

Calculation of volume of standing trees

The volume of standing trees can be calculated by using the following formula.

$V = \pi r^2 h \times$ form factor (Form factor have to be determined separately and if already available it can be used)

Where V is the volume of standing trees

r is the radius calculated from diameter

h is the height of the tree assessed through altimeter

44. How age of the tree can be determined in which annual rings are not clearly observed. (2007)

Age determination is necessary to calculate the rate of growth of species or individual tree and also it is required to compare the financial outturn of a particular land use. The following methods can be used to determine the age of the tree without annual rings.

(i) **Through past plantation records:** For trees without annual rings the reference of existing plantation records play a significant role in determining age of trees. In several plantations, the maintenance of plantation register is mandatory which can act as a reference material for determining age of trees.

(ii) **Through general appearance of tree:** The appearance of the tree plays a vital role in determining age of trees when it has not growth rings. The appearance includes colour and condition of the bark, crown, size and shape and the tapper of the stem. In early years the bark will be green in colour and when it ages it turned into brown in colour. Similarly the crown in early stages will be linear and after certain period it is horizontal and wide spread. Trees at early ages have higher tapper and when it matures it really equates with cylindrical bole.

Example:	Teak	–	Crown shape
	Acrocarpus	–	Bark colour
	Melia	–	Stem tapper

(iii) Based on the number of whorls of branches: Some tree species like *Anthocephalus cadamba, Bombax ceiba, Sterculia fortida* etc. produces one whorls of branches annually and the age can be determined by counting the number of whorls by adding proportionate number of years for the portion wherein such whorls are not visible.

(iv) By formulae method: In case of trees without annual rings the age can be determined through periodic measurement of diameters and the age can be determined by the following formulae.

$$Age = \frac{1}{P_1 S}$$

Where, $S = \frac{\log P_{1-} \log P_2}{\log d_{2-} \log d_1}$

$$P_1 = \frac{\text{Difference in diameter in first two measurement}}{\text{Initial diameter} \times \text{Difference in year of first two measurement}}$$

$$P_2 = \frac{\text{Difference in diameter in last two measurement}}{\text{Mid diameter} \times \text{Difference in year of last two measurement}}$$

(v) Through age diameter curve: The age of trees without annual rings may also been determined through age diameter curves. However these curves have to be prepared for the species of interest.

(vi) Establishing relationship between age and growth parameters: Age can also been determined by establishing relationship between age and other growth attributes like height, diameter, volume, biomass etc. For this purpose suitable equations have to be prepared considering age as dependant variable and other growth attributes as independent variables.

Chapter 11

Surveying and Forest Engineering

1. Describe methods of using prismatic compass in forest surveys. (2010)

(i) Finding One's place on a survey plan or a map:

(a) **By compass alone** - *Q*uite often one may be required locate his position in the forest, on a plan or a map the area. From the observer's station, take bearing to two (preferably three, for check) prominent features, such as boundary pillars, which are also plotted on the map. From these points observed to, protract their back bearings corresponding to observed fore-bearings. Inter section of these two gives the observer's station on the plan. Another bearing to a third object/feature may taken for a check. This process of finding one's place in a survey is called interpolation.

(b) **By Compass and Chain**- Take bearing to a well defined object/feature from the observer's station, and measure distance to it. On the plan, at the object observed, protract the calculated back-bearing and scale off the distance on it. The point so fixed is the observer's station.

(ii) Finding the position of an inaccessible or a distant object to which it is not easy to chain: Select two or more known points - stations, or other points on the chain line or other fixed points inside the traverse from where the object is visible. Any two of these stations and the object is visible. Any two of these stations and the object should form, as nearly as possible, an equilateral triangle (for accurate intersection). Take bearings from these points to the object, and plot the same. Intersection of any two of them flues the position of the object; others serve as check.

(iii) Surveying a winding stream : finding its width: For surveying a meandering stream/river, open traverse with compass and chain is the best method. To locate bends, etc., offsets may be taken from traverse lines to various points on the stream banks, on either side. Fix one point on the other side of the river, either by intersection of two bearings from two stations to a flag on the opposite bank and proceed with open traverse on the other bank. To find the width of a stream, select a base line on one bank and fix a flag on the other bank. Take two bearings from two ends of the base line on to the flag and plot the bearings on the plan. Intersection of two bearings locates the flag on the opposite bank. Width of the stream can be scaled off on the plan. A third bearing from an intermediate point on the base line would be advisable as a check.

(iv) Ranging out a straight Hoe between two points In a forest: This operation is often required in forestry works, e.g., cutting a straight fire line overgrown with tall weeds, cutting a straight boundary line between two pillars, etc. This may be done by any of the following two methods

- **Open Traverse Method:** Let P and T be two points in a forest between which a straight line is to be ranged and cut. Run an open traverse from P to T, *via* Q, R and S keeping as close to the approximate direction of T as possible, After reaching T, plot the open traverse data ; join T & P. Now read the bearing of PT or TP on paper, and lay out the line on the ground by setting out with a compass from P or T end, the corresponding bearing as read from the plan
- **Random Line Method:** A straight line is to be cut between A and B. Proceed from A at a bearing of K towards B', estimating the direction of B from A as best as possible. Stop at B', so that B is visible from B' and BB' is perpendicular to AB'. Measure BB', and plot A, B', B on any convenient scale. Join AB on plan and measure angle B'AB, i.e., 0. To strike B from A, move at a bearing K+ !6 (i.e. angle NAB), or at a reverse bearing of (K+ !6) ± 180° from B to strike A.

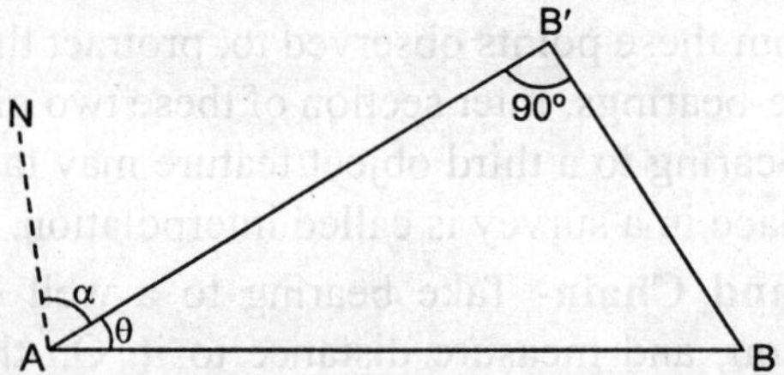

(v) Laying out a Forest Coupe on the ground:

(a) When a Map with the coupe shown on it is available : Suppose PQRS (Fig.) is the outline of the coupe shown on the map.

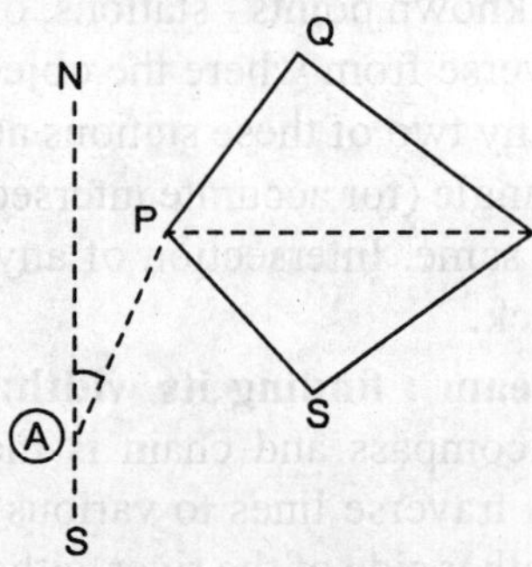

Fix P on the ground with respect to a known feature, say a boundary pillar A (also shown on the map), by measuring the bearing and distance of P from A on the map and running it on the ground from A to P. Points Q, R and S of the Traverse can be fixed by traversing, i. e., setting out bearings find distances successively, from P, Q and R.

(b) When no map is available: Sometimes in case of fuel coupes only the area of the coupe is specified in the Working Plans and the demarcation is left to the discretion of the territorial staff, subject to a certain prescribed sequence. In such cases suitable shape of the coupe should be decided - whether square, m tangle, parallelogram or a trapezoid and its sides (and angles) worked out, corresponding to the given area. When the plan is made, it can be laid out on the ground by compass and chain, as Ascribed in the preceding paragraph ("laying out of a coupe when limp is given")

(vi) Demarcation and checking of Boundaries:

(a) Demarcation: Proceed as in problem 5(a) above.

(b) Checking of Boundaries : In case boundary record giving Mir bearings and distances of successive boundary pillars is available (in, what is known as Boundary Register), set off from a known pillar the bearings and distances to successive pillars, and check discrepancies, if any.

If only a map showing forest boundaries is available, first of all measure distances and bearings (and offsets to curved boundaries, if necessary) between the boundary pillars as shown on the map and proceed to check on the ground.

2. Write down the chemical constituents of earth (Chemical formulae and percentage range of contents) for manufacture of good quality of bricks. (2013)

Chemical properties: Clay basic ingredient giving plasticity – excess of clay causes bricks to shrink and warp when drying.

Bricks are an important materials for the constructions of buildings. They are being increasingly used for building constructions even in forest areas as nowadays bricks can be transported to most sites in and near forests viz., division range block or beat headquarters.

Compositions:

Constituents	Chemical formula	% composition
Sand (Silica)	Si	50–60
Clay (alumina)	Al	20–30
Lime;Magnesia;Manganese	$CaCo_3$, Mn	Reset
Oxide of Iron, sodium and potassium	Feo, Na, K	

Sand:

- Produces resistance to heat
- Provides durability
- Preserve the shape of the bricks
- Prevent shrinkage and wrapping
- Provides the relative degree of softness to the brick

Clay:

- Imparts the required plasticity for moulding the bricks
- Helps the bricks to maintain their shape and size

Lime:

- Reduce shrinkage in raw bricks
- Fushing and binding the particles

Iron oxide: A very small quantity of iron oxide in the brick imparts it with a pleasant red colour. This is also done by magnesia.

3. Distinguish between Irish bridge, suspension bridge and cantilever bridge (2010 & 2016)

Irish Bridge: The Irish bridge is a paved dip surface of the road having a span of not more than 7 mts. The formation of the road is dropped to follow the bed of the shallow water course. A cross slope of about 1 in 12 inches on hill roads and 1 in 30 inches on flat ground is proved for.

Irish bridges are located at re-entrant curves and never at salients. Edges are rounded off to avoid sharp dips in the road. The bed is paved with dry stones or stone in lime or cement mortar lime or cement concrete.

These structures are preferred to culverts where there are chances of debris blocking the road, particularly in hilly areas.

The advantages of Irish bridges are:

- Cheap to construct
- A high level of technical know-how is not required
- Material that is available locally can be used.

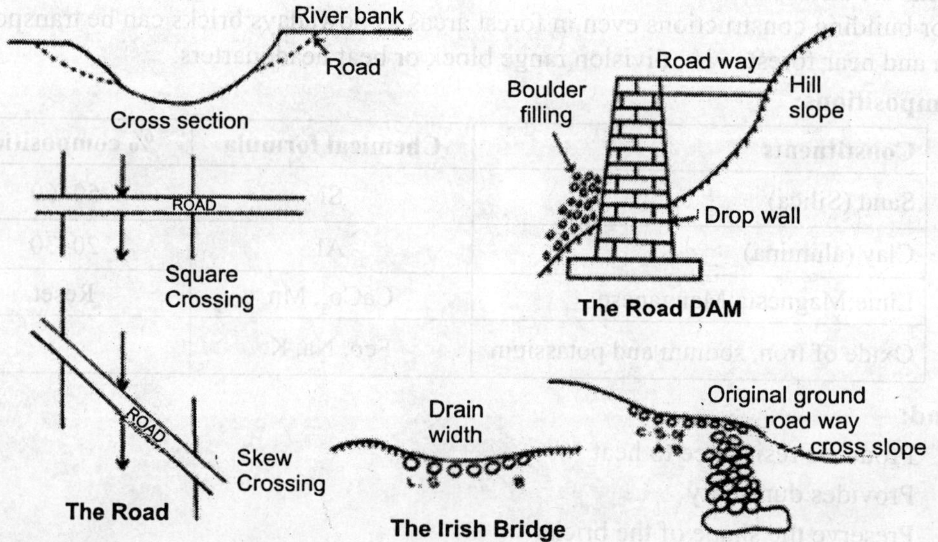

Suspension Bridge:

Introduction – To be erected across obstacles

1. More than 50 feet wide
2. Not admitting any intermediate support due to
   (i) Deep gorge
   (ii) Unstable foundation bed
   (iii) High velocity of flowing water

Suspension bridges of forest requirements carry generally laden men and mule traffic; of such loads, bridge suspended form:

- Two longitudinal ropes
- Two ropes pass over tops of high uprights (towers) on either side of obstacles.
- The free ends of ropes fixed to firm anchorages buried beneath the ground on either side of obstacle.
- Small ropes called 'suspender' or suspensors hang from longitudinal ropes.
- Hanging ends of suspenders carry cross-pieces
- Cross –pieces carry longitudinal beams
- Longitudinal beams carry the roadway decking

Cantilever Bridge: Bridge is independent of the nature of obstacle; Consists of two systems of counter-balanced beams built out from either side of the obstacle to be spanned; for stability of bridge, centres of gravity of the system of counter-balanced beams should fall well within the base of abutment over which the system of beams rests.

- Cantilever bridge is made up of two system of counter poised beams built out from the abutments supporting the road bearers in the middle.
- The beams are placed one above the other so that one end of each beam projected beyond the end of lower beam.
- The span gap is to be short so that it is to be connected by a single road bearer.
- The stability of bridge is depends upon the length of the part of the beam projecting out from the bank, counter weight, center of gravity of the counter weight and its distance from the abutments and size of the beams and their transverse strength.
- Cantilever and road bearers of similar size are used for the construction because each bears the same stress.
- The length of projection of beam should be 116th to 117th of the bridge span.
- Cantilever arms are connected by wall plates and transom to provide lateral stability. Abutments are constructed at a height so that the super structure is kept at above flood level.

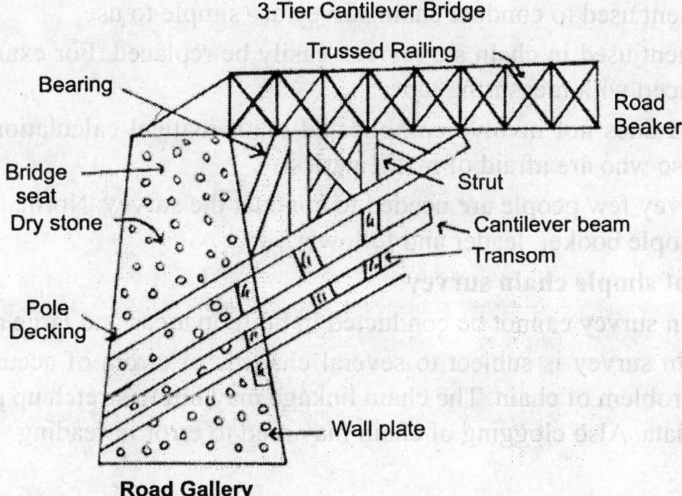

It is very difficult and cumbersome to construct besides huge wastage of materials. In additions, this, high-level technical knowledge is required for construction.

4. Describe all the five kinds of chains used in survey and advantages and disadvantages of the chain surveys. (2013)

Various kinds of Chains:

(i) Gunter's or Surveyor's Chain: It is 22 yds or 66 feet long; named after its inventor in1620. Each of its 100 links measures 0.66 foot or 7.92 inches. Formerly it was mainly used for land surveying. It is very convenient for measuring distances in miles and furlongs and areas in acres, because of its simple relation to these units; 10 chains = 1 furlong; 80 chains = 1 mile; 10 square chains = 1 acre.

Distances measured with Gunter's chain are recorded in chains and decimals, e.g. 3.63 chains; or chains and links, e.g. 3 chains and 63 links.

(ii) Engineer's chain: It is 100 ft. long chain; each one of its 100 links measures 1ft. Till recently it was exclusively used for leveling and engineering surveys, because of facility of computation of quantities of work. Levelling staffs were also graduated in feet, tenths and hundredths of a foot. Now metric chains are being used.

(iii) 50-ft Chain: It was mainly used in hilly country along steep slopes on account of its lightness.

(iv) 60 ft chain: Mainly used for mining surveys; the chain being ten fathoms long, each fathom being six feet. The tallies in such chains mark each fathom.

(v) Metric chain: In countries where unit of length is a meter (as now in India also) 10, 20, 25 and 30 meter chains are used; 20 meter or 2 – dekameter chain (65.617') is the most commonly used chain. It is divided, as usual, into 100 links, each 2 decimetres in length, and brass tallies are attached at every 2 meters from each end.

Advantages of chain survey

- Chain survey is simplest and commonest method used in surveying exercises
- The equipment used to conduct chain survey are simple to use,
- The equipment used in chain survey can easily be replaced. For example measuring rods can be replaced with measuring tape.
- This method does not involve complicated mathematical calculation. I know this is the relief to those who are afraid of mathematics
- In chain survey few people are needed to conduct the survey. Normally chain survey team has three people booker, leader and follower.

Disadvantages of simple chain survey

- Simple chain survey cannot be conducted in built up areas and large areas.
- Simple chain survey is subject to several chances of errors of accumulation which may caused by problem of chain. The chain linkage may fail to stretch up properly and result in inaccurate data. Also clogging of chain may read to error in reading.

- It is time consuming
- It may not be conducted in areas with steep slopes or water logged areas. Chain survey is usually conducted in dry areas with gentle slopes. It becomes more complicated when survey is conducted in areas that are too wet.
- Chain survey becomes more complicated method when there are raised points (obstacles) in between areas to be surveyed

5. Topographic surveys are found useful in forest management. Discuss. (2015)

Topographical survey is used in preparing a contour map. This is useful in depicting the topographical features in an area. Distinguishing feature of a topographical features in an area. Distinguishing feature of a topographical survey is the determination of location of the objects, both in plane as well as elevation.

Contouring over large areas is known as *Topographical survey.* Distinguishing feature of a topo-survey is the determination of location of the points and objects, both in plan and in elevation, necessary for plotting of contours and preparation of a topographic map, on which the relief features of the surface can be interpreted. Topo-survey, therefore, collects necessary field data required to prepare a contour map depicting the topographical features of the area. Topographical map is a graphic representation of the results obtained from such a survey; and as it portrays the third dimension of the ground features as well, it is of great utility.

Aerial photogrammetry is increasingly used for topographic surveying of extensive areas. Even for ground surveys, the surveyor should study aerial photographs of the area, whenever available; examination of over lapping pairs of photographs by means of simple stereoscope affords vision as in three dimensions, and is of great help.

A forester needs topographical survey for planning, silvicultural treatment purposes, soil conservation, timber extraction, road alignment, etc. From the topographical survey one can obtain the following useful information:

(i) It gives detailed topographic information of the terrain such as relief, slope, etc.

(ii) Laying, determination of gradients and alignment of roads are easier from the topographical maps.

(iii) It helps in the selection of sites for forest plantation.

(iv) It facilitates lay out of fire lines and irrigation channels for forest nurseries.

(v) It helps planning of stock mapping and enumeration operations.

(vi) It facilitates computation of earth work for cutting and filling

(vii) Inter-visibility of two points can be ascertained from a topographical map.

(viii) The watershed area, capacity of a reservoir and water supply can be estimated.

Uses of Topographic (Contour) Map:

(i) Contour maps give all important information regarding the relief features and terrain of the tract, whether flat, undulating, hilly, etc.

(ii) They help the forester, engineer and a surveyor in selection of suitable sites for layout of roads, irrigation channels, reservoirs, nurseries, soil conservation works, plantations, etc.

(iii) Quantities of earth work, cut and fill, involved in various engineering projects can be computed from a contour map.

(iv) Intervisibility of two points can be ascertained from a topo-map.

(v) With the help of a contour map, area of the drainage basin (watershed) and the capacity of a reservoir for irrigation purposes and water supply can be estimated.

(vi) Watershed line separating one drainage basin from another is traced on the map, say, along the ridge RR' and through the saddle S. Extent of the drainage basin area can be estimated by measurement (with a planimeter or any other method), taking care that the entire area draining into a valley is included.

(vii) A Contour map is useful in tracing a given contour gradient through a point of purposes of road alignment, layout for irrigation channels, etc.

6. Write the simple design and construction of timber bridge across river in forest. (2009)

Simple Wooden Bridges: Simple wooden bridges are the most commonly constructed form of cross-drainage for forest roads, particularly in the interior areas.

The main components of simple wooden bridges are:

(i) Abutments made of masonry or timber poles to support the earth pressure.

(ii) Road bearers are longitudinal beams to span the gap between the abutments.

(iii) Planks or ground poles are laid side by side on the road bearers for forming the road surface.

(iv) A layer of brushwood and earth or other suitable material is laid on these poles or planks.

The maximum span of such bridges is about 6 mt. This ensures that the bridge is not dangerous and unsafe. However, the length of the span can be more than 6 mt. If piers are constructed in between the abutments.

There are a number of modifications of the simple wooden bridge:

(i) Strutt and straining beam bridge: In case bearers of the required length are not available, 2 or 3 lengths can be made use of, supported by a straining beam and two inclined struts built into erodiable banks. The former should have a length of atleast half span.

The struts are as wide as the straining beam though depth depends on the lad and angle of inclination. A 45 degree inclination is commonly adopted in case of strutts

(ii) Strutted Beam Bridge: This bridge is constructed for spanning more than 6 mts by a wooden bridge. Long logs are used as longitudinal beams or road bearers. However, there is need for intermediate support so that the beams do not sag. An inclined strut is provided for giving this support. The upper end of the strut abuts against a block of wood bolted to the road bearer while its food end rests against a food block of stone slab.

A strutted beam is provided for each road bearer. Its size is equal to that of the road bearer.

This type of bridges is used for spans of upto 15 mts. All the strutts are fastened together by cross pieces to prevent them from spreading out.

Super Structure: The super structure of simple wooden bridges consist of the following:

(a) Road bearers: Timber beams of cross section upto 35 × 35 cm or 40 × 40 cm are used as road bearers. They must be of a durable timber species. The road bearers are fixed on the bridge seats in a longitudinal direction with the help of drift bolts.

(b) Decking: The main decking is formed of planks having a thickness of 8 cms. They must be strong enough for supporting the load over the span between the road bearers. The planks are nailed, bolted or spiked to the road bearers. A wearing layer of 3 to 5 cms thick plants is laid horizontally or diagonally to the axis of the bridge over the main decking on bridges carrying heavy loads.

This wearing layer can be replaced as frequently as desired; without the need to re-lay the main decking.

(c) Wheel Guides: Wheel guides or ribands with splayed approaches are provided on the bridge for guiding the wheels and helping drivers in negotiating the bridge.

These are made of beams with a cross-section of 12 x 12 cm. Guard posts are added to the wheel guides.

(d) Hand Rails: Handrails are provided for preventing traffic, human beings and cattle passing over the bridge from falling off.

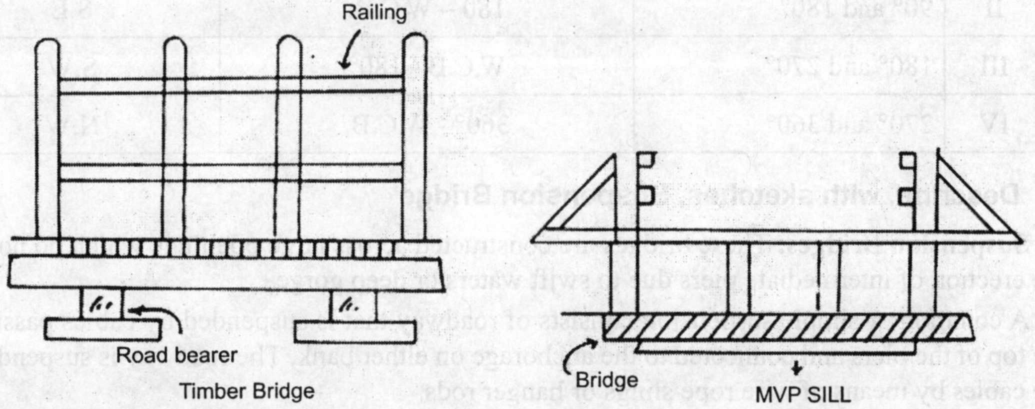

7. Explain the terms (2011)

(i) Whole circle bearing, and (ii) Quadrantal bearing

(i) Whole Circle System: In this system the bearing of a line is reckoned Clockwise from the north pint of the reference meridian, right round the circle. The angle thus measured is called the Whole Circle Bearing (W.C.S) which may be defined as the angle which a line makes with the meridian (N/S line) measured form the north point in clockwise direction. It may have any value between 0 ° and 360°. In this system, W.C.B. completely specifies the direction of the line. Prismatic compass indicates whole circle bearings.

(ii) Quandrantal System: In this system the bearing of a line is measured clockwise or anticlockwise, from the north or the south point, whichever is nearer. The angle so measured is called the Quadrantal Bearing (Q.B) or Reduced Bearing (R.B.) which may be defined as the angle which a line makes with the North / South line, in clockwise or anticlockwise direction, from the north or the south point (but never from East / West pints), whichever is nearer. It may also be defined as the minimum angle which a survey line makes with the North / South line, with quadrant specified. It is obligatory to specify the quadrant, alongwith the magnitude of the angle. Numerical value of quadrantal bearings cannot be more than 90°, which is definitely advantageous in finding the values of trignometrical functions (sines, cosines, etc.) as most trigonometrical tables deal with angles from 0° to 90° only.

When the whole circle bearing of a line exceeds 90°, it must be reduced to the corresponding angle less than 90° which has the same values of trignometrical functions. This angle is known as the Reduced Bearing (R.B.)

The method of obtaining R.B from W.C.B. is tabulated below

S.No.	W.C.B. lies between	Reduced bearing is equal to	Quadrant
I	0° and 90°	W.C.B.	N.E
II	90° and 180°	180 – W.C.B.	S.E
III	180° and 270°	W.C.B - 180°	S.W
IV	270° and 360°	360° - W.C.B.	N.W

8. Describe, with sketches, Suspension Bridge (2011)

Suspension Bridges: These bridges are constructed across large openings which do not permit the erection of intermediate piers due to swift waters or deep gorge

A common or simple suspension consists of roadway that is suspended by cables passing over the top of the piers and connected to the anchorage on either bank. The roadway is suspended from the cables by means of wire rope slings or hanger rods.

Parts: The main parts of suspension bridges have been described in the following text :

(a) Cable

(b) Towers or Piers

(c) Anchorage

(e) Suspenders

1. More than 50 feet wide
2. Not admitting any intermediate support due to
   (i) Deep gorge
   (ii) Unstable foundation bed
   (iii) High velocity of flowing water

Suspension bridges of forest requirements carry generally laden men and mule traffic; of such loads, bridge suspended form:

- Two longitudinal ropes
- Two ropes pass over tops of high uprights (towers) on either side of obstacles.
- The free ends of ropes fixed to firm anchorages buried beneath the ground on either side of obstacle.
- Small ropes called 'suspender' or suspensors hang from longitudinal ropes.
- Hanging ends of suspenders carry cross-pieces
- Cross –pieces carry longitudinal beams
- Longitudinal beams carry the roadway decking

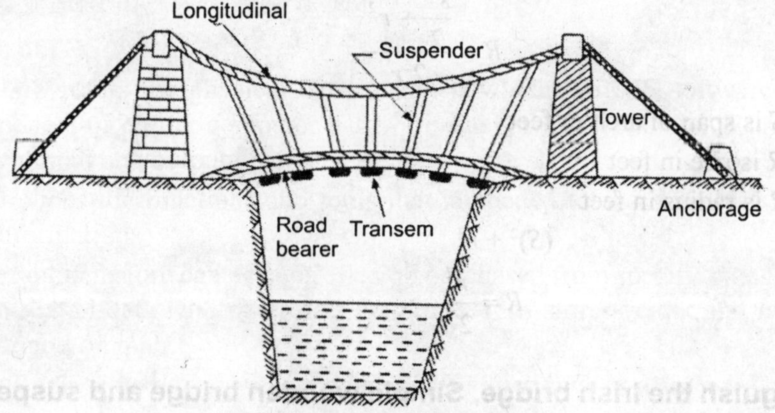

Fig. Suspension bridge

9. Define arches used in construction. What are the di-erent characteristic required in an arches? (2013)

Arches: Arches are an arrangement of wedge shaped blocks mutually supporting each other constructed to form a curve and supported at both ends by abutments or pillars. Arches are designed to support the load of any superstructure over them.

Characteristics required in an arch: A well constructed arch should have the following main characteristics:

1. The abutments must be strong enough to resist thrust from the arch. It should be able to counter the overturning moment of the arch thrust.
2. The depth or thickness of the arch should be able to bear the load coming over it.
3. The size of the arch ring must be such that it does not fail by crushing.
4. Thickness of various arches is calculated by the different formulae given below:

   (a) For segmental arches

$$T = \sqrt{0.12 \times \text{radius in feet}}$$

(b) For segmental arches in series

$$T = \sqrt{0.2 \times \text{radius in feet}}$$

(c) For segmental arches in series

$$T = \sqrt{0.17 \times \text{radius in feet}}$$

(d) For semi circular arches in series

$$T = \sqrt{0.25 \times \text{radius in feet}}$$

Where *T* is thickness of the arch in feet. *R* is radius of intrados.

It is calculated by :

$$R = \frac{\frac{s^2}{2} + r^2}{2\,r}$$

Where, *S* is span of arch in feet

R is rise in feet

R is radius in feet

$$(S)^2 + r^2$$

$$R = \frac{r^2}{2\,r}$$

10. Distinguish the Irish bridge, Simple wooden bridge and suspension bridge with details and neat sketches? (2009)

Irish Bridge: The Irish bridge is a paved dip surface of the road having a span of not more than 7 mts. The formation of the road is dropped to follow the bed of the shallow water course. A cross slope of about 1 in 12 inches on hill roads and 1 in 30 inches on flat ground is proved for.

Irish bridges are located at re-entrant curves and never at salients. Edges are rounded off to avoid sharp dips in the road. The bed is paved with dry stones or stone in lime or cement mortar lime or cement concrete.

These structures are preferred to culverts where there are chances of debris blocking the road, particularly in hilly areas.

The advantages of Irish bridges are:

- Cheap to construct
- A high level of technical know-how is not required
- Material that is available locally can be used.

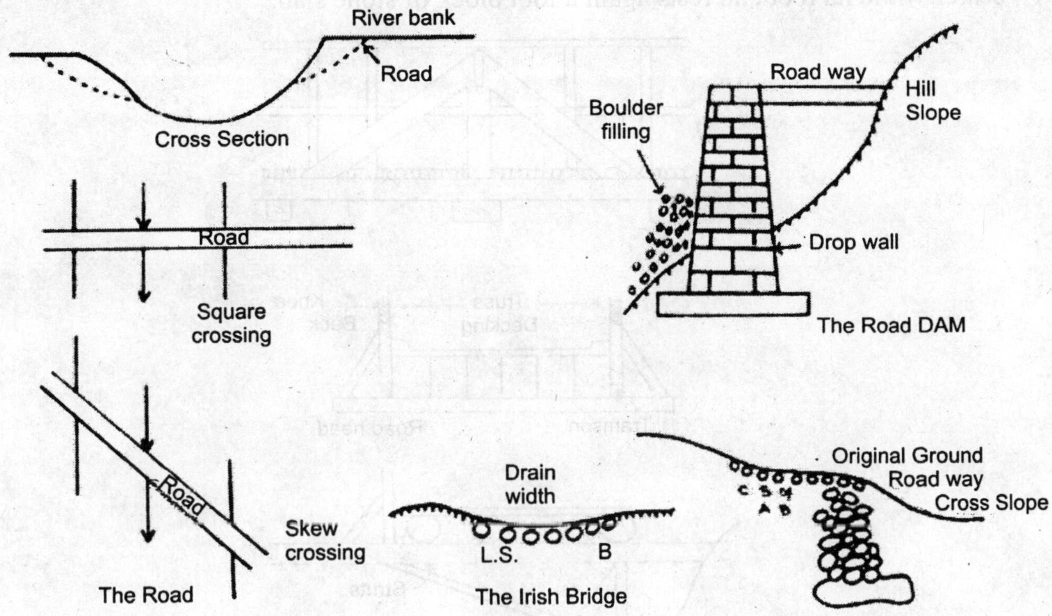

Simple Wooden Bridges: Simple wooden bridges are the most commonly constructed form at cross-drainage for forest roads, particularly in the interior areas.

The main components of simple wooden bridges are:

- Abutments made of masonry or timber poles to support tit earth pressure.
- Road bearers are longitudinal beams to span the gap between the Abutments.
- Planks or poles are laid side by side on the road for forming the road surface.
- A layer of brushwood and earth or other suitable material is laid on these poles or planks.

The maximum span of such bridges is about 6 mt. This ensures that the bridge is not dangerous and unsafe. However, the length of the span can be more than 6 mt. if piers are constructed in between the abutments.

There are a number of modifications of the simple wooden bridge:

(i) Strutt and Straining Beam Bridge: In case bearers of the required length are not available, 2 or 3 lengths can be made use of, supported by a straining beam and two inclined strutts built into erodiable banks. The former should have a length of at least half span.

The strutts are as wide as the straining beam though depth depends ca the load and angle of inclination. A 45 degree inclination is commonly adopt in case of strutts.

(ii) Strutted Beam Bridge: This bridge is constructed for spanning more than 6 mts by a wool bridge. Long logs arc used as longitudinal beams or road bearers. However there is need for intermediate support so that the beams do not sag. An inclines: strutt is provided for giving this support. The upper end of the strutt abuts against a block of wood bolted to the road bearer while its foot end rests again a foot block of stone slab.

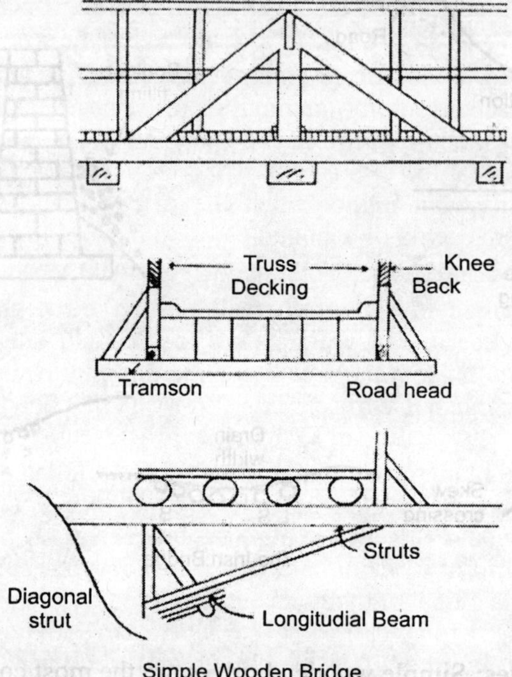

Simple Wooden Bridge

A strutted beam is provided for each road bearer. Its size is equal that of the road bearer. This type of bridges are used for spans of upto 15 mts. All the strutts are fastened together by cross pieces to prevent them from spreading out.

Suspension Bridges: These bridges are constructed across large openings which do not permit the erection of intermediate piers due to swift waters or deep gorge

A common or simple suspension consists of roadway that is suspended by cables passing over the top of the piers and connected to the anchorage on either bank. The roadway is suspended from the cables by means of wire rope slings or hanger rods.

Parts: The main parts of suspension bridges have been described in the following text :

(a) Cable

(b) Towers or Piers

(c) Anchorage

(d) Suspenders

More than 50 feet wide

Not admitting any intermediate support due to

(i) Deep gorge

(ii) Unstable foundation bed

(iii) High velocity of flowing water

Suspension bridges of forest requirements carry generally laden men and mule traffic; of such loads, bridge suspended form:

- Two longitudinal ropes
- Two ropes pass over tops of high uprights (towers) on either side of obstacles.
- The free ends of ropes fixed to firm anchorages buried beneath the ground on either side of obstacle.
- Small ropes called 'suspender' or suspensors hang from longitudinal ropes.
- Hanging ends of suspenders carry cross-pieces
- Cross –pieces carry longitudinal beams
- Longitudinal beams carry the roadway decking

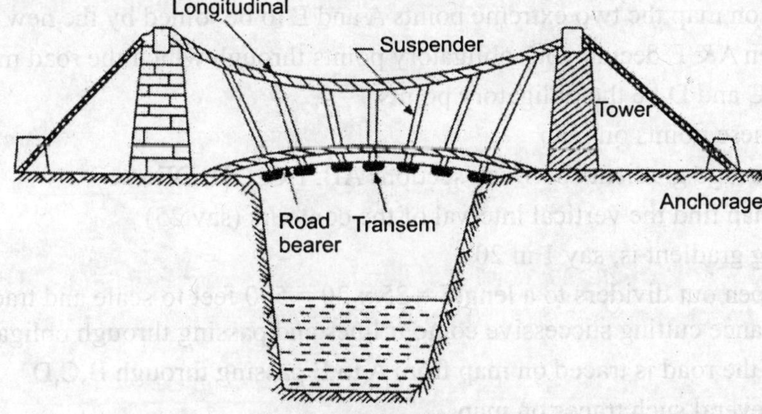

Fig. Suspension bridge

11. What are the major principles of designing roads in remote forest area? Provide diagrams wherever necessary (2009)

Broadly road in and adjoining forest areas which are constructed and maintained by the forest department are termed as forest roads. The practice of constructing roads in forest areas is fairly old and the work is carried out by the forest department.

Requirements of good forest roads

The basic requirements of good forest roads are listed below:

- The road surface should be impervious, hard, durable non slippery and neither too wet nor too dry (in wet and dry weather respectively).
- The gradient should not be too steep
- The road surface must have least co-efficient of friction
- The forest road must be easy to clean and carry out repairs

Principle of designing a road

- Good alignment, directness, visibility
- Easy gradients

- Sufficient width
- Properly designed curves
- A stable sub-soil (sub-grade) to carry foundation adequately drained.
- Proper drainage
- An adequate foundation (base)
- Wearing course i.e. a suitable surface for general character of the traffic
- Correct camber and super-elevation
- Economical construction in materials, labour and transport

Road Alignment in Hills for Ghat Road

- Have a good map of the area in which the new alignment is to be made
- Locate on map the two extreme points A and E to be joined by the new alignment.
- Between A& E decide your obligatory points through which the road must pass.
- Let B, C and D be the obligatory points
- Mark these points on map
- Find average gradients between sections AB, BC, CD, DE
- From map find the vertical interval of the contours (say 25)
- If ruling gradient is, say 1 in 20
- Then open out dividers to a length = 25 x 20 = 500 feet to scale and tract out on map with the distance cutting successive contour lines and passing through obligatory points.
- Finally the road is traced on map from A to E passing through B,C,D
- Draw several such traces on map
- Reconnoitre these traces actually on ground
- Select the best trace
- Align on the best trace selected

Hill roads: The section of hill roads varies from place to place depending upon the topography, geology and quantity of rainfall that is received.

The following points are to be kept in mind while designing the section of hill roads:

- The slope may be uniform either towards the hill side or outwards as the case may be. In either case, the transverse slope of the road should not exceed 1 in 20 as otherwise there will be heavy erosion on the surface.
- An outward slope is preferred in areas having rainfall less than 50 cms per year. The reason for this is that there will not be much cross drainage. The cost of constructing the inner gutter and a number of culverts will also be saved. There will also be an increase in the effective width of the road on account of the absence of the inner gutter.
- An outward slope is not suited in areas receiving heavy rainfall as the road surface will be severely eroded and damaged by the run off flowing from the uphill side. An inward slope is preferred with an inner gutter so as to intercept the rainwater flowing in from the uphill side and to drain away the surface run off.

- A general cambered cross section is preferred for permanent roads. Two drains are provided on either sides of the road. On steep gradients, the camber is increased so as to confine the water running down the slopes to the side drains.

Gradient / Slope: Gradient refers to the longitudinal slope of the centre of the roadway. It is usually expressed in terms of the horizontal distance in which the road rises or falls by one unit. It is the degree of elevation or depression of the road surface with reference to the horizontal plane.

General Dimensions: The general dimensions recommended for an ordinary single lane forest road of different classes is given in the following table:

S.No.	Class	Width (in mts.)	Camber	Gradient	
				Max.	Ruling
1.	Main motorable	3.5	1/2 to 1/25	1 in 15	1 in 20
2.	Branch Jeepable	3	1/5 to 1/20	1 in 10	1 in 15
3.	Bridle path	2 to 2.5	-	1 in 5	1 in 8
4.	Inspection	0.6 to 1	-	1 in 2	1 in 5

Road Alignment in the Plains: Alignment of forest roads in the plains is relatively easier than in hilly country. The main considerations in road alignment in the plains are

- Main roads are laid out in the central part of the area to be covered with there being feeder roads to connect different parts.
- Unnecessary crossings of rivers, streams and natural drainage channels are avoided
- Wherever possible, the alignment of the road is on high ground to keep the road dry.
- The monotony of very long and straight stretches on the road is broken by reverse curves; planting trees or digging wells.
- A small deviation of upto ten percent of the total length of the road may be made in order to connect a village, inn, railway station or even rest house. This will help to accelerate the growth of the locality.
- A detour can be made to negotiate an isolated hill or block of hills between terminal stations if the avoidable rise is saved by it.
- As far as possible, the existing roads and paths are used.
- Private land holdings are avoided in order to reduce the overall costs.
- For small streams, bridges may be located wherever they cross the general alignment of the road. However, the road may be deviated for approaching a good bridge site for fording larger rivers.

Road alignment in the hills: Alignment of a forest road in hilly terrain involves a considerable level of expertise. The main considerations for such alignments are

- An extraction road should follow the valleys and should be as near as possible to the lower boundary of the areas being worked. This will facilitate the carriage of timber or other forest produce down to the road head.

- A main communication road, should as far as possible, follow the ridges.
- The work of road alignment may be started from the higher obligatory points to the lower obligatory points on either side.
- Unnecessary ascents and descents area avoided. Every meter of elevation gained is maintained.
- If possible, the road is aligned on the sunny side of the valley or on the side exposed to winds. This will help to keep the road dry.
- The slopes may be within the limits of the ruling gradient, unless it is absolutely unavoidable. To begin with the gradient is slightly flatter than the ruling gradient, easing it at all reentrants, culverts, bridges and curves.

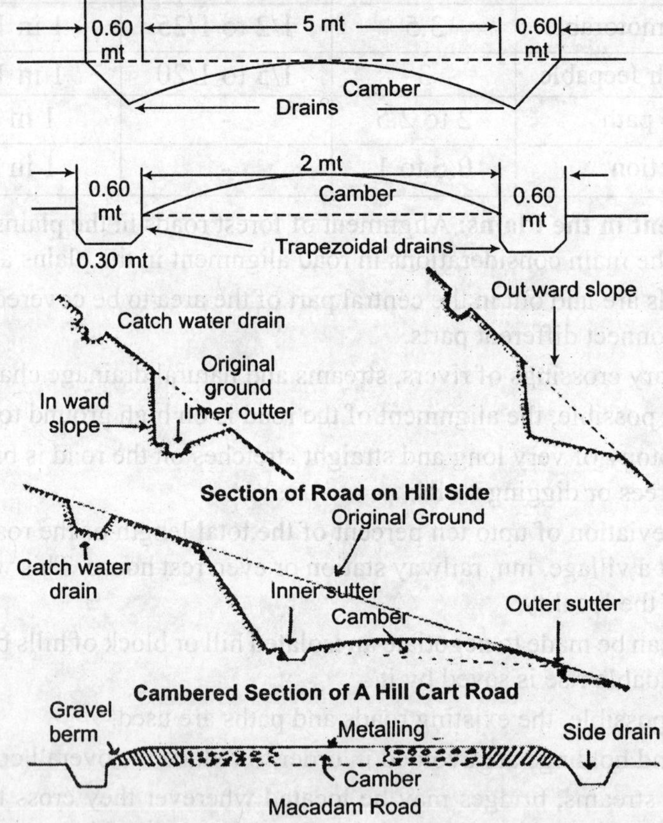

- Approaches to bridges may be selected keeping in view the following points.
  - ♦ The approach should be straight for 50 mtrs. on either side of the bridge. In case it absolutely necessary to provide a curve, the road may be widened with provision for adequate sighting.
  - ♦ The gradients may be eased off if it is not possible to have a level stretch of road over the bridge.

- Heavy rock cuts and locations likely to be affected by landslides are avoided. The cost of road construction may also be reduced by detours to avoid deep cuttings and high embankments.
- Wherever possible, hairpin bends and zig-zags are avoided as these are both difficult to construct and maintain.
- Halting places are provided at regular intervals in long continuously rising roads.
- Road should be aligned on that side of the hill where the strata are dipping into the mountain side and locations where the reverse in the case are avoided.

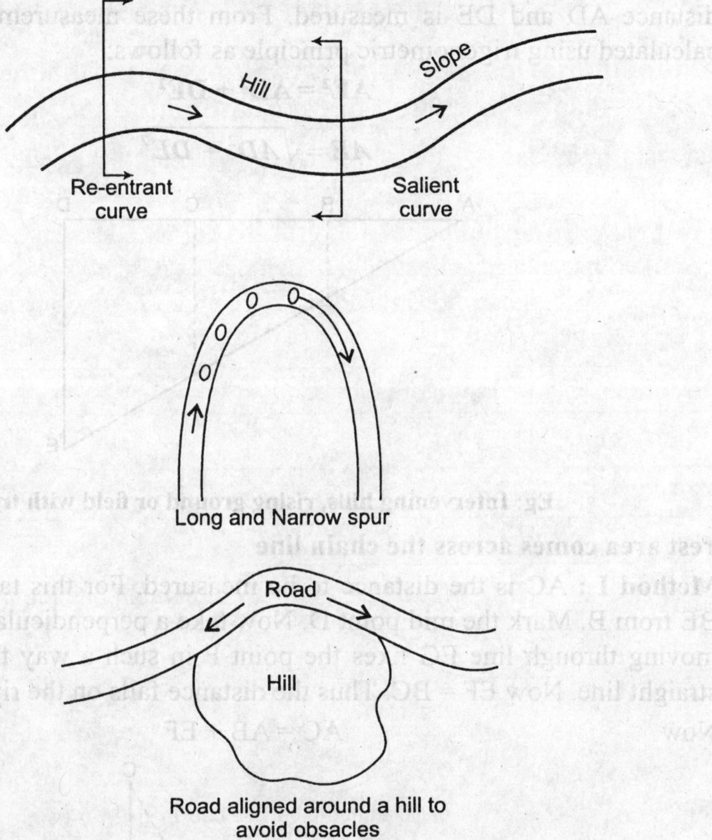

Road aligned around a hill to avoid obsacles

- It is preferable to align roads on forested slopes than on bare mountain sides; as such alignments are stable and make the roads less costly to maintain.
- Long stretches of dead level roads are avoided as they tend to hold up surface runoff and are difficult to drain.
- A downward gradient (reverse gradient) is provided from both sides of the valley while aligning a road across a deep ravine on ruling gradient.

12. Describe how you would continue the line with the chain only, when (2011)

(i) Large river interrupts the chain line and

(ii) A forest area comes across the chain line

(i) Large river interrupts the chain line

- **Both ends are visible from intermediate points**: This can be suitably measured by using reciprocal ranging methods.
- **Both ends are not visible from intermediate points**: AE is the distance to be needed. For this a random line AD is laid. From which a perpendicular DE drawn. The ground distance AD and DE is measured. From these measurements, the distance AE is calculated using trigonometric principle as follows:

$$AE^2 = AD^2 + DE^2$$

$$AE = \sqrt{AD^2 + DE^2}$$

A B C D

E

Eg: Intervening hills, rising ground or field with trees

(ii) A forest area comes across the chain line

- **Method I :** AC is the distance to be measured. For this take a perpendicular offset BE from B. Mark the mid point D. Now take a perpendicular offset EG from E. Now moving through line EG fixes the point F in such a way that F, D and C falls in a straight line. Now EF = BC. Thus the distance falls on the river measured indirectly.

Now $$AC = AB + EF$$

C

River

E B

D

F A

- **Method 2:** AC is the chain line and part of this line BC is obstructed by a river. Take a perpendicular offset BD. Make another perpendicular offset at point D by looking point A and C side by side. Now the triangles ABD and BCD are similar and hence the distance BC is calculated using trignometrical principles.

$$\frac{BC}{BD}=\frac{BD}{AB},\ BC=\frac{BD^2}{AB}$$

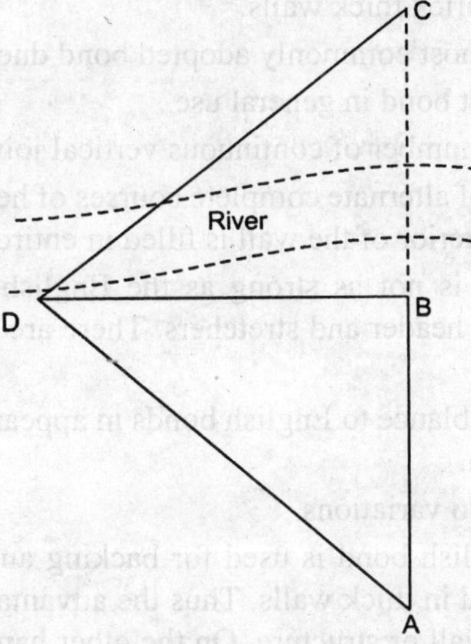

13. Describe di-erent types of mortar and its function in building construction.(10). (2005)

Mortar is a mixture of lime or cement with sand, surkhi or cinder and water used in a plastic state for binding individual bricks, stones, tiles etc. into one solid mass. It is a mechanical mixture consisting of .

- A binding or cementing material. This may be lime or cement,
- An inert material. This may be sand, crushed stone, brick or boulder

Functions: Mortar performs the following functions:

- Due to its plasticity, mortar forms an even bed or layer on which bricks, stones or even metal is laid in a structure. As a result, the superincumbent weight is uniformly distributed.
- Mortar serves as an adhesive cementing material for binding bricks, stones and metal into a solid mass of the desired strength and dimensions.
- Mortar is used for providing a hard, smooth covering for walls, roofs and floors in the form of a plaster. This plaster may either be grey, white or of some other colour.
- For facilitating moulding
- For filling up joints and making the structure resistant to the influence of weather.

14. Describe types of bonds in brick masonry works and explain general rule for good brick bonding. (20). (2005)

Types of Bond: The following types of bonds are usually used in forest engineering.

Heading bond: This bond is made up entirely of headers with a lap of 6 to 8 cms. The heading bond is used for footings and in turning small radius curves.

Stretcher Bond: This bond consists entirely of stretcher courses with a lap of 10 to 12 cms. These bonds are used for half brick thick walls.

English bond: This is the most commonly adopted bond due to the following reasons.

- It is by far the strongest bond in general use.
- These are a minimum number of continuous vertical joints in thickness of walls.

The English bond consists of alternate complete courses of headers and stretchers on the face of the brick or stone work. The interior of the wall is filled in entirely with headers.

Flemish bond: This bond is not as strong as the English bond described above elevation consists of alternate courses of header and stretchers. There are more continuous vertical joints in the interior of the wall.

This bond has a close resemblance to English bonds in appearance. It is used where appearance is more important than strength.

Flemish bonds may have two variations

Single Flemish bond: English bond is used for backing and Flemish bond for the wall face where extra strength is required in thick walls. Thus the advantages of both these bonds are made use of while constructing the wall or structure. On the other hand their disadvantages are more or less eliminated.

Double Flemish bond: In this arrangement, the Flemish bond is made use of on both the faces of the wall or structure.

Rules for Good Bonding: Good bonding in brickwork is obtained in the following manner

- Bricks are in an uniform arrangement
- The number of bats used in as less as is possible
- The vertical joints in every course should be perpendicular in line in the external and internal face of the wall.
- Stretches are used for the faces of the wall while the interior is made up only of headers.
- Dimensions of the width is such that after they have been bedded, the length is two times the width plus one mortar joint.
- Mortar should be thoroughly filled up in both the bed joints and vertical joints
- Load or pressure must be distributed over the entire masonry work. For this straight joints should be avoided and each brick should break vertically with its neighbours.
- The transverse joints must be continuous.

  The compressive strength of bricks depends on the following variables :

  - Strength of the mortar

- Type of bond being used
- Strength and nature of the bricks.

15. What is the importance of Prismatic Compass in Forest Surveys? (10). (2006)

Prismatic compass: The prismatic compass is one of the magnetic compass in which there is a prism for taking observations. The prismatic compass is generally smaller in size than a surveyor compass. The prismatic compass is used to determine the whole circle bearing of the lines. It consists of circular box, which box is made up of brass or a non metallic material. At the centre of the box there is a hard steel pivot which supports the magnetic needle, on focussing shows the magnetic bearing of the object.

Applications of Prismatic compass in forest surveys:

(i) Finding one's place on a survey or plan

(a) By compass alone: Quite often one may be required to locate his position in the forest, on a plan or a map of the area. From the observers bearings take bearings to two or three permanent features, such as pillars, which are also plotted on the map. From these points observed to, protract their back bearings corresponding to observed fore bearings. Intersection of these two gives the observers station on the plan. Another bearing to a third object/ feature may be taken for a check. The process of finding one's place in a survey is called interpolation.

(b) By Compass and chain: Take bearing to a well defined object/feature from the observer's station, and measure distance to it. On the plan, and measure distance to it. On the plan, at the object observed, protract the calculated back bearing and scale off the distance on it.

(ii) Finding the position of an inaccessible or a distant object to which it is not easy to chain: Select two or more known points, or other stations on the chain line or other fixed points inside the traverse from where the object is visible. Any two of the stations and the object should form, as early as possible an equilateral triangle. Take bearings from these points

To the object, and plot the same.Intersection of any two of them fixes the position of the object; others serve as check.

(iii) Surveying an winding stream; finding its width: For surveying a meandering stream / river open traverse with compass and chain is the best method. To locate bends, etc. offsets may be taken from traverse lines to various points on the stream banks, on either side. Fix one point on the other side of the river, either by intersection of two bearings from two stations to a flag on the opposite bank and proceed with open traverse on the other bank. To find the width of a stream, select a base line one bank and fix a flag on the other bank. Take two bearings from two ends of the base line on to the flag and plot the bearings on the plan. Intersection of two bearings locates the flag on the opposite bank. Width of the stream can be scaled off on the plan.

(iv) Ranging out a straight line between two points in a forest: This operation is required in forestry works, e.g., cutting a straight fire line over grown with tall weeds, cutting a straight boundary line between two pillars. This may done either by open traverse method or random line method.

(v) Laying out the forest coupe on the ground:

(a) When a map with a coupe show on its is available: Fix a point on ground with respect to a known feature by measuring the bearing and distance by traversing i.e., setting out the bearings and distances successively from neighboring points.

(b) When no map is available: In such cases suitable shape of the coupe should be decided, whether square, rectangle, or trapezoid and its sides worked out corresponding to the given area.

(vi) Demarcation and checking of boundaries: If a map showing forest boundaries is available, measure distances and bearings between the boundary pillars as shown on the map and proceed to check on the ground.

16. How measures is carried out in chain survey when an obstacles coming in chain line can be chained across but not seen across. (2005)

While performing chain survey operations in the field, a Surveyor may come across various kinds of obstacles such as ponds, rivers, buildings, woods, spurs, etc., which obstruct ranging and/ or chaining along the survey lines. In order to continue chaining, special methods (based on simple geometrical or trigonometrical principles) are used for measuring distances across such obstacles.

Some typical cases of various categories are discussed below:

(a) Where obstacles can be chained across but not seen across, i.e., chaining free, vision obstructed – Examples of such obstacles are an intervening hill, rising ground or wooded fields with trees or brushwood. So far as the ends are concerned, they are not visible from each other. Two cases arise in each of such obstacles, *viz :-*

(i) Both ends are visible from intermediate points on the chain line: in such cases the difficulty is got over by *Reciprocal Ranging.*

(ii) When both ends are not visible from any intermediate point, e.g., wooded fields with brushwood, etc. In such cases *Random Line* method is the most suitable. This is also applicable for cutting straight lines (say, firelines) between two end points invisible from each other, through brushwood area.

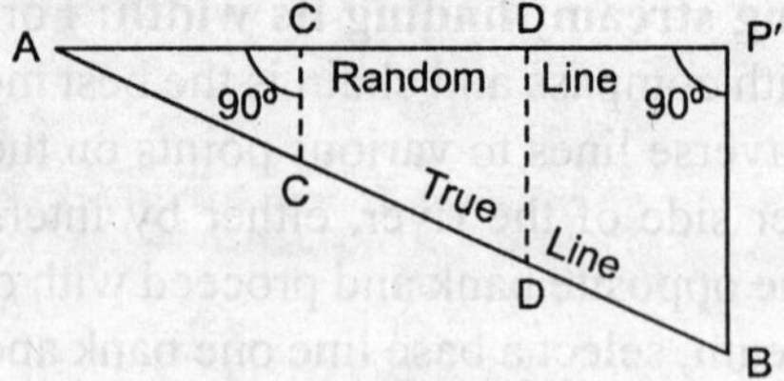

Fig. Random Line Method

Refer to Fig. AB is a line which is to be ranged and chained. From the point A run a straight line AB', known as *Random Line,* in a direction as near to that of B as can be estimated, and station B is visible from B'. Fix station B' such that BB' is perpendicular to AB'. Measure BB'.

Then $(AB)^2 = (AB')^2 + (BB')^2$

or $AB = \sqrt{(AB')^2 + BB'^2}$

If line AB is to be ranged and it is desired to fix intermediate points C and D on this line, we can do so by the following relationship:

$$CC'/BB' = AC'/AB'$$

or $CC' = (AC'/AB') \times BB'$

Similarly, $DD' = (AD' \times AB') \times BB'$

Lay out perpendiculars *CC'* and *DD'* on *AB'* at *C'* and *D'* which will fix points *C* and *D* on the line *AB*. A sufficient number of points may be located in this way and then the straight line *AB* can be cleared and measured.

17. Describe the limitations and advantages of Plane Table Survey (2012)

Advantages of Plane Table Survey

- It is suitable for location of details as well as contouring for large scale maps directly in the field.
- As surveying and plotting are done simultaneously in the field, chances of getting omission of any detail get less.
- The plotting details can immediately get compared with the actual objects present in the field. Thus errors as well as accuracy of the plot can be ascertained as the work progresses in the field.
- Contours and specific features can be represented and checked conveniently as the whole area is in view at the time of plotting.
- Only relevant details are located because the map is drawn as the survey progresses. Irrelevant details get omitted in the field itself.
- The plane table survey is generally more rapid and less costly than most other types of survey.
- As the instruments used are simple, not much skill for operation of instruments is required. This method of survey requires no field book.

Limitations of Plane Table Survey:

- The plane table survey is not possible in unfavorable climates such as rain, fog etc.
- This method of survey is not very accurate and thus unsuitable for large scale or precise work.
- The method requires large amount of time to be spent in the field.
- Quality of the final map depends largely on the drafting capability of the surveyor.
- Plane table is cumbersome and heavy to carry.
- It needs many accessories.
- This method is effective in relatively open country where stations can be sighted easily

18. Discuss the "Two-Point Problem" – the special case of resection in Plain Table Survey (2012)

Two-Point Problem: Two point problem helps in locating the instrument station (I.S) on the plan, by means of observations to only two well defined points, which have already been plotted on the plan. The problem may be enunciated thus: "Given two points A and B, visible from a station P over which the table is set; a and b being their plotted positions. Required to plot p".

Two instruments stations are required to solve the problem. The problem appears indeterminate unless we effect orientation by declinator in the process. The most convenient solution is as follows:

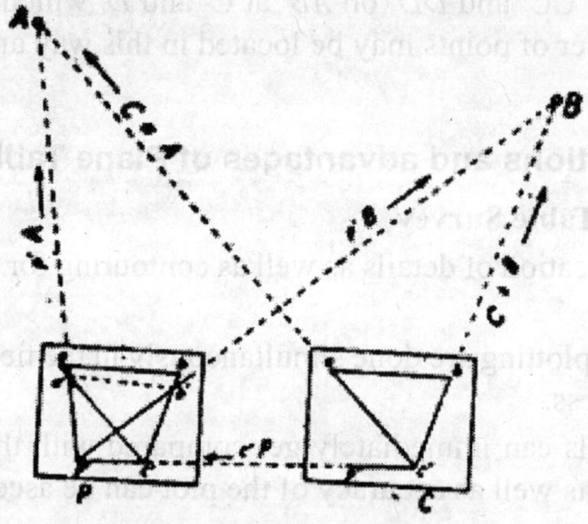

Fig. Two-Point Problem

(i) Select another auxillary point C, such that angles PAC and PBC are not very acute for good intersections at A and B. Set the P.T. at C, and orient it approximately by estimation (by judging ab to be parallel to AB) or by declinator. With alidade centred at a and b sight A and B respectively and draw rays intersecting at c', which is the exact or approximate plot of C, according as the orientation at C was exact or approximate. With alidade touching c' sight P and draw a ray c' P.

(ii) Shift the table to P and set it up; assume a point p' on the last ray drawn to represent P, by estimating the distance PC. Bring the table to the same orientation as at C by placing the alidade along p'c' and sighting C; clamp the table. Now with the alidade touching p', sight A and B, and draw rays intersecting c'a and c'b in a' and b' respectively.

(iii) These points have been obtained by ordinary intersection from C and P, so that a'b' is parallel to AB. But a'b' does not coincide with ab, the correct plot of AB, because: (1) the orientation at C and P, although the same, are only approximate : (2) p'c' represents PC by estimation only. In other words, a'b'c'p' is similar to ABCP, but is of incorrect size and is incorrectly located on the sheet. The error of orientation is the angle between ab and a'b'; which is to be corrected.

(iv) To orient the table correctly it must be rotated through this angle unit lab become parallel to the present direction of a'b'. The best method to do this is to place the alidade against a'b' and to fix a ranging rod at a sufficient distance, aligned on the line a'b'. Now shift the alidade and place it along ab, and rotate the table until the ranging rod is aligned. Clamp the table. Now ab is parallel to AB and the table is correctly oriented.

(v) To find the correct location of P on the plan: centre the alidade on a, sight A and draw a ray a A and similarly, centering the alidade on b sight B and draw a ray bB. The intersection of these two rays, p, gives the plot of observer's station P on the map.

19. Under which circumstances are chain and plane table survey used in forestry? Write in brief on procedure for any one survey. (2006)

Chain surveying is the system of surveying in which sides of various triangles are measured directly in the field and no angular measurements are required. The underlying principles are to divide the whole area to be surveyed into a framework of triangles. Since triangle is a only figure which is determinate in size and shape if the lengths of sides are known. the location and disposition of the triangle will depend on configuration of the ground, natural obstacles and extinct and nature of the details desired to be plotted. Chain survey deals with only linear measurement in the field. it is convenient method suitable for small areas and open grounds with simple details.

Plane tabling is a method of surveying, the peculiar feature of which is that the field observation, measurement and plotting proceeds. It is admirably suited for filling in interior details of a closed traverse, surveyed by prismatic compass, the theodalite, or triangulation. It is also extensively used for small scale mapping, topography and engineering surveys, fire maps etc.,

20. Enumerate the methods of direct linear measurements? Discuss in brief the application of prismatic compass in forest survey. (2016)

(i) **By pacing:** For approximate results, distances may be taken by pacing. One has to walk over the line to be measured, keep on counting the number of paces, multiply the same by the length of a pace (average is taken as 21 ft. or 0.75 meter). An experience pacer should know the length of his pace, fairly accurately, and he can measure with a precision of 1 in 100. This method is quite rapid and alright for shorter distances, but monotonous for long distances, for which it is not advisable. It also provides a rapid means of approximately checking more precise measurements of distances.

(ii) **By ocular estimation:** It is rough method of determining distances.

(iii) **By speedometer**: The speedometer of an automobile (a car, etc.) may be used to give approximate distances. It gives better result than pacing, if the route is smooth. Some of the speedometers can read upto 0.01m.

(iv) **The wheel pedometer**: It consists of a wheel, whose circumference is a multiple of a meter. It is rolled with a bandle along the line to be measured, its rotation being communicated to a cyclometer arrangement, the dial on which registers the distance traversed. It is mainly used by P.W.D for fixing positions of milestones.

(v) **Odometer:** It is a simple device which can be attached to the wheel of any vehicle, such as a cart, cycle or a carriage. It directly records the number of revolutions of the **wheel** which, multiplied by circumference of the wheel, gives the distances traversed.

(vi) **Passometer:** It is a pocket instrument, resembling a pocket watch, which automatically records the number of steps taken in pacing a given distance. It mechanism is operated by motion of the body. It saves the surveyor from monotony of counting the steps in pacing long lines.

(vii) **By recording time:** By recording time interval of travel and the average walking speed of the surveyor, the distance transverse can be determined approximately.

(viii) **By chaining:** Of all the methods, the most common and accurate is the measurement with a chain or a tape, called chaining.

Application of prismatic compass survey:

- The area to be surveyed little accuracy is required.
- Area where chaining is tedious and liable to inaccuracy. E.g., hilly or broken terrain
- Area where triangulation is not possible due to obstrucles
- It is quicker and quite satisfactory in case of linear surveys such as road, canal, railway track or a stream.
- Compass survey is unsuitable for magnetic areas like iron posts, pipes, bridges etc.,

21. Find out the maximum and minimum pressures of a wall which is 60 ft long and 4.5 ft wide at the base of its footing, carrying loads at the following distances from the left hand side: 20 tons at 10 ft, 30 tons at 25 ft, 40 tons at 29 ft, 48 tons at 50 ft and 12 tons at 55 ft. (2013)

Let x_1 = Length, x_2 = wide, x_3 = height

Surface area of

$$S = x_1x_2 + 2x_2x_3 + 2x_3x_1$$

$$S = 60\, x_1 \times 4.5\, x_2$$

The distance from origin to any point

$$d = \sqrt{(10x_1 - 20)^2 + (25x_2 - 30)^2 + (28x_3 - 40)^2 + (50x_4 - 48)^2 + (55x_5 - 12)^2}$$

$$F = d + S$$

$$S = 270\, x_1x_2$$

$$F = (10x_1 - 20)^2 + (25x_2 - 30)^2 + (28x_3 - 40)^2 + (50x_4 - 48)^2 + (55x_5 - 12)^2 + 270\, x_1x_2$$

$$\frac{\partial F}{\partial x_1} = 2(10x_1 - 20) + 270\, x_2 = 20\, x_1 + 270x_2 - 4 \quad ...(1)$$

$$\frac{\partial F}{\partial x_2} = 2(25x_2 - 30) + 270\, x_1 = 270\, x_1 + 50x_2 - 60 \quad ...(2)$$

$$\frac{\partial F}{\partial x_3} = 2(28x_3 - 40) = 56\, x_3 - 40 = 0.7142 \quad \left(x_3 = \frac{40}{56}\right)$$

$$\frac{\partial F}{\partial x_4} = 2(50x_4 - 30) = 10\,x_4 - 60 = 6 \quad \left(x_4 = \frac{60}{10}\right)$$

$$\frac{\partial F}{\partial x_5} = 2(55x_4 - 12) = 110\,x_5 - 24 = 0.2181 \quad \left(x_5 = \frac{24}{110}\right)$$

$$\frac{\partial F}{\partial x} = 270\,x_1 x_2$$

Solve (1) and (2)

$$20\,x_1 + 270\,x_2 = 40 \qquad (1)$$

$$270\,x_1 + 50\,x_2 = 60 \qquad (2)$$

$$(1) \times 13.5 \rightarrow 270\,x_1 + 3645\,x_2 = 540 \qquad (1)$$

$$270\,x_1 + 50\,x_2 = 60 \qquad (1)$$

$$3595\,x_2 - 480$$

$$x_2 = \frac{480}{3595} = 0.13351$$

Sub $x_2 = 0.13351$ in (2)

$$270\,x_1 + 50 \times 0.13351 = 60$$

$$270\,x_1 + 0.0178 = 60$$

$$270\,x_1 = 59.9822$$

$$x_1 = 0.222156$$

Sub $x_{1,} x_{2,} \ldots\ldots x_5$ value is d

$$d_1 = \sqrt{(10\times0.22156-20)^2 + (25\times0.13351-30)^2} + (28\times0.714-40)^2 + (150\times6-48)^2 + (55\times0.218-12)^2$$

$$d_1 = 254.8165$$

Chapter 12

Forest Ecology and Ethnobotany

1. Write short notes on Riparian bu-ers. (2011)

Riparian bu3er: A riparian buffer is a vegetated area (a "buffer strip") near a stream, usually forested, which helps shade and partially protect a stream from the impact of adjacent land uses.

It plays a key role in increasing water quality in associated streams, rivers, and lakes, thus providing environmental benefits. With the decline of many aquatic ecosystems due to agriculture, riparian buffers have become a very common conservation practice aimed at increasing water quality and reducing pollution.

Benefits of Riparian bu3er: Riparian buffers act to intercept sediment, nutrients, pesticides, and other materials in surface runoff and reduce nutrients and other pollutants in shallow subsurface water flow. They also serve to provide habitat and wildlife corridors in primarily agricultural areas. They can also be key in reducing erosion by providing stream bank stabilization.

- **Provide habitat** - Riparian buffers can act as crucial habitat for a large number of species, especially those who have lost habitat due to agricultural land being put into production.
- **Increase biodiversity** - By adding this vegetated area of land near a water source it becomes a prime location for species that may have left the area due to non-conservation land use to re-establish. With this re-establishment the number of native species and biodiversity in general can be increased.
- **Bu3ers acting as corridors** - Buffers also serve a major role in wildlife habitat. The habitat provided by the buffers also double as corridors for species that have had their habitat fragmented by various land uses.
- **Shading water** - The large trees in the first zone of the riparian buffer provide shade and therefore cooling for the water, increasing productivity and increasing habitat quality for aquatic species.
- **Large woody debris** - When branches and stumps fall into the stream from the riparian zone, more stream habitat features are created. Carbon is added as an energy source for biota in the stream.

2. Write short notes on Forest decline. (2011)

Forest Decline: Forest dieback is a condition in trees or woody plants in which peripheral parts are killed, either by pathogens, parasites or due to conditions like acid rain and drought. Two of the nine tipping points for major climate changes, forecast for the next century, are directly related to forest diebacks. *Phomopsis azadirachtae* is a fungus of the Phomopsis genus that has been identified as the fungus responsible for dieback of *Azadirachta indica* (Neem) in India.

Forest dieback refers to the phenomena of a stand of trees losing health and dying without an obvious cause. This condition is also known as forest decline, forest damage, canopy level dieback, and stand level dieback. This usually affects individual species of trees, but can also affect multiple species. Trees and woody plants are killed by a complex group of factors that include parasites, such as fungus and beetles, pollution, such as acid rain and organic compounds, and drought. Diseases and pests can kill large groups of trees very easily, but the premature and progressive loss of clusters of trees without an obvious or specific cause is known as forest dieback. Dieback is an episodic event. Dieback takes on many locations and shapes. It can be along the perimeter, at specific elevations, or dispersed throughout the forest ecosystem.

Forest dieback has many symptoms: falling off of leaves and needles, discoloration of leaves and needles, thinning of the crowns of trees, dead stands of trees of a certain age, and changes in the roots of the trees. Forest dieback has many dynamic forms. A stand of trees can exhibit mild symptoms, extreme systems, or death at the same time. Forest decline can be viewed as the result of continued, widespread, and severe dieback of multiple species in the forest. Current forest decline is described by rapid development on individual trees, occurrence in different forest types, a long duration (over 10 years), and occurrence throughout the natural range of affected species.

3. List merits of Bentham and Hooker system of plant classification used in Dendrology. (2010)

Merits of Bentham and Hooker System

(i) Bentham and Hooker's classification is the most natural system, based on actual examination of specimens.

(ii) The description of plants is quite accurate and reliable.

(iii) As it is easy to follow, it is used as a key for the identification of plants in Kew herbarium and several other herbaria of the world.

(iv) Although this system is natural, most of the aspects of this system show affinity to modern concepts of evolution. For example, theorder Ranales, which is the first order in the arrangement of plants, has been given a primitive position in this system. Recent taxonomic findings also indicate that the members of Ranales are the most primitive living angiosperms.

(v) The placement of monocotyledonae after the dicotyledonae also appears to be in accordance with the evolutionary trends.

(vi) The description of families and genera is very accurate.

(vii) The system is very handy for identification purposes.

(viii) The system is of great practical convenience.

(ix) Each family had a synopsis at the beginning which is very useful in identification.

(x) The gamopetalae placed after polypetalae is justified since union of petals is considered to be an advanced feature.

4. Basis of forest classification and why there is need for such classification. (2012)

Basis of Classification: The forests can be classified into forest types on the basis of the following factors:

(i) Physiognomy: Physiognomy means the general appearance of a forest community and therefore, forms an easy basis for rough differentiation of very broad classes. It is described by dominant growth-form (e.g., trees, shrubs, grasses, etc.), the seasonal changes (e.g., evergreen and deciduous habit) and such other features as may be associated with very dry or very wet sites.

(ii) Structure: Structure of a forest is described by stratification (i.e., the way in which different species are aligned in different layers of the forest) and dimensions of trees including height and spacing. It is generally observed that more favourable the site to tree growth the greater is the number of strata and the less favourable the site, the lesser is the number of strata in which the forest is divided. Therefore, structural stratification gives good basis for classifying forest types.

(iii) Function: Function refers to the most common morphological characters of the species such as leaf characters, leaf size, stem and root characters, e.g., buttress formation, development of stilt roots, etc., which form the basis of classification.

(iv) Floristics: Floristics refers to the species present in a particular forest. While this forms an important basis for delimiting a forest type, there is a great difference of opinions as to whether the frequency of the species should be used as a basis or not. However, this can be used to distinguish subtypes.

(v) Dynamics: As a result of interaction between vegetation and the site, there is continual change between the two. This results in succession and development of climax communities. Though the general view favours Whittaker's theory of vegetational gradients, it is convenient for the time being to classify the relatively stable types as climax, those still developing as seral, the stable community resulting from the special soil peculiarities as edaphic climax and that resulting from the biotic interference as biotic climax.

(vi) Habitat: Habitat refers to the effective environmental conditions in which a forest community exists. Thus, climate and edaphic factors often form the basis of classifying forest vegetation.

(vii) Physiography: Physiography refers to the natural features of the earth surface. As it modifies the microclimate and results in different vegetation occurring in the same climate on different aspects of the hill slope, it forms a good basis for classifying vegetation.

(viii) History: History refers to past biotic influences on a site and its vegetation. Though these are very important in determining the present condition and future potentialities in vegetational communities, it is often difficult to assess these factors correctly.

Need for Classification: The main purpose of classification of forests into forest types is to find out correct silvicultural techniques and management practices for the development of forests. These techniques and practices cannot have universal application because forests vary from place to place. Therefore it becomes necessary to classify the forests into forest types, so that suitable silvicultural techniques and management practices may be evolved for each type to be applied to similar types in the field. This avoids waste of money and effort as well as disappointment from the failure resulting from the application of wrong techniques.

5. Importance of plant succession in forestry practices (2012)

The plant succession is very important in forestry practices due to the following factors.

(i) Classification of Forests: The succession forms the basis of classification of forests There are many bases of forest types which can be easily studies during the course of establishment of succession irrespective of their types. The possible bases of forest classification could be as follows:

(a) Physiognomy

(b) Structure

(c) Floristics

(d) Dynamics

(e) Habitat and

(f) Physiography

All these characteristics are generally studied during the process of succession. Therefore, succession is helpful tool in classifying the forests.

(ii) Species choice for a3orestation: At present, the forest cover is low in India (say 24.16%) which is much below the recommended cover of 33%. There is a lot of setback to the forests due to deforestation activities with the virtue of man (a biotic factor), a greatest destructor of mother nature. Not only this, there is another point of anxiety i.e., rapid increase in-population. Our natural resources, particularly the forests (forest resources, are getting depleted day-by-day. It is difficult to meet out the growing demands of the people. Hence, there is no alternative but to go for affforestation wherever possible.

There are several considerations for afforestation e.g., site conditions, species availability, water sources, labour force availability, etc. To achieve targets of afforestation, there is an urgent need to study the local sites. Therefore, succession provides an ample opportunity for such conditions.

(iii) Evaluating the sites: According to BCFT (1953), site is complex of physical and biological factors of an area that determine what forest or other vegetation it may carry. The sites are classified as:

(a) Qualitative sites: These are influenced by the factors like, climate, soil, vegetation, etc. These are called Site types.

(b) Quantitative sites: Potential wood production into site classes.

The locality factors, particularly the climatic ones, are influenced in a number of ways by vegetation. The succession plays a crucial role in this regard.

(iv) Delineating economic species: Plant succession is helpful in determining most promising the economical tree species. During this phenomenon, the successional stage of these species can be determined. In this process, the performance of a particular species can be judged for their preference with regard to climate / soils, etc.

6. Write short notes on Source-sink relationship with respect to carbon cycle. (2012)

Sources - Sinks Relationship (Carbon Cycle): Compounds that contain the element carbon are referred to as "organic." They are present in all living things. Carbon is continually moving among Earth's lithosphere, hydrosphere, biosphere, and atmosphere in various forms: as carbon dioxide (CO_2) in the atmosphere, sugars or carbohydrates ($C_nH_{2n}O_n$) in living organisms, and calcium carbonate ($CaCO_3$) in rocks and minerals, to name just a few. The movement of carbon among Earth's spheres, as diagrammed below, is known as the carbon cycle.

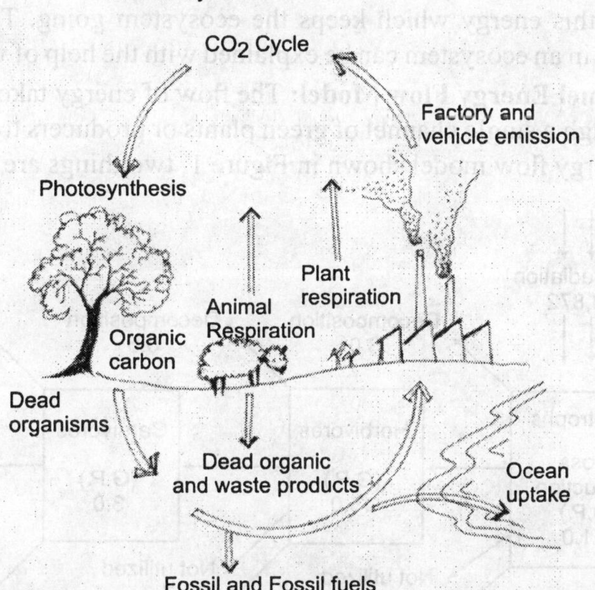

The black numbers in the diagram indicate how much carbon is stored in carbon sinks (areas of storage) in billions of tons (gigatons – GTC). The arrows show how carbon moves among Earth's spheres.

Green plants play a very important role in the carbon cycle. They absorb carbon dioxide (CO_2) from the atmosphere and produce carbon-containing sugars. This process is called photosynthesis. There are two main steps in photosynthesis. First, plants trap the sun's light energy in a compound called chlorophyll. This energy is converted to a chemical form called adenosine triphosphate (ATP). In the second step plants use the energy from ATP to produce sugar ($C_6H_{12}0_6$). The process of photosynthesis requires water (H_2O). It also produces water as well as oxygen (O_2). The net chemical reaction for the process of photosynthesis is

$$6CO_2 + 6H_2O + \text{sunlight} \longrightarrow C_6H_{12}O_6 + 6O_2.$$

Animals eat plants to obtain the energy trapped during photosynthesis. As the animals' bodies break down the carbohydrates in the plant tissue, CO_2 is released to the atmosphere. This process is called respiration. The net chemical reaction for the process of respiration is the exact opposite of photosynthesis: $6O_2 + C_6H_{12}O_6 \longrightarrow 6H_2O + 6CO_2$. Plants respire also as they break down the organic molecules in themselves in order to release the stored energy. Plants and animals also release CO_2 to the atmosphere when they decay.

Hence the plants act both as a source and sink to balance the carbon cycle. During photosynthesis it absorbs carbon-di-oxide and releases oxygen and during respiration it takes oxygen and release carbon-di-oxide thereby it acts both as a source and sink and balances the carbon cycle.

7. Write short notes on Energy flow model in ecosystem (2012)

Energy flow models in Ecosystem: Flow of energy in an ecosystem lakes place through the food chain and it is this energy which keeps the ecosystem going. The flow of energy through various trophic levels in an ecosystem can be explained with the help of various energy flow models.

1. **Single Channel Energy Flow Model:** The flow of energy takes place in an unidirectional manner through a single channel of green plants or producers to herbivores and carnivores. From the energy flow model shown in Figure 1, two things are clear:

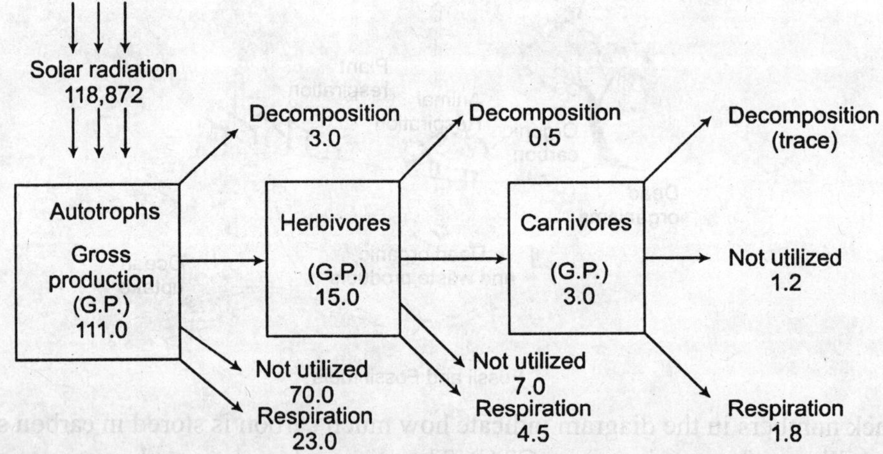

(i) There is unidirectional flow of energy. The energy captured by autotrophs does not revert back to solar input but passes to herbivores; and that which passes to herbivores does not go back to the autotrophs but passes to consumers. Due to one way flow of energy, the system would collapse if the primary sources of energy (i.e., sun) were cut off.

(ii) At each trophic level, there occurs progressive decrease in energy. This is accounted largely by the energy lost as heal in metabolic reactions (respiration) coupled with unutilized energy.

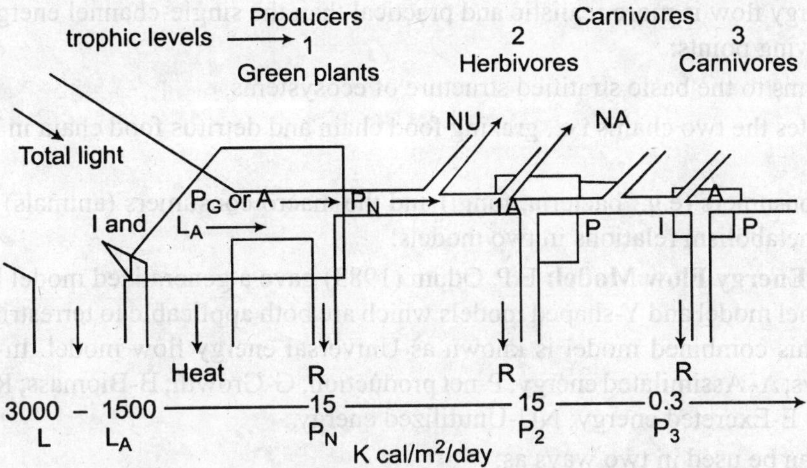

Fig. 2. A simplified energy flow diagram depicting three trophic levels (boxes numbered 2.3) in a linear food chain. 1-total energy input; LA-Light absorbed by plant cover, PC-gross primary production; A-total assimilation, PN-net primary production; P-secondary (consumer) production; NU-energy not used (stored or exported); NA-energy not assimilated by consumers (egested); R-respiration Bottom line in the diagram shows the ordr of the magnitude of the energy losses expected at major transfer points. Starting with a solar input of 3,000 Kcal per square metre per day.

2. Y-Shaped or Double Channel Energy Flow Model: Y-shaped energy flow models as pioneered by H.T. Odum in 1956. This model shows a common boundary, light and heat flows as well as the import, export and storage of organic matter. Decomposers is placed in a separate box as a means of partially separating the grazing and detritus food chains. In terms of energy levels, decomposers are, in fact, a mixed group. The significant part in Y-shaped model is that the two food chains are not isolated from each other.

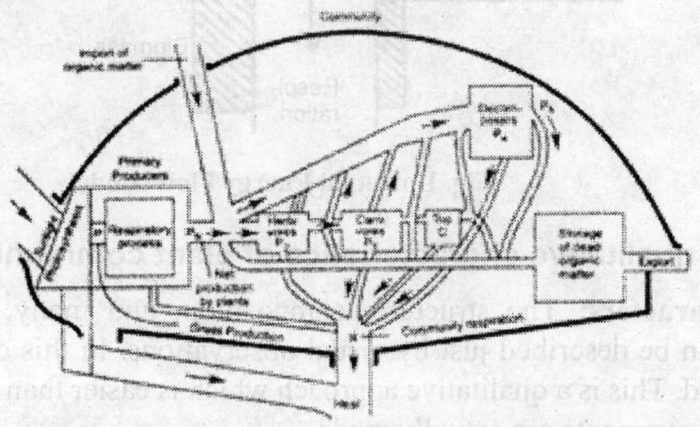

Fig. The relationship between flow of energy through the grazing food chain and detritus pathway.

Y-shaped energy flow is more realistic and practical than the single-channel energy flow model because of following points:

(i) It conforms to the basic stratified structure of ecosystems.

(ii) It separates the two chains i.e., grazing food chain and detritus food chain in both time and space.

(iii) Micro consumers (e.g.. bacteria, fungi) and the macro consumers (animals) differ greatly in size-metabolism relations in two models.

3. Universal Energy Flow Model: E.P. Odum (1983) gave a generalized model by combining both single channel model and Y-shaped models which are both applicable to terrestrial and aquatic ecosystem and this combined model is known as Universal energy flow model. In this model I-Incident solar rays; A-Assimilated energy; P-net production; G-Growth; B-Biomass; R-Respiration; S-Stored energy; E-Excreted energy; NU-Unutilized energy.

This model can be used in two ways as:

- It can represent a species population in which case the appropriate energy inputs and links with other species would be shown as a conventional species oriented food levels
- The model can represent a discrete energy level in which case the biomass and energy channels represent many populations supported by the same energy source.

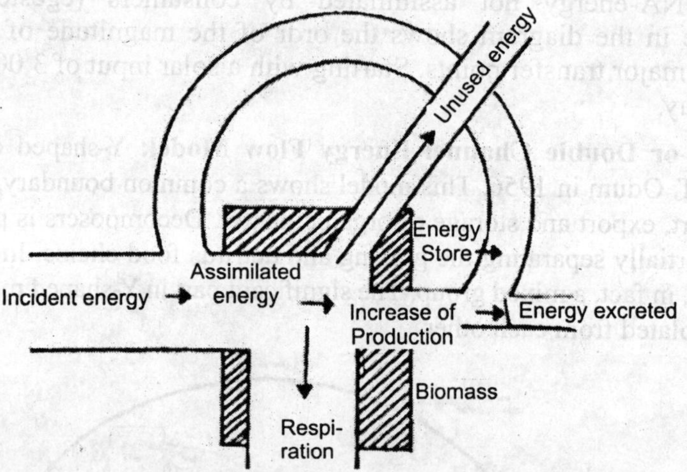

Fig. Universal Energy Flow Model

8. Describe the qualitative characteristics of plant community (2013)

Qualitative characters: The structure, composition, and many other characters of plant community (ies) can be described just by visual observations. In this case, the measurements, of any type are not mad. This is a qualitative approach which is easier than the quantitative vegetation analysis where measurements are actually made.

The following are some qualitative characters of community mainly the plant communities

- **Kind of species:** This is very important factor here the study of flora or kind of species present in an area is considered. This requires knowledge of taxanomy which may be obtained by observation of flora with regards to period, collection of plant material, systematic studies, herbarium preparation etc. for whole of the year.

- **Stratification of vegetation:** The life forms like shrubs, trees, climbers, herbs, forbs, grasses etc., should be recorded for each species individually. It is true that with a change in season, the appearance of vegetation also changes. For this the periodicity and phenology should be recorded for different seasons and years.
- **Sociability:** This is also called as 'nature of grouping'. It is estimated / observed that whether the species grow singly, in patches, in colonies, or evenly intermixed.

9. Describe flora and distribution of Group: 16/C_1 of Champion & Seth's Forest Type (2012)

This type of forest is called as Dry alpine scrub and is distributed in a small area. This group is small area. The group has only one forest type i.e., Type 16 / C_1. This type has been given the name as **'Dry alpine scrub'**. The plant growing in the stony soil are subjected to high wind and high intensity. However, certain drought resistant species, e.g., grasses, sedges, mosses and lichens are very common. Moreover, only few short-lived plants can manage to grow through crevices, such as *Oxyria digyna, Ranuculus hyperboreus, Saxifrage sibirica* etc., And this type occurs in drier alpine areas.

The common species are *Juniperus communis* and *J. wallichiana* in western Himalayas and *J.recurva* in eastern Himalayas. The trees are bushy generally never reaching height more than 2 meters. These scrub forests may be found right upto permanent snowline.

10. Write about species diversity and centre of origin of Willows (Salix species). Describe its various uses under short rotation forestry. (2013)

Species Diversity: Species diversity is the number of different species that are represented in a given community. The effective number of species refers to the number of equally abundant species needed to obtain the same mean proportional species abundance as that observed in the dataset of interest (where all species may not be equally abundant). Species diversity consists of two components: species richness and species evenness. Species richness is a simple count of species, whereas species evenness quantifies how equal the abundances of the species are.

Centre of Origin of Salix

(i) The distribution of Salix species among the continents: There are about 526 species of Salix in the world, most of which are distributed in the Northern Hemisphere with only a few species in the Southern Hemisphere. In Asia, there are about 375 species, making up 71.29 percent of the total in the world, including 328 endemics; in Europe, about 114 species, 21.67 percent with 73 endemics; in North America, about 91 species, 17.3 percent with 71 endemics; in Africa, about 8 species, 1.5 percent, with 6 endemics. Only one species occurs in South America. Asia, Europe and North America have 8 species in common (excluding 4 cultivated species). There are 34 common species between Asia and Europe, 14 both between Europe and North America and between Asia and North America, 2 between Asia and Africa.

According to the Continental Drift Theory, the natural circumstances which promoted speciation and protected newly originated and old species were created by the orogenic movement of the Himalayas in the middle and late Tertiary. Besides, the air temperature was a little higher in Asia than in Europe and North America (except its west part) and the dominant glaciers were mountainous in Asia during the glacial epoch in the Quaternary Period. Then willows of Europe moved southwards to Asia. During the interglacial period they moved in opposite direction. Such a to-and-fro willow migration between Asia and Europe and between and North America occurred so often that it resulted in the diversity of willow species in Asia. Those species of willows common among the continents belong to the Arctic flora.

(ii) **The multistaminal willows are of the primitive group in Salix:** Asia has 28 species of multistaminal willows, but Europe has only one which is also found in Asia. These 28 species are divided into two groups, "northern type" and "southern type", according to morphology of the ovary. The boundary between the two forms in distribution is at 40°N. The multistaminal willows from south Asia, Africa and South America are very similar to each other and may have mutually communicated between these continents in the Middle or Late Cretaceous Period. The southern type willows in south Asia are similar to the North American multistaminal willows but a few species. The Asian southern type willows spreaded all over the continents of Europe, Asia and North America through the communication between them before the Quaternany Period. Nevertheless, it is possible that the willows growing in North America immigranted through the middle America from South America. The Asian northern type multistaminal willows may have originated during the ice period. The multistaminal willows are more closed to Poplars in features of sexual organs. They are more primitive than the willows with 1-3 stamens and the most primitive ones in the genus.

(iii) **The center of origin and development of willows:** Based on the above discussion it is reasonable to say that the region between 20°and 40°N in East Asia is the center of the origin and differentiation of multistaminal willows. It covers Southern and Southwestern China and northern Indo-China Pennisula.

Uses under Short Rotation Forestry: The concept of short rotation forestry aims at reducing the harvest period without comprising the productivity through classical and modern breeding tools coupled with precision silvicultural technologies. Willows are one such species which are amenable for short rotation forestry and suitable for multifarious industrial utility due to the following characteristics.

- The wood is tough, strong, and light in weight, but has minimal resistance to decay.
- The stems (withies) from coppiced and pollarded plants are used for basket-making.
- Charcoal made from the wood was important for gunpowder manufacture.
- The bark tannin was used in the past for tanning leather.
- The wood is used to make cricket bats. *S. alba's* wood has a low density and a lower

transverse compressive strength. This allows the wood to bend, which is why it can be used to make baskets.

- Willow bark contains Indole-3-butyric acid, which is a plant hormone stimulating root growth; willow trimmings are sometimes used to clone rootstock in place of commercially synthesized root stimulator.
- Willows are excellent coppicer and hence highly deployed in Short Rotation Forestry towards generating raw material for several wood based industries like plywood, particle board, packing cases, sports goods, charcoal, tannin etc.

11. Discuss about the di-erent kinds of food chains in di-erent habitats and ecosystems. (2013)

A series of organisms each dependent on the next as a source of food. The series of processes by which food is grown or produced, sold, and eventually consumed. For an ecosystem to work there has to be a flow of energy within it. The organisms of the ecosystem need energy in the form of food.

The ultimate source of this energy is the sun. Producers like green plants trap solar energy and convert it into the chemical energy of food. When a primary consumer eats the producer, a part of this energy is passed on to it. The primary consumer is then eaten by a secondary consumer. And the secondary consumer may be eaten by a tertiary consumer, and so on. In this way energy gets transferred from one consumer to the next higher level of consumer. A series of organisms through which food energy flows in an ecosystem is called a food chain. It may also be defined as follows.

A food chain in an ecosystem is a series of organisms in which each organism feeds on the one below it in the series.

(i) **Food chain in Forest Habitat:** In a forest ecosystem, grass is eaten by a deer, which in turn is eaten by a tiger. The grass, deer and tiger form a food chain (Figure). In this food chain, energy flows from the grass (producer) to the deer (primary consumer) to the tiger (secondary consumer).

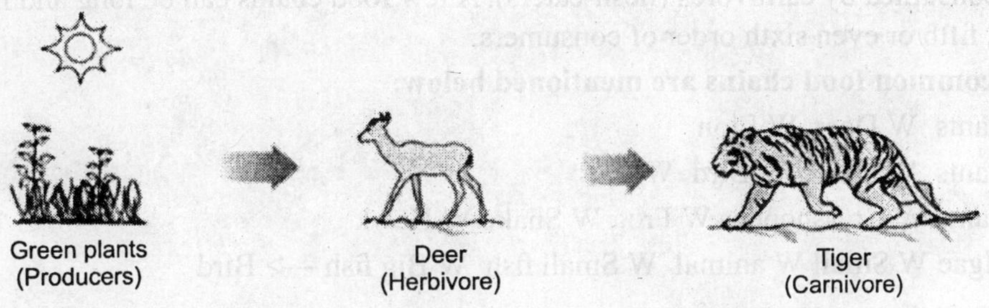

Fig. A Food chain in a forest Ecosystem

(ii) **Food chain in Grasslands:** A food chain in a grassland ecosystem may consist of grasses and other plants, grasshoppers, frogs, snakes and hawks (Figure).

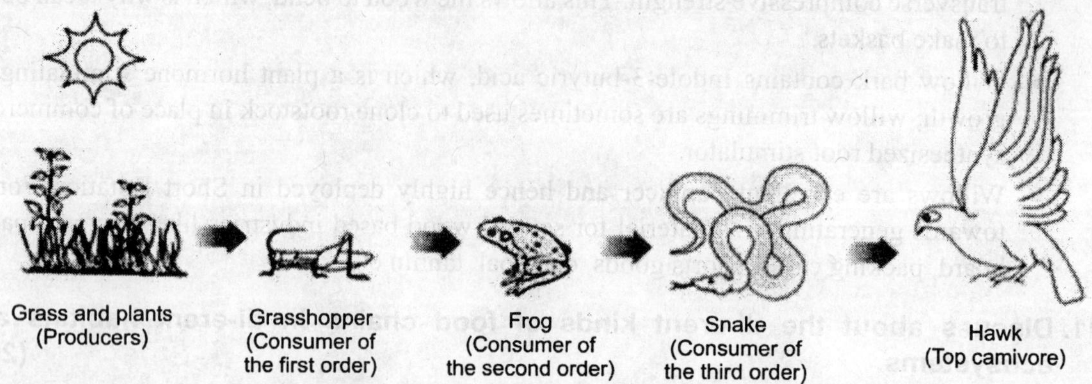

Fig. A Food chain in a forest Ecosystem

(iii) Food Chain in Fresh Water Ecosystem: In a freshwater aquatic ecosystem like a pond, the organisms in the food chain include algae, small animals, insects and their larvae, small fish, big fish and a fish-eating bird or animal (Figure).

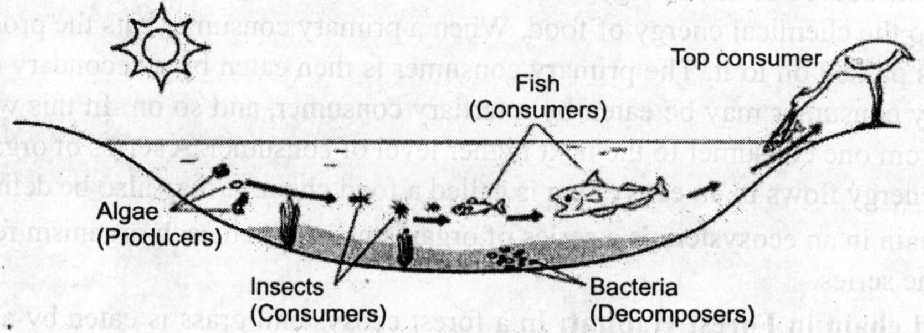

Fig. A Food chain in a freshwater pond

A food chain always begins with producers. Herbivores (plant-eaters) come next in the chain. They are consumed by carnivores (flesh-eaters). A few food chains can be long and may extend to the fourth, fifth or even sixth order of consumers.

Some common food chains are mentioned below:

- Plants W Deer W Lion
- Plants W Worm W Bird W Cat
- Plants W Grasshopper W Frog W Snake W Hawk
- Algae W Small W animal W Small fish W Big fish —> Bird

12. Describe the role of Clonal parks in Forestry with examples. (2009)

Clones are genetically identical plants derived from a common parent which are predominantly multiplied through vegetative means. The clones are multiplied and deployed in operational forestry improvement programme. All the intended operational forestry plantation programme are aimed at deploying superior clones. These clones need to be assembled in a decentralized as well as centralized manner in the form of clonal parks. This clonal parks involve clones from centers of origin, clones isolated from provenances, seed sources and progenies and clones developed from hybrid derivatives.

Advantages of Clonal Park

- It helps to conserve the rare and endangered plant for further utilization and breeding programme. Example- *Shorea macrophylla*
- The clone has an ability to utilize rapidly a greater proportion of the additive and non additive genetic variation than through the seed based progenies.
- It has an ability to select and utilize greater genetic diversity than is normally found in a single progeny.
- It has a greater simplicity and flexibility of managing sets of stock plants than in seed orchards.
- The clones are early flowering and hence facilitate rapid seed production through the clonal parks.
- The clones have increasing superiority over seed based genetic resources due to its multiple-trait selection characters.
- It ensures genetic uniformity among the multiplied populations.
- It ensures higher productivity due to increased genetic uniformity.
- It is possible to develop and breed clones resistant for biotic and abiotic stress.
- It is possible to reduce the rotation cycle and increase the productivity.

Examples of Clonal Parks

Poplar clonal park: Poplar has been one of the potential fast growing tree species amenable for plywood and matchwood industries in Northern India. Hence the wood based industries particularly WIMCO, Forest departments of Punjab, Haryana and Uttar Pradesh and Institutes like Punjab Agricultural University, Central Agroforestry Research Institute, University of Horticulture and Forestry, Solan have assembled wide range of Poplar clones and established in the form of clonal garden or park. This park acts as a base breeding population for immediate utilization and incorporation in future breeding programme.

Eucalyptus Clonal park: Eucalyptus is one of the largest planed species deployed across the country due to its multifarious utility in pulp and paper, plywood, firewood etc. and gained significant attraction among the wood based industries and research organizations. One of the major wood based industry M/s. ITC, Andhra Pradesh has developed wide range of clones and assembled clonal park deploying over 200 clones of Eucalyptus. Similarly the Tamil Nadu Agricultural University has established Eucalyptus clonal park with 175 clones and Institute of Forest Genetics and Tree breeding has assembled clonal park with over 100 clones. Besides, the Forest Plantation Corporations of Tamil Nadu, Karnataka, Andhra Pradesh, Haryana and Punjab have assembled wide range of clones in the form of clonal park and deploying them both for utilization and for further breeding programme.

Casuarina Clonal Park: Casuarina is one of the fast growing species exhibited wide range of genetic variability in two species viz., *Casuarina equisetifolia* and *Casuarina junghuhniana*. The state forest department of Tamil Nadu and Andhra Pradesh have assembled wide range of Casuarina clones suitable for domestic and industrial utility besides their utility in coastal afforestation projects. The Tamil Nadu Agricultural University has assembled over 60 clones and Institute of Forest Genetics and Tree Breeding has assembled over 100 clones in the form of clonal garden or parks and deploy them for direct utility and for incorporation in future breeding programme.

Melia Clonal Park: Melia is one of the fast growing species amenable for plywood, pulp and paper, matchwood and timber industries. It is naturally distributed in moist deciduous forest and has now been domesticated through systematic tree improvement programme. This has resulted in establishment of over 55 clones in the form of clonal park by Tamil Nadu Agricultural University, Coimbatore which is deployed by paper and plywood industries both for commercial adoption and for incorporation in future breeding programme.

Multi species Clonal Garden: The Forest College and Research Institute of Tamil Nadu Agricultural University has identified 30 different tree species suitable for agroforestry activities in the state of Tamil Nadu and established clonal parks deploying these 30 species. Similarly the forest department of Tamil Nadu, Karnataka, Andhra Pradesh and Haryana has established clonal garden for the various species.

13. What are ecosystem energetic? Describe the energy flow in a forest ecosystem (2009)

Ecological energetics is the quantitative study of the flow of energy through ecological systems. It aims to uncover the principles which describe the propensity of such energy flow through the tropic or energy availing levels of ecological networks. In ecology, the principles of ecosystem energy flows or ecosystem laws are considered formally analogous to the principles of energetic.

- **Energy flow:** Energy flow refers to the flow of energy through a food chain.
- **Energy flow in forest ecosystem:** Living organisms can use energy in two forms, radiant and fixed energy. Radiant energy is in the form of electromagnetic waves, such as light. Fixed energy is potential chemical energy bound in various organic substances which can be broken down in order to release their energy content.
- Organisms that can fix radiant energy utilizing in organic substances to produce organic molecules are called autotrophs. Organisms that cannot obtain energy from abiotic source but depend on energy rich organic molecules synthesized by autotrophs are called heterotrophs. Those which obtain energy from living organisms are called consumers and those which obtain energy from dead organisms are called decomposers.

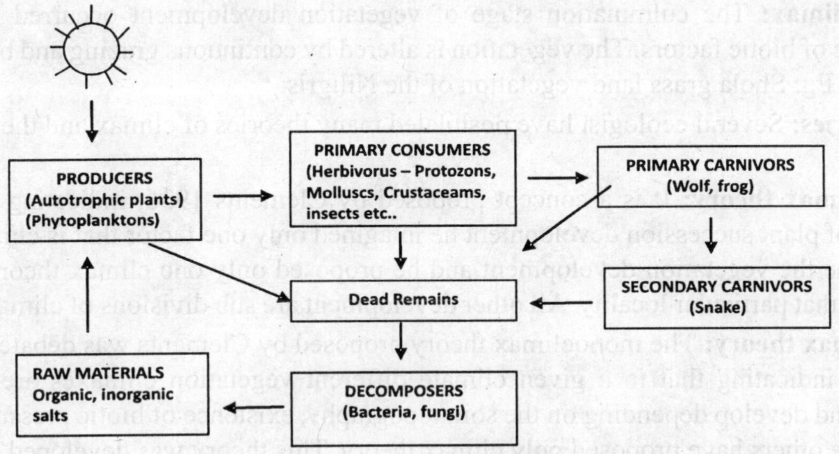

Fig. Flow of energy at di3erent levels of forest ecosystem

When the light energy falls on the green surfaces at which is stored in various organic products in the plants. When the herbivores consumes plants as food and convert chemical energy accumulated in plant products into kinetic energy, degradation of energy will occur through its conversion into heat. When herbivores are consumed by carnivores of the first order (secondary consumers) further degradation will occur. Similarly when primary carnivores are consumed by top carnivores, again energy will be degraded.

14. Define Climax. what are the di-erent climax types that are found in plant succession? State di-erent theories of Climax. (2009)

Climax: It is a stage wherein the succession ends with a community in which the species stabilize themselves through reproduction. The climax has been defined by various authors and they are furnished below.

Climax is the culmination stage in plant or community succession for a given environment, vegetation being conceived as having reached a highly stable condition.

Di3erent types of Climax: Based on the climax theories, the following climax vegetation have occurred and reported by many authors.

(i) **Climatic climax:** The culmination stage of vegetation development occurred due to the climate of the locality. Eg: Dry teak types in southern tropical moist deciduous forests, Himalayan sub tropical pine forests and Moist teak in tropical moist deciduous forests

(ii) **Edaphic climax:** The culmination stage of vegetation development occurred due to the influence of edaphic factors like soil. Eg: Bamboo and Cane plantations in tropical moist deciduous forests, Occurrence of Xylia in lateritic soils in semi evergreen forests, Occurrence of Hardwickia on shallow gravelly soils etc.

(iii) **Topographic climax:** The culmination stage of vegetation development occurred due to the influence of topographical factor like altitude. Eg: Lower Chirpine forests (*Pinus roxburghii*), Oak forests (*Quercus spp.*) etc.

(iv) **Biotic climax:** The culmination stage of vegetation development occurred due to the influence of biotic factors. The vegetation is altered by continuous grazing and browsing of animals. Eg: Shola grass land vegetation of the Niligris.

Climax theories: Several ecologist have postulated many theories of climax and the prominent theories are

(i) **Monoclimax theory:** It is a concept proposed by Clements 1936, indicating that in the serious of plant succession development he imagined only one factor that is climate which determine the vegetation development and he proposed only one climax theory which is stable in that particular locality. All other development are sub divisions of climatic climax.

(ii) **Polyclimax theory:** The monoclimax theory proposed by Clements was debated by many workers indicating that in a given climate different vegetation climaxes are known to appear and develop depending on the soil, topography, existence of biotic pressure etc. and hence the others have proposed poly climax theory. This theory was developed by Tansley in 1935 and supported by other workers.

(iii) **Information theory:** This theory was proposed by Fosbergy (1967) and Odum (1969) which indicated that succession is a function of energy balance and nutrient cycles.

15. What morphological, anatomical and physiological features are suited in xerophytic plants? (2010)

(A) Morphological Features

- **Modified stem:** The stems of xerophytic species of plants become somewhat fleshy and are changed into a leaf-life structure. This is known as phylloclade. This is a green coloured, flat, or a bit rounded stem. It performs the function of a leaf, the true leaf is modified into spines, which are present all over the fleshy modification. The rate of respiration is, therefore, considerably reduced and the plants are able to pass through extreme drought conditions.
- **Small leaf-size:** The leaf blades, pinnules, etc., are small in size. This adaptation helps to reduce transpiration, e.g., *Larrea*. However, sometimes, the leaves are transformed into a structure known as phyllode. The leaves may become fleshy and hardened, e.g., agave, aloe, so that there may be enough water stored, to pass through the plants during extreme drought period.
- **Long-root system:** In xerophytes, a long-root system is developed which helps the plant to tap the moisture from below strata. In some conditions, a superficial root system may be developed, which may absorb moisture from the soil surface after a brief shower.
- **Curling and rolling:** Due to dry conditions, some stem fruit stalks and some other parts are curved inwards. This condition is known as hygrochasy. The disposal of seed occurs only in the moist season. However, the flowering takes place in the dry season. The exciting example is *lepidium spinosum*. Similarly, in some cases, the plants are broken loose from the soil and are scattered which are called as rolling plants. For example, some lichens (*Parmelia esculenta*) are detached from the soil by the effect of speedy winds and scattered over large distance.

(B) Physiological Features:

- **Respiration:** The respiration in xerophytic species takes place at night. The stomata open after daylight and organic acids are produced. The Respiratory Quotient (RQ) in most of the species is lower than one and often zero (0) or say negligible.
- **Consumption of Co_2:** The stomata open only during night in most of the species, hence, there is less loss of transpirational moisture. The Co_2 consumed in photosynthesis during day time is derived partly from the organic acids which are produced in the anaerobic respiration. The process of hydrochasy is also a physiological adaptation.
- **Osmotic pressure:** The osmotic pressure is high in xerophytes, and the sap density of a particular species, varies from time to time. The high osmotic pressure helps in the following ways :
    - To check loss of water;
    - To maintain rigidity of plant;
    - To continue the process of photosynthesis for longer period.

(C) Anatomical Features

- The epidermis of Xerophytic species is covered with a layer of cuticle. This may control the transpiration to a great extent.
- The cork is produced in some more quantities. This is a tool to reduce transpiration and generally seen in deserts and alpine conditions.
- The resins and gums help to check the loss of moisture and to increase moisture absorption. Similarly, wax reduces the loss of moisture from the plant, e.g., *Capparis spinosa*, etc.
- Mucilage, produced in some plants, helps to reduce transpirational loss of water. Lignification acts as a barrier against extreme temperature. Also increases the power of plants against mechanical force.
- In some species, small hairs may develop (mainly on the lower leaf surface, e.g., *Populus alba* (white poplar). These hairs protect the leaf surface from the sun's rays by providing shade.
- The stomata are adapted in such a manner so as to reduce loss of moisture from some plants. Hydathoes are special features which help to increase the loss of injurious salts.
- The stems do not droop under extreme drought. The hypodermal cells consist of several layers of masses of dignified fibres. The air in the intercellular space is less as compared to hydrophytes.
- There is development of water storing tissues which is utilized in the deficiency of water in the roots.
- In xerophytes, relatively, small cells and vacuoles are present. This helps to reduce the condition of dehydration.

16. How is damage due to teak defoliator and stem borer managed in the plantations? (2010)

Teak defoliator –*Hyblaea puera*

Biology of insects

- Larva cause damage and attack commences with onset of monsoon rain
- Larva almost dark greyish-green with faint longitudinal lines or black with a broad dorsal orange red band
- Adult moth with reddish brown forewings and dark brown hindwings having orange band

Nature of damage

- Defoliation
- Entire leaf is eaten leaving midrib

Management of Pests

- Clip badly affected leaves with larva
- Spraying of Neem Seed Kernel Extract (NSKE) 5%
- Setting up of light trap to trap adult moths

Stem borer – *Batocera rufomaculata*

Symptoms of damage

- Grub tunnels in the sapwood on the trunk or branches
- Grub bore into the sap wood and macking irregular tunnels.
- Feeding the vascular tissues
- interruption of nutrient and water transport on the tissue
- Drying of terminal shoot in early stage
- Frass comes out from several points and some times sap oozes out of the holes
- Wilting of branches or entire tree

Identification of pest

- Grub - Linear, fleshy, apodous
- Adult - Grayish beetle with two pink dots and lateral spine

Management of pest

- Remove and destroy dead and severely affected branches of the tree
- Remove alternate host, silk cotton and other hosts
- Grow tolerant mango varieties viz., Neelam, Humayudin.
- Swab Coal tar + Kerosene @ 1:2 or Carbaryl 50 WP 20 g / l (basal portion of the trunk - 3 feet height) after scraping the loose bark to prevent oviposition by adult beetles.
- If infestations are severe then apply the copper oxychloride paste on the trunk of the tree.
- One celphos tablet (3 g aluminum phosphide) per hole
- Apply carbofuran 3g or 5 g per hole and plug with mud.

17. How are ectomycorrhizal fungi beneficial in managing soil borne diseases of forest nurseries? Give examples. (2010)

Among the most beneficial root-inhabiting organisms, ectomycorrhizal fungi can cover plant roots, forming what is known as a fungal mat.

The Ectomycorrhizal fungi protect soil borne diseases in several ways:

- By providing a physical barrier to the invading pathogen. A few examples of physical exclusion have been reported. Physical protection is more likely to exclude soil insects and nematodes than bacteria or fungi. However, some studies have shown that nematodes can penetrate the fungal mat.
- By providing antagonistic chemicals. Ectomycorrhizal fungi can produce a variety of antibiotics and other toxins that act against pathogenic organisms.
- By competing with the pathogen.
- By increasing the nutrient-uptake ability of plant roots. For example, improved phosphorus uptake in the host plant has commonly been associated with Ectomycorrhizal fungi. When plants are not deprived of nutrients, they are better able to tolerate or resist disease-causing organisms.
- Protection from the pathogen *Fusarium oxysporum* was shown in a field study using a cool-season annual grass and Ectomycorrhizal fungi. In this study the disease was suppressed in mycorrhizae-colonized grass inoculated with the pathogen.

18. Write causal pathogens of important diseases of Poplar and *Gmelina arborea*. Write integrated management of any one disease in each species. (2010)

Disease	Casual organisms	Symptoms	Control measures
Gmelina arborea			
Leaf spot	*Cercospora ranjita*	❖ The infection starts as water-soaked lesions, which later turns to minute grey spot with dark margins. As the disease progresses, the spots become thin and papery and drops off leaving shot hole symptoms.	❖ Affected plants must be removed and burnt
Root rot	*Poria rhizomorpha*	❖ Brown cuboidal rot develops in the root of the trees causing die-back and subsequent death of the affected individuals	❖ Soil mixed with 0.1%, Carbendazim (or) Mancozeb @ 0.2% or chlorothalonil @ 0.2%.
Poplars			
Leaf spot	*Cylindrocaldium quinqueseptatum*	❖ Minute grayish black water soaked lesions on young and old leaves. Several spots coalesee leading to blighting.	❖ Affected plants must be removed and burnt

Leaf blight	*Alternaria alternate*	❖ A brown water soaked lesion develop on the leaf and spreads fast to occupy major portion of the leaf leading to complete drying of the lamina. The disease can easily identified by presence of concentric rings in the blighted area showing the sporulation of the fungus.	❖ Removal of diseased crop debris reduce the disease.
Rust	*Puccinia psidii*	❖ The disease develops as minute yellow dots on the lower surface of the leaves with chlorotic spots on the upper surface. The yellow dots represent the uredia (uredosori). As the disease advances thousands of black raised rough spots develop on the lower surface representing telia (teliosori). In the most advanced stage of infection defoliation occurs leaving plants barren.	❖ Foliar spray with wetable-sulphur @0.2% or Carbendezim @0.1% or tridemorph @ 0.1% removal and destruction of infected plants debris

Integrated Disease Management in Leaf spot of Gmelina

- Removal of affected trees and destruction of infected trees is important in reducing the primary source of infection.
- Crop rotation is of primary importance in avoiding early season infection.
- Keep weeds under control.
- Spacing are important considerations.
- The early and late leaf spots are effectively controlled by the 2-3 spray application of Carbendazim 1gm or Mancozeb 2gm or Chlorothalanil 2g, hexaconazole 2 ml/litre of water at 15 days interval.

Integrated Disease Management in Rust of poplar

- Crop rotation and field sanitation.
- Strict plant quarantine regulations should be enforced to avoid the spread of rust on the seeds to disease free areas.
- Destroy reservoir hosts and debris to reduce the primary source of inoculums.
- Intercropping of pearl millet is useful in reducing the intensity of rust.
- Foliar application of aqueous neem extract @ 2-5 percent is useful and economical for the control of rust.

- 3-4 Spray of Tridemorph 2gm/lit of water at 15-20 days interval gives good management of rust.
- Spray application of Chlorothalonil 2g/lit or Mancozeb 2g/lit or Wettablesulphur 2g/lit of water on 35 and 50 days after sowing.

19. Name five aromatic grasses with their uses and methods of extraction (2010)

Aromatic Grasses

(i) Lemongrass (*Cymbopogon citratus*)
(ii) Palmarosa (*Cymbopogon martini var. martinii*)
(iii) Citronella (*Cymbopogon nardus*)
(iv) Ginger grass (*Cymbopogon fleuxosus*)
(v) Vetiver (*Vetiveria zizanoides*)

Uses

- Lemongrass essential oil is used for headaches, muscle pain, indigestion, acne and fevers. In addition, lemongrass is used as an insecticide
- Palmarosa is used in aromatherapy for problems such as acne, dermatitis, skin care, stress, and intestinal infections
- Ginger grass is used to treatment of digestive problems, fever, intestinal problems and menstrual difficulties
- Citronella is commonly used as an insect repellent, although it is also used in aromatherapy for colds, flu, headaches, and oily skin
- Vetiver essential oil is extensively utilized in perfumery, including the creation of perfumes for the body, room fresheners and coolers, as well as soaps, cosmetics, and oils. It is also a flavoring agent in beverages, sorbets and other foodstuffs. Its aromatic effect on the mind is grounding, calming and balancing, while its other notable actions are antiseptic, antispasmodic, immune-stimulating, warming and sedative to the nervous system, and stimulating to the circulatory system.

Methods of Oil Extraction

(i) Enfleurage
(ii) Expressed Oils
(iii) Steam Distillation
(iv) Solvent Extraction
(v) Carbon Dioxide Extraction
(vi) Phytonic Process

(i) **Enfleurage:** Enfleurage is one of the oldest methods of extracting essential oils and is rarely used these days because of its high cost. It involves placing the flower petals on a layer of glass that is first spread with a thin layer of fat called «chassis». The volatile oil diffuses into the fat, then the fat is collected and the oil is extracted from the fat using alcohol. Once the alcohol evaporates what is left behind is called the absolute. As you can imagine, this is a very time consuming process.

(ii) Expression of Oils: When oils are taken from the rind of fruits they are called "expressed oils". This method is cold and does not involve any solvents or heat of any kind. These are not technically considered essential oils for this reason, but they are still definitely therapeutic oils. The most important thing to consider with cold pressed oils is the source of the crop.

(iii) Steam distillation: Steam distillation has been used for hundreds of years and today remains one of the most favorably methods of extracting essential oils. There are actually three different steam processes. But in all of the methods, steam is used to rupture the oil membranes in the plant and release the essential oil. The steam carries the essential oil to a condenser and then as it re-liquefies the lighter essential oil floats on top. The water and oil is then separated and the water portion is referred to as the hydrosol, hydrolat, or flower or floral water.

(iv) Solvent Extraction: Solvent extraction is a method of extracting essential oils that is dominated by the perfume industry. And technically does not produce therapeutic grade oils because chemicals such as hexane, acetone, di-methylene-chloride and others are used in the process. The oil particles are released when the concrete substance is then mixed with alcohol. Once it is separated out, the chemicals remain in the oil and the oil is used by the perfume industry.

(v) Super Critical Fluid Extraction (Carbon Dioxide Extraction): Super critical fluid extraction or carbon dioxide extraction is also a fairly new solvent extraction process that uses carbon dioxide at very high pressure. The carbon dioxide is injected into the tank where the plant material is contained and acts as a solvent to extract the oil.

(vi) Phytonic Process: The Phytonic process is a one of the recent methods of extracting essential oils using non-CFCs (non-chlorofluorocarbons). It is also called Florasol Extraction and the oils are referred to as phytols.

20. List ten main bamboo species of commerce in India with their scientific names and the State of origin (2011)

S.No	Species	States/UTs
1.	***Bambusa arundinacea***	Arunachal Pradesh, Karnataka, Orissa, Maharashtra, Himachal Pradesh, Andhra Pradesh and Gujarat
2.	***Bambusa balcooa***	Arunachal Pradesh, Mizoram
3.	***Bambusa pallida***	Arunachal Pradesh, Nagaland, Mizoram, Tripura
4.	***Bambusa tulda***	Arunachal Pradesh, Assam, Mizoram, Nagaland, Tripura
5.	***Bambusa polymorpha***	Tripura
6.	***Dendrocalamus hamiltonii***	Arunachal Pradesh, Assam,Mizoram, Nagaland
7.	***Dendrocalamus longispathus***	Mizoram

8.	***Dendrocalamus strictus***	Andhra Pradesh, Assam, Gujarat, Maharashtra, Himachal Pradesh, Madhya Pradesh, Manipur, Orissa, Karnatak, Uttar Pradesh, Rajasthan
9.	***Melocanna bambusoides***	Assam, Mizoram, Nagaland, Tripura, Manipur, Meghalaya
10.	***Neebenzia balcooa***	Nagaland
11.	***Oxytenanthera nigrociliata***	Tripura, Assam
12.	***Oxytenanthera parviflora***	Assam
13.	***Pseudostachyus polymorphium***	Arunachal Pradesh
14.	***Polystachia pargracile***	Orissa

21. Give legal definition of forests in India. Write about the major groups of forest types of India. (2011)

Definition: In legal terminology, forest is defined as an area of land proclaimed to be forest under a forest law. This definition describes the forest not as a biological unit but as property having a owner and with rights or certain people. This definition is useful only in law courts, where cases pertaining to offences committed are tried. From the point of view of silviculture, the first two definitions are more important.

Major groups and Type groups of forest in India

I. Tropical forests

- Wet evergreen forests
- Semi-evergreen forests
- Moist deciduous forests
- Littoral and swamp forests
- Dry deciduous forests
- Thorn f orests
- Dry evergreen forests

II. Montane sub-tropical forests

- Subtropical broad leaved forests
- Subtropical pine forests
- Subtropical dry evergreen forests

III. Montane temperate forests

- Montane wet temperate forests
- Himalayan moist forests
- Himalayan dry forests

IV. Sub-alpine forests

- Sub-alpine forests

V. Alpine scrub

- Moist alpine
- Dry alpine

22. Describe the initial causes of secondary succession. Write various seral stages of succession leading to the development of *Shorea robusta* forests. (2011)

Initial causes of Secondary Succession

(a) Climate: Climate is the initial cause when the vegetation is destroyed by the action of drought, wind, snow or frost. For instance, a fair portion of a forest may be killed by drought. It left to self, secondary succession will start on this bare area. Wind may lay bare an area by uprooting the original crop. Snow may, similarly, destroy forest by sliding.

(b) Physiography: Physiography is the initial cause when configuration of the land surface is responsible for the destruction of vegetation in combination with some other factors. For instance, landslide may take place on a steep slope, destroying the forest.

(c) Biotic factor: Biotic factor is the initial cause where a forest is destroyed as result of the activity of man, his animals, or even wild animals. For instance, a forest may be destroyed by reckless cutting, clearing, burning, indiscriminate heavy grazing, etc.

Seral stages of succession in sal forest: The various stages of this succession leading to the establishment of a sal forest are given below:

(i) *Saccharum spontaneum, Tamarix dioica*

(ii) *Acacia catechu, Dalbergia sissoo.*

(iii) *Acacia catechu, Holoptelia, Adina, Albizzia procera*

(iv) *Holoptelia, Adina, Lagerstroemia parviflora, Bombax, Terminalia belerica*

(v) *Shorea, Lagerstroemia, Terminalias, Adina*

If the soil is moist,

- *Trewia nudiflora* and *Toona ciliata* appear in stage (iv),
- *Terminalia tomentosa* in stages (v) and (vi) with *Syzygium cumini* below the top canopy.

23. Explain dynamics of Forest Vegetation giving example of evolution of Sal Forest in Uttaranchal (2012)

The dynamics of forest vegetation is very importance for evaluation of plant succession. The plant community is not a static unit but changes over a period of time as a result of its interaction with the surrounding environment. The growth and dynamics of forest vegetation depends on climatic, edaphic and biotic factors. The edaphic factor plays a vital role because in a given area as the organic layer of soil deepens, the water retaining capacity of soil increases under the influence of newly formed vegetation. This results in replacement of pioneer plant community by a new plant community and the sequence of vegetational and the associated ecosystem change continues till a community is evolves which attains equilibrium with the environment. This change in forest dynamics towards creation of a biotic community in equilibrium with environment is termed as climatic climax. This is the result of both horizontal and vertical expansion of plant community which is the success of dynamics in forest vegetation.

The very classical example is evolution of sal forests in Uttranchal wherein the pioneer plant associations of *Acacia catechu* and *Dalbergia sissoo* in the gangetic alluvium. In this ecosystem, the pioneer species of *Acacia catechu* and *Dalbergia sissoo* were replaced with *Shorea robusta* forests and lead to climax vegetation. The following composition of crop and the various stages of dynamics of forest vegetation in Uttranjal area is furnished.

(i) *Acacia catechu – Dalbergia sissoo*
(ii) *Acacia catechu – Holoptelea – Adina – Albizzia procera*
(iii) *Holoptelea – Adina – Lagerstroemia parviflora – Salmalia – Terminalia*
(iv) *Adina – Lagerstroemia parviflora – Terminalia – Shorea.*
(v) *Shorea – Lagerstroemia – Terminalia – Adina*

24. Based on the objectives, what are the di-erent classifications of the 'Forestry'? (2013)

Forest is an area set aside for the production of timber and other forest produce, or maintained under woody vegetation for certain indirect benefits which it provides, e.g., climatic or protective. Forestry is also defined as the theory and practice of all that constitutes the creation, conservation and scientific management of forests and the utilization of their resources.

Objectives of Forestry: The main objectives of forestry are

(i) To protect the existing forest and natural resources from all biotic and abiotic factors.
(ii) To produce the maximum possible wood and non wood for domestic and commercial utility.
(iii) To develop forestry to benefit the society.
(iv) To protect the environment and to create Clean Development Mechanism.

Classification of Forestry: Based on the objective, forestry is classified as

(a) Protection forestry: The Protection forestry ensures

(i) Protecting lands whether those upon which the forest is situated or those at a distance from it, against wind and water erosion.
(ii) Conserving water supplies for human consumption, fish culture etc.
(iii) Reducing hazards from flood damage to human life and property and
(iv) Amelioration of adverse climatic effects.

(b) Production Forestry: The objective is to produce timber and other industrial wood to meet the requirements of people, wood based industries and other stake holders.

(c) Social forestry: The Social forestry is forestry for the people, by the people and to the people and the primary objective is to meet the fuel, fodder, fibre and food requirements of the rural and urban communities.

Special types of forestry

(a) Farm forestry: Farm forestry is the practice of forestry on farms lands in the form of block plantations or sporadic distribution of trees in the farm lands.

(b) Agroforestry: Agroforestry is the sustainable land use system which integrates agriculture and allied sectors of agriculture with forestry. This includes silviagriculture, silvihorticulture, silvipasture, wind breaks, shelter belts, alley cropping, home gardens etc.

(c) Extension forestry: Extension forestry is the practice of forestry in areas devoid of tree growth and other vegetation and situated in places away from the conventional forest areas with the object of increasing the area under tree growth. It includes within its scope the following:

   (i) Mixed forestry: Mixed forestry is practice of forestry for raising fodder grass with scattered fodder trees, fruit trees and fuel wood trees on suitable wastelands, Panchayat land and village commons.

   (ii) Shelterbelts: Shelterbelt is defined as a belt of trees and/or shrubs maintained for the purpose of shelter from wind, sun, snow drift, etc. They are generally more extensive than the wind breaks covering areas larger than a single farm and sometimes whole regions on a planned pattern.

   (iii) Linear strip plantations: These are plantations of fast growing species on linear strips of land on the sides of public roads, canals and railway lines.

(d) Recreational forestry: Recreational forestry is the practice of forestry with the object of raising flowering trees, ornamental trees, fruit trees and shrubs mainly to cater the recreation needs of urban, semi urban and rural population. This kind of forestry is otherwise called as Aesthetic forestry or Urban forestry or Semi urban forestry.

25. Describe flora and distribution of group: Type 14/C2 East Himalayan sub-alpine birch fir forests. (2013)

Type 14/C2: This type of forest is called East Himalayan Sub-Alpine birch/fir forest. This type occurs in eastern parts of Himalayas above 3000 m elevation.

The examples are, *Abies pindrow, Juniperus wallichiana, Larix griDthiana, Rhododendron wightii, Betula utilis, Acer caesium, Padus cornuta,* some bushes of *Viburnum nervosum, Rosa webbiana*, etc.,

The following seral and degraded types of sub alpine forests are also found:

- Hippophae / Myricaria brakes ($1s_1$)
- Deciduous sub-alpine scrub ($1s_2$)
- Sub- alpine blue pine forest ($2s_1$)
- Sub alpine pastures (Ds_1)

The first two sub-types have already been described. Sub-alpine blue pine forest (*Pinus wallichiana*) occurs in parbati valley (Himachal Pradesh). They are similar to seral blue pine forest occurring in temperate region. Sub-alpine pastures consists of grasses – *Agropyron semicostatum, Brachypodium sylvaticum, Bromus asper, Dactylis spp, Oryzopsis spp, Poa spp.*

26. Explain how the latitude influences the forest types of the earth. (2015)

The latitude plays a vital role in influencing the forest type in the earth which ranges from tropical forest to tundra forest. The latitudinal influences in association with climatic factors towards creating forest types are discussed below.

Latitude and Solar radiation: Latitude and solar radiation have an inversely proportional relationship that is latitude increase the intensity of solar radiation decrease solar radiation constitute. Light and temperature so here photoperiod at higher latitude is less. Many characteristics in plants (growth, seed dormancy, germination, leaf fall, flowering and fruiting) are affected by Photo-period

Temperature is the function of solar radiation. Equator zone receives more heat. Near the equator the rays fall almost vertically upon the earth at noon whereas in higher latitudes, the noon day rays strike the surface obliquely. The change from day to night cause fluctuation in temperature (more in higher latitude, less in equator). The difference between highest and lowest temperature during 24 hours called diurnal range. So the air temperature influence various physiological activity i.e., transpiration, photosynthesis and respiration. Temperature is perhaps one of the most important single factors which determines the distribution of the vegetation.

The ice deserts of Antarctic zone have a temperature of -40° C to -60°C, while temperature in arid desert may be high as 50°C. The species which grow well in low temperature are not likely to grow in sites having constantly high temperature.

Latitude and Precipitation: Precipitation is limited by the latitude. The major from precipitation is rainfall, which is maximum at the equator. From equator to pole the precipitation will be decreased. There are several factors play in it. (Ocean, currents, temperature etc). As a result of the above factor latitude influence informing the different type of vegetation which in turn constitute different type of vegetation which in turn constitute different type of forest from tropical to tundra.

27. How are the forests classified in India? Discuss its significance in forest management? (2016)

Forest and Its Classification

Definition: Forest is defined as- "An area set aside for the production of timber and / or other forest produce, or maintained under woody vegetation for certain indirect benefits which it provides e.g., climatic or protective"

Classification of forests

Forests can be classified on the basis of

- Method of regeneration
- Age
- Composition
- Objects of management
- Ownership and legal status and
- Growing stock

(i) **Classification based on method of regeneration:** Based on regeneration, forests can be classified as

- High Forest – Forest regenerated through seeds
- Coppice Forests – Forest regenerated through coppice

(ii) **Classification based on age:** Based on age, forests are classified as

- **Even aged forests:** Forests consisting of trees which are approximately of the same age (differences allowed up to 25%).
- **Uneven aged forests:** Forests which vary widely in age and contain trees of all ages (the range of age difference being usually more than 20 years)

(iii) **Classification based on composition:** Based on composition, forests are classified into

- Pure Forest – a forest composed of almost entirely of one species, usually to the extent of not less than 80%. It is also called pure crop or pure stand.
- Mixed forest - a forest composed of trees of two or more species intermingled in the same canopy; in practice, and by convention, at least 20% of the canopy must consist of species other than the principal one.

(iv) **Classification based on objects of management:** Based on the objects of management, forests are classified into

- **Production Forest:** It is a forest managed primarily for its produce viz., timber, non-timber and other industrial utility.
- **Protection Forest:** It is an area wholly or partly covered with woody growth, managed primarily to regulate stream flow, prevent erosion, hold shifting sand or to exert any other beneficial influence on the physiographic features of the site
- **Farm Forest:** It is a forest raised on farms and its adjoining areas either as individual scattered trees or as woodlots to meet the domestic and / or industrial wood requirement.
- **Recreation forest:** It is a forest managed to meet the recreational needs of the urban and rural populations.

(v) **Classification based on ownership and legal status:** On the basis of ownership, forests are classified into

- **State forest:** It is a forest owned by state
- **Community forest:** Forest owned and generally managed by a community such as a village, town, tribal authority or local government, the members of which share the produce or proceeds of sale
- **Panchayat forest:** It is a forest where management is vested with a village panchayat (i.e., a body of people elected by the villagers from among themselves for specific administrative or other purposes pertaining to the village).

On legal basis, forests are classified into

- **Reserved forest:** It is an area so constituted under section 3 in chapter II of the Indian Forest Act 1927.
- **Protected forest:** It is an area subject to limited degree of protection under the provisions of Chapter IV of the Indian Forest Act, 1927.
- **Village forest:** A village forest is a state forest assigned to a village community under chapter III of the Indian Forest Act, 1927.

(vi) Classification on the basis of growing stock: On the basis of growing stock, forests are classified into

- **Normal forest:** It is a forest with normal growing stock, normal series of age gradation and normal increment.
- **Abnormal forest:** It is a forest with abnormal growing stock, i.e., growing stock in deficit or in excess or in which the relative proportions of the age or size classes are defective resulting in abnormal increment.

Significance in Forest Management: The forest classification has direct significance in forest management because forest management involves practical application of scientific, technical and economic principles of forestry. In simple terms forest management aims to achieve certain specific objectives by applying the principles of silviculture, ecology, mensuration, finance, economics and etc. All these activities depend on the type of forest and the objectives of forest management. Hence classification of forests plays a vital role in forest management.

Forests are classified based on regeneration pattern as high forest and coppice forest. This is very important for forest management particularly the high forest developed through seeds are managed differently compared to the coppice forests regenerated through coppice shoots. Hence the forest type classification is very significant towards forest management.

Forests are also classified based on the objective of forest management as productive, protective and social forests. When the forest is planned for production oriented purposes like timber, pulpwood, plywood, matchwood, firewood, fodder, gums and resins etc, these forests need separate and scientific management system in order to ensure sustained yield. At the same time if the forest is planned for water conservation / soil conservation / climate amelioration then the management option may be multiple cropping systems with different canopy architecture. Hence forest classification is essential for forest management.

For administrative purpose, the forests are legally classified into reserved forests and protected forests. This ensures full control on forest flora and fauna, legal control of all resources which will ensure conservation and protection of forest resources. No forests unless prescribed under working are allowed for felling. Hence legal classification of forests is very well significant towards over all control and management of forests.

Similarly the forest management lies on the nature of crop composition like growing stock. Accordingly the forest with normal growing stock, normal series of age gradation and normal increment demand separate management strategies compared to abnormal forests with abnormal growing stock, abnormal and sub normal increment and varying age classes. Hence forest classification is very vital for forest management.

28. a. Garg (1988) described the ecological pyramid as a graphic representation of relationship between the trophic levels of an ecosystem. Explain with the help of neat diagrams. (2016)

b. What are ecological Pyramids? Describe any one of the them. (2009)

Ecological pyramid is a graphical representation of some relationship between trophic levels of an ecosystem. Generally, pyramids of numbers, biomass, and energy transfer are drawn in the study of ecosystem as a rough representation of numerical, biomass, or energy relationship among trophic levels of the ecosystem.

Types of pyramids: The ecological pyramids are of three types:

(i) Pyramid of Numbers

(ii) Pyramid of Biomass

(iii) Pyramid of Energy

(i) Pyramid of numbers: This pyramid deals with the relationships between the numbers of primary producers and consumers of different order. At the base of such pyramid is always the number of primary producers. The top represents the number of top carnivores in an ecosystem. In figure, two types of up-right pyramids of numbers have been shown. A grassland ecosystem on the left side and cropland ecosystem on the right where the grass and weeds or the crop plants are in very large numbers. The number of rabbits in the grassland or grasshoppers in the crop field is generally less than the number of grasshoppers. Similarly, figure b, depicts an instance where the number of primary producers (a tree) is less than that of herbivore birds feeding upon the tree fruits. This is the case of inverted pyramid of numbers in a tree-ecosystem.

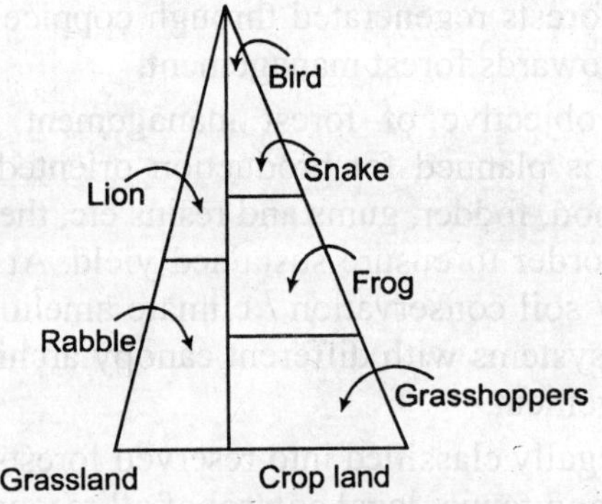

Fig a. Upright Pyramid of number in herbaceous ecosystem

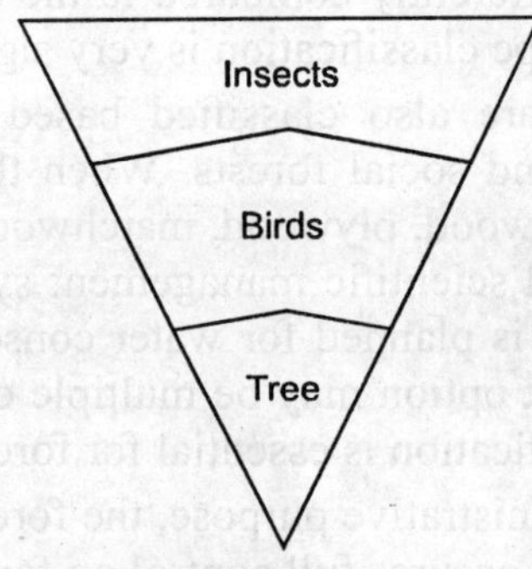

Fig b. Inverted pyramid of numbers in a tree-ecosystem

(ii) Pyramid of biomass: The biomass of one tree is very high. Moreover, the biomass of a number of Birds feeding upon the tree is far less than the tree itself. In the same fashion, the biomass of even a very large number of parasites in and on the body of birds is very less. The pyramid of biomass become upright (Figure a). But there can be instances where the pyramid of biomass also gets deformed or inverted (Figure b).

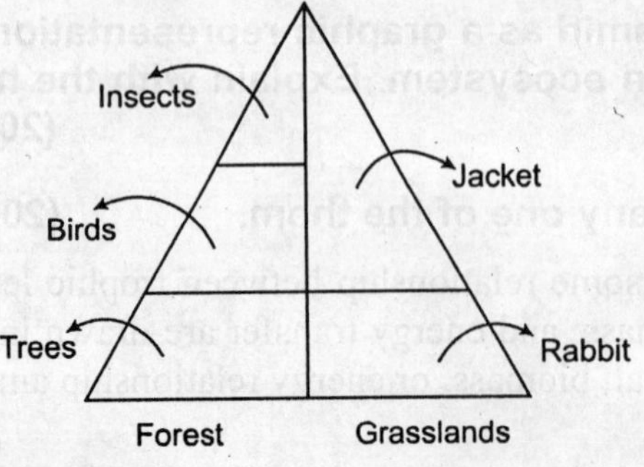

Fig a. Upright Pyramid of biomass

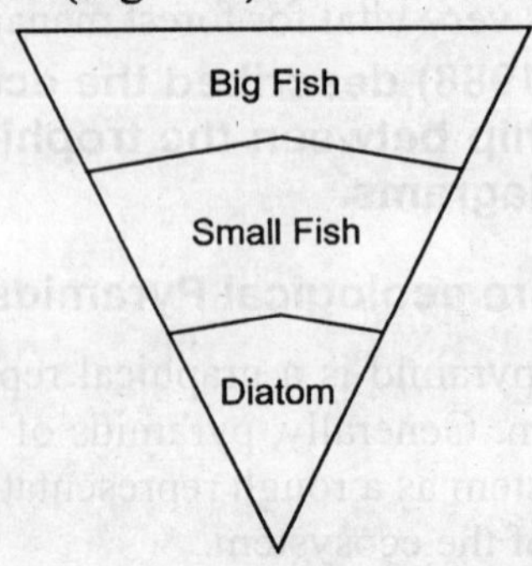

Fig b. Pyramid of biomass in an aquatic ecosystem

(iii) Pyramid of energy: The shape of the pyramid of energy is always upright. This is due to the factor of time. It represents the total quantity of energy utilized by different trophic level organisms of an ecosystem per unit area over a set period of time. The cumulative energy content that generation of phyto plantation trap in course of a year is much more than a few generations of herbivore fishes. This is in the corresponding time and space. Thus, the energy content of top carnivores is the least (Fig. a). The energy contents in K Cal/m^2/ yr from 20,810 in primary producers decrease to 3,368 to 383 to 21consumers of the first (C_1), second (C_2), and third (C_3) orders (Fig. b).

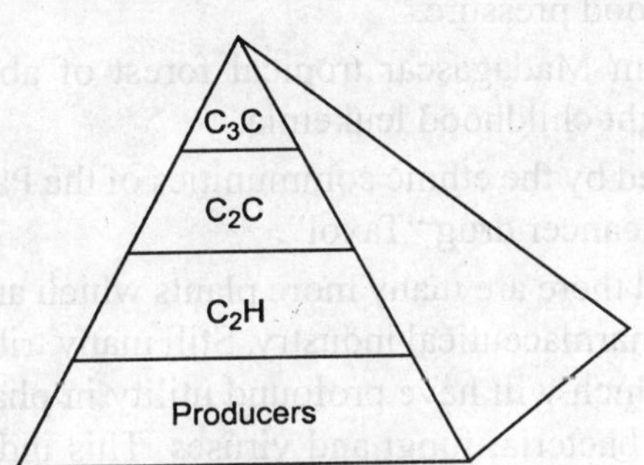

Fig a. Pyramids of Energy

21 C_3
383 C_2
3,368 C_1
20,810 P
Silver Spring (U.S.A.)

Fig b. Quantitative Values of Energy

29. The knowledge of Ethnobotany helps to identify viable medicinal plants for pharmaceutical industries. Explain. (2016)

Ethnobotany, the largest sub-discipline of ethnobiology, is generally defined as the "science of people's interaction with plants". This circumscription includes the study of plants that have therapeutic applications. While the primary objectives of modern ethnobotany are neither to develop new pharmaceuticals, nor to discover new bioactive chemical moieties, elucidating the pharmacological activities of a particular plant is part of some ethno botanists' research. While bio-prospecting (exploring biodiversity for new sources of natural products) is still the objective of some pharmacologists, the field of ethno botany is generally more concerned about the cultural meaning of the relationship between humans and plants than in mining for plant pharmaceuticals and information about plants. However the knowledge of ethnobotany helps to identify viable medicinal plants for pharmaceutical industries as indicated below.

- The plant *Ancistrocladus korupensis* obtained from the indigenous people of Cameroon forest in South Africa had the capacity to inhibit the HIV (AIDS) virus due to the presence of an active compound called Michellamine – B which is obtained from the leaf extracts of the plant. This active ingredient can play a vital role in pharmaceutical industry towards combating the AIDS virus.
- A rare medicinal plant *Trichopus zeylanicus* is used by the Kani tribe of Kerala, India is a health food to maintain vitality and increase resistant to the diseases.
- The winged bean *Psophocarpus tetragenolobus,* a crop native to South East Africa has great potential for feeding many in the third world.

- During the last few decades, several wonder drugs such as a resarpine, quinine, ephedrine, cocaine, Kkhellin, colchicines, digexin, D-tubercurarine, artemisinine and gugulipid have been discovered from plants with rich ethnobotanical role in tribal societies.
- The May apple (*Podphyllum pelatum*) has a toxin resin, Podophyllotoxin, formerly used by Amerindians to remove warts; its cytotoxic action has led to its modern used in treating uterine warts which formerly required surgery.
- The tranquillisers, rescinnamine and reserpine have been obtained from the roots of *Rauvolfia serpentina* used in India for more than a thousand years in folk medicine for snake bite, insanity, epilepsy and high blood pressure.
- The Rosy Periwinkle (*Vinca rosea*) from Madagascar tropical forest of aborigines has provided the million dollar weapon to fight childhood leukemia.
- The Pacific yew (*Taxus baccata*) preserved by the ethnic communities of the Pacific Islands has provided he million dollar worth anticancer drug "Taxol".

These are only few examples and citations and there are many more plants which are associated with ethnic groups will have profound utility in Pharmaceutical industry. Still many tribal groups in the country has wider ethnobotany knowledge which will have profound utility in pharmaceutical industry towards controlling the diseases due to bacteria, fungi and viruses. This indigenous and ethnobotanical knowledge of ethnic group will have significant role in pharmaceutical industries.

30. Enumerate the classification of tropical dry deciduous forests given by Champion and Seth (1964). Mention two species for each forest type.

Group	Forest type	Species
5A – Southern tropical dry deciduous forests	C1 Dry teak bearing forest 1a Very dry teak forest 1b Dry teak forest	*Anogessus latifolia,* *Boswellia serrata*
	C2 Dry red sanders bearing forests	*Pterocarpus santalinus,* *P.marsupium*
	C3 Southern dry mixed deciduous forest	*Boswellia serrata,* *Terminalia chebula*
5 B – Northern tropical dry deciduous forests	C1 Dry sal bearing forest 1a Dry Siwalik sal forest 1b Dry plains sal forest 1c Dry peninsular sal forest	*Aegle marmelos,* *Terminalia alata*
	C2 Northern dry mixed deciduous forest Ds1 Dry deciduous scrub Ds2Dry savannah Forest Ds3 Euphorbia forest Ds4 Dry grassland	*Butea monosperma,* *Acacia catechu*

Degradation seral types of dry deciduous forests	E1 *Anogeissus pendula* forest Ds1 *Anogeissus pendula* forest	*Anogeissus pendula,* *Acacia catechu*
	E2 *Boswellia* forest	*Boswellia serrata,* *Tectona grandis*
	E3 Babul forest	*Acacia nilotica,* *A.leucophloea*
	E4 *Hardwickia* forest	*Hardwickia binata,* *Lannea coromandalica*
	E5 *Butea* forest	*Butea monosperma,* *Acacia nilotica*
	E6 *Aegle* forest	*Aegle marmelos,* *Diospyros melanoxylon*
	E7 Laterite thorne forest	*Terminalia chebula,* *Careya arborea*
	E8 Saline / alkaline savannah 8a Phoenix savannah 8b Babul savannah 8c Salvadora-tamarix scrub	*Phoenix sylvestris,* *Acacia nilotica,* *Salvadora persica*
	E9 Dry bamboo brake	*Dendrocalamus strictus,* *Sterculia urens*
General seral types of dry deciduous forests	1s1 Dry tropical riverain forest 1s2 Khair-sissoo forest 1s3 Inundation babul forest	*Acacia catechu,* *Dalbergia sissoo*
	2s1 Secondary dry deciduous forest	*Prosopis cineraria,* *Grewia tilieafolia*

31. a. Defind ecological amplitude. Describe the classification and characters of plant communities. (2016)

b. Discuss "ecological amplitude" and the "law of tolerance" in the context in which they are generally used

Ecological Amplitude: It is the degree of adaptation of a living organism to changes in its environment. A species trait and ecological amplitude are expressed quantitatively as the range of environmental changes within which a given species is able to carry on its normal vital activities.

Classification:

(a) The amount of heater present in the habitat, determines the different types of communities. Thus, the communities may be divided into the following groups :-

- **Hydrophytic:** Predominantly aquatic habitats;
- **Mesophytic:** Moderately moist soils; and
- **Xerophytic:** Arid or dry conditions.

(b) The communities growing in conditions of abundant light are called heliophytic, and, those growing in shade are called sciophytic.

(c) The communities may be similarly classified on the basis of other environmental considerations. In terms of their general shape and growth form with a particular environment, communities have been classified into:

- Forest communities; and
- Desert communities.

(d) Clements (1916) recognized the fact that communities are not always the same at any place and he classified the communities on two parallel lines:

- Seral communities - the process of change
- Stable/climax communities - remain stable.

Characters of Communities: The communities are judged in two ways, viz., (1) Qualitative characters, (2) Quantitative characters.

Qualitative characters: The structure, composition, and many other characters of plant community (ies) can be described just by visual observations. In this case, the measurements, of any type are not mad. This is a qualitative approach which is easier than the quantitative vegetation analysis where measurements are actually made. The following are some qualitative characters of community mainly the plant communities:

- Kind of species
- Stratification of vegetation
- Sociability

Quantitative characters: By qualitative characteristics, the vegetation could be studied in field without the help of any special sample technique. Elsewhere in the world some systems of classification and description of communities have been developed. This aspect of ecology is known as 'phytosociology'. But the structure of sociological order in any community or set of communities cannot be studied by observing each and every individual in an area. Therefore, some sort of vegetation sampling is required in such cases. Some quantitative characters are:

- Density
- Frequency
- Dominance

Law of Tolerance: Any organism can tolerate the environmental factors which will be variable due to the intensity of the environmental factor. The organism has a tolerance which varied between lower and upper limit in any environmental range and between which the organism functions efficiently. The capacity of a organism to adopt to minimum quantity or a condition of a any one factor is termed as threshold level and the response of organism both above and below threshold level will be variable.

V.E. Shelford in 1913 incorporated the idea that factors could be limiting at the maximum as well as minimum quantity as the Law of Tolerance.

Each ecological factor to which an organism responds has maximum and minimum limiting effect between which lies a range of gradient that is known as limits of tolerance.

Moderate rise or decrease can lead a condition of physiological stress for that organism and further rise or decline can cause death of the organism as it reaches the zone of intolerance or zone of coma.

Shelford's Law of Tolerance

"The presence and success of an organism depend upon the completeness of a complex of conditions. Absence or failure of an organism can be controlled by qualitative or quantitative deficiency or excess with respect to any one of the several factors, which may approach the limits of tolerance for that organism".

Between the lower and upper limit of tolerance lies a broad sector of a gradient, which is called, as zone of compatibility or zone of tolerance or zone of biokinetic or zone of capacity adaptation. The region at either end of zone of compatibility is called the lethal zone or zone of resistance or zone of intolerance. The zone of compatibility too includes broad range of optimum and narrow zone of physiological stress in between the range of optimum and lethal zones.

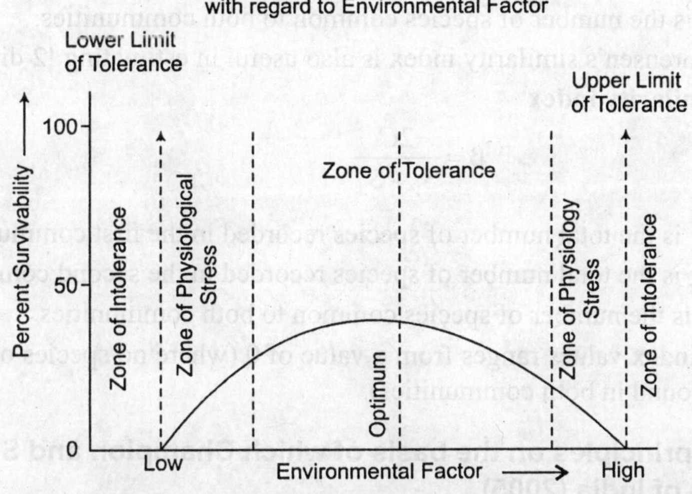

Fig. Zone of tolerance and resistance for an organism with regard to environmental factor

32. What do you mean by alpha and beta diversity, and how they are measured? (10). (2005)

Alpha diversity: This is also known as point diversity. This denotes the number of species in a specified area. Alpha diversity refers to the diversity within a particular area or ecosystem and is usually expressed by the number of species (i.e species richness) in that ecosystem. This can be measured by counting the number of taxa (distinct groups of organisms) within the ecosystem. However, such estimates of species richness are strongly influenced by sample size. Simpson diversity index is more useful in estimating the alpha diversity which is given below:

$$D = 1 - \frac{\sum_{i=1}^{s} n_i (n_i = 1)}{N(n_i = 1)}$$

Where, S is the number of species,
N is the total number of organisms,
n_i is the number of organisms of species i.

Beta diversity: This refers to the degree to which species composition changes along the environmental gradient. This is represented by turnover of species across a space. Beta diversity is a comparison of diversity between ecosystems or along environmental gradients. It is usually measured as the amount of species change between the ecosystems. It gives the number of species that are unique to each ecosystem. It is the rate of change in species composition across habitats or among communities.

Beta diversity can be represented by the following equation

$$\beta = (S_1 - C) + (S_2 - C)$$

Where, S_1 is the total number of species recorded in the first community,
S_2 is the total number of species recorded in the second community, and
c is the number of species common to both communities.
Sorensen's similarity index is also useful in estimating !2 diversity.

Sorensen's similarity index

$$\beta = \frac{2C}{S_1 + S_2}$$

Where, S_1 is the total number of species recorded in the first community,
S_2 is the total number of species recorded in the second community, and
c is the number of species common to both communities.

The Sorensen index values ranges from a value of 0 (where no species overlap) and 1 (exactly same species are found in both communities).

33. Explain the principles on the basis of which Champion and Seth have classified forest types of India (2005).

Principle of Champion and Seth Classification: The Champion and Seth classified forest types based on the principle of ecosystem as indicated below.

Classification based mainly on ecosystem: This systems gives sufficient weightage to both vegetation and climate. The most important classification of this system have been proposed by Burtt-Davy, Swain, Gaussen and by Champion and Seth in the revised classification of forest types of India.

Revised Classification of Forest Types of India: On the basis of additional information available Champion and Seth revised the preliminary classification of forest types of India prepared in 1935. In this classification, the forests of India have first been divided into the following five major groups:

(i) Tropical forests
(ii) Montane subtropical forests

(iii) Montane temperate forests

(iv) Sub-alpine forests and

(v) Alpine scrub

The major groups have, as described below, been further divided into type groups or, simply, groups on the basis climatic data and vegetation. The tropical forests have been differentiated into seven groups, the montane subtropical forests into three groups, montane temperate forests into three groups, sub-alpine forests into one group and alpine scrub into two groups. Most of these groups or type groups have been further differentiated into two subgroups describing southern and northern forms.

Each subgroup is again divided into types in which climax formations have been designated by letter C, edaphic climax formation by letter E, primary sere by IS and secondary sere by 2S, the tropical swamps by letter TS and fresh water swamps by FS. Within each subgroup, the types are given in serial number as C_1, C_2, E_1, E_2 and so on. These types are differentiated into subtypes by suffixing letter a, b, c and so on for each subtype. Subtypes are sometimes further classified into varieties by suffixing number, e.g., i, ii, iii. Thus a variety of type C_1 is designated as $C_{1a(i)}$.

Major groups (5)	**Type groups (16)**
I. Tropical forests	1. Wet evergreen forests
	2. Semi-evergreen forests
	3. Moist deciduous forests
	4. Littoral and swamp forests
	5. Dry deciduous forests
	6. Thorn forests
	7.Dry evergreen forests
II. Montane sub-tropical forests	8.Subtropical broad leaved forests
	9. Subtropical pine forests
	10.Subtropical dry evergreen forests
III. Montane temperate forests	11. Montane wet temperate forests
	12.Himalayan moist forests
	13.Himalayan dry forests
IV. Sub-alpine forests	14. Sub-alpine forests
V. Alpine scrub	15. Moist alpine
	16. Dry alpine

34.Name di-erent types of nutrient cycling which occur in tropical forest ecosystems. Give examples. (2006)

(i) Oxygen Cycle: Oxygen (O_2) is very essential for life and it makes up 21% (by volume) of the atmosphere. However, it occurs in a bound form as carbonates and oxides in rocks and water. In the process of photosynthesis, the plants also release some amount of oxygen into the atmosphere such s

$$6CO_2 + 6H_2O = C_6H_{12}O_6 + 6O_2$$

(Carbon dioxide) (Water) (Sugar) (Oxygen)

The oxygen cycle is given as under:

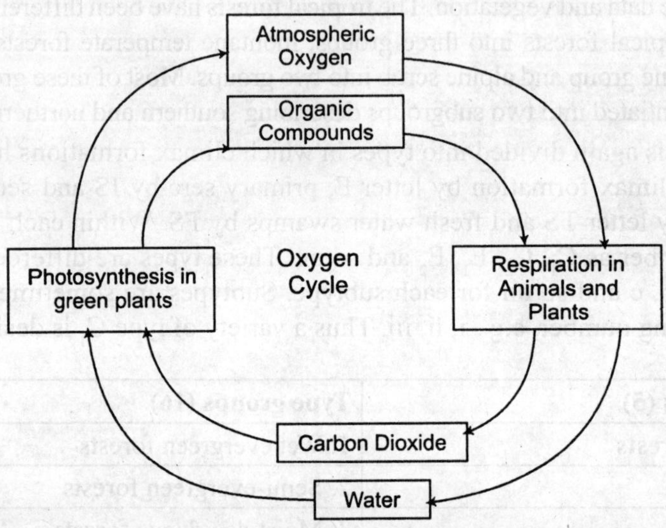

Further, the oxygen in gaseous form is utilized by all organisms during the process of respiration. Oxygen (O_2) is used in the process of conversion of energy.

Formation of Oxygen: The formation of oxygen (O_2) in the atmosphere is shown as under:

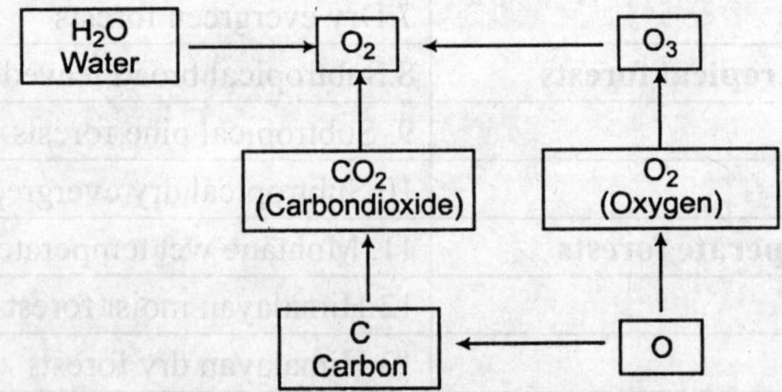

The oxygen also helps in the oxidation of organic matter.

(ii) Carbon cycle: The carbon dioxide (CO_2) is reduced during the process of photosynthesis, whereas, it is incorporated in various organic compounds. It is observed that a portion of these substances are oxidized during the process of respiration and carbon dioxide (CO_2) is released back into the atmosphere. The death of plants and animals leads to the formation of petroleum and coal, which form a complex combination of carbon dioxide (CO_2)

The carbon cycle is as follows:

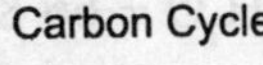

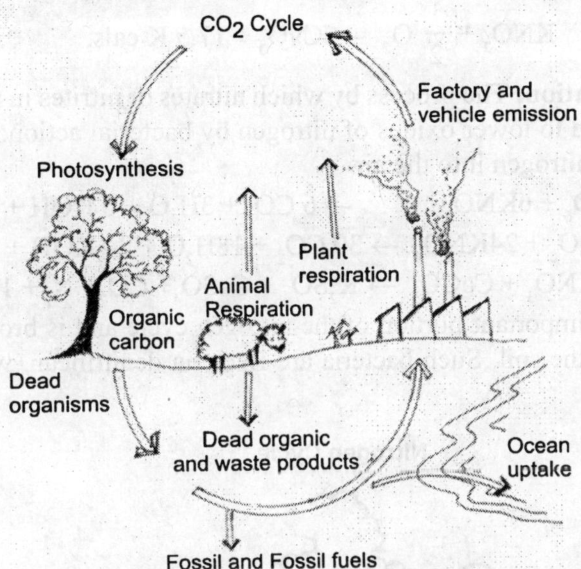

The reserves of carbon (Coal and petroleum) are drawn up continuously for their use as fuel and carbon is fed back into the atmosphere. In addition to the earth's surface, some portion of carbon cycle operates in seas and oceans. The oceans contain about 50 times more CO_2 than the atmosphere, in the form of carbonates. The CO_2 dissolves in water and carbonic acid is formed, converts carbonates to bicarbonates. The following are the reactions:

$$H_2O + CO_2 \quad W \quad H_2CO_3$$
$$H_2CO_3 \quad W \quad H_3O \; + \; HCO_3$$
$$HCO_3 \quad W \quad H_3O \; + \; CO_3$$

The carbonate deposition is hastened due to the excess of calcium in sea water (Generally alkaline)

(iii) Nitrogen Cycle: The atmosphere contains 78% nitrogen (approx.) It is estimated that there are some 35,000 tons of nitrogen per acre of land. For utilization of nitrogen, it must be combined with hydrogen or oxygen. This process is called nitrogen fixation. This is, however, fixed by various soil organism. Lightning also fixes smaller amounts of nitrogen which is carried in the soil by the rain. The plant and animal wastes are returned to the soil, carrying nitrogen along with them.

The nitrogen may be lost to atmosphere by denitrification or by volatilization where ammonia is produced at or near the soil surface and leaching of nitrates may move nitrogen below the root zone where it cannot be utilized by the crops. This continuous recycling of nitrogen is knows as nitrogen cycle.

Reactions: Following are the reactions involved:

(a) Nitrification: The formation of nitrates and nitrites from ammonia (or ammonium compounds), as in soils by micro-organisms.

$$NH_3 + \frac{3}{2}O_2 \rightarrow HNO_2 + H_2O + 66 \text{ K cals}$$

$$KNO_2 + \frac{1}{2}O_2 \rightarrow KNO_3 + 17.5 \text{ K cals.}$$

(b) Denitrification: The process by which nitrates or nitrites in the soil or organic deposits are reduced to lower oxides of nitrogen by bacterial action. The process results in the escape of nitrogen into the air.

$$C_6H_{12}O_6 + 6KNO_3 \rightarrow 6\ CO_2 + 3H_2O + 6\ KOH + 3\ N_2O + 545 \text{ K cals}$$

$$5C_6H_{12}O_6 + 24KNO_3 \rightarrow 30\ CO_2 + 18H_2O + 24\ KOH + 12\ N_2 + 570 \text{ K cals}$$

$$S + 2KNO_3 + CaCO_3 \rightarrow K_2SO_4 + CaSO_4 + CO_2 + N_2 + 132 \text{ K cals}$$

This is an important portion of the nitrogen cycle and is brought about by the bacteria present in the soil. Such bacteria are Bacteria denitrifican, which live under anaerobic conditions.

Nitrogen Cycle

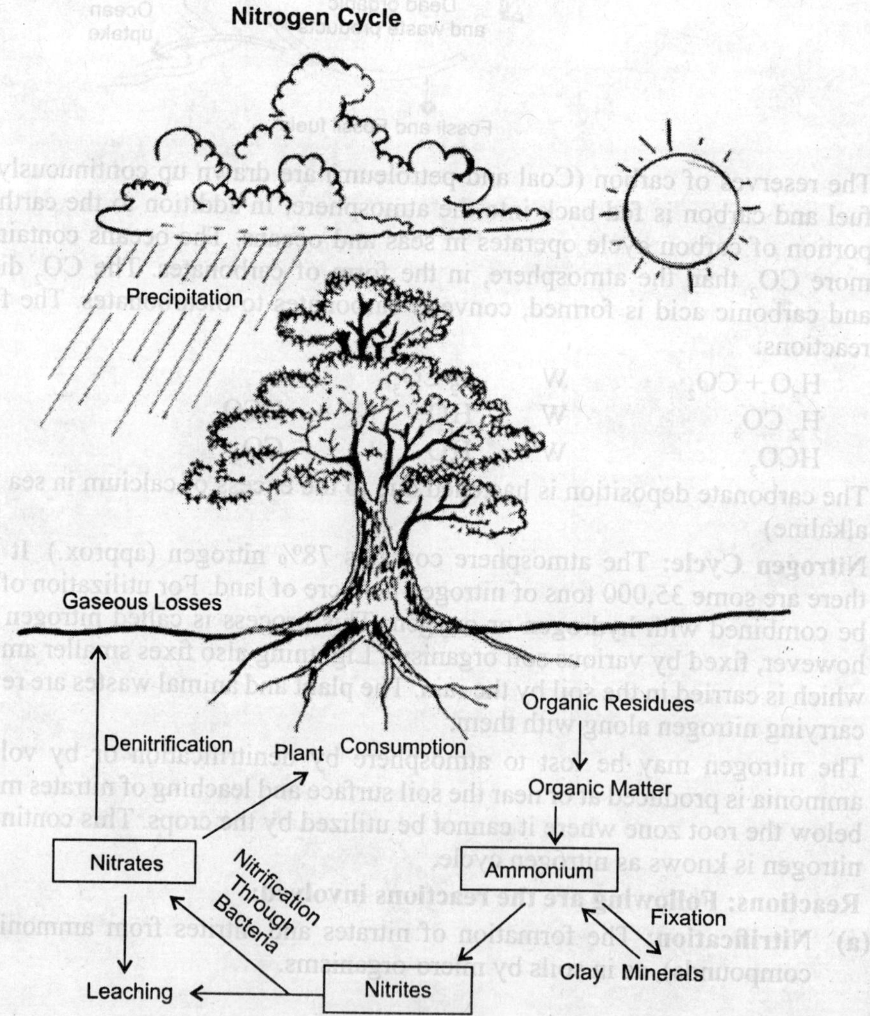

(c) **Ammonification:** Refers to "the formation of ammonium compounds or ammonia", when sufficient raw organic matter is not present in the soil, which could suitably serve as food material for the soil micro-organims, the bacteria feed upon soil humus. This brings about the conversion of complex lingo proteins into simple amino acids and then into ammonia, carbon dioxide and water.

(iv) **Sulphur Cycle:** The uptake of sulphur by plants is in the form of sulphate ions (SO_4). This may also be absorbed from the air through leaves in areas where the atmosphere has been enriched with sulphur compounds from industrial sources. Sulphur is a constituent of three amino acids (e.g., cystine, methionine, and cysteine) and therefore essential for protein synthesis. It is also essential for nodule formation of legume roots.

The sulphur cycle also resembles nitrogen cycle as it also involves the oxidation (SO_2) and reduction phases (H_2S). It is found in abundance in the soil and only a limited amount of it is available in atmosphere. The sulphur bacteria has the capacity to utilize elemental sulphur. As a part of the cycle, sulphur dioxide and hydrogen sulphide are returned to the soil as sulphates along with rain. There are some lichens that utilize sulphur dioxide.

The sulphur cycle is shown as follows.

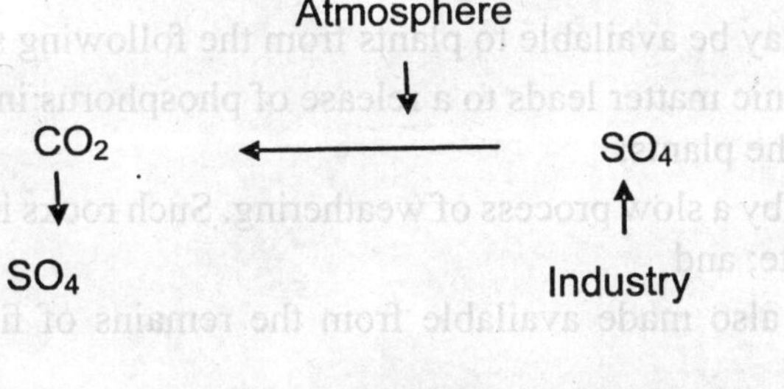

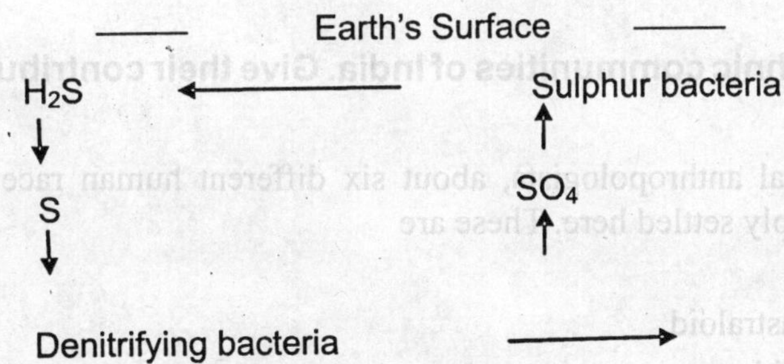

The decomposition may be brought about by the following:

(a) 'Anaerobic Conditions, e.g., Escherichia sp., *Proteus sp., etc.*

(b) Aerobic Conditions, e.g., *Aspergillus* sp., *Neurospora sp., etc.*

However, in waterlogged and submerged conditions (soils), there may be formation of H_2S (Hydrogen sulphide). The bacteria use up the oxygen present in the sulphate molecule and may lead to the release of energy on oxidation of CHO (carbohydrates) arid-other organic compounds.

The reactions may be expressed as under:

$RCH_2OH + SO_4$ W $RCOOH + H_2O + S$

$CH_2NH_2COOH + H_2O + SO_4$ W $H_2S + HS + HCO_3 + NH_4$

(v) **Phosphorous cycle:** Phosphorous is absorbed by plants as H_2PO_4, HPO_4 or PO_4 depending upon the soil pH. In neutral to alkaline soils, calcium phosphate is formed whereas in acid soils, iron and aluminum phosphates are produced. The available soil phosphorus may be only 1% or less of the total amount present.

The phosphorus cycle is a relatively simpler one, yet it is very significant as phosphorous is the main carrier of energy in the form of ATP. Phosphorus is present in all living cells. It is utilized by plants to form nucleic acids (DNA and RNA). It is used in storage and transfer of energy through energy-rich linkages (ATP and ADP). It has been known to create many environmental and ecological implications.

Phosphorous stimulates early growth and root formation. Its supplementation is required most by crops under the following conditions:

(a) Fast top growth;

(b) Growth in cold weather; and

(c) Limited root growth.

However, the phosphorus may be available to plants from the following sources:

(i) Decomposition of organic matter leads to a release of phosphorus in the soil. This is again available for reuse by the plants;

(ii) From phosphatic rocks by a slow process of weathering. Such rocks include rock phosphate or phosphorite or apatite; and

(iii) A little phosphorus is also made available from the remains of fishes and bird excreta (guano)

34. Describe the ethnic communities of India. Give their contribution to ethno botany. (10) (2007)

According to racial anthropologists, about six different human races have migrated to our country and have deeply settled here. These are

(i) Negrito,

(ii) The Proto-Australoid

(iii) The Mongoloid

(iv) The Mediterranean,

(v) The western Brachycephals, and

(vi) The Nordics

Generally the aborigines of Indian subcontinent are known as tribes. However some of the primitive tribal societies of our country are now facing the threat of extinction. There are some tribal societies in India which are considered as most primitive human societies on the land surface on earth depending totally on the forest areas for their survival or smooth running of life. These tribals are observed and led their life by having complete harmony with nature. Such ethnic societies include Onges, Jarawas, Sentineles and Shompens of Andaman and Nicobar islands. Their mode of survival completely depends on hunting, food gathering and fishing. These tribals strictly observe and follow several taboos and totems for ensuring the sufficient regeneration of wild plant species (Sinha, 1996). But unfortunately with the gradual and continuous decline of these tribals the rich repository of biodiversity are eroding at rapid pace.

According to the informations available their population in 1858 was estimated 4800 which has now deteriorated to 500 or may be even lesser in numbers. The Onges and Jarawas almost inhabit in the stone age culture and rapidly approaching to the extinction. The number of Jarawas estimated to 200 in 1911 which has now declined to 31 only in 1992, while the numbers of Onges from 200 has slide down to mere 80, during this time period. The Jarawas are said to be hostile towards outsiders, while the Santin else inhabit in complete isolation. They were not at all having any idea of world beyond the seas till recently. The population numbers of the Totos of Bengal has been also declining in the similar manner. They are now only 675 in number (Anonymous 1994).

The comprehensive but brief account of major ethinic groups are furnished below.

(a) Bhotiyas
(b) Jaunsari Tribe
(c) Tharu Tribe
(d) Bauxa Tribe
(e) Kinnaur Tribe
(f) Garos Tribe
(g) Khasi Tribe
(h) Naga Tribe
(i) Abors
(j) Mizo
(k) Khamtis
(l) Kukis
(m) Dabla Tribe
(n) Santhals
(o) Gond Tribe
(p) "Ho" Tribe
(q) Khond Tribe
(r) Toda Tribe
(s) Urali Tibe

- Some Rare Tribes of India on the Verge of Extinction
  - ♦ Raji Tribe
  - ♦ Lepcha Tribe
- The Tribal Societies of Andamaan& Nicobar Islands

- The Promment Tribal Groups of Rajasthan
  - Bhils
  - Meena
  - Sahariya
  - Garasia
  - Damor
  - Kathodia
  - Bhil-Meena
- Nomadic and Denotified Tribes of Rajasthan
  - GadiaLohar
  - Banjaras
  - Kanjars
  - Kalbelias
  - Sansis and Nats
  - Baurias & Baghri

Contribution to Ethnobotany

- The invaluable role of ethnic people in the conservation of plant genetic resources is very significant. The people of different ethnic communities in village keep various kinds of seeds for fulfillment of their needs about domestic consumption and use for cultivation.
- The tribal people and the traditional farmers are still adopting cultivation practices of their traditional cultivars and land races which constitute valuable germplasm bank.
- The ethnic societies have provided several miracle plants of immense food and medicinal value to the modern civilization.
- The primitive cultivars and wild relatives of crop plant preserved by the indigenous ethnic people may act as a genetic key for many valuable crops for use in the near future.
- A rare medicinal plant *Trichopus zeylanicus* is used by the Kani tribe of Kerala, India is a health food to maintain vitality and increase resistant to the diseases.
- The winged bean *Psophocarpus tetragenolobus,* a crop native to South East Africa has great potential for feeding many in the third world.
- During the last few decades, several wonder drugs such as a resarpine, quinine, ephedrine, cocaine, Kkhellin, colchicines, digexin, D-tubercurarine, artemisinine and gugulipid have been discovered from plants with rich ethnobotanical role in tribal societies.
- Nutritional evaluation of about 200 wild edible plants collected from tribal areas in India have so far been carried out.
- Certain psuedocerals (*Amaranthus polygamus* and *Fagopyrum esculentum*) are traditionally used by the tribals of the Himalayas for edible purposes are considered exceptionally rich in good quality protein with the essential amino acid "Lysine" which is usually deficient in most cereals.

The capability of many tribals to identify the plants even at the variety level has surprised many plant taxonomists. While botanists sometimes feel difficulty even in distinguishing genera and species without consulting the floras, the tribals are capable of identifying these plants even by the vegetative parts. Deep investigations and thorough studies about such knowledge can lead to the discovery of new relationships between modern taxonomy and primitive knowledge. The traditional knowledge and resource management practices of the ethnic people should be applied in modern development strategies. Hence the ethnic groups have contributed very significantly in ethnobotany.

35. Describe the importance and future scope of medicinal and aromatic plants in India. (10). (2007)

Importance of Medicinal Plants: Medicinal and Aromatic plants form a numerically large group of economically important plants which provide basic raw materials for medicines, perfumes, flavors and cosmetics. These plants and their products not only serve as valuable source of income for small holders and entrepreneurs but also help the country to earn valuable foreign exchange by way of export. Medicinal plants are those plants which are rich in secondary -metabolitcs and arc potential source of drugs. These secondary metabolites includes alkoloids, glycosides, coumarins, flavonides, steroids etc.

- Among various plants there is great demand in the country and abroad are as opium poppy, tropane alkaloid bearing plants, sapogenin bearing yams, senna, cinchona and ipecae.
- The ancient Indian system of medicine is mainly plant based matesiamedica making use of most of our native plants. It caters the needs of rural population of our country.
- ISM (Indian System of Medicine) offers most appropriate or first line theorapy against many diseases like jaundice, asthama, diabetes etc.
- India has about 2,000 species of medicinal plants and a vast geographical area with high production potential and varied agroclimatical conditions.
- India is already a major exporter of medicinal plants. It is estimated that Rs, 86 crores worth of raw materials and drug from medicinal plants are exported from India.

Importance of Aromatic Plants: Aromatic plants are from a numerically large group of economically important plants. These are increasing demand for essential oils, aroma chemicals drugs and pharmaceuticals in the world market since two decades. Aromatic compounds are present in plants i.e. in root, wood, bark, foliage, flower, fruit, seed etc.

- Aromatic plants produce essential oils, perfumes and flavours are in use with our civilization since several thousand years.
- Due to Vast area and varied agro-climatic condition, it can be commercial cultivated in different part of India successfully.
- Essential oils and aroma chemicals are indispensable in various human activities.
- They are adjuncts of cosmetics, soaps, pharmaceutical preparation, perfumer confectionery, ice-cream, aerated waters, disinfectants, agarbatti etc.

- Some of the important aromatic plants like Lemon grass, Citronella, Palmarose, Vetiver, Geranium, Lavender, Dawana etc. have great demand in our country.

Future Scope

- The demand for essential oils is increasing day by day with the advancement of education and prosperity in the country.
- Essential oils are now a basic raw material for consumer products meant for mass consumption.
- The search for natural resources has been intensified as synthetics have failed to provide versatility and a situation of saturation and later stagnation started developing.
- The aroma therapy is gaining momentum across the world and the interest in aromatics for their therapeutic value is also increasing due to the worldwide scare of the side-effects of synthetics and the revival of interest in herbs.
- Natural essential oils have the potential of being very safe insecticides. One good example in this regard is of the essential oil obtained from Acoruscalamus which has ß-asarone as an active principle, and produces sterility among a variety of insects of either sex. It has been found very effective and safe for the storage of food grains.
- The growing and processing of aromatic crops is labour-intensive and, hence, generates lot of employment.
- Thus, they provide a natural asset to these countries for the export trade.
- The World Health Organization (WHO) has emphasized the need for better utilization of the indigenous system of medicine, based on the locally available medicinal plants in the developing countries.

36. Why Herbaria and Arboreta are prerequisites for dendrology. (2007)

Dendrology is a division of botany the deals with taxonomy of trees and other woody plants including nomenclature, classification, identification and distribution. For this purpose, the plants have to be established and preserved in the form of herbaria and arboreta. This will help to understand how trees are named, including scientific names, common names and the code of botanical nomenclature. It also helps to classify the tree based on families, genera and other groups according to their arrangement. The herbaria and arboreta are also important to identify the trees and conserve them for future utility. Hence herbaria and arboreta are prerequisites for dendrology. The specific prerequisites amenable for herbaria and arboreta are furnished below.

Herbaria

- It is an invaluable conservatory of plant material and data.
- It is storehouses of collection including the valuable type specimens. The herbaria greatly aid in all kinds of taxonomic researches.
- Serves as a fundamental resource for identification of all plants of the world.
- It serves as a source for collections biodiversity. Most estimates on global biodiversity today are based on herbarium collection only.

- It serves as a repository of voucher specimens on which varieties Botanical researches are carried out.
- Aids in assessment of conservation status of a taxon.
- Vast collection of a particular species in a herbarium aids in assessing the diversity or variations exhibited by a species in its distributional ranges helping in population biology studies.
- It serves as a source for search of new genetic material for improvement of cultivated stock.
- The tags of herbarium carry all the information about habitat, habit, local name, flower colour and other characters of the plant, use of plant, frequency and abundance of species etc. It also includes the morphological description, range of distribution, variation and uses. In this way it provides data for botanical, ethnobotanical and phyto geographical studies etc.
- It aids in assessment and cataloguing of all species of economic potential, as commercial species, medicinal herbs, food plants etc.
- Herbarium is a source of knowledge about the flora of a region or a locality or a country.
- It provides materials for taxonomic and anatomical studies.
- Typical pollen characters have been well emphasized in taxonomy.

Arboreta

- An arboretum is a plot of land devoted to the planting of trees and shrubs.
- They are great places to educate the public in how to use plants best in their own landscape.
- Arboretums are important in conservation efforts to save endangered trees and the wildlife that depends on these trees.
- Trees receive protection within their borders, thus, protecting biological diversity.
- They offer a beautiful place for people to visit and enjoy.
- Scientists use arboretums as tools for scientific and educational studies related to trees.

37. Which are the six major attributes of an ecosystem? Explain in details. (2014)

Six Major Attributes of Ecosystems: The term ecosystem is a concept with six major attributes:

(i) **The attribute of Structure:** Ecosystems are made up of biotic and abiotic subcomponents. At the very least, a terrestrial ecosystem must have green plants, a substrate, and an atmosphere, and in most ecosystems there must be an appropriate mixture of plants, animals and microbes if the ecosystem is to function. Terrestrial ecosystems normally of a complex biotic community together with soil and atmosphere, a source of energy (generally the sun), and a supply of water.

(ii) **The attribute of function:** The constant exchange of matter and energy between the physical environment and the living community. Because living and non-living things are both composed of energy and matter, and because it is often difficult to define when organic material is alive and when it is dead, there are considerable advantages in looking at an ecosystem in terms of physical-chemical entity. Within this entity there is a constant exchange of matter and energy between different components, some of which have the characteristics of life and some of which do not. This way of looking at ecosystems in no way denies the importance of the more traditional genetic view of life; it is complementary to it.

(iii) **The attribute of complexity:** This results from high level of biological integration that is inherent in an ecosystem. All events and conditions in ecosystems are multiply determined. They are therefore difficult to predict without a considerable knowledge of the structure and functional processes of the system.

(iv) **The attribute of interaction and interdependency**: So complete is the interconnectedness of the various living and non-living components of the ecosystem that a change in any one will result in a subsequent change in almost all the others.

(v) **No Inherent Definition of Spatial Dimensions:** The term ecosystem focuses on the structure, the complexity of organization, the interaction and interdependency, and the functioning of the system, and not on the geographical boundaries of the system.

(vi) **Temporal change.** Ecosystems are not static, unchanging systems. In addition to the continuous exchanges of matter and energy, the entire structure and function of an ecosystem undergoes change over time.

38. How is the site quality important in timber production? Describe the methods used to measure the site quality (2014)

Site quality is the important factor in timber production because it is the relative productive capacity of site for particular species or group of species. If the site factors contribute optimally the timber produced may be of very good quality and it has been very well witnessed for species like *Tectona grandis, Dalbergia sissoo* and *Cedrus deodara.*

If one of the site factors is not present in optimal condition it may affect the growth and development of species. Several site factors are responsible for root production. Any change in site factor may cause the change in composition and volume of wood. For example if the soil is poor in nutrition and reduced water which may affect total volume of wood. Hence, site quality is very important in timber production.

Example: Teak produced from site quality I is more superior to Site Quality II and Sit Quality III. This Superiority in Site Quality I is due to the optimal availability of all site factors.

Methods used to measure site quality

(i) Soil Evaluation
(ii) Climatic Evaluation
(iii) Topographic Evaluation
(iv) Tree dimensions / size
(v) Species Composition
(vi) Site Index
(vii) Modelling using GIS

(viii) Remote Sensing

(ix) Site Quality Maps

(i) **Soil Evaluation:** Classification of a site using soil characteristics Individual or collective soil properties that correlate with site quality are chosen and measured. Permanent (soil depth and texture, etc.) and dynamic (nutrient levels, structure, pH, etc.) soil characteristics can both influence site quality.

Measurement of soil factors

$$\text{Log } H = b_o + b_1 (1/A) + b_2 (B) + b_3(C) + \ldots\ldots + b_n (N)$$

Where

H = height

A = age

B, C.......N = soil factors

b_0, b_1.....bn = constants to be determined by least squares

(ii) **Climatic Evaluation:** The climatic factors like rainfall, temperature, length of growing season, wind and light all affect plant survival and growth and are therefore related to site quality.

Climatic evaluation

$$\text{Climate, vegetation and productivity index (CVP)} = \frac{Tm \times P \times G \times E}{Ta \times 12 \times 100}$$

Where,

Tm = Maximum temperature

Ta = Minimum temperature

P = Precipitation (mm)

G = Growing period (month)

E = Evapotranspiration

(iii) **Topographic Evaluation:** Based on the assumption that growth can be correlated with landform variables like aspect, gradient, position on the slope, length of slope and the geometric shape of the site. The Terrain Shape Index (TSI) is one example of using topographic measurements to correlate with site.

(iv) **Tree dimensions / size:** Based on the assumption that trees grown on a good site will be bigger *(eg.* taller, larger average diameter, greater natural basal area, *etc.*) than those on a poor site of the same age or structure.

(v) **Species Composition:** Identification of native vegetation or even weed species may act as an indicators of site characteristics.

(vi) **Site Index:** The mean height of a specified number of dominant (and co dominant) trees at an index age is referred as site index. The number and selection of trees and the index age varies.

(vii) **Modelling using GIS:** Use the data handling and modelling capabilities of a GIS to extrapolate site quality measurements from areas with similar environmental conditions.

(viii) **Remote Sensing:** Determine the amount of intercepted photo synthetically active radiation. The more radiation intercepted and used efficiently, the higher the productive capability or site quality of the land.

(ix) **Site Quality Maps:** Accurate and reliable site maps are essential under intensive forest management.

These maps are the basis of:

- Evaluation of land for purchase.
- Siting of species.
- Forecasting the productivity capabilities of planted stands (yield forecasting and yield regulation).
- Forecasting thinning, pruning, fertiliser application and other tending and protection strategies.

When site maps are used to determine thinning and related logging priorities, they must be made before the time of first thinning, *i.e.* at age approximately 9-10 years.

39. Explain "adiabatic lapse rate". (2011)

Lapse Rate: It is the rate at which atmospheric temperature decreases with increase in altitude. Fall in temperature implies that the pressure also falls. With increase in elevation, the concentration of GHGs decrease (water vapor and CO_2 fall sharply in elevation). Hence the heat absorption capacity of atmosphere will also decrease.

This temperature decrease in elevation called Temperature Lapse and the rate at which it happens is called Temperature Lapse rate.

Adiabatic lapse rate: Lapse Rate is the Rate of fall in temperature of atmosphere with elevation. Adiabatic Lapse Rat is the rate of fall in temperature of increase in elevation "adiabatically".

Adiabatically or Adiabatic: Heat does not enter or leave the system. All the temperature change are internal. Adiabatic Lapse Rate is governed by "Gas law".

Gas Law: According to Gas law, Pressure 'P' is directly proportional to temperature 'T' when volume 'v' is constant.

Adiabatic change refers to the change in temperature with pressure. On descent through atmosphere, the lower layers are compressed under atmospheric pressure. As a result, the temperature increases. On ascent, the air expands as pressure 'decreases'. This expansion reduces the temperature and aids condensation of water vapor. Condensation of water vapor releases the Latent Heat of Condensation in the process. This latent heat of condensation is the major driving force behind tropical cyclones, convectional rains. Adiabatic lapse rates are usually differentiated as dry or wet (moist)

The Dry Adiabatic Lapse Rate

- The Dry Adiabatic Lapse Rate (DALR) is the rate of fall in temperature with altitude for a passage of dry or unsaturated air (air with less moisture) rising under adiabatic conditions.
- Unsaturated air has less than 100% relative humidity

  [Saturated air = The air that cannot hold any more moisture.

Unsaturated air = It can accommodate some more moisture]

- When a rising air passage has little moisture, condensation during upliftment is low, the latent heat of condensation released is low [Less additional heat from inside]. As a result, the fall in temperature with height is greater compared to Adiabatic Lapse Rate (normal passage of air)
- The dry adiabatic lapse rate for the Earth's atmosphere equals 8° C per kilometre.
- Dry Adiabatic Lapse rate is mainly associated with stable conditions [because it has less moisture]

The wet Adiabatic Lapse Rate

- When an air passage that is saturated with water vapor rises, some of the vapor will condense and release latent heat. This process causes the passage to cool more slowly than it would if it were not saturated.
- The moist adiabatic lapse rate varies considerably because the amount of water vapor in the air is highly variable. The greater the amount of vapor, the smaller the adiabatic lapse rate [because the condensation process keeps on adding more latent heat of condensation]. On an average it is taken as 4° C per kilometre.
- Wet Adiabatic Lapse rate is mainly associated with unstable conditions [because it has more moisture].
- As an air passage rises and cools, it may eventually lose its moisture through condensation; its lapse rate then increases and approaches the dry adiabatic value.

Chapter

13

Forest Resources and Utilization

1. Di-erentiate between NTFPs and MPTs. (2012)

NTFPs

- Non-timber forest products (NTFPs), (also known as Minor Forest Produce) special, non-wood, minor, alternative and secondary forest products, are useful substances, materials and/or commodities obtained from forests which do not require harvesting (logging) trees.
- They include gums, resins, tannins, dyes, game animals, fur-bearers, nuts, seeds, berries, mushrooms, oils, foliage, medicinal plants, peat, fuelwood, fish, spices and forage.
- NTFPs in particular highlight forest products which are of value to local people and communities but have been overlooked in the wake of forest management priorities (for example, timber production and animal forage).
- In recent decades, interest has grown in using NTFPs as alternatives or supplements to forest management practices.
- In some forest types, under the right political and social conditions, forests can be managed to increase NTFP diversity and, consequently, to increase biodiversity and potentially economic diversity.

MPTs

- Multipurpose trees are defined as all woody perennials that are purposefully grown to provide more than one significant contribution to the production and/or service functions of a land-use system.
- They are so classified according to the attributes of the plant species as well as to the plant's functional role in the agroforestry technology under consideration.
- In this system multipurpose tree species like *Azadirachta indica, Moringa oleifera, Cocos nutifera, Bamboo, Acacia species*. Mango are scattered haphazardly or according to certain pattern on bunds terraces or bounders.
- The primary role of this system is to meet various needs of grower for food, fodder, timber, soil conservation or protection of farm.
- *Gliricidia sepium* – the most common tree used for living fences in Central America, firewood, fodder, fixing nitrogen into the soil.

2. What are the keys to identify timbers for construction purpose? (5) (2010)

Keys to identify timber for construction purposes: The identification of timber may seem to be very simple at the first sight. However it is extremely difficult to identify timbers extremely due to wide variability in species and the associated wood quality. Though several species are available for timber utility not all the part were used for construction purposes. The distinguishing character available for timber identification is not numerous. But it should be based on examination of features that are known to be reliable. The following characteristics are generally used for identification of timber suitable for construction.

(i) **The colour of the sample:** The colour of the sample plays a vital role in identifying the wood samples. The dark coloured wood is always heart wood and the light coloured wood is soft wood. Based on this the wood can be identified as a soft wood or a hardwood.

For example: Teak – Harwood – Yellowish brown
Melia – Softwood – White colour

(ii) **Mass and density of the sample:** Density and specific gravity of the wood plays an important role in timber identification. The species with over 950 kg of weight / m^3 is termed as extremely heavy wood and the species with 450 – 600 kg/m^3 is termed as moderately heavy.

(iii) **Distinctive smell on the wood:** The smell of the wood is also identified as the one of the key factor in timber identification.

For example: Sandal is known for its fragrant smell

(iv) **Presence / absence of vessels:** This parameter is also identified as one of the key factor in timber identification process but it has to be tested through image analysis system.

(v) **Texture of wood on the planed surface:** The surface texture is also used as one of the key factor in determining timber identification process. The smooth texture is preferred in wood panelling and rough texture is preferred in floor panelling.

(vi) **Presence / absence of Conspicuous rays:** The presence of rays also used as one of the key factor in timber identification process as present in rosewood, terminalias and imported teak.

(vii) **Grain pattern and structure:** The grain pattern viz., straight, wavy grain, spiral, interlocked etc. is also one of the significant factors for timber identification. For example wavy grain wood in Red sanders (*Pterocarpus santalinus*) is highly preferred in the international market than the straight grain wood.

3. Name five important tree species used for axle and wheels of di-erent carts.
(5mark) (2010)

Species suitable for Axle: A very hard and toughwood is essential for the purpose.

Examples:

Common Name	Species
Sandan	*Ougeinia oojeinensis*)
babul	*Acacia nilotica*
axlewood	*Anogeissus latifolia*
sundri	*Heritiera minor*
Sal	*Shorea robusta*
sissoo	*Dalbergia sissoo*
khair	*Acacia catechu*
anjan	*Hardwickia binata*
hopea	*Hopea parviflora*
mesua	*Mesua ferrea*
kusum	*Schleichera oleosa*
satinwood	*Chloroxylon swietenia*

Species suitable for Wheels: Wheels, are composed of three main parts (i) the nave or hub (ii) the spokes and (iii) the felloes. Each of these parts requires certain distinct qualities in the woods used.

(i) Hub: This parts is subjected to great stress and has to be hard enough to prevent the spokes from working loose. A very hard timber is desirable. The timbers are babul (*Acacia nilotica*), khair (*Acacia catechu*), trincomalee wood (*Berrya ammonilla*), satin-wood (*Chloroxylon swietenia*), sundri (*Heritiera minor*), hopea (*Hopea parviflora*), mesua (*Mesua ferrea*), sandan (*Ougeinia oojeinensis*), kusum (*Schielechera oleosa*), sal (*Shorea robusta*) and irul (*Xylia xylocarpa*).

(ii) Spokes: The wood should be straight grained, strong, elastic, free. from defects and not liable to warp or shrink. The best woods are sissoo (*Dalbergia sissoo*), dhaman (*Grewia tiliaefolia*) and bijasal (*Pterocarpus marsupium*).

(iii) Felloes: The woods are subjected to crushing and alternations of wetting and drying. The wood must be hard, strong, elastic and durable. The best woods are babul (*Acacia nilotica*), sissoo (*Dalbergia sissoo*), dhaman (*Grewia tiliaefolia*), bijasal (*Pterocarpus marsupium*) and sins (*Albizia* spp.)

4. Write short notes on Heartwood and sapwood (5 mark) (2012)

Sap wood: Sap wood is the younger softer living or physiologically active outer portion of wood that lies between the cambium and the heartwood and is more permeable, less durable, and usually lighter in color than the heartwood

Characteristics of Sapwood

- The outer region of the old trees forms the sap wood
- It is also called as alburnum
- It is soft and not durable
- It is light coloured and formed of living cells
- Vessels are not blocked by tyloses
- The function of this region is conduction of water and nutrients and also storage of food

Heartwood: Heartwood is the older harder nonliving central wood of trees that is usually darker, denser, less permeable, and more durable than the surrounding sapwood

Characteristics of Heartwood

- The central region of the old trees forms the Heart wood
- It is also called as duramen
- It is hard and durable
- It is dark coloured due to the deposition of various substances
- Vessels are blocked by tyloses with various deposits
- The function of this region is mechanical support

5. Briefly comment on the extent of wastages (in approximate percentage of total timber content) from harvesting to marketing stages and give the reasons for such wastages. (5mark) (2012)

Wastage is one of the plantation residues commonly obtained in any plantation operation from harvesting and transportation to the consuming industries. The wastages are normally felling wastages, conversion loss and sometimes transportation loss.

S.No	Type of wastage	Species	Approximate percentage
1	Pulpwood plantations	Casuarina spp.	5 -10 tons out of 100 tons of wood produced
2.	Plywood	*Melia dubia*	30% loss of wood due to branches, less than 18 inches girth
		Eucalyptus	20% waste due to lower girth below the plywood requirement of 18 inches
3.	Timber	Teak	25% felling loss due to branches, twigs and lower girth class
4.	Match wood	Ailanthus	30% due to branches

Conversion Loss

S.No	Type of wastage	Species	Approximate percentage
1	Pulpwood plantations	Eucalyptus	8-12% loss due to debarking
2.	Plywood	*Melia dubia*	35% loss
		Eucalyptus	50% loss
3.	Timber	Teak	18% loss
4.	Match wood	Ailanthus	60% loss

Transportation loss: The transportation loss is ranged between 2% and 6% depending on the species and the moisture content.

Reason for wastages

(i) Size di3erentiation: Normally wood is used for various utility viz., timber, paper, ply, match splint etc., which demands a prescribed size class.

For example, timber is preferred with 30"and above girth class and the tree with stem portion upto 30" girth class is used for timber and the remaining wood is considered as waste.

Similarly for plywood industries, the minimum girth class of 18 inches is preferred and only 60% of the tree height will have 18 inches girth and hence the tree portion less than this size is considered as waste.

(ii) Tree taper: Tapering is decreasing in diameter from base upwards. It is very clearly evident that only the portion of the tree is useful for industrial utility and the remaining wood is considered as waste

(iii) Moisture content: Wood moisture content is very significant factor which decide the wastages. Some industries utilize the tree with 40% moisture content and some use only 12% moisture content which is one of the deciding factors for wood wastages

(iv) Pest and diseases: The incidence of pest and diseases is also one of the factors for occurrence of wastage.

For example the presence of borer and heart rot will increase the amount of wastages

(v) Wood discolouration: The percentage of wastages increased due to discolouration of wood. The discolouration is possible due to temperature, rainfall, humidity, occurrence of pest and diseases etc which will increases the wastages

(vi) Stump height: One of the major wastages is contributed by stump and stool height in felling operation. The basal portion is always has a higher girth which is left as such in the field to facilitate felling operation. This stump is the major contribution of wastages of timber.

6. Write critical notes on Plants of sacred groves with two examples well-known in the country. (5 mark) (2013)

Critical notes on Plants of Sacred Groves

- The general floristic composition and physiognomy of vegetation of the sacred groves are typically like the low level evergreen forest.
- The vegetation in undisturbed groves is luxuriant and comprises several stories of trees mixed with shrubs, lianes and herbs.
- The ground is humus laden and covered with litter.
- The macro-fungus are abundant coupled with the populations of ferns.
- The water bodies in and around the sacred groves attracts the algae and water plants which grow very gregariously.
- Floristic variations have occurred in many sacred groves exposed to human and animal interferences and climatic and edaphic changes.
- Some characteristic morphological adaptations accompany the members of the sacred groves. Broad prominent buttresses are seen in most of the trees like Ficus spp, Hopea parviflora etc. which attain enormous heights with tall bole, the branches confining to the top most region

Composition of Sacred groves: The composition of sacred grove involve tall trees in the top canopy followed by co dominance in second story and small trees herbs and shrubs in the ground layer.

(a) Tropical sacred groves – Kerala: Top canopy is represented by species like *Artocarpus hirsutus, Vateria indica, Hopea parviflora, Hopea ponga, Alstonia scholaris* etc.

Second storey consists of Macaranga peltata, Mimusops elengi, Hydnocarpus wightiana, Holigarna arnottiana etc. Common lianas are Strychnos colubrina, Anamirta cocculus, Tetracera akara, Acacia intsia etc. Chassalia curviflora, lxora nigricans, Ixora brachiata are constituents of the shrubby layer.

The ground layer is usually formed of seasonal members as, Centella asiatica, Aerva lanata, Andrographis paniculata, Biophytum sensitivum.

Common climbers are Calamus spp, Anamirta cocculus. Strychnos colubrina, Calycopteris floribunda etc. Other climbers commonly found in these groves belong to Vitaceae. Menispermaceae. Asclepiadaceae and Apocynaceae.

Total parasites like Cassytha filiformis and semiparasites like Loranthus spp, are also common. Common constituents of the shrubby layer include species like Memecylon umbellatum, Pavetta indica, Chassalia ophioxyloides, lxora spp etc.

The ground layer is usually thickly populated with species which prefer humus and love shade. Along with a few angiosperms, ferns, Selaginellas and many macrofungi like species of Agaricus, etc. occur.

Dead trunks of fallen trees harbor a variety of Polyporales especially species of Fomes and Polyporus as is common in Iringole Kavu.

(b) Evergreen sacred groves: Typical evergreen elements of the tree layer of these sacred groves are Artocarpus hirsutus, Hopea parviflora, Mesua ferrea, Vateria indica, Ficus spp etc. There are also a few semi evergreen representatives like Fagraea ceylanica, Murraya exotica, Samadera indica etc. A few deciduous and semi deciduous members also are seen.

Semideciduous representatives are species of Cinnamomum, Syzygium cumiini, Litsea spp, Mangifera indica etc. Mallotus philippensis, Dillenia pentagyna and Trema orientalis are some of the deciduous species.

Semi-evergreen representatives are found among the undergrowth also, common members being Pavetta indica, Curcuma aromatica, Nephrolepis spp etc. Epiphytic orchids like Vanda spathulata, Vanda tessellata, and Bulbophyllum spp are also common.

(c) Mangrove sacred groves: A very interesting constituent of the sacred groves is represented by members of the typical mangroves swamp found in poorly drained regions of South Travancore with a very long rainy season. Common members represented are Myristica magnifica with prominent dense stilt roots, Myristica malabarica. Hydnocarpus spp. Eugenia spp. etc. These members are known to develop a high profile of humidity in the surroundings their presence in certain groves is justified as they promote luscious growth of undergrowth mostly members of Araceae, Urticaceae. Zingiberaceae, Acanthaceae etc The exact physiological implication behind this high humidity is not experimentally proved, though it may be ascribed to very high transpiration rates of the leaves of these trees.

7. Write critical notes on Importance of *Butea monosperma*. (2013)

Butea monosperma commonly called as palas or flame of forest is a medium sized deciduous tree grows to a height of 12 m and a girth of upto 2m. The tree is found in greater parts of India in the tropical and sub tropical zones. The tree is a multipurpose tree and the wood is moderately dense with the density of 545 kg/m^3 at 12 % moisture content. The species is important for its multifarious utility which are indicated below.

(i) Host for lac

- The most important use of this tree is as a host of a lac insect for production of *Rangini* lac. It is cultivated for this purpose in Madhya Pradesh, Bihar, and in West Bengal.

(ii) Raw material for pulp and paper

- The species is suitable for raw material for the production of newsprint either alone or in mixture with bamboo pulp in proportion of 60:40

(iii) Raw material for other uses

- The leaves are lobed for fodder for buffalo and are used for making leaf plates and cups and beedi wrapping.
- Flowers yield orange dye for colouring cotton fabrics but has little permanency.
- Seeds used as vermifuge for elephants and horse. Kernels yield 80% of oil in expeller and 16 to 17% by solvent extraction. The oil is comparable to groundnut oil or sesame oil for use in soap industry. The seed and flower extract have contraceptive value. When the seeds are pounded with lemon juice, they act as a powerful rubefacient and have been successfully used as a cure for a form of herpes called dhobi's itch.
- The bark particularly of roots yields a coarse brown fibre used for rough cordage and caulking boats. Incision in the bark gives a clear red gum known as Butea gum or bengal kino substituted for kino of commerce obtained from *Pterocarpus marsupium.*

- Wood used mainly for well- curbs and water -scoops; also employed as a cheap board wood and for structural work.

Hence *Butea monosperma* is a tree of importance in tropical and sub tropical regions of India due to its multifarious utility to meet the domestic and the industrial demand.

8. Write critical notes on Siddha system of medicines. (2013)

Siddha system of medicine: Siddha system of medicine is the most primitive medical system. This system was formulated and established about more than 25000 years back by the emminent powers called Siddhars and hence the name Siddha Medicine.The medicines were prepared by the various research work done by the Siddhars on herbs,minerals and animals.The father of Siddha Medicine is the primordial Guru, Agasthiar. There are also 18 prime Siddhars who are the followers of the primordial Guru, contributed their valuable knowledge and experiences in this field.

Siddha medical system doesn't consider treatment and prevention separately. The main aim of this system is prevention of disease, as it is well said that "Prevention is better than cure"! This system was then carried over generations after generations by the Traditional practioners under Gurukula system.

This system consists of both internal and external medicines which are today available in market, manufactured by various companies. However there are traditional practioners still practicing with their self prepared drugs.

External medicines like Thokkanam (massage), Ottradam (formentation),Vedhu (steam therapy), Varmam (stimulation of vital points of body therapeutically), Kaya-karpam, Attanga Yogam (8 limbs of Yoga) etc are very effectively used for even diseases like hemiplegia, parkinsonism, arthritis etc.

Today, the treatment like physiotherapy, panchakarmam, marmam, accupressure, naturopathy and yoga are well known separately but it can be confirmed that all these are evolved from Siddha system of Medicine or Siddham. These methods are present in this single system of medicine.

Siddha system emphasizes not only a healthy body but a peaceful mind and pure soul. Hence, it is unique when compared to any other medical system.

9. Write critical notes on Drugs of alkaloids. (5 mark) (2013)

Alkaloids: Alkaloids are naturally occurring chemical compounds containing basic nitrogen atoms. The name derives from the word alkaline and was used to describe any nitrogen - containing base.

Alkaloids are produced by a large variety of organisms, including bacteria, fungi, plants, and animals and are part of the group of natural products (also called secondary metabolites). Many alkaloids can be purified from crude extracts by acid - base extraction. Many alkaloids are toxic to other organisms. They often have pharmacological effects and are used as medications, as recreational drugs, or in entheogenic rituals. Examples are the local anesthetic and stimulant cocaine, the stimulant caffeine, nicotine, the analgesic morphine, or the ant malarial drug quinine. Some alkaloids have a bitter taste.

Alkaloids are generated by various living organisms, especially by higher plants – about 10 to 25% of those contain alkaloids. Therefore, in the past the term "alkaloid" was associated with plants. The alkaloids content in plants is usually within a few percent and is inhomogeneous over the plant tissues. Depending on the type of plants, the maximum concentration is observed in the leaves (black henbane), fruits or seeds (Strychnine tree), root (Rauwolfia serpentina) or bark (cinchona). Furthermore, different tissues of the same plants may contain different alkaloids.

The different alkaloids isolated along with their action or furnished in table.

S.No	Alkaloid	Action
1.	Ajmaline	Antiarrhythmic
2.	Atropine, Scopolamine, Hyoscyamine	Anticholinergic
3.	Caffeine	Stimulant, Adenosine receptor antagonist
4.	Codeine	Cough medicine, Analgesic
5.	Colchicine	Remedy for gout
6.	Emetine	Antiprotozoal agent
7.	Ergot alkaloids	Sympathomimetic, Vasodilator, Antihypertensive
8.	Morphine	Analgesic
9.	Nicotine	Stimulant, Nicotinic acetylcholine receptor agonist
10.	Physostigmine	Inhibitor of acetylcholinesterase
11.	Quinidine	Antiarrhythmic
12.	Quinine	Antipyretics, Antimalarial
13.	Reserpine	Antihypertensive
14.	Tubocurarine	Muscle relaxant
15.	Vinblastine, Vincristine	Antitumor
16.	Vincamine	Vasodilating, Antihypertensive
17.	Yohimbine	Stimulant, Aphrodisiac

10. Write critical notes on Biological evaluation of drugs. (2013)

Biological evaluation of drugs: When physical or chemical methods of evaluation or quantity of sample are very less are not able to produce satisfactory result in drugs then the drugs are evaluated by biological methods of evaluation. These methods are performed on living animals, isolating living organ and tissue, animal preparation and micro-organism. The biological methods of evaluation is performed on living organism are called bioassay or biological assay.

Following methods are used for biological evaluation of drugs:

(i) Anti-inflammatory activity of drugs Eg. *Pinus roxburghii, Erythrina variegata*

(ii) Antipyretic activity of drugs Eg. *Acacia catechu, Terminalai arjuna*

(iii) Anti-diabetic activity of drugs Eg. *Syzygium cumini, Ficus bengalensis*

(iv) Analgesic activity of drugs Eg. *Toona ciliata, Ficus glomerata, Polyalthia longifolia*

(v) Antiulcer activity of drugs Eg. *Ocimum sanctum, Morus alba*

(vi) Anthelmintic activity of drug Eg. *Carica papaya, Mimusops elengi, Punica granatum*

(vii) Cardiac activity of drug Eg. *Digitalis purpurea, Thevetia peruviana,* Apocynum spp.

(viii) Microbiological methods Eg. Eugenia spp., *Myristica fragrans, Morinda citrifolia*

11. Describe the Boucherie process of wood preservation with its advantages (2012)

Boucherie process of wood preservation: Developed by Dr. Boucherie of France in 1838, this approach consisted of attaching a bag or container of preservative solution to a wood, thereby injecting the liquid into the sap stream. The modified Boucherie process consists of placing freshly cut, unpeeled timbers onto declining skids, with the stump slightly elevated, then fastening watertight covering caps or boring a number of holes into the ends, and inserting a solution of copper sulfate or other waterborne preservative into the caps or holes from an elevated container. Preservative oils tend to not penetrate satisfactorily by this method. The hydrostatic pressure of the liquid forces the preservative lengthwise into and through the sapwood, thus pushing the sap out of the other end of the timber. After a few days, the sapwood is completely impregnated; unfortunately little or no penetration takes place in the heartwood. Only green wood can be treated in this manner. This process has found considerable usage to impregnate poles and has experienced a revival of usage to impregnate bamboo in countries such as Costa Rica, Bangladesh, India and the state of Hawaii.

The equipment and facilities required are:

- Container able to take pressure of 1.5 kg/cm^2
- Rubber pipes
- Stopcocks
- Metallic clamps to seal the culm-pipe joints
- Electric or gas-powered pump (a bicycle pump would be suitable on a small scale)
- Pressure gauge

Procedure

- The cylinder is ¾ filled with preservative. (boron compound).
- The cylinder is pressurized (up to 20-25 psi) using a simple manual pump.
- Valve in the nozzle is open for a split second to let the air out.
- Nozzle is connected to the bamboo, which is made airtight using rubber tube.
- Sap starts dripping from the branch in almost 5 minutes. It takes about half hour for the preservative to come out from the opposite end)
- Treatment is done for atleast an hour so that the preservative can reach all parts of the bamboo.
- Bamboo is then stored horizontally in a rain-protected area till it dries.

Advantages

- Affordability
- Versatility
- Durability
- Ease of repair

- Insect repellency
- Natural

12. Give the classification of timbers based on air dry weight with suitable examples. (2013)

Classification of Indian Timber Based on Seasoning: Based on the extensive observation of Indian timbers, they have been classified into three classes according to the ease or difficulty in seasoning and recovery of defect free seasonal material.

Class A – Highly refractory wood: These are slow drying timber and difficult to season without any defects, especially cracks and splits. This type of wood requires good deal of protection against rapid drying to avoid defects originating while drying.

Ex: Sal (*Shorea robusta*), Laurel (*Terminalia tomentosa*), Axelwood (*Anogeissus latifolia),* Hopea (*Hopea species)*.

Class B – Moderately refractory wood: Timber in this class shows a moderate tendency to crack and split while drying. This type of wood requires little protection against rapid drying.

Ex: Sissoo (*Dalbergia sissoo)*, Teak (*Tectona grandis)*.

Class C – Non refractory wood: These can be rapidly seasoned without defects. This can be dried even in the open air. If not dried rapidly, they are prone to develop discolouration and mould on the surface.

Ex: *Semul* (*Bombax ceiba*), Mango (*Mangifera indica)*, Gutel *(Trewia nudiflora*).

13. Write in detail the processing and uses of Tannins in India. (2013)

Tannin: Tannins are organic compounds, mostly glucosidal, which have an acid reaction and are very astringent. Most plants contain some tannin, but only a few species have a sufficient quantity to be of commercial importance. Tannins are found in the cell sap or in other definite areas in bark, wood, leaves, roots, fruits and galls. Such structures are of little value for other purposes, so that the extraction of tannin is usually incidental to other industries.

Processing of tannin from bark

Harvesting the bark: Harvesting consists of peeling and drying.

Season for peeling: Peeling should be effected as soon as the buds begin to shoot i.e., at the beginning of the growing season; first, on account of comparative ease with which peeling can be done early in the season and secondly, because it is probable that there is more tannin in the bark during that period.

Method of peeling bark: Bark can be peeled either after the stems have been felled or from standing trees. Generally bark is obtained from trees felled under the ordinary annual cut. The operation of peeling is done differently in different places. Generally all wood fit to be peeled are cut into round billets of the length customary in the locality; the workmen then takes each billet and removes the bark with a knife in longitudinal pieces (opens out the bark along a certain line and separated from the wood along this line) as far as possible without tearing it. These strips of bark are then rolled together into bundles.

Drying the bark: The chief object or drying is to lighten the bark and reduce its volume. The best conditions for drying are to isolate the bark from the moisture of the ground, to expose it fully to air currents and protect it from rain. The degree of dryness attained may vary considerably. Bark is said to be green bark, when fresh striped from the tree; air dried when, on bending, it breaks easily and meal dried, when it has lost all flexibility and becomes brittle.

Preparation of tannin extracts: There are various methods for preparing tannin exact the general principles being to heat the bark with water and continue heating the liquor obtained with successive fresh relays of chips. The liquor is leaches out of barks and purified and mixed with other extracts to give any required colour or quality of leather. Tannin extracts is said to complete the tanning process in a much shorter period.

Uses of tannins:

- Tannins are an important ingredient in the process of tanning leather. Tanbark from wattle, Terminalia, oak and emblica has traditionally been the primary source of tannery tannin.
- Tannins produce different colours with ferric chloride according to the type of tannin. Iron gall ink is produced by treating a solution of tannins with iron(II) sulfate.
- *Pinus radiata* tannins has been investigated for the production of wood adhesives.
- Tannins can be used for production of anti-corrosive primer, sold under brand name-Nox Primer for treatment of rusted steel surfaces prior to painting, rust converter to transform oxidized steel into a smooth sealed surface and rust inhibitor.
- The use of resins made of tannins has been investigated to remove mercury and methyl mercury from solution. Immobilized tannins have been tested to recover uranium from seawater.

14. Explain the term logging, its purposes and stages. (2014)

Logging: Logging includes felling of timber and its conversion into various forms which can be used by end users. Conversion can be done both in the forest site and outside the forest areas in saw mills.

Purpose of logging

- To fell the matured, dead, drying and diseased trees towards final utilization and clearing the sites.
- To convert trees into logs of various length to be suitable to cater the needs of various utility like timber, plywood, matchwood, firewood, pulpwood etc.
- To transport the logs from forest to temporary storage areas.
- To transport the logs from site of felling to the site of destination.
- To efficiently handle the tools in order to reduce the log and improve the harvest efficiency.
- To reduce the loss due to felling conversion and transportation.
- To fell and convert the trees suitable for varied utility.

Stages of logging: The following are the major stages of logging operation.

- **Felling:** This is felling of trees in the identified forest sites which is predominantly done by axe, saw, billhook, power chain saw, tractor mounted circular saw, wedges, pickaroons, canthooks etc. For felling, scientific rules and regulations have to be followed taking it to consideration the slope and gradient of the site in order to have reduce the impact logging.

- **Conversion:** Felled timber is converted into various usable forms either in the forest site or outside the forest. Conversion can be carried out manually or through mechanized implements in a saw mill. The underlying principle of conversion is to keep the wastage as minimum as possible. The main purpose of converting the felled trees in the forest is to facilitate them for easy transport and extraction. Such type of conversion is done in remote and inaccessible forests too. The simplest form of conversion is to saw the felled trees into logs. Conversion generally leads to sawn material of various sizes and nomenclature as follows: Scantlings, Beams, Reepers, Poles, Planks, Sleepers and Squares.
- **Transport:** Transportation of timber is generally categorized into two types *viz.* Minor transportation and Major transportation. Transportation from the felling site in the form of logs or converted material at a common point within the forest/plantation is termed as minor transportation while major transportation comprises of transporting such logs/ converted material to the final destination.

Methods of transport: Major methods involving transport of timber are as follows:

- Land transportation
- Water transportation
- Overhead transportation

Factors deciding the method of transport are:

- Cost
- Possibility of loss/damage to the produce
- Total volume of timber available for transport
- Topography of the terrain concerned
- Existing transportation facilities in the site under consideration
- **Storage:** A depot is a place where wood or other forest produce is stored, pending its disposal. Depots may be for minor or major forest produce, for log scantlings, sleepers, posts, piles, grass or any minor forest produce.
  - The depots can also be classified according to their situation as land and water depots.
  - A **land depot** is the one used for storing wood transported by land.
  - A **water depot** is one where timber borne by water is stored. Water depots are situated on a river in still water or on a sea shore.
  - **River depots** are attended with the risk of flood damage.
  - Both land and water depots may be transit or sale depots.

The depots can be classified as,

- Forest depots
- Transit depots
- Sale depots

Forest depots: These depots are usually of temporary nature intended chiefly for collection and checking of produce before despatch from forest. The timber converted in the forest and lying scattered near the conversion sites is also accounted against the forest depots.

Transit depots: These are the intermediate depots inside or outside the forest. Transit depots are used mainly for collection, checking and examination of land borne timber, particularly where the sale is by royalty of system, and for catching, collecting, examining and rafting of water borne timber.

Sale depots: Sale depots are those depots where the timber or other forest produce is pending its disposal by sale. Sale depots are generally near the markets or from where the timber can be easily transported.

15. There are various processes involved in the expression and extraction of oilseeds available in forests. Elaborate on the processes involved and their utilization (2015)

Oil Seeds: There is a very large number of forest plants which bear seeds yielding oils of varying commercial importance and they are categorized as Tree Borne Oil Seeds (TBOs). Indian forests are known to contain over 100 species with oil yielding potential of commercial importance. However, tree borne oil seeds due to scattered nature of trees cannot compete with oil seeds produced as field crops. The oils are chiefly used for cooking, lighting and adulteration of more expensive oils.

Oil extraction process: There are two main methods of extracting oil from seeds. These are expression and extraction.

Expression or expelling method: In expression the clean seeds are pressed between chilled iron rollers, the oil is forced out by the pressure between the rollers. A higher yield is obtained by heating the seeds. The crushed seeds are placed in a kettle provided with a steam jacket and a mechanical stirrer to ensure uniform heating. The oil is expressed from the heated mass in hydraulic presses. The country ghani is commonly used for the purpose.

Extraction or solvent method: The other method of obtaining oil from seeds is extraction with volatile `solvents which give a higher yield of oil but the resultant cake is unfit for consumption by cattle because traces of solvents remain behind and are difficult to remove. The crushed seeds are placed in an extractor and are subjected to the solvent, the oil is separated by distillation and the solvent used again. The solvents commonly used are petroleum, ether and carbon disulphide.

Important tree borne oil seeds

(i) ***Azadirachta indica* (Neem):** The trees are found to a certain extent in forests. The oil is used in medicine, as an illuminant and for manufacture of soap.

(ii) ***Calophyllum inophyllum*:** The tree is found in the coastal regions and locally in the hilly tracts of peninsular India. The oil is used for burning and for soap making.

(iii) ***Garcinia indica*:** The tree yields a valuable edible fat known as `Kokam Butter'.

(iv) ***Madhuca indica* (Mahua):** The tree grows wild in forests throughout the country. The seed oil is known as `Mahua Butter' and is the most important of tree seed oils. It is used for cooking purposes, hair oil, soap making and as an illuminant. The oil finds use in medicine also.

(v) ***Mesua ferrea* (Mesua):** The tree is found in abundance in Assam, West Bengal and - Western Ghats. The oil is used for burning, lubricating, soap-making and applying on sores.

(vi) ***Pongamia pinnata* (Karanj):** The tree is found in central and southern India. The seeds yield an oil which is used in medicines, soap-making and leather tanning.

(vii) ***Schleichera oleosa* (Kusum):** This tree occurs in deciduous forests throughout central and peninsular India. It produces seeds yielding the kusum oil of commerce. The oil is used for cooking, lighting and soap-making.

(viii) ***Shorea robusta* (sal):** The tree has a wide distribution in the country. The cotyledons of its seed yield the well known 'Sal Butter' used for cooking, lightning and for adulterating ghee. It is a suitable, confectionery fat and is used for soap-making.

16. Give the scientific names of a least five drug yielding plants with their use in pharmaceutical industry. (2015)

S.No.	Botanical Namc	Common Name	Uses
1.	*Panax schinseng*	Ginseng	Stimulant and stomachic
2.	*Urginea maritime*	Squills	Expectorant and Stimulants
3.	*Valeriana oDcinalis*	Valerian	Relieve nervous affictions, pain, coughing and hysteria
4.	*Chondrodendron tomentosum*	Curare	Cardiac lailure, Chromic spastic condition
5.	*Barosma betulina*	Buchu	Treatment of indigestion and urinary disorders
6.	*Eucalyptus globulus*	Eucalyptus	Treatment of throat, nose disorder, malaria and fever
7.	*Marrubium vulgare*	Hoar hourd	Arthristics, dyspepsia and Ailments
8.	*Cassia acutifolia*	Senna	Purgative
9.	*Datura stramonium*	Stramonium	Narcotic effect
10.	Piper cubeda	Cubebs	kidney stimulant
11.	*Strychnos nux-vomica*	nux-vomica	Treat nervous disorder and paralysis

17. a. Write procedure of extraction of Katha and Lac. (2009)

b. Write the Processing and Sale procedure for Katha and Lac. (2006)

Extraction of katha

(a) Country method

- Katha is prepared from the heartwood showing white specks of kheersal.
- As like above cutch preparation, syrup of desired consistency is obtained.

- This is poured into mould dugs containing fine sand.
- Fine sand absorbs the tannic acid and leaves catechin.
- The residue of catechin is crystallized and sold as katha.
- In this, all tannic acid (cutch) is wasted while preparing katha.

(b) Modified method of extraction of katha

- The heartwood chip is boiled in tinned copper vessels.
- The boiled liquid is further concentrated in another vessel.
- Meanwhile, fresh water is added to the boiled chips and boiled again.
- Cutch is soluble in both hot and cold water but katha is soluble in hot water but sparingly soluble in cold water.
- Thus cooling of solution causes crystallization of katha first.
- Seeding of katha while cooling encourages crystallization process.
- Crystallized katha is separated by filtering the liquid through a muslin cloth where cutch is passes away.
- The retained katha is washed with water to remove the traces of cutch.
- This is dried and marketed as katha.

(c) Improved method of katha

- Wood is chipped by chippers and extracted in copper vessels with hot water under pressure.
- The extract is concentrated in vacuum and cooled in refrigerated tank for a week.
- In this katha crystallizes out.
- This is speared by fitter presses yielding cutch which forms 10 to 12 percent by weight of the wood.
- The yield of katha is at least 4 to 4.5 percent.

Extraction of Lac: Lac cut from the host plant is called as "stick lac". Lac can be scraped from the twigs before or after the emergence of larvae. If it is used for manufacturing before the emergence of larvae, the type of lac produced is called as "Ari lac" and if it is used for manufacturing purpose after swarming of larvae has occurred, the lac is said to be Phunki lac".

The scraping of lac from twig is done by knife, after which they should not be exposed to sun. The scraped lac is grinded in hard stone mills. The unnecessary materials are sorted out In order to remove the finer particles of dirt and colour, this lac is washed repeatedly with cold water.

Now at this stage it is called as "Seed lac" and is exposed to sun for drying. Seed lac is now subjected to the melting process. The melted lac is sieved through cloth and is given the final shape by molding. The final form of lac is called "Shellac". Colour or different chemicals may be mixed during melting process for particular need.

Processing of Lac: The following points highlight the three main stages involved in processing of lac. The stages are:

1. Production of Stick Lac
2. Production of Seed Lac
3. Production of Shellac from Seed Lac.

Stage 1. Production of Stick Lac: During harvesting, the lac encrusted twigs are cut from the host plants. Then the lacs are removed from the twigs by scrapping. The lac thus obtained is known as stick lac or scrapped lac or crude lac or raw lac, that contains insect body, lac dye, sand and twig debris, in addition to resins.

If harvesting is done before emergence of larvae, then the lac obtained is referred to as Arilac. The stick lac, if required, can be stored on the floor in layers, less than 30 cm in height and racked frequently.

Stage 2. Production of Seed Lac: Stick lac, if stored for long time, may form lump due to its high moisture content and may invite other storage pests, and thereby causes substantial loss and deterioration in quality of desired industrial demands. So seed lac is produced from the stick lac by further processing.

Stick lac is first crushed in a machine and sieved to remove sand and dust. Once the sieving is done the crushed stick lac is fed by basket into a Barrel Washing Machine. The loading time taken is generally 15-20 minutes. Capacity of loading at a time of the barrel is its (diameter x length)/2.

This washing removes the dye from the sick lac and also removes other materials like sand, wood impurities and insect body. The process usually takes around two to three hours to complete. It uses a lot of water and the rotary device inside the barrel, helps the churning of the stick lac and helps in the cleaning process.

The water that comes out of the barrel is often taken onto a field and is allowed to dry down. The contents are sometimes used as pesticides and fertilisers. However, the use is much restricted nowadays. The washing time and the methods adopted to do this improves the quality of the seed lac.

Different types of washing are done for different grades single wash for low quality, double wash for medium quality and triple wash for superior quality:

(a) **Single Wash:** This involves simple washing without or with negligible amount of soda for some 45 minutes.

(b) **Double Wash:** Soda cleaning (750 gm for 500 kg) with low quantity of water for 20 minutes is carried out. Water is poured into the barrel from the top through pipe for about 25 minutes and stopped when colour of water passing through a 60 no. net at the back side of the barrel appears white.

(c) **Triple Wash:** Water cleaning is for 20 minutes. Then Soda washing (1 kg for 500 kg material) for one hour (superior quality) or 45 minutes (medium quality) has to be carried out followed by water cleaning for another 30 to 35 minutes. This is done till the water coming out of the barrel shows no pigment.

After the completion of washing, the bottom part of the barrel is opened and the total mass is poured into a circular cemented basin (size of the basin depends upon the capacity of the washer). The masses thereafter are filtered through cloth (different quality of cloth is used for different grade). After removing the lighter impurities ('Khari' or 'Pati') from the mass, seed lac is collected.

The seed lac is then spread into the open cement floor for drying. The layer of seed lac is continuously rolled over the floor from one side to the other side with the help of a 'Patri' (a wooden flat strip having a long handle) for facilitating the drying.

The process of drying would take between an hour or two. Usually the washing should not go beyond 2 p.m. so that the finished product is dried by the end of the day.

After drying, the material is sieved through a number of Chaluni' (circular hand operated sieving tools. Each of them is of different size and helps sieve of the seed lac based on the size of the sieve. The residue so left out at this stage is called the 'Kuni' and 'Molamma'.

Then all materials so sorted out are gathered and remaining impurities are cleaned using the hand operated winnowing baskets (bamboo made instrument called 'Kulo') usually done by women.

18. a. What are Gums? List some important gums of India along with names of respective sources trees. (2009)

b. What are gums? List some important gums of India along with the names of source tree. (2005)

Gums: Gum is a more or less viscous substance which exudes from cracks or wounds in the bark of many trees shrubs and climbers. True gums are formed as the result of the disintegration of internal tissues, for the most part from the decomposition of cellulose, through a process is known as gummosis. Gums chiefly occur in the cortex tissues. They contain a large amount of sugars and are closely allied to the pectin. They are soluble in water, either dissolving completely are swelling up to form mucilage but insoluble in alcohol and other. Gums are specially common in plants of dry regions.

Process of gum secretion: True gums are formed as the result of the disintegration of internal tissues, for the most part from the decomposition of cellulose, through a process known as gummosis. Gums chiefly occur in the cortex tissues. They contain a large amount of sugar and are closely allied to the pectins. They are soluble in water, either dissolving completely or swelling upto form a mucilage, but are insoluble in alcohol and other (unlike resins). They exude naturally from the stems especially when bark cracks with prolonged heat or in response to wouding. Commercial gums reach the market in the form of dried exudations. Gums are specially common in plants of dry regions.

Uses of gums

- Gums are predominantly used in adhesive industry
- They find profound utility in calico printing
- The gums are used in paints and other bonding materials
- Gums are also used in food industry
- Gums are used in pharmaceutical industry.

Important gums and the species

S.No	Source Name	Botanical Name
(i) True gum		
1.	**Indian Gum Arabic** (Babul gum)	*Acacia arabica*
2.	**true gum arabic**	*Acacia senegal*

3.	substitute gum arabic	*Macaranga beltata*
4.	'Jhingan' gum	*Lannea coromandelica*
5.	Neem Gum	*Azadirachta indica*
6.	'Khair' gum	*Acacia catechu*
7.	'Dhaura' gum	*Anogeissus latifolia*
		Aegle marmelos
		Anacardium occidentale
	–	*Feronia limonia*
(ii) Pseudo-gums (Gum which is not soluble in water but absorbs water readily and swell up into a jelly mass.)		
8.	Gum Tragacanth (Katira gum)	*Astragalus heratensis*
9.	karava or Kullu gum	*Streculia urens*
		Bombax ceiba
		Morinea oleifera
		Cochlospermum religiosum
		Saccopetalum tomentosum
(iii) Astringent gums		
10.	Gum Kino	*Pterocarpus marsupium*
11.	Bengal Kino	*Butea monsperma*

19. Write on the present status of the composite wood industry in India. (10 mark) (2009)

Status of composite wood industries in India: Despite an enormous and varied forest resource, India is timber-deficient. Over- exploitation and the lack of effective forest management have resulted in virtual logging bans in natural forests. India has long been a net importer of forest products and wood fibre, and these imports are increasing rapidly. In 2006 imports were valued at US$ 2.75 billion with exports valued at US$ 0.55 billion, a deficit in forest products trade of US$ 2.2 billion (Flynn *et al.*, 2007). Khanduri and Mandal (2005) has estimated that the wood supply deficit will exceed 90 M m3 by 2020.

As the demand for wood products shows no sign of abating, alternative materials have be developed. Wood composites are appropriate alternatives to meet the rising demands. Any wood material made of smaller pieces and glued together is known as wood composite. Some common wood composite materials are plywood, block board, particle board, fibre board, medium density fibre boards, and modified wood. With the development of wood composites it has become possible to utilize various secondary species and lingo-cellulosic wastes as a substitute for conventional timbers.

The advancement in knowledge and improvement in technology has led to development of new wood composites. The latest development is 'Engineered Wood': a form of composite materials that are engineered to precise design specifications.)

Engineered Wood Products (EWP): Engineered wood, also called as composite wood, man-made wood, or manufactured board; which includes a range of derivative wood products and are manufactured by binding the strands, particles, fibers, or veneers of wood, together with adhesives, to form composite materials. Engineered wood products are used in a variety of applications, from home construction to commercial buildings to industrial products. The products can be used for joists and beams that replace steel in many building projects.

Typically, engineered wood products are made from the same hardwoods and softwoods used to manufacture timber. Saw mill scraps and other wood waste can be used for engineered wood composed of wood particles or fibers, but whole logs are usually used for veneers, such as plywood, Laminated Veneer Lumber (LVL). Alternatively, it is also possible to manufacture similar engineered cellulosic products from other lignin-containing and lignocellulsoic material such as rye, straw, wheat straw, rice straw, hemp stalks, kenaf stalks, or sugarcane residue, in which case they contain no actual wood but rather vegetable fibers. Furniture such as flat pack and design furniture are typically made out of man-made wood due to its low manufacturing costs and its low eight, making it easy to transport.

Composite wood industry is less discussed at national level. Now a day, these industries have tremendous potential and growing at a phenomenal pace. Plywood and laminates are likely to play a more prominent role in future due to the fast growth of housing sector. Indian plywood industry is as big as Rs 10,000 crore and the laminate industry is almost Rs 5,500 crore. The industry is growing at a rapid pace of 10-20 percent per annum. Approximately 750 units are currently functioning all over the country. There is tremendous growth potential as the players are yet to penetrate majority of the market. In composite wood industry, organized industries contribute around 10-15 % market share and the rest is with the unorganized players. The people are not accepting the higher quality and most of them prefer cheaper and inferior varieties. Currently the market demand is very huge and it will be difficult to cover everyone.

20. How is Forest Certification done in developed countries? Comment on its present status in India. (2010)

Certification: It is a process by which written assurance is given that a product, process or service is in conformity with certain standards of good practices.

History of Forest Certification: The certification has a history of over four centuries and traced back in the year 1637 wherein wood products were labelled. However the use of forest certification as a potential tool to deal with deforestation rates is of recent origin that too only after 1980s. After 1992, UNCED, the certification has gained significance based on the principle of sustainable forest management.

The certification has emerged from a group of stakeholders in California during 1990 which recognized the need for a credible system to identify acceptable sources of forest products. Then the rain forests alliance launched its smart wood programme as a tool to curb deforestation. Later the WWF activated the FSC and help tropical timber producers to avoid environmental boycotts and meet the demand from importing countries for environmentally sensitive wood products from well managed forests. The FSC has a set of principles for certifying forests.

Forest Certification in developed countries: The certification concept was emerged in US during 1941. At that the American Tree Farm System (ATFS) was created to draw public attention towards active management and investment made in private forests. This certification was based on the set of forest management principles and practices which have been reviewed once in every five years.

After that in the year 1990 the American Forests and Paper Association initiated its own certification system. In 1996, the Canadian Standards Association (CSA) developed sustainable forest management standards which later updated in 2002. The Sustainable Forestry Initiative (SFI) which is an industry self regulated programme recognized the CSA system as the functional equivalent of SFI system.

In Europe some forest owners created an alternative to FSC and they have named it as Programme for the Endorsement of Certification Scheme in 2003 which was again renamed as Programme for the Endorsement of Forest Certification Scheme (PEFC).

Forest certification is a mechanism for forest monitoring and labelling timber, wood and pulp products and non-timber forest products where quality of management from environmental, social and economic perspectives is judged against a series of agreed standards.

Certification Agencies

- Forest Stewamdship council (FSC)
- Pan European Forest Certification (PEFC)
- International Organization for standardization (ISO)
- Sustainable Forestry initiative (SFI)

Leading Regional & National Certification Programs

- Canadian standards association
- Keurhat foundation
- American tree farm systems
- Malaysian timber certification council

Types of Certification: There are two types of certification

(i) Sustainable Forest Management Certification which confirms that forests are managed on a sustainable basis

(ii) CoC Certification which confirms forest products are made with wood harvested from sustainably managed forests.

Status of Forest Certification in India: Forest Certification in India is still in its inference stage. India has secured two or three forest management certification (Rubber plantation in Tamil Nadu, TNPL plantation in Tamil Nadu and ITC in Andhra Pradesh) and a wide range of CoC certification by paper and other wood based industries in order to meet export demand. The number of FSC certification in India is growing and several companies are in the pipe line. As on 2009,

India has around 21 CoC certificates and the demand is increasing in India due to export demand. Some of the industries which have got certification are listed below.

- Tamil Nadu Newsprints and Papers Ltd (Management and CoC)
- BILT Graphic Paper Products Ltd
- ITC Ltd
- Thomson Press India Ltd
- Gopson Papers Ltd
- Manipal Press Ltd
- Kraftile International of Santhosh Group
- Khanna Paper Mills Ltd
- Repro India Ltd
- Seshasayee Papers and Boards Ltd

21. What are the di-erent types of water soluble wood preservatives? (2010)

Water Soluble Type: These preservatives may further be of two types:

Leaching Type: These are inorganic or organic salts that are soluble in water. They are suited only for timber which has to be used indoors as the preservatives any readily leach out when in contact with rainwater. Thus have a limited application. Important preservatives in this category are:

Zinc Chloride: Highly toxic to fungi and most insect. Also acts are a fire retardant.

Boric Acid and Borax: Both these preservatives are highly toxic to all types of organisms which may tend to occur in building timbers. Thus they are ideally suited for wood and wood panel products like plywood, primarily because of their penetrating properties.

Sodium Pentachlorophenate: The water soluble salt of pentachlorophenol shows a high degree of effectiveness against sap stain.

Benzene Hexachloride: This preservative is suited for protecting the timber from borers and termites.

Fixed Type: The salient features of fixed type of preservatives are

Timbers treated with these preservatives can be used even under exposed conditions. They are mixtures of various salts that form insoluble salts in the presence of wood and get permanently fixed. The timber treated with fixed type of preservatives take time to dry and about 1 to 2 weeks for the preservative to have its full effect. They can be applied even at room temperatures. Timber treated with this type of preservatives is very clean c n be painted, polished or waxed. Important fixed type of preservatives finding wide use in India has been discussed below:

(A) Copper-Chrome-Arsenic Composition (CCA Or ASCU): This is one of the best preservatives being used in many parts of the world. It was developed by the forest research institute and is commonly known as ASCU. The composition of this preservative is furnished.

Chemical	% by weight
Copper sulphate	35 to 37.5
Sodium dichromate or Potassium dichromate	47.5 to 50
Arsenic pentoxide	10 to 12.5

The total solid content is not more than 95%. The basic formulation is a uniform and homogenous mixture of the above salts reacted together at controlled temperatures. Over the years, ASCU has become one of the most widely used preservatives all over the world. Some minor modifications have been made to suit local conditions.

(B) acid-cupric-chrome composition (ACC): Developed by the FRI the composition of this preservative is furnished.

Chemical	% by weight
Copper sulphate	47.5 to 50
Sodium dichromate or Potassium dichromate	42.5 to 45
Chromic acetate	4.75 to 5

The total solid content is not more than 95% and the pH of the solution for use varies from 2.7 to.0

(C) Chromated Zinc Chloride (CZC): The composition of this preservative is given is furnished.

Chemical	% by weight
Zinc Chloride	81.5%
Sodium dichromate or Potassium dichromate	18.5

(D) Copper Chrome Boric Composition (CCB): This preservative was developed by the FRI in order to substitute the expensive arsenic by boron. Its composition is furnished.

Chemical	Part
Copper sulphate	3
Sodium dichromate or Potassium dichromate	4
Boric acid	1.5

22. "There is no replacement for rattans in forest based industry". Justify this statement and list five important species reported from di-erent parts of the country. (2010)

Yes. There is no replacement for Rattons in forest based industry because of the following reasons.

- Raw rattan is used in rattan baskets, plant containers and other decorative works. Processed into several products to be used as materials in furniture making. Rattan accepts paints and stains like many other kinds of wood, so it is available in many colours; and it can be worked into many styles. Moreover, the inner core can be separated and worked into wicker.

- From a strand of rattan, the skin is usually peeled off, to be used as rattan weaving material. The remaining "core" of the rattan can be used for various purposes in furniture making.
- Rattan is a very good material mainly because it is lightweight, durable, suitable for outdoor use, and to a certain extent flexible.
- Many of the properties of rattan that make it suitable for make it a popular choice for handicraft and art pieces.
- The inner core of the Rattans as well as the shoots of some of ratton species are rich source of proteins and hence used for edible purposes.
- For several industries rattan supplies sustained and long term based raw material and ensures industrial sustainability and rural livelihood.
- By providing a source of income to rural people, it allows them to become stewards and guards of their forests and biodiversity.
- Helps to protect forests from degradation
- Provides habitat for endangered and threatened species

Important Rattan species

S.No	Genus	Distribution
1.	*Calamus* L.	Tropical Africa, India and Sri Lanka, China, south and east to Fiji, Vanuatu and eastern Australia
2.	*Calospatha* Becc.	Endemic to Peninsular Malaysia
3.	*Ceratolobus* Bl.	Malay Peninsula, Sumatra, Borneo, Java
4.	*Daemonorops* Bl.	India and China to westernmost New Guinea
5.	*Korthalsia* Bl	Indo-China and Burma to New Guinea
6.	*Laccosperma* (Mann & Wendl.) Drude	Humid tropical Africa
7.	*Myrialepis* Becc.	Indo-China, Thailand, Burma, Peninsular Malaysia and Sumatra
8.	*Plectocomia* Mart.	Himalayas and south China to western Malaysia
9.	*Plectocomiopsis* Becc.	Laos, Thailand, Peninsular Malaysia, Borneo, Sumatra

23. Write common and scientific names of ten tree species yielding resins and oleo resins. (2011)

Resins

Resin represents oxidation products of various essential oils and are very complex and varied in their chemical composition.

Sl.No	Common name	Botanical name
1.	White dammar	*Vateria indica*
2.	Black dammar	*Canarium strictum*
3.	Sal dammar	*Shorea robusta*
4.	Rock dammer	*Hopea odarata*
5.	Green dammar	*Shorea tumbuggaia*
6.	Salai tree	*Boswellia serrata*
7.	Ailanthus, perumaram	Ailanthus triphysa
8.	Asafoetida	*Ferula asafoetida*
9.	Siamese gamboges tree	*Garcinia hanburyi*
10.	Gamboges tree	*Garcinia gambugea*

Oleo Resins: The oleo-resins contain a considerable amount of essential oils in addition to the resinous materials. Consequently they are more or less liquid in nature. The essential oils are secreted in the intercellular spaces where they combine with resin formations to produce the oleo-resins which exude from the stem. They have a distinct aroma. They are also known as wood oils.

Sl.No	Common name	Botanical name
1.	Gurjun oil	*Dipterocarpus turbinatus*
2.	Gurjun	*Dipterocarpus tuberculatus*
3.	Gurjun	*Dipterocarpus indicus*
4.	Malabar mahogany	*Hardwickia pinnata, Kingiodendron pinnatum*
5.	Maritime pine	*Pinus pinaster*
6.	Slash Pine	*Pinus elliottii var elliottii*
7.	Dragon's blood	*Agathisdammara*
8.	Caribbean pine	*Pinus caribaea*
9.	Chir pine	*Pinus roxburghii*
10.	Radiata pine	*Pinus radiata*

24. Write short notes on the following medicinal plants: (2013)

(i) ***Aconitum heterophyllum***

(ii) ***Orchis latifolia***

(iii) ***Podophyllum emodii***

(iv) ***Morchella esculanta***

(i) ***Aconitum heterophyllum:*** Ativisha (Aconitum heterophyllum wall) of family Ranunculaceae is an Ayurvedic herb which is known for its important medical properties.

It is extensively discussed in the ancient ayurvedic texts written by Charaka and Sushruta. Ativisha grows in colder parts of India. The roots of the plant are the source of most of the alkaloids obtained from the plant such as Atidine, hetisine, heteratisine, Diterpene alkaloids, heterophylline, heterophylline, heterophyllidine heterophyllisine, hetidine, atidine & Atisenol, and a new entatisene diterpenoid lactone from roots.

(ii) ***Orchis latifolia*:** Orchis latifolia Linn. Is commonly known as Panja, Salam-Panja, Salep, Hath-Panja or Hatajari. It got all these local names due to its hand like roots. The appearance of its root is like a palm with fingers. Salam-panja is found in Himalaya at altitudes of 2, 500-5000m in damp places. It is an orchid plant and bears purple flowers. It has fleshy tuberous roots. These roots are used for medicinal purpose. In Unani, the roots of plant are considered aphrodisiac and nervine tonic. The roots are cooling, emollient, aphrodisiac, rejuvenating and tonic.

(iii) ***Podophyllum emodii*:** P. emodii is an herbaceous perennial found in damp, open woodlands and wet meadows and the family of Berberidaceae. A variety of pharmacologically active lignans are present in podophyllum. These lignans (podophyllotoxin being the best known) act as antimitotic agents, with the greatest effects in tissues undergoing rapid cell division. Podophyllum has been used topically in the treatment of genital warts and hairy leukoplakia. Semisynthetic derivatives of podophyllotoxin are used to treat RA and various cancers, including refractory testicular tumors and small-cell lung cancer; however, limited information is available. Use as a cathartic is considered unsafe.

(iv) ***Morchella esculenta*:** *Morchella esculenta*, (commonly known as common morel, morel, yellow morel, true morel, morel mushroom, and sponge morel) is a species of fungus in the *Morchellaceae* family of the Ascomycota. It is one of the most readily recognized of all the edible mushrooms and highly sought after. Each fruit body begins as a tightly compressed, grayish sponge with lighter ridges, and expands to form a large yellowish sponge with large pits and ridges raised on a large white stem. *Morchella esculenta*, like all morels, are among the most highly prized of all edible mushrooms. Raw mushrooms have a gastrointestinal irritant, hydrazine, but parboiling or blanching before consumption will remove it. Old fruit bodies that show signs of decay may be poisonous. *M. esculenta* fruit bodies have several medicinal properties, including anti-tumor effects, immune regulatory properties, fatigue resistance, and antiviral effects. Extracts from the fruit bodies have antioxidant properties.

25. Indian forests are found to be rich in di-erent kinds of non-timber forest products (NTFPs). Can you list at least 10 NTFPs. (2015)

The Indian forests housed with plants and animal species are equally valuable and sometimes much more important than timber products of forests. Such important products other than timber are classified as non timber forest products.

Sources of NTFP in Forest: Indian forests supports over 400 species of edible nature 5000 species of medicinal important, over 500 species of fodder importance over 1000 species with various social, cultural and religious utility, 60 species have been recorded as insectivorous plants,

400 plants are recorded to yield gums, resins and dyes and 150 species are source for essential oils used in perfumery industries.

Importance of NTFP: These Non Timber Forest Products commonly contribute to meet the food and other based needs, as source of income and of inputs to the agriculture, help households control exposure to various kinds of diseases, and constitute an integral part of the habitat and of the social and cultural structure of those living within that environment.

- The women folk usually depend on NTFP for household and income generation.
- The collection and processing of NTFP provide employment during lean seasons of agriculture and provide buffer against risk and household emergencies.
- NTFP provide essential food and nutrition, medicine, fodder, fuel, thatch, construction materials, mulch and non farm income.

Major ten NTFPs

S.No.	NTFPs	Example
1	Fibres and flosses	*Sterculia villosa, Agave spp., Ceiba pentandra,*
2	Grasses, bamboos and canes	*Vetiveria zizanoides,Cymbopogan martini, Bamboo spp., Calamus spp.,*
3	Essential oils	*Santalum album, Eucalyptus, Cedrus deodara*
4	Oilseeds	*Azadirachta indica, Jatropha, Madhuca indica*
5	Tans and dyes	Terminalia chebula, Acacias,Bixa orellana, Butea monosperma
6	Gum	*Acacia senegal, Sterculia urens, Anogeissus latifolia, Pterocarpus marsupium*
7	Resins and Oleo-resins	*Pinus gerardiana, Boswellia serrata, Dipterocarpus turbinatus*
8	Drugs, poisons, spices and insecticides	*Piper longum, Saraca asoka, Azadirachta indica*
9	Edible products	*Mangifera indica, Syzygium spp., Ziziphus spp., Embilica oDcinalis*
10	Animal, mineral and miscellaneous products	*Laccifera lacca, Apis dorsata, Morus alba,* Hides, ivory and horns

26. Enumerate the role and importance of the Forest Survey of India (FSI) in forest management systems. (2015)

Forest Survey of India (FSI): Forest Survey of India (FSI) is a government organization in India under the union Ministry of Environment and Forests for conducting forest surveys and studies. The organization came into being in June, 1981. It is headquartered at Dehradun, Uttarakhand, India. The objective of the organization is monitoring periodically the changing situation of land and forest resources and present the data for national planning; conservation and management of environmental preservation and implementation of social forestry projects.

The Role and Importance of Forest Survey of India

- To prepare State of Forest Report biennially, providing assessment of latest forest cover in the country and monitoring changes in these.

- To conduct inventory in forest and non-forest areas and develop database on forest tree resources.
- To prepare thematic maps on 1:50,000 scale, using aerial photographs.
- To function as a nodal agency for collection, compilation, storage and dissemination of spatial database on forest resources.
- To conduct training of forestry personnel in application of technologies related to resources survey, remote sensing, GIS, etc.
- To strengthen research & development infrastructure in FSI and to conduct research on applied forest survey techniques.
- To support State/UT Forest Departments (SFD) in forest resources survey, mapping and inventory.
- To undertake forestry related special studies/consultancies and custom made training courses for SFD's and other organisations on project basis.

Forest Survey of India assesses forest cover of the country every 2 years by digital interpretation of remote sensing satellite data and publishes the results in a biennial report called 'State of Forest Report'(SFR).Beginning in 1987, 13 SFRs have come so far. Since 2004, FSI has been monitoring forest fires across the country using MODIS (Moderate-Resolution Imaging Spectrometer) and GIS based technology.

27. What are the major fibres and flosses which are obtained from forests in India? Citing at least ten species, discuss their commercial importance. (2015)

Fibres and Flosses: Fibres derived from plants have been used by human beings since time immemorial. Flax was the earliest of such fibres. It was used for spinning threads and cordage. Coarse fabrics were weaved from such fibres. Cotton came later. Today it is the mainstay of modern textiles. Even though thousands of plant fibres are known to mankind, only a few are of commercial importance. The rest are put to local use.

Fibre yielding species:

S.No	Species	Commercial importance
1.	*Acacia leucopholea*	Making fishing nets, lines and coarse cordage
2.	*Marsdenia tenacissima*	Making Fishing nets, fishing lines, strings of bows
3.	*Bauhinia vahlii*	Making ropes and for domestic purposes
4.	*Butea monosperma*	Making ropes and cordage
5.	*Ficus bengalensis*	Making coarse rope
6.	*Tremna orientalis*	Making ropes, twine and coarse cloth.
7.	*Careya arborea*	Making coarse cordage, for tying carts and rafters
8.	*Kydia calycina*	Making ropes and cordage

9.	*Agave americana*	Ropes and cordage
10.	*Caryota urens*	Making fishing nets and lines
11.	*Pandanus tectoris*	Cordage, fishing nets, fishing lines, and sacks

Flosses: Silky flosses are produced in the fruits of many trees and shrubs found in different forests of India. Flosses are used for a wide variety of purposes which include manufacture of life belts, and as a stuffing in mattresses, pillows and cushions.

Floss yielding species

S.No	Species	Commercial importance
1.	*Bombax ceiba*	Stuffing life belts, Mattresses, Pillows and Cushions
2.	*Calotropis gigantea*	
3.	*Ceiba pentandra*	
4.	*Cocholospermum religiosum*	
5.	*Beaumontia grandiflora*	
6.	*Cryptolupis buchanani*	
7.	*Hoiarrhena antidysentrica*	
8.	*Populus ciliata*	
9.	*Salix daphonides*	
10.	*Wrightia tomentose*	

28. Clonal forestry approach proved successful in meeting the industrial wood dem and particularly pulpwood. Substantiate your view and perception in this regard. (2016)

Trees are reverend in our mythological scriptures and farmers have been maintaining sporadic trees of useful local species in their farm lands and around their farm houses since ages. National needs of timber, industrial round wood for processing and to a large extent fuel-wood have been traditionally met from forest resources. However, with the economic development of our country, demand for timber and wood based products have been growing continuously, while sustainable supply of wood from forests has decline because of low growing stock, poor annual increment and degradation of nearly 42 per cent of 69.09 million ha of our country's forest cover. As per national forestry action programme timber requirements of India during 2006 were 82 million m^3, whereas the domestic availability was just 27 million m^3, Timber and wood based products worth US$ 2.75 billion were imported into India during 2006. Estimated domestic wood supply deficit in India will exceed 90 million m^3 by 2020 – a substantial increase from the round wood deficit of 21 million m^3 in 1999. Growing shortages of industrial round wood have hampered modernization and growth of wood based industries.

Need for clonal plantation: As the Government owned forests cannot meet growing requirements of industrial round-wood, national forest policy 1988 suggests that wood based industries should source their future wood supplies through persuading farmers to raise agroforestry plantations.

Substantial improvement in productivity of forest resources on sustainable basis and promotion of large scale agroforestry plantations are most essential for meeting the national needs for fuel-wood, timber & wood products on sustainable basis and conservation of the forest resources. Productivity and net returns from seed based agroforestry plantations have been generally very low mainly because of extremely poor genetic quality of seed most often used and non availability of adequate quantities of genetically improved planting stock.

Development and large scale deployment of tested genetically superior clones of Eucalyptus and Poplars supplemented with improved scientific package of practices have redefined productivity standards making agro-forestry plantations an attractive lan use option with net returns significantly higher than those from traditional crops. Average productivity of commercial poplar and eucalyptus clones is ranged between 20 and 25 $m^3ha^{-1}yr^{-1}$. However, many farmer have achieved trend setting high productivity 50 m^3 $ha^{-1}yr^{-1}$ both for poplars and clonal eucalyptus plantations.

(a) Clonal *Eucalyptus* based pulpwood Plantations: Even though large scale seed based *Eucalyptus* plantations have been raised on degraded forest and marginal lands and along railway tracks/road strips, canals and drainage channels in India, yet most of these past plantations have low productivity ranging from 4-10 m^3 $ha^{-1}yr^{-1}$ and large genetic variation because of poor genetic quality of seed used. ITC Bhadrachalam Paperboards Limited launched an ambitious programme in 1989 for development, testing and commercial scale deployment of fast growing, high yielding and disease resistant clones of eucalypts in Andhra Pradesh state. Productivity and profitability of eucalypt plantations has been revolutionized with the deployment of genetically improved clonal planting stock of *Eucalypts* since 1992. Average productivity of commercial eucalyptus clones is around 20 to 25 m3 ha-1 yr1 and many farmers have achieved record growth rates of 50 m^3 ha^{-1} yr^1 even under rain-fed conditions making clonal *Eucalyptus* plantations an economically attractive land use option both for reforestation projects and agro forestry plantations.

Eucalyptus clones have been adopted by many integrated pulp and paper mills in India for promoting agro- forestry plantations around their mills. Current planting level of clonal eucalypt plantations in India during 2010, mostly under agro-forestry system, is estimated to be around 55 to 60 million plants covering 22000-24000 ha area annually. Andhra Pradesh still leads in clonal eucalypt plantations.

Wood based pulp and paper mills in Orissa, Karnataka, Tamil Nadu, Madhya Pradesh, Gujarat and Maharashtra are also promoting clonal *Eucalyptus* plantations. Amongst the Government agencies, Andhra Pradesh Forest Development Corporation and Haryana Forest Department are followed by a few other State Forest Departments in raising/ promoting clonal Eucalyptus plantations.

(b) Casuarina clone based pulpwood plantation: Similar to Eucalyptus, Casuarina clonal plantations have made significant impact in the raw material availability of pulp and paper industries in Tamil Nadu. The development of high yielding clones like MTP 2 has a tremendous potential of yielding 150 tonnes / ha within a short period of 3 years. This has resulted in establishment of over 40000 ha of clonal plantation in Tamil Nadu and is able to generate over 2 lakh tonnes of pulp annually and satisfy the demands of pulp and paper industries in Tamil Nadu.

Hence clonal forestry has made significant impact in improving the productivity to the levels over 20 – 25 m^3/ha/annum compared to less than 10 m^3/ha/annum from the unimproved seedling based plantations. Similarly the rotation or the harvesting period has brought down from 7 years to 4 years in case of Eucalyptus and 5 years to 3 years in case of Casuarina clonal plantations. Hence the productivity improvement coupled with reduced rotation extended by the clonal forestry approaches is proved lucrative and successful in meeting the raw material requirement of pulp and paper industries in India.

(c) **Melia clone based pulpwood plantation:** *Melia dubia* has been identified as a potential pulpwood species with a pulp yield of over 50 percent which is far superior to Eucalyptus and Casuarina. The productivity of Melia is over 40m^3/ha/annum which is highest among all pulpwood species. The Tamil Nadu Agricultural University has identified potential clones (MD26) exclusive for pulpwood plantation with the productivity of 150 tonnes / ha in a short duration of two years. This clone based pulpwood plantation has created significant attraction towards tree husbandry.

29. Dry deciduous forests are rich in Non-Timber Forest Products (NTFPs). Justify the statement with examples. (2016)

India is endowed with 5 major forest types and 16 sub types distributed across the country and among them the predominance of tropical forests are very well witnessed. Among the tropical forests the potential of tropical dry deciduous forests is very significant due to its rich biological resources leading to the production of timber and non timber forest products. It is estimated that more than wood the non timber forest products obtained from dry deciduous forest of India have ensured income and employment generation activities of the tribal people and the villages near by forests.

The dry deciduous forests extend wide range of non timber forest products which are classified as oil seeds, gums, resins essential oils, dyes, lac, silk, honey, bamboos, medicinal plants etc. It has been observed that the dry deciduous forests incorporated over 4000 edible species 500 species with fodder value, 400 plants with gums, resins and dye yielding potential, 150 species with essential oil utility. Over and above, over 5000 species of medicinal plants are known to occur in the dry deciduous forests and among them nearly 1000 species have a potential to yield drugs for pharmaceutical industries.

The dry deciduous forests are known for honey production with economic values. The honey holds high plays in the esteem of mankind and it is believed to be a sacred substances. It is used in all religious ceremonies, as a food and also as a medicine. It contains, proteins, amino acids, enzymes, organic acids and minerals and hence consider potential for food industry. Several species of dry deciduous forests are housed with beehives which are potential yielder for honey collection.

The dry deciduous forests are housed with over 1000 plants with accredited medicinal value. The original dwellers and the tribal communities traditionally use herbal medicines. The medicinal plants are acting as a source of income, source of medicinc and also as a source of livelihood. The species like aloe, atropa, cassia, centella, digitalis, commiphora, panas, ocimum etc., are traditional plant based medicines supporting the livelihood activities of forest dwellers.

The dry deciduous forest comprises trees which are known to secrete gums and resins. The species like Terminalias, Sterculia, Boswellia, Acacias, Lannea etc., are known to produce gums and resins and they are playing significant potential in socio-economic status of the local people. Similarly the tamarind seed gum is tapped from dry deciduous forests of India which is also playing significant role.

The bamboos distributed and spread in dry deciduous forests of India have excellent potential of raw material for cottage and corporate industries. The dry deciduous forests comprise two major bamboo species, viz., *Bambusa bambos* and *Dendroclamus strictus* which are excellent raw material for cottage industries and livelihood opportunities for local dwellers.

The dry deciduous forests are known to contain over 500 species of fodder utility which supports the livelihood opportunities of tribal people and the local villagers. The component of cuminat animals, sheeps and goats are increasing at an alarming rate due to the availability of wide range of fodder trees in dry deciduous forests.

Hence dry deciduous forests are one of the potential resources for wide range of non timber forest products and play a vital role in socio-economic development the weaker sections of the society.

30. Give an account of mechanical properties of wood and suggest the standard tests for the same. (2016)

The mechanical properties of the timber are very essential for its suitability towards various utility. The following are the major properties which decide the suitability of wood.

(i) Hardness: It signifies the resistance offered, by a wood specimen, to penetration by another body, i.e., resistance to indentation and not to cutting as is commonly understood, which depends on nature of grain, presence of silica and other substances. The hardness of wood depends on anatomical structure, specific gravity, moisture content, presence of resinous material and soundness. Based on hardness the wood is classified into

a. **Extremely hard** : *Schleichera oleosa, Mesua ferrea*
b. **Very hard** : *Terminalia alata, Dalbergia sissoo*
c. **Hard** : *Chukrasia tabularis, Lagerstroemia spp.*
d. **Soft** : *Lannea coromandelica, Cedrus deodara*
e. **Very soft** : *Boswellia serrata, Pinus roxburghii*
f. **Extremely soft** : *Bombax ceiba, Cryptomeria japonica.*

(ii) Flexibility: Flexibility is the capacity of wood to bend out of shape freely without rupture. Resistance to bending is stiffness. Brittleness is the property of allowing little deformation without failing. Steaming of wood increases its flexibility. Soft light woods are more flexible than hard heavy woods because their loose structure allows fibres some play. Young trees are more flexible than old. Coppice shoots are more flexible than seedlings. Climbers are most flexible because of long straight loose fibres.

(iii) Elasticity: Elasticity in wood is the property which enables it to regain its original shape after the stress has produced deformation. Elastic wood always possesses some flexibility whereas a flexible wood need not be elastic. Comparative dryness increases elasticity while moisture decreases it. Based on elasticity the wood is classified into

a. **Very elastic** : *Grewia tiliaefolia, Anogeissus latifolia,*
b. **Elastic** : *Mimusops elengi, Chloroxylon swietenia,*
c. **Moderately elastic** : *Mangifera indica, Tectona grandis,*

(iv) Fissility or Fissibility: This is the capacity of wood for being split along the fibres, by means of implements such as an axe or a wedge. This is an important requisite for split fuel. Based on fissility the wood is classified into

a. **Extremely fissile** : All Himalayan conifers. *Tectona grandis,*
b. **Easily fissile** : *Acacia catechu, Adina cordiafolia*
c. **Moderately fissile** : *Bombax ceiba, Anogeissus latifolia*
d. **Di3cult to split** : *Dalbergia sissoo, Shorea robusta*
e. **Very di3cult to split** : *Dalbergia latifolia, Ougenia oojeinensis*

(v) Strength: The strength of wood is its ability to resist external forces tending to alter its shape. These forces, depending on the direction of application, may be compressive, tensile or shearing, hence there is the necessity of qualifying the "strength" in compression, tension or shear. A timber specimen strong in one respect may not be equally strong in another, other things being equal, strength varies with density, conditions of seasoning, soundness, and absence or presence of defects. Based on strength the wood is classified into

a. **Exceptionally strong** : *Mesua ferrea, Schleichera trijuga*
b. **Extremely strong** : *Acacia nilotica, Grewia tiliaefolia*
c. **Very strong** : *Tectona grandis, Anogeissus latifolia*
d. **Strong** : *Dalbergia latifolia, Dalbergia sissoo,*
e. **Moderately strong** : *Athena cordifolia, Cedrus deodara*
f. **Somewhat weak** : *Abies pindrow, Bischofia javanica,*
g. **Weak** : *Picea smithiana, Cedrela toona,*
h. **Very weak** : *Hymenodictyon excelsum, Bombax ceiba*

Standard timber tests

- Stiffness and elasticity
- Tensile strength
- Compressive or crushing strength
- Shearing strength
- Transverse or bending strength
- Toughness
- Hardness
- Cleavability
- Resilience

31. Bamboos become the major raw material for many cottage as well as corporate industries. Explain. (2016)

Bamboos are tall perennial grasses, belonging to Bambusae, a tribe under Graminae, There are about 140 species in India. Bamboos are characterised by woody stems, commonly called culms, which arise from the rhizome. The number of culms produced annually is very variable. There is also a great variation in the length of culms. The tulles of bamboos are generally round and smooth. The growth of bamboos is very rapid. The most common, most valuable and most universally used of all bamboos is *Dendrocalamus strictus*, the 'male bamboo' of commerce which is found in deciduous forests throughout the country. The culms in this bamboo are solid. *Bambusa arundinacea* is the next in importance, occurrence and utility, which is found mainly in Orissa, Assam, Southern and western India.

Bamboos in cottage industries

- The strength of culms, their straightness, lightness combined with hardness, their hollowness, the facility with which they can be split, and the range in size in which available, their abundance, ease of propagation and exploitation, make bamboos a usable commodity for a variety of purposes, ranging from house posts to fountain pens.
- The common uses being roofing, rafters, walling, flooring, matting, spear and lance shafts, lathis, masts, spars, tent poles, furniture, water pipes, cart shafts, basket making, musical instruments, bows and arrows. Bamboo leaves are much valued as fodder. Bamboo seeds are eaten.
- The rhizomes and young shoots are pickled and eaten.
- Bamboo is now an important raw material for the paper industry in the country. Many species are suitable for paper and rayon pulp.

Some of the important bamboos are as under:

- ***Arundinaria falcate***: This grows gregariously forming an an undergrowth in forests of oak and deodar, usually in moist shady places at situations between 1,200-2,000 m elevation. The culms are 1.84.0 in high and are used for basket making, fishing rods and hookah pipes.
- ***Arundinaria spathiflora***: A small bamboo found growing in Western Himalayas at 2,000-2,700 m in elevation as an under growth of fir, spruce and deodar forests.
- ***Bambusa arundinacea***: A thorny bamboo, having culms 24-30 m in height and 15-18 cm in diameter. It occurs wild, especially in the hill forests of western and southern India, ascending upto 900 m. The bamboo is used in paper manufacture.
- ***Dendrocalamus giganteus***: This is the biggest of the Indian bamboos cultivated in parts of West Bengal, Assam and Malabar. It is 24-30 m high and 20-25 cm in diameter.
- ***Dendrocalamus strictus***: This occurs throughout India, in deciduous forests often of a dry type, on hill-slopes upto 1,000 m elevation. The calms are usually quite solid, 6•15 m high and 2.5-7.5 cm in, diameter. It is most universally used of all the bamboos. It is put to a great variety of uses.

- ***Dendrocalamus hamiltonii***: This occurs in the sub-Himalayan tract upto 900 m. The culms are upto 25 m long, 10-18 cm in diameter, and thin-walled.
- ***Dendrocalarnus longispathus***: This occurs in parts of West Bengal and Assam, usually along streams upto 1,200 m. The culms are from 1248 m in height and 7,5;10 cm in diameter.
- ***Melocanna bambusoides***: This bamboo is found in Garol Khasi and Lushai Hills in Assam. The culms are upto 2 m in length and 9.5 cm in diameter. It is typically gregarious.
- ***Ochlandra travancorica***: This is found in the mountains of Kerala, Karnataka and Tamil Nadu. It is an erect reed-like, gregarious bamboo, which covers large areas.
- ***Oxytenanthera nigrociliata***: This is a tufted bamboo with culms 9-15 m high and 1.3-10 cm in diameter, found in Assam, Orissa and Andaman Islands.
- **Other Bamboos** *Bambusa balcooa, Bambusa tulda, Cephalostachyum pergracile, Gigantochloa macrostachya, Ochlandra scriptoria, Oxytenanthera ritcheyi, Pseudostachyum polymorphum* are of importance.

Bamboos in corporate industries

Pulp and paper: Bamboo is the main suitable fibrous raw material, found growing extensively in the forests of the country. There is a very large number of species but the main which are used for paper making are: *Bambusa arundinacea, B. nutans, B. balcooa, B. tulda, B. vulgaris, Dendrocalamus strictus, D. hamiltonii, Melocanna barnbusoides and Oxytenanthera monostigma, Ochlandra rheedii and 0. travancoria.* Bamboo is used for almost all the common varieties of paper such as writing, printing, wrapping paper and also as part of furnish in newsprint.

Furniture industry: Bamboos are the major raw material for furniture industries across the world and chineese are leading the business. The bamboo based furnitures are preferred by the consumers than the wood based due to its appearance, easy handling and plasticity. Wide range of furnitures like chairs, tables, stools, doors, huts etc., are made using wide range of bamboo species. The species like *Dendrocalamus strictus, Bambusa bambos, Bambusa balcooa, Dendrocalamus brandsii* etc. are playing a vital role in furniture industry.

Bamboo based biomass generation: There is growing interest towards biomass based power generation particularly using bamboo as a raw material. For every megawatt of biomass power generation, it is estimated that 1.5 tonne of bamboos are required. Bamboo is amenable for power generation both through combustion and gasification technology. One of the bamboo species like *Bambusa balcooa* has been used for this purpose due to high calorific value, combustion efficiency and low ash content. Hence the corporate involved in biomass power generation are deploying bamboo as a raw material due to its fast growth and coppicing ability.

Bamboo based activated carbon: The industries involved in activated carbon production started deploying bamboo as a raw material. The activated carbon generated from bamboo has found multifarious utility from filter paper, paints and varnishes, pharmaceutical industries, colouring industries etc., and play a significant role in corporate sector.

32. What are the general principles and advantages of wood seasoning? Describe the solar kiln seasoning with neat sketches. (2016)

Wood seasoning: Wood seasoning the process of systematic removal of moisture from the wood in order to dry the wood to the equilibrium moisture content before it is put into commercial utility.

Principles of wood seasoning: The seasoning of wood involves drying which starts from surface zones in advance of interior zone because of direct evaporation of moisture from the surface area. During drying a moisture gradient is setup which causes the movement of moisture from interior surface to exterior surface. This movement of moisture depends on the following factors.

- The rate of movement of moisture through the wood.
- Externally applied temperature.
- Equilibrium moisture content conditions.
- Moisture gradient
- Drying stresses
- The Drying process is also influenced by the interaction of the external drying conditions such as temperature, relative humidity, rate of air circulation and the rate of moisture diffusion within the wood.

Seasoning process: The seasoning of wood is carried out by air seasoning or kiln seasoning using the principles indicate above.

Advantages of wood seasoning:

- Wood seasoning aims to bring about maximum attainable dimensional stability from and timber during use.
- It also minimizes losses through cracking, splitting and warping, which take place when green timber is put to use.
- Adequate seasoning also protects the timber against primary decay, fungal stain and attack by insect pests.
- Seasoned timber is lighter, thus being easier and cheaper to handle.
- It has better strength properties than wood in a green condition.
- Preservative treatment of wood is best done if the timber is in a dry condition.
- The finishing and gluing qualities of seasoning timber is better than unseasoned wood.
- Seasoning improved the electrical and thermal insulation properties of wood.

FRI Solar Kilns: The FRI Solar Kiln is developed to satisfy the demands of low cost industries. The FRI Solar Kiln has a charging capacity of 7 cubic meter for 25 mm thick planks with a maximum length of 5.5 meter. This kiln has a super structure of timber or metal frame double sheathed on the roof, south east and west walls with 5.5 mm thickness as indicated in the figure. A clear transparent glass is fixed on the outside and 0.4mm thick glasses on the inside and with the 40 mm wide intervening air gap. The northern side wall is sheathed with 9.5 mm BWR grade plywood. The FRI Solar Kiln is oriented towards east west direction along its length. The roof of the kiln is tilted towards the south at an angle to the horizontal equal to 0.9 times the latitude for maximum absorption of solar energy throughout the year.

The Kiln also has the timber stack which is built under a horizontal false ceiling consisting of a corrugated galvanized iron sheet which acts a solar absorber. The Kiln also provided with two 91 cm dia, 650 rpm and 1 kilowatt electrically driven propeller fans to maintain the air circulation through the stake in a recirculating system in which the heat is continuously passed through the air. The vents have been provided at the bottom of north and south walls for exchange of air. The entire structure is painted black in colour to maximize absorption of solar energy. The fundamental principles of FRI Solar Kiln is based on glasshouse principle.

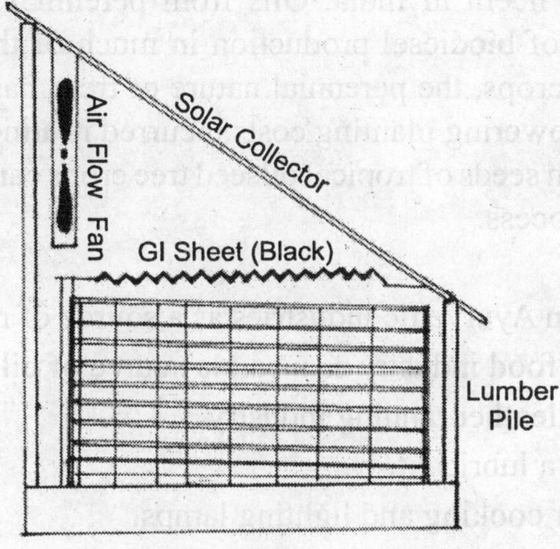

Fig. FRI Solar Kiln

33. Write critical notes on the following: (2016)

(i) Gums

(ii) Oilseeds

(iii) Katha

(iv) Bidi leaves

(i) **Gums:** Gum is a translucent and amorphous substance partly dissolved in water to produce viscous solution. These are the group of plant products related to sugar and carbohydrates and consist of polysaccharides and their derivatives.

Source and significance of gum: This is a more or less sticky substance with a plastic consistency. It is not soluble in alcohol or other organic solvents. However it dissolves or softens in water. These are also known as true gums. Gums are a group of plant products. They are formed due to disintegration of plant tissue; primarily cellulose. This process is known as gummosis. The gums are found utility in food, pharmaceutical, paints and adhesive industries and hence they are valued as a potential raw material and extend great employment opportunities through their collection and processing.

Important gum yielding species: *Acacia nilotica, Acacia catechu, Acacia modesta, Acacia Senegal, Anogeissus latifolia, Bauhinia retusa, Cohlospermum religiosum, Lannea coromandelica, Pterocorpus marsupium, Sterculia urens.*

(ii) **Oilseed:** The trees that yield oil from the seeds are termed as oils seeds which are now pronounced as tree borne oil seeds. The Indian forests are housed with over 100 species of different nature with oil bearing capacity. The oil content of these species varied from 25% to 60%. The important oil seeds are sal, mahua, karanj oil, kusum, neem, mango and khakan. Marketing of sal seed alone is organized. The seeds are processed and brought to depots of forest department or corporations for sale. There is a potential for production of 200,000 tons of kusum, 50,000 tons of pilu, 1000 tons or pisa, 11,0000 tons of karanj and 4,00,000 tons of neem in India. Oils from perennial tropical oilseed bearing trees are potential sources of biodiesel production in much of the insular areas of the Pacific. Compared to oilseed crops, the perennial nature of trees has the advantage of minimizing soil disturbance and lowering planting costs incurred routinely for annual crops. Biodiesel from oil extracted from seeds of tropical oilseed tree crops can be produced by an established trans-esterification process.

Utility of Oil

- The oil is useful in Ayurvedic industries as a source of medicine.
- The oil is used in food industry as a edible source of oil.
- The oil is used in leather tanning industry.
- The oil is used as a lubricant.
- The oil is used for cooking and lighting lamps.
- The oil is used for generating biodiesel through transesterification process.
- Some of the oil seeds are acting as a pesticide, fungide and nematicide.

Important oil yielding species: Jatropha (*Jatropha curcas*), Karanj (*Pongamia pinnata*), Wild Apricot (*Prunus armeniaca*), Cheura (*Diploknema butyracea*), Simarouba (*Simarouba glauca*), Mahua (*Madhuca indica*), Kokum (*Garcinia indica*), Jojoba (*Simmondsia chinensis*), Neem (*Azadirachta indica*), Tung (*Aleurites fordii*), Olive (*Olea europaea*)

(iii) **Katha:** Katha is obtained by boiling of heartwood of *Acacii catechu* in water. Catechin is called as katha.

Extraction of Katha

- Katha is prepared from the heartwood showing white specks of kheersal.
- As like above cutch preparation, syrup of desired consistency is obtained.
- This is poured into mould dugs containing fine sand.
- Fine sand absorbs the tannic acid and leaves catechin.
- The residue of catechin is crystallized and sold as katha.
- In this, all tannic acid (cutch) is wasted while preparing katha.

Uses of Katha

- Katha is ingredient in the preparation of chewing pans.
- It is used in medicines as an astringent and digestive.
- It is applied as a cooling agent to ulcers and eruptions of the skin.

(iv) **Bidi Leaves:** The leaves obtained from *Diospyros melanoxylon* are commercially called as bidi leaves. Tendu leaves are one of the most important non-wood forest produce, which fetches huge revenue. These leaves are used for making bidi wrappers. The leaves of coppice shoots are prepared for bidi wrapping because of its bigger size. The leaves of mature trees are shorter, thicker and brittle. Larger leaves with inconspicuous veins are considered as superior quality leaves.

The tree adapted to varying soil and climatic conditions. Tendu tress are found in forests, wastelands and cultivated lands in Madhya Pradesh, Orissa, Maharastra, Andhra Pradesh, Bihar, Rajasthan, Uttar Pradesh, Gujarat, Tamil Nadu and West Bengal.

Important oil yielding species: *Diospyrous tomentosa, Bauhinia racemosa, Butea monosperma, Ficus religiosa, Shorea robusta*

34. Explain the recent development in mechanization of logging operations and their impact on the e-ciency and wastage. (2016)

Introduction: Forest management concepts have changed over years from production forestry to conservation forestry and sustainable forest management. Yet logging is a very important aspect in forestry and forest management. There is one school of thought that the only way to prevent forest depletion is to discard all forms of timber extracting practices. Other feels that if the timber production stops, the economic value of forests diminishes and forest land would be diverted for more profitable and productive options. In between these two are some who believe that forests could be sustainably utilized. Logging is a very important economic activity in the temperate regions and tropical regions. Timber is an important economic product that contributes to the State exchequer. Hence it is very essential to understand logging operations and how best they could be carried out without resulting in depletion and degradation of our forests and forest resources.

Logging generally refers to the process of felling and extracting timber from the forests. In forestry, the term logging is sometimes used in a narrow sense where it refers to the log making only and consists of trimming off the branches, debarking and cutting into sections. In some instances, logging is used to refer to the whole logistics of moving wood from the site of felling to somewhere outside the forest, usually a saw mill or a depot from where it is further disposed. However, the term may be used to indicate a range of forestry or silviculture activities including the felling, on-site processing, extraction and transportation to saw mills or storage depots. Logging is distinct from timber harvesting in the sense that the latter includes all the stages of harvesting starting from pre-harvest planning to harvest, technical supervision and post harvest assessment and even the impact of the operations on the non-timber forest resources and the state of forests.

Recent developments in logging operations: Due to international compulsions and the implementation of reduced impact logging has necessitated advancements in logging operations. This has compelled deployment of semi mechanized harvesting system using power saws. This has increased felling efficiency and reduced the human drudgery besides reducing the logging impacts. In case of larger operational area the introduction and implementation of feller buncher machine has revolutionized felling operations in terms of man power reduction, felling efficiency, conversion at in-situ level, rapid felling operation and reduced costs. This besides the logging impact is very well reduced. Hence almost all the paper industries in Brazil have deployed this kind of mechanization system for rapid and efficient harvest operation. As far as India is concerned manual harvest is slowly replaced with mechanized harvesting and improving the efficiency and reducing the logging loss.

Impacts of Logging operations: No doubt logging and timber operations help to meet the ever increasing demand for timber and firewood. It adds to the revenue of the country and aids economic and industrial (forest based industries) development. It provides income and employment to people, especially those living in and near vicinity of the forests. But conventional logging is a cause of great concern as they cause significant damages to the forest ecosystems. Concerns about the depleting forest cover are the most dominant factors that evoke the feeling of absolute non interference with the forests and zero timber operations. The major impacts of logging on the forests and environment are:

- Loss of biodiversity
- Soil erosion and siltation
- Damages to regeneration and residual trees
- Effect on watershed
- Opening up of forest land and conversion to agriculture
- Effects on non-wood forest product
- Climate change and carbon sequestration
- Effect on scenic beauty and aesthetics

35. Do you think that the controlled fire is beneficial to forest vegetation? Discuss (2014)

Yes controlled fire is beneficial to forest vegetation due to the following reasons.

- Reduces hazardous fuels, protecting human communities from extreme fires;
- Minimizes the spread of pest insects and disease;
- Removes unwanted species that threaten species native to an ecosystem;
- Provides forage for game;
- Improves habitat for threatened and endangered species;
- Recycles nutrients back to the soil; and
- Promotes the growth of trees, wildflowers, and other plants;

Benefits of controlled Fire

- **Hazardous fuel reduction:** Forest fuels accumulate rapidly in pine stands and pose a serious threat from wildfire. Prescribed fire is the most practical way to reduce dangerous accumulation of combustible fuels. Wildfires that burn in areas where fuels have been reduced by prescribed fire cause less damage and are much easier to control.
- **Wildlife habitat improvement:** Prescribed fire is highly recommended for wildlife habitat management where loblolly, shortleaf, longleaf, or slash pine is the primary over story

species. Periodic fire tends to favour under story species that provide browse for wildlife. Deer, dove, quail, and turkey are some of the game species that benefit from prescribed fire.

- **Insect and disease control:** Prescribed fire is the most effective and practical means of controlling brownspot disease in longleaf pine seedlings and cone insects such as the white pine cone beetle.
- **Aesthetic appearance enhancement:** Prescribed fire improves recreation and aesthetic values by increasing occurrence and visibility of flowering annuals and biennials and maintains open spaces for vistas.
- **Native vegetation improvement:** Use of prescribed fire encouraging the new growth of native vegetation, and maintaining the many plant and animal species whose habitats depend on periodic fire.
- **Grazing improvement:** Prescribed fire improves grazing in open pine stands on the Coastal Plain by increasing availability, palatability, quality, and quantity of grasses and forbs.
- **Seeding and planting preparation:** Prescribed fire is useful when regenerating southern pines. On open sites, prescribed fire can expose mineral soil and control competing vegetation until seedlings become established.

36. Explain properties of good wood preservatives and classify the various wood preservatives with the help of a flowchart. (2012)

Preservatives and Their Properties: The chief characteristics of an ideal wood preservative are :

(i) High toxicity, against fungi, insects and or marine organisms.

(ii) High permanency under all service conditions i.e., resistance to leaching by water or to rapid evaporation due to heat or to chemical transformation due to oxidation or reduction.

(iii) High penetrability, i.e., amenability to impregnation over the entire cross section of the wood to be treated.

(iv) High stability during treatment i.e., resistance to decomposition at temperatures normally used during treatment and to precipitation due to reducing action of chemicals in timber.

Further the preservative:

(i) Should not unduly increase the inflammability of wood but in fact should aim at increasing the natural resistance of timber to fire.

(ii) Should not impair the health of labour during treatment or the workmen during assembly of treated timber or the residents of a house either.

(iii) Should be economically cheap and commercially available if not produced from indigenous sources.

(iv) Should be capable of being transported to distant - places without excessive charges.

(v) Should not corrode metals used in manufacture of treating equipment or metals used along with treated timber during actual service.

(vi) Should allow treated timber to be painted over or varnished.

It is not always possible to obtain such an ideal preservative but care is 'necessary before a compromise is made and a suitable preservative is chosen for any particular use ; the deciding factor being proven efficiency."

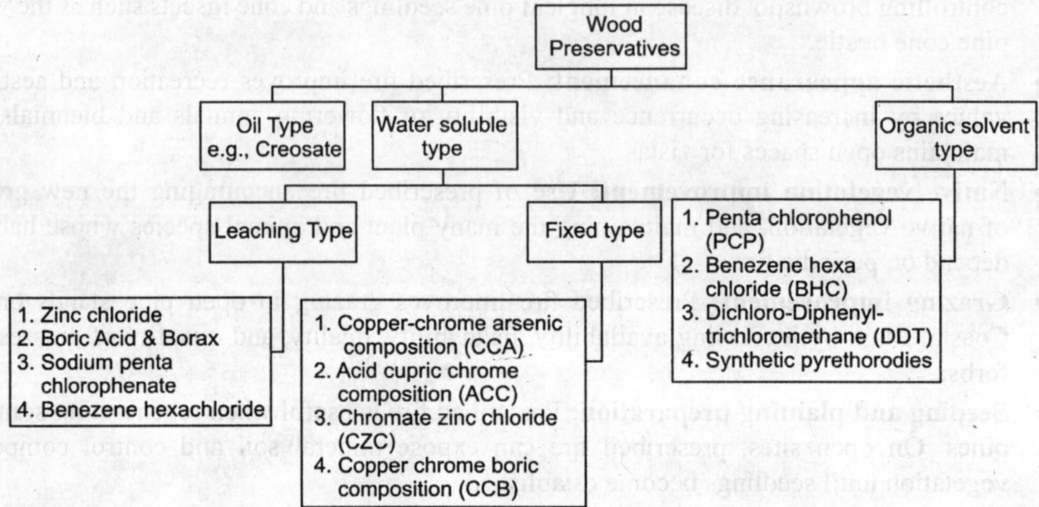

37. What are advantages of wood seasoning? Describe the various methods of wood seasoning and mention classification of timbers for seasoning. (2013)

Advantages of Wood Seasoning

- Seasoning increases timber value and versatility. Many of the properties of well dried timber are more superior to those of increased or poorly dried timber.
- Seasoned timber will have dimensional stability while in service. Wood products made of unseasoned timber will shrink considerably and lose their shape as the wood dries.
- Seasoned timber is about double the strength of unseasoned timber.
- Timber at less than 20% MC has no risk of fungal activity that leads to decay, stain and mould besides reducing the susceptibility of attack by insects viz. termite, borer etc.
- Gluing, machining and finishing of seasoned material are much easier than unseasoned timber.
- Dried timber exhibits better electrical resistance and thermal insulation properties.
- Seasoned timber holds fasteners far better than unseasoned timber, especially when the unseasoned material subsequently dries.
- Drying reduces weight and thereby handlings and transport costs.
- During wood preservative treatment, dryer and quick penetration can be achieved in the seasoned wood than unseasoned wood.

Methods of Seasoning: The fundamental principle behind deciding a proper method of seasoning is to obtain maximum amount of defect free wood with desired moisture level in minimum time. The chief difficulty in wood seasoning is the tendency of outer layers to dry out more rapidly than the interior. If the outer layers are allowed to dry much below FSP while the interior is still saturated, stress can develop leading to rapture or evaporation of moisture from outer surface and the outward diffusion of moisture from inside. In most species, except highly permeable ones, the diffusion is much slower than evaporation and therefore the seasoning process should be controlled by increased RH at the initial stages.

The seasoning methods fundamentally consist of circulating air over the surface of timber to allow evaporation and removal of moisture. The air serves both as a carrier of heat required for evaporation of moisture from the timber and as a vehicle for removal of evaporated moisture from the timber surface. The most important methods of seasoning are air-seasoning and kiln seasoning.

Air Seasoning: Air seasoning is the process of placing stacks of timber in the prevailing natural condition so that they dry evenly to suitable moisture content. This method basically consists of making a good stack of the timber preferably under shade and allows drying by atmospheric air. The drying agent is the atmospheric air. No artificial drying controls are employed in this type of drying. The only way by which the drying is controlled is by stacking the material in the proper way. As a general rule, when seasoning non refractory material in a mild climate, care should be taken to provide maximum circulation and on the other hand, in case of refractory species in a hot climate the rate of drying must be kept as low as possible. Air drying is used to dry the timber initially before it is placed into a kiln dryer. Usually no definite moisture target is set for this. Air drying is a low capital and low energy method of drying timber.

Kiln Seasoning: This consist primarily drying wood using artificial heat. Timber is stacked in chambers called "Seasoning Kilns". Seasoning kilns are fitted with equipments for manipulation and control of temperature, relative humidity of the drying air and the air circulation.

In air seasoning the drying is completely due to the effect of climate. In kiln-seasoning, on the other hand, the drying conditions are completely controlled. It is a faster method and perfect control at the most critical periods of drying possible.

Classification of Indian Timber Based On Seasoning

- Based on the extensive observation of Indian timbers, they have been classified into three classes according to the ease or difficulty in seasoning and recovery of defect free seasonal material.

Class A – Highly refractory wood

- These are slow drying timber and difficult to season without any defects, especially cracks and splits. This type of wood requires good deal of protection against rapid drying to avoid defects originating while drying.
- Ex: Sal (*Shorea robusta*), Laurel (*Terminalia tomentosa*), Axelwood (*Anogeissus latifolia*), Hopea (*Hopea species*).

Class B – Moderately refractory wood

- Timber in this class shows a moderate tendency to crack and split while drying. This type of wood requires little protection against rapid drying.
- Ex: Sissoo (*Dalbergia sissoo*), Teak (*Tectona grandis*).

Class C – Non refractory wood

- These can be rapidly seasoned without defects. This can be dried even in the open air. If not dried rapidly, they are prone to develop discolouration and mould on the surface.
- Ex: *Semul* (*Bombax ceiba*), Mango (*Mangifera indica)*, Gutel *(Trewia nudiflora*).

38. What is wood preservation? Why is preservation of wood necessary? Mention main types of wood preservation methods. (2014)

Wood Preservation: Wood preservation is defined as the process of improvement of wood's natural durability by treatment with chemicals that are toxic to insects, fungi and other decaying agents.

Need for wood preservation

- To increase the life span of wood.
- To decrease the cost of product
- To avoid the frequent replacement of woods
- There are numerous species of wood-destroying insects and fungi.
- Hot and humid weather facilitates and speeds up any decomposition of wood.
- Only a limited number of the great variety of species are naturally durable.
- The lines of supply from the forest to the final destination of the timber usually extend over comparatively long distances and the risk of damage by insects or fungi is a constant threat even after the finished product has arrived at its final destination.

Major Wood Preservation Methods: There are two major wood preservation methods viz., Non pressure and pressure methods.

Non-Pressure Methods: This is a general term used for preservative techniques/methods in which pressure is not used for treatment of the wood. The main non-pressure methods used for the treatment of wood are described below:

(i) Steeping: In this method the preservative solution is either applied with a brush or sprayed on the surface of the timber being treated. The timber may also be dipped in the preservative solution for a short period. The wood being treated should be near or above the fibre saturation point if water soluble preservatives are being used. The period of protection offered by these preservatives is usually short.

The moisture content may be around 15% in case of organic solvent type of preservatives. In this case the timber may be dipped for 3 to 5 minutes and this protects the timber from degradation for a longer period of time. Use of organic solvent preservatives is considered an acceptable alternative to pressure methods, if the latter are not available.In steeping, a water repellent such as paraffin wax may be added to the preservative solution. This helps in the following manner:

- Ensuring dimensional stability of the treated wood
- Affording protection against rain water

(ii) Di3usion processes: Diffusion processes are used in the treatment of green timber with the help of water soluble preservatives. The free water occurring in the cell aids in the penetration of the preservative through a concentration gradient. The four diffusion processes that may be used for wood preservation are:

(a) Normal di3usion process: The wood is sawn and then soaked for fairly long periods in preservative solutions of 4 to 10% concentration. As a result, the preservative is soaked by the timber. The rate and level of absorption depends on:

- Nature and type of wood being treated
- Moisture content and thickness of the wood
- Nature of the preservative being used

This process is best suited for the treatment of wood which become refractory to treatment under pressure after drying. Bamboos give good results when treated by this process.

(b) Hot-soak process: In this process, the green timber is immersed in open tanks in hot solutions of boron compounds (usually boric acid). It is not used in India.

(c) Steam quenching followed by di3usion: In this process, the green or partially green wood or bamboos to be treated are first steamed under atmospheric pressure for periods varying between 2 to 3 hours. Thereafter, the heated material is transferred to the preservative solution at ambient temperature. This is known as quenching and the period for which quenching is done varies with the size of the material being treated, species and concentration of the preservative.

The diffusion period under non-drying conditions follows quenching. This enables the preservative to penetrate into the wood. Even bamboos may be treated by this process.

(d) Momentry dip process: In this process, the wood being treated is dipped for brief periods in the preservative solution which comprises of a combination of boric acid and borax. The treated wood is stacked for some time so that the preservative solution moves to the inner layers by diffusion.

(iii) Sap displacement processes: The preservative treatment by this process involves flow of the preservative solution along the length of the wooden member along the flow of the sap stream. The timber being treated is placed vertically or partially inclined in solutions of water soluble preservative for 3 to 4 days after which they are inverted and allow to stand for similar periods. Various sap displacement processes have been discussed below:

(a) *Boucherie and modified boucherie*: This process involves treating timber in a green condition with or without bark by using inorganic water soluble preservatives. The butt end of the timber being treated is held in an inclined position, usually at angles of about 45 degrees and attached to a reservoir containing the preservative at a higher level. This enables the preservative to move into the wood. Bamboos can also be treated by this method. In a modified version of the boucherie suitable air pressures may be used over the preservative for ensuring better and quicker penetration.

(b) *Hot and cold process*: This process is used for timbers that can be readily treated and the preservative is easily absorbed. In the first stage, the timber being treated is heated in the preservative solution at about 90°C for a period of 2 to 3 hours. As a result, the air inside the wood escapes partially due to expansion. Thereafter, both the timber and preservative at left to cool.

After, it has cooled, the preservative is absorbed by the wood due to the partial vacuum created by contraction of the air. The following variables are taken into consideration while using this process to treat timber

- Density
- Treatability class
- Heating time
- Temperature
- Cooling time
- Absorption of dry salt

Pressure Methods: Various pressure methods/processes/techniques for the treatment of timber are discussed below:

(i) **Full cell process:** The full cell process is adopted when it is desired to have high levels of absorption of the preservative. It involves filling the cells and saturating the cell walls with the preservative. The timber to be treated is placed in the pressure chamber. It is commonly termed as the charge. The door of the chamber is tightly closed and a vacuum of atleast 56 cm of mercury is created and maintained for about 30 minutes. The aim is to force out as much air from the cells as is possible.

Thereafter, the preservative is introduced into the cylinder with the vacuum working. The vacuum pump is stopped when the preservative has completely filled the cylinder. It is then subjected to an antiseptic pressure whose intensity depends on

- Species
- Size of the material being treated
- Refractoriness of the material

Due to this pressure, the preservative is introduced into the timber. It is maintained till absorption of the desired extent is obtained after which the preservative is withdrawn from the cylinder. The dripping preservative is removed from the timber by applying a vacuum of 38 to 56 cm of mercury for a period of about 15 minutes.

This process is fairly effective if the correct technique has been adopted and the right preservative used in the treatment.

(ii) **Empty cell processes:** These processes involve maximum penetration of the preservative into the timber with least net absorption. The following two processes may be adopted:

(a) **Lowry process:** The pressure cylinder is first loaded with the charge and then closed. After this, it is filled with the preservative and an antiseptic pressure is applied till the required level of absorption is obtained. The quantum of pressure applied depends on the species and size of the material being treated.

The pressure is then released. A portion of the preservative is expelled as a result of the entrapped air within the cells. Then the cylinder is drained off and a vacuum is applied for cleaning the surface.

(b) **Rueping process:** The material to be seasoned is placed in the cylinder and an air pressure varying from 2 to 5 kg per square cms is applied for a specific time period, depending on the sapwood content. This is maintained when the cylinder is being filled up with the preservative.

Once the cylinder has been filled up an antiseptic pressure is applied till absorption of the desired level is obtained. After this, the vacuum is generated as in the other processes. The rueping process is suited for treating mixed species and also for timber that contains both sapwood and heartwood.

(iii) **Alternate pressure method (APM):** This process for timber seasoning has been recently developed. The advantages of this method are:

(a) The timber need not be pre dried. It is possible to treat even freshly sawn timber at moisture per cent ranging from 100 to 200, by applying a cyclic pressure schedule.

(b) It is convenient, cheaper and faster than other pressure methods.

The preservative used is of a higher concentration as is required to expel/replace water from the cells. The treated wood may subsequently be dried to the desired moisture content and further processed.

(iv) **Boulton process:** This process combines seasoning and preservation of woods. The green timber is boiled with a gradual vacuum being applied in the presence of suitable oil. In this way, the boiling point of water in the wood is reduced and it evaporates at temperatures below 100°C. Timber can thus be dried at lower temperatures causing less damage to it. The preservative treatment comprises of creosoting.

39. a.Classify medicinal plants on the basis of plant parts used and explain general rules of sustainable collection of medicinal plants, with suitable examples. (2009)

b. Curiosity medicinal plants on the basis of plant parts used and explain general rules for collection of medicinal plants. (2005)

The medicinal plants are classified based on the plant part used in various medicinal utility. Accordingly, they are classified based on

(a) Leaf based
(b) Bark based
(c) Wood based
(d) Flower based
(e) Fruit based
(f) Root based
(g) Seed based

Classification of medicinal plants

Leaves: The leaves of following plants are accredited with medicinal utility and are furnished.

(i) *Azadirachta indica*

(ii) *Aloe vera*
(iii) *Ocimum sanctum*
(iv) *Phyllanthus amarus*
(v) *Cymbopogon martni*
(vi) *Eucalyptus*

Bark

(i) *Saraca asoca*
(ii) *Cinnamomum zeylanicum*
(iii) *T.arjuna*
(iv) *T.bellarica*
(v) *Careya arborea*

Wood

(i) *Santalum album*
(ii) *Pterocarpus santalinus*
(iii) *Pterocarpus marsupium*
(iv) *Caesalpinia sappan*
(v) *Acacia catechu*

Flower

(i) *Syzygium aromaticum*
(ii) *Madhuca latifolia*
(iii) *Michelia champaca*
(iv) *Phyllanthus emblica*
(v) *Calophyllum inophyllum*

Fruit

(i) *Aegle marmelos*
(ii) *Piper longum*
(iii) *Emblica oDcinalis*
(iv) *Terminalia chebula*
(v) *Terminalia bellirica*

Root

(i) *Rauwolfia serpentine*
(ii) *Vetiveria zizanioides*
(iii) *Withania somnifera*
(iv) *Asparagus*
(v) *Gmelina arborea*

Seed

(i) *Terminalia bellerica*
(ii) *Gloriosa superb*

(iii) *Strychnos nux-vomica*
(iv) *Terminalia chebula*
(v) *Strychnos nux-vomica*

Guidelines for root harvesting: In many medicinal plants, the medically effective substances are found in the root. Thus, in many cases, the whole plant gets uprooted by the gatherers and thus can never grow again. If this technique is used with a high number of plants in a short time, the plant will eventually go extinct In order to ensure sustainable harvest of root material for follow these rules:

- Dig the root at a considerable distance, at least 30 cm, from the main stem or tap root.
- Avoid severing of the tap root Collect only the lateral roots.
- After digging cover the hole to ensure protection against infection and invasion by pests.

Guidelines for bark harvesting: Traditionally, bark is harvested with an axe or bush knife. If a particular species is in high demand and intensively used, the plant can become endangered by this technique. The most common unsustainable practice is ring-barking where entire rings of bark are removed around the tree, inevitably leading to death of the tree. In order to ensure sustainable harvest of bark material the following recommendations should be followed:

- Peel the bark from the tree in small pieces leaving most of it intact on the trunk on the East and West side of the tree as has been the tradition. Remove the bark in long vertical strips using a thin flexible blade/bush knife.
- Do not practice ring barking, which is the cut of off entire rings around the tree.
- Remove the bark in small sections and leave some inner bark to protect the wood. Do not cut the edges of the strip with an axe as this cause the remaining bark to lift from the wood and dry out.
- Possibly use 'tree seal', e. g. apply a piece of wet cow-dung to the bark wound. This will prevent the wound from drying out, though it does not facilitate bark recovery or prevent the development of insect infestation or development of infections on the wound.

Guidelines for leaf harvesting: In most cases, leaf harvesting is regarded as least destructive form of harvesting to the plants. However, studies revealed that "business as usual" harvesting sometimes resulted in the collection of all the leaves from the tree including cutting down the branches and twigs. In order to ensure sustainable harvesting of leaves it is recommended that one can:

- Pluck individual leaves instead of leaf striping and avoid use of sharp pruning shears for leaves.
- Regularly prune branches to improve the quality and quantity of leaves.
- Injure a limited number of roots to encourage sprouting of new root suckers.

Guidelines for fruit harvesting: The commercial harvesting of fruits can affect not only species regeneration but also the quality of the resource in an area. This is particularly the case if the fruits and seeds of an inferior quality, such as small and rotten ones, are left to regenerate. Thus, fruits also need to be extracted in a sustainable manner so that the biotic integrity of forests and woodlands is supported. In order to ensure sustainable harvesting of leaves it is recommended to:

- Do not collect all good looking, high quality fruits but leave some on the ground so that more plants of good quality can germinate.
- Only collect fruits from some trees and leave others completely.

40. a. What are the wood composites? How they are prepared? What are the common glueing agents used in wood composites? (2011)

b. Give an account of composite wood products and their utilities. (2016)

c. What is Composite Wood? Write raw material, manufacturing process, marketing and uses of the most commonly used composite wood in India. (2006)

Composite wood: It is the general term for built up bonded product consisting either wholly of natural wood or of wood in combination with metals, plastics, etc.

Plywood: It is a glued wood construction built up of veneers in such a manner than the grain of each veneer is at right

Raw Materials

- *Dipterocarpus indicus*
- *Eucalyptus*
- *Grevillea robusta*
- *Melia dubia*
- *Acrocarpus fraxinifolius*
- *Mangifera indica*
- *Albizzia lebbeck*
- *Lannea coromandelica*
- *Anthocephalus cadamba*
- *Holoptelea integrifolia*
- *Artocarpus heterophyllus*
- *Azadirachta indica*
- *Melia azadarach*
- *Boswellia serrata*

Composite wood: It is the general term for built up bonded product consisting either wholly of natural wood or of wood in combination with metals, plastics, etc.

The most common composite wood used in India is plywood. The raw material used in plywood industries are

(i) Poplars
(ii) Eucalyptus
(iii) Silver oak
(iv) *Melia dubia*
(v) *Acrocarpus fraxinifolius*
(vi) *Maesopsis eminii*

(a) Plywood: It is a glued wood construction built up of veneers in such a manner than the grain of each veneer is at right.

Plywood manufacturing process

- **Step 1: Felling and Transportation:** *Melia dubia* plantations grown after 5 years are eligible for felling. The minimum girth requirement for felling *Melia dubia* tree is 20 inches girth at breast height. The plantations with the girth of 20 inches and above are felled into bellets of 4.3 ft length uniformly using power chain saw or two men saw. The converted logs are then transported to Ambi Ply factory located at Mettupalayam, Tamil Nadu.
- **Step 2: Peeling of Logs:** For peeling logs, the 4 feet peeling lathe is used. The logs with 51 inch length (4 feet & 3 Inch) logs are kept in the Lathe for peeling. While peeling, the core veneers are produced into two different thickness viz., 1.8 mm or 2.0 mm as per the standard requirement. The core veneers are then cut into 2 feet length through the chopping machine. The logs are normally in green state with moisture content ranged between 30 and 44% and are directly deployed for peeling and hence all the peeled veneers are with very high moisture content.
- **Step 3: Drying of Veneers:** The veneers produced have higher moisture content and for making plywood the optimal moisture requirement is 8-12%. Hence the veneers have to be dried for which a drier is used. The core veneers are fed into the dryer wherein the temperature is maintained at 120°C. The drying process ensured the moisture content of around 12%.
- **Step 4: Gluing:** After drying, the veneers are sized into uniform sizes and are intended for gluing. For this purpose, two types of resins are used. Melanine Formaldehyde resin (MF resin) is used for gluing for the process of commercial plywood production and phenol formaldehyde resin (PF resin) is used for gluing towards the production of water proof plywoods. For the purpose of gluing, a glue spreader machine is used. Using the glue spreader machine, all the core veneers are glued either by MF Resin or PF Resin.
- **Step 5: Construction of Plywoods:** After gluing, the plywoods are constructed using both face and core veneer. In Ambi Ply Panels and Doors, the face veneer is used from *Dipterocarpus turbinatus*. In the plywood construction process, first one face veneer is laid and upon the face veneer, glued core veneers are kept horizontally and vertically to the required thickness and again another face veneer is laid on the assemble of core veneers. After the assemblage of face veneer at two ends with core veneers in the middle and the whole process is called plywood construction.
- **Step 6: Pressing Plywoods in the Hot Press:** After construction, the plywood construct has been placed in a hot press for pressing. It is done through a hot press having a hydraulic compressing capacity of about 1000 tonnes. In the hot press, the temperature of about 140° C is maintained to ensure uniform pressing. The assembled plywood is kept inside the hot press for about 20 minutes. During this process, the glue applied on the core veneers is spread uniformly and bonding of the core veneers takes place very solidly due to the hydraulic compression and the hot press plywood is ready for further process.

- **Step 7: Cutting Process:** The hot dried plywood constructs have uneven sizes which need to be sized in all directions. For this purpose, the DD saw cutting machine is employed and the four edges of plywoods are cut in order to make the required sizes of plywood. This cutting process is preferred to ensure uniformity and to remove the uneven veneer edges.
- **Step 8: Sanding:** After the cutting process, the plywood needs to be sanded to convert rough surface into smooth surface. For this purpose, a sanding machine is used in which both sides of plywoods are sanded. During this process, the plywood becomes very smooth on both faces of the plywood and are ready for supply to consumers.
- **Step 9: Labelling:** After the sanding process, the plywoods are labelled and are ready for despatch.
- **Step 10: Loading and Transportation:** Ambi Ply Panels and Doors have established very good network of dealers and retailers across South India. The required grades of plywood are loaded and transported to the dealers and retailers through trucks on a regular basis

Composite woods

(a) **Laminated wood:** These are a build-up product made of wood layers (laminate) in which all layers laid with their grain parallel and glued or otherwise fastened together. The glued laminated wood is called as glulam. Lamina refers the thin board of veneers. Lamina are seasoned in kilns and cut into uniform width and length. These are arranged in a proper order. This feed into a glue spreader and assembled into a required shape. Then pressure is applied in a clamp fitting assembly.

(b) **Core boards:** These are composite board built up of a core composed of strip of woods of various dimensions glued together to form a slab. This is intern glued between two or more outer veneers with the direction of grain of the core strips running perpendicular to that of adjacent veneer.

Board	**Width**
Batten board	7.5cm
Block board	<2.5 cm
Lamin board	<7 cm

(c) **Sandwich board:** It is the general term for built up boards having a core of light material, faced on both sides with relatively thin layers of woods having high strength properties. The thin faces are called as skins which are strong enough, have high strength and more stiffness. It is used in manufacturing parts of air craft, motor boats, table tops and containers.

(d) **Fibre board:** It is a sheet of material made from fiber of wood. Fiberboards are made from all wood wastes and agricultural wastes. First the wood is defibriated or pulped inter-felted into a mat. Bonding agents and supplementary agents added at the felting stage to improve the mechanical properties. This is consolidated by pressured and heat. It is used as a core material in sandwich and core boards, building boards, insulation boards and wallboards. Based on densities, these are classified as i) Hardboard ii) Semi hardboard iii) Medium hard board iv) Soft board.

(e) **Particle board:** It is a board or sheet constituted from the fragments of wood and other lignocellulosic materials bonded with organic binders with the help of one or more agents like heat, pressure, humidity, catalysts, etc. This is made from chips, flakes, splinters. The main types are: 1. Chip board 2. Flake board and 3. Shaving board

Waste materials are chipped in chippers and screened where unwanted materials are removed. This is dried to suitable moisture content in a drier. Mechanical mixers were added and flat pressed by hot process method.

(f) **Improved wood:** These are also called as modified wood. It is a general term of wood that has been specially treated in various ways to reduce or retard working or to increase strength.

The wood properties such as hygroscopicity, strength, density etc., are modified by impregnating with resins or by compression or by heating.

(g) **Impregnated wood:** It is a wood impregnated with impregnants. The common impregnants used in this process are wax, paraffin, resins and oils. The main objectives are to make the wood dimensionally stable and prevention of water absorption. Now commercial water soluble, thermosetting formaldedyde resins is available. In the manufacturing process, first woods are impregnated through vacuum impregnation with or without pressure. This is thermosetted and impregnated again. After curing, individual materials are bonded together as one.

(h) **Heat stabilized wood:** This is obtained by simple heating. In this, woods are passed beneath the surface of a molten metal or fused salt. It is subjected to 260 – 315 ° C for few minutes under non oxidizing conditions.

(i) **Compressed wood:** In this, woods are compressed without impregnation. It increases specific gravity and strength of the wood so that it can withstand high wear and tear. Compressed wood is formed from solid or laminated wood under the pressure of 90 – 140 kg/cm^2. Material obtained after compression have a specific gravity of 1.2 to 1.3.

(j) **Compregnated wood:** This designates the wood which are subjected to impregnation and compressed. Woods are spliced into veneers and impregnated with urea formaldehyde by compression. This increases tensile strength and improves insulation property. Because of this treatment, wood becomes 20 times harder than normal woods.

(k) **Heat stabilized wood:** In this woods are subjected to heat to improve wood property.

(l) **Chemically modified wood:** In this wood, OH group of cellulose is replaced. Chemicals used for this are acetic anhydride influences of pyridine as catalysts. Urea hydroxylin and sulphuric acid also used. It decreases the hygroscopicity and also of swelling and shrinkage.

Common gluing agents: Synthetic resign glues replaced protein and starch glues for plywood manufacture, with urea formaldehyde predominating because of price, properties and ease of application. Australian mills once used tannin formaldehyde resigns or casein and soybean glues.

- **Urea formaldehyde** glues are extensively used for interior and intermediate grade bonding, which covers the majority of hardwood plywood produced. Formulas differ somewhat and can be adapted to meet specific conditions.
- **Polyvinyl acetate** glues and also polyvinyl acetate/urea formaldehyde glues are used for edge jointing and veneering. Polyvinyl acetate glues resistant to boiling water are available.

- **Melamine formaldehyde** glues are not extensively used for plywood gluing but are used where a high grade bond is required, and where black phenolic glues cannot be tolerated. They are used to fortified urea-formaldehyde glues to increase the weathering resistance of the bond. The largest application of the melamine formaldehyde resin is in the production of decorative overlays.
- **Phenol-formaldehyde** glues are standard for exterior bonds. With a glue properly formulated for exterior use and suitably employed, no exposure conditions or laboratory testes are known which will degrade hot pressed phenolic glue bonds without destroying the adjacent wood layers. Phenolic glues are also used for impregnating veneers and paper overlays for plywood.
- **Resorcinol-formaldehyde** and **phenol/resorcinol-formaldehyde** glues are similar to phenolic glues in quality of performance but being more reactive, can be cured at room temperature. They are more expensive than phenolic glues, and therefore limited in use to special applications.

Utility of composite woods

- Composite woods are useful in construction industries
- They are flexible, uniform and consistency in performance and they are used in all interior decorations.
- In multistory buildings the composite wood are preferred due to its light weight and durability.
- Some of the composite woods are stronger and more dimensionally stable and hence they are used in construction of wood houses.
- Composite woods are also used to make headers, beams, girders, columns and also for heavy trusses.

41. What is the role of Land depots and Water depots in commercial storage of forest products? (2006)

Depots: A depot is a place where wood or other forest produce is stored, pending its disposal. Depots may be for minor or for major forest produces, for logs, scantlings, sleepers, posts, piles, grass or any minor forest produces. The depots can be broadly classified as

1. Forest depot,
2. Transit depots and
3. Sale depots.

The depots can be classified according to their situation as land and water depots. A land depot is the one used for storing wood transported by land. A water depot is one where timber borne by water is stored. Water depots are situated on a river, in still water or on sea shore. River depots are attended with the risk of flood damage. Both land and water depots may be transit or sale depots.

42. Write properties and uses of Turpentine oil and Lac. (2007)

Properties and uses of turpentine oil

- Oil of turpentine is a colourless oily liquid with a strong odour and a hot disagreeable taste.
- Its boiling point is 155°C, specific gravity between 0.85 to 0.88.
- It is insoluble in water, is miscible with absolute alcohol, ether and dissolve Sulphur, phosphorous, resins and caoutchous.
- On exposure to air and absorbing oxygen, gives off ozone. It is mixture of terpenes.
- Turpentines is chiefly used in preparation of paints and varnishes, shoe and other polishes, chemicals and pharmaceuticals.
- It is employed for dissolving fats, resins and for a variety of domestic uses.
- Its products are used in manufacture of synthetic rubber, waxes, synthetic camphor, insecticides and germicides.

Properties of Lac

- Lac is insoluble in water and readily soluble in cheap solvents.
- It is comparatively hard among gums
- It is highly elastic
- It has a power of adhesion to smooth wood and metal surface
- It helps to assume the shape of mould
- It is an electrical resistance

Uses of lac

- It is used for moulding and gramophone industries
- In varnishes, polishes, sealing waxes, hats, emery wheel, ammunition (cartriduges and shells), electric instruments of all kinds, it find its greatest use

43. Describe general scenario of marketing and Trade of timber and Non-Timber Forest Products (NTFP) in India. (2007)

Marketing and Trade of timber: Marketing is a process encompassing all the steps involved in flow of goods and services from the producer to the consumer. It is a system operating within a network of channels where middleman and institutions not only influence but also play significant roles in production, distribution and exchange of products. The timber products include timber and the associated wood and wood products. The following are the marketing channel prevailing in India as far as forest products are concerned.

(a) Direct Marketing (Through the contract farming system)

Producer → Consumers/Self Consumption

The direct marketing is gaining significance through the introduction and implementation of contract tree farming. This contract tree farming facilitated linking of growers with consumer industries through organized and pre-determined support price mechanism. This system is gaining significance in few forest products and the successful direct marketing and trade mechanism prevailing in Tamil Nadu is furnished.

Direct Marketing Mechanism of timer prevailing in Tamil Nadu

S.No.	Species	Utility	Price support Rate/ mt	Industry Associated
1	Teak	Timber	Rs.20000	Suresh Timbers, Elumalai, Madurai Dt.
2	Gmelina	Timber	Rs.8000-10000	
3	Casuarina species	Paper	Rs.4900	Tamil Nadu Newsprints and Papers Limited (TNPL), Kagithapuram, Karur Seshasayee Paper and Boards Ltd., (SPB), Pallipalayam, Erode
4	Eucalyptus	Paper	Rs.3900	Tamil Nadu Newsprints and Papers Limited (TNPL), Kagithapuram, Karur Seshasayee Paper and Boards Ltd., (SPB), Pallipalayam, Erode
5	Eucalyptus	Plywood	Rs.5500	Century Plyboards (I) Ltd, Chennai Ambiply Panels and Doors, Mettupalayam,
6	Leucaena leucocephala	Paper	Rs.3900	Seshasayee Paper and Boards Ltd., Pallipalayam, Erode
7	Leucaena leucocephala	Energy	Rs. 3000	Auromira Energy Company Private Limited, Chennai
8	Ailanthus excelsa	Match Splints	Rs.6500	Vasan Match Works, Vellore The Ideal Splints and Veneers, Coimbatore
9	Melia dubia	Paper	Rs. 4400	Tamil Nadu Newsprints and Papers Limited (TNPL), Kagithapuram, Karur Seshasayee Paper and Boards Ltd., Pallipalayam, Erode
10	Melia dubia	Plywood	Rs. 7500	Century Plyboards (I) Ltd, Chennai Ambiply Panels and Doors, Mettupalayam, Sri Ranga Ply Industries, Mettupalayam Asian Timber Industries, Pollachi

Similarly there are few industries like ITC in Andhra Pradesh, WIMCO in Central India practices direct marketing mechanism. However this system is not practiced widely in the country.

(b) Through Middlemen (Widely present)

Producer → Local Contractor (Middlemen) → Wholesaler → Consumers

This marketing and trade mechanism is widely prevalent across the country. Here the growers are not getting adequate prices for their product and the actual benefit goes to the local contractor and the wholesaler. Several wood based industries are also not interested in establishing sufficient linkages with growers due to the field level problems and hence predominantly depending on the middlemen

(c) Through Organized institutions

Producer → Felling groups → Consumer industries

This model exists in Tamil Nadu through an institution called Consortium of Industrial Agroforestry (CIAF). This institution has established felling groups and consumer industries for major forest products like timber, plywood, particle board, match splints, pulp and paper, biomass power, packing cases, pencil, industries etc. The purpose of this arrangement is to facilitate the farmer in terms of easy felling by resolving the issues in felling and transportation of timber. This model is gaining significance across South India which may find utility across the country in the near future.

(d) Marketing and Trade of Non Timber Forest Products: The forests have primarily been seen as wood-producing units and other products have conventionally been referred to as "minor forest products" which has now been renamed as Non Timber Forest Products and their supply was not given high priority till the recent past. It is only in recent years that proper attention has been paid to the importance of these products, especially in their local context where they may be considerably more valuable than the wood obtained from the forest. It is now becoming accepted that in many cases recognition of the economic and social importance of NTFP may be the key to the active involvement of people in forest management.

Major NTFPs gaining significance in India: There are nine different non timber forest products gained significant attraction due to their potential towards meeting the livelihood security of original dwellers of forests and they are

(i) Fibres and flosses
(ii) Grasses
(iii) Essential oils
(iv) Oil seeds
(v) Tans and dyes
(vi) Gums, resins and oleoresins
(vii) Drugs, spices, poisons and insecticides
(viii) Edible products
(ix) Animal, mineral and miscellaneous products

Marketing and Trade: In general, the NTFPs are collected by the tribal people to meet their domestic requirement and livelihood options. But due to developments in NTFP utilization the collection and processing of NTFP has significant attraction. The tribal people particularly the women tribals have been involved in collection of the NTFP. However the marketing and trade of NTFP is not organized as it involves wide range of products. In most cases, the local contractors and middleman collect the materials from the tribal community without paying the actual revenue to the collectors. The following institutions are involves in marketing and trade of NTFPs across the country.

(i) Primary Cooperatives

(ii) Large Area Multipurpose Societies (LAMPS),

(iii) Tribal Development Federations,

(iv) MFP Trade & Development Federations and

(v) TRIFED

(vi) TRIFED, NAFED and STC are carrying out export of NWFPs.

The trade of NTFPs and the price mechanism is still not very well established due to the inherent nature of the NTFP supply chain process.

In general the following marketing channels are witnessed in marketing and trade of NTFPs.

(e) Marketing channel for medicinal plants

(i) Producers (Farmers) W Small traders W Wholesaler (Trader – II) W Exporter / Manufacturing Unit.

(ii) Producers (Farmers) WNGOs WManufacturing Unit

(iii) Producers (Famers) WContract farming traders W Exporter / Manufacturing Unit

(iv) Farmers W Siddha Practitioners or Traditional healers

(v) Producers (Farmers) W Farmers Association W Manufacturing Units

(f) Marketing channel for flosses

Producer W Wholesaler W Retailer W Consumer

(g) Marketing channel for tannis and dyes

Tribal collector W Local contractor W Wholesaler W Consumer

(h) Marketing channel for TBOs

Local collector W Middlemen W Wholesaler W Consumer

Most NTFPs are sold without any processing or value addition. Unlike timber and timber products, the producer access to consumers in NTFP supply chain is very limited and the sales are made in local villages and weekly markets. A major portion of their product is sold to intermediaries like local contractors, middlemen and commission agents who operate in the area. Hence the marketing and trade of non timber forest products is highly unorganized and needs suitable interventions to resolve the issues from each and every stage of Production to Consumption System.

44. List important modern harvesting practices adopted in forest plantations. (2007)

Harvesting in forest plantations of India is still done by semi mechanized harvesting system and not through the mechanized harvesting system practiced in most of the developed and developing countries like Brazil. However, the following are the modern mechanized harvesting methods practiced in major forest plantations.

Modern mechanized felling operations: Mechanized harvesting systems have taken over the conventional systems especially where the growing stock is uniform and even aged. The level of mechanization varies from low level to high level where felling and processing are done using highly specialized equipments and machines. High efficiency is the often quoted advantage of using machines in logging.

There are three basic types of mechanical felling machines.

- **Single-function machines:** They are only capable of directionally felling a tree.
- **Dual-function machines – Feller-Bunchers:** These machines are able to fell a tree and move the cut stem to a pile or a site of interest. It is a motorized vehicle which runs on tracks or wheels, with an attachment that can rapidly cut and gather several trees before felling them. It is a type of harvester with a cutting head that can hold more than a single tree at a time. The cutting head is not only used for cutting but also for holding and placing the stems on the ground but feller bunchers cannot do any kind of processing. A feller buncher consists of a standard heavy equipment base with a tree-grabbing device furnished with a chain-saw or circular saw or shear (shear is a device designed to cut small trees off at the base). The machine keeps the cut trees on a stack facilitating the further removal of the loads by a skidder or forwarder or other means of transport.
- **Multifunction machines – Harvesters:** These machines can fell, delimb, buck and place the processed pieces. It is the first of a pair of machines, the second being a forwarder, that jointly makes the mechanized "cut-to-length" (CTL) system. It is a 4 or 6 wheeled articulated machines used to fell trees, remove branches and cut the logs to required lengths which help to maximize the value of the end product. It is a heavy machine powered by diesel engine and is usually fitted with a hydraulic loader onto which a harvester head is mounted. A harvester head consists of a hydraulically powered chain saw, two or more curved de limbing knives which reach around the trunk to remove branches, feed rollers to grasp the tree and to move the tree through the head, diameter sensors to calculate the volume of timber harvested and a measuring wheel to measure the length of the stem as it is fed through the harvesting head. The whole harvesting process is controlled by an on-board computer and an operator sitting in the cab of the vehicle. Harvesters are mostly employed for cut-to-length logging operations, for felling, delimbing and bucking trees.

Advantages of Mechanized Logging

- High Efficiency in harvesting trees especially smaller trees
- Wide applicability in commercial thinning
- Reduced residual stand damage due to controlled felling and proper placement of felled stems.
- Higher wood utilization due to lower stump height
- Safely used in weather and lighting conditions that are unacceptable for manual cutters
- Improved safety
- Use of improved and modern techniques

Disadvantages of Mechanized Logging

- Very high initial and operational costs
- Soil disturbances
- Potential wood damage with shear-type cutters
- Stability limitations on steep slopes
- Equipment utilization needs technical knowhow and skill upgradation of the persons
- Tree stem size and terrain could be major limitations
- Mechanization may not be feasible in all types forests

Other harvesting equipments

Chain Saw: Chain saws are the most commonly used equipment for felling and delimbing. It is powered by diesel, hydraulic or electric motor and has a saw with a loop of chain that is pulled around a bar which rotates at a high speed. The chain contains cutting teeth that are capable of sawing wood. A chainsaw can be used to fell and process trees. Chains come in a variety of pitches (the distance between the rivets in the chain), gauges and cutter teeth and a range of power sizes and bar lengths. Though the operations are faster, proper safety measures have to be ensured due to the inherent dangers associated with the working of power chain saws. The greatest advantage of chain saws is that felling is generally not limited by slope or soil conditions.

45. Timber Identification is pre-requisite to check adulteration in the timber market. Give reasons. (2007)

Yes timber identification is pre-requisite to check adulteration in the timber market due to the availability of wide range of wood varieties in the market. If not identified correctly, the consumers will be cheated both in terms of quality and price of the wood.

In general, the carpenters, builders, cabinet makers are familiar with timbers and they can identify the timbers based on colour, odour, texture, figure and density. These properties vary due to the species and hence only experienced people can be able to identify the wood. Hence the timber identification is very essential for a common man in order to check the adulteration and to buy the perfect wood.

Timber identification is very pre-requisite particularly if the consumer looks for both hardwoods and softwoods. Normally the hardwoods and softwoods have entirely different distinguishing features which need to be identified before purchasing the material.

Some of the woods may be artificially stained due to infections. The common people or the consumer may be confused due to the staining or colouring of the wood. If the fungus stained wood is purchased it will affect the quality and durability and hence timber identification is very essential.

Some of the timbers may exhibit odour naturally. But the commercial wholesalers artificially spray the odour using chemicals and adulterate the wood. Hence identification of wood is very very pre-requisite.

The identification towards certification of wood is another major breakthrough in the recent past due to international compulsion. This can also be checked which will ensure the origin of the species and through which it can be identified.

Under such circumstances timber identification is pre-requisite in order to avoid the adulteration of the products and to purchase the quality and durable product.

Features for identifying softwoods

- Woods from regions with strong seasonal differences had noticeable difference in tracheids that form the latewood and early wood.
- Softwoods from tropical regions have same tracheid size.
- Presence of resin canals.
- Presence of bordered pit arrangement.

Features for identifying hardwoods

- Hard woods from regions with strong seasonal differences had different vessel (pore) size.
- In woods from tropical regions has same vessel (pore) size.
- Vessels can be solitary or can be clustered into radial multiples.
- Pore groups can be distributed evenly or arranged in radial diagonal lines, or tangential lines.
- Parenchyma cells exhibit many different patterns, which are useful for identification.

46. Discuss types of preservatives available for preservation of timber. (2005)

Preservatives are chemical substances which have high toxity, permanency, permeability and stability which will help to protect the timbers from all biological organisms. These preservatives are of three types and are furnished below.

Types of preservatives

(a) **Oil type:** This type includes creosote and coal tar. Creosote is a brownish black oily distillation from coal. It is used in admixture with petroleum oils. This is effective against borer, marine organisms and termites. These are leaching resistant and not corrosive. The depth of penetrations is easily visible. But it is relatively dirty and emits bad odour thus by gives disagreeable look to wood. It increases the inflammability of wood and hence it is not useful for inside timber works in home. Creosote with 2% PCP (Penta Chloro Phenate) is effectively prevents fungal decay.

(b) **Water soluble preservatives:** These are the water soluble chemicals which effectively protects the wood from various decays. As it is liable to washed away with water, it is not suitable for outdoor works. This may be of following two types.

(i) **Non-fixed water soluble type**: This preservative is easily washed by water and hence it is useful for woods used for above ground level works. This is applied as dip diffusion method.

Example: Borax, Boric acid, Sodium PCP, Copper sulphate, Mercuric chloride, Sodium fluoride and zinc sulphate.

(ii) **Fixed water soluble type**: This preservative is not liable to leaching through water and is retained in the wood for longer time. These are colourless, odourless and non corrosive one.

Example: CCP, Chromated Zinc Chloride, Zinc Meta Arsenate (ZMA), Ammonical Copper Chromate (ACA) and Acid Copper Chromate (ACC)

(c) **Organic solvent:** This type of preservative consists of copper and zinc salts of organic acids like naphthenic acid, stearic acid, abietic acid and chlorinated phenols and benzens like pentachlorophenol and hexa chloro-benzene. These preservatives are used after dissolving them in suitable organic solvents with low boiling point which facilitate easy drying of timbers.

Example: Hardi proof and cuprinol are the proprietary preservatives

47. Explain general rules followed during felling of trees in forests? (2005)

Felling rules: Scientific felling lays emphasis on the production of the maximum quantity of timber / industrial wood besides reducing the risk of damage to the surrounding vegetation in the felling series / coupe. In order to achieve this, the following rules are to be strictly followed.

- Felling of trees should be done as close to the ground as possible to avoid valuable timber/ industrial wood lying in the forest / plantation as wastage
- Care should be taken to undertake felling in such directions where the damage caused to the trees being felled as well as the surrounding vegetation is kept at the barest minimum. In other words, scientific and sound felling should reduce the impact caused to the surroundings
- In hilly terrain, fellings should be done uphill
- Felling should be avoided when it is not possible to extract the timber from the felled area
- Lopping should be done first in the trees to be felled when there is a potential danger to the surrounding regeneration / young crop; this will help in reducing the damage to the surrounding crop
- On slopes, felling should begin at the top and proceed downwards.
- Fellings are to be conducted under strict supervision to ensure that the rate of conversion keeps pace with felling

48. Name five aromatic grasses with their uses and method of extraction. (2010)

Grass oil: This group of oils are obtained from tropical grasses that are rich in aromatic essential oil. Andropogon and cymbopogon are two genus of grasses that yield grass oils

Important grass oils of India and its uses

(i) Citronella oil – *Cymbopogon fleuxosus*

- It is used in low cost perfumes, soaps, floor waxes, insecticides and other household products
- It is mixed with cedar oil to produce a popular mosquitoe repellent.
- This essential is the basic source from which chemicals like geraniol, citronella etc are produced.

(ii) Lemon grass oil – *Cymbopogon martini var motia*

- This oil is used for the manufacture of perfumes, soaps and cosmetics in different flavours.
- Lemongrass oil is the source material for the production of ionoes, methyliononis and vitamin A.
- Citral is isolated from the oil. It is in turn used for making perfumes and flavours or is process into other perfume making material.
- It is also used to prepare pharmaceuticals such as pain balms, disinfectants and mosquitoe repellent.

(iii) Palmarosa oil – *Cymbopogon martini var sofia*

- It is used as a base for perfumes and cosmetics.
- It is also used for adding flavour to tobacco.
- This oil is used to mosquitoe repellent.
- This oil forms an important constituent of many medicines.
- It is geranoil which is usually of high quality.
- It blends well with other soap and perfume materials.

(iv) Ginger grass oil – *Cymbopogon nardus*

- It is sweet or pleasant smelling oil.
- It is used for making scents and cheaper varieties of soaps.
- A number of chemicals are also obtained from this oil.

(v) Vetiver oil - *Vetiveria zizanioides*

- It is widely used in the manufacture of perfumes, soaps and cosmetics. This oil gives a pleasing note to the perfume and serves as a valuable fixative.
- This oil blends well with ionones, mimosa, cassia, lavender, sandalwood, patchouli and rose oil.
- It is also used for flavouring beverages.

Extraction of essential oils: In grasses, the oil is extracted predominantly by stream distillation but the following methods are employed both for commercial and academic importance.

Stream distillation: In this method, the grasses is pulverized and stream distilled to extract oil. In this process, the stream is generated at 140 – 212 ° F and the stream is passed through the sample which releases the oil from the grasses that is locked within the cellular structure of the leaves. The oil with the stream is transferred to the condenser wherein it gets cooled and the oil and water are separated in two different layers. The oil is separated and collected in a vessel.

Hydro distillation: It is a traditional method wherein the sample is soaked in water and boil through open fire. On boiling, the oil is expelled and will float on the surface of the water. The floating oil can be separated and collected. It is a crude method and involve huge volume of fire wood cost and time.

Solvent extraction: In this method, the sample is extracted using solvents like hexane, petroleum ether and ethanol are used. In this method, the oil is extracted in thick and highly concentrated form. This extract is mixed with alcohol to extract only the oil.

49. a. Discuss the various common defects encounter after harvesting of the wood during stacking. (2011)

b. Which are the important defects observed in stored timber? (2006)

c. Timber encounters many di-erent problems during the process of drying. What are those problems? (2015)

The following are the Common drying defects:

I. Mechanical failure

(a) **Knots:** The portion of the branch within the wood as a knot. Different forms of wood knots are pin knots, small knots, large knots, round knots and live knots.

(b) **Shakes:** This is a separation of fibres along the grain of standing or freshly felled trees. A crack or fissure is usually formed which is restricted to the interior of the timber.

(c) **Cross Grain:** This is a term used to denote the deviation of wood fibres along the direction of the longitudinal areas of the tree. These may be diagonal, spiral, and interlocked grains.

(d) **Reaction Wood:** An automatically distinctive wood may be formed in the parts of leaning trees and its branches formed due to compression of tissues.

(e) **Compression Failure:** These are defects caused by the budding or failure of fibres as a result of compression. i.e.) the end compression.

Resin pockets and Resin streaks, bark pockets may be included in the main stem.

(f) **End-splits:** It is the complete separation of the fibres with the grain starting from one end and extending along the length of the wood, caused by more rapid drying of the end selections than of the other surfaces. A certain amount of end cracking and splitting is difficult to avoid even in best drying process. The serious end cracking can be avoided by employing high EMC condition in early stages of drying.

(g) **Split-knot:** It is rupture and complete separation of the fibres in and around a knot which is originally present in a piece of wood. This defect is caused by unequal shrinkage of the woody tissues at that point in and around a knot during the process of seasoning.

(h) **Surface cracking:** This defect is caused by the rupture of fibres in and through the surface of a piece of timber usually across the annual growth rings i.e. in the direction of the medullary rays. It is caused by the more rapid drying of the surface than of the inside portion of the piece.

(i) **Warping:** Any distortion of wood from its original shape is called warping. The following terms denote the various kinds of distortion that occur in practice.

- **Cup:** When the distortion is by a curvature across the width of a piece of timber it is known as cup. It usually occurs due to the more rapid drying of one face of a plank than the other.
- **Bow:** This is longitudinal distortion of a board in such a manner that one of the faces becomes concave along the length. It can be caused by some inherent defect in the wood structures, such as cross or twisted grain or may be due to irregularity in conversion,

but in most cases it is the result of poor piling.

- **Twist:** It is identified as the spiral distortion of the edges of a piece of timber so that the four corner of any face are not in the sample planks. It is measured by the degree of twist and in generally expressed qualitatively as slight, moderate and heavy.
- **Spring or Crack:** This is edgewise distortion of a piece of wood and can be detected by laying a straight - edge along one edge of a board from end to end. It is general due to a difference in the longitudinal shrinkage between sapwood and heartwood.

(ii) Defects observed in stored timber: The main causes of timber defects/timber deteoriation during stacking are fungi, insects, mechanical failure, fire and weathering.

Wood decaying fungi: Fungal decay may be proceeding quite rapidly in timber on which no signs of fungal growth are visible.

(i) Stain: Fungi causing stain in wood, when it feeds only on food material stored in the sapwood. In this case, fungi do not attack the heartwood. Stain defect does not affect strength properties of wood. Blue stain is common in pines. A bluish gray discolouration on the woods surface is caused by mould. Some other stains are like brown stain, mineral stain, oxidation stains.

Eg: Ceratocystis.

(ii) Decay or rot: This is observed due to wood destroying or wood rotting fungus of wood. These fungi nourish cell wall material and breakdown the cell structure and enzymatic activities. Decay fungi attack both sapwood and heartwood. This defect reduces the strength properties of wood. White rot mainly concerned with heartwood of hardwood and Brown rot mainly concerned with sapwood conifers. When the wood is stacked in highly humid or moist weather soft rot occurs on the surface of wood due to the breakdown of soft rot. Wood rot fungi frequently causes these types of rots in addition to Heart rot, root rot, But rot, in the stacked timber.

(iii) Insects: Several group of insects attacks the staked timber because of the moisture content. The major insects under the order beetles, termites, moth are groups attacking timber and starch feeders, fungus feeders, etc.

- Pin holes and Shot holes borers. Eg: Xylenborus, Cyctus, Aeolesthes
- Worm holes, Grub holes and Borer holes. Eg: Hoplocerambyx
- Damage by flat headed borer. Eg: Chrysobothris
- Damage by dry timber borer. Eg: Stromatium
- Carpenter bee holes. Eg: Xylocopa
- Termite damage. Eg: Heterotermes, Odonototermes
- Marine wood type borer. Eg: Teredo, Limnoria

Ambrosia beetles attack freshly felled logs, more serious problem in the tropics than they are in temperate countries. In dry or converted timber, damage by Ambrosia beetles is recognized by the black staining of the tunnels. Common furniture beetles called *Anobium punctatum* is normally found in temperate climates. Death watch beetle are normally found in temperate climates. It is primarily a pest of timbers in old buildings in which damp has been allowed to react the wood work.

(iv) Weathering: Any unprotected wood exposed to alternate weathering and rapid drying will soon suffer deterioration at the affected surface, which swell and shrink in response to moisture changes and develop case hardening, checking and lifting of the grain.

(v) Fire: Timbers vary greatly in the ease with which they can be ignited, and in the rate which they will burn.

50. Discuss briefly about the role of NTFP in Indian economy. (2005)

- The Non Wood Forest Products in India contribute significantly to Indian economy.
- Today the NTFP contributes over 75% of total forest export revenue in India.
- Nearly 400 million people live in and around forest depend on non wood forest product for sustenance and supplementing income.
- Non wood forest product contributes significantly the income of 30% rural people.
- More than 80% of forest dwellers depend on NWFP to meet their basic necessities. Over 17% landless labourers and another 39% wage labours are directly depending on NWFP for their occupation.
- The occupational activities involved collection, cleaning, grading and to some extend of processing of NTFP's.
- Traditionally NWFP has received only little attention hence the activities at significance NWFP are not witnessed but due to economic progress of the country associated with the technological advancement as attracted several non wood products as a source of economic returns.
- Indian forests support over 1500 plants which are accredited with medicinal values. Several pharmaceutical companies have emerged to meet the market demand for natural medicine. Such a development pose a threat to many medicinal plants through demand based over exploitation.
- Similarly the significance of gums, resins, dyes, beedi leaves, edible fruits, fibres, flosses are increasing at an alarming rate which are contributing significantly towards Indian economy.
- The institution like Forest Development Corporation, Co-operatives and Tribal societies are playing vital role in bringing economic prosperities through non wood forest products and the associated value addition process.
- It is estimated that several non timber products have been demanded on large scale and the annual growth is increasing at an alarming rate of over 15% in many of the products like Amla, Brahmi, Guggal, Kalihari, Mulethi, Senna, Thulsi etc.
- These products are contributing significantly towards Indian economy hence non wood forest products are vital source of socio-economic development and contribute directly to Indian economy.

Forest Protection and Wildlife Biology

1. What are fire-prone areas? How are these detected? (2010)

Fire prone areas are those which are liable for occurrence of wild fire due to the forest type (deciduous forest), failure of monsoon, availability of flammable materials and large spell of dry season.

Detection of fire prone areas: The fire prone areas are detected through the following approaches.

(a) **Fire - watchers:** Allotment of fire watchers in drier and arid localities will help to identify and assess the fire occurrences either periodically or seasonally. This will help to identify and locate the areas as fire prone which will help to facilitate early control.

(b) **Fire watch towers:** Establishment of fire watch tower in each and every location is very opt in order identify the fire occurrence. The permanent watch towers will help to identify whether the area is fire prone continuously or seasonally.

(c) **Look-out points:** The look-out points is also essential in order to identify and locate the fire prone areas.

(d) **Aerial patrols:** Aerial patrol though not common in India but very well deployed in developed nations in order to identify the fire prone areas. This is normally to identify the dryness of the forests, colour changes in leaf and other illegal or intentional fire incidences.

(e) **Local communities:** It is very very opt to always discuss and get the ideas of local communities towards the identification of fire prone areas. The local communities or the aboriginals have traditional knowledge of fire occurrences and hence they may act as a potential source for identifying the fire prone areas.

(f) **Real time warning:** The real time warning system is also one of the easiest way of detecting fire occurrences. A smoke detector or a smoke alarm is a device that detects smoke and issues an alarm to alert the nearby people. The wireless communication enables transfer of data or signals over part of the entire communication network. The wireless implementation of sensor network ensures safety in terms of saving life and property. The successful realtime warning has been implemented in Andhra Pradesh using ARM 7 based micro controller based wireless sensor network system. The Forest Survey of India also alarms through real time warning system.

2. a. Comment on the view that after deforestation forest fires are most important cause of forest destruction. Also give di-erent types of forest fires and their causes, and preventive measures of forest fires.

b. There are measures which are taken to protect the forests from forest fire damage. Which are those and how e-ective have they been (2015)

After deforestation, the debris and other residues lying on the ground act as a source material for occurrences of fire and they are the most important cause of forest destruction. Normally after deforestation, the leaves, twigs, branches are left as such in the field and on drying they become highly inflammable. Similarly after deforestation, the area is invaded with grasses which may also act as a source material for fire occurrences.

These forest residues and the new grasses due to seasonal changes, lightning, thunder, heavy winds etc. get sparking of fire which rapidly spreads across the resources and act as a major cause for forest destruction. The human beings involved in deforestation activities also through lights after cigarette smoking will also play a major role in fire occurrences and cause a major destruction to the forests. The human beings involved in deforestation activity also establish temporary settlements and cook their own food. This may also results in spreading of forest fires and act as a major cause of destruction. Hence after deforestation the exposure of debris and other residues to extreme climatic conditions cause occurrence of forest fire which may lead to rapid destruction of the forests.

The forest fires are classified into various types based on

(a) Nature of occurrence: Based on nature of occurrence the forest fires are classified into natural fires, accidental fires and deliberate or intentional fires.

- **(i) Natural Fires:** The fire occurs due to natural causes like lightening, rolling of stones, rubbing of bamboos, etc. and this may comprise about 5% of the total fire occurrence in the country.
- **(ii) Accidental Fires:** The fires that occur due to carelessness of human being and are not intentionally done. This occurs due to burning of match sticks, bedi, cigarette, wooden torches etc. Sometimes the throwing of glowing coal pieces from the train may also cause fie. The forest camping and fire camping of tourist people may also create accidental fires.
- **(iii) Deliberate or intentional fires:** It is a type of fire caused deliberately with the intention of certain objectives like collection of honey, preparation of charcoal, scaring wild animals, enemity with the forest department, collection of non timber forest products and induction of new shoots of grass after fire.

(b) Place of occurrence: Based on place of occurrence and action the forest fires are classified into

- **(i) Creeping fire:** It is a fire which spreads slowly over the ground with low flame. This fire normally occurs in forest flow covered with the layer of dry leaves which burns slowly in the absence of strong wind.
- **(ii) Ground fire:** It is a type of fire which occurs on the ground and burn the carpet of herbaceous plants and low shrubs which covers the soil.
- **(iii) Surface fire:** It is a type of forest fire which burns the not nearly the ground cover but also undergrowth.
- **(iv) Crown fire:** It is a type of forest fire which spreads through the crowns of trees and damages all or part of upper branches and foliages. This type of fire is more predominant in coniferous forests.

Measures to prevent forest fires

(a) **Direct measures**

(i) **Establishment of Shelterbelts and fire breaks:** The establishment of shelterbelts and fire breaks along with silvicultural measures can help to prevent the occurrence of forest fires. A vegetated fire break is a strip of land about 100-300 metres wide which is overgrown with fire retarding trees, shrubs and fire resistant groundcover plants will help to prevent the occurrence of fire.

(ii) **Fire fighting reservoirs:** Water is still the main way to extinguish forest fires. Therefore in large, contiguous and fire endangered forested areas it is necessary to have, or build and maintain, a fire fighting water supply system within suitable water courses or to create artificial reservoirs for water extraction. It is important that these extraction points are sufficiently identified and easily accessible by fire engine

(iii) **Forest Fire Monitoring:** Since the introduction of automatic, camera supported forest fire observation systems, the number of forest fires has not reduced but their extent has. The early and exact identification of forest fires and a fast notification has ensured that the technical team can be onsite quickly and can quickly control the fire.

(iv) **Controlled burning:** It is one of the silviculture activity which will help to destroy the leaves and under growths through controlled burning practices. This can be done at the beginning of dry season before the leaves and undergrowth gets dried completely. Through this activity all the inflammable material in the forest get destroyed which will help to prevent spreading of fire.

(v) **Slash disposal:** Fire normally occurs due to the presence of unusable residues after logging like branches, lops and tops, leaves, uprooted stumps, roots etc. which are lying on the ground and may act as a debris to initiate fire. Hence slash disposal is one of the silvicultural treatment which will help to remove the unutilized slash and help to prevent the occurrence of forest fire.

(vi) **Creation of fire lines:** Fire line is the cleared permanent fire break intended to prevent fires from crossing one area to another. It is already an existing practice in most forest area but due to non maintenance it also acts as a cause for fire occurrence and spread. Hence new fire lines have to be created and the existing firelines have to be maintained in order to prevent the occurrence of fire.

(b) **Indirect measures:** There are certain indirect measures which will help to reduce the risk of fire on a sustainable basis. The following are the indirect measures which will help to prevent the occurrence of forest fire.

(i) Obtain always the goodwill of the people living within the forests and forest boundaries

(ii) Educating the people regarding the causes and effects of forest fire will help to prevent the occurrence of fire in the long run

(iii) Imposing restrictions on collection of some of the forest products like horns and bones of wild animals, collection of honey in unscientific manner, charcoal production etc. will help to prevent the occurrence of forest fire.

(iv) Extending rewards to the people associated with forest fire monitoring and control will also help to prevent the occurrence.

(v) Fixing sign boards and extending notices regarding fire prone areas causes and effects will help to prevent the occurrence of forest fire.

3. a. Describe the extent, method of cultivation and e-ects of shifting cultivation. Suggest some suitable alternatives to shifting cultivation (2010)
b. What is shifting cultivation? How it is practiced in hilly areas? (2005)

Shifting cultivation is a land use system in which the forested land is cleared and cropped for a few years, and then left for the regeneration of the bush (fallow period), the duration of the fallow period being much longer than that of the cropping period.

Extent of shifting cultivation: Shifting cultivation is practiced in the entire north eastern states viz., Assam, Mehalaya, Manipur, Nagaland and Tripura and the two union territories of Arnunachala Pradesh, Mizoram. It is also practiced in by Madhya Pradesh, Uttar Pradesh, Andhra Pradesh, Bihar, Orissa and Karnataka states. It is commonly called as Jhum cultivation in north eastern states and podu in Andhra and Orissa. It is estimated that an area of 2.8 million hectares of land is available for shifting cultivation in the north eastern region in comparison to that of 6.76 million hectares of land available in the whole country. Over 5.4 lakhs tribal families of north eastern region are involved in this shifting cultivation process. In Orissa an area of 5.29 lakh hectare and in Andhra Pradesh and area of 1.03 lakh hectare were utilized shifting cultivation in a year.

Method of shifting cultivation

(i) Traditional shifting cultivation: This has been practiced traditionally by tribal people whose lives and religions have been closely tied to the natural systems around them.

(ii) Shifting cultivation imposed by necessity: The cultivator is not linked to a tribal community and usually owns no land or not enough land to support permanent agriculture. Therefore, the cultivator often clears forests for cultivation and starts a cycle of shifting agriculture. This type of shifting cultivation can also be called land, hunger shifting cultivation.

Impact of Shifting cultivation

(a) Degradation of Soil Fertility: Shifting cultivation results in huge lose of soil and soil fertility and contributes a large volume of carbon to the atmosphere. It ultimately sustains on natural regeneration of soil fertility. It is an important period since the natural processes takes a minimum period to restore the cultivated land to its original fertility level.

(b) Deforestation: It has a profound influence on the structure and function of human habitat locally as well as globally and also the largest reservoir of flora and fauna. The cause of deforestation is the extension of shifting cultivation which has been identified as one of the principal causes of deforestation in most of the tropical regions. Loss of natural forest ecosystems always results in loss of biodiversity and often in a range of environmental impacts.

(c) Loss of Biodiversity: Environmental protection, sustainable development, good management of natural resources and conservation of biodiversity are some of the emerging subjects of growing public concern in the world. Increasing attention is drawn on the issues of human impact on the fast depletion of biological diversity. The earth is a home of rich diverse and array living organisms where genetic diversity has relationships with one another and with their physical environment constitutes rich biodiversity. Moreover, the ethical, aesthetic, spiritual, cultural and religious values of human societies are the integral part of these complex resources.

The inception of modern civilization and increase of population and continuing practice of traditional method of cultivation bring changes in the environment. The physical alteration of forest and jungles is one of the sole factors for the great loss of biodiversity. The felling of trees and forest permanently has brought great loss and the recovery would take many years.

(d) **Forest Fire:** Forest fire is common in all the areas under shifting cultivation as burning of Jhum fields sets in during the dry seasons which are susceptible to fires. There is considerable loss of forest cover and its product due to frequent burning of forest. It has a wide range of negative environment impact. These immense fires are visible from space, disrupt air travels and increase carbon-dioxide adds to the atmosphere, treasure house of biodiversity. The burning of vegetation in forest cover release large quantities of smokes into the atmosphere which impact the ecological environment by arising in climatic variations of an area.

(e) **E3ect on Hydrological Cycle:** Deforestation that results from the cause of shifting cultivation interference with the rain as vegetation plays a vital role in the rain cycle. The mass destruction of forest cover with forest canopy gaps lead to deforestation and this limit in rain formation due to low limited evapo-transpiration. It is undertaken as one of the primary sources for water vapour during the formation of cloud. The water catchment areas also dry out when forest cover are cleared which results into droughts.

(f) **Land Degradation:** The productivity of land deteriorates easily in the areas where shifting cultivation is prominent. In India, people of the Eastern and Northeastern regions heavily practiced Jhum on hill slopes. With the increase land use, the cycle of cultivation is affected. It has been observed that, Jhum cycle has been reduced from 10-15 years now to 8-10 years. In some ranges in the district it has come down to 5-7 years. Due to the reduction in cycle, the resilience of ecosystems has been broken down and the land falls into deteriorating condition. Under this, the land is deteriorated with more vulnerable to soil erosion and loss of soil fertility.

Alternatives to Shifting Cultivation

(i) **Livestock based livelihood system:** Shifting cultivation is normally practiced to meet the livelihood requirements of food, fuel, fodder etc. Hence if a suitable and alternate livelihood system is introduced then people may not practice shifting cultivation. For this purpose, the integration of livestock department will help to establish livestock based livelihood system which will help to control shifting cultivation.

(ii) **NTFP based livelihood option:** The forest dependants have to be educated scientifically toward collection, processing and value addition of various non timber forest products. Forest harbours over thousands of species accredited with wide range of products like fibre, floss, edible fruits, oils, gums and resins, medicinal plants etc. If these products are collected scientifically which will facilitate sustained availability of these resources which will help to extend significant income and employment generation activities and the people avoid practicing shifting cultivation.

(iii) **Promotion of agroforestry models:** The shifting cultivation involves clearing forests and then cultivation of annuals. Instead of this system, the local dwellers have to be trained to adopt agroforestry models in order to generate both wood and food from the same unit area which will help to provide food to the people and wood for commercial activity. Hence promotion of agri-silviculture, silvihorticulture and silvi-pasture models will act as an excellent alternate land use system and curtail the activities of shifting cultivation.

(iv) **Introduction of high yielding varieties:** The introduction of high yielding varieties in the agroforestry system will help to increase the productivity and profitability. The tribal communities should be exposed to the state of art technologies of new varieties and the benefit released will help to resolve the issues in shifting cultivation.

(v) **Promotion of seed villages:** The shifting cultivators have to be trained with adequate skill and expertise in order to produce certified seeds without depending on agriculture through shifting cultivation. The identification and practice of seed village concept will help the shifting cultivators to shift to new land use pattern which may require less area with more income generation activities.

4. Discuss a-orestation of inland sand dunes by giving their distribution, site con - ditions, planting techniques and species suitable in such areas. (2012)

Sand dunes: It is a specific type of bed form created when a fluid moves across a deformable surface of non-cohesive sediments.

Geographically sand dune is a hill of sand built by aeolian process.

Distribution: Sand dunes are distributed along the coastal and dessert areas. It is estimated that, in India the sand dune occupies to an area of 22.7% of the total geographic area especially in the four states of Rajasthan, Gujarat, Punjab and Haryana. The drifting and shifting of sand dunes cover the adjoining the states as well.

Site Conditions:

- It is often dry and lacks organic matter, has very low water retention properties, and as soon as it stop raining, it rapidly dries out.
- It is sometimes affected by seawater, at least at high tide when the sea is rough. The salts in the seawater cause a desiccating effect as a result of osmosis.
- It has an unstable surface which makes seedling establishment difficult.
- It is nutrient deficient because it lacks the organic matter which is present in most soils and provides a pool of nutrients that is continually liberated by micro-organisms.
- It also has low cation exchange capacity (the ability to retain cations) which, in most soils is provided by organic matter and clay.

- Areas in which sand dune occurs is characterized by severe wind erosion, high temperature, scanty & erratic rainfall, frequent droughts, high rate of evaporation, devoid of perennial river, sparse vegetation.

Planting Techniques

Suitable Species for sand dune a3orestation

	Rainfall zones of 15 cm to 30 cm	**Rainfall zone of 30 cm to 40 cm**	**Rainfall zone of 40 cm and above**
Trees	*Prosopis juliflora* (Vilayathi babool), *Acacia tortilis* (Israeli babool), *Acacia Senegal* (Kumat)	*Acacia tortilis*, *Acacia Senegal*, *Prosopis juliflora*, *P. cineraria* (Khejri), *Tecomelta undulata* (Rohida), *Dichrostachys glomerata* *Casuarina equisetifolia.*	*Prosopis juliflora*, *Prosopis cineraria*, *Acacia tortilis*, *Acacia Senegal*, *Acacia nilotica sub-sps indica* (Kikar). *Tecomella undulate* *Parkinsonia aculeate* *Albizzia lebbek* (Siris), *Dalbergia sissoo* (Shisham) *Dichrostachys glomerata* *Colophospermum mopane*
Shrubs	*Calligonum polygonoides* (Phog) Zizyphus sps. *Calotropis* spp.	*Zizyphus sps.*, *Calligonum polygonoides*, *Ricinus communis* (Arandi) *Cordia gharaf*	*Ricinus communis* *Zizyphus sps.*, *Cassia auriculata etc*
Grasses	*Lasiurus sindicus* (Sewan)	*Lasiurus, sindicus*, *Cenchrus ciliaris* (Dhaman), *Cenchrus setigerus* (Kala Dhaman), *Saccharum rnunja* (Moonj).	*Cenchrus ciliaris* *Cenchrus setigerus* *Panicum antidotale* (Gramna), *Saccharum munja*

5. Briefly explain the Monteith formulas for biomass estimation of plants (2012)

Monteith established a robust linear relationship for estimation of biomass of plants. He has used the concept of radiation use efficiency (RUE), which can be empirically estimated as the slope value from the linear regression of accumulated biomass over accumulated radiation intercepted (RI).

RUE represents the conversion of radiation energy into biomass and it indicates the amount of dry biomass (DBM) produced per unit of radiation energy intercepted by the crop canopy. The RUE value for crops under non limiting condition is about 1.4 g MJ^{-1} or approximately

2.8 – 3.2 gram MJ^{-1} of photo-synthetically active radiation (PAR). As per Monteith framework the accumulated dry biomass over a period of time interval is written as

$$DBM_1 = \int_0^t RI_t \, x RUE_t x \, dt$$

Where *DBM* – Accumulated dry biomass

RI – Radiation intercepted

RUE – Radiation use efficiency

T – Time interval

6. a. Discus in detail the importance; success and limitations of the 'Project Tiger' (2014)

b. Tiger census raises question mark in Indian wild life management. Give your view points. How can you e-ectively tackle the problem. (2007)

Importance of Project Tiger:

- Project Tiger scheme has been under implementation since 1973 as a centrally sponsored scheme of government of India.
- Project tiger was started with the objective of protecting tiger.
- Tiger is at the apex of the food chain and it should be ensured that safety of tiger is protected which in turn will protect like its preys and other carnivores like leopards, wild dogs, hyenas as these animals share the same prey base.
- The healthy herbivores population is an indication of healthy ecosystem. So tiger was chosen to protect the entire ecosystem. The decision to protect tiger was a wise decision as it shall automatically protect the entire ecosystem of a forest.
- To ensure a viable population of tiger in India for scientific, economic, aesthetic, cultural and ecological values and to preserve for all time, areas of biological importance as a natural heritage
- This project will help the people in terms of recreational and esthetic nature of habitat besides educating the people and extends enjoyment to the people.

Success and achievements of Project Tiger

- The project initially with 9 tiger reserves, covering an area of 16,339 sq.km., with a population of 268 tigers.
- At present there are 27 tiger reserves covering an area of 37761 sq.km., with a population of 1498 tigers.
- This represents to almost 1.14% of the total geographical area of the country. Project Tiger is undisputedly a custodian of major gene pool.
- Tiger Reserves are constituted on a 'core-buffer' strategy. The core area is kept free of biotic disturbances and forestry operations, where collection of minor forest produce, grazing, human disturbances are not allowed within.

- However, the buffer zone is managed as a 'multiple use area' with twin objectives of providing habitat to the wild animals from the core conservation unit, and to provides specific eco development inputs to surrounding villages.
- The project tiger has resulted in increase in population, habitat improvement and the associated ecosystem development.

Limitations of Project Tiger: The Project Tiger experienced the following limitations.

(i) Demarcation of core and bu3er zone: The major limitation of the project tiger programme is the demarcation of core and buffer zone. In most of the project tiger areas the delineation of core and buffer zone is a major problem and in several cases buffer zone activities are carried out within the core zone area and affect the habitat.

Eg: Mudumalai Tiger Reserve is one of the classical example of this issue wherein tourism activities are still carried out within the core zone area and hence the major objectives envisaged in project tiger is destowed.

(ii) Relocation and compensation of forest dwellers: Another major limitation experienced in the project tiger is relocation of the forest dwellers and distributing the compensation. In several cases, the original forest dwellers are resistant towards relocation and they are not vacating the project tiger areas. This conflict is existing in almost several project tiger areas. Similarly the compensation both in terms of quantity and dispersement is still a problem because the compensation offered by the Forest Department has not been accepted by the forest dwellers and hence conflict has arisen.

Eg: Sathyamangalam Tiger Reserve in Tamil Nadu

Sariska Tiger Reserve in Rajasthan

In both these tiger reserves relocation and compensation is a major issue and because of that the people have not been relocated and conflict exists.

(iii) Tiger census method: The tiger census method followed in most project tiger areas are still questionable and the scientific validation of the methodology followed has not been well established which lead to wrong projection of tiger population. This can be tackled only by total direct count or enumeration assistant by Camera trap method. Deploying the unmanned drones are also possible to exactly quantify the populations. Some tiger reserves have low population recorded and some tiger reserve have high population.

Example like Sariska Tiger Reserve in Rajasthan has recorded low population and Ranthambore tiger reserve has recorded high population.

Relocation of over populated tiger to less populated area with due care on protection part and 24 × 7 monitoring through radio caller is advisable to tackle the tiger population.

7. a. What is the role of rotational and controlled grazing to overcome damage due to grazing animals? (2009)

(i) In the present context, what is the role of rotational and controlled grazing to overcome damage due to grazing animals? (2006)

(ii) Does rotational grazing have an advantage over controlled grazing in the smooth management of forest ecosystems? Give your reasons for and against both (2015)

Rotational Grazing: Rotational grazing is the use of different parts of the grazing area or pasture in an orderly sequence based on the carrying capacity of the grazing land.

It is the practice of moving grazing livestock between pastures (often called *paddocks*) as needed or on a regular basis. There are many approaches and types of grazing that fall under the broad umbrella of rotational grazing. The simplest is moving livestock between paddocks every set number of days viz., two days, one week and one month depending on the carrying capacity of the grazing land.

Well-managed rotational grazing means that the area should have the required nutritional and forage needs of the animal based on forage quality and quantity and also involve regulation and controlling of grazing animals.

Controlled Grazing: It is one of the grazing system wherein the conservative grazing through controlling the period and incidence by the movement of livestock in different parts of the area in a prescribed sequence. This is again based on the carrying capacity of grazing land and the associated management objectives prescribed.

Advantages of rotational grazing over controlled grazing

- It helps to allow the regenerations very easily and hence animals can be rotated in a period time.
- It helps to improve the soil quality and thereby enhances productivity
- It helps to allow more number of species under regeneration and helps to act as a alternate species for grazing.
- It helps to plan the possibility of increasing or reducing the number of animals based on the productive and carrying capacity.
- It also helps the animals to find wide range of species amenable for fodder quality.

Disadvantages of rotational grazing over controlled grazing

- Several areas have to be managed simultaneously which involve wide range of managerial and financial resources.
- Managing the animals in different areas through rotational system is very difficult to control particularly due to the involvement of both domestic and wild animals.
- The rotational grazing also facilitates regeneration of non fodder trees which may sometimes eliminate the native fodder species due to competition and fodder species may not be available to the animals intended through rotational grazing.
- During the process of translocation there are chances of animals getting new diseases and may result in death of animals.
- There are chances of spreading diseases across biotic system through spreading of diseases from one location to other location.

8. What are the causes of the declining vulture population in India and what are the likely adverse ecological impacts of such decline? (2009)

Vultures are one of the major components of food chain and they are recognized potential omnivores and manages and stabilizes the food chain. But due to urbanization, industrialization coupled with the recent developments in science and technology have caused serious decline in vulture population and pose a threat to the ecosystem.

Causes of declining vulture population

(i) **Electrocution:** Due to urbanization and industrialization, electrification has gained significant attraction and almost all the villages and cities are now well connected. But this electrification and the associated electrocution from overhead electricity wires is one of the major cause for declining vulture population. Physically the vultures are large bird and has expanded wings which during flight movement are caught under electrocution and the population has declined.

(ii) **Food Shortage:** One major reason for decline of vulture population is shortage of food and food resources. Normally the vultures are omnivores and depends on dead animals for their food. However the government policies and directions like Gram Swachata Abhiyan and Swatch Bharath have made non availability of dead animals. The livestock carcasses have to be buried and they should not be left outside which ushered in lack of food material. This has led to a major reduction in vultures population.

(iii) **Diclofenac:** Use of the chemical may be one of the causes for the vulture deaths. The chemicals Diclofenac poisoning is one of the cause for vulture population declining. Though the chemical has been banned but still it is acting as one of the cause for death of vultures. This Diclofenac poisoning causes a disease called Gout which lead to organ disfunction and ultimately death of vultures.

(iv) **Malaria:** Vultures are also infected due to Malaria and causes death of vultures.

(v) **Pesticides:** Many farmers spray cattle carcasses with the pesticide such as Organo-chlorine and Organo-phosphorous to prevent them from spreading foul odour. This pesticide infested carcass may be eaten by the vulture leading to its death.

(vii) **Lack of Nesting Trees:** Some species of vulture like the Oriental White-backed Vulture prefer to make their nest on coconut trees. But when the trees are chopped off, the vultures lose their home. Some trees may also wither away due to the frequent vulture droppings on the trees, and therefore many times they are cut even though they are healthy.

(viii) **Establishment of Mobile towers:** It is also believed that increasing number of mobile towers and the associated radiation emitted from these towers are also acting as one of the reasons for declining of avian population in general and vultures in particular.

(ix) **Increasing air tra3c:** It is also believed that the increasing number of air travel across the country resulted in more accidents of vulture populations which have lead to death of these birds. The growing number of aeroplanes coupled with increasing air traffic and the associated noises also contributed towards housing, breeding and populations of vultures.

Ecological impact of Vulture decline: Vultures are nature's most successful scavenger and provide an array of ecological, economical and cultural services. The decline in vulture population will have serious implication on the ecology and the associated ecosystem.

(i) The loss of vultures can allow other scavengers to flourish which could bring bacteria and viruses from carcases into the human cities.

(ii) The decline in vulture population may change the community composition of scavengers at carcases and an increased potential for disease transmission between mammalian scavengers.

(iii) The decline in vulture population may spurt the population of feral dogs and the chances of increasing the intensity of rabies are high. There is also a high risk of rabies outbreak in the absence of vultures.

(iv) Wild life Scientists have reported that absence of vultures may create resurgence in anthrax among human population. Animal anthrax remains endemic in India due to inadequate vaccination and lack of cooperation among farmers which have kept livestock susceptible to the infection. The association of physicians of India reported that the incidence of human anthrax is on the raise which is linked towards the decline in vulture population.

Hence the decline of vulture population is a very big threat to the scavenging activity and may change the composition of the community which causes serious disease to both humans and animals.

9. a. What are the functions of "Indian Board for Wildlife" with regard to conservation of wildlife? (2010)

b. Write the constitution of National Board of Wild life. Explain the provisions of the sections of wildlife act, 1972 used to declare an area as sanctuary. (2011)

Functions of Indian Board of Wildlife

- The National Board for Wild Life is a "Statutory Organization" constituted under the Wildlife Protection Act, 1972.
- The board is "advisory" in nature and advises the Central Government on framing policies and measures for conservation of wildlife in the country.
- It is a very important body because it serves as apex body to review all wildlife-related matters and approve projects in and around national parks and sanctuaries.
- The primary function of the Board is to promote the conservation and development of wildlife and forests.
- It has power to review all wildlife-related matters and approve projects in and around national parks and sanctuaries.
- No alternation of boundaries in national parks and wildlife sanctuaries can be done without approval of the NBWL.
- The board ensures the protection and conservation of wildlife in the country through necessary directions, policy and legal issue.

Composition of the board: The National Board for Wildlife is chaired by India's Prime Minister and its vice chairman is Minister of Environment. Further, the board is mammoth body with 47-members including Parliament Members, NGOs, eminent conservationists, ecologists and environmentalists, Government secretaries of various departments, Chief of the Army Staff, Director General of Forests, tourism etc. The members include 15 non-government members, 19 ex-officio members and 10 government officials such as secretaries. There is a standing committee of the board which approves all the projects falling within protected wildlife areas or within 10 km of them; but notably, law does not specify the formation of the standing committee. The rules under the wildlife law allow the board to take decisions with half the members but are silent about the standing committee.

Declaring an area as Sanctuary as per 1972 Act: The 1972 wild life protection act under the provisions of the chapter IV (protected areas) in the section 18, indicated that the state government may by notification, declare its intention to constitute any areas other than an area comprised within any reserve forests or the territorial waters as a sanctuary if it considers that such area is of adequate ecological, faunal, floral, geomorphological, natural or zoological significance, for the purpose of protecting, propagating or developing wild life or its environment.

The notification referred to the sub section (1) shall specify, as nearly as possible, the situation and limits of such areas.

For the purpose of this section, it shall be sufficient to describe the areas by rods, rivers, ridges or other well-known or readily intelligible boundaries.

10. How severe is the damage of Ganoderma in Indian Forests? Discuss with some case history. (2011)

Indian Forestry has witnessed severe damages due to *Ganoderma* and the issues related to damages are furnished.

The fungus *Ganoderma* has been recorded in almost 144 tree species of Indian forests and causes major damage like root rot and reduce the productivity which ultimately lead to death of the plants. The major pathogens in *Ganoderma* are:

(a) ***Ganoderma lucidium***: This organism has been recorded on 91 species and causes most wide spread diseases.

(b) ***Ganoderma applanatum***: It is recorded in about 30 different tree species.

Precautions in Control of *Ganoderma*

- A thorough understanding of the etiology and epidemiology of the diseases on different hosts
- Clarifying current ambiguity in species names
- Assessing the inter-relationships between populations of *Ganoderma* on different hosts
- Developing tools for early detection of diseases in important crops.

Case Studies

(i) **Root rot in Khair:** The root rot caused by *Ganoderma lucidium* is a very serious disease in several Khair plantations which includes coppice plantations as well. Generally *Ganoderma* is endemic in natural forests, it infects the plantations grown species as well. There is a poor development of lateral roots in the early stages of infection. The infected plants are killed in the first few years of plantation and their small mass of roots disintegrates quickly with no effects in the further spread of disease. The increasing age has the chances of lateral growth contact and the infection may occur through root and injure the plantations.

Control of Khair Ganoderma

- Mechanized soil working and removal of residual roots and stumps are the potential preventive and control measures.
- Creation of isolation trenches may help to prevent root contact and arrest the spread of diseases. The trench with 1 m deep, 1.5 m long and 0.5 m wide are preferred.

(ii) **Root rot in Eucalyptus:** Several species of Eucalyptus are also infected due to *Ganoderma lucidum*. Since Eucalyptus is a coppiced crop once the plantation is infected the organism will be available across the coppiced rotations. The residual stumps, stools and roots act as infection centers. To control the disease the infected trees may be isolated by digging isolated

11. How grazing is managed for the browsing animals? Discuss various management options for sustainable grazing in forest. (2011)

Browsing animals are those animals which feed on twigs or shoots with or without leaves of shrubs, trees and woody climbers.

Eg: Elephants, camel, goat etc. The grazing management of browsing animal is different from grazing animals and hence suitable methods have to be established for sustainable grazing in forest area. The following management options are essential towards managing browsing animals on a sustainable basis.

Grazing Management: Grazing management involves manipulation of the soil-plant-animal complex to achieve the desired results. Thus, it is important to recognize that it is the system that is being managed, not just the livestock or the forages alone. The pieces must act in concert with each other.

Grazing management can be made more intensive or more extensive. Intensive grazing management uses additional inputs of resources, labor, or capital to increase livestock production per acre or per head or to improve forage production and utilization. Extensive grazing management uses lower optimum inputs of labor, resources, and capital to achieve profitable and sustainable plant and animal production.

Grazing Management Options: Grazing management is very import and the browsing animals can be managed through the following options viz., continuous grazing, development of tree fodder bank, rotational grazing, deferred grazing, creep grazing, strip grazing and limit grazing which are discussed below.

(i) **Continuous stocking:** This is the simplest grazing method and is almost certainly the most commonly practiced for grazing animals. However for browsing animals incorporation of native and wild fodder trees will be essential and continuously stocking the fodder trees will be significant towards sustaining the grazing management system.

(ii) **Development of tree fodder bank:** Unlike grazing animals, the browsing animals do not graze on the grassland and they always get the feeds from the trees. Hence establishment of tree fodder bank deploying major fodder trees like *Ficus bengalensis, Ficus religiosa*, Acacia, Melia, Hardwikia, Subabul, Glyricidia, Sesbania etc. which are amenable for lopping are essential for sustainable grazing.

(iii) Rotational stocking: This method is commonly referred to as "rotational grazing" although animals are actually stocked on the pasture on a rotational basis. Under this system the grazing area is divided into several small "paddocks". Animals are concentrated on these paddocks for relatively short periods of time with the ultimate goal being uniform and efficient utilization of forage species. The number of paddocks can vary from 2 to over 40. Large numbers of paddocks improve control of grazing and animals, but increases input costs and labor. In general, 8 to 12 paddocks provides sufficient utilization efficiency and rest periods for most forage and animal systems. Some operations may benefit from more paddocks- particularly when multiple forage species or herds are grazed.

Grazing period varies according to number of paddocks and range from 14 days to less than 1 day. Following the grazing period animals are moved to another paddock for grazing and the previously grazed paddock is allowed to rest and re-grow. This system minimizes the amount of individual animal diet selection and can reduce individual animal performance. However, the improved forage utilization normally allows increased stocking rates and increased animal gain per acre. In addition, the rest periods enable less grazing tolerant species like endophyte-free tall fescue, orchard grass and native species to persist for longer periods of time. This method also allows a large amount of producer flexibility. During periods of rapid forage growth some paddocks can be deferred from grazing and used for hay production. This can be an excellent system for beef cow-calf producers, particularly when cool season perennials are grazed.

(iv) Creep grazing: Creep grazing is essentially identical to traditional creep feeding young animals, except that forages are grazed in place of grain feeding. This method allows young animals with high nutrient requirements to access high quality forages like pearl millet, chicory, grazing tolerant alfalfa or winter annuals. Access to these high quality paddocks is provided either underneath electric fences or through a creep opening (See Southern Forages for gate design). Dams are maintained on traditional perennial base forages like tall fescue or Bermuda grass and prevented from grazing high quality forages. This is a system that offers excellent potential to improve weaning weights and should be utilized more often in beef cattle operations

(v) Strip grazing: This is a self-descriptive grazing term where animals are held in small areas by a movable electric fence graze and normally graze a one or two day forage supply. Once this 'strip' is utilized, the front fence is moved forward in the pasture. A back fence may or may not be used depending on circumstances. Due to lack of forage selectivity, performance of animals with high nutrient requirements will likely be depressed when strip grazing is used. Labor requirements can also be high for strip grazing. This method works particularly well when dry cows are grazing stockpiled forages as it typically forces high forage utilization rates.

(vi) Limit grazing: Where animals are allowed limited time in certain paddocks. This method is typically practiced when animals are grazing a base paddock containing low quality forages (like dormant Bermuda grass or low quality hay) and are allowed periodic access to high quality and high cost pastures (like winter annuals). This is an extremely effective practice where animals limit graze a pasture for a few hours per day or on an alternate day basis. Advantages include decreasing intake of high quality forages to more effectively "balance" animal nutrient requirements (particularly with mature animals grazing winter annuals). Decreased pugging or trampling of winter annuals also improves forage utilization. This method can be used with summer annuals to improve cow condition in mid-summer

12. Describe the importance of pug marks and illustrate sex di-erentiation based on pug marks. How do these help in forest management system? (2012)

Importance of Pug marks

- Pugmarks are easy to find indirect evidence of an animals presence.
- With some basic training and an honest approach, the information contained in the pugmarks can be easily recorded through traces and plaster casts for analysis at a higher level.
- If analyzed skillfully and honestly, pugmarks can provide reliable data of:
  - ♦ Presence of different species in the area of study.
  - ♦ Population of large cats.
  - ♦ Sex ratio of large cats.
- Identification of individual animals. The greater the skill, the greater the accuracy. This forms the basis for estimating tiger and leopard populations estimating tiger and leopard populations.

Sex di3erentiation of pug marks

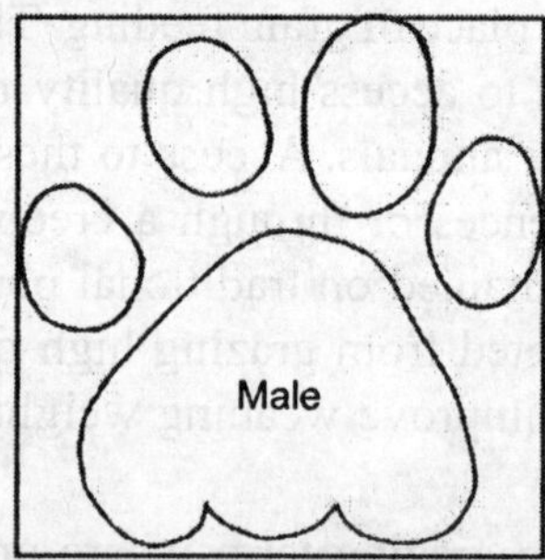

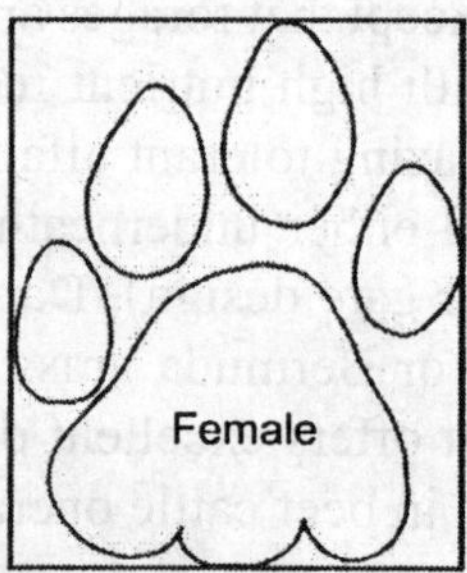

Use of pugmark in forest management system

(i) Estimation of the number of individuals of a species in a population is an important function in the field of ecology and wildlife conservation.

(ii) Pugmarks are also used for tracking rogue animals which may be a danger to mankind or even to themselves because of injuries etc.

(iii) It is possible to make an accurate identification of species, sex, age and physical condition of an animal by those trained in the field.

(iv) Compared to any other method of obtaining data on population of large carnivores, 'Pugmark tracking' is quick, involving about 10 days of ground preparation, 6 days of rigorous data collection, and about two to four weeks to analysis the data.

(v) It is very cost-effective or economic, and all money spent in the process goes to local tribal people who act as assistants as they possess the skill to track animals in Indian jungles.

(vi) It results in data which shows which forest beat possess how many, of what sex/age and which type of large carnivore.

(vii) This brings a sense of responsibility among the Guards, as none of the animals is 'virtually' generated through statistical interpretations. Like any study technique, Pugmark tracking also calls for sincerity for true reflection of structure and spatial distribution of the population of large carnivores.

13. Explain with the help of suitable examples the various kinds of population interactions during their growth period and give di-erence between Commensalism and Amensalism of plant relationship. (2012)

In nature, organisms do not exist in absolute isolation but co-occur in a matrix of many species. Thus every organism must interact with the environment and other organisms. In order to survive, an organism must interact with its biotic and abiotic environment and thus it becomes functional part of the ecosystem. Species interaction range from once in a generation (e.g. Pollination) or completely residing within the host (e.g. endosymbiosis). Similarly the effects also varies ranging from consumption of another individual (predation, herbivory, cannibalism), to mutual benefit (mutualism). Interactions can be direct or other indirect through intermediaries such as shared resources or common enemy.

Kinds of Population Interaction: Broadly the population interactions can be divided into three main types

(i) Neutral

(ii) Negative

(iii) Positive

These three type of interactions also depend on the fact whether the two populations are coexisting or not.

(i) Neutral Interaction

Neutralism: The two populations are not affected by each other whether they are coexisting or not. In this interaction the fitness of one species has absolutely no effect on the other. However in nature true neutralism is nonexistent and hence unlikely find any suitable example.Ecosystem itself characterize complex network of interactions so one cannot assert positively that there is absolutely no interaction: positive or negative. Since true neutralism is rare the term is often used to describe situations where interactions are merely insignificant or negligible.

(ii) Negative population interactions: These are the interactions between the two populations in which there is a negative effect of one on the other population. These interactions include competition, amensalism, predation, parasitism and parasitoidism.

(a) Competition: Competition is a type of negative population interaction which arises whenever at least two parties strive for resources which cannot be used in sharing and cooperation. Competition occurs naturally between living organisms which co-exist in the same environment.

(b) **Predation:** Predation is a type of negative interaction which describes the act of hunting (predator) and being eaten up or consumed (prey). Predators may or may not kill their prey prior to feeding on them, but the act of predation often results in the death of its prey. Predators are larger than their prey, whereas parasites are smaller than their host.

(c) **Amensalism:** Amensalism is a relationship in which a chemical compound produced by one organism as part of its normal metabolism has a negative effect on growth or survival another organism when it comes in its contact. It is specifically a population interaction in which one organism is harmed, while the other remains unaffected. Amensalism is converse of commensalism.

There are two basic modes:

(a) **Act of competition:** In this competition, a larger or stronger organism eliminates a smaller or weaker one by denying living space or food near its vicinity. It even prevents the growth of its own seed close to itself.

(b) **Act of antibiosis:** One organism secretes a chemical compound which is detrimental to the other. This is also known as allelopathy. This property has been greatly exploited by humans in the field of medicine. Production of pathogen specific antibiotics and other medicine is the result this biological interaction.

Examples: The bread mold *penicillium*: The bread mold *Penicillium* secretes *penicillin*, a chemical that kills bacteria. This discovery of the bacteria-killing property of penicillin led to its use as an antibiotic in medicine. However, the *Penicillium* itself is not benefitted from the death of bacteria.

(d) **Parasitism:** Parasitism may be defined as the relationship between two organisms of different species in which the smaller one, called parasite has the potential to harm the larger individual called host. In the association, the parasite relies on the host for nutrition or for niche. Parasitism is an interaction between the two species which live together but one drives its nourishment at the expense of other. Parasite is benefitted but the host growth rate declines. A parasite may gradually cause the disease or death of its host.

Eg. Sandal Root Parasite

(e) **Parasitoidism:** A parasitoid is an organism that acts initially as a true parasite and spends a significant portion of its life history in the host. During this stay it may not kill the host. But, gradually it becomes predatory and ultimately kills the host. They are also called necrotrophs. This is different from the typical parasitic relationship in which the parasite and host live side by side without lethal damage to the host. Typically, the parasite takes enough nutrients to thrive without killing the host.

About 10% of the insects are parasitoids. A well-known example of parasitoids is the ichneumon wasps, which live a free life as an adult. They lay eggs on or in another species such as a caterpillar of a butterfly or moth. The eggs hatch into larvae and start feeding on the growing host causing not much harm. But soon they start devouring the internal organs until finally resulting in prey death. By this stage the young wasps are developed sufficiently to move to the next stage in their life cycle. Because of this feature, these wasps are used as agents of biological control of insect pests.

(iii) Positive Interaction: These interactions between the two populations result in a beneficial effect of one population on the other. It may include a series of evolutionary steps of beneficial effect on one population (commensalism) to a non-obligatory beneficial association (proto- cooperation) finally resulting in an obligatory association mutualism

Symbiosis: The symbiosis is used for a relationship in which two or more organisms live together such that both are more successful within the partnership than they would have been if they were living on their own. This term is more commonly used only where both organisms are benefitted from the presence of the other.

Proto-cooperation

a. Commensalism

b. Mutualism

Proto-cooperation: It refers to involuntary or automatic interaction of two species with each other in order to benefit mutually. The interaction is not necessary for their growth and survival but they interact purely for the gain that they receive from doing this. This interaction occurs between different phyla or Kingdom, which are distantly related.

Example:

(a) Ants and aphids: The ants and aphids interact to derive mutual benefit from each other. The trees and shrubs are hosts to honeydew secreting species such as aphids. The ant gathers the honey dew produced by aphids as their food. In turn, ants protect them from their natural predators. Ants also stimulate the honeydew secretion from aphid.

(i) Commensalism: In commensal relationships, one member derives benefit while the other is unaffected. The organism, which derives the benefit, is called the commensal while other is the host. The word commensalism is derived from commensal meaning "eating at the same table" in human social interaction. Originally, the term was used to describe the consumption of leftover food by animals, like the carcass eaters that follow hunting animals. The commensal derives the benefit by obtaining nutrients, shelter, support, or locomotion from the host species but without harming it in the process. The commensal is may be larger or smaller than the host. The host organism is unmodified, whereas the commensal species may show great structural adaptation in accordance with its habits.

(ii) Mutualism: Mutualism is relationship between two or more species, where both the species derive a mutual benefit in terms of increasing their carrying capacity. Similar interactions within a species are known as cooperation. Mutualism may be classified in terms of the closeness of association, which is often confused with symbiosis. Mutualism can be contrasted with interspecific competition where one species benefits at the expense of the other. Each species experiences reduced fitness and exploitation or parasitism. It is an obligatory association known as obligatory symbiosis.

14. Describe the altitudinal variations in flora of Eastern and Western Himalayas through a schematic diagram. (10 mark) (2012)

The altitudinal variation in flora of eastern and western Himalayas is depicted below.

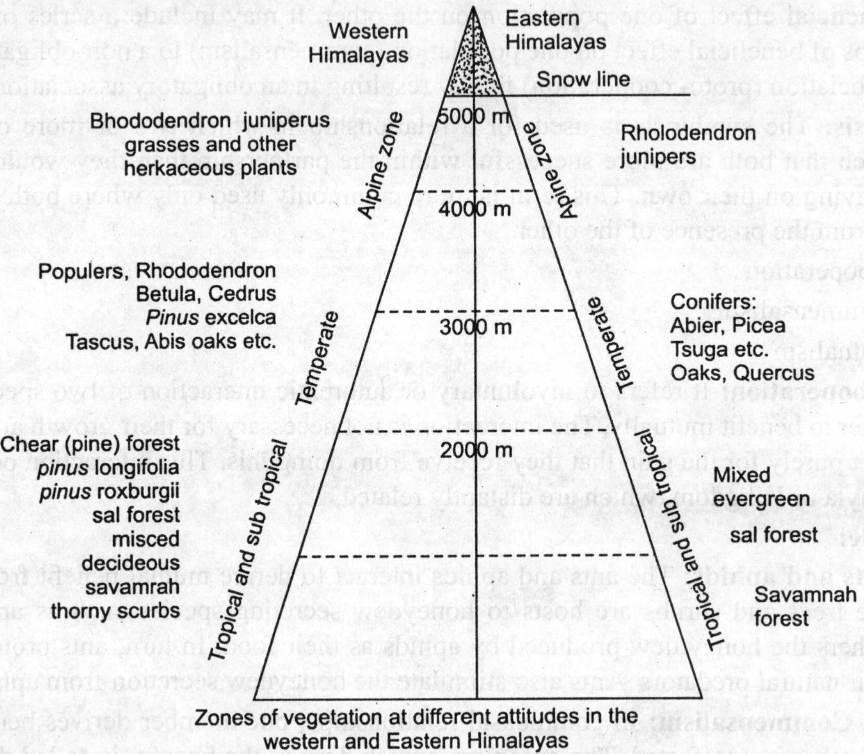

Zones of vegetation at different attitudes in the western and Eastern Himalayas

15. What do you understand by wildlife census? Write di-erent methods of census in brief. (2013)

Wildlife census is the estimation of population of the wild animals which may bring out the following type of data or information:-

- Population size of actual numbers.
- Number per unit of the habit viz population density. These figure may be for the whole area or a part of the protected area such as near water holes and salt licks.

Di3erent methods of Census

(i) Total Counts or Census: A total count or census involves estimation of the population in the entire area of the reserve or protected area. It involves a thorough search of the entire tract and all animals/marks seen are tallied. However this does not mean that all animals in the area have been or are to be counted in totality though this is usually the inference/ assumption that is drawn.

Total counts have the following draw backs or disadvantages.

(a) It is a very difficult method census particularly if it has to be carried out in a large reserve.

(b) When census is being carried out for a small animal it is very difficult to search and count them viz hog dear, porcupine or jungle fowl etc.

(c) The time and manpower required for total count is extraordinarily high. This may lead to the problem of under/over counting dues to the movement of animals during the census operations.

(d) Interpretation of the results of total counts may pose a problem as there is usually no way in which it can be accurately determined whether all animals of a particular species have been counted.

(ii) Road side index for population trends: This is simple technique for monitoring the population trends of animals such as deers, antelopes, goats and sheep. It is designed for a preliminary population survey by vehicles on motorable roads and on foot/ cycle on small tracts and trails. A combination of both vehicle and foot / cycle survey may be used for more detailed results.

(iii) Dung surveys: Another reliable indicator of the presence of animals is their dung or droppings. This had frequently been used for estimating their population or as an index of population abundance.

The following steps are involved in dung surveys for census:

- The observer must be able to correctly recognize the dung of different species. Individual pellets of species like sambhar, nilgai and cheetal are akin. The best way for correct identification is by keeping sample of pellets of different species likely to be found in the area being surveyed in plastic bags. The pellets found on the ground may then be compared with the samples for correct identification.

(iv) Waterhole survey: This method is based on the assumption that in dry areas all animals come to drink water at waterholes which are usually limited in number. The large animals are counted from a hide. The waterhole survey method is best suited for dry and open areas e.g. Sariska Ranthambor and Gir protected area.

(v) Calls for census: Calls of birds and sounds emitted by other animals are used for estimation of their population. Sounds indicate the presence of a particular animal in the locality. However, it cannot be said that if there is no sound, there are no animals in the areas or locality. This technique is primarily used for the estimation of the population of bird species though it is also helpful in the census of other animals. A minimum estimate of the number of individuals of a particular species of birds in an area may safely be arrived at by this method.

(vi) Pug mark count method: This is a widely accepted method for the census of large carnivores in India. This technique has largely been developed in India during the seventies. It involves the use of pug marks to identify individual members of thc cat family so as to ascertain their actual population in a given area or reserve.

(vii) **Single team block census:** This technique is based on the sample count of animals using actual sightings. The entire area is divided into blocks and actual counting is done in only a few sample blocks which may be a forest compartment or unit that is defined by distinctive physical or vegetative features. The area of the sample block should not usually be more than 2 sq. kms. The total area of the sample blocks may be upto 5 to 10% of the total area of the reserve.

(viii) **Beat or line drive count:** In this method a sample block or area is systematically covered by a group of people walking abreast in a line. The sample area is selected and earmarked as described for the single team block census method. However one essential pre-requisite of this method is that the start line should be clearly defined and should be along the long axis of the counting block.

(ix) **King's census method:** This is a simple, flexible and relatively unbiased method for determination of population density. The advantages of this technique are outlined below: It has a considerable flexibility.

- It can be adopted in almost all types of terrain except perhaps in the steep mountains where visibility and movement along a fixed transect is restricted.
- This method is best suited for determining and population density of any animal which does not hide in order to avoid detection.

(x) **Ratio method:** This is a simple method for calculating the population of one animal species based on the reliable population data of another species of similar size and ecology. The number of cattle in a forest area is usually known. This can be used for determining the population of an animal like nilgai in the same area.

(xi) **Total count:** This pertains to a detailed to a detailed count of each and every individual of the target species in the entire reserve and not in a sample block.

(xii) **Aerial survey or census:** The population of many animal species is being monitored by aerial survey in the developed countries.

(xiii) **Camera trap method:** Camera trap is a method of recent origin which helps in animal census. It is a globally preferred method for tracking down the movement of animals by installing cameras deep inside the jungle. This method is reliable that the pictures are automatically taken when the animals cross the camera.

16. Describe the benefits and limitations of chemical and biological control of diseases in forest nurseries and plantations. (2016)

Various management strategies have been used to control plant diseases with various degrees of success. Each method has its own merits and demerits.

Benefits of chemical control

- Chemical pesticides/fungicides are highly effective against pests/diseases that feed directly on crops over a very large area,
- The use of chemical is an effective method that gives immediate control over the target organism

- Immediate knock out of the disease causing organism is achievable
- Different methods of application can be adopted based on the area to be covered

Limitations of chemical control of diseases

- Chemical are highly effective, but must be used safely and economic problems of modern pesticide use must be taken into consideration greatly.
- Use of chemicals is becoming more expensive and undesirable due to environmental hazards.
- Most chemical nematicides are highly toxic to humans and animals and have negative effects on the environment.
- Pesticides kill not only the pests but also the natural enemies of these pests, a problem known as resurgence
- The use of pesticides produces alterations in the ecosystem
- A waiting period is required for about a couple of weeks for sowing or harvesting
- Toxicity or residual effect is found in the economic plant parts.

Benefits of Biological Control

- Biological control is recognized as being more effective against pests that are tolerable enough to only need to be suppressed, not destroyed entirely
- This method is friendlier to the environment and generally more profitable than chemical pesticides.
- It is a long term process which involves the application of naturally occurring organisms viz., *Bacillus subtilis, Pseudomonas fluorescence* and *Trichoderma viride*
- It is less expensive and operational cost when compared to the chemical control
- Endophytic biocontrol organisms persist inside the host plant and provide longstanding control over the pathogens and attack the pathogens in the target site which is not possible in the chemical method.
- Various mechanisms of actions like competition for space, effective root colonization, promising ability to sporulate, plant growth promotion, production of lytic enzymes viz., chitinase and glucanase, production of antibiotics viz., surfactin, iturin and fengycin are potentially involved in the biocontrol process.
- Combination of two or three biocontrol agents is recommended to provide a stable antagonism towards the plant pathogens as one will take care when other fails.
- There are few or no detrimental effect on other organisms or even humans.
- Pests rarely resist biological control.
- Often the results are permanent or for a long time.
- There are no problems of intoxication.

Limitations of Biological Control

- Although less expensive use of biocontrol agents is less effective when compared to chemicals

- They require repeated or inundated application at different time intervals s as to maintain the recommended inoculums level.
- More expertise is required on the biology and the host pathogen interaction
- Application requires more monitoring

An integrated approach on judicious application of chemicals and proper selection of biocontrol agent would benefit in a better crop production

17. Discuss the phytosociological analysis describing formulae. (2009)

In phytosociological analysis, the frequency, relative frequency, density, relative density, relative dominance and important value index are very essential parameters which will help to understand the diversity of species and the population within a species. These factors are described as follows.

Frequency: This indicates the dispersion of a particular species in a community.

The chance of an individual of a given species to be present in a randomly placed quadrats

No. of quadrats in which a species occurs

$$F = \frac{\text{No. of quadrats in which a species occurs}}{\text{Total No. of quadrats sampled}} \times 100$$

where, F = Frequency.

Relative frequency: No. of occurrence of a species

$$\text{R.F.} = \frac{\text{No. of occurrence of a species}}{\text{No. or occurrence of all species}} \times 100$$

where, R.F. = Relative Frequency.

Density: Density is the quantitative expression of abundance. It is expressed as number of organisms per unit area. It is expressed in numbers per sq.m.

Total number of individuals of a species in all quadrats

$$D = \frac{\text{Total number of individuals of a species in all quadrats}}{\text{Total number of quadrats sampled}}$$

Where, D = Density.

Relative Density: The study of numerical strength of a species in relation to total number of individuals of all species and it can be calculated by the following formulae:

$$\text{R.D.} = \frac{\text{No. of individuals of the species in all quadrats}}{\text{No. of Individuals of all species in all quadrats}} \times 100$$

where, R.D. = Relative Density

Relative dominance: Relative dominance is the coverage of the rest of the species in the area.

$$\text{R.D} = \frac{\text{Total basal area of the species in all the quadrats}}{\text{Total basal area of all the species in all the quadrats}}$$

Where, R.D. = Relative dominance

Importance Value index (IVi): The sum of the above three relative measures for a species is an index called the Importance Value (IVi)

$$IVi = RDi + Rfi + RCi$$

The value of IV may range from 0 to 3.00 (or 300%). Dividing IV by 3 will result in a figure that ranges from 0 to 1.00 (or 100%). This value is referred to as the importance percentage. The importance value, or the importance percentage, gives an overall estimate of the influence of importance of a plant species in the community.

18. What are major ecological considerations in a-orestation. (2009)

a. Climate based consideration

- It brings proper rain.
- It reduces Global warming.
- It brings in beauty to the nature.
- It avoids acid rain.
- It is useful in medicine field.

b. Soil based consideration

- It restores ecological balance of all eco systems.
- It act as catchments for the soil and water conservation
- Plating tress can stabilize soils by increasing interception
- Prevent floods and safeguard future of the tribal people.
- Bring soil together and prevent soil erosion.
- Afforestation act as windbreaks

c. Biodiversity based consideration

- It maintains biological diversity
- Afforestation causes fundamental changes in ecosystem structure and functioning, including changes in shading, the micro-climate, production, nutrient cycling and the water balance, all of which may affect biodiversity.
- Conversion from treeless landscapes to forests can have negative impacts on organisms adapted to open habitats
- In general, the cover of vascular plant species, mosses and lichens decreases following afforestation.
- At the same time there is a significant change in the relative proportions of many of the species present.
- These changes following afforestation were mainly found to be due to decreased light at the level of the forest floor.
- The surface invertebrate fauna that characterize heath lands disappear as conifers are introduced and new invertebrate species colonize the area.

19. Distinction between Rotational and Deferred Grazing. (2006)

S.No.	Rotational Grazing	Deferred Grazing
1.	Rotational grazing means that animals are moved from one pasture to another on a schedule basis. The shifting of livestock to different units of a pasture or range in regular sequence to permit the recovery and growth of the pasture plants after grazing.	Deferred Grazing means that the vegetation is not grazed until seed maturity is nearly complete or assured that it is grazed after seed maturity. Deferment in the second year permits establishment of seedlings.
2.	Rotational grazing is perhaps the most rotational form of grazing.	Deferred grazing pertains to delayed grazing.
3.	One unit is being grazed, the other units of the same forest are closed to grazing and are able to put on vigorous growth.	The season of the beginning of grazing may vary from pastures to pastures.
4.	A rest of grazing is beneficial for forest plant.	It is immensely beneficial for plant species which reproduce by vegetative means. It helps in the production of rhizomes and other vegetative parts.
5.	A large number of animals per unit area is possible in rotational grazing	Number of animals depends on the forage availability.
6.	The forage is more uniform.	The forage is not uniform. The beginning of grazing can be delayed once every few years then the palatable species have a better chance of reproducing.
7.	Rotational grazing is beneficial both for animals and for ensuring sustainable fodder production.	Deferred grazing is not always very beneficial for the pastures. It must be adopted carefully and only those measures which are absolutely necessary should be taken up.

20. What are the causes of forest fire? Discuss in brief the damage caused to forest by fire along with its control. (2010)

Forest fires is the most common type of hazard in forests which leads to a tremendous loss of flora and fauna. Various causes of this are:

(i) Natural causes: Lighting sets dry trees or leaves on fires which spread due to wind. High temperature with low humidity and presence of dry leaves increase the chances of spread of forest fire. Friction of bamboo swaying due to high wind velocity

(ii) Man made causes: Cigarettes, beedies thrown in forests, Shifting cultivation, Use of fire to ward off wild animals, lighting fire for recreation in places near to forests, for getting good fodder for cattle etc.

Damage caused to forests

(a) **Damages to habitat of flora and fauna:** Forest fires take away homes, wildlife, as well as vegetation. All of the inhabitants of the wildlife environment now are found homeless. People often lose their houses as well if the fires are close enough to human housing. Vegetation is now obsolete if this area is near a farm or near the food of other inhabitants. Millions of dollars are spent repairing these damages and re-building homes and areas of vegetation.

(b) **Damages to soil:** The soil in the area of the forest fire has been completely destroyed. The soils in the forest are made with decaying nutrients and debris that have a lot of natural ingredients that help make the earth what it is. When a wildfire hits this soil it becomes too hot and all of those nutrients are gone for good.

(c) **Damages to Animals:** Animals lose their lives. It is sad but true fact that birds, squirrels, rabbits, and other wildlife animals are no longer a part of this great earth.

(d) **Damages to Productivity of Trees:** The productive trees will be eliminated due to forest fire. The timber and non timber forest product producing species will be destroyed and hence the total productivity of the forests is reduced.

(e) **Damages to protective power of forests:** Forests in general has an excellent role to protect the soil, water and the associated ecosystem. The forest fire damages the trees and plants and hence the role played by the tree in protecting the soil and water will be a major issue. The soil will be exposed to all degradation and the possibility of water conservation is very meager. Hence the protective capacity of tree species is damaged due to forest fire.

(f) **Damage to environment:** The forest fire will create large amounts of smoke which will be released into the air which makes it difficult to breathe and also causes air pollution. This may in turn cause climate change and the chances of global warming and acid raining will be at alarming rate.

(g) **Damages due to recreational and aesthetic value:** The absence of forests due to forest fire will eliminate the chances of recreational and aesthetic value. This in turn will reduce the ecotourism value generated through this.

Control of forest fires: Forest fire can be controlled by the following methods

(i) **By Water:** Water is still one of the robust method of controlling forest fire. Though in India forest fire control through water is applicable only to small areas, its application in larger areas through water bomb is practiced in most of the developed countries.

(ii) **By Earth:** Forest fire is also controlled by application of earth but it is a time consuming process along with requirement of huge money for excavation of soil. Therefore this method has minimal utility.

(iii) **By Beating:** It is one of most practicable method used to control mild forest fire. For this purpose the brooms are made by cutting branches of shrubs and are used for controlling forest fire through continuous beating.

(iv) **By Counter firing:** This is one of the method widely practiced in Indian forests which is an attempt to extinguish an advancing fire by deliberately burning the forests from the opposite direction. When the fire is advanced from the direction towards the advancing fire, the two fires meet at a point and gets extinguished.

21. State what are the plant parasitic nematodes associated with nursery forest plant species. (2010)

Plant parasitic nematodes: Nematology is an important branch of biological science, which deals with a complex, diverse group of round worms known as Nematodes that occur worldwide in essentially all environments. Forest crops are not exceptions from the noninvasion of nematodes. A good example to substantiate the statement is the prevalence of most serious tree disease, pine wilt disease in the world.

Plant parasitic nematodes are usually known to damage root and above ground plant parts either by sucking cell sap, modifying the cells or rough inter and intracellular movement in plant system for their development and multiplication. These parasites are recognized as causal agents for heavy monetary and plant stand loss in forest nurseries, plantations and natural forest stands by way, of loss of transplanted seedling which leads to the problem of replanting often. Plant parasitic nematodes have a wider distribution on tree crops and have a definite negative bearing on forest productivity.

Symptoms of damage: Infected seedlings show yellowing, stunting and loss of vigour due to lack of water and nutrient transportation. Below ground symptoms include root galling or formation of course roots. As the progress of the symptom occurs gradually wilting and death of trees happen especially in the years of productive age of the crop. The symptom of the nematode even though it is not an observable in a tree of year 19th year, the death of the tree will happen on 20th year" Hence, the study on nematodes affecting forest tree crop is essential besides studying insects and disease.

1. **Root-knot nematode, *Meloidogyne* sp.:** Root-knot nematodes (RKN) are most important parasites of tree crop seedlings in the nursery. Many crops are known to be attacked by either of the four major species of *Meloidogyne* such as *M. incognita*, *M. javanica*, *M. arenaria* and *M. hapla*. They are easily detectable by their characteristic symptoms in the form of galls on the roots of host crops.

Damage Symptoms: Stunted growth, yellowing of foliage and wilting appearance in patch areas are the most common symptoms.

Biology: *Meloidogyne* spp. are sedentary endo-parasites. Second stage juveniles are infective and establish a feeding site (syncytium) involving vascular tissues, endodermis and pericycle cells. The feeding site comprises of 8-12 multinucleate 'giant cells'. The second stage juveniles start swelling and develop to third and fourth stages. After final moult, males acquire vermiform shape and come out into the soil. Females attain saccate shape and lay eggs in a gelatinous matrix (egg mass) on the root surface. Reproduction is generally by parthenogenesis and nematode completes 7- 8 generations in a year

Mode of spread: Seedlings or planting material, irrigation water, farm implements
Management:

- Prophylactic measures like selection of nematode free site for nursery preparation.
- Nematode infested seedling should not be used for planting.
- Nursery beds can be treated with carbofuran 3G @ 3gai/m2. Deep summer ploughing
- Solarization of nursery beds

- Destruction of weeds
- Organic amendments with neem/mustard/ castor cake @ 1 kg/ sq. m.
- Seedling root dip with phorate/aldicarb/carbofuran @ 500 ppm for 1 hr.
- In main field, carbofuran or phorate can be applied spot treatment @ 1.5 kg ai/ha
- Bio-agents such as *Pasteuria penetrans*, *Glomus fasciculatum*, *G. mossae*, *Paecilomyses lilacinus* and *Pseudomonas fluorescens* can be applied as seed treatment, seedling root dipping or soil application in nursery or in plantation.

2. Reniform nematode, *Rotylenchulus reniformis*

Damage symptoms: Reniform nematode causes general growth reduction. It also affects germination and seedling emergence. The affected plants show stunting, decline in vigour, reduction in leaf size, marginal drying of leaves and perceptible reduction in yield.

Management

- Summer ploughing
- Application of organic amendments
- Seed treatment with Pseudomonas flourescens @10 g/kg seed
- Soil application of Pseudomonas flourescens @ 2.5 kg/ha
- Application of carbofuran 33 kg/ha.

3. Burrowing nematode, *Radopholus similis*: Burrowing nematode is the most important plant-parasitic nematode associated with tree crops. Larvae and adults of these nematodes freely move inside the host roots, feeds on root cortical cells and kills the cortical cells by their metabolites. Cell death results in brown lesions on the roots. Lesions begin on one side, but may encircle a root and thereby girdle it. The overall effect is a weak, shallow root system with a lot of dead areas.

Damage symptoms: The nematode infested roots are found with reddish brown lesions. These lesions are clearly seen when an affected root is split longitudinally and examined immediately.

Management

- Deep plough during summer (April –May) kill nematodes
- Selecting nematode free seedlings
- Application of VAM or Pseudomonas fluorescens or Paecilomyces lilacinus in nursery soil mixture.
- In the main field the plants can be treated once in 6 months with carbofuran @ 45 kg/ha.
- Application of neem cake @ 0.5 to 1 kg / plant also reduces the nematode damage.

Other nematode genera: *Longidorus spp*, *Pratylenchus spp*, *Helicotylenchus spp*, *Xiphinema* spp, *Tylenchulus spp*, and *Rotylenchulus reniformis* are found to occur in the plantations of *Melia dubia*, *Pterocarpus santalinus*, *Casuarina* and *Bixa orellana*.

22. a. Discuss di-erent methods of disease control in forest nursery with examples

b. Enlist insect pests and disease in forest nursery methods to prevent and control both of them. (2006)

a. Nursery diseases are very serious in forestry which will decide the success and failure of plantations. The following are the different methods of disease control in nursery.

(i) Cultural / Silvicultural method

(ii) Chemical method

(iii) Biological method

(i) Cultural / Silvicultural method

- Continual monitoring of the seedling crop is very essential in order to identify the disease at the early stage and to take efforts for the controlling disease. The sanitation is an important aspect of managing disease in forest nurseries. The potting media may act as a source for many pathogens and hence the potting media handled properly in order to reduce the disease occurrence.
- Many of the most important pathogens can reside saprophytically on many types of organic matter that may be present within seedling-growing areas.
- Removal of organic matter within greenhouses that may harbor pathogens is very important in reducing disease losses.
- The care and maintenance of propagation chambers, periodical, fumigation will help to develop sterile environment and avoid disease occurrence.
- Seedlings with disease symptoms should be periodically removed from both bare root and container stock to reduce chances for secondary spread of pathogens.
- The seeds used should be free from pathogens and if possible necessary seed production has to be given before sowing the seeds. Always the seeds have to be washed with clean running tap water in order to eliminate the pathogens harbouring on the seed coat.
- Irrigation is another source of infection which has to be judicially used. Some of the diseases caused due to Cylindrocarpon, Pythium, and Phytophthora spp are due to excess application of water. This will lead to rotting of roots and ultimately death of plants.
- Irrigation should be applied only in the morning to allow foliage to dry quickly during the day to help control Botrytis blight.
- Air circulation within greenhouse is important so that seedling foliage can rapidly dry after irrigation to help reduce losses from Botrytis and Fusarium spp., both of which may attack aboveground tissues.

(ii) Biological method

- The available biocontrol formulations (biofungicides) developed for agricultural crops show promise in forest nurseries. These are made up of either fungi or bacteria that are antagonistic towards pathogens.
- Biocontrol formulations are usually applied early in the growing cycle.They are either incorporated into soilless growing media, which is made up of mixtures of peat moss with other organic or non-organic materials, or they can be applied directly adjacent to seed during sowing. Some biocontrol agents are applied to directly to seed prior to sowing.
- A few ectomycorrhizae are also antagonistic toward pathogens. Commercially-available mycorrhizal preparations are available and can be applied several times during the seedling stages.

(iii) Chemical method

- In most nursery disease control chemicals are predominantly used and they are specific to some organisms. For example, metalaxyl only controls oomycete pathogens (*Pythium, Phytophthora*) and is not effective against other pathogens such as *Fusarium*.
- Chemicals should be applied according to label instructions for timing and dosage rates. It is best to rotate the chemical fungicides in order to reduce chances for pathogens to develop resistance.
- Most fungicides are not effective therapeutically; they help prevent pathogen infection and colonization rather than kill pathogens that are already colonizing hosts.
- Fungicidal applications just prior to lifting may be important to preclude fungal development during cold storage.

(b) Major nursery pests and their control

Nursery pests	Control measure
Defoliators	Monochrotophos 0.036%, Endosulfan or Malathion (0.1 – 0.2%)
Grasshoppers	Physical killing, Spraying of Neem oil
Crickets	Malathion (0.25%), Thiodan 35 EC at 1%
Sapsuckers	Dimethoate 30 EC (0.01 to 0.02%), Monocrotophos 36 EC (0.01%)

Major nursery diseases and their control

Nursery disease	Control measure
Damping off	Application of Bavistin 0.1 to 0.2% solution as foliar drench
Leaf spot	Blitox 0.2%
Leaf blight	Blitox 0.2%
Powdery mildew	Sulphur based fungicides or calixin 0.1%
Leaf rust	Diathane M-45 0.5%
Seedling wilt	Application of Bavistin 0.1 to 0.2% solution as foliar drench
Root rot, collar rot and Stem rot	Dithane M-45 0.5% or Bavistin 0.1 to 0.2%, Dithane Z-78 0.2%

23. Describe the mathematical expression for Biotic potential and Environmental resistance. (2010)

(a) Biotic potential (*r*) is capacity of a population to increase. The maximum rate of increase of a population living under ideal conditions.

Ideal conditions exist when conditions are uncrowded.

- It usually is potential and not actual.
- It is also called the per capita rate of increase or intrinsic rate of natural increase.
- It is a measure of the instantaneous rate of change of population size per individual.
- It is the difference between the instantaneous birthrate and the instantaneous death rate.

(b) Quantitative expression

- The biotic potential is the quantitative expression of the ability of a species to face selection in any environment.
- The main equilibrium of a particular population is described by the equation:

  Number of Individuals = Biotic Potential/Resistance of the Environment (Biotic and Abiotic)

(c) Environmental resistance: Environmental resistance refers to the limitations placed on the biotic potential due to unfavorable environmental conditions.

Eg. Overcrowded and predation.

These are the limiting factors which limit the growth of a population. The biotic potential and the associated environmental resistance determine the carrying capacity (K): the maximum population of a given species that a particular habitat can sustain indefinitely without being degraded.

The growth rate of a population decreases as its size nears the carrying capacity of its environment because resources such as food, water, and space begin to declined.

The sum total of the environmental limiting factors, both biotic and abiotic, which together act to prevent the biotic potential of an organism from being realized.

Such factors include the availability of essential resources (e.g. food and water), predation, disease, the accumulation of toxic metabolic wastes, and, in some species, behavioural changes due to stress caused by overcrowding.

The equation summarizes the interaction of biotic potential with environmental resources, as seen in populations showing the S-shaped growth curve, as:

$$dN/dt = rN(N - K)/K$$

where, N is the number of individuals in the population,

t is time,

r is the biotic potential of the organism concerned, and

K is the saturation value or carrying capacity for that organism in that environment.

The resulting growth rate or logistic curve is a parabola, while the graph for organism numbers over time is sigmoidal.

24. Describe the damage caused to the forest by grazing and browsing animals. (2005)

Animals play a major role in damaging forests and the associated ecosystem through their grazing and browsing activities. The damages are attributable due to both domestic as well as wild animals. Hence the damages caused by domestic and wild animals are described separately.

Damages caused by domestic animals

- Uncontrolled grazing in the forest ecosystem affects the crop composition of the locality.
- Continuous and uncontrolled grazing affects the regeneration of young crop thus resulting in poor regeneration and establishment of new crop.
- Continuous and indiscriminate grazing compacts the soil and reduces the soil porosity,

water holding capacity, volume expansion and makes a soil susceptible to erosion and the associated soil loss and poor fertility.

- Some of the domestic animals debark the standing trees and causes death of trees.
- Browsing of apical shoots by some of the domestic animals alters the apical dominant growth of trees which resulting in changes in tree farm and structure.
- Trampling by most of the domestic animals affects damages to the seeds and young regenerates and cause poor establishment.
- Over grazing causes physiological disturbances to the trees resulting in reduction in growth attributes like height, diameter and reproductive structures.
- Uncontrolled grazing causes changes in plant community composition thereby creates dominance of one plant community.
- Invasion of new and non palatable species is another major problem caused due to over grazing.

Damage caused by wild animals

- Most wild animal damages the tree and forest cover by trampling the young regenerations, seedlings and saplings of favored and unflavored species. This is very important in case of elephants which damages the plants in and around the corridor and cause serious damages.
- Most elephants feed on trees by uprooting them or damaging them which causes serious damage.
- Many wild animals eat the bark of species and arrest the growth of tree crops and cause damage lead to death of trees. Eg. Elephants eat the barks of Artocarpus, Shorea, Deers eat the barks of Melia, Bison strip the barks of Acrocarpus etc.
- The young regeneration of many plants are eaten away by bison, wild buffalos, wild pigs, deers, rodents and monkeys which arrests the establishment of new forest cover.
- Many wild animals like monkeys and birds cause severe damage to the reproductive part particularly the flowers, fruits and seeds and thereby cause damages to seed production and the associated seedling and plantation establishment.
- The wild pigs readily eat the roots and tubers of wide range of trees and shrubs and causes not only damage to the plant communities but also expose the soil resulting in loss of soil cover and fertility.
- All wild animals particularly elephants and bisons destroy the nursery and plantations, institutional establishment and even fences which may affect all the forestry operations.

25. What is control burning? Under Indian condition how it is advantageous to prevent occurrence of wild fire in a forest? During which season it is followed. (2007)

Control Burning: It is a process of deliberate burning which includes planned application of fire to the natural fuels including slash, with inherent to confine it to a predetermined area / site. Or otherwise it is a deliberate application of a fire where fire control is exercised in some particular respects.

Under Indian conditions, control burning is practiced in some of the areas particularly in parts of north eastern states and dry deciduous forests of South India. Deliberate burning to curtail wide spread of forest fire has been in practice in the Nilgris. In India, sophisticated and modern equipments are not available to control the natural fires. Moreover the fire occurrence particularly in dry deciduous forests is very fast and hence control burning is one of the effective and efficient tool in order to spend the fire.

Under Indian condition, particularly in dry deciduous forests the presence of huge volume of inflammable materials which facilitate the natural fire. Removal of such large volume of inflammable material in the inaccessible areas is not possible. This will involve huge managerial and financial resources. Under such circumstances, controlled burning is one of the alternate and acceptable methods of controlling fire.

Under Indian conditions the following factors viz., temperature, wind, humidity, topography etc., will play a significant role in expansion of fire. Hence the control of these factors to arrest forest fire is very tedious process under Indian conditions. Hence control burning is one of the viable method under Indian weather and terrain conditions.

This controlled burning can be practiced before the beginning of summer season in order to reduce the quantum of inflammable materials present in the forest ecosystem.

26. Write in brief on important conservation project for wild life animal in India. (2007)

The following are the important conservation projects for wild animals operating in India.

(i) Project Tiger: Project tiger launched in 1973-74, is one of our most successful conservation projects in the recent times. The project aims at tiger conservation in specially constituted tiger reserves, which are representative of various bio-geographical regions falling within our country. The project was launched in 1973 and various tiger reserves were created in the country on a core buffer strategy. The core areas were freed from all sorts of human activities and the buffer areas were subjected to conservation oriented land use. Management plans were drawn up for each tiger reserve, based on the principles outlined below:

- Elimination of all forms of human exploitation and biotic disturbance from the core area and rationalization of activities in the buffer zone
- Restricting the habitat management only to repair the damages done to the ecosystem by human and other interferences, so as to facilitate recovery of the ecosystem to its natural state
- Monitoring the faunal and floral changes over time and carrying out research about wildlife.

(ii) The Gir lion project: The Asiatic lion is localized in the Gir forests of Gujarat. In the past, the lion was distributed in the northern and central India. The lion population has declined owing to the shrinkage of its habitat at the turn of the century, coupled with uncontrolled shikar. The Government of Gujarat, in 1972, launched the Gir lion project, which apart from the conservation of lions in their habitat, also provided for the social upliftment of the pastoral maldhari tribes, whose cattle frequented the lion habitat.

(iii) Crocodile breeding project: This project was undertaken from the 1st of April 1975 for all the three crocodilian species found in our country, based on the advice of an FAO expert. The crocodilian population in India declined considerably due to unabated persecution of these animals for their skin. At present, crocodile hunting is legally banned in our country. The Wildlife Protection Act of 1972 has listed all the crocodilian species under schedule I, thereby affording total protection to these animals.

(iv) Project Elephant: Elephant is an animal included in Schedule – I of the Wildlife (Protection) Act, 1972. Project Elephant (PE) was launched by the Government of India in the year 1992 as a centrally sponsored scheme to conserve the elephants and its habitat. The project seeks to address (a) the threats to the survival of elephant in our country on account of loss of its habitat, migration path and poaching for tusks; and (b) issues arising out of elephant depredation in villages. Project elephant also seeks

(i) To maintain an appropriate size of the population in any habitat to safeguard it against extinction; and

(ii) To maintain genetic continuity of isolated populations, which are otherwise susceptible to inbreeding.

(v) Project Leopard: The Indian government launched 'Project Snow Leopard' on January, 2009 to safeguard and conserve India's unique natural heritage of high-altitude wildlife populations and their habitats by promoting conservation through participatory policies and actions. This project covers Jammu and Kashmir, Himachal Pradesh, Uttarakhand, Arunachal Pradesh and Sikkim. It aims at promoting a knowledge-based and adaptive conservation framework that fully involves the local community.

27. Do you think that the controlled fire is beneficial to forest vegetation. Discuss. (2014)

Yes, controlled fire is beneficial to forest vegetation due to the following reasons.

(i) Enhancing natural regeneration: The controlled burning facilitates rapid regeneration of forest species during the ensuing monsoon and facilitates complete natural regeneration. This enriches the flora and fauna of the forest ecosystem and hence improves the whole vegetation.

(ii) Removal of inflammable materials: The early removal of inflammable materials due to controlled fire will help to expose the seeds and rhizomes present on the forest floor. This will help to increase germination of this vegetation.

(iii) Control of soil borne pest and disease: The controlled fire helps to kill the soil borne pest and disease there by helps to prevent the pest and disease outbreak.

(iv) Prevention of natural fire: Controlled fire also helps in preventing the natural fire and ultimately protects the forest from heavy damage

Control burning has some negative e3ects and they are as follow:

(i) Soil development and nutrient cycling: The control burning leads to increase in soil erosion due to negative in vegetative cover. Due to controlled burning all the organic matter will be lost which ultimately change the soil profile and affect the fertility status of soil.

(ii) **Water retention:** The controlled burning changes the rate of penetrate ability of water inside soil which leads to both soil and water erosion.

(iii) **Changes in animal species:** Due to controlled burning animals may temporary shift or migrate to new places to meet food and shelter needs.

(iv) **Control burning is not possible in all localities:** In inaccessible areas, control burning is very difficult to carry out and manage. In sloping ground, the chance of conversion of controlled burning into disasters fire is high. Hence, controlled burning is not possible in all localities

28. How is the forest damage report framed in encroachment cases having nearby human settlements. (2009)

Forest damage report is a written document of damage done to the forest land after the encroachment taken place and it reflect the picture of illegal activities and it is prepared for the purpose to realize the compensation for the damage done to the forest in addition to the punishment under forest acts.

It is prepared by the concerned Beat Officer/ Forest Guard. There is a format to be filled by the Forest Guard in that the following information are recorded

- Date
- Time of encroachment
- Name of the compartment, boundary marks/pillars, legal status-RF,PF, WLS,NP or Tiger Reserve any damages to those boundary marks are recorded
- Details of damage caused

(i) **Trespass:** The duration, size, purpose, any permanent structures are enumerated and this will give the assessment of the rent for the unauthorised occupation of the land

(ii) **Removal of forest produce:** Wood and non wood produce- based on compartmental records and beat map the damage is assessed the value of the forest produce is based on the market value of the forest produce removed from the land after due measurement of their quantity if there is available in that land before encroachment and during the time of illegal occupation. The market value is fixed by the PCCF for various forest produces is applicable for calculation of the value in terms of money to be paid as compensation.

(iii) Any rights are affected to any right holder on the forest land- is valuated based on the records of the forest settlement (No. of rights, forest produces in the rights).

- Name, age and address of encroacher
- Name and address of witness
- Any other case or previously booked case and damage report information

The report prepared in triplicate copies. Two copies immediately send to Forest Range Officer, and the FRO as early as possible or within two weeks verify the damage report sent by FG and with his comments, FRO sent the damage report beat wise and serial wise to DFO concerned.

DFO concerned maintains the Register of Cases of encroachments and based on the cases, DFO decides the case to be compounded, case to be prosecuted and case to be dismissed as error of judgement of FG.

29. To protect the wild animals venturing out from their core areas, some administrative and management arrangements are made. What are those discuss. (2015)

- Management measures such as demarcation of buffer zone around core areas are most important.
- Administrative measures are more protection around core zones by intensifying the patrolling activities
- Creation of fodder reserves inside the core zone area will help to stop the migration of animals from core area to other areas.
- Another major cause of migration is shortage of water for the wild animals. Hence adequate provisions have to be made to augment the water resources inside the core area.
- The disturbance due to wild animals has also occur due to the influence of local inhabitants. Hence the local inhabitants should have to be rehabilitated and their compensation should be settled in tune with the provision.
- The animals are also moving out due to the activities of tourism. Generally the tourism activities are restricted inside the core area barring few exceptions like Mudhumalai Tiger Reserve (MTR) wherein tourism activities influences the movement of animals from core area to the buffer zone.

In the recent development the wildlife conservation in particular tiger conservation is having the different kinds of administrative and management arrangement to protect tiger or wild animals venturing out of core area

The whole protected area is divided into three different components:

(i) Core area/critical wildlife habitat : this area is dedicated to reproduction of the animals with adequate prey base, having low human impact by village relocation from the source area, ensure no development activities, ensure the population surplus in the landscape for population maintenance for long period and regulate the tourism as per govt. Guidelines.

(a) The area is managed on various zone plans and theme plans-

(b) The animal movement is due to poor prey base should be analyzed and the cause of poor prey base is due to the competition of livestock and the intervention is focus to reduce the livestock presence and augment the prey base with proper habitat improvement measures such as ensure proper fodder availability and water sources in core area and the proper monitoring of the prey base health However, such interventions should not change the ecological character of the forest habitat. The zoning also focus the importance of animal status endemic, endanger and its unique habitat, the human settlement may be relocated on basis of the govt. Guidelines to make the area inviolate and give more space to the animals, proper tourism plans regulation as per govt. Guidelines.

(c) The theme plans covers the general intervention such as protection against poaching, monitoring regular basis of the movement of animals, capacity building of staffs to handle various issues and habitat intervention.

(d) For protection of animals in core area, there are measures adopted by management authorities as follows 1.The Tiger Cell 2. The Strike Force. 3 Strategy for Protection and Communication- putting camera trap, radio collars and alert the movement of animals and proper patrolling 4. Relocation of human settlements, 5. Reintroduction of animals in the earlier occupied area after necessary improvement of habitats and 6. Intelligence Gathering and Coordination

(ii) Bu3er Zone: This zone is for the spill over population which are venturing out of core area and is used as a "multiple use area", it may encompass adjoin forest area under territorial wing, conservation or community reserves, apart from revenue lands, private holdings, villages, towns and other production sectors. This area is legally notified and the wildlife protection act is covered for the area notified as buffer. The existing land uses in the buffer zone can continue with due mainstreaming of wildlife concerns and all the activities are regulated by zone and theme plans which integrate various subjects and stakeholders in the buffer area for the protection of the animals and its movements.

(iii) Corridors: Connecting habitats, apart from facilitating animals movements also act as refuge for spill over populations from the core areas. It is facilitating re-colonization in habitat patches through restorative management and providing opportunity to tiger for colonizing new areas through patches of habitats (stepping stones) between isolated populations.

In addition to the above, the following components are also suggested to protect the wild animals from venturing out from the core areas.

Creation of conservation reserve: usually found around the Protected Area Networks like National Parks and Wildlife Sanctuaries.

Creation of community reserve: On land of private or community ownership, if a group of people are doing conservation work and Government give it a legal recognition.

Locally the PA management has the EDC and the Foundations for conservation and protection of the wild animals straying out of PA and

Infrastructures created to monitor the animal movement by establishing total survivelance stations along the boundaries of PA and putting radio collar on the animals for their movement pattern study, created various monitoring team on ground to find out the animals venturing out the Pas and PA management is having quick response team to deal the animals venturing out equipped with all instruments for capturing and releasing in the wild again. The PA management is also having the compensation provisions to the loss or damage by the venturing out animals in their respective areas.

30. Enumerate and discuss the factors responsible for restricting tribal populations in the national parks. (2015)

The following factors are responsible for restricting the tribal population in the National Park.

A. Legal and Policy Factors: The tribal population in the National Parks is legally delt by the provision of Scheduled Tribes and other Traditional Forest Dweller (Recognition of Forest rights) Act, 2006 and Supreme Court Order on ban on collection of non wood forest produce from NP and WLS.

- **Rights on forest land:** The forest right act restrict their occupation on only to 4ha of land maximum or the actual holding if it is less than 4 ha and the claims are eligible by the cut off date 13 dec 2005.
- Various agency involved in processing of claims of Forest Rights of Tribals and Claims of right can be rejected for any false information to Gram Sabha and different level Committees by objections from Forest Department
- In National Park area, there is a provision of creating Critical Wildlife Habitat in which the tribal have to relocate from the CWH to make the area inviolate for wildlife.
- Relocation of tribals from CWH involves
  - (i) It is volunteer in nature - harsh condition of the habitation inside NP-Trust between FD and Tribals to be built and
  - (ii) Consent of gram sabha
  - (iii) Package – option 1 - land (2ha)package or option 2- money package(10 Lacs)
  - (iv) Different agencies involved- FD, Tribal department, revenue and others
  - (v) Availability of budget, land and other resources- better resources availability help to move the tribals from core area.
  - (vi) Rehabilitation process - quick administrative process and proper co-ordinations between line agency- of infrastructures and other facilities for the livelihood

B. Socio-economic conditions: Though the communities residing in the park are allowed to collect wood and other materials from the forest, the sale of the forest products is strictly prohibited. Lack of viable alternatives compels the people to rely heavily on the park resources for their livelihood. The illiteracy among them makes them unequipped to look for alternative sources of income. Therefore, better education facilities up to higher levels should be provided to the local communities, so that their dependency on forests resources is reduced. Eco-development programmes must be initiated, including educational program, awareness towards major conservation issues and importance of wildlife training program for alternative livelihood sources both for men and women. Changing attitude of tribes is possible towards mainstreaming with proper education and employment by constant efforts of forest department and other NGOs in the protected areas.

C. Conflicts with wildlife: Degradation of resources leads to conflict with wildlife for their livelihood such as fodder collection, fuelwood, Non-wood forest produce and grazing inside core areas, loss of life and properties by predation of wild animals and no compensation inside core area for any loss. To avoid the loss, the tribes are willing to move voluntarily for the conflict free zone with the relocation packages.

31. Large-scale mortality has been noticed in *Dalbergia sissoo* and *Acacia nilotica*? What could be the possible reason for this mortality? (2012)

Reasons for mortality of *Dalbergia sissoo*: Shisham (*Dalbergia sissoo* Roxb.) is an important rough barked deciduous timber tree species of multipurpose utility and distributed throughout India. In the recent past Shisham trees are reported to dye due to various symptoms which included reduction in leaf size, change in foliage colour and dye back of twigs from top downwards. These defects lead to death of plants within 3-4 years depending on the age and site. It has been reported that mortality is not due to only single factor but due to cumulative action of several climatic and edaphic factors followed by infestation due pest and diseases.

(a) **Environmental causes:** The mortality is site specific and is due to the dominance of either environmental or miscellaneous factors of physico-chemical properties of soil. The soil with stiff clay and water logged areas are more prone to high mortality. Similarly high mortality also coincided with dry spell prevalent in the northern state and abnormal climatic conditions during winters such as prolonged foggy weather would have cost mortality. These physical factors have weeken the tree which attracted several pest and diseases which lead to huge mortality.

(b) **Biological causes:** The species experienced wide range of mortality due to pest and diseases. This species is attacked by over 125 insect pests and of which 10 are known to cause significant mortality. Among them, the damage due to *Plecoptera reflexa* is very important which causes heavy defoliation and repeated attacks causes large scale mortality in plantations.

The catterpillars of *Leucoptera sphenograpta* mine into the leaf destroying the palisade and large veins of leaves and causes mortality. The sap suckers viz., *Drosicha mangiferae, Destebbingii* are very important because they feed on the sap of foliage, shoots and twigs of the plants and causes heavy mortality in the plantations.

Similarly this species is heavily damage due to diseases and among them the fusarium wilt caused by *fusarium solani* is very important. It is a facultative parasite inhabiting soil and attacks the tree through dyeing or week roots. The clay soil or water logged conditions favour the development of disease. This disease cause severe mortality in the plantations. The root rot due to Ganoderma is also one of the reasons for severe mortality.

Reasons for mortality of *Acacia nilotica*: *Acacia nilotica* is a multi-purpose tree species and widely grown across the country. It has been reported in Haryana that over 9 lakh trees have been died in 2000 different land use system. The mortality of *Acacia nilotica* was not due to single factor but it is due to cumulative action of several factors for the mortality. The mortality is site specific and the intensity is governed by either environmental or miscellaneous factors like physico-chemical properties of soil. No mortality was observed in mixed plantations. The mortality was maximum from plantations facing acute biotic pressure.

(a) **Environmental causes:** Erratic rainfall which leads to heavy water logging during 2001 in North India played a major role in mortality of *Acacia nilotica.* The soil properties particularly the salinity, alkanity, water logging followed by aridity would have caused maximum mortality. These factors weaken the trees followed by infestation by pest and diseases.

(b) **Biological causes:** *Acacia nilotica* is heavily infested with *Celosterna scabrator* and

Psiloptera fastuosa which bore the stem and root of young plants, make them hallow which lead to death of plants. Two fungi viz., *Fomes badius* and *Ganoderma lucidum* causes spongi rot and heart rot which causes serious damages and lead to death of the plants.

32. Willow is the life line in dry temperate region (Lahaul – Spiti) but its large scale drying is causing great concern. Give your view point. (2013)

Yes. Willows are very adaptable to dry temperate region particularly in the cold desert areas of Lahaul – Spiti. But Salix species gets dried due to extreme temperatures. Wide range of factors are responsible for drying and the predominant factors are furnished below.

- The temperature of dry temperate region is less than zero degree centigrade for most of the period which inhibit water absorption by willows.
- The atmosphere of the region is very dry and the mean annual rainfall is less than -100 mm which creates desert like influences. But willows grow very well in moist temperate region and due to lack of precipitation the willows are drying.
- This dry temperate region experiences heavy snowfall which occurs between late November and carly April. During this pcriod willows arc hcavily damagcd.
- This dry temperate region is experiencing heavy cold wind erosion which leads to rapid degradation of land and ultimately drying of willows.
- The soil of the region has poor organic matter content and low in water retention capacity and it is not supporting the growth and development of willows. The willows require perpetually moist, well aerated, deep, fertile, rich loamy soil which is never subjected to drought and this may be reason for drying of willows.
- Because of extreme cold conditions, the growth of willows is slow and sometimes it experiences stunted growth and further extreme in weather condition leads to death of willows.

33. What are live fences? Name five plant species most commonly used as live fences. How do these di-er from other types of fences? (2010)

Live fences: A live fence is made of living trees and shrubs, made from thorny or non thorny plants. it is also known as green fence or hedge. Live fences are established by planting trees around the edges are aerable field or area of pastures. Fodder and hedge trees are planted along the boundaries which serve as a live fence in addition to provide fodder.

Important Species: Agave sisalana, Albizia amara, Caesalpinia pulcherrima, Calliandra calothyrsus, Euphorbia tirucalli, Gliricidia sepium, Lantana camara, Pithecellobium dulce, Prosopis spp., and Protium cordatum.

Characteristics and usefulness of live fences

- Most of the species indicated are robust, shrubby and sometimes thorny which facilitates protection against entry of biotic animals.
- Most of the species indicated are amenable for lopping and hence they can be easily managed as hedges.

- They are amenable for high density hedges due to their easy adaptation.
- Most of the species are fast growing and amenable for multiple branches and hence their management as hedges will prevent the entry of animals.
- It can act as wind break and also an ornamental plant.
- Control movement of cattle
- Some of the species have nitrogen fixing ability and helps to improve the soil in terms of soil fertility.
- The leaves of fence species have high nitrogen and hence their mixing with the soil will help to improve the soil, binds the soil and prevent soil erosion.
- Live fences are protecting the land, various products such as fodder, firewood, medicines, timber, herbs and etc.,
- The live fences save money
- Interplant with fruits and timber trees for economic diversification crops are protected again harmful plants.
- It provides shade

Chapter

15

Forest Economics and Legislation

1. Give the important features of demand and supply curves. What are the factors responsible for demand and supply of forest produce? (2007)

Salient Features of Demand Curve

(i) The demand curve illustrates the variation in the demand of a particular commodity with its change in its price.

(ii) If other variables remain constants the demand curve will have a negative slope as depicted in figure.

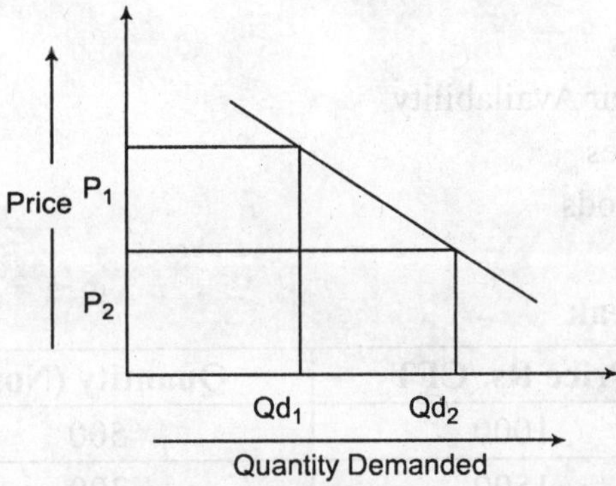

(iii) For some essential commodities, the demand would remain the same irrespective of the prices.

(iv) In some cases, the demand curve could be positive slope which indicates the increase in price and the demand condition goes up. This could happen in a condition is a particular hoarded.

(v) The demand curve shifts only if there is a change in factors other than price.

Factors responsible for Demand

- Price Control and Rationing
- Black Market
- Rent Control
- Minimum Support Price
- Incidence of Indirect Taxes
- Price and Income Stabilisation
- Buffer Stock Operations
- Income, Price of related commodities

Salient Features of Supply Curve:

(i) The supply curve depicts the variation in supply of commodity or goods with change in price.

(ii) The normal supply curve has a positive slope which indicates the supply and price of commodity increases or decreases simultaneously.

(iii) A shift in supply curve indicates a change in the quantity of commodity that will be supplied at each possible price. This shift is due to involvement of change in one or more factors.

(iv) The shortage occurs when at a given price, the quantity of a commodity demanded exceeds the quantity which is supplied.

Factors responsible for Supply

- Price
- Cost of Production
- Natural Conditions
- Technology
- Transport Conditions
- Factor Prices and their Availability
- Government's Policies
- Prices of Related Goods

Examples:

Hypothetical model in teak

Price Rs./CFT	Quantity (Nos.)
1000	500
1500	300
3000	220
5000	100

2. Explain the concept of demand with the help of demand curve and its application in forest trade. (2016)

Demand: Demand is the quantity required by the purchase or buyer which requires particular commodity at a particular price, in a particular market, in a particular period of time.

Demand Variables

(i) Income

(ii) Preference (or) taste

(iii) Price of commodity

(iv) Price of other commodities

Demand Function: It refers to the relation between the quantity of commodity that has been demanded and all the factors which determine the demands viz., Income, taste (or) Preference, Price of Commodity, Price of related commodities. The demand function is

$$D_x = f(P_x, P_y, t, i, E, N, U)$$

Or

$$D_x = f(P_x, \textit{Cetrisperibus})$$

Where D_x – Demand for a commodity X

P_x – Price of commodity X

P_y – Price of an related commodity Y

t – refers to a taste or preference of the consumer

i – income of the consumer

E – Expectation of the consumer

N – Number of sellers

U – Unknown factors

Cetrisperibus – Means that all other factors remain the same.

Demand Curve (DC): Demand curve is the fundamental and important tool of an economics which illustrates the numerical and graphical relation between the quantity of demand and the price of the demand. During this, analysis it is assumed that all other variables are associated.

Salient Features of Demand Curve

(i) The demand curve illustrates the variation in the demand of a particular commodity with its change in its price.

(ii) If other variables remain constants the demand curve will have a negative slope as depicted in figure.

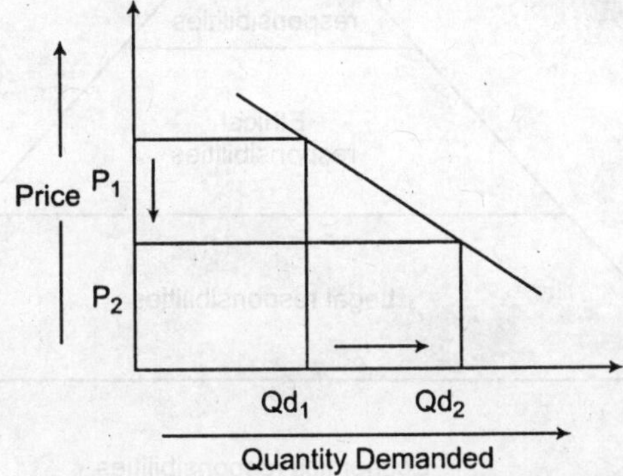

(iii) For some essential commodities, the demand would remain the same irrespective of the prices.

(iv) In some cases, the demand curve could be positive slope which indicates the increase in price and the demand condition goes up. This could happen in a condition is a particular hoarded.

(v) The demand curve shifts or changes only if there is a change in one or more factors ·

Examples

Hypothetical model in teak

Price Rs./CFT	Quantity (Nos.)
1000	500
1500	300
3000	220
5000	100

3. Describe corporate social responsibility towards sustainable forest production through public private partnership approach.

Corporate Social Responsibility (CSR, also called corporate conscience, corporate citizenship or responsible business) is a form of corporate self-regulation integrated into a business model. CSR policy functions as a self-regulatory mechanism whereby a business monitors and ensures its active compliance with the spirit of the law, ethical standards and national or international norms.

CSR entails the simultaneous fulfilment of a company's economic, legal, ethical, and philanthropic responsibilities. In other words, a company should strive to make profit, obey the law, be ethical, and be a good corporate citizen. The pyramid of corporate social responsibility is depicted below.

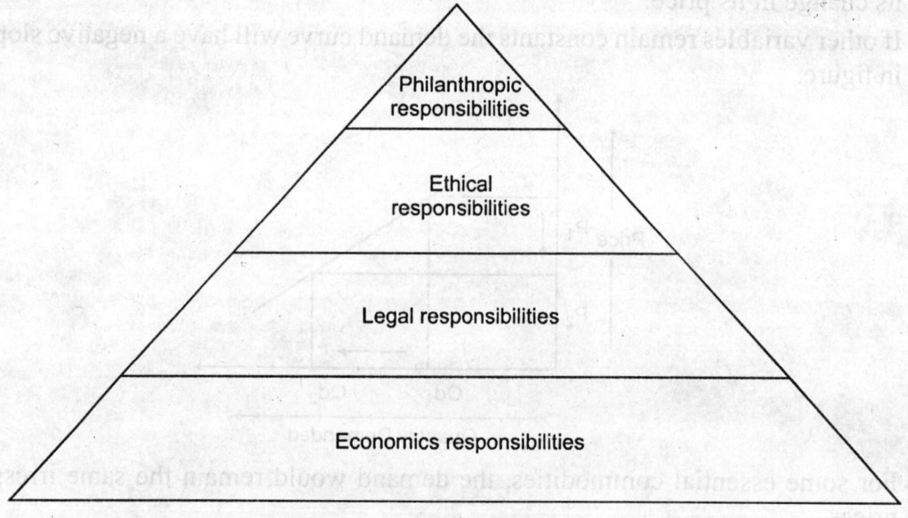

Pyramid of corporate social responsibility

The pyramid depicts the four components of CSR, beginning with the main functions of every company i.e. economic performance. Apart from carrying on economic activities, the company is also expected to obey the law which codifies acceptable and unacceptable behaviour in the society. In addition, the company also needs to be ethical i.e., it is obliged to do what is right, just, and fair, to avoid or minimize harm to stakeholders (consumers, employees, environment, and others). Lastly, the company is also expected to be a good corporate citizen which is patent in its philanthropic acts of contributing financial and human resources to improve the quality of life of the community.

Public-private partnership (PPP) is broadly defined as a government service or private business venture which is funded and operated through a partnership between government and one or more private sector companies.

PPP arrangements are defined as contracts between a private sector entity and the government that call for the private partner to deliver a desired service and assume the associated risks. In return for agreeing to provide the service, the private partner receives payment (in the form of a fee, tariff or user charge) based on pre determined standards of service and other criteria, as specified in the contract. While the government is relieved of the financial and administrative responsibilities of service provision, it retains and focuses on the role of regulating and monitoring the performance of the private partner(s).

Examples of forestry PPPs from around the world demonstrate that the resource can be managed sustainably, with assured livelihoods for forest dependent communities. Countries including Ghana, Nepal, Bolivia and Ethiopia have successfully implemented PPPs in carbon sequestering, ecologically sustainable cropping, marketing of non-timber forest products and wild coffee3 production. While Pakistan falls short in concept and practice, a number of forestry PPPs and Corporate Social Responsibility (CSR) initiatives which constitute important precedents in alternative forms of forest management are already in place.

Approaches to Forestry PPPs: Countries around the world have employed several approaches that divide into four broad categories.

Ownership: Outright privatization is the least common approach and has not gathered any sizable momentum. A much more common approach has been the incorporation of a private sector role in afforestation on private and state land as part of a planting or reforestation drive

Forest Use: Private companies within the forestry industry extract timber/forest products from state forests in one of three broad ways. Concessions or lease arrangements entail the firm acquiring land without affecting the land title. During the lease period, private operators act as owners and are bound by the conditions of the lease agreement. In volume permits or standing timber sales the state sells standing timber in a specifically marked area, which is then logged by the timber company. Finally, in contractual arrangements, the forest authority awards a contract to a professional logging company and out sources logging activity to that entity, while it retains the marketing and retail functions.

Forest Management: In some situations where public sector actors are incapable of managing the forest resource, the responsibility is devolved to private entities. Resource management tasks such as silviculture, inventories, planning, reforestation and monitoring are all carried out by the private sector actor.

Sustainable Forest Product Models through PPP mode: The following are the CSR projects operating in Tamil Nadu through a Public Private Partnership mode which ensure sustainable availability of raw material resources and providing income and employment generation activities to the society.

Product	Type of deal
High-quality timber	**Joint venture – timber** Forest communities manage timber in partnership with private company
	Concessions leased from communities Forest communities lease concessions to private industry, communities retain substantial control
	Outgrower schemes – timber Small farms or communities participate in outgrower or crop share schemes with private companies to establish plantings of improved high-value timber
	Corporate social responsibility project Company contributions to local development in return for access or 'social license to operate'
Industrial pulpwood	**Outgrower schemes – pulp** industry assists farmers to establish and manage pulpwood plantings, in guaranteed supply contracts
	Farm forestry support – pulp Farmers establish plantings with technical support from industry, and sell output without purchase contracts
	Farm forestry crop-share – pulp Plantings established on farmer's land with support from industry, and crop profits shared
	Joint ventures – pulp Communities and companies share equity in pulpwood production venture
	Land leased from farmers/communities Forest owners lease to private companies for pulpwood production
	Corporate social responsibility project Company contributions to local development
Commodity wood	**Outgrower scheme/farm forestry support – commodity wood** Schemes that directly link producers with commodity wholesalers or final users
	Contracts from communities – commodity wood Contracts or agreements for wood-using or logging companies to harvest wood from community forests
Non-timber forest	**Co-management for NTFPs** Communities manage/benefit from NTFPs in company products controlled areas producing wood or pulp
	Outgrower scheme – NTFPs Small-scale farmers grow and sell NTFPs through out-grower schemes with private industry
Commodity wood	**Outgrower scheme/farm forestry support – commodity wood** Schemes that directly link producers with commodity wholesalers or final users

The improvement towards sanitation, health care, availability of protected drinking water, education, protecting cultural and religious interest of local community are also inculcated in the

CSR project activities. The corporates like ITC in Andhra Pradesh, TNPL in Tamil Nadu, Century Ply in South India, Green Ply in North India etc., are involved in CSR project activities through PPP mode and establishes huge forestry and agroforestry plantation programmes.

4. Sustainable forestry accounts for the ecological, economics, social and cultural values of forests. Give your opinion. (2011)

Sustainable Forest Management (SFM) entails the management of forests to maintain their full range of environmental, social, economic and cultural values. Sustainable forest development, therefore, means recognizing the limits of forests to withstand environmental change, individually and collectively, and in managing human activities to produce the maximum level of benefits obtainable within these limits. It recognizes three critical parameters *viz.*, productive capacity, renewal capacity and species and ecological diversity.

Sustainable forestry for economic values: The sustainable forestry aims at producing maximum benefits not only for the current generation but also taking into consideration the future needs. The forests are known for wide range of economic products like timber, pulpwood, plywood, particle board, panel board, MDF, biomass and other non timber forest products like gums, resins, silk, lac, fibre, floss, medicinal plants, dyes, bamboos, oils etc. The timber and non timber forest products are obtained through sustainable forest management by evolving suitable criteria and indicators for managing product based forest utility.

The paper industries in the country have developed sustainable forest management activities to generate pulpwood requirement through deployment of clonal technology coupled with precision silvicultural packages. This has increased the productivity and assured self reliance in raw material security hence sustainable forestry accounts for economic value through production of economic wood and wood products.

Similarly the plywood industries have evolved industrial agroforestry plantations to ensure sustained production of raw materials.

Sustainable forestry for ecological values: The concept of sustainable forest management has aimed at providing all ecological services in order to mitigate and adapt to growing climate change risks. This besides, the forests have played a key role in soil and water conservation and thereby help to augment the ecological cycle. The forest due to their continuous litter fall, stem and through fall contribute towards soil fertility improvement besides ameliorating the soil and ecosystem. The forests play following key roles in protecting the ecological values.

The ecological services of forests are those environmental or ecological processes, which directly benefit humans. Some of the key ecological services are:

(i) Carbon storage and sequestration,

(ii) Preservation and protection of hydrological function and conservation of biodiversity.

(iii) Forests provide some of the most bio-diversity rich ecosystem on earth and are supposed to provide habitat for an estimated 90 percent of the threatened and endangered species.

Social and cultural values: Forests provide a wide range of social and cultural values. The people who depend on forests particularly the dwellers enjoy and benefit through the forestry activities. The forest development projects, tribal development plan, collection, processing and value addition of non timber forest products, rehabilitation of degraded sites etc., create wide range of income and employment generation activities thereby help to protect the social values of forest dependence. It also provides wide range of products which are used by the local dwellers both for domestic utility and for livelihood security. Hence the forest management through sustainable concept help to protect the social values of the dependence.

Similarly the sustainable forestry helps to protect the cultural values of the local society. The forest being supported with wide range of ethnic communities and the cultural values of each and every ethnic society is different. The sustainable forest management helps to protect the cultural interest of the society and thereby contributes significantly towards sustaining the activities. Several tree species are culturally integrated in the ethnic groups, rituals and festivals and the sustainable forestry helps to protect these tree species and ultimately protects the cultural interest. The wide range of sacred grooves and the heritage sites present within the forest ecosystem help the cultural values through sustainable forest management.

5. a. Why it is necessary to take up Cost: Benefit analysis of Forest Plantations? Describe the main points to be taken into consideration while conducting Cost: Benefit analysis of short rotation commercial species and traditional timber species. (2009)

b. Discuss the role of cost benefit analysis in forestry sector. (2005,2006)

Cost Benefit Analysis: Cost benefit analysis is one of the basic tools used in economics for analysing the costs that may be incurred in a project and comparing it with the benefits that may accure. This technique can also be readily used in forestry, particularly while examining the costs and benefits of plantations.

General Benefits: A Whole range of benefits may be obtained from forests. These would include the tangible and intangible benefits. The main benefits that can be valued are:

(i) Timber for construction, small timber and also that for other uses.
(ii) Fuel wood
(iii) Fodder including grasses
(iv) Food
(v) Fibres, flosses and coir
(vi) Essential oils
(vii) Fatty oils
(viii) Resins and oleo resins
(ix) Medicines and drugs

In addition to the above benefits which can be directly measured, there are a number of other indirect or intangible benefits which are difficult to measure in money terms though they are very important. These include:

(i) Regulation of the water cycle
(ii) Carbon sequestrating
(iii) Ecological stability and security

(iv) Soil and water conservation
(v) Mitigation of environmental degradation
(vi) Ecosystem services

General Costs: The general costs in forestry enterprises would include::

(i) Land expectation value or land rent
(ii) Capital value or costs
(iii) Costs of nursery
(iv) Cost of raising the plantation
(v) Costs of silvicultural practices such as tending and thinning
(vi) Establishment costs

Costs-Benefit Analysis: The salient features of cost-benefit analysis in forestry have been brought out in the following points:

(i) It does an appraisal of the costs and benefits of a project or programme, usually in monetary terms.
(ii) Cost-benefit analysis is not restricted to marketable goods alone, rather it also takes into account non-marketable goods and services. The cost of all those inputs required for providing the goods and services are identified while producing the goods and services are taken into account. Any distortions that may be prevailing in the market are also considered.
(iii) Intangible cost and benefits such as diversion of land for a specific purpose and the ecological benefits of forests are taken into consideration.
(iv) This method of analysis is primarily a tool of economic planning brought into use for analysis a specific course of action.
(v) It helps in taking the right decision about a plan of action in respect to forests and plantations.
(vi) It is possible to analyse if the benefits from a specific line of action are higher or lower than the costs involved and if so by how much.

Limitations in BCA:

(i) Human attitude is not considered in this analysis.
(ii) The rate of interest is assumed as same for the whole period. In real it is not so.
(iii) Sustained supply of raw material in future is questionable.
(iv) In forestry, environmental pollution is not considered earlier. Now pollution is highly considered in all analysis.

Cost-Benefit Analysis of Plantations: The following are the main points to be taken into consideration while conducting cost-benefit analysis of a forest plantation.

Costs: The cost include

- Preparation of the plantation project
- Establishment costs
- Nursery establishment, management and cost of seedling production
- Preparation of site including pit digging and fencing

- Planting operation
- Replacement of failures
- Tending and thinning
- Harvesting and conversion
- Transport and marketing
- Cost of the land or land rent
- Protection costs

Benefits: The benefits accrued from plantations would include:

- Yields from thinning
- Final yields of timber
- Non-timber products that may be obtained
- Environmental and ecological benefits
- Employment generation, both direct and indirect
- Socio-economic upliftment.

Often, it may be difficult to put a price tag on benefits such as ecological and environmental and also the socio-economic benefits that would accure over a long time period. Nevertheless these should be taken into consideration and some monetary value may assign.

The ratio of cost to benefits determined the level to which the plantation is likely to succeed. It is necessary to take up the cost-benefit analysis of all forest plantation programmes.

6. Discuss objectives and methods of forest valuation. (2005)

Forest valuation: Valuation is the process of estimating the value of something. Forest valuation refers to the estimation of the value of a forest property. It includes both tangible and intangible benefits derived from the forest. Both biotic and abiotic components of the forests are taken into considerations while assessing the value of forest.

Objectives of forest valuation

(i) To determine the capital value of forest/tree crops.

(ii) To determine the rate of forest capital formation.

(iii) To transfer the forest land to other government or private agencies.

(iv) To acquire a private forest land by government and give compensation for that.

(v) To analyze the past and present facts of forest capital for forest valuation studies.

(vi) To monitor the economic returns from the investments made in forests.

(vii) To predict the value of forest in future.

Forest valuation refers to the valuation of both biotic and abiotic value of the forest and the land involved. It considers both measurable and non-measurable properties. Land includes both animate and inanimate sense except man. The value depends upon the use and user.

The value of the land is estimated based on the use of land and output obtained per unit area. In the same way, the value of a forest is calculated by estimating the output/product from the forest. It calculates values of animals, plants and inanimate such as minerals and other valuables in the area. For example for a forest, its value is the average annual income from that area.

Basis for valuation

1. **Cost value:** It is based on historical cost, replacement cost, restoration cost and present land value cost. This is poor base for valuation due to the non availability of information regarding the historical costs and the increase in the forests land value.
2. **Income value:** Income value is the estimated present net worth of all future earnings or other returns expected from a forest property. It depends on the rate of interest. This is also not a good measure due to the non availability of yield tables for various species and market prices for all products.
3. **Market value:** It is an excellent indicator of the value of a product or commodity, parameters affecting market value. This considers a) site quality, b) topography, c) composition, density, age, d) road accessibility, e) water availability and f) mode of appraisal. This gives better results. This is useful for acquiring a private land for public use.

   Faustman obtained the soil expectation value based on the assessment of output from the forest. He considers a cleared land and which is waiting to be planted. From this land, he assumed intermittent yield of thinning and final yield throughout the infinite series of rotation. He expressed this in a mathematical equation form. The principle involved here is "the whole value is equal to the sum of all intermittent and final yields"

   Soil expectation value refers to the discounted present value of all future returns from a property minus the discounted present value of all future expenses necessary to earn these returns, the discounting done at a given rate of interest at compound interest. The value is calculated from the forecasted future net incomes and forecasted future expenses.

Valuation Methods: The methods adopted for the economic valuation of forest goods and services generally include direct methods, which determine the value a person is willing to pay for the products or goods through a resource survey instrument. Indirect methods are also used to determine the value of forest goods and services.

(i) **Indirect valuation techniques:** The travel cost method is an indirect valuation technique that was designed to model recreation behaviour. This method calculates a value based on the fact that the price paid to travel to the site is the ultimate value of that site. It should be considered that no fees may be imposed on the use of the resource. The costs associated with travelling to the resource (fuel, mechanical maintenance of vehicle, time spent travelling there) become the variables to be used to determine the value of a resource. The weakness of this method is that, it only deals with single destination trips and assumes that travel is a means, rather than an end in itself.

These variants of the travel cost method can be used to analyse the effect of the quality of the site characteristics rather than the gross value.

The Hedonic price models are based on a hypothesis that goods are aggregations of characteristics and that the demand for these goods is interrelated to characteristics it passes. A good example would be to consider that the price of a house in a city includes the contribution of certain market goods (e.g., size, and design of the house, number of rooms, etc.) and the neighbourhood environmental conditions (e.g. air quality when near a sewage, noise pollution if near an airport, etc).

The household production function model involves situations where individuals purchase private inputs at market prices and combine them with their time and natural resources and environmental attributes to produce out-door recreation experiences. This model has two stages. Firstly, the household reduces the cost of producing a given level of experiences. Secondly, the recreationist maximizes their utility subject to their budget constraint to determine the level of recreation experiences to consume.

(ii) **Direct valuation techniques:** Contingent valuation is a direct way of capturing consumer surplus by means of eliciting the willingness to pay value for the preservation of a resource or opportunity in a simulated market. This method comprises a number of techniques to elicit valuation responses including a bidding game, the payment card, open-ended Questions and close-ended questions. Another value that can be elicited through contingent valuation is a willingness to accept value. A willingness to accept provides an estimate of the amount of money an individual would like to be compensated for to forgo an opportunity. This value is estimated based on the fact that the payment is equal to the benefits that an individual would enjoy through salvaging that opportunity.

Experimental economics approach is another direct method of deriving un-priced values of environmental goods and services. High profile experiments can be put in place to elicit individuals' valuation for environmental amenities. However, conducting such meaningful experiments is generally difficult and expensive.

7. Describe Section (2) of Forest Conservation Act of 1980. (2012)

Forest Conservation Act of 1980 section (2) deals with Restriction on the dereservation of forests or use of forest land for non-forest purpose.

Not with standing anything contained in any other law for the time being in force in a State, no State Government or other authority shall make, except with the prior approval of the Central Government, any order directing-

(i) That any reserved forest (within the meaning of the expression "reserved forest" in any law for the time being in force in that State) or any portion thereof, shall cease to be reserved;
(ii) That any forest land or any portion thereof may be used for any non-forest purpose;
(iii) That any forest land or any portion thereof may be assigned by way of lease or otherwise to any private person or to any authority, corporation, agency or any other organisation not owned, managed or controlled by Government;
(iv) That any forest land or any portion thereof may be cleared of trees which have grown naturally in that land or portion, for the purpose of using it for reafforestation.

Explanation: For the purpose of this section, "non-forest purpose" means the breaking up or clearing of any forest land or portion thereof for-

(a) The cultivation of tea, coffee, spices, rubber, palms, oil-bearing plants, horticultural crops or medicinal plants;

(b) Any purpose other than reafforestation; but does not include any work relating or ancillary to conservation, development and management of forests and wildlife, namely, the establishment of check-posts, fire lines, wireless communications and construction of fencing, bridges and culverts, dams, waterholes, trench marks, boundary marks, pipelines or other like purposes.

8. Describe procedure of declaration of reserved forests mentioning important section of the Indian Forest Act, 1927. (2005, 2013)

Chapter II, Section 4. Notification by State Government.

(i) Whenever it has been decided to constitute any land a reserved forest, the State Government shall issue a notification in the Official Gazette

(a) Declaring that it has been decided to constitute such land a reserved forest;

(b) Specifying, as nearly as possible, the situation and limits of such land; and

(c) Appointing an officer (hereinafter called "the Forest Settlement-officer") to inquire into and determine the existence, nature and extent of any rights alleged to exist in favour of any person in or over any land comprised within such limits or in or over any forest-produce, and to deal with the same as provided in this Chapter.

Explanation: For the purpose of clause (b), it shall be sufficient to describe the limits of the forest by roads, rivers, ridges or other well-known or readily intelligible boundaries.

(ii) The officer appointed under clause (c) of sub-section (1) shall ordinarily be a person not holding any forest-office except that of Forest Settlement-officer.

(iii) Nothing in this section shall prevent the State Government from appointing any number of officers not exceeding three, not more than one of whom shall be a person holding any forest-office except as aforesaid, to perform the duties of a Forest Settlement-officer under this Act.

Chapter II, Section 20. Notification declaring forest reserved.

I. When the following events have occurred, namely:

(i) The period fixed under section 6 for preferring claims have elapsed and all claims (if any) made under that section or section 9 has been disposed of by the Forest Settlement-officer;

(ii) If any such claims have been made, the period limited by section 17 for appealing from the orders passed on such claims has elapsed, and all appeals (if any) presented within such period have been disposed of by the appellate officer or Court; and

(iii) All lands (if any) to be included in the proposed forest, which the Forest Settlement-officer has, under section 11, elected to acquire under the Land Acquisition Act, 1894 (1 of 1894), have become vested in the Government under section 16 of that Act, the State Government shall publish a notification in the Official Gazette, specifying definitely, according to boundary-marks erected or otherwise, the limits of the forest which is to be reserved, and declaring the same to be reserved from a date fixed by the notification.

II. From the date so fixed such forest shall be deemed to be a reserved forest.

9. What are the factor responsible a-ecting forest policies? Discuss. (2013)

Factors a3ecting forest policy: The following factors are significant in designing the forest policy.

(a) The national priorities such as:

(i) **Conservation of forests:** The objective of forest management plays a significant role in designing the forest policy. If the major objective is towards conservation, the policy should be designed incorporating more conservation issues than the production functions.

(ii) **Bio-diversity conservation:** It also plays a vital role in forest policy formulation. India is one of the hotspot country for biodiversity and hence all the policy guidelines and directions may be oriented towards conservation of biological resources. The 1988 forest policy has incorporated biodiversity conservation as one of the key strategy.

(iii) **Wildlife conservation and protection**: India is known for diverse faunal population which resulted in sustained food chain in the forest ecosystem. The wildlife management and conservation issues are of great significance in deciding the policy strategies.

(iv) **Ecological security**: Forest in general and the forest ecosystem in particular is gaining significant attraction towards conservation of all natural resources. This will play a major role in forest policy. The 1988 forest policy has mandated ecological security as one of the major factor.

(v) **Research and training needs:** It is also one of the major factors which forms a part in policy strategy. One of the key issues in 1988 forest policy is research and training needs particularly to train and develop professional foresters to manage the complex ecosystem. Hence it plays a significant role and affects the policy issues.

(vi) **Survey of forest resources:** India's forests are rich in resources which are categorized into timber resources, non timber resources and other ecosystem resources. These resources significantly affect the decision on forest policies.

(b) Regional priorities such as:

(i) **Generating revenue:** Any forest policy always take into consideration the revenue generated through the forest activities. All the developmental activities are associated closely with the revenue generation principle and hence it affects the policy decisions.

(ii) **Maintaining the water cycle:** Another major significant factor is restoration of water cycle in order to cater the needs of the region and the locality. Based on this the policy strategy is liable for changes.

(iii) **Wildlife conservation and protection:** The regional wildlife conservation is also one of the significant factors which can decide the policy issues.

(c) Local Priorities:

(i) **Meeting local timber and non-timber needs:** The policy though mandated for resolving national issues but takes into consideration the local demand. The growing concern on timber and non timber requirement will also affect the policy decision.

(ii) Rights and concessions: The forest policy should aim to satisfy the demands of rights and concessions of the ethnic groups and the local dwellers. Any policy is highly influenced by the rights and concession issues of ethnic society.

(iii) Tribals: The populations of tribes are intrinsically associated with the forest and the associated ecosystem. Hence any policy decision will directly or indirectly have influence both on the tribe and the policy decisions.

(iv) Generation of employment: Several policies are aimed to generate income and employment generation and hence they play a vital role in policy issues.

10. Explain the application of the Indian penal code in forestry. (2006, 2009)

The IPC is applicable both directly and indirectly to forest offences that are listed below;

(i) Theft of timber from government sale depot.
(ii) Receiving of stolen timber
(iii) Thefts of forest department property (other than forest produces)
(iv) Criminal beach of trust
(v) Assault on the staff of the forest department
(vi) Unlawful assembly inside the forest
(vii) Omitting to give information about a forest offence
(viii) Giving false information to forest officials
(ix) Usage of false property marks
(x) Tampering of property marks

11. Mention the salient features of the National Forest Policy, 1988 and Discuss its superiority over National Forest Policy, 1894. (2005, 2006, 2013)

The aim of the policy is to ensure environment stability and maintenance of ecological balance for the sustenance of all life forms, humans, animal and plant. This national forest policy is enunciated due to the following need:

- To meet the increasing energy demands of the populations
- To obey the new directive principles and duties given in the Indian constitution
- Constitutional amendments of 1976 and its stress on environment
- Marked growth of forest based industries in this era
- Need for the guidelines on conservation of wildlife

Basic objectives

- To maintain the environment stability and ecological restoration through preservation
- To check soil erosion and denudation in the catchment areas and mitigation of floods
- To checks the sand the dune extension in desert and costal lands
- To increase the forest and tree cover in the demand and degraded lands
- To meet the requirement of the tribal peoples
- To conserve the natural heritage

- To increase the productivity of the forest to meet the various needs
- To increase the substitution of wood and reduction of wood usages
- To create a massive people movement and minimize the pressure on existing forest

Essential of forest management

- Existing forest should be protected and its productivity improved
- Diversion of agricultural lands to forestry should be discouraged
- Conservation of biodiversity should be strengthened through national park, sanctuary and biosphere reserves
- A forestation programme encouraged to meet the local requirement of fuel wood

Management strategy

- The area under forest should be a minimum of 1/3rd of the total geographical area. In which hills and mountainous region should have 2/3rd cover
- Afforestation, social forestry and farm forestry programme are encouraged
- Rights and concessions of the local peoples are suitability addressed
- Diversion of forest lands for non- forestry purposes is discouraged
- Wildlife conservation is given more importance
- Shifting cultivation practices are discouraged and cultivation of perennial crops through horticulture and tree forming is encouraged
- Fire and grazing management is given special consideration
- Forest based industries are advised to meet their raw materials requirements from outside the forest
- Forest extension, research and personal management are more emphasized

Advantages over NFP 1894

- Conservation of natural heritage, biodiversity and genetic resources
- Symbiotic relationship between forest department and tribals
- Massive afforestation programme emphasized through various measures
- 33% under forests cover area with 60% tree cover in hills
- Grazing is allowed according to carrying capacity
- Discourage shifting cultivation
- Personal management for professional foresters
- Detailed guidelines on training and scientific forestry research is given for increasing productivity, afforestation, wildlife and wood substitution
- Focus on people's participation in forest management.

12. Describe the salient features mentioned in new draft forest bill, 1994? (2013)

New Draft Forest Bill: In 1994, the new draft forest bill was circulated for discussions, views and observations of all concerned. The aim is to solicit the opinion of the state governments, non-governmental organizations, legal experts, general public and all concerned on this proposed bill.

This bill is proposed for replacing the existing Indian Forest Act of 1927.

The salient features of the new draft forest bill, as circulated, have been discussed in brief:

(i) The proposed title is – The Conservation of Forests and Natural Ecosystems Act.

(ii) The proposed bill aims to provide for the restoration, conservation and management of forests and natural ecosystems and matters connected therewith and incidental thereto.

(iii) A detailed procedure is prescribe to be followed for the constitution of reserved forests in which the Forest Settlement Officer inquires into and settles the rights, if any. The appeal against rejection of claims by the Forest Settlement Officer is proposed to lie with the district courts and also subsequently with the High court.

(iv) The state governments are proposed to be empowered to re-define the limits of a reserved forest, within a period of one year from the date of the original notification.

(v) The state government is proposed to be empowered to notify, after clearance form the central government, that a forest is no longer reserved.

(vi) The state government is empowered to declare any tree or class of trees or forest produce in a protected forest as reserved.

(vii) The state government may also take over the management of any forest or natural ecosystem which not is property.

(viii) The owners of forests or natural ecosystems may request the government to protect and manage them.

(ix) There is a proposal for registration of tree growers in a prescribed manner.

(x) The Divisional Forest Officer may provide assistance to any owner who wishes to raise plantations on his land.

(xi) Detailed regulations and provisions are proposed for the following other matters:

(a) Control of trade, possession and transit of timber and other forest produce and its processing

(b) Collection of drift and stranded timber.

(xii) Various procedures have also been laid down in the draft new forest bill. Penalties are proposed for different offences and violations of the provisions.

13. Mention the salient features of the National Forest Policy, 1988. How does it di-er from 1952 forest policy. (2014)

The aim of the policy is to ensure environment stability and maintenance of ecological balance for the sustenance of all life forms, humans, animal and plant. this national forest policy is enunciated due to the following need:

- To meet the increasing energy demands of the populations
- To obey the new directive principles and duties given in the Indian constitution
- Constitutional amendments of 1976 and its stress on environment
- Naked growth of forest based industries in this era
- Need for the guidelines on conservation of wildlife

Basic objectives

- To maintain the environment stability and ecological restoration through preservation
- To check soil erosion and denudation in the catchment areas and mitigation of floods
- To checks the sand the dune extension in desert and costal lands
- To increase the forest and tree cover in the demand and degraded lands
- To meet the requirement of the tribal peoples
- To conserve the natural heritage
- To increase the productivity of the forest to meet the various needs
- To increase the substitution of wood and reduction of wood usages
- To create a massive people movement and minimize the pressure on existing forest

Essential of forest management

- Existing forest should be protected and its productivity improved
- Diversion of agricultural lands to forestry should be discouraged
- Conservation of biodiversity should be strengthened through national park, sanctuary and biosphere reserves
- A forestation programme encouraged to meet the local requirement of fuel wood

Management strategy

- The area under forest should be a minimum of 1/3rd of the total geographical area. In which hills and mountainous region should have 2/3rd cover
- Afforestation, social forestry and farm forestry programme are encouraged
- Rights and concessions of the local peoples are suitability addressed
- Diversion of forest lands for non- forestry purposes is discouraged
- Wildlife conservation is given more importance
- Shifting cultivation practices are discouraged and cultivation of perennial crops through horticulture and tree forming is encouraged
- Fire and grazing management is given special consideration
- Forest based industries are advised to meet their raw materials requirements from outside the forest
- Forest extension, research and personal management are more emphasized

Di3erentiate National Forest Policy of 1988 & 1952

S.No	Parameters	National forest policy of 1952	National forest policy of 1988
1.	Objectives	Balanced and complementary use of forest produce	Conservation of natural heritage, biodiversity and genetic resources
2.	Forest Classification	Protection forests National forests Village forests Tree lands	No classification

3.	Tribal people and forest	Not much guidelines	Symbiotic relationship between forest department ' and tribal
4.	Scope for increasing tree cover	Increasing vegetative cover through planting on railways, PWD lands.	Massive afforestation programme emphasized through various measures
5.	Control of private forest	Few guidelines given	Detailed guidelines given
6.	**Proportion of forest area**	1/3 rd of the area, 60% in hills and 20% in plains	33% under forests cover anc 66% tree cover in hills
7.	**Protection to wild life**	Protection of rare and endangered species only	Network of protection measures strengthened
8.	**Grazing**	Grazing is allowed with nominal fee	Grazing is allowed according to carrying capacity
9.	**Shifting cultivation**	No coercive measures to stop shifting cultivation	Discourage shifting cultivation
10.	**Sustainable yield**	Limited scope	Detailed measures are given
11.	**Forestadministration**	Trained foresters are recommended	Personal management for professional foresters
12.	**Forestry legislation**	Enactment of forest acts	Appropriate legislation is given
13.	**Forestry training education and research**	Few guidelines given through rangers college and universities	Detailed guidelines on training and scientific forestry research is given for increasing productivity, afforestation, wildlife and wood substitution
14.	**Good will of local peoples**	Instructs need for peoples cooperation	Focus on people's participation in forest management.

14. Discuss the important role of the Wildlife Protection Act, 1972 in forest management. Give salient points only. (2015)

The wildlife protection act comprised of 6 chapters and 66 sections which dealt predominantly with protection and conservation of wild animals and ensures significant forest management. The following provisions in the wild life act help to manage the forests.

(i) Hunting of wild animals: This act prohibits hunting of any wild animals specified in schedule I to IV expect provide under Sec. 11 and Sec. 12. But under special conditions the C WW can permit hunting of scheduled I animals 'Sec. 11).

Special conditions necessitating animal killing

- If the animal is dangerous to human life

- If the animal is disabled or injured one
- If the animal is diseased which is beyond the recovery stage

(ii) Collection or capturing of wild animals: The CWW with the permission of Central Government can permit collection or capturing schedule I animals under special conditions which are as follows (Sec. 12);

- Education
- Scientific research
- Scientific management (Translocation and population management)
- Collection of specimen for zoological gardens

(iii) Protection of specified plants (Chapter - IIIA, Sec. 17): This act prohibits picking, uprooting, damaging, destroying and acquiring of specified p from forests. This prohibits collection of specified plants from any forestland. Possession, transfer and sale of these plants are prohibited except bonafied personal use by the scheduled tribes.

The CWW by the prior approval of state government can permit collection of these plants specific conditions. Cultivation of specified plants without license is also prohibited. The dealing these specified plants requires specific license from the forest department.

Special plants are those plants that are denoted in schedule VI

- *Cycas beddomi*
- Red vanda
- Blue vanda
- Pitcher plant (*Sasuaria lappa*)
- Ladies slipper orchid

(iv) Sanctuary: Sanctuary is an area declared under section 18 or 38 or deemed under sub section 3 of section to be declared as Wildlife Sanctuary If the central government is satisfied that the conditions specific under section 18 are fulfilled, the area by notification is declared as sanctuary (Sec. 28 (1)).

The following are restricted inside the sanctuary. Entry inside the sanctuary is restricted one. Damage of boundary work is restricted (Sec. 27). But the CWW can permit the entry inside the sanctuary for the purpose of photography, scientific research, tourism and transaction of any useful business with person residing inside the sanctuary (Sec. 28). Destruction to forest property (Sec. 29), causing fire c. 30), use of rigorous substances (Sec. 32), grazing (Sec. 33), and possession of arms (Sec. 34) inside the sanctuary is prohibited. All livestock in and around the sanctuary region should be properly immunized (Sec. 33A).

(v) National park: Under sec 35(1), the state government may declare any area as a national park with the importance of its ecological, faunal, floral or geomorphologic association for the purpose of protecting, propagating or developing wild life in its environment. The provisions of sec 19-26 are applicable in declaring a national park.

No alteration of the boundaries of a national park shall be made except on a resolution passed by the state legislature. No person shall destroy, exploit or remove wildlife or its habitat from the national park. No grazing of any cattle shall be permitted inside the national park.

Hence the issues in wild life protection act from hunting of animals to establishment of sanctuaries and national park will help to establish strong protected area network and they play a vital role in managing the forest resources.

15. Describe the role of permanent forest policy after enactment for regulation of forest management in India. (2015)

Essential of forest management

- Existing forest should be protected and its productivity improved
- Diversion of agricultural lands to forestry should be discouraged
- Conservation of biodiversity should be strengthened through national park, sanctuary and biosphere reserves
- A forestation programme encouraged to meet the local requirement of fuel wood

Management strategy

- The area under forest should be a minimum of 1/3rd of the total geographical area. In which hills and mountainous region should have 2/3rd cover
- Afforestation, social forestry and farm forestry programme are encouraged
- Rights and concessions of the local peoples are suitability addressed
- Diversion of forest lands for non- forestry purposes is discouraged
- Wildlife conservation is given more importance
- Shifting cultivation practices are discouraged and cultivation of perennial crops through horticulture and tree forming is encouraged
- Fire and grazing management is given special consideration
- Forest based industries are advised to meet their raw materials requirements from outside the forest
- Forest extension, research and personal management are more emphasized

16. Compare and Contrast the Indian Forest Act, 1927 and Forest Conservation Act, (1980). (2016)

S. No.	Indian forest act 1927	Forest conservation act,1980
1.	The act contains : 13 chapters 86 sections	The act contains : 5 chapters
2.	The act is to consolidate the then existing laws relating to forest, the transit of forest products and duties.	The act of Parliament with a view to provide for the conservation of forest and for matters connected therewith or ancillary or incidental thereto.

3.	The act deals with reserved forest, village forest, protected forest, control over forests and lands not being property of government, the duty on timber and other forest produce, regulation transit of forest products, collection of timber, penalties and procedures, cattle – trespass, forest officers and other miscellaneous provisions	The act makes a provision of a prior approval of the Central Government necessary before a State Government or any other authority issues direction for dereservation of reserved forests (which have been reserved under the Indian Forest Act 1927), use of forest land for non – forest purpose, assigning forest land by way of lease or otherwise to any private person or to any authority, corporation, agency or any other organization not owned, managed or controlled by the government and clear felling of naturally grown trees
4.	The act establishes three categories of forests, Reserve forest, Protected forest and Village forest.	The term "forest" shall not be applicable to the plantation raised on private land except notified private forest. Tree falling in such plantation would however be governed by state acts and rules.
5.	The Indian Forest Act 1927 prohibits a number of activities including making fresh clearings, tree felling, lopping, burning, grazing, quarrying, manufacturing activities, hunting, shooting, etc. in the forest. Violation of provisions of Section 26 specifically with regards to creating fire, felling, girdling, lopping, etc. of trees, quarrying and manufacturing operations or clearing breaking up of any land for cultivation is punishable with imprisonment for a term which may extend to two years or with fine which may extend to Rs.20,000 but which shall not be less than Rs. 5,000.	The regional empowered committee shall decide the proposal involving diversion of forest land upto 40 ha. Proposals involving forest land of more than 40 ha. and all proposals relating to mining and encroachments irrespective of the area of forest land involved, shall be forwarded by the concerned State Government / Union Territory along its recommendations to the MoEF Government of India.

17. Bring out the background and need for forest policy of 1988 with its basic objectives. (2012)

Background and Need for 1988 Forest Policy: The National Forest Policy of 1952, the Government of India in the erstwhile Ministry of Food and Agriculture enunciated a Forest Policy to be followed in the management of State Forests in the country. However, over the years, forests in the country have suffered serious depletion. This is attributable to relentless pressures arising from ever-increasing demand for fuel-wood, fodder and timber; inadequacy of protection measures; diversion of forest lands to non-forest uses without ensuring compensatory afforestation and essential environmental safeguards; and the tendency to look upon forests as revenue earning resource. The need to review the situation and to evolve, for the future, a new strategy of forest conservation has become imperative. Conservation includes preservation, maintenance, sustainable utilisation, restoration, and enhancement of the natural environment. It has thus become necessary to review and revise the National Forest Policy which resulted in the establishment of National Forest Policy 1988.

The aim of the policy is to ensure environment stability and maintenance of ecological balance for the sustenance of all life forms, humans, animal and plant. This national forest policy is enunciated due to the following need:

- To meet the increasing energy demands of the populations
- To obey the new directive principles and duties given in the Indian constitution
- Constitutional amendments of 1976 and its stress on environment
- Make growth of forest based industries in this era
- Need for the guidelines on conservation of wildlife

Basic Objectives

- Maintenance of environmental stability through preservation and, where necessary, restoration of the ecological balance that has been adversely disturbed by serious depletion of the forests of the country.
- Conserving the natural heritage of the country by preserving the remaining natural forests with the vast variety of flora and fauna, which represent the remarkable biological diversity and genetic resources of the country.
- Checking soil erosion and denudation in the catchment areas of rivers, lakes, reservoirs in the "interest of soil and water conservation, for mitigating floods and droughts and for the retardation of siltation of reservoirs.
- Checking the extension of sand-dunes in the desert areas of Rajasthan and along the coastal tracts.
- Increasing substantially the forest/tree cover in the country through massive afforestation and social forestry programmes, especially on all denuded, degraded and unproductive lands.
- Meeting the requirements of fuel-wood, fodder, minor forest produce and small timber of the rural and tribal populations.
- Increasing the productivity of forests to meet essential national needs.
- Encouraging efficient utilisation of forest produce and maximising substitution of wood.
- Creating a massive people's movement with the involvement of women, for achieving these objectives and to minimise pressure on existing forests.

18. Give a comparative and appropriate statement on the National forest policy of 1988 with that of 1952. (2015)

S.No	Parameters	National forest policy of 1952	National forest policy of 1988
1.	Objectives	Balanced and complementary use of forest produce	Conservation of natural heritage, biodiversity and genetic resources
2.	Forest Classification	Protection forests National forests Village forests Tree lands	No classification
3.	Tribal people and forest	Not much guidelines	Symbiotic relationship between forest department ' and tribal
4.	Scope for increasing tree cover	Increasing vegetative cover through planting on railways, PWD lands.	Massive afforestation programme emphasized through various measures
5.	Control of private forest	Few guidelines given	Detailed guidelines given
6.	Proportion of forest area	1/3 rd of the area, 60% in hills and 20% in plains	33% under forests cover and 66% tree cover in hills
7.	Protection to wild life	Protection of rare and endangered species only	Network of protection measures strengthened
8.	Grazing	Grazing is allowed with nominal fee	Grazing is allowed according to carrying capacity
9.	Shifting cultivation	No coercive measures to stop shifting cultivation	Discourage shifting cultivation
10.	Sustainable yield	Limited scope	Detailed measures are given
11.	Forest administration	Trained foresters are recommended	Personal management for professional foresters
12.	Forestry legislation	Enactment of forest acts	Appropriate legislation is given
13.	Forestry training education and research	Few guidelines given through rangers college and universities	Detailed guidelines on training and scientific forestry research is given for increasing productivity, afforestation, wildlife and wood substitution
14.	Good will of local peoples	Instructs need for peoples cooperation	Focus on people's participation in forest management.

19. Which are the defects in National Forest Policy of 1894 and 1952. (2007)

Defects in the National Forest Policy of 1894

- This policy was enunciated by the Britishers which was formulated to suit the needs of British regime.
- The policy has mandated public benefit as a sole issue and did not consider other conservation issues.
- This policy has indicated conversion of forest land for agriculture purposes which resulted in massive conversion of forests for agriculture purposes.
- This policy did not taken into consideration the growing demands of society and the industry which lead to increased deforestation.
- The policy did not taken into account the professional man power needs of the country through systematic forestry education.
- It was also silent on systematic research on forestry programme for production as well as conservation.
- Wildlife is a major part in forest management and the policy did not make any provision for conservation and management of wildlife population which have started decreasing due to more human interaction.
- The policy did not consider the regulation of water and other ecological resources which resulted in rapid loss of these resources.

Defects in the National Forest Policy of 1952: After the enunciation of National Forest Policy 1952, tremendous development have happened across the country in terms of utilization of wood and wood products, mushrooming of industries, diversion of forest land for non forestry purposes, development of infrastructure like roads and communication facilities, construction of dams and reservoirs etc. which resulted in decline of forest cover at an alarming rate. It has been estimated that during the period between 1950s and 1980s the country has lost over 1.5 million ha per annum. This has necessitated a change in forest policy which has not been addressed by the National Forest Policy 1952.

(i) **Production Vs Conservation:** The 1952 forest policy was aimed more towards production lead forestry programme which resulted in large scale deforestation. However during the period between 1960s and 1980s the emphasis was shifted from production forestry to conservation forestry and forestry had undergone a massive change.

(ii) **Mushrooming of Industries:** After 1952 mushrooming of industries have been established across the country like saw mills, pulp and paper, plywood, veneer, particle board, match splints, sports goods, agriculture implements and construction industry. These industries have taken heavy toll of Indian forests and the 1952 forest policy have not directed any issues to meet the industrial raw material.

(iii) **Government Vs. Participatory Management:** Traditionally Indian forests were managed by forest department with the stringent rules and regulations. The 1952 forest policy has not taken into consideration the role of people in managing the forests through a participatory mode which is a major drawback in the policy.

(iv) **Maintenance of Ecological Balance:** The 1952 forest policy has not taken into consideration the maintenance of ecological balance. As indicated earlier due to continuous exploitation to meet domestic and industrial demand the forests have been degraded which lead to imbalance in ecosystem. This has not been taken care of by the 1952 National Forest Policy.

(v) **Rights and Concessions:** The 1952 National Forest Policy has not taken to consideration the Rights and Concessions of the local dwellers. Though it has given claims of neighbouring communities but there is no clarity on rights and concessions.

(vi) **Forestry Education:** The 1952 forest policy dealt only with in service forestry education and does not taken into consideration development of professional forestry education.

(vii) **Forestry Research:** Similarly to education the policy does not taken into consideration the research outside the forest department. It has not given any direction on agroforestry research which is one of the important land use systems to satisfy the demands of domestic and industrial requirement.